MERRILL

Focus On Earth Science

AUTHORS

Dale T. Hesser
North Syracuse Central Schools—North Syracuse, New York

Susan S. Leach
Upper Arlington Schools—Columbus, Ohio

CONSULTANT

Dr. Berry Sutherland
University of Texas at San Antonio—San Antonio, Texas

CONTENT CONSULTANTS

Dr. Allen A. Ekdale, Department of Geology
University of Utah, Salt Lake City, Utah

Dr. Robert Howe, Department of Geography and Geology
Indiana State University, Terre Haute, Indiana

Dr. G.H. Newsom, Department of Astronomy
The Ohio State University, Columbus, Ohio

Dr. James B. Phipps, Department of Marine Geology and Environmental Science
Gray's Harbor College, Aberdeen, Washington

Dr. Jeffrey Rogers, Department of Geography
The Ohio State University, Columbus, Ohio

Dr. Russell O. Utgard, Department of Geology and Mineralogy
The Ohio State University, Columbus, Ohio

MERRILL
PUBLISHING COMPANY
Columbus, Ohio

A MERRILL SCIENCE PROGRAM

Focus on Earth Science: Student Edition
Focus on Earth Science: Teacher Edition
Focus on Earth Science: Teacher Resource Package
Focus on Earth Science: Review and Reinforcement Guide
Focus on Earth Science: Review and Reinforcement Guide, Teacher Annotated Edition
Focus on Earth Science: Laboratory Manual
Focus on Earth Science: Laboratory Manual, Teacher Annotated Edition
Focus on Earth Science: Overhead Transparency Package
Focus on Earth Science: Chapter Review Software
Focus on Earth Science: Test Generator Software
Earth and Space Science Skillcards
Safety Card Package
Focus on Life Science Program
Focus on Physical Science Program

Dale T. Hesser currently serves as the Assistant Superintendent of Schools in North Syracuse, New York. A past recipient of the Outstanding Earth Science Teacher award from the National Association of Geology Teachers, Mr. Hesser received his B.S. in earth science from Buffalo State College, New York, and holds an M.S. and Certificate of Advanced Studies in Science Education from Syracuse University. He has over 20 years of classroom teaching experience in the earth sciences ranging from junior/senior high school through college astronomy, and numerous pre-service and in-service teacher training institutes. Mr. Hesser is co-author of a number of textbooks and publications related to the earth sciences.

Susan S. Leach is a teacher of earth science at Jones Middle School, Upper Arlington School District, Columbus, Ohio. She serves on the Board of Trustees of North American Astrophysical Observatory and has served on the Boards of Directors for state and national science organizations. Ms. Leach received a B.S. in Comprehensive Science from Miami University, Oxford, Ohio, and a M.S. in Entomology from the University of Hawaii. She has 15 years of teaching experience and is author of various educational materials. Ms. Leach, in addition to receiving Exemplary Earth Science and Career Awareness in Science Teaching Team awards from NSTA, was the 1987 Ohio Teacher of the Year, and one of four finalists for the 1987 National Teacher of the Year.

Reading Consultant
David R. Urbanski, Reference Librarian, Dublin Branch Library, Dublin, Ohio

Special Features Consultants
Julie Herold, Science Teacher, Westerville North High School, Westerville, Ohio
Laurel Sherman, Affiliate Scholar, Oberlin College, Oberlin, Ohio
Nancy Von Vrankeken, Teacher's Clearinghouse for Science and Society Education, New York, New York

Reviewers
Sr. Johanna Danko, Science Teacher, Most Blessed Sacrament School, Franklin Lakes, New Jersey
Linda E. Delano, Science Coordinator, Dodgen Middle School, Cobb County, Georgia
Terry Dyroff, Science Teacher, St. Albans School, Washington, DC
Nancy E. Greenwood, Earth Science Teacher, Northwood Junior High, North Little Rock, Arkansas
John F. Hartnett, Earth and Marine Science Teacher, Wilmington High School, Wilmington, Massachusetts
Norman E. Holcomb, Science Teacher, Marion Elementary School, Maria Stein, Ohio
Mazie R. Lunn, Science Teacher/Coordinator, Busbee Middle School, Cayce, South Carolina
Susan T. Roberts, Science Teacher, Wirt County High School, Elizabeth, West Virginia
Lawana M. Scoville, Science Teacher, Laurel County Junior High, London, Kentucky
Dennis W. Sterner, Science Supervisor, Warwick School District, Lititz, Pennsylvania
Timothy W. White, Earth Science Teacher, Cimarron Middle School, Edmond, Oklahoma
Luigina B. Yerino, Earth Science Teacher, Dayton High School, Dayton, Kentucky

Series Editor: **Joyce T. Spangler;** *Project Editor:* **Mary Dylewski;** *Editors:* **Nancy F. Gore, Jane L. Parker, Greg A. Shannon;** *Book Designer:* **Kip M. Frankenberry;** *Project Artist:* **Catharine L. White;** *Illustrators:* **Bill Robison, Charles Passarelli, Dennis Tasa, David German;** *Photo Editor:* **Mark Burnett;** *Production Editor:* **Joy E. Dickerson**

Cover Photograph: Waterfall, Arizona: Harold Sund/The Image Bank

ISBN 0-675-02671-7

Published by
MERRILL PUBLISHING COMPANY
Columbus, Ohio 43216

Printed in the United States of America

To the Student

Most students enjoy their study of earth science because they are curious about the world around them and the planet on which they live. In using ***Focus on Earth Science,*** you will increase your knowledge of topics such as Earth history, the solar system, and the universe beyond. You will find out how natural resources are recovered and used and how we can conserve resources. You will learn how the movement of the continents has affected life on Earth. You will learn about earthquakes, volcanoes, and the weather. Thus, earth science is a study of the planet Earth and its place in space.

Many jobs and careers require a background in earth science. Careers in mining, teaching, ceramics, wastewater treatment, astronomy, and map making involve earth science fields. As you read ***Focus on Earth Science,*** *Career* features and *Biographies* will help you explore your interests in these fields and in many others.

Scientists use certain methods to solve problems and find answers. In this textbook, you will learn about scientific methods and how to use them to solve problems. *Investigation, Skill,* and *Problem Solving* activities will help you discover how you can use these same methods. Using scientific methods will increase your success in solving everyday problems and accomplishing tasks.

Focus on Earth Science contains many features that will help you learn. Each unit begins with a photograph and a brief introduction to the theme of the unit. A *Time line* points out important discoveries and historical events. The photograph and introductory paragraph at the beginning of each chapter describe the major theme of the chapter and relate it to your everyday life.

Each chapter has several major divisions. At the beginning of each major division, a list of *Goals* identifies what you will learn as you study the short, numbered sections. *Margin questions* printed in blue emphasize the main ideas of each section. Use these questions as self-checks to evaluate your progress. Major terms are highlighted in boldface type. At the end of each major division, *Review* questions provide another means of self-evaluation.

At the end of each chapter are study and review materials. The *Summary* provides a list of the major points and ideas presented within the chapter. *Vocabulary* lists important new terms, and contains a ten question vocabulary review. *Main Ideas* contains questions that are useful as a review of the chapter's concepts and questions that require you to apply what you have learned. *Skill Review* includes questions about and applications of the skills learned in this chapter and previous chapters. *Projects* provides thought-provoking problems and ideas for projects. Sources of more information are listed under *Readings.*

Several special features have been included to make your study of earth science more interesting. *Technology* features provide exciting information on new technological developments in earth science. A *Science and Society* feature located at the end of each unit offers you the opportunity to explore the interactions and effects of earth science on society.

At the end of the textbook are the *Appendices, Glossary,* and *Index.* The Appendices contain tables, charts, and safety information. The Glossary contains definitions of the major terms presented in the textbook. The complete Index will help you quickly locate specific topics within the textbook.

This textbook has been written and organized to help you succeed in your earth science class. As you do your classwork and complete your assignments, you will gain the satisfaction of understanding earth science and its application to everyday life.

Table of Contents

UNIT 1
EARTH SCIENCE FOUNDATIONS

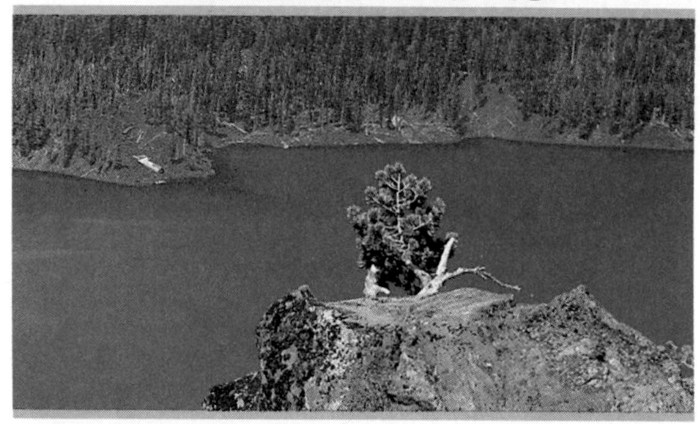

UNIT 2
EARTH IN SPACE

Chapter 5 Earth-Moon System 97

Chapter 6 Exploring Space 117

Chapter 7 The Solar System 139

Chapter 8 Stars and Galaxies 159

UNIT 3

EARTH'S AIR AND WATER

Chapter 9 Air 185

Chapter 10 Weather and Climate 207

Chapter 11 Earth's Ocean 233

Chapter 12 Oceanography 251

UNIT 4
SURFACE PROCESSES

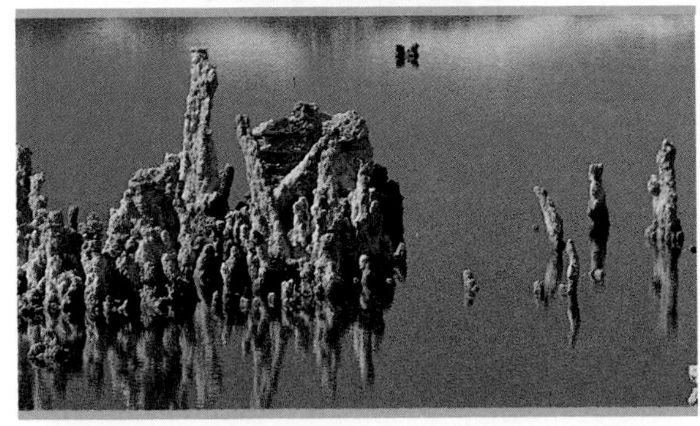

Chapter 13 Weathering and Erosion 279

Chapter 14 Water Systems 305

Chapter 15 Glaciers 325

UNIT 5
ROCKS AND MINERALS

Chapter 16 Minerals 349

UNIT 7

EARTH'S HISTORY

Chapter 21 Clues to Earth's Past 459

Chapter 22 Geologic Time 479

UNIT 8

EARTH'S RESOURCES

Chapter 23 Renewable Resources 507

Chapter 24 Mineral Resources 529

Chapter 25 Energy Resources 551

APPENDICES

Skills and Investigations

UNIT 1

Earth is a dynamic planet. Some changes occur within minutes or hours. Other changes span millions of years. Streams change Earth every day by eroding and depositing sand, gravel, and pebbles. Mountains form over millions of years as the result of forces within Earth. What kinds of processes and forces produce these changes?

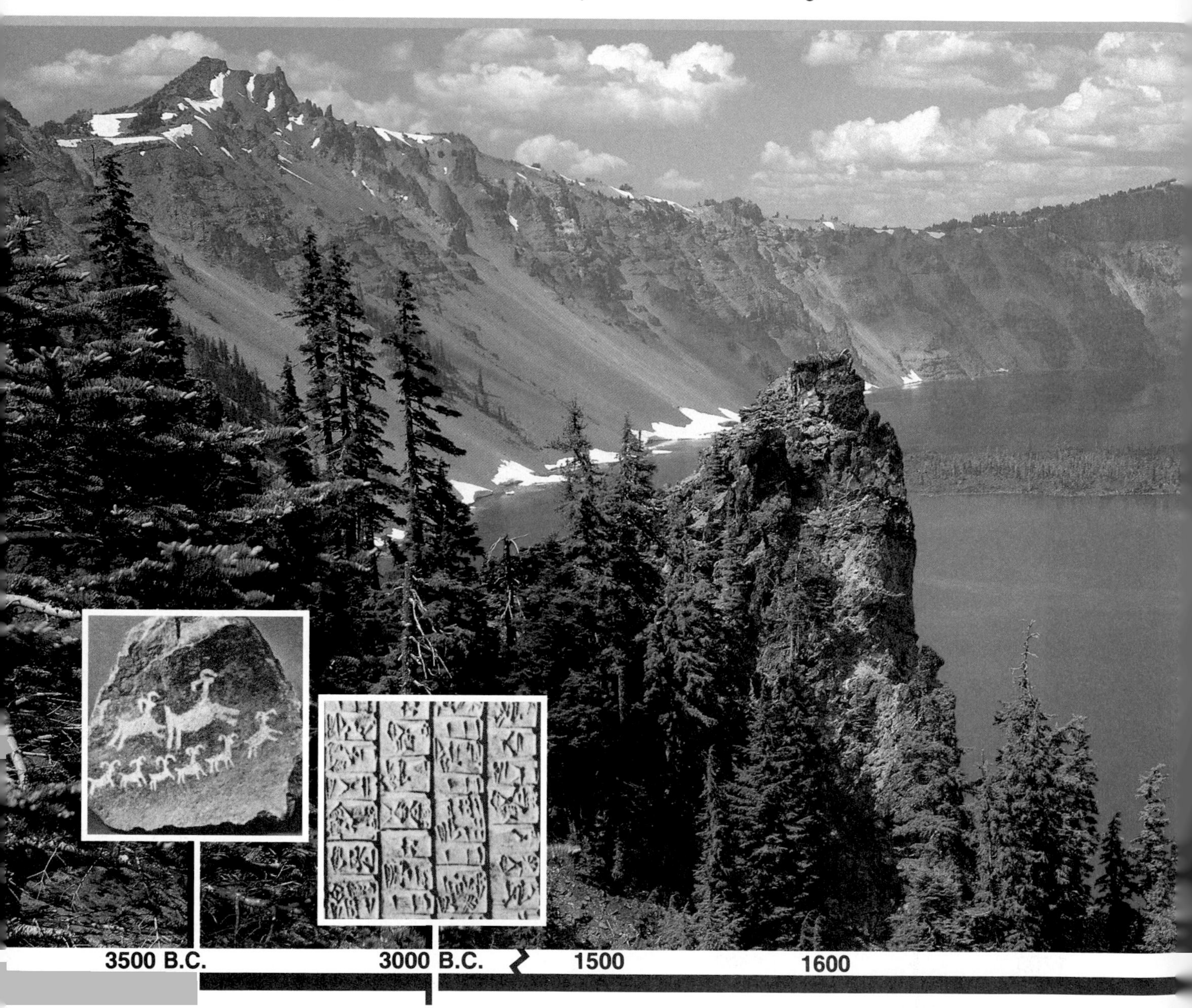

3500 B.C. | 3000 B.C. | 1500 | 1600

~3490 B.C.
Petroglyphs are carved into cave walls.

~3000 B.C.
Hieroglyphics are a form of writing.

1617
Triangulation method of land measurement used.

EARTH SCIENCE FOUNDATIONS

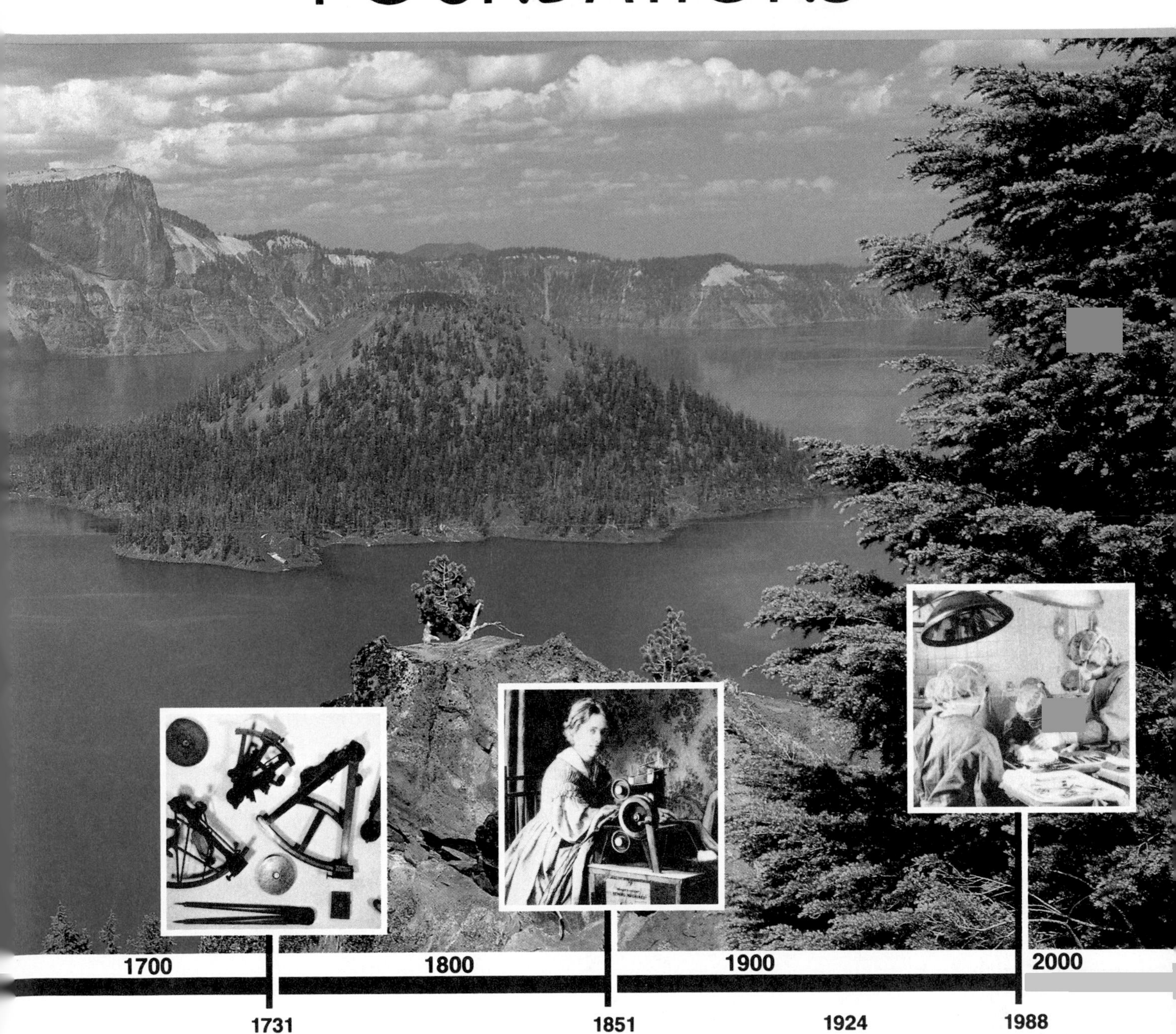

1731
Navigational sextant is invented.

1851
Isaac Singer manufactures first sewing machine.

1924
Walt Disney makes *Alice in Wonderland.*

1988
Modern medicine increases average life expectancy.

PROBLEM SOLVING AND SCIENCE
SCIENTIFIC PROCESSES
WHAT IS SCIENCE?

The Nature of Science

It has been predicted that by the time you are 50 years old, 95% of all human knowledge will have been developed during your lifetime. Throughout the years, people have taken what is known and tried to expand their knowledge. Science is learning and doing. Science means "having knowledge." Much of our scientific knowledge has come from a process of observing and studying problems occurring in the world.

PROBLEM SOLVING AND SCIENCE

GOALS

1. You will learn that solving a problem involves an organized process.
2. You will learn how to solve problems by applying some problem solving strategies.

In order to solve a problem, it is important to approach it in an organized manner. In solving problems, the same path may not be used each time. However, there are methods that are common to most problem solving. Scientists work on problems in a logical way. This is called the scientific method.

1:1 Hot Tub Mystery

After a long jog around the neighborhood, John and Mary Doe decided to relax in their hot tub. While in the water, their eyes began to water and become irritated. After 15 to 20 minutes in the swirling water, they got out to dry off. They noticed that their skin was dry and irritated. They were puzzled about why this occurred. The previous night, they had cleaned the tub and had put in fresh water and new chemicals. John and Mary suspected that the chemicals were irritating their eyes and skin. They had always used the same brand of chemicals and were reasonably sure the amount was correct. However, they took a sample to a laboratory to be tested to see if the chemical composition had changed. When the report

What was the first suspected cause of John and Mary's irritation?

came back, they knew the irritation was not due to a change in composition.

The next time the hot tub was used, eye irritation and skin burning occurred again. Now, John and Mary suspected the water, the amount of the added chemicals, or possibly an allergic reaction. But, Mary doubted she and John would both develop an allergy to the chemicals at the same time. They called the water company to report the incident and asked to have the water analyzed.

Upon contacting the water company, they discovered that there had been other complaints about eye and skin irritation when other people in their area had bathed and showered. Thus, John and Mary were able to rule out an allergic reaction. They then asked the water company to test the water to determine the percent of chemicals in the water from the hot tub, wondering if perhaps they had added too much. They took the chemicals they had used along with a sample of water from the hot tub to the water company. The tests showed that the correct amounts had been used, but 50 parts per billion of benzene were found in the sample. Benzene is a cancer-causing chemical found in gasoline. Dilute concentrations of some chemicals enter the skin far more easily than the pure chemical. Thus, this amount of benzene presented a health problem.

The source of the benzene had to be determined. They thought about possible sources and realized that there was a gasoline station on the corner near their housing development. Mary had recently read an article in the newspaper about leakage from underground storage tanks and wondered if this could be the source of the problem. They called authorities to check the underground storage

FIGURE 1–1. Warm, swirling water in a hot tub is relaxing to tired muscles.

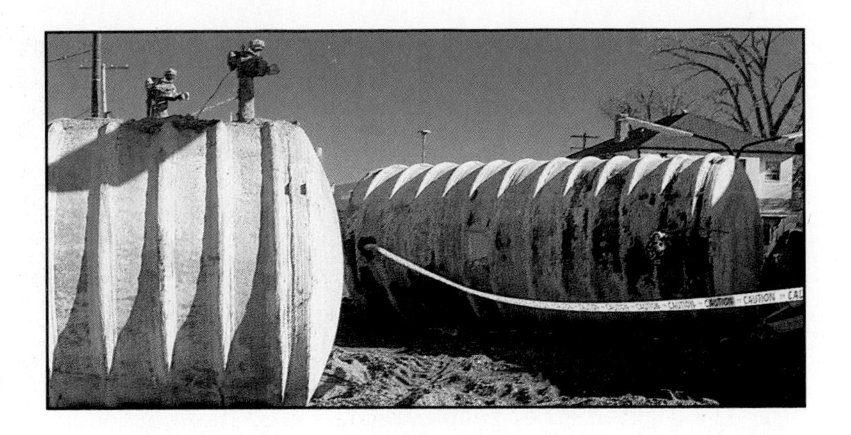

FIGURE 1–2. An underground tank can be used to store gasoline.

tank. The steel tank had been in the ground about 15 years. Upon completion of the test, the tank was found to be leaking. Gasoline is composed of nearly 300 different chemicals, some of which can cause anemia, nervous system disorders, kidney disease, or cancer. Upon further testing, it was determined that the gasoline had seeped into the groundwater source for the town. It was contaminated groundwater that was the cause of the irritation.

What was the cause of John and Mary's irritation?

1:2 Problem Solving

People are faced with problems to solve on a daily basis. How can we solve our pollution problems? How can we find a cure for cancer? How can we find enough time to be involved in extracurricular activities, accomplish home chores, and do homework for school classes? These and other questions are not of equal importance. They do not have direct solutions immediately available. What is now known, and what we need to know are different. Still, a problem exists that needs to be solved. **Problem solving** is a process for finding solutions to problems that face us daily. You have been identifying and solving problems all of your life. Now it is time to identify the processes you have been using.

What is problem solving?

Problems vary in complexity. There is no right or wrong way to come up with a solution to a problem. There is often more than one way to solve a problem. You may choose one or several ways. Sometimes people think they are traveling down paths that lead nowhere but find they learn from their mistakes. In the case study, John and Mary used the strategy of eliminating possibilities. Sometimes it is easier to find out what does not work. They finally determined that the water was contaminated. These various methods are called problem solving strategies.

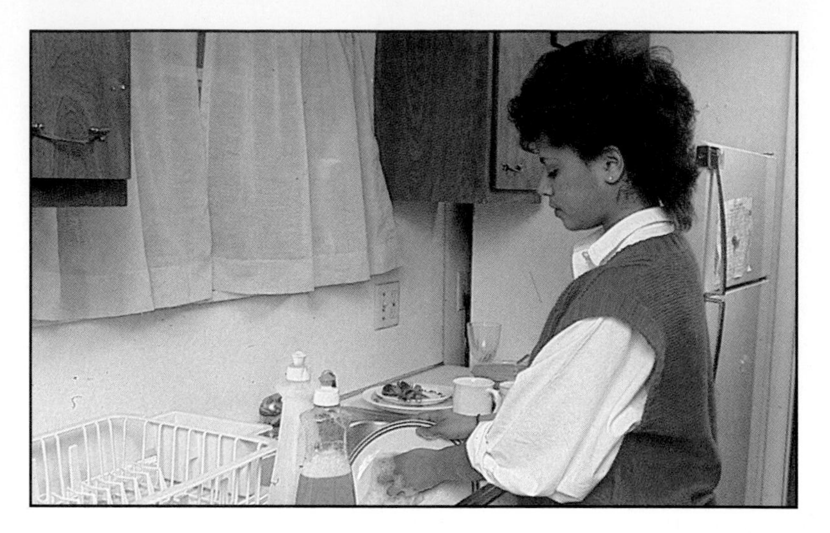

FIGURE 1–3. Which liquid soap can dissolve the most grease from dishes?

1:3 Problem Solving Strategies

To solve a problem, you must have a plan. You must determine the order in which you will move from the known to the unknown. Identifying the problem is the first step. Many times, the problem is stated as a question, such as, "Which brand of soap cleans the most dishes?" Problem solving strategies can be used to solve this problem and many other everyday problems.

One strategy to solve a problem would be to guess and check it out. If you cannot find the solution, make another guess. In the case study, Mary and John first thought their eye and skin irritation might be due to the chemicals added to the water in the hot tub. They checked this possibility and discovered the guess was wrong. Next they guessed the irritation was due to something else in the water. This guess checked out and they were right. If their guess had been wrong, they would have needed to check other possibilities.

Finding answers often depends on the problem, your knowledge, and your approach. It might be easier to work backwards, or solve a simpler, related problem. Looking for a pattern is another way to solve problems. If there seems to be a pattern, predict what will happen next and check your prediction. Another possibility is to make and use a model or drawing, or construct a table or graph. Both physical and mental models can aid us in understanding unfamiliar ideas.

In order to solve problems, you must understand the problem and perhaps restate it in your own words. Determine the goal you are looking for and decide on a strategy. You must have the facts to find your goal. You may have to test your strategy many times to determine the answer. If your strategy does not work, keep trying different strategies. When you find a solution, check to see if you think it is a reasonable answer to the problem.

Review

1. Explain problem solving.
2. Name some problem solving strategies.
★3. Acrylic windows on international airplanes were clouding up with pits and scratches until they became opaque. What are some possible explanations for the problem?

SCIENTIFIC PROCESSES

GOALS

1. You will learn how to use the scientific method to solve problems.
2. You will learn the differences among a hypothesis, a theory, and a law.
3. You will describe the difference between a physical model and mental model.

Have you ever had a problem to solve, but you did not know where to begin? Scientists are constantly trying to solve problems. To do this, they work on the problem in a very logical way. The steps they follow are called the scientific method. You will learn how the scientific method can help you to solve problems.

1:4 Scientific Methods

Solving problems in science involves these four basic steps: (1) determining the problem, (2) testing, (3) analyzing the results, and (4) drawing conclusions. These processes are parts of the **scientific method.** Each of these steps is made up of several processes.

What are the steps of the scientific method?

In order to solve their problem, Mary and John had to first determine what it was that they wanted to know. They made observations about the situation. An **observation** is an act of gathering information using the senses. In the "Hot Tub Mystery," the couple experienced skin and eye irritation. This is an observation. John and Mary also made inferences. An **inference** is a conclusion that is based on both observations and knowledge of the subject.

Define observation.

After they determined the problem, "What is causing the irritation of the skin and eyes?", they did some research. Their observations helped them propose an answer to the problem. A **hypothesis** is a proposed answer

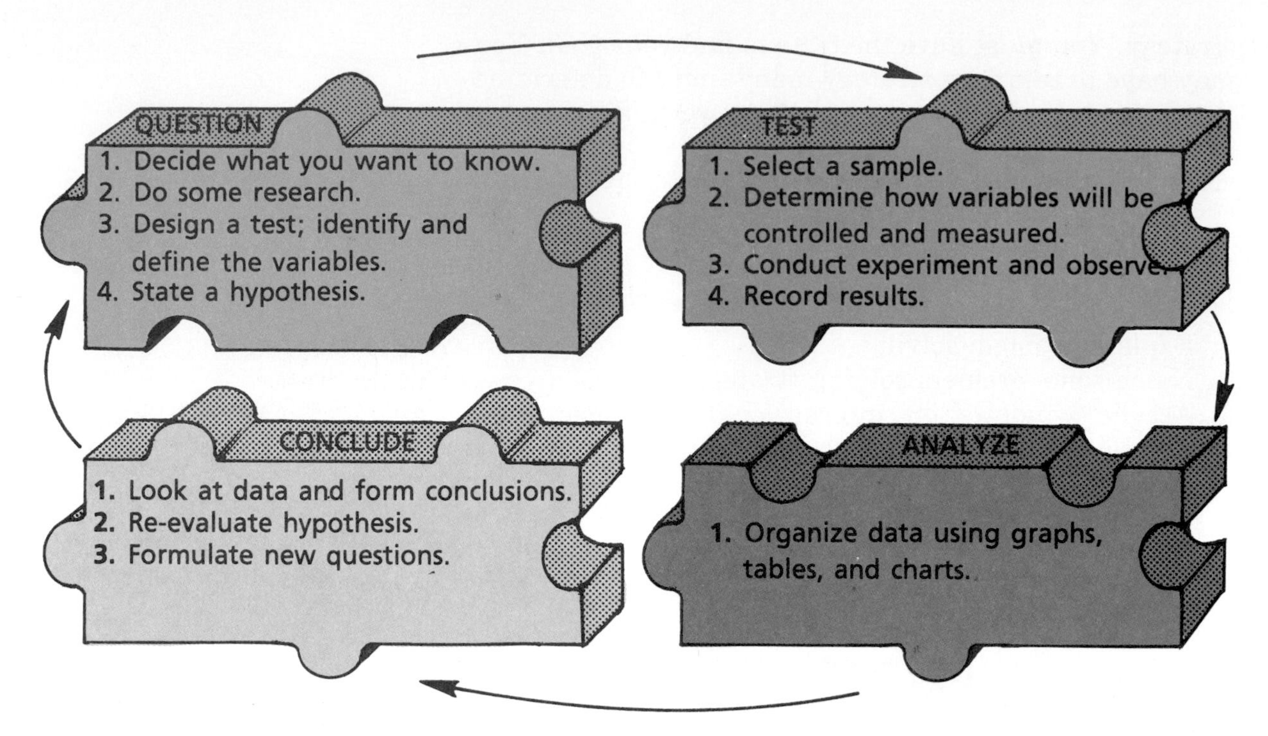

FIGURE 1–4. The processes involved in the scientific method generate both information and questions.

to a problem that can be tested. The couple hypothesized that the water in the hot tub was the source of the irritation. Experiments are often repeated to determine if the same results occur each time. If the results of the repeated experiment support the hypothesis, it is more certain that the hypothesis is correct.

F.Y.I. Aristotle's hypothesis that Earth is the center of the universe was proven wrong by Copernicus 19 centuries after it was made.

Designing an experiment that will test the hypothesis is important. An experiment can either prove or disprove a hypothesis. Water experts tested the water from the hot tub and determined that it was contaminated with benzene. The results of the tests were analyzed and conclusions were drawn. A **conclusion** is an answer to a question which is based on analyzing data and observations gathered in an experiment. The observation of the skin and eye irritation led John and Mary to form hypotheses, eliminate possibilities, guess and check, and finally come to the conclusion that the water was contaminated.

What is a variable?

What is a control?

Experiments involve variables and controls. Changeable factors in an experiment are called **variables.** It is usually not possible to test all variables in an experiment. The best experiments test only one variable. A **control** is a standard for comparison. A control is used in many experiments to show that results are not due to some other condition. An example of a control would be to perform an experiment or test a hypothesis at a controlled, exact temperature.

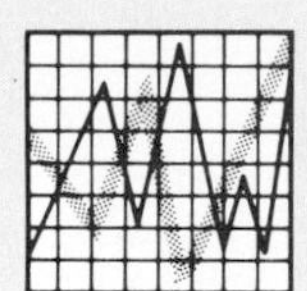

SKILL 1–1

Observations and Inferences

Problem: How do observations differ from inferences?

Materials

identical sheets of paper (6)
paper
pencil

Procedure

1. With a partner, take two sheets of paper, place them together, and fold them in half one time.
2. Take the remaining four sheets and place them together, and fold them in half one time.
3. Now drop both sets of paper from eye level at the same time. Drop them several times and then record your observations and inferences in a data table. You and your partner should record your own observations and inferences separately.
4. Next, fold the two sheets in Step 1 in half again. Repeat Step 3. Record your observations and inferences.
5. Fold the two sheets in Step 1 in half again. They should now be folded three times. Repeat Step 3. Record your observations and inferences.
6. Fold the two sheets in Step 1 in half again. They should now be folded four times. Repeat Step 3. Record your observations and inferences.

Data and Observations

Trial	Observations	Inferences
1		
2		
3		
4		

Questions

1. Did your observations change as you tried new tests with the papers?
2. Did your inferences change as you tried new tests with the papers?
3. Was it easier to make observations or to make inferences?
4. When you kept folding the papers, what factors did you change?
5. Can you make inferences without first making observations?
6. What else do you need to have before you can make an inference?
7. Supposc you see little black lines in a rock. Is this an observation or an inference?
8. Suppose that you conclude that the lines in the rock were made when it broke away from a larger rock, or that someone scratched them into the rock with a pencil. Are these observations or inferences?
9. Why might your inferences be quite different from another person's?
10. Compare your data chart with your partner's. What did you both observe?
11. What did you both infer?
12. How are observations and inferences important in solving a problem?
13. List an observation that you made today.
14. List an inference that you made today.

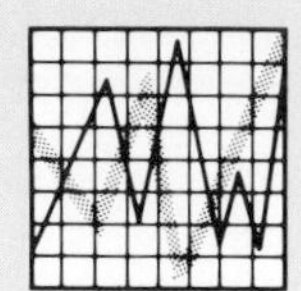

SKILL 1–2

Determining Variables and Controls

Problem: How can you determine variables and controls?

Materials  

large beaker (1)
small beakers (3)
water
sugar
hot plate
stirring rods (3)
spoon
paper
pencil

Procedure

1. Bring water to a gentle boil in the large beaker on the hot plate.
2. Half fill the first beaker with this boiling water.
3. Half fill the second beaker with warm tap water.
4. Half fill the third beaker with cold water.
5. At this time, you and two lab partners should each put one spoonful of sugar into the beakers at the same time.
6. Stir and observe each beaker. Record your observations.

Questions

1. What was the variable in the experiment above?
2. What was the control?
3. How did the variable affect the experiment?
4. Why was it important to put the same amount of sugar in each beaker?
5. Why was it not necessary to record the exact temperature of the water?

Use the information in Paragraphs A, B, C, and D to answer the questions.

A. Suppose you want to know at which temperature a particular dishwashing liquid dissolves the most grease.

6. What would be your variable?
7. What would be your control?

B. Sometimes an experiment will have a control sample that is just like the test sample except for the one factor that is being tested. For example, if you want to know if a stream has been polluted by a particular factory, you would collect water both downstream and upstream from the factory.

8. Which sample would be the control?
9. Why is a control sample necessary in this experiment?
10. What other controls should you have in order to make this experiment valid?

C. Another way of controlling an experiment is to repeat it several times to see if the results are the same each time. Also, you should test many samples, not just one or a few, so you will know that your results are representative.

11. If you want to determine the effect of atmospheric pressure on how high a tennis ball bounces, why should you test more than one ball?
12. If you want to know if the air in your city was polluted, why would you test several sites and repeat your testing several times?

D. Your science teacher asks you to determine the effect of light on plant growth. You are given a light and 5 small plants.

13. What is the variable?
14. How could you control the variable in this experiment?
15. Would it be possible to have exactly the same control for each plant used?

During an experiment, observations must be made and results must be recorded. After the test has been completed and the data recorded, the results are analyzed. Conclusions are made based on observations. The hypothesis can be reexamined in light of the test results. Scientists constantly revise their ideas as they gather data through experiments and research. Often, the discovery of more data shows that a hypothesis is wrong and needs to be changed. Thus, people must keep an open mind as they experiment and search for answers.

1:5 Theories and Laws

All scientists have a common goal, to find out about something in the universe. Scientists are always testing their hypotheses. When there are more data from tests to support the explanation, a scientist is more convinced that the hypothesis is correct. An explanation backed by results obtained from repeated tests or experiments is a **theory.** A hypothesis may become a theory when all new data gathered over a long period of time support it.

How does a hypothesis differ from a theory?

If a theory has been sufficiently tested and validated, it becomes a **law.** An example of a law is Newton's first law of motion. It states that an object continues at rest, or in motion, until acted upon by an outside force. This law began as Newton's hypothesis 300 years ago. He and others tested and retested it, and it became a theory. This theory has withstood the test of time and is now considered a law. However, if future scientists find that it is not always true, then the law will no longer be accepted.

F.Y.I. You will learn about Newton's other laws in Unit 2.

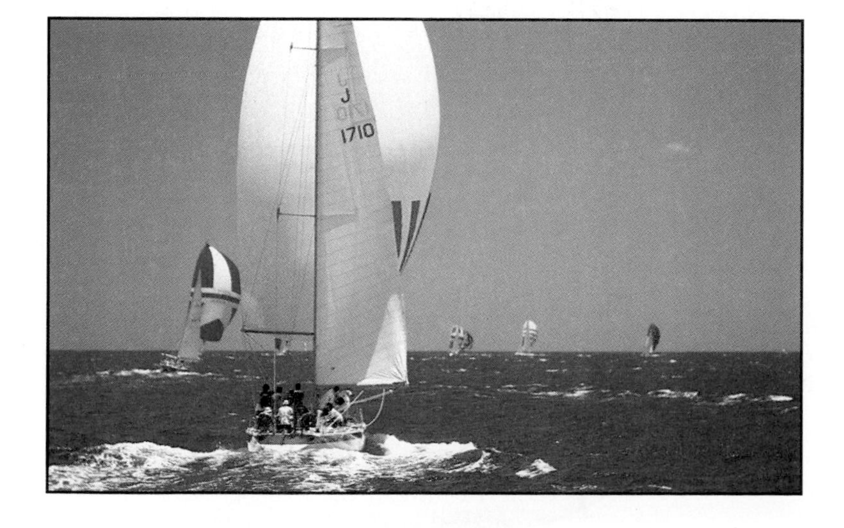

FIGURE 1–5. Sailboats in motion demonstrate Newton's first law.

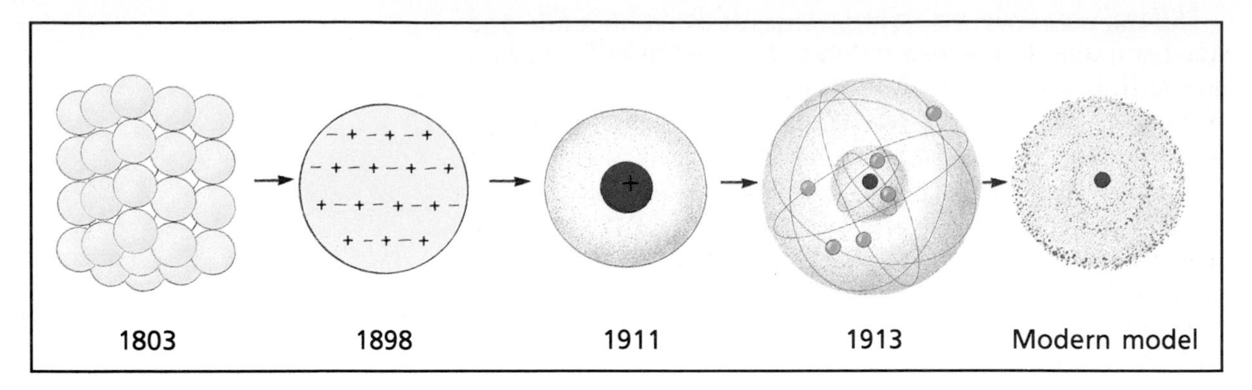

FIGURE 1–6. The model of the atom changed as new facts were gathered.

1:6 Models

What is a model?

Often, scientists make models that help them test a hypothesis or demonstrate a theory. A **model** is a representation of an actual object or an idea of how an object looks. A scale model represents a real object reduced in size. The **scale** is a fixed ratio between the size of the model and the size of the real object. For example, if a model river is 3 meters deep and the actual river is 30 meters deep, the ratio is 3:30 or 1:10. One unit of measure of depth on the model equals 10 units of the same measure on the actual river.

Scientists use two types of models. **Physical models** represent actual objects. Scale models of airplanes, ships, buildings, and maps are physical models. Maps are good examples of physical models that have practical uses. Imagine how hard it would be to find your way in a large, unfamiliar city without a map.

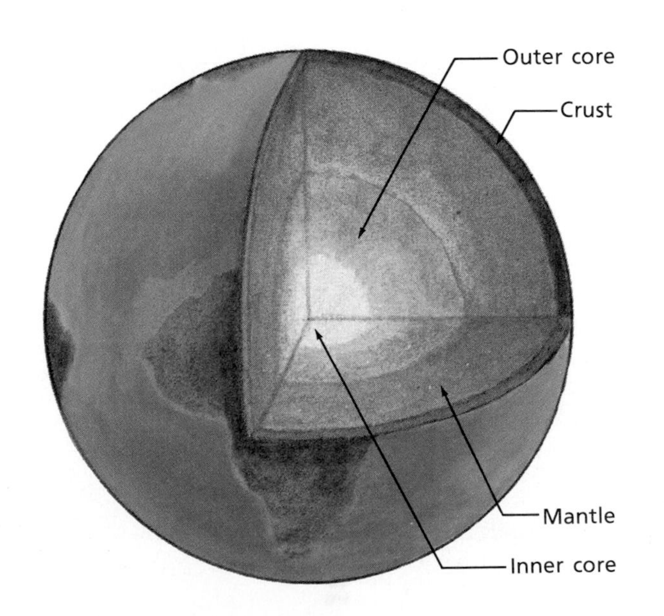

FIGURE 1–7. This model of Earth has been tested and is currently considered accurate.

Mental models are ideas of how objects look. No one has been able to see much detail of an atom because it is so small. But scientists do have a good idea of its structure from observing the behavior of atoms. Thus, scientists have constructed a mental model of the atom. Niels Bohr proposed a model of the atom in 1913. Over time, Bohr's model has been replaced by newer models as data obtained from experiments showed that Bohr's idea was not correct (Figure 1–6).

Today, computers are widely used in making both physical and mental models. Computers make the task of updating changes easier by making the changes faster and more accurately. A physical model of Earth that was made with the aid of a computer is shown in Figure 1–8.

Review

4. What are four steps in the scientific method?
5. What is a control? What is a variable?
6. How does a theory become a law?
★ **7.** To build a scale model of your house, what measurements must you know?

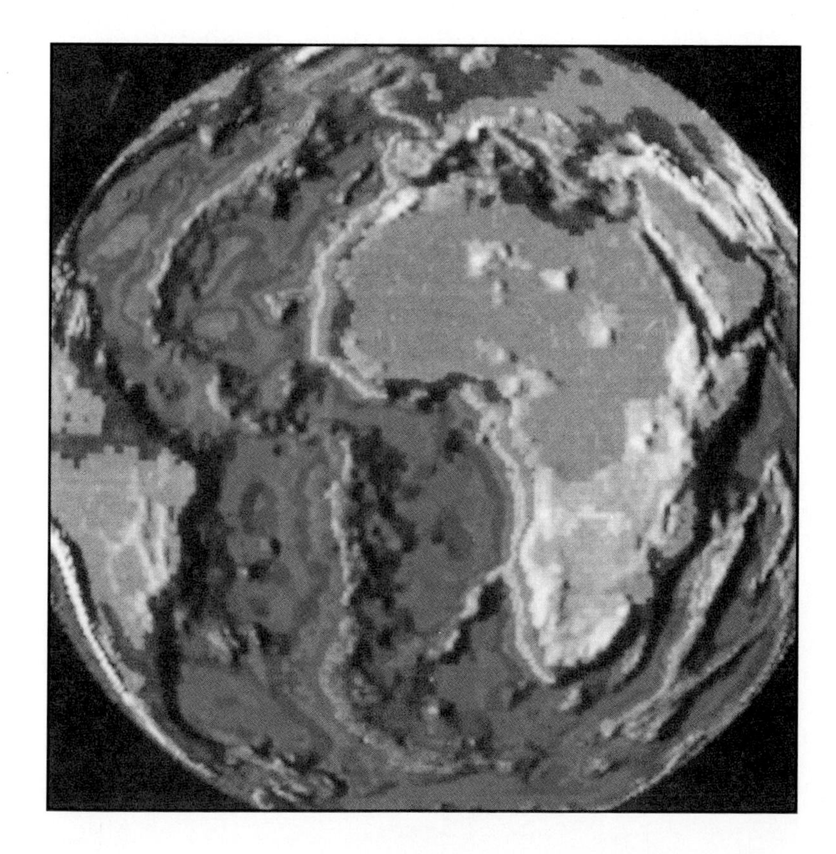

FIGURE 1–8. This model of Earth was generated by a computer using data collected by ocean- and land-based surveys.

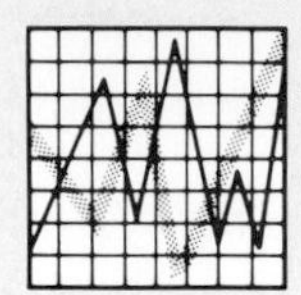

SKILL 1–3

Constructing Models

Problem: How do you construct a scale model?

Materials

Figure 1–9
pencil
metric ruler
paper
mathematical compass
modeling clay

Procedure

1. You will be constructing a scale model of a hill. Determine the height that you want your model to be. Record this value on your paper.
2. To determine the scale, divide the proposed height of your model into the height of the actual hill. See Figure 1–9. Record your answer.
3. Now you must decide how wide to make the base of your model hill. To do this, divide the width of the actual hill, shown in Figure 1–9, by the number you computed in Step 2. Round off your answer to the nearest tenth of a centimeter. Record your answer.
4. Use the metric ruler to draw a line on your paper that equals the distance you calculated in Step 3.
5. Use your compass to bisect the line. Then set the point of your compass on the center of the line. Adjust the pencil end of the compass until it touches the end of the line. Make a circle with this radius. This circle represents the base of your model hill.
6. Now begin constructing your hill with clay. First fill in the circle, and then build up the hill to the height you determined in Step 1. Shape your model as shown in Figure 1–9.
7. When you have completed your model, compare it to Figure 1–9. Do they look the same? If not begin again at Step 2.

Questions

1. How high would your model have to be if the scale were 1:10?
2. How wide would the base of your model have to be using the 1:10 scale?
3. Suppose you wanted your model to be exactly 10 centimeters high. What scale would you use?
4. How do you construct a scale model?

FIGURE 1–9. Side view (a) and top view (b) of a hill

a

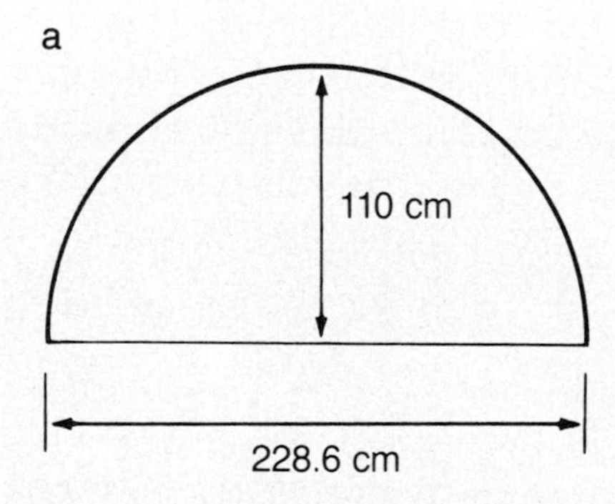

b

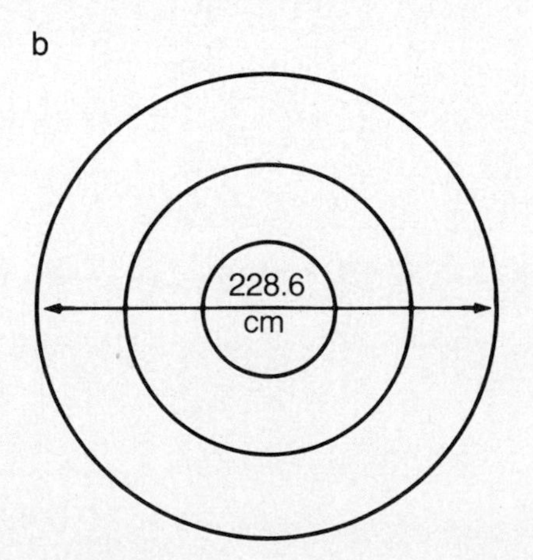

WHAT IS SCIENCE?

GOALS

1. You will learn about the relationship between science and technology.
2. You will learn how earth science is related to chemistry, physics, and biology.

Science is a process that produces knowledge about the physical world. Many natural events can be explained by using scientific methods. Using methods of science, people arrive at solutions to problems or are able to answer questions about the natural world. Is science finding a new star or proposing a new theory to explain gravity? Is it inventing a new telescope? Is science discovering a living fish that was thought to have disappeared millions of years ago? Science aids in discovering answers to many questions.

What is science?

1:7 Science and Technology

With each generation, data about how the universe works are gathered. The concepts you are learning in science are based on evidence collected over many years. People have progressed from making observations with the unaided eye to using complex instruments. Microscopes magnify small objects. Some satellites take photos of Earth and other planets. Telescopes probe the depths of space. These instruments allow people to make detailed observations of the natural world. Yet, many processes taking place in the universe still are not fully understood.

a

b

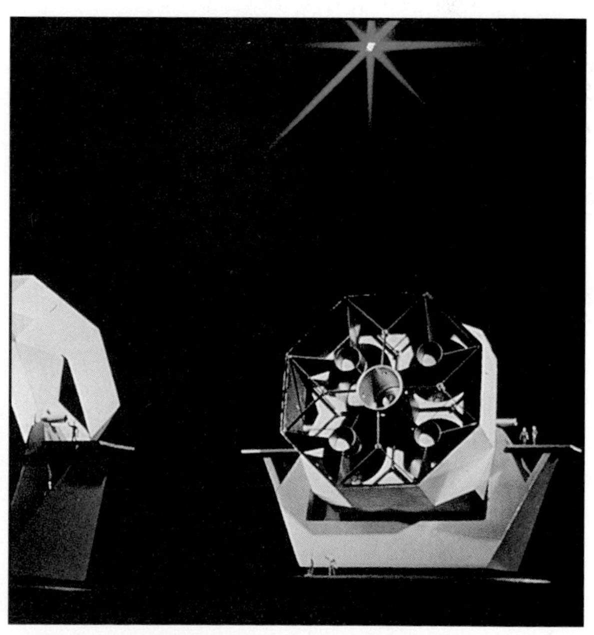

FIGURE 1–10. Throughout time, people have used different tools to gather knowledge. Stonehenge (a) was built as an observatory centuries ago. The National New Technology telescope will help scientists probe space (b).

BIOGRAPHY

Katherine Johnson
1918–

Katherine Johnson is an aerospace technologist for NASA. One of her projects is the Earth Resources Satellite, used to locate minerals and other Earth resources. In 1977, Johnson received the Group Achievement Award for work on the Lunar Spacecraft and Operations Team.

A **scientist** is a person who uses scientific methods to learn about and explain natural events. People working in each branch of science gather data about that science. The relationship of each to earth science is described in the paragraphs that follow.

Chemistry is the study of the properties and composition of matter. Everything on Earth and in the universe is composed of matter. You will need to understand some basic concepts about matter in order to understand rocks and minerals.

Physics deals with forces, motion, and energy and their effects on matter. You will learn some of the principles of physics as you study the motions of Earth and the moon, the properties of starlight, and the circulation of ocean currents.

Biology is the study of living organisms. Life exists on Earth and perhaps at other places in the universe. You will consider some biological concepts as you learn about Earth's history, resources, and the environment.

Earth science is the study of planet Earth and its place in space. As you will see, earth science includes data from other branches of science. Biologists, chemists, and physicists all contribute to our knowledge of Earth and its history.

Technology is the use of scientific discoveries. It is applied science. Many of the things we take for granted are the results of technology. Modern health care, food processing, jets, telephones, satellites, and many other conveniences are all technologies resulting from scientific knowledge.

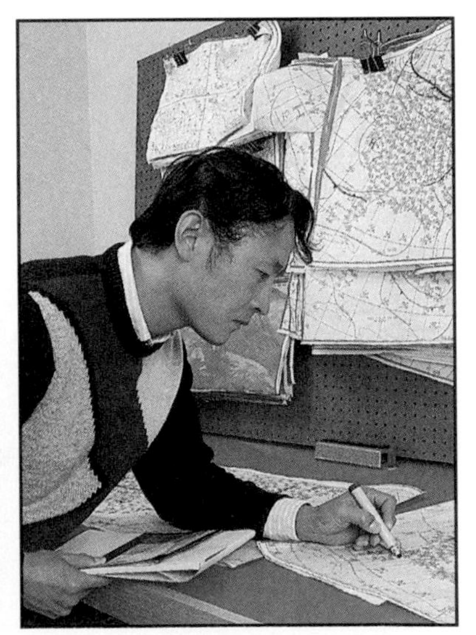

FIGURE 1–11. Scientists gather data to learn about and explain planet Earth.

TECHNOLOGY: APPLICATIONS

Road Building

The first roads were built by the Romans to allow armies and horses to have speedy access between cities. Queen Elizabeth I of England was the first to charge tolls, or fees, which were used for repair of local roads. It was not until the end of the 18th century that scientific principles were applied to road construction.

Today, before a road is constructed, a surfacing material must be chosen. A thorough study of the soils and rocks is made by boring holes into the ground to collect samples. Climatic conditions, such as annual amounts of fog, frost, and rain, are also researched. The land is carefully surveyed and detailed sketches are drawn.

Typically, roads today are made of a layer of gravel or crushed rock that has been covered with either blacktop or concrete. Blacktop, also called asphalt, is made from heavy oil residues and crushed rock, sand, or gravel that are heated and mixed. To improve skid resistance on blacktop, a final coating of oil-residue-coated rock chips is spread over the top. When concrete is used, it has joints at intervals of 4.6 meters to enable it to expand and contract with temperature changes. To make it more skid resistant, wet concrete is brushed or grooved to a shallow depth.

Scientists are researching road surface materials that are porous. This will allow rain to drain into the surface, instead of staying on top where it can cause tires to skid. The Strategic Highway Research Program is presently studying salt corrosion problems, pothole prevention, and pavement failure. They are experimenting with plastics, fibers, and chemicals that will prolong road life. The goal of this program is to develop better tests and specifications for asphalt products, so that product strength and durability are assured.

CAREER

Earth Science Teacher

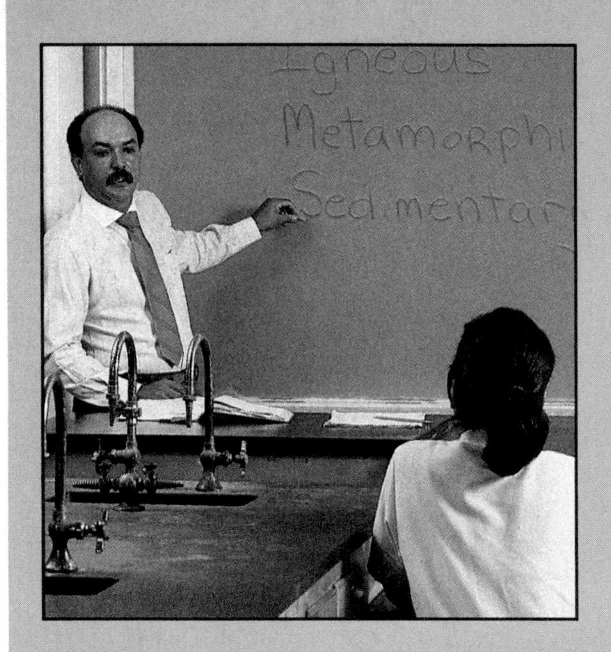

Samuel Lawson teaches earth science in a public school. His goal is that each of his students be successful in learning about Earth. He uses lectures, discussions, demonstrations, lab investigations, and audiovisual aids to instruct students about astronomy, meteorology, geology, and oceanography. Mr. Lawson prepares a teaching outline of his course of study, assigns lessons, and corrects student papers.

His duties also include administering tests in order to evaluate student progress, recording results, and sending progress reports to parents or guardians. Mr. Lawson keeps attendance records, maintains discipline, and participates in faculty and professional meetings, educational conferences, and teacher workshops. He performs extra duties, including helping to coordinate the school science fair, advising the science club, and monitoring the school computer lab.

For career information, write:
The National Science Teachers Association
1742 Connecticut Ave., N.W.
Washington, DC 20009

1:8 Earth Science

Many different types of scientists study Earth, its history, and its place in the universe. These scientists have a common goal. They strive to understand Earth and the processes that are constantly changing it. Earth scientists are working to predict earthquakes and volcanic eruptions, to warn people of severe weather, and to better use our resources. In general, there are four major areas of specialization in earth science.

What is geology?

Geology is the study of Earth, its matter, and the processes that form and change it. Among other tasks, geologists search for oil, study volcanoes, identify rocks and minerals, study fossils and glaciers, and determine how mountains form.

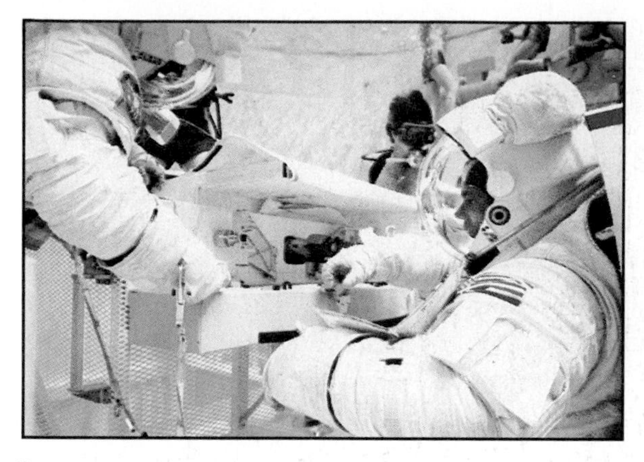

a

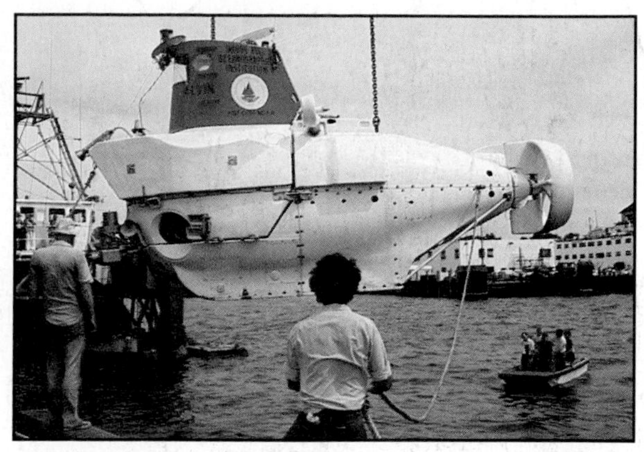

b

FIGURE 1–12. Astronauts must have knowledge in many areas of science (a). Oceanographers use submersibles like *Alvin* to study the ocean floor (b).

Astronomy is the study of objects in space, including stars, planets, and comets. Before telescopes were invented, this branch of earth science mainly dealt with descriptions of the positions of the stars and planets. Today, scientists who study space objects seek evidence for the beginning of the universe. The study of astronomy helps scientists understand Earth's origin.

Meteorology is the study of weather and the forces and processes that cause it. A meteorologist is a scientist who studies storm patterns and climates in order to predict daily weather. Some meteorologists are trying to determine how to change certain weather patterns and to predict the results of such changes.

What is oceanography?

Oceanography is the study of Earth's oceans. Scientists who study the oceans conduct research on the physical and chemical properties of ocean water. Oceanographers also study the processes that occur within oceans and the effects humans have on these processes.

Some earth scientists work in fields that are combinations of the areas above. Table 1–1 on page 22 lists just a few of these specialized areas of earth science.

Review

8. What is technology?
9. What are two of the topics that would be studied by a meteorologist?
10. What is earth science?
★ **11.** If you were a scientist specializing in meteorology, how could you use technology to help other people?

What is geomorphology?

Table 1–1

Specializations in Earth Science	
Earth science area	**Subject studied**
astrogeology	geology of the moon, planets, meteorites and other cosmic objects
paleontology	fossils and Earth history
geomorphology	Earth's surface features and their origins
petrology	rocks
mineralogy	minerals
paleoclimatology	ancient climates and weather patterns
geophysics	Earth's internal structure and processes
tectonics	effects of internal processes on Earth's surface, including ocean formation and mountain building
geochemistry	Earth's composition

PROBLEM SOLVING

The Pepper Chase

It was Gerome's turn to wash the dinner dishes. He filled the sink with water and slowly added the plates and silverware. His family had had peppered beef, mashed potatoes, and green beans for dinner. As the plates soaked, pepper from the beef came loose from the plates and began to float on top of the water.

Gerome's little sister, Dawn, came in from playing outside and began washing her hands over his clean dish water. The bar of soap slipped from her hands and dropped into the sink. Gerome looked into the sink and saw that the pepper was rapidly moving away from the soap. He hypothesized that soap repels floating grains of material. Was Gerome's hypothesis correct? Explain.

Chapter 1 Review

SUMMARY

1. Using science, people find solutions to problems and questions about the natural world. 1:1
2. There is no right or wrong way to solve a problem. 1:2
3. Problem solving strategies include checking possibilities, solving related problems, or using models. 1:3
4. Determining the problem, testing, analyzing the results, and drawing conclusions are parts of the scientific method. 1:4
5. A hypothesis becomes a theory when results from repeated tests support the hypothesis. A theory becomes a law when it has been sufficiently tested and validated. 1:5
6. Scientists use models to represent an actual object or an idea of how an object looks. 1:6
7. Technology is applied science. 1:7
8. Geology, astronomy, meteorology, and oceanography are the four major areas of specialization in earth science. 1:8

VOCABULARY

a. astronomy
b. biology
c. chemistry
d. conclusion
e. control
f. earth science
g. geology
h. hypothesis
i. inference
j. law
k. mental models
l. meteorology
m. model
n. observation
o. oceanography
p. physical models
q. physics
r. problem solving
s. scale
t. scientific method
u. scientist
v. technology
w. theory
x. variables

Matching

Match each description with the correct vocabulary word from the list above. Some words will not be used.

1. four basic steps used by scientists to find answers to problems
2. changeable factors in an experiment
3. a proposed answer to a question or problem
4. person who uses scientific methods as an approach to learning about and explaining natural events
5. a standard for comparison
6. applied science
7. results when a hypothesis has been supported by repeated tests
8. representation of actual object or idea of how an object looks
9. study of objects in space
10. fixed ratio between size of a model and size of the real object

Chapter 1 Review

MAIN IDEAS

A. Reviewing Concepts

Choose the word or phrase that correctly completes each of the following sentences.

1. The branch of science that studies the composition of matter is *(chemistry, physics, biology)*.
2. A scientist who predicts and studies storms is a(n) *(astronomer, meteorologist, geologist)*.
3. When observations do not support a hypothesis you should *(use the hypothesis and look for other observations that will support it; make a theory; alter your hypothesis so it is supported by the observations)*.
4. An explanation backed by results from repeated tests is a *(hypothesis, theory, control)*.
5. A model of an idea of how something looks is a *(physical model, mental model, theory)*.
6. If you want to determine how long you need to soak dishes in order to remove grease, time is a *(control, variable, constant factor)*.
7. If you want to determine how long you need to soak dishes in order to remove grease, the amount of water you use is a *(control, variable, hypothesis)*.
8. The ideal number of variables in an experiment is *(one, two, five)*.
9. Sally sees that Jeff has purple spots on his shirt. This is a(n) *(inference, observation, hypothesis)*.
10. After seeing Jeff's spotted shirt, Sally guesses that his pen broke. This is a(n) *(inference, observation, hypothesis)*.
11. A process for finding solutions to problems is called *(conclusions, hypotheses, problem solving)*.
12. The first step in the scientific method is *(running the test, defining the variables, determining the question)*.
13. Charts, graphs, and tables are used when *(determining the problem, testing, analyzing the results)*.
14. An example of a physical model is *(a model car, a model of the universe, a model of an atom)*.
15. A geologist is most interested in studying *(Earth's oceans, the laws of motions, Earth and its processes)*.

B. Understanding Concepts

Answer the following questions using complete sentences.

16. Describe how a hypothesis becomes a theory.
17. How does a theory become a scientific law?
18. Explain why an earth scientist needs to know information from other branches of science.
19. What is the difference between a variable and a control?
20. Explain the difference between a physical model and a mental model.
21. Explain why an experiment should test only one variable.
22. Describe the four basic steps of the scientific method.
23. Explain how technology is related to science.
24. What is the first thing that needs to be done in order to solve a problem?
25. What is the difference between an observation and an inference?

C. Applying Concepts

Answer the following questions using complete sentences.

26. Why is a hypothesis that is unsupported by observations rejected by scientists?
27. How does the scientific method help scientists in their research?
28. How would you determine which flashlight battery lasts the longest?

29. State a problem and determine a plan to find the solution.
30. Why would scientists who are studying flood control of a particular river want to make a scale model of the river and its valley?

SKILL REVIEW

1. The scale on a map is 1:62 500. If two cities are 2 centimeters apart on the map, how many centimeters are they apart on Earth?
2. You suspect that it is raining outside when your dog comes in with wet fur. Is your guess that it is raining an observation or an inference?
3. If you wanted to study the reaction time of people, would it be more scientific to test a few people many times, test many people a few times, or test many people many times? Explain your answer.
4. If you wanted a model train to be 20 centimeters long, and the actual train is 1740 centimeters long, what scale will you use?
5. Suppose that you want to know whether cold water causes fish to breathe slower than warm water. How would you set up your experiment? What would be your variables? What would be your control?

PROJECTS

1. Use polystyrene, cardboard, balsa, or clay to make a scale model of your home, apartment, or school.
2. Devise a simple experiment and perform it. Be sure to use all steps of the scientific method.

READINGS

1. Diamond, Jared. "Soft Sciences Are Often Hard." *Discover*. August, 1987, pp. 34-39.
2. Medawar, P. B. *The Limits of Science*. New York: Harper & Row, 1984.
3. Weiss, Anne. *Seers and Scientists*. San Diego: Harcourt Brace Jovanovich, 1986.

MEASUREMENT
USING SI
LABORATORY AND SCIENCE SKILLS

Chapter 2

Earth Science Skills

In earth science, you will be using many skills, including measuring, laboratory skills, and graphing. Reporting of measurements and other data must be understood by other people. The International System of Units is a standard system of measurement used by all scientists. You will learn about this system and how to use it.

MEASUREMENT

How did scientists determine that the temperature on the bright side of the moon is about 120°C? They investigated and measured. When making measurements, scientists use the International System of Units (SI), which is a modern form of the metric system.

GOALS

1. You will learn to recognize measurements in terms of two parts—units and numbers.
2. You will learn to identify SI units and understand prefixes.

2:1 Scientific Measurement

When making a measurement, an object is compared to a standard. All **measurements** include a unit of measure and a number stating how many of the units are present. For example, if someone asked how far you lived from school, you would use both a unit and a number to reply. If you live three kilometers away, the unit of measure is kilometers, and three is the number of these units.

What two components make up a measurement?

Measurements include length, mass, weight, volume, temperature, and time. The length of your classroom can be measured with a meter stick. Temperature can be measured with a thermometer. Mass can be measured with a balance or a scale.

A standard is an exact quantity people agree to use for comparison when measuring. For example, if you and a classmate each use a meter stick to measure the distance from the bulletin board to the teacher's desk, you each should get close to the same results. Standards must be the same all over the world.

CAREER

Precision-Instrument and Tool Maker

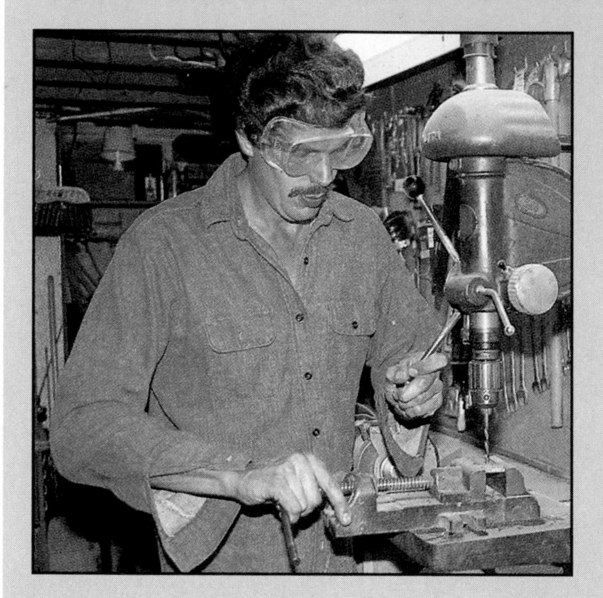

Jim Snyder is a precision-instrument and tool maker. He makes, modifies, and repairs mechanical instruments or mechanical assemblies of electrical or electronic instruments such as timing devices, thermostats, and seismographs. He applies knowledge of mechanics, metal properties, shop mathematics, and machining procedures in his work. Sometimes Mr. Snyder must measure, mark, and scribe such materials as silver, nickel, and plastic.

In order to make some instruments, he must install wiring and electrical components to certain specifications. For all of the instruments and tools he makes, he must verify the dimensions of parts and the installation of components. To do this, he uses measuring instruments such as micrometers, calipers, and electronic gauges.

For career information, write:

International Association of Tool Craftsmen
1915 Arrowline Ct.
Bettendorf, IA 52722

What is SI?

2:2 International System of Units

Since 1670, several versions of the metric system have existed. The original metric system used three basic units of measurement: meter, gram, and second. In 1960, after 12 years of work, the General Conference of Weights and Measures adopted the International System of Units. The **International System of Units (SI)** uses units and standards that are agreed upon around the world. This system is a modern version of the metric system.

FIGURE 2–1. The length of a guitar is about one meter.

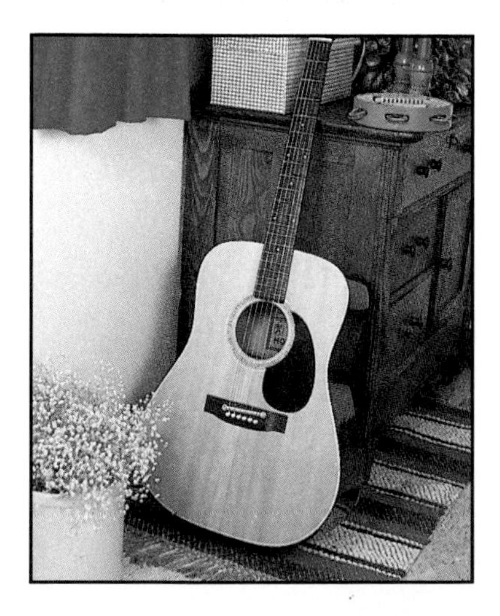

SI is based on a decimal system similar to the money system of the United States, which uses 10 as a base unit. Metric units such as meter, kilogram, second, and kelvin are also used. These are units of measurement for length, mass, time, and temperature. Smaller or larger units are identified with prefixes.

You have seen many examples of SI units of measure. A milk carton gives the volume in milliliters. A bread wrapper lists mass in grams. Some highway signs give

a

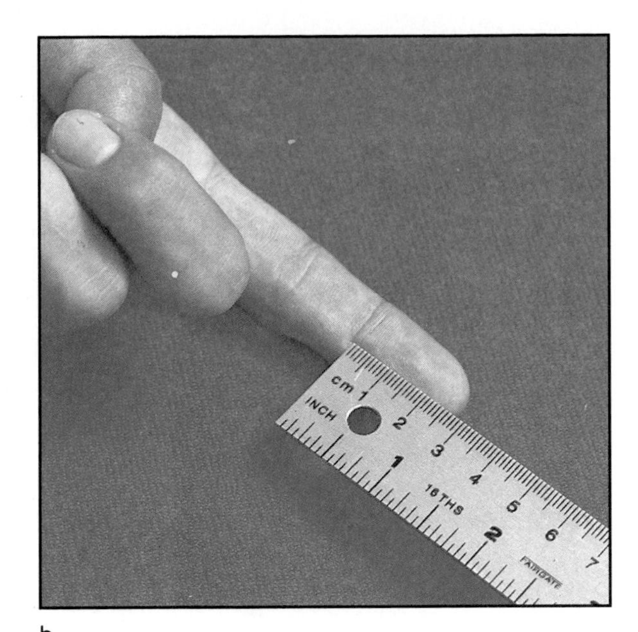

b

FIGURE 2–2. Twenty lengths in an Olympic-sized pool is one kilometer (a). The tip of an index finger is about 2.54 centimeters in length (b).

distances in kilometers. The speedometers of most cars give the speed in kilometers per hour.

The standard unit of distance or length in SI is the **meter** (m). A guitar is about one meter long. Smaller distances and lengths are measured in centimeters (cm). There are 100 centimeters in one meter. Larger distances are measured in kilometers (km). A kilometer is 1000 meters. The standard unit of liquid volume is the **liter** (L). The **gram** (g) is the standard unit of mass. Smaller or larger units of measure in SI are identified by their prefixes. Study Table 2–1 to learn to identify the prefixes.

F.Y.I. The first meter was equal to one twenty-millionth of the length of the meridian that passes through Paris, France. The official standard meter today is based on a certain wavelength of light.

F.Y.I. One kilometer is about five city blocks long.

Table 2–1

Important SI Prefixes

Prefix	Symbol	Multiplying Factor
kilo-	k	1000
deci-	d	0.1
centi-	c	0.01
milli-	m	0.001
micro-	μ	0.000 001

Review

1. Why is it necessary to label all measurements with the units?
★2. Convert one kilometer to centimeters. Convert ten milliliters to deciliters.

GOALS

1. You will learn how length, area, volume, mass, weight, and density are measured.
2. You will learn about time and temperature.

USING SI

Making accurate measurements is an important part of the scientific method. If accurate measurements are not made during an experiment, the results and conclusions are invalid and other scientists cannot duplicate them.

2:3 Length, Area, and Volume

Throughout this course, you will be asked to make measurements of length, area, mass, and volume during skills and investigations. To measure lengths in centimeters (cm) and millimeters (mm), a metric ruler is used. Longer lengths are measured with a meter stick. **Area** is the amount of surface included within a set of boundaries. Area can be found by measuring its length and width, then multiplying the length times the width. Area is expressed in square units such as square centimeters (cm^2).

What is area?

All objects take up space. The space that an object occupies is its **volume.** A volume unit must represent the length, width, and height of an object. Volume units are based on units of length. The cubic meter (m^3) is the basic unit of volume. Because this unit is very large, smaller units such as the cubic centimeter (cm^3) are often used. If an object has a regular shape, an equation can be used to determine its volume.

$$\text{length} \times \text{width} \times \text{height} = \text{volume}$$

Liquid volume measurements are made using graduated cylinders and beakers. These volumes are usually expressed in liters (L) or milliliters (mL). One milliliter of a liquid will just fill a container with a 1 cm^3 volume. Thus, milliliters can be expressed as cubic centimeters.

a

b

FIGURE 2–3. The area of this box can be found by measuring its length and its width (a). Graduated cylinders and flasks are used to measure liquid volume (b).

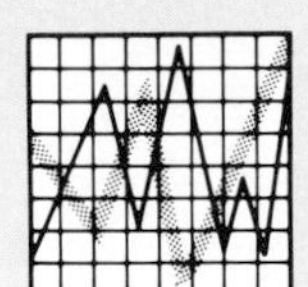

SKILL 2–1

Determining Length, Area, and Volume

Problem: How are length, area, and volume determined?

Materials

shoe box
coffee can
small rock
metric ruler
graph paper
string
graduated cylinder (100 mL)
water

Procedure

1. Measure and record the length, width, and height of the shoe box using the metric ruler.
2. Calculate and record the area of the top, side, and end of the shoe box using the equation: *area* = *length* × *width*.
3. Calculate and record the volume of the shoe box using the equation: *volume* = *length* × *width* × *height*.
4. Measure and record the distance across the top of the coffee can. Be sure that you measure the diameter of the circle.
5. Measure and record the height of the can.
6. Calculate and record the area of the top of the can using the equation: *area* = π × *radius* × *radius*. The radius of a circle is one-half the diameter. Pi, π, is a constant used by scientists approximately equal to 3.14.
7. Calculate and record the volume of the coffee can by using the equation: *volume* = π × *radius* × *radius* × *height*.
8. Trace the outline of the rock on a piece of graph paper. Determine a surface area of the rock. Explain your answer.
9. Fill the graduated cylinder half full of water and record the volume in mL.
10. Tie a piece of string around the rock and lower it into the cylinder. Record the volume reading. Remember to express each volume measurement in cm^3 or mL.
11. Remove the rock. Check to make sure that the cylinder has the same volume of water in it as when you started.
12. Subtract the volume of the cylinder with water only from the cylinder with water and the rock. Record the volume of the rock.

Questions

1. How did you determine the volume of the coffee can?
2. Which was easier to measure, the area or the volume of the rock?
3. Why did you need to know the volume of water in the cylinder before you added the rock?
4. How could you determine the volume of an oddly shaped object that floats in water?
5. What area does a house 10 m wide and 15 m long cover?

Data and Observations

Object	Length	Width	Height	Area	Volume
shoe box					
coffee can					
rock					
graduated cylinder					

2:4 Mass, Weight, and Density

What instrument and unit are used to measure mass?

Mass is a measure of the amount of matter in an object. Mass depends on the number and kinds of atoms that make up an object. Mass can be measured using a balance. It can be read directly from the triple beam balance shown in Figure 2–4. On a pan balance, mass is determined by adding known masses to the pan opposite the object being studied. Mass is measured in grams (g) or **kilograms** (kg). The mass of an object remains the same no matter where it is measured in the universe.

What is weight?

Weight is a measure of the gravitational force exerted on an object by another body. Isaac Newton described gravity as a force that pulls every particle of matter toward every other particle. According to Newton, there are two factors that affect the amount of gravitational attraction between two objects. These are the mass of the objects and the distance between them. The greater the mass, the greater the gravitational force. The shorter the distance between two objects, the greater the force.

When you weigh yourself on a scale, it is the force of Earth's gravitational pull on your mass that is being measured. In SI, gravitational force is measured in **newtons** using a spring scale. Weight on Earth varies slightly from place to place depending on the strength of the gravitational force.

F.Y.I. 1 newton = 0.807 kg

Gravitational force is stronger at sea level than at the top of a mountain because an object at sea level is closer to Earth's center. Gravity is stronger at the North Pole than at the equator, because Earth's rotation on its axis affects the gravitational force.

FIGURE 2–4. Weight is measured in newtons using a spring scale (a). A balance is used to measure mass (b).

a

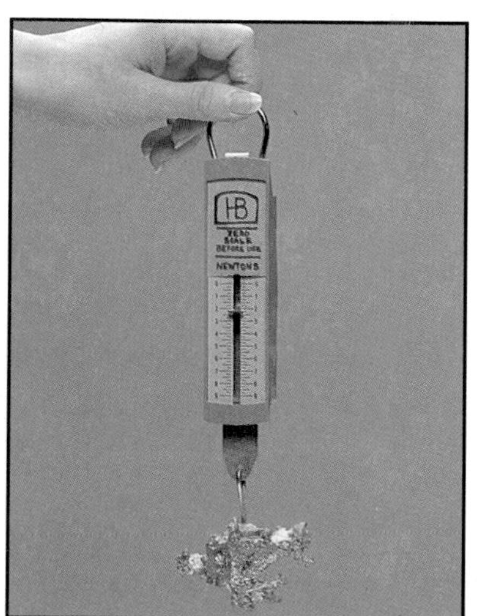

b

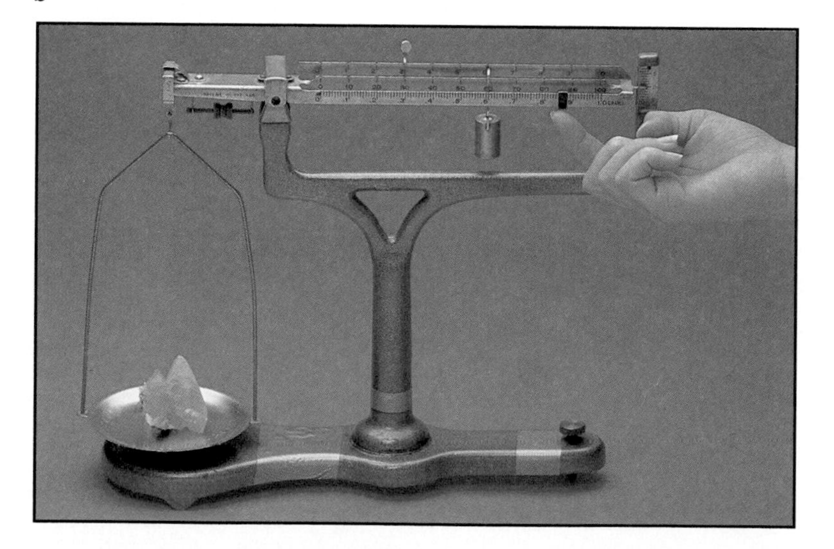

FIGURE 2–5. Time and temperature are important measurements made in science.

The expression of some measurements requires the combination of SI units. These units are called derived units. In order to determine volume, the length, width, and height of an object must be measured. Units for volume are derived from units of length.

Density is the amount of matter in a unit volume of any substance. The density of a material is found by dividing the mass of an object by its volume.

$$\text{density} = \frac{\text{mass}}{\text{volume}} \qquad D = \frac{m}{v}$$

Grams per cubic centimeter (g/cm^3) is a unit often used to express density.

2:5 Time and Temperature

Time is the measurement of the span between two events. Time and change are related because changes occur during the passing of time. The SI unit for time is the second (s). A stopwatch or clock is used to measure how long an event lasts or is observed.

Temperature is measured with an instrument called a thermometer and is recorded in units called degrees. Most scientists use the Celsius temperature scale. The symbol for the Celsius degree is °C. The Celsius temperature scale is based on the freezing and boiling points of water. The freezing point of pure water is 0°C. The boiling point is 100°C. There are 100° between the freezing and boiling points of water on the Celsius scale. A comfortable room temperature is 21°C. The average human body temperature is about 37°C. These two temperatures can be used as reference points when you encounter other Celsius temperatures.

The SI unit for measuring temperature is Kelvin. On the Kelvin scale, absolute zero is 0, which is the coldest

BIOGRAPHY

Garrett A. Morgan
1875-1963

In 1916, Garrett Morgan invented the first breathing device that could be used by people attempting to rescue trapped miners. Before his invention, rescuers were prevented from entering the mines because of the large quantities of smoke, natural gases, and dust. The device, later called the Morgan inhalator, is also used by many manufacturers and fire departments.

possible temperature. The symbol used for Kelvin is K. Although most laboratory thermometers are read in Celsius units, these units can be changed to Kelvin. The conversion to Kelvin can be made by adding 273.16 to Celsius degrees.

$$\text{degrees Celsius} + 273.16 = \text{Kelvin}$$

Review

3. What are examples of SI units of measurement of area and volume?
4. What is the difference between mass and weight?
5. What instrument is used to determine the mass of an object?
★ 6. Why is the force of gravity unimportant when you use a pan balance to measure mass?

PROBLEM SOLVING

Mass Hysteria

Tim wanted to lose weight. He tried several diets, but he found it very difficult to stick to them. He began a rigid exercise program, going to the health spa every day. But, his weight did not change.

After several weeks of dieting and exercise, Tim decided to leave his beach home in Alaska and fly to the equator for a vacation. He thought if he could just quit worrying the problem might go away. His hotel was on the top of a gorgeous mountain with a great view of the rain forest below. He began to relax and enjoy himself. But he was still concerned about his weight, so he weighed himself on the hotel scales. To his delight, he had lost a half of a kilogram.

Upon his return home, he reweighed himself. The half of a kilogram was back. He was perplexed. Was Tim really trying to lose weight or mass? Why did his weight vary in the two different locations?

LABORATORY AND SCIENCE SKILLS

GOALS

1. You will learn safe procedures to use in the science laboratory and classroom.
2. You will learn about first aid.

Laboratory investigations are interesting to perform, and usually follow the scientific method discussed in Chapter 1. When performing investigations, safe practices and methods must be used. Scientific equipment and chemicals need to be handled safely and properly. You will be observing and gathering data while performing investigations and skill activities. You will learn about laboratory safety and graphing.

2:6 Safety Precautions

Being aware of possible hazards and taking sensible precautions can prevent accidents in the science classroom and especially in the laboratory. Safety begins with you. Think about safety at all times. Most injuries in the science laboratory are due to heated objects or splatters and broken glass. The safety rules that follow will help you protect yourself and others from injury.

1. Always obtain your teacher's permission before beginning an investigation or skill activity.
2. Study the procedure. If you have questions, ask your teacher. Be sure you understand any safety symbols shown on the page.
3. Use the safety equipment provided for you. Goggles and a safety apron should be worn when any investigation or skill calls for heating, pouring, or using chemicals.
4. Always slant test tubes away from yourself and others when heating them.
5. Never eat or drink in the lab, and never use laboratory glassware as food or drink containers. Never inhale chemicals. Do not taste any substance or draw any material into a tube with your mouth.
6. If you spill any chemical, wash it off immediately with water. Report the spill to your teacher.
7. Know the location and proper use of the fire extinguisher, safety shower, fire blanket, first aid kit, and fire alarm.
8. Keep all materials away from open flames. Tie back long hair and loose clothing.
9. If a fire should break out in the classroom, or if your clothing should catch fire, smother it with the fire

What should you do if you spill any chemical?

FIGURE 2–6. Know the location and proper use of laboratory safety equipment.

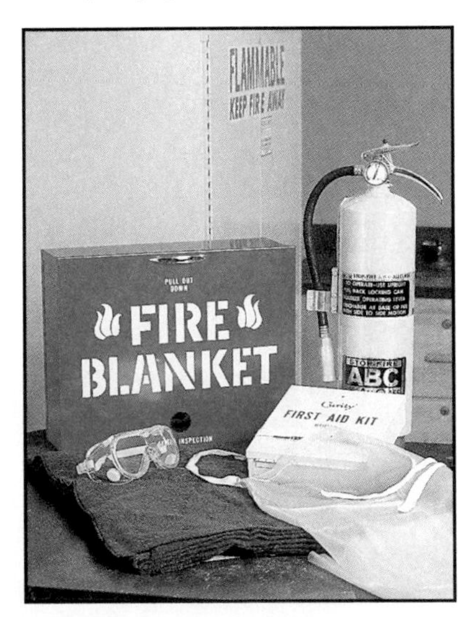

blanket or a coat or get under a safety shower. NEVER RUN.

10. Report any accident or injury, no matter how small, to your teacher.

It is very important to be cautious when you are cleaning up after an investigation or skill. Follow these procedures as you clean up your work area.

1. Turn off the water and gas. Disconnect all electrical devices.
2. Return all materials to their proper places.
3. Dispose of chemicals and other materials as directed by your teacher. Place broken glass and solid substances in the proper containers. Never discard materials in the sink.
4. Clean your work area.
5. Wash your hands thoroughly after working in the lab.

2:7 First Aid

What is first aid?

First aid is emergency care or treatment given to an ill or injured person before regular medical aid can be obtained. Your teacher or school nurse will usually be able to administer first aid if needed. However, you need to be aware of safe responses so that you can help yourself or a classmate during an emergency. Any injury, no matter how small, should be reported to your teacher at once. Appendix B lists safe responses to injuries that may occur in a science classroom. Learn this information.

FIGURE 2–7. Following safety rules while performing an investigation (a) and cleaning up (b) is necessary. Why?

a

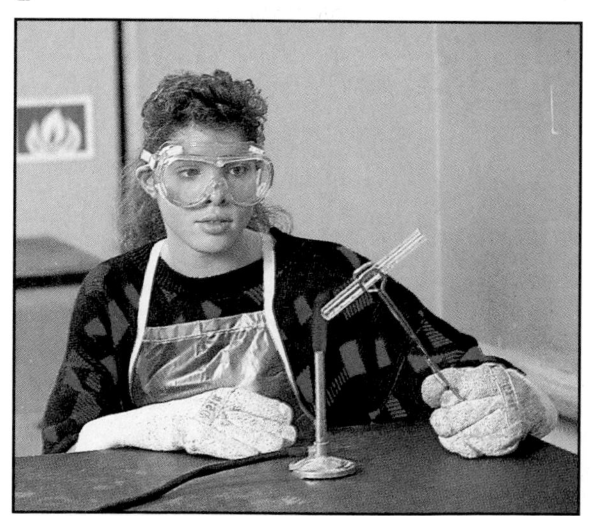

b

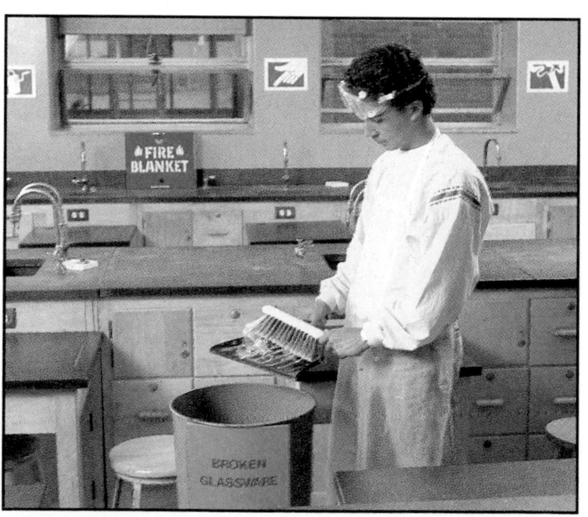

Table 2–2

Safety Symbols			
	DISPOSAL ALERT This symbol appears when care must be taken to dispose of materials properly.		**ANIMAL SAFETY** This symbol appears whenever live animals are studied and the safety of the animals and the students must be ensured.
	**BIOLOGICAL HAZARD** This symbol appears when there is danger involving bacteria, fungi, or protists.		**RADIOACTIVE SAFETY** This symbol appears when radioactive materials are used.
	**OPEN FLAME ALERT** This symbol appears when use of an open flame could cause a fire or an explosion.		**CLOTHING PROTECTION SAFETY** This symbol appears when substances used could stain or burn clothing.
	**THERMAL SAFETY** This symbol appears as a reminder to use caution when handling hot objects.		**FIRE SAFETY** This symbol appears when care should be taken around open flames.
	**SHARP OBJECT SAFETY** This symbol appears when a danger of cuts or punctures caused by the use of sharp objects exists.		**EXPLOSION SAFETY** This symbol appears when the misuse of chemicals could cause an explosion.
	**FUME SAFETY** This symbol appears when chemicals or chemical reactions could cause dangerous fumes.		**EYE SAFETY** This symbol appears when a danger to the eyes exists. Safety goggles should be worn when this symbol appears.
	**ELECTRICAL SAFETY** This symbol appears when care should be taken when using electrical equipment.		**POISON SAFETY** This symbol appears when poisonous substances are used.
	**PLANT SAFETY** This symbol appears when poisonous plants or plants with thorns are handled.		**CHEMICAL SAFETY** This symbol appears when chemicals used can cause burns or are poisonous if absorbed through the skin.

This textbook uses **safety symbols** to alert you to possible laboratory dangers. These symbols are explained in Table 2–2. Be sure you understand each symbol before you begin an investigation or skill.

2:8 Performing Investigations and Skills

Scientific knowledge is valuable only if it is learned and used. By doing investigations, you will have the opportunity to use knowledge. Also, you will be able to formulate new questions to be answered, to develop concepts, to explain observations, and to establish hypotheses. The chapters in your textbook contain investigations. These follow the steps of the scientific method.

TECHNOLOGY: APPLICATIONS

Watches

Over the years, watches have become nearly perfect mechanical devices to measure time. If a watch loses or gains 20 seconds in a day, the error is only 0.023%. Few machines are this accurate.

The time keeping of today's watches is controlled by a balance wheel that turns in alternate directions at a fixed rate. Early watches had two lengths of stiff hog bristle that acted as a balance wheel. Early watches had only an hour hand.

A coiled spring was the first source of power for watches. For some of these watches, the stem is twisted back and forth to wind the spring. For others, wrist movement automatically winds it. The first electronic watches, powered by small batteries, were produced in the 1950s.

Another type of watch is the quartz watch. There are several types of quartz watches. One uses a quartz oscillator that produces an alternating current to drive a small motor. Another type of quartz watch has no moving parts. A third uses solar power. The accuracy of a well-made quartz watch is usually better than a loss or gain of one minute per year.

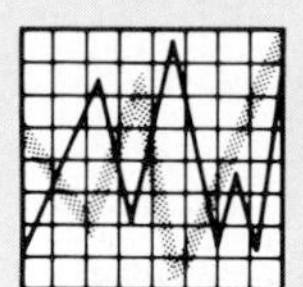

SKILL 2–2

Performing Investigations

Problem: How do you perform investigations?

Background

Recall from Chapter 1 that solving problems in science involves four basic steps.

1. Determining the problem
2. Testing
3. Analyzing the results
4. Drawing conclusions

Every chapter of this textbook contains investigations and skills. These follow the four steps of the scientific method. By performing these activities, you will have the opportunity to use knowledge, to formulate new questions to be answered, to develop concepts, to explain observations, and to establish hypotheses.

Procedure

1. Before performing an investigation, you should always read the entire set of directions. Previewing the investigation in this way will enable you to ask your teacher for help if there are any procedure steps that you do not understand.
2. Read Investigation 9–1. Note the parts of the investigation and the order they follow: Problem, Materials, Procedure, Data and Observations, Analysis, and Conclusions and Applications.
3. First, a question is asked. What is the question in Investigation 9–1?
4. Note that the testing requires having the equipment that is listed under "Materials." It is a good idea to collect all materials that you will need to complete the investigation before you begin. This step will save you time and may keep you from missing a procedure step.
5. A method of testing the question is then suggested in the section entitled "Procedure." It is important to note any safety symbols and to be cautious when performing investigations. Why is the diagram of the apparatus included with the procedure?
6. As the investigation is performed, observations are made. Some observations require making accurate measurements. All observations must be immediately and precisely recorded. What is the title of the section where you will record these data? Will you always record data in the same manner? Explain your answer.
7. Finally, the data are analyzed and the conclusions are drawn. Your text helps you to analyze your observations by having you answer questions in the "Analysis" section. What types of questions would you expect to find in the "Analysis" section?
8. Your text helps you draw conclusions by providing questions for you to answer in "Conclusions and Applications."
9. As you examined this sample investigation, you probably made some conclusions about how a student should perform an investigation in order to get the most accurate results. Answer the following questions.

Questions

1. Why should all directions be read first?
2. What safety symbols are used in Investigation 9–1? What do they mean?
3. A student performs the investigation, but decides not to write down the results until the next day. What problems might occur in drawing conclusions?
4. Refer to Chapter 4, Investigation 4–2. What observations do you predict you will be recording about the masses of the wood block, metal block, clay, and 10 mL of water?
5. Refer to Chapter 10, Investigation 10–1. How do you think wetting the gauze on one thermometer, as instructed in Step 3, will affect the temperature?

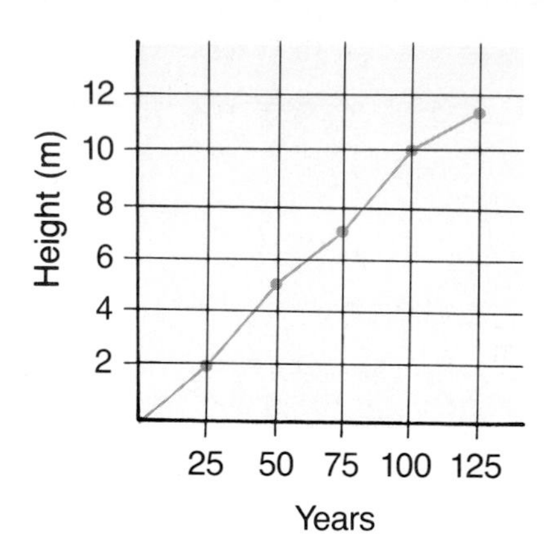

FIGURE 2–8. This line graph shows the height in meters of a certain cactus during its lifetime.

What is a line graph?

2:9 Graphing

Scientists often arrange their data into charts and tables. Charts and tables are used to classify and organize large amounts of scientific data. Data may also be analyzed by making graphs. Graphs show data in picture form. They make it possible to make quick interpretations of a great deal of data. Graphs also make it easier to make comparisons and draw conclusions. Knowing how to make and interpret graphs is an important skill to learn.

A **line graph** consists of a vertical axis, or y-axis, which is perpendicular to the horizontal, or x-axis. A line graph shows the relationship between two types of data. A point is plotted that represents two pieces of information. The points are then connected with a line. For example, if you are making a graph to show the size of a certain cactus during its lifetime, you would need to know the age and the height. Study the data in Table 2–3.

Table 2–3

Age *(years)*	**Height** *(m)*
0	0
25	2
50	5
75	7
100	10
125	11

In this example, time is plotted on the x-axis. Height is shown on the y-axis. Study Figure 2–8. Run your finger along the x-axis until you come to 25. Keep one finger on this number. Then move your other hand to the y-axis and find 2 meters. Move your fingers toward each other. A point is plotted on the graph where your fingers intersect. Now repeat the procedure with 50 years on the x-axis and 5 meters on the y-axis. The point plotted there represents the height of the cactus at 50 years. After all the points have been plotted, they are connected with a line. The graph illustrates the growth pattern for a certain cactus.

A **bar graph** uses thick bars to display information. It is similar to the line graph since it, too, has an x-axis and a y-axis. The difference is that instead of plotting points, bars are used. You begin plotting data in the same way that you do with the line graph, but instead of using a

FIGURE 2–9. This bar graph shows the amount of snowfall in centimeters in certain cities for a particular November day. How much fell in city C?

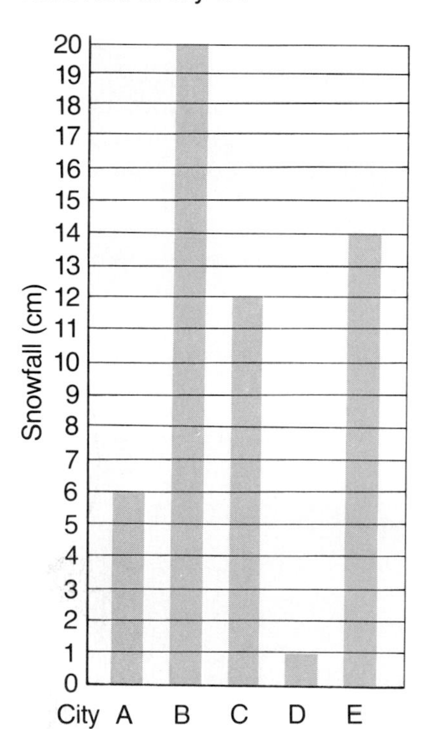

point, you plot a straight line parallel to the x-axis. Then, two vertical lines are drawn to connect it to the x-axis, which forms the bar. A completed bar graph showing the amount of snowfall in centimeters in certain cities for a day in November is shown in Figure 2–9.

A **pie graph** uses parts of a circle to show information. This type of graph shows how each part is related to the total. Each part of a pie graph is called a section. Each section is a certain percentage of the whole. When all sections of the pie are added, they equal 100 percent. In order to make a pie graph, you need to calculate the fractional amount of the total quantity for each part.

For example, if you want to make a pie graph that shows where Earth's water supply is found, you need to know the total amount of Earth's water supply. Earth's total supply is 1398 million km^3. Therefore, the total circle will represent 1398 million km^3. There are three main locations of this water: (1) in the oceans, (2) in glacial ice, and (3) in freshwater, which includes lakes, rivers, and groundwater. The volume of water in Earth's oceans is 1356 million km^3. Glacial ice contains 28 million km^3. Lakes, rivers, and groundwater contain 14 million km^3.

Each of these three locations will be represented by a section of the graph. To find the fraction or the percent for each section, divide the amount of water in each location by 1398. Then multiply the fraction by 360 degrees to determine the angle for the section. For example, the ocean section would be determined by the following calculation: $1356 \div 1398 \times 360$. Round the answer to the nearest whole degree. The angle would be 349 degrees. To make a pie graph, use a compass to draw a circle. Then draw a line from the edge of the circle to the center. Place your protractor on this line and use it to mark a point on the edge of the circle at 349 degrees. Connect this point with a straight line to the center of the circle. Repeat this procedure for the other two sections. Label all sections of the graph. The completed graph would look like the one shown in Figure 2–10.

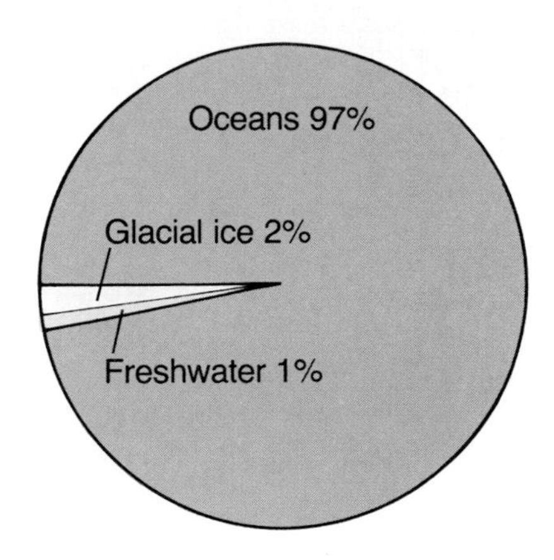

FIGURE 2–10. This pie graph shows the locations of Earth's water supply.

What does a pie graph show?

Review

7. What is first aid? What should you do if your clothing should catch fire?

8. What is a pie graph?

★ **9.** Why is it necessary to wash your hands after performing an investigation?

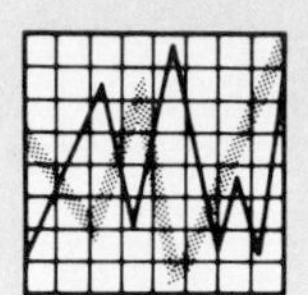

SKILL 2–3

Constructing Graphs

Problem: How do you construct graphs?

Materials

graph paper
metric ruler
mathematical compass
protractor
colored pencils

Procedure

Part A

1. Make a line graph that shows the daily tide for San Diego, California, on January 19, 1988. Label the x-axis with the time of day in hours. Have each horizontal block represent 1 hour. Label the x-axis starting on the left with 12:00 midnight.
2. Label the y-axis with the height of the tide in meters. Let each vertical block represent 0.2 meters. Start at the bottom of the graph with 0 meters.
3. Label each axis and write a title for your graph.
4. Plot the information that is given in the table below.
5. Use a red pencil to connect your points.

Data Table

Time	Tide (Meters)
12:00 midnight	2.3
6:00 A.M.	0.2
12:00 P.M.	2.5
6:00 P.M.	0.4
12:00 midnight	2.6

Part B

6. Use the graph shown in Figure 2–11 to answer the following questions.
   a. What kind of graph is this?
   b. What is shown on the x-axis?
   c. What is shown on the y-axis?
   d. What is the approximate area of the Pacific Ocean?
   e. Which of the oceans has the least area?
   f. How much larger in area is the Indian Ocean than the Arctic Ocean?
   g. What is the total area of the four oceans?

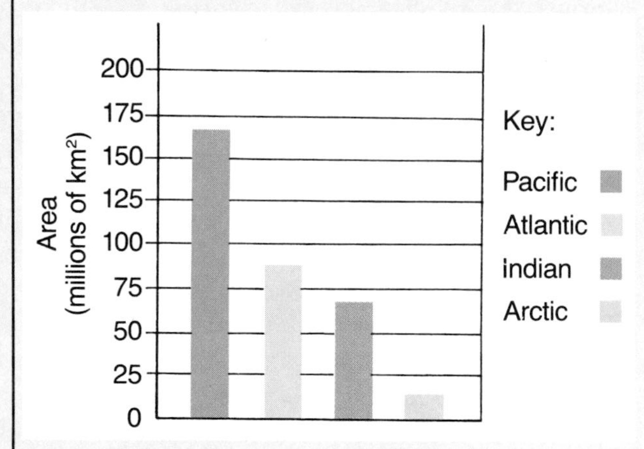

FIGURE 2–11.

Part C

7. Make a pie graph that shows the amounts of dissolved elements in seawater. In every kilogram (1000 g) of seawater, there are approximately 35 g of dissolved substances. Of this 35 g, 19.0 g are chlorine, 10.6 g are sodium, 2.6 g are sulfate, and 2.8 g are all other elements.
   a. Have four sections in your graph to represent (1) chlorine, (2) sodium, (3) sulfate, (4) other elements.
   b. Refer to Section 2:9 for the procedure needed to graph this information.
   c. Be sure to label each section with the percentage and the chemical name.
   d. Color the sections and write a title for your graph.

Chapter 2 Review

SUMMARY

1. Measurements include a unit of measure and a number indicating how many units are present. 2:1
2. The International System of Units (SI) is based on the decimal system. 2:2
3. Area is the amount of surface included within a set of boundaries. The space that an object occupies is its volume. 2:3
4. Mass is a measure of the amount of matter in an object. 2:4
5. Weight is the measure of the gravitational force exerted on an object by a more massive object. 2:4
6. Density is the amount of matter in a unit volume of a substance. 2:4
7. Time and temperature are measurements often made in science. 2:5
8. Safety rules should be followed to protect people from injury. Safety symbols alert laboratory users to possible dangers. 2:6
9. First aid is emergency care given before medical treatment can be obtained. You should know safe responses to injuries. 2:7
10. Investigations follow the steps of the scientific method. 2:8
11. Line, bar, and pie graphs are used to make comparisons and draw conclusions. 2:9

VOCABULARY

a. area
b. bar graph
c. density
d. first aid
e. gram
f. International System of Units (SI)
g. kilograms
h. line graph
i. liter
j. mass
k. measurements
l. meter
m. newtons
n. pie graph
o. safety symbols
p. time
q. volume
r. weight

Matching

Match each description with the correct vocabulary word from the list above. Some words will not be used.

1. standard unit of length in SI
2. emergency care or treatment
3. space that an object occupies
4. standard unit of volume in SI
5. measure of amount of matter in an object
6. measure of the force of Earth's gravitational pull on an object
7. standard unit of weight in SI
8. the amount of matter in a unit volume of any substance
9. graph showing how each part is related to the total
10. warn students of possible laboratory dangers

Chapter 2 Review

MAIN IDEAS

A. Reviewing Concepts

Choose the word or phrase that correctly completes each of the following sentences.

1. Temperature is measured with an instrument called a *(meter stick, pan balance, thermometer)*.
2. A kilogram is equal to *(1000 grams, 100 grams, 10 grams)*.
3. A unit of length in SI is the *(gram, centimeter, milliliter)*.
4. *Milli-* means *(0.1, 0.01, 0.001)*.
5. You should always *(dispose of solid materials in the sink; wear goggles when working with chemicals; rinse broken glassware with water)*.
6. The *(meter, kilogram, gram)* is the standard unit of mass in SI.
7. Chemicals should never be *(used, inhaled, poured)*.
8. A graduated cylinder is used to measure the *(volume of liquids, area of oddly shaped solids, mass of oddly shaped solids)*.
9. Area is often expressed in *(cm^2, mm^3, km^3)*.
10. The horizontal axis on a line graph is called the *(y-axis, x-axis, z-axis)*.
11. A graph that shows how each part or section is related to the total is a *(line graph, bar graph, pie graph)*.
12. If you spill acid, you should wash it off immediately with *(alcohol, water, vinegar)*.
13. If your clothes should catch fire, you should first *(walk to the safety shower and put it out with water; run to the safety shower and put it out with water; turn on the fire alarm)*.
14. Every injury should be *(flushed with water; attended to by a nurse; reported to the teacher)*.
15. The area of a rectangular surface can be found by measuring its length and width and *(dividing length by width; dividing width by length; multiplying length times width)*.

B. Understanding Concepts

Answer the following questions using complete sentences.

16. Describe how mass is measured with a pan balance.
17. Describe how weight is measured with a spring scale.
18. What instrument is used to measure temperature? What SI unit is used?
19. Describe how line and bar graphs are similar.
20. Describe how line and bar graphs are different.
21. Why should you report any accident or injury, no matter how insignificant, to your teacher at once?
22. Explain the meaning of the SI prefix *centi-*.
23. What safety equipment should you use when you are heating, pouring, or using chemicals?
24. Why should you always slant test tubes away from yourself and others when heating them?
25. Describe what you should do if one of your classmates faints. See Appendix B.

C. Applying Concepts

Answer the following questions using complete sentences.

26. Why would your weight be less on the moon, but your mass be the same as on Earth?
27. How does the distance between two objects affect the gravitational force between them?
28. How is SI similar to the money system of the United States?
29. Why is there more gravitational attraction for an object at sea level than for an object on top of a mountain?

30. When performing investigations, why should you not draw any material into a tube with your mouth?

SKILL REVIEW

1. Explain how you can measure the volume of a small, oddly shaped object using string, a graduated cylinder, and water.
2. How do you calculate the volume of a cylinder like a coffee can?
3. In a laboratory investigation, what does the procedure section tell you?
4. Find the area of this page in SI units of measurement.
5. Calculate the area and volume of three sizes of boxes. Then graph your data by making a bar graph with area on the x-axis and volume on the y-axis.

PROJECTS

1. Design an experiment to determine the effects of water on small plants. Use three plants, and give each a different measured amount of water each day. Record the amount of water used and plant growth in SI units.
2. Perform an experiment to study the expansion of heated air. Partially fill a balloon with air. Tape two wooden sticks parallel to the sides of the balloon. Tape the balloon to a ring stand with a heating lamp above it. Record the temperature near the balloon, the size of the balloon, and the time at set intervals. Use SI units. Record this information in a data table.

READINGS

1. Ardley, Neil. *Making Metric Measurements.* New York: Franklin Watts, 1984.
2. Cole, K. C. "Small Differences." *Science Digest.* June, 1985, pp. 42, 79, 81.
3. White, Jan V. *Using Charts and Graphs.* New York: R. R. Bowker, 1984.

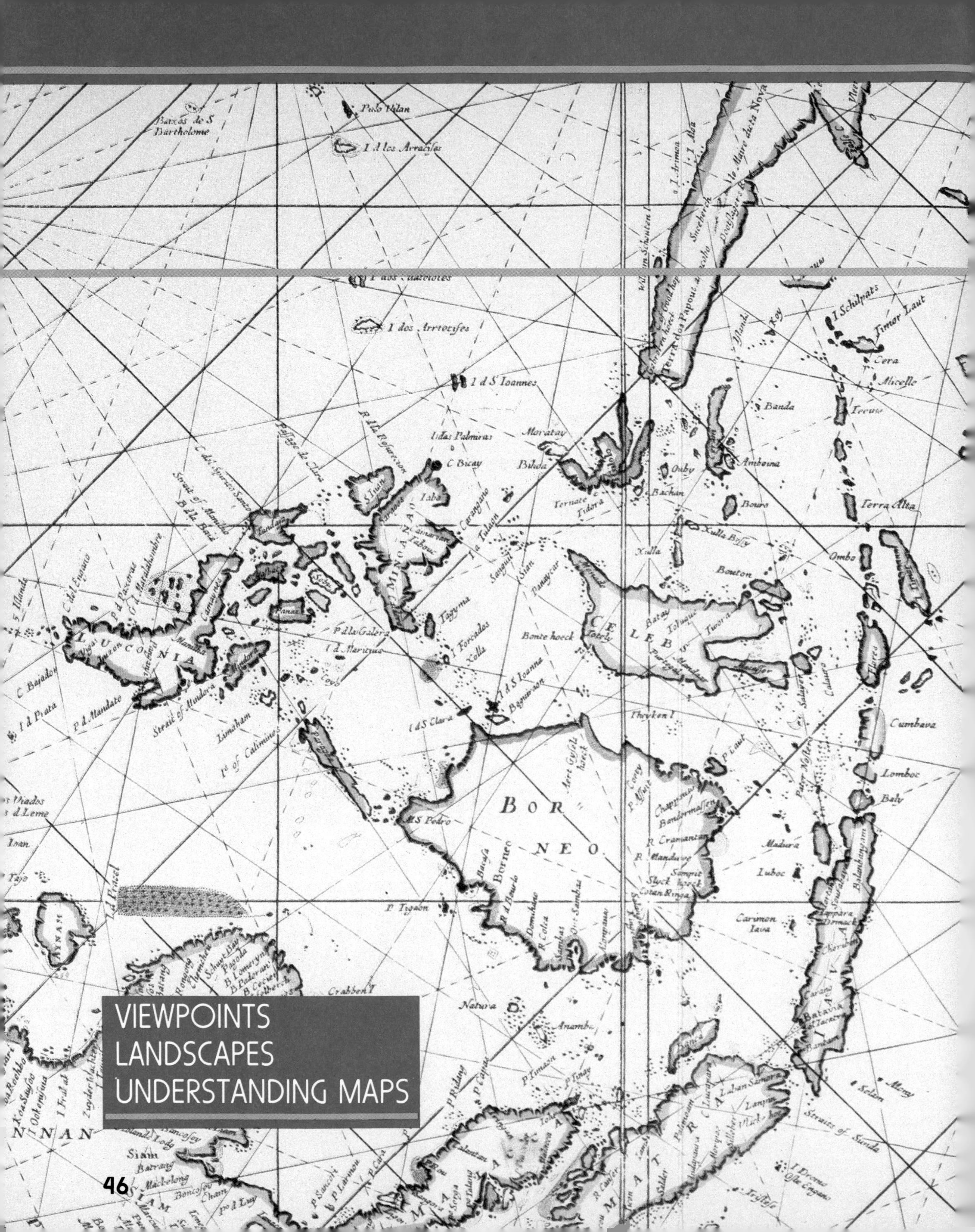

VIEWPOINTS
LANDSCAPES
UNDERSTANDING MAPS

Views of Earth

When viewed from space, Earth appears as a water- and cloud-covered sphere. A closer view of Earth's surface reveals a number of large landmasses surrounded by the water. There are no lines, however, that show boundaries of countries or states. In order to be able to locate certain points on Earth, models that represent Earth must be used. These models are globes and maps. Knowing how to use these models of Earth will help you learn more about our planet and how to get around on it.

VIEWPOINTS

Maps and globes help us locate certain points or positions on Earth's surface. Every point on Earth's surface has its own latitude and longitude. Each point is different from any other. By knowing how to use latitude and longitude, it is possible to identify the exact location of any point on the surface of Earth.

GOALS

1. You will use latitude and longitude to locate features on Earth's surface.
2. You will see how Earth models, such as globes and maps, can be used to describe differences in time and location.

3:1 Latitude and Longitude

How would you describe the location of Cairo, Egypt? You might answer that it is located along the southeast coast of the Mediterranean Sea. Or, you might say it is located on the northeast tip of Africa. Both answers are correct, but there is a more precise way to locate places on Earth. Lines of latitude and longitude form an imaginary grid system that enables points on Earth to be located exactly.

What is a precise way of locating a point on Earth's surface?

Look closely at a globe or map of the world. First, find the equator, which lies halfway between the North and South Poles. Notice the other lines, both north and south of the equator. These parallel lines are lines of latitude. **Latitude** identifies locations or distances north or south

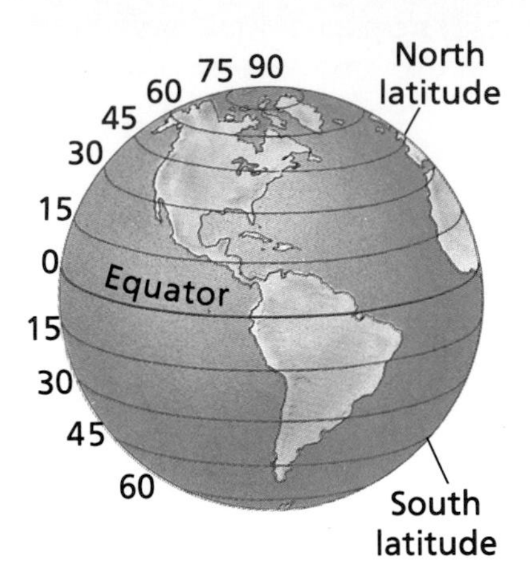

a

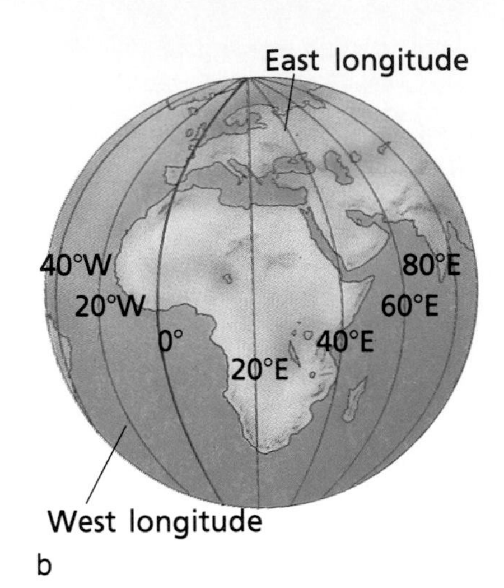

b

FIGURE 3–1. Lines of latitude are parallel to one another (a). Lines of longitude identify locations east or west of the prime meridian (b). All lines of longitude pass through the poles.

of the equator. Each of these lines is numbered in degrees either north or south of the equator. The equator is numbered 0° (zero degrees) latitude. The poles are each numbered 90°. Thus, latitude is measured from 0° at the equator to 90° at the poles. Locations north of the equator are referred to by degrees north latitude. Those south are degrees south latitude. Observe that Ottawa, Canada, is located at 45° north latitude. What is the latitude of Cairo, Egypt?

Where is 0° latitude?

Other lines on a globe or map are perpendicular to the equator and pass through the poles. These lines are meridians, or lines of longitude. **Longitude** identifies locations or distances east or west of the prime meridian. They divide Earth into 360 degrees, like a circle. Unlike lines of latitude, meridians are not parallel. They are farthest apart at the equator, and get closer together as they near the poles, where they meet. In order to divide Earth into 360°, there must be a starting point. The **prime meridian** runs through Greenwich, England, and represents 0° longitude. Points west of the prime meridian have west longitude. Points east of the prime meridian have east longitude. The east and west longitude lines meet at the 180° meridian, the **International Date Line**. The International Date Line is directly opposite the prime meridian on the other side of Earth. Warsaw, Poland, has a longitude of about 20° east. What is the longitude of Cairo, Egypt?

F.Y.I. Because the International Date line must be located only in a nonpopulated area, there are several points along the 180° meridian where it shifts either to the east or west to avoid islands or countries. For most of its length, however, the 180° meridian is the IDL.

As you have seen, lines of latitude and longitude form a grid system that can be used to locate places on Earth's surface. Look at Figure 3–2. What is the latitude and longitude of point P? Now, using latitude and longitude, find your approximate location on Earth's surface.

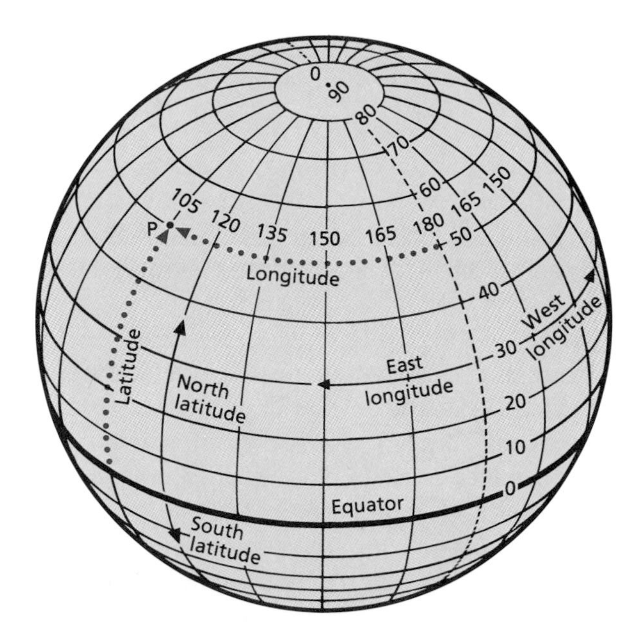

FIGURE 3–2. Latitude and longitude are used to locate points on Earth's surface.

CAREER

Cartographer

Robert L. Stewart is a cartographer at the Ohio Department of Natural Resources, Division of Geological Survey. He received his degree in architectural drafting and worked for an engineering firm prior to joining the Survey. Historically, engineers were this country's first mapmakers. Many cartographers like Mr. Stewart acquired their jobs because of their engineering training. Cartography students today often fulfill an internship in addition to classes in technical and mechanical drafting, design, technical writing, trigonometry, calculus, and cartography. These courses teach students how to use mapmaking tools such as pens, templates, and scribers.

Although Mr. Stewart is an office cartographer, mapmakers also work in the field from airplanes and helicopters, in trucks, and on foot. While cartographers have been working for more than 100 years on mapping the United States, there are still "white spaces" for which precise mapping has not been done. Also, many older maps must be revised.

For career information, write:
American Congress on
Surveying and Mapping
210 Little Falls St.
Falls Church, VA 22046

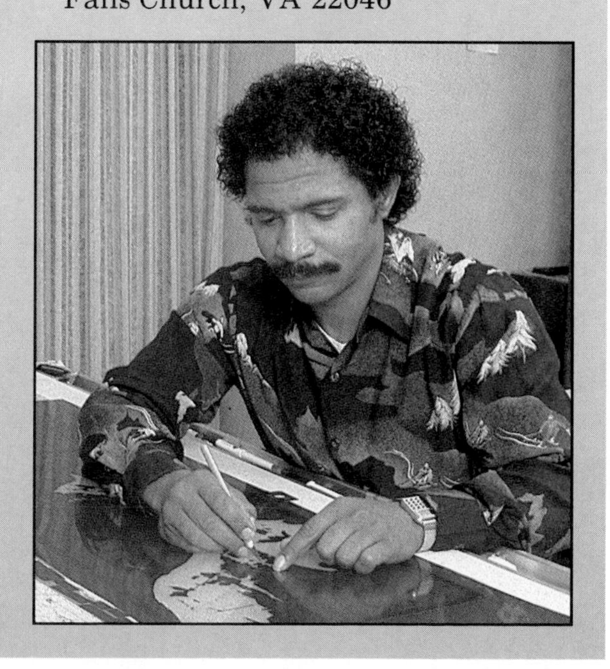

3:2 Earth Time

The time of day or night at any point on Earth's surface depends on the relationship of the longitude position of that point to the position of the sun. You know that Earth's surface is divided by lines that represent latitude and longitude. Lines of longitude divide Earth into 24 units, each unit being 15° (fifteen degrees) wide (15° × 24 = 360°).

How wide is each time zone?

For each 15° of longitude, there is a one-hour difference in time from the previous meridian. Thus, each 15° unit represents a single time zone. The 180° meridian, or International Date Line, is 12 time zones from the prime meridian. One day is skipped going west across the line. One day is added going east across the line. Can you tell why? Assume that you leave the prime meridian at midnight on May 2. You are flying west at 1600 kilometers per hour. If you fly along a latitude line where Earth's circumference measures 38 400 kilometers, how many hours will it take you to return to your starting point? On what date and at what time will you arrive?

Daylight saving time is a plan in which clocks are set one hour ahead of standard time for a certain number of months. As a result, darkness comes one hour later than on standard time. A state may stay on standard time if

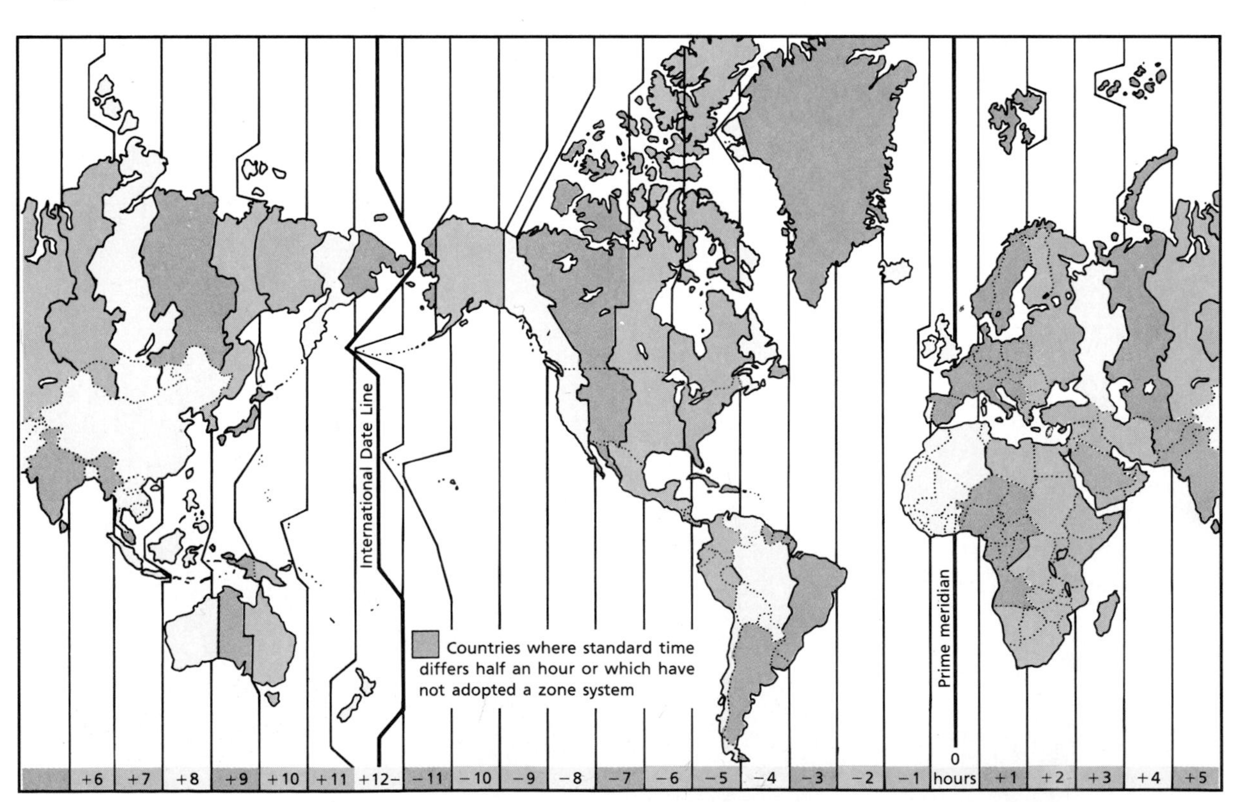

FIGURE 3–3. Time zones are roughly determined by lines of longitude.

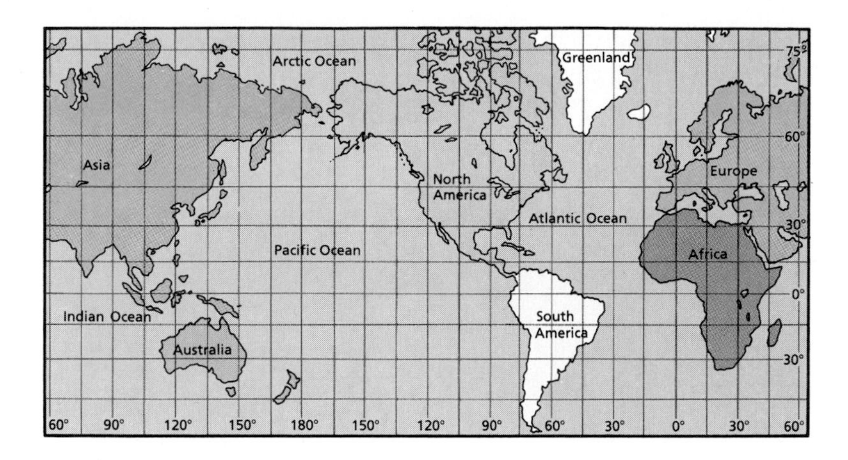

FIGURE 3–4. A Mercator projection exaggerates the size of areas near the poles.

it wishes to do so. Daylight saving time usually begins on the first Sunday of April and often ends on the last Sunday in October.

3:3 Map Projections

You have learned that maps are models of Earth's surface. A globe, however, is the best Earth model because its shape is similar to Earth's. Places or areas on a globe closely correspond to their actual locations and relative sizes on Earth. But globes are not easy to carry around or store. Maps are flat pieces of paper. They can be used easily, hung on a wall, or folded into small packages. How can Earth's curved surface be represented on a flat piece of paper?

Where does the most distortion occur on a Mercator projection?

Map projections are used to produce convenient representations of Earth's surface. Using map projections, points and lines on the globe's surface are projected onto a flat piece of paper. On a Mercator (mur KAYT ur) projection, Figure 3–4, both latitude and longitude lines are parallel and intersect at right angles. Mercator projections produce maps that are distorted near the poles. The land and sea areas near the poles appear much larger than they actually are. This is because the longitude lines are spread apart to make them parallel on the map. Because both latitude and longitude lines are straight, Mercator maps are widely used for surface navigation.

In a polar projection, a flat piece of paper is placed above either pole. Again, projections of points and lines are made from the globe's center. A polar projection map has longitude lines that extend outward from the pole like spokes of a wheel. Lines of latitude appear as circles, with the smallest circles near the poles. Polar projections are useful for studying polar areas, because the landmasses near the poles on these maps are not distorted.

FIGURE 3–5. Points and lines are projected from the center of a globe to make a polar projection.

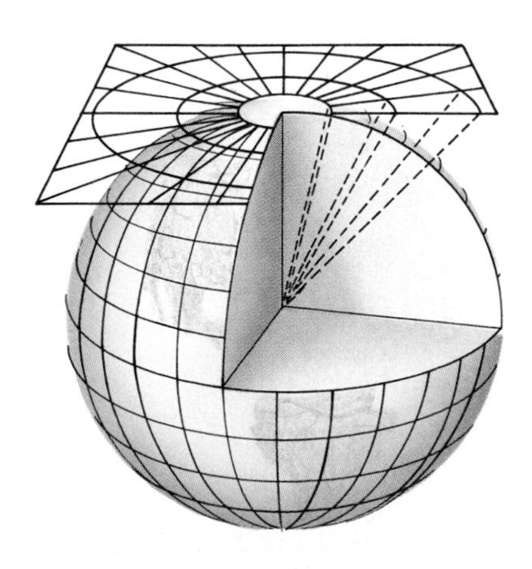

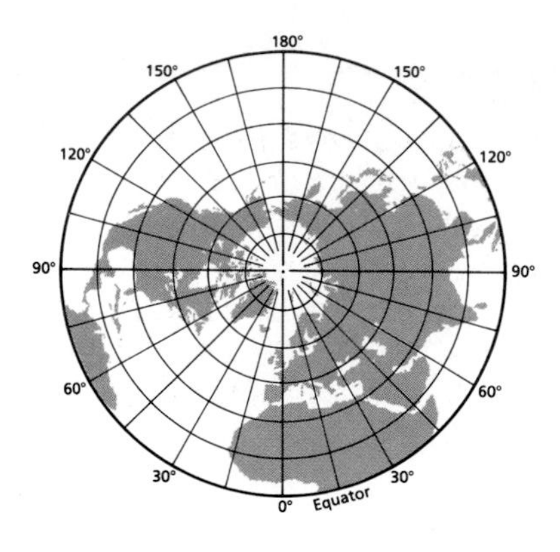

FIGURE 3–6. Lines of latitude appear as concentric circles on a polar projection.

Conic projections are sometimes used to produce accurate maps of small areas. A flat piece of paper is rolled into the shape of a cone. The cone is placed over a globe and contact is made along a line of latitude. Projections of points and lines are made like those of the Mercator and polar maps. Road maps and weather maps are often conic projections.

Review

1. Identify the locations of these three cities on a globe using latitude and longitude: Tokyo, Japan; Paris, France; Rio de Janeiro, Brazil.
2. On a globe, how many degrees apart are lines of latitude? Lines of longitude?

Use Figure 3–3 to answer questions 3 and 4.

3. If it is 9:00 A.M. in Florida, what time is it in Utah?
4. If it is 8:00 P.M. on Monday in Minneapolis, what time and day is it in London, England?
5. How is the arrangement of lines of latitude and longitude on a Mercator map different from that of a globe?
6. ★ When crossing the International Date Line, a traveler either gains or loses one day. Is there ever a situation where this is not true? Explain why or why not.

FIGURE 3–7. Some modern maps, such as this Dymaxion map by Buckminster Fuller, are designed to show the correct proportions of all landmasses.

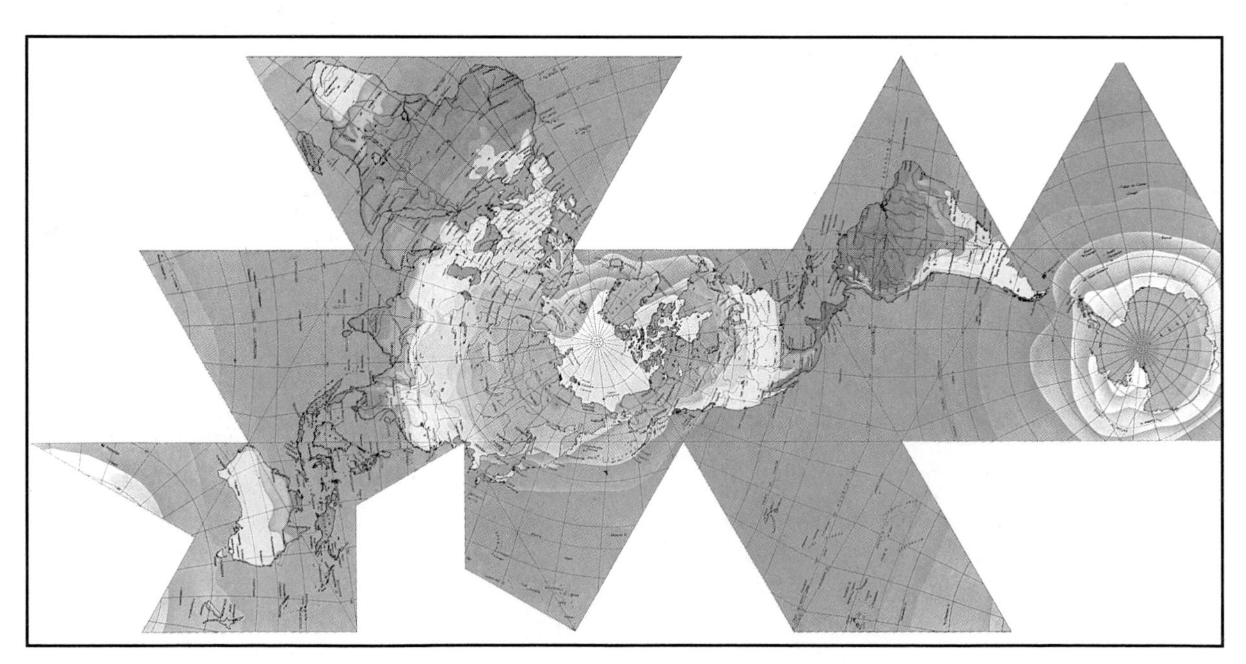

LANDSCAPES

Earth is a dynamic planet. Forces within Earth have produced the widely varied and complex arrangement of rocks and soil on Earth's crust. The features that make up the landscape at Earth's surface are landforms. Landforms include beaches, bars, and spits found along coastlines. They also include the high peaks and steep canyons of the mountains, the low hills and shallow valleys of the plains, and the deep canyons of the high plateaus. If you traveled across the United States, you would see three basic types of landforms that cover large areas—mountains, plains, and plateaus. In this section, you will learn about some of the landforms that exist throughout the United States. Locate the major landform that encompasses your area.

GOALS

1. You will be able to distinguish between different landforms on Earth's surface.
2. You will learn how to describe some of the geologic history of an area by identifying the landforms in that area.

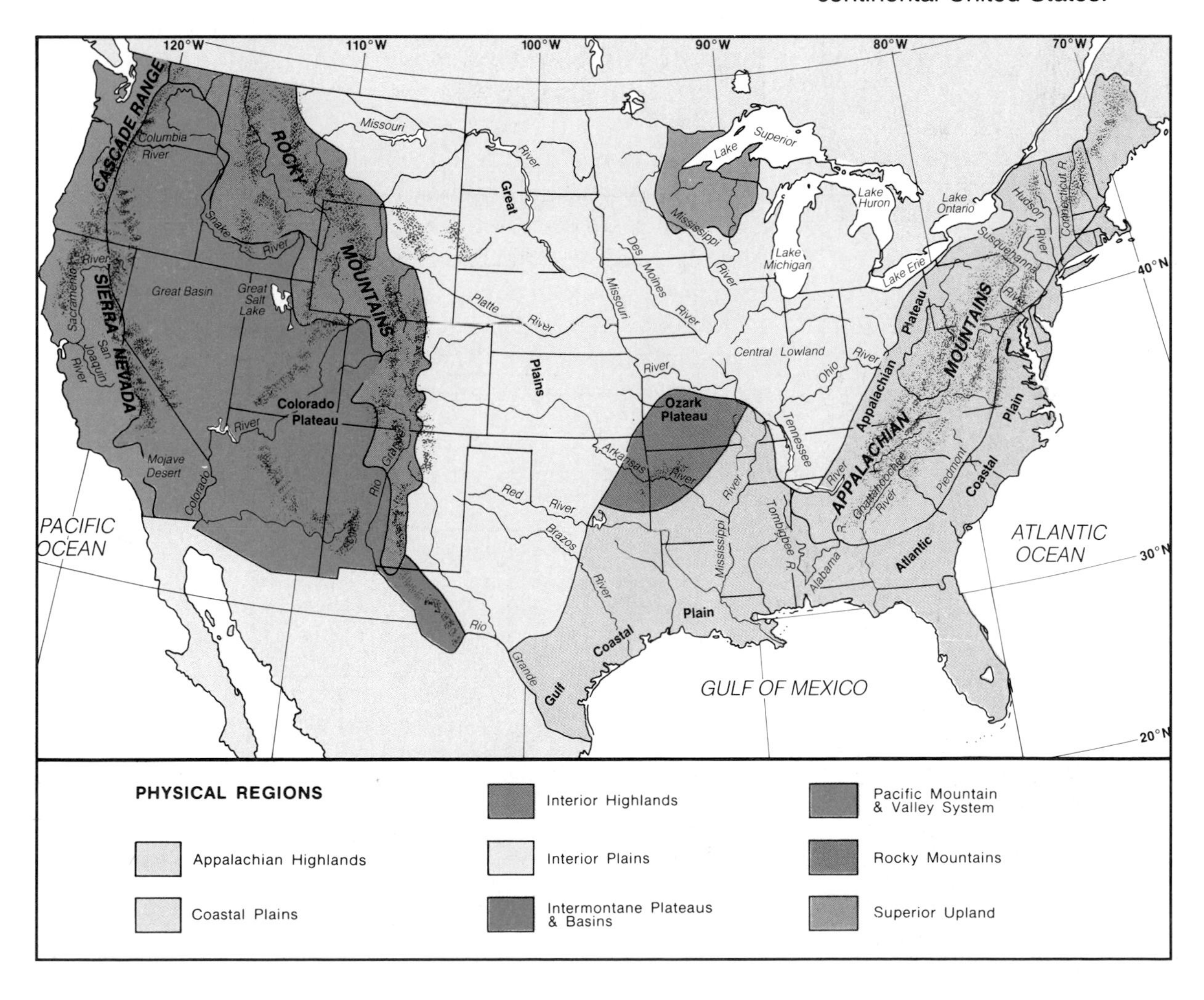

FIGURE 3–8. Plains and plateaus cover large areas of the continental United States.

3:4 Plains and Plateaus

Plains are vast flat areas that encompass about one-half of the landforms of the United States. Two examples of plains shown on the map in Figure 3–8 are the Atlantic Coastal Plain and the Great Plains. The Atlantic Coastal Plain is the exposed edge of North America that has emerged from the sea during the past several million years. It is a lowland that is made of rock covered by soft, loose materials, called sediments, such as sand, silt, and clay. Low hills and valleys occur only where the plain, has been slightly uplifted. Wide river valleys have cut through the plain, producing low hills between river systems. Much of the area is covered by swamps and lakes.

Where did the sediments come from that form the Great Plains?

The Great Plains is a large flat region in the interior of the continent. It is a young landform overlain by loose materials that have been eroded from the Rocky Mountains. Streams have deposited these materials in nearly horizontal layers during the last few million years. The Great Plains' elevation ranges from 1500 meters at its western boundary to 350 meters at its eastern border. Thus, rivers that cross the plain flow to the east.

Plateaus are high, relatively flat areas next to mountains. They are uplifted regions in which rocks remain nearly horizontal. Two examples are the Appalachian Plateau and the Colorado Plateau. The Appalachian Plateau is composed of old, nearly horizontal rock layers. The rocks have been cut by major streams to form steep valleys. The northern end of the plateau is covered by eroded material deposited when massive ice sheets melted.

FIGURE 3–9. The low, swampy Everglades of Florida lie on the Atlantic Coastal Plain (a). The Great Plains of the central United States is a farming area (b).

a

b

a

b

FIGURE 3–10. Bryce Canyon (a) and the Grand Canyon (b) are erosional features of the Colorado Plateau.

The Colorado Plateau is higher in elevation than the Appalachian Plateau. Its elevation averages more than 1500 meters. This area, which has been recently uplifted, also is composed of nearly horizontal rock layers. Rivers have cut deep into these rock layers, forming spectacular features such as the Grand Canyon. Because the plateau is located in an arid region, only a few river systems have developed on its surface. Many of the landforms on the Colorado Plateau are typical of a desert landscape.

How did the Grand Canyon form?

3:5 Mountains

Any part of Earth's surface that rises high above the surrounding land may be called a mountain or hill. Mountain building involves many processes. Some processes occur deep within Earth. Other processes take place at or near Earth's surface. The processes of mountain building are called **orogeny**. These processes include folding, faulting, upwarping, and volcanic activity. Mountain ranges are formed as the result of one or more of these processes.

What is orogeny?

The Appalachian Mountains extend from Quebec, Canada, south to Alabama, a distance of about 2400 kilometers. This mountain range is the oldest prominent range in North America. These mountains formed between about 450 and 200 million years ago. The Valley and Ridge province of the Appalachian Mountains are folded mountains. **Folded mountains** are made of rocks that have been squeezed from opposite sides, forming folds.

The Black Hills in western South Dakota are **upwarped mountains.** They were formed millions of years

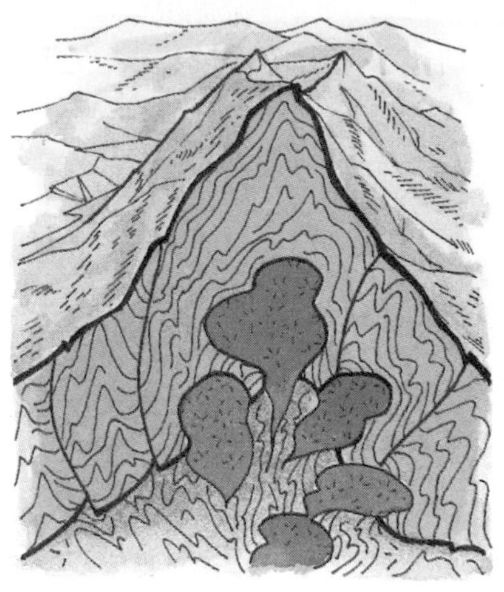

a

b

FIGURE 3–11. Folded mountains (a) result from compression of Earth's crust from opposite sides. The Valley and Ridge province of the Appalachians (b) is an example of folded mountains.

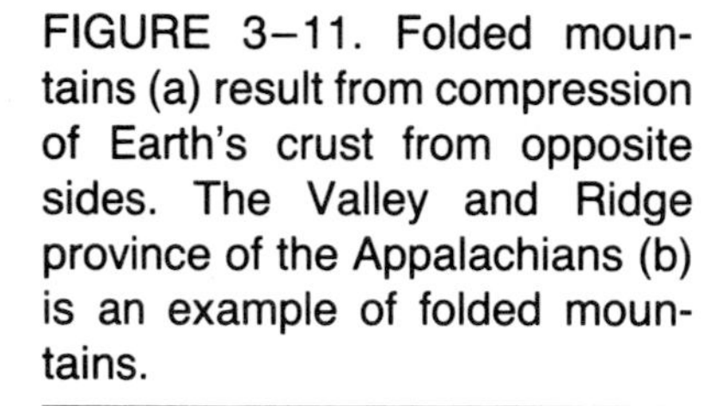

ago as the result of crustal uplifting. Hard rocks form the core of these mountains. As the region was upwarped, much of the softer rocks were worn away, exposing the hard core. Other examples of upwarped mountains include the Adirondacks in New York state, the Front Range of Colorado, and the Bighorn Mountains of Wyoming.

How are fault-block mountains formed?

Fault-block mountains are bounded on at least one side by faults. A fault is a fracture along which there has been movement. During orogeny, melted material called magma flows upward into the crust. This can cause uplifting and fracturing of the crust. The resulting blocks may move vertically to form mountains. The Grand Teton Mountains of Wyoming are fault-block mountains.

FIGURE 3–12. Upwarping of Earth's crust (a) formed the Black Hills (b).

a

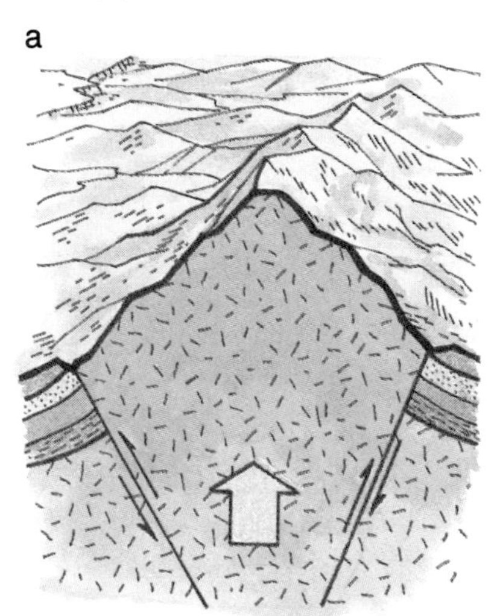

b

a

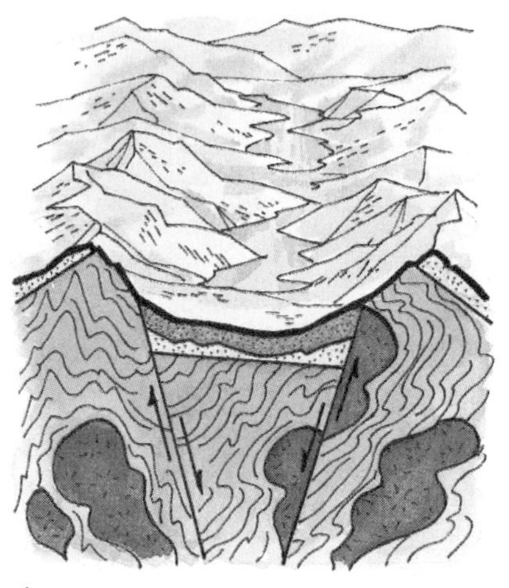

b

FIGURE 3–13. The Sierra Nevada Mountains (a) are the result of vertical movement of blocks of rock (b).

A fourth type of mountain is called a **volcano**. A volcanic mountain forms by the building up of molten material, which has been forced out of its interior onto its surface. Volcanoes are formed in several different ways and can be found in many locations on Earth. Mount St. Helens in Washington state is a volcanic mountain at Earth's surface. Volcanoes also occur on many ocean floors. Mauna Loa, the largest volcano on Earth, rests on the ocean floor 5000 meters below sea level. Its summit reaches a height of over 4100 meters above the water. Volcanoes appear to occur in patterns. A more complete description of volcanoes is found in Section 17:4.

How are volcanic mountains formed?

FIGURE 3–14. Mount St. Helens (a) is a volcanic mountain formed by the accumulation of molten material from Earth's interior (b).

a

b

Review

7. What are three basic types of landforms?
8. Compare and contrast plains and plateaus.
9. Describe how folded mountains are formed.
★ 10. How are the Appalachian and Colorado Plateaus different? Why do they differ so much?

UNDERSTANDING MAPS

GOALS

1. You will learn how to use a topographic map and identify various surface features on these maps.
2. You will construct a map.

If you were asked to make a model of Earth's surface as you saw it in Figure 3–15a, how would you do it? What information would you need to complete the model?

One of the most useful tools to represent Earth's surface features is the topographic map. In the next several sections, you will construct and use topographic maps to identify and describe landforms.

3:6 Topographic Maps

What is shown on a topographic map?

Topography (tuh PAHG ruh fee) describes the surface features of an area. Topography refers to the "lay of the land." Mountains, seacoasts, and prairies are different types of topographies.

Topographic maps show the topography of an area in detail. These maps show the location, landscape, and cultural features of a small part of Earth's surface. Among the landscape features shown are mountains, hills, plains, lakes, and rivers. Cultural features include roads, cities, dams, and other structures built by people. Thus, topographic maps are scale models of some part of Earth's surface.

FIGURE 3–15. Mapmakers translate features of a landscape (a) into a topographic map (b).

a

b

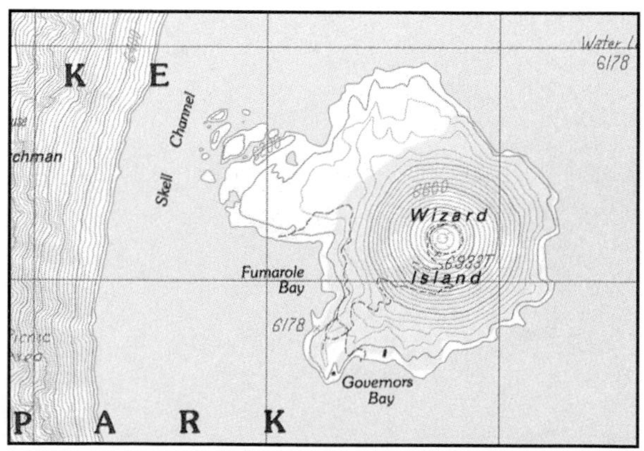

Recall that the scale of the model river in Chapter 1 was a fixed ratio. A map scale is also a fixed ratio between a unit of measure on the map and the distance that it represents on the surface. For example, a map of Maine may have a scale that reads "1 cm = 80 000 cm," which means that one centimeter on the map represents 80 000 centimeters on land.

A map scale can be shown in other ways. It may be a ratio such as 1:62 500 or 1/62 500. The ratio, however, is always expressed in the same units of measure. That is, 1 cm = 62 500 cm or 1 m = 62 500 m. A map scale also may be in the form of a small bar graph that is divided into a number of units. The scale can be used to measure the distance between two points on the map.

Look carefully at the topographic map in Figure 3–17. A number of different colors appear on the map. Blue is used for rivers, lakes, and streams. Green identifies vegetation. Observe the small symbols or designs. These symbols may include solid and dotted lines, and numbers,

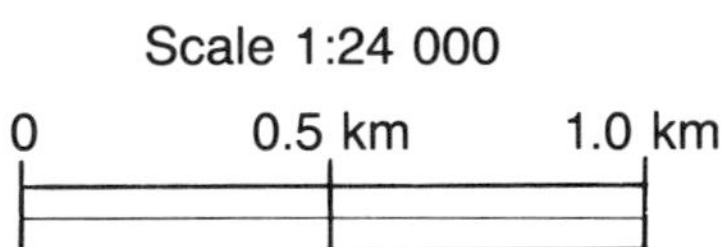

FIGURE 3–16. A map scale is a fixed ratio between a distance on the map and the distance it represents.

FIGURE 3–17. Kingston, RI

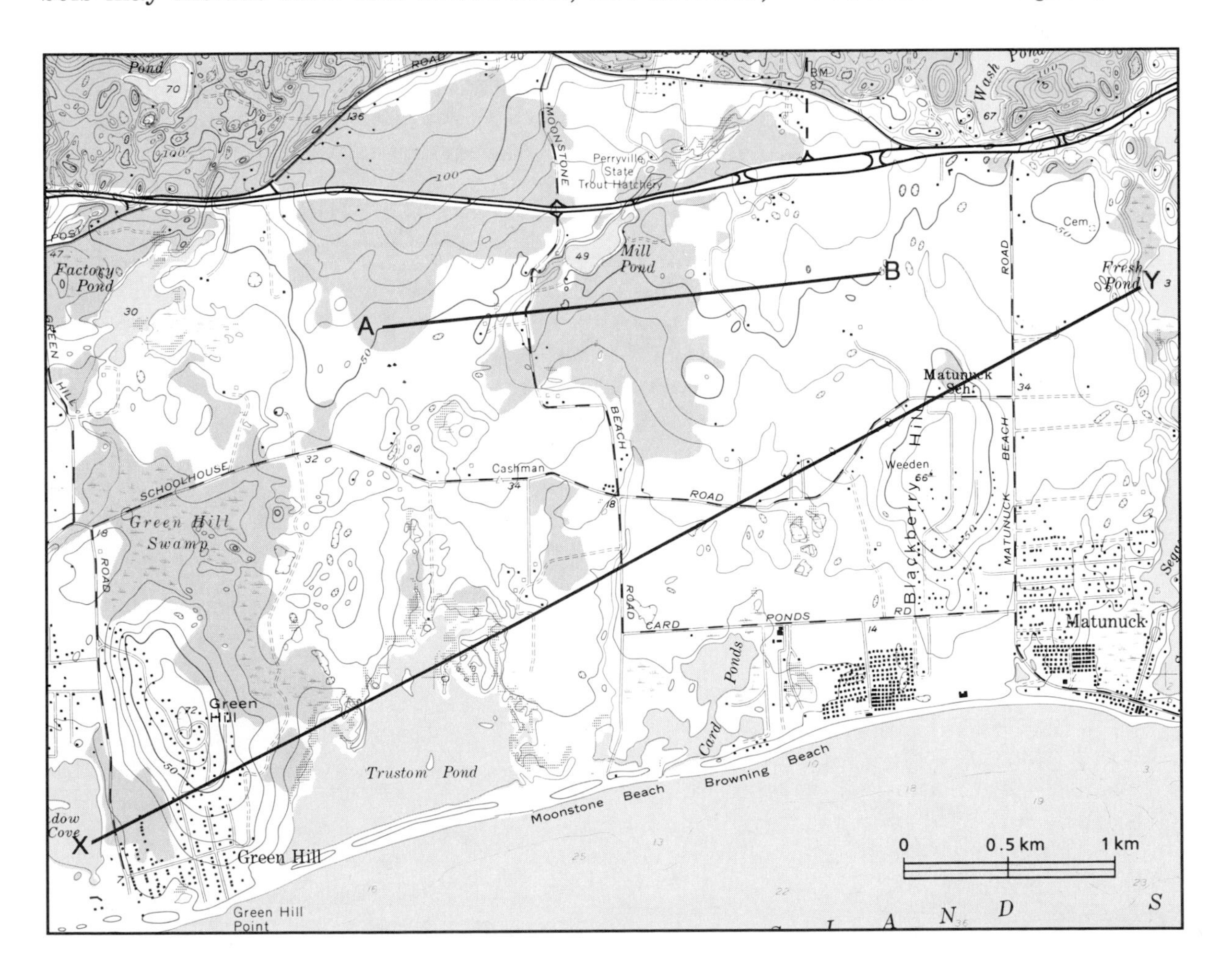

What is a legend?

triangles, circles, and dots. Some symbols look like the features they represent. In order to understand what other symbols mean, you need to look at a map legend. A **legend** explains each symbol used on the map. Legends vary depending on the type of map, but most maps have a legend. Some frequently used symbols for topographic maps are shown in Appendix E.

The most outstanding feature of a topographic map is that it shows the **elevation**, or height above sea level, of various features on the map. Thus, topographic maps are three-dimensional models of parts of Earth's surface. How is this third dimension, elevation, shown on a map?

TECHNOLOGY: APPLICATIONS

Electronic Road Maps

An electronic road map is a navigation device that calculates the position of your automobile by using dead reckoning, which is an ancient Polynesian way of plotting the course and distance traveled by using a previously known position. This navigation device contains a display screen mounted on the dash, a mini computer in the trunk, a compass near the rear window, sensors mounted on the nondrive wheels, and a cassette player complete with tapes of every street and address for an area twice the size of an ordinary paper map.

With these systems, a person can program a destination into the computer, and the route will be shown on the screen. In the future, auto manufacturers hope to offer navigation systems that will speak to drivers and tell them which way to turn, so the drivers do not need to watch a screen. If this device could be linked to radio data systems, the navigation system could change your route according to changes in road conditions, traffic jams, construction, icy roads, and other hazards.

Compact discs can hold 150 000 sheets of paper information, each containing 50 lines of 80 characters each, which is a pile about 5 stories high! This would dramatically increase the address information that is presently contained on the cassette tapes.

3:7 Using Topographic Maps

To show the elevation of various features on a topographic map, contour lines are used. A **contour line** is a line drawn on a map to join all points of the same elevation. Elevation is expressed as a distance either above or below sea level. Sea level has 0 (zero) elevation. Look at the topographic map below. Notice that not all of the contour lines are numbered. Each contour line on this map represents a vertical change of 80 feet.

What point is considered zero elevation?

FIGURE 3–18. Point Reyes, CA

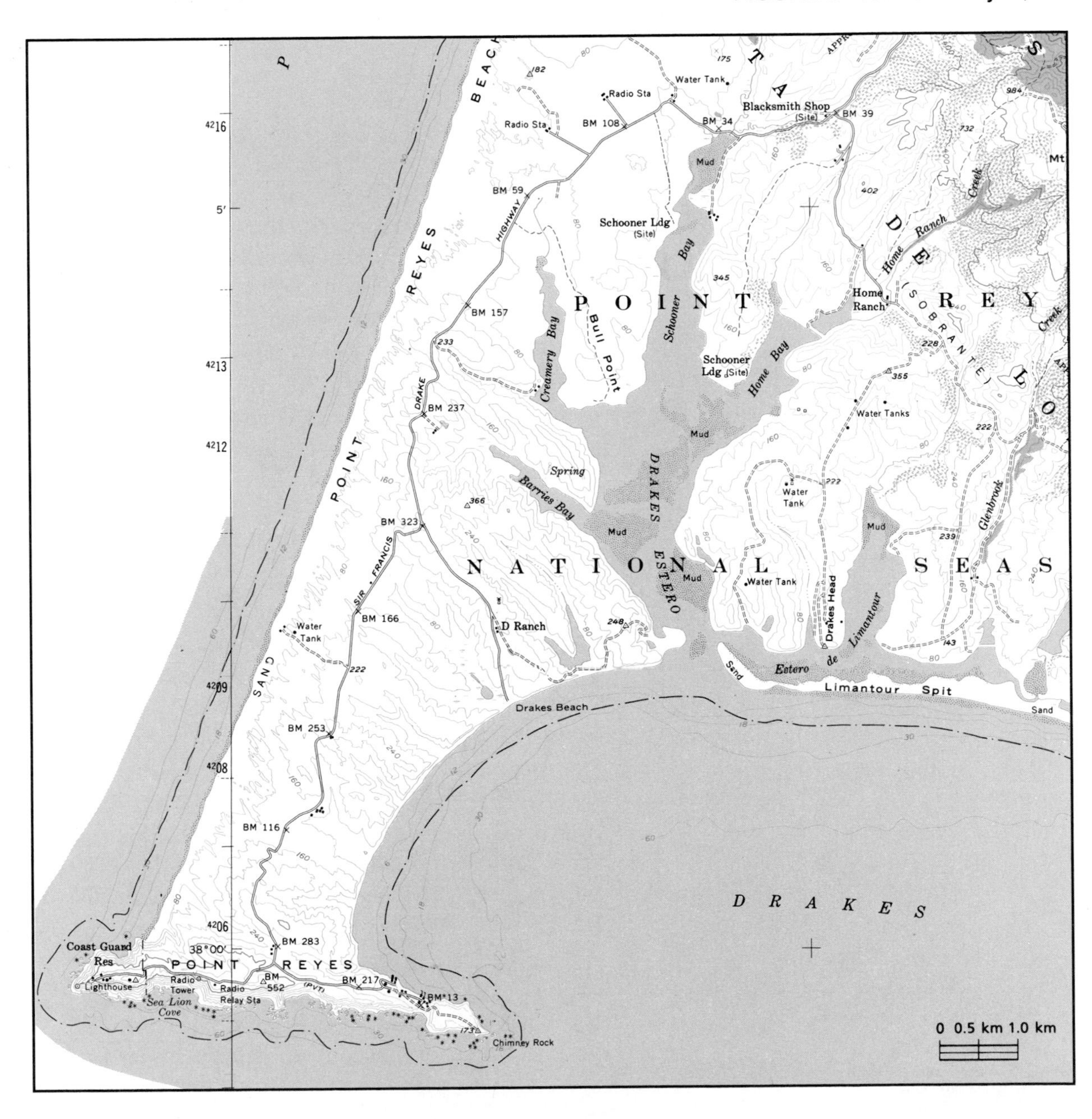

The difference in elevation between two adjacent contour lines is the **contour interval**. Contour intervals usually are given in even numbers or as a multiple of five. The size of the chosen interval depends on how much the elevation changes in the area mapped. In mapping mountains, a large contour interval is used. Otherwise, there would be so many lines you could not read them. Small contour intervals may be used where there are small differences in elevation. **Relief** describes how much variation in elevation an area has. Rugged or high relief, for example, describes an area of many hills and valleys. Gentle or low relief describes a plain area or a coastal region. Five general rules should be remembered in studying contour lines on a map.

F.Y.I. Often the relief of a geographic area will determine the contour interval that must be used on the topographic map of that area. For instance, a mountainous region of high relief might be mapped with a 50 m or 100 m contour interval, while a contour interval of 5 m or 10 m usually indicates relatively low relief.

What are hachures?

1. *Contour lines close around hills and basins or depressions.* To decide whether you are looking at a hill or a basin, you must read the elevation numbers. Hachure (ha SHOOR) lines are used to show depressions. Hachures are short lines placed at right angles to the contour line, and they always point toward the lower elevation. Examine area A on Figure 3–19.
2. *Contour lines never cross.* Contour lines are sometimes very close together. See areas B and C. But, each contour line represents a certain height above sea level. Thus, one site cannot be at different elevations.
3. *Contour lines appear on both sides of an area where the slope reverses direction.* Contour lines show where an imaginary horizontal plane would cut a hillside or both sides of a valley. See area C on either side of Stink Lake.

Why do contour lines "V" upstream?

4. *Contour lines form V's that point upstream when they cross streams.* Streams cut beneath the general elevation of the land surface, and contour lines follow a valley. In which direction does the stream in area D flow?
5. *All contour lines either close or extend to the edge of the map.* No map is large enough to have all its contour lines close. See area E and all edges of the map in Figure 3–19.

FIGURE 3–19. Voltaire, ND

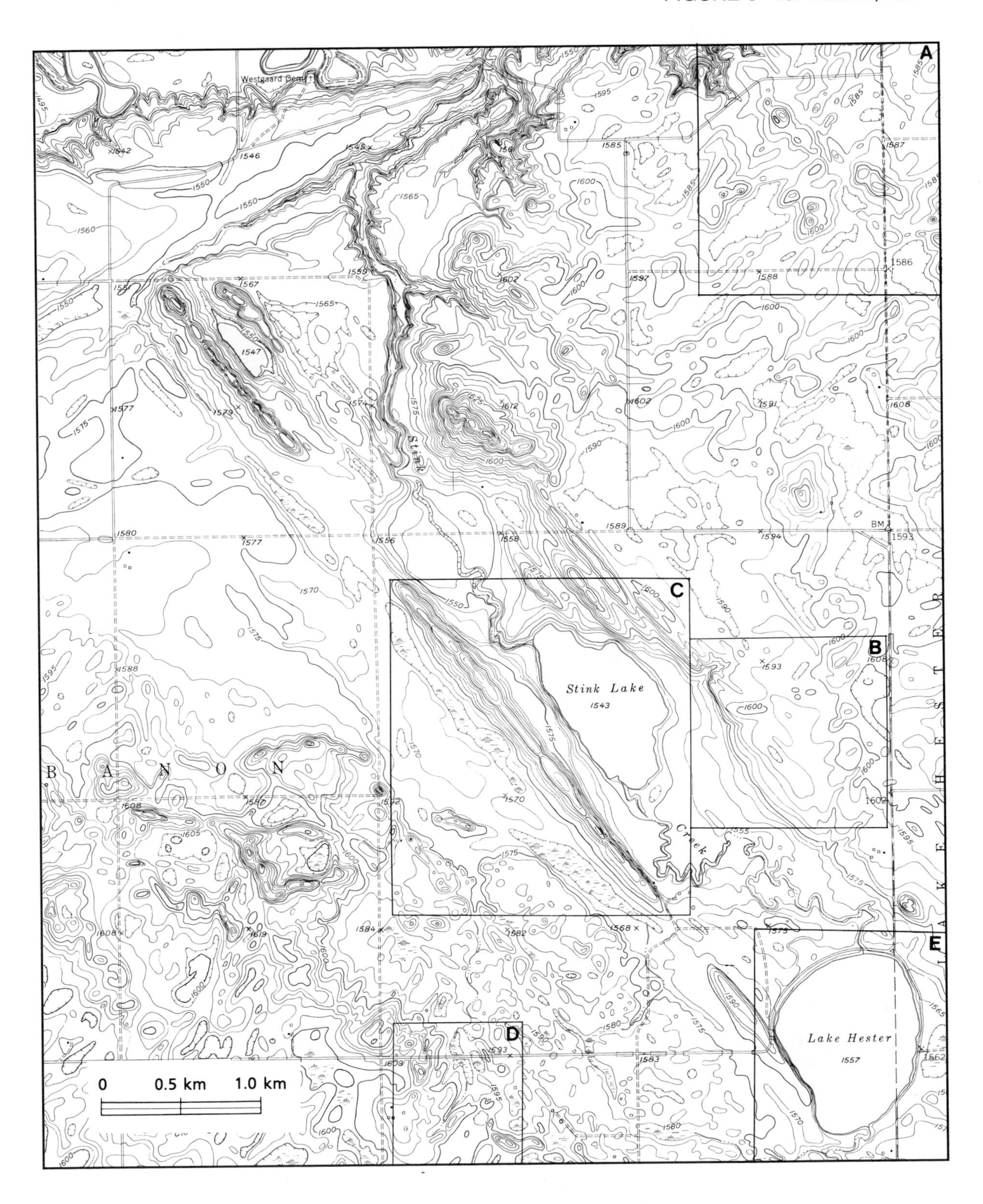

INVESTIGATION 3–1

Determining Elevation

Problem: How is elevation indicated on a map?

Materials 

plastic model landform
water
transparency
clear plastic storage box with lid
beaker
metric ruler
tape
glass marker

Procedure

1. Using the ruler and the glass marker, make marks up the side of the storage box 2 cm apart (Figure 3–20).
2. Secure the transparency to the outside of the box lid with tape.

FIGURE 3–20.

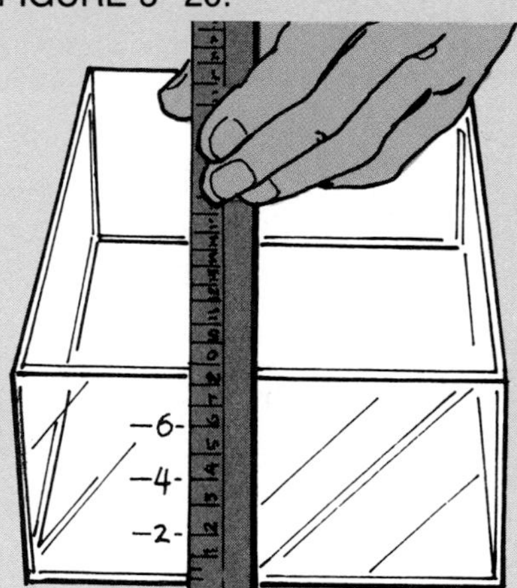

3. Place the plastic model in the box. The bottom of the box will be zero elevation.
4. Using the beaker, pour water into the box to a height of 2 cm. Place the cover on the box.
5. Trace the top of the water line on the transparency. See Figure 3–21.
6. Using the scale 2 cm = 10 m, mark the elevation on the line.
7. Remove the lid and add water until a depth of 4 cm is reached.

FIGURE 3–21.

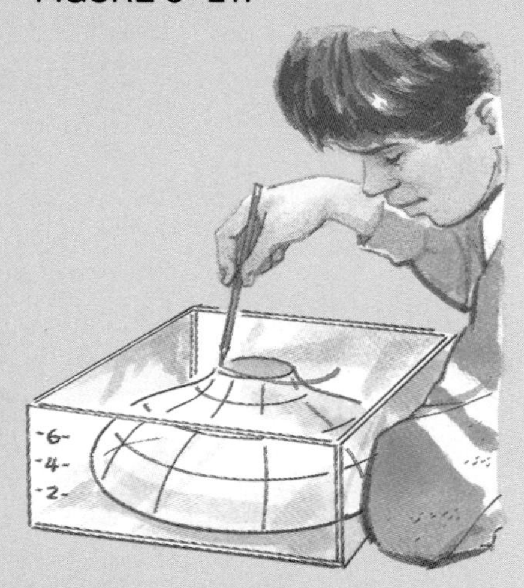

8. Map this level on the lid and record the elevation.
9. Repeat the process of adding water and tracing until you have the hill mapped.
10. Transfer the tracing of the hill onto a white sheet of paper.

Analysis

1. What is the contour interval of this topographic map?
2. How does the distance between contour lines on the map show the steepness of slope on the landform model?
3. What is the total elevation of the hill?
4. How was elevation represented on your map?

Conclusions and Applications

5. How are elevations shown on topographic maps?
6. Must all topographic maps have a 0 m elevation contour line? Explain.
7. How would the contour interval of an area of high relief compare to one of low relief on a topographic map?

INVESTIGATION 3–2

Reading Topographic Maps

Problem: How can distances and features be determined on a topographic map?

Materials

Figure 3–17
paper
Appendix E

NOTE: The contour interval on Figure 3–17 is expressed in feet above sea level.

Procedure

1. Lay a piece of paper along a straight line between Matunuck and Green Hill. Use the first letter of each name for the measurement.
2. Make a mark on the paper where the two towns are located.
3. Move the paper to the scale and determine the distance.

Questions

1. What is the distance between the two towns? Is this measurement the same distance you would cover if you traveled by car between the two cities? Explain.
2. How would you measure this same distance using a ratio scale?
3. How can you determine the distance between two points on a topographic map?
4. What is the contour interval of this map?
5. What is the approximate elevation of the Matunuck School?
6. How many closed contour lines would Green Hill have if the contour interval were 50 feet?
7. Would a contour interval of 50 feet give more or less detail? Explain.
8. What is the distance between the Matunuck School and the intersection of Moonstone Beach Road and Schoolhouse Road?
9. If sea level rose 50 feet, would the State Trout Hatchery be covered by water?
10. What is the relief of the area south of Post Road?
11. Locate the closed contour line just northeast of the intersection of Moonstone Beach Road and Card Ponds Road. Is this feature a hill or a depression? Explain.
12. What is the approximate elevation of Mill Pond?
13. What is the approximate elevation of Green Hill Swamp?
14. What is the elevation of Schoolhouse Road where it intersects Green Hill Road?

FIGURE 3–22.

FIGURE 3–23. How would topographic maps of these two areas differ?

3:8 Interpreting Topographic Maps

The United States contains many kinds of landforms. Recall that these features formed as a result of many forces acting on rocks over varying lengths of time. To better understand landforms, we need to look at them in more detail. One way to study landforms is by using topographic maps.

Both erosion and deposition along shores can be seen on topographic maps. Some coastlines are irregular because the sea has risen and covered former hills and valleys. The hills now appear as islands; the valleys are bays. Features of a coastline, such as spits, bars, and beaches, indicate that currents are active in shaping the boundary between land and water. Figure 3–18 is a topographic map of a shoreline.

Regions once covered with ice also may be recognized from topographic maps. Figure 3–19 is a topographic map of a glaciated area. A glacier is a massive body of ice in motion. As glaciers move over Earth's surface, they deposit materials, as well as erode the bedrock over which they move. Note the long, narrow ridges on the map. These are valley glacial deposits. Clusters of small lakes are common in regions that were once covered by continental glaciers.

How can the presence of limestone be recognized on a topographic map?

A topographic map can indicate what types of rocks are present at the surface. Look at Figure 3–24. This map shows a "negative" topography. That is, many sinkholes are present on the map. Many of these depressions are filled with water. The rock, limestone, was weathered by rainwater to produce this type of topography. While many

FIGURE 3–24. Horse Cave, KY

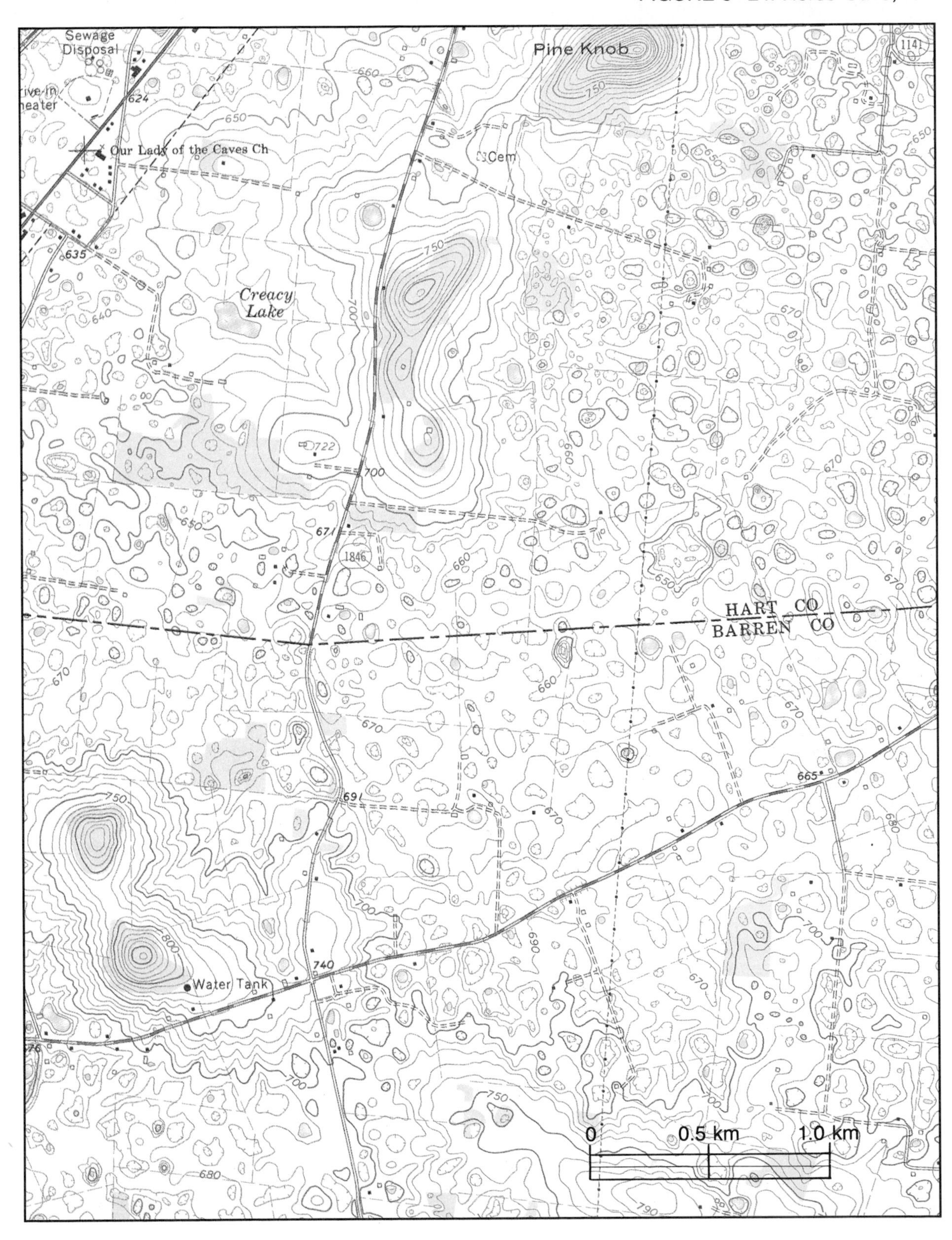

BIOGRAPHY

Gerhardus Mercator
1512–1594

Mercator is remembered for his map projection of Earth. By drawing lines of longitude and latitude at right angles, he was able to make a flat map of the spherical Earth. His projection is a very good approximation of Earth's surface near the equator, but stretches the polar areas so that they are quite distorted.

of the holes are filled with water, others drink up water and stay open. Streams disappear into the ground at some of these holes.

Figure 3–25 is a map of a highly eroded landscape. What basic type of landscape is shown on this map? What is the direction of flow of the stream in Timber Canyon? How can this direction be determined?

Review

11. What is topography?
12. What does a map scale indicate?
13. How does the spacing of contour lines indicate the relief of an area?
★ **14.** What is the highest possible elevation on a topographic map with a contour interval of 10 meters when the highest contour line is labeled 460?

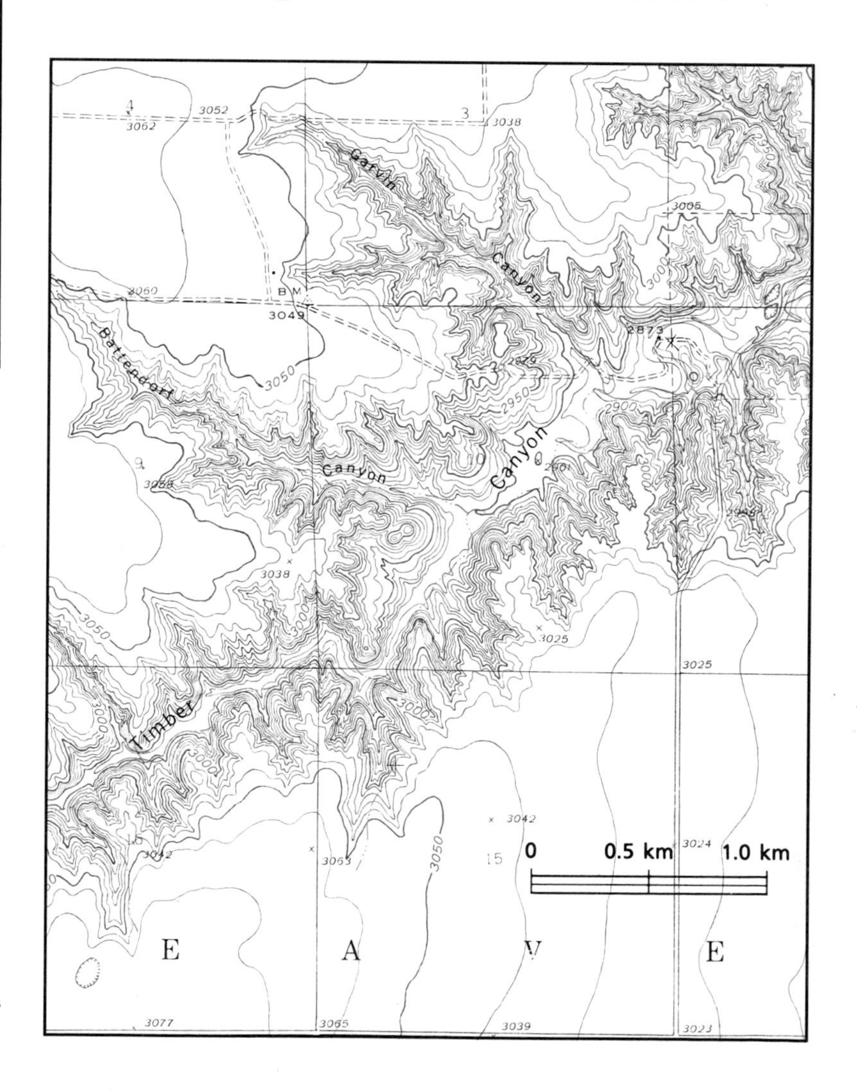

FIGURE 3–25. Lake McBride, KA

PROBLEM SOLVING

A Climb to the Top

The map below is a topographic map of an area in California. One sunny day, three hikers started from the point marked with the + in the center of the map. One hiker headed for the peak of Cedar Mountain, another for the peak of Orr Mountain, while the third intended to climb to the top of Garner Butte.

All three would travel at the same rate on flat or gentle slopes. The climb would be slower on the steeper slope. If each hiker could choose any route to take to the top of the intended goals, which one do you think would reach the top first? Explain. Could the three hikers see each other at the top? Why or why not?

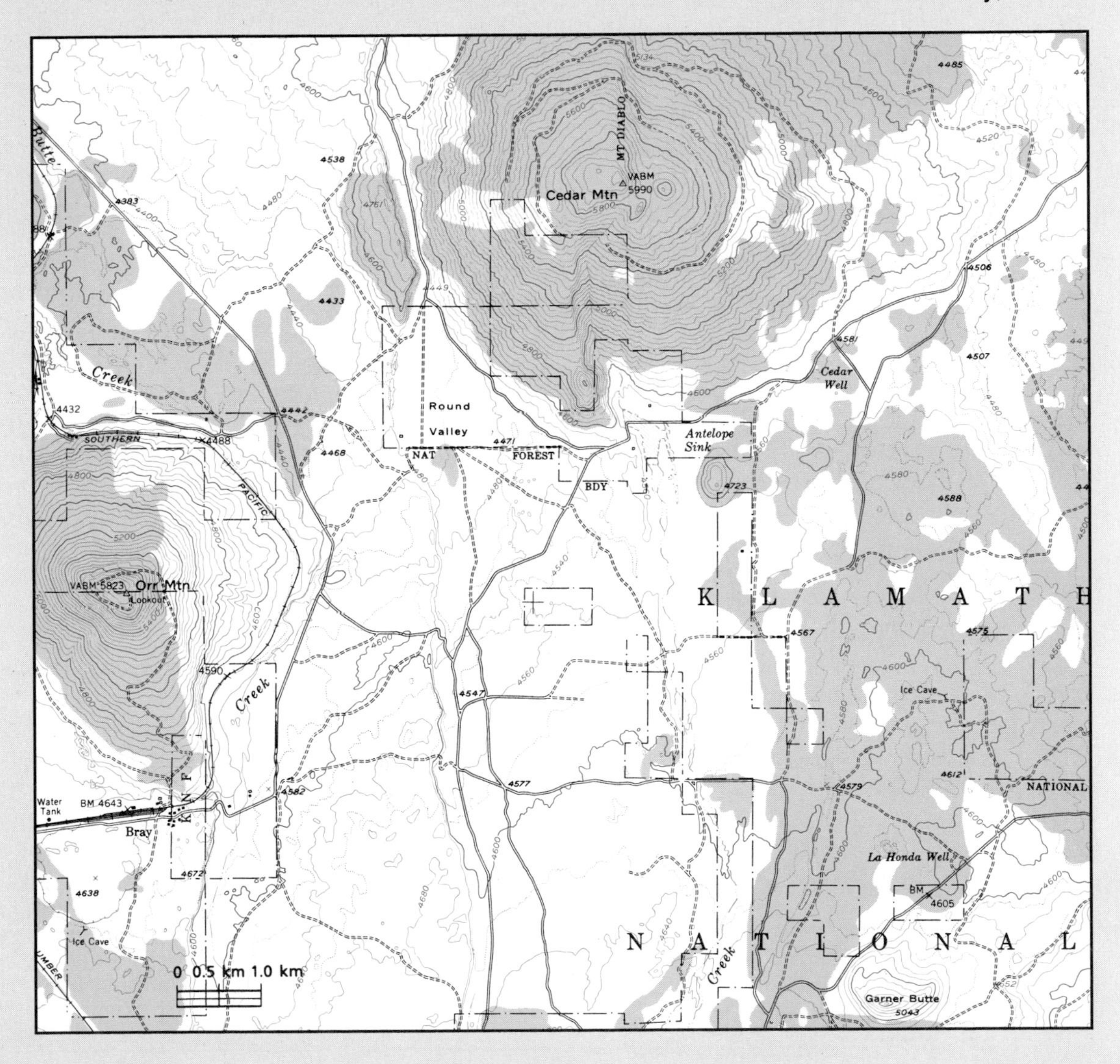

FIGURE 3–26. Bray, CA

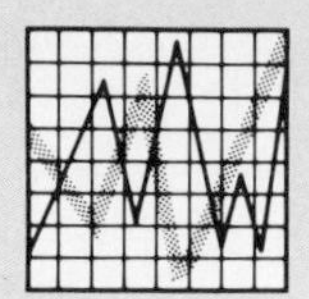

SKILL

Making a Map

Problem: How can you make a map?

Materials

sextant
hammer
meter stick
string, about 20 m
stakes, about 1.5 m long (2)
wrapping paper, about 1 m

Procedure

1. Select an area outside, no larger than 100 m^2, that you wish to map. Hammer the stakes into the ground.
2. With two stakes (labeled A and B) and a string, mark and measure a baseline along one edge of your area. The baseline should be no longer than 20 m. Record the length of your baseline.
3. Hold the sextant on top of stake A with the base of the sextant in line with the string (Figure 3–27).
4. Sight the object that is to be mapped.
5. In the data table, record the object sighted and the angle indicated by the sextant. See Figure 3–27.
6. Go to stake B and sight the same object and record this angle in the table. The sextant should be attached to the stake.
7. Repeat Steps 3–6 for all objects you wish to map.
8. At the bottom of the wrapping paper, draw your baseline using a scale of 1 cm = 1 m.
9. Mark the ends of the baseline A and B to correspond to stakes A and B.
10. Place your sextant at each end of the base line and draw the angles recorded for the first object sighted. The point where the two lines meet is the location of the object. Use this same method to locate and map all of the objects sighted.
11. Measure and record the map distance, in cm, between some of the objects. Now, go outside and measure the actual distance in meters. Record the actual distance.

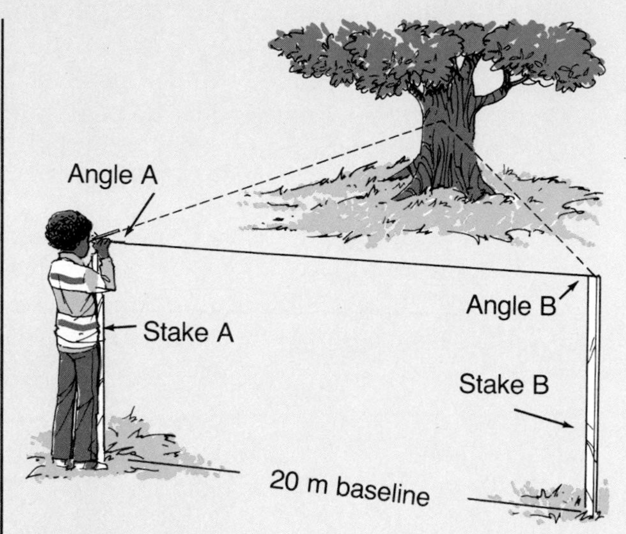

FIGURE 3–27.

Questions

1. Why was it necessary to measure your outside baseline?
2. If you had mapped a larger area, would your map have been more or less accurate? Explain your answer.

Data and Observations

Object	Angle at Stake A	Angle at Stake B	Distance on Map (cm)	Actual Distance (m)

Chapter 3 Review

SUMMARY

1. Latitude lines are parallel to the equator and are measured in degrees either north or south from the equator. 3:1
2. Longitude lines, called meridians, divide Earth into 360°. Each longitude line passes through both poles. 3:1
3. Earth time is determined by the relationship between the longitude position of a point on Earth's surface and the location of the sun in the sky. 3:2
4. Mercator, polar, and conic are three types of map projections. 3:3
5. Plains are vast, flat lowlands. Plateaus are high flatlands usually adjacent to mountains. 3:4
6. Mountain building processes include folding, upwarping, faulting, and volcanic activity. 3:5
7. A map scale is a fixed ratio between a unit of length on a map and the actual distance it represents on Earth's surface. 3:6
8. All points on a given contour line have the same elevation. 3:7
9. Topographic maps show the location, landscape, and cultural features of a small part of Earth's surface. They provide clues to the geologic history of an area. 3:6, 3:8

VOCABULARY

a. contour interval
b. contour line
c. daylight saving time
d. elevation
e. fault-block mountains
f. folded mountains
g. International Date Line
h. latitude
i. legend
j. longitude
k. orogeny
l. plains
m. plateaus
n. prime meridian
o. relief
p. topography
q. upwarped mountains
r. volcano

Matching

Match each description with the correct vocabulary word from the list above. Some words will not be used.

1. the difference in elevation between the highest and lowest points in an area
2. refers to the "lay of the land" or surface features
3. high areas of flatlands adjacent to mountains
4. 0° longitude
5. lines that describe north-south locations in degrees
6. connects points of equal elevation
7. the processes of mountain building
8. the 180° meridian
9. the difference in elevation between two adjacent contour lines
10. the height above sea level of various features shown on a map

Chapter 3 Review

MAIN IDEAS

A. Reviewing Concepts

Choose the word or phrase that correctly completes each of the following sentences.

1. Lines parallel to the equator represent *(longitude, latitude, meridians)*.
2. Exaggeration of land areas near the poles results from a *(Mercator projection, polar projection, globe)*.
3. The time and day in Philadelphia is *(1 A.M. Wednesday, 1 A.M. Tuesday, 7 P.M. Tuesday)* when it is 10 P.M. on Tuesday in Los Angeles.
4. Topographic maps show elevations above sea level by *(latitude lines, contour lines, meridians)*.
5. A large contour interval is needed *(along coastal plains, in flat desert country, in the mountains)*.
6. Closed contours on a map suggest *(rivers, steep cliffs, hills)*.
7. As contour lines cross streams, they *("V" downstream, "V" upstream, become hachured)*.
8. Most road maps are *(conic, Mercator, polar)* projections.
9. High relief is typical of a *(mountainous, hilly, flat)* landscape.
10. The Valley and Ridge province is an example of *(upwarped, folded, fault-block)* mountains.
11. Locations east or west of the prime meridian are identified by *(contour lines, latitude, longitude)*.
12. A hill is indicated on a topographic map with a contour interval of 20 m. The highest elevation indicated by the innermost contour line is 160 m. The highest possible point on the hill is *(160 m, 179.9 m, 180 m)*.
13. By moving either east or west on a globe or map, a change will occur in *(relief, latitude, longitude)*.
14. An airplane departs from New York at 9 A.M. for New Orleans. The flight time is 3.5 hours. What time (local time) will the plane arrive in New Orleans? *(12:30 P.M., 11:30 A.M., 10:30 A.M.)*
15. Which landform will probably have the lowest relief? *(plain, plateau, mountain)*

B. Understanding Concepts

Use Figure 3–18 to answer questions 16 through 20.

16. What is the distance in kilometers between the coast guard lighthouse and Chimney Rock on Point Reyes?
17. In what general direction does Home Ranch Creek flow? Explain your answer.
18. What is the contour interval of this topographic map? *(Note: These units are given in feet above sea level.)*
19. What is the approximate relief of this map?
20. What is the approximate elevation of the base of the radio tower on Point Reyes?

Answer the following questions using complete sentences.

21. Define latitude and longitude. Draw a map to indicate the directions of latitude and longitude.
22. Define elevation. How are contour lines used to represent elevation?
23. How can you tell the difference between a stream and a contour line?
24. How do you know the distance represented by each contour interval? What interval is represented by each contour line in Figure 3–25?
25. Explain why a polar projection gives a more accurate view of the poles than a Mercator projection.

C. Applying Concepts

Answer the following questions using complete sentences.

26. Why is a map considered to be a model?

27. Compare a Mercator projection map of Australia with a polar projection map showing the same area. How do the shapes of the landmasses vary on the two maps? Explain the differences.

28. Find a variety of maps that show rugged relief. In what ways is rugged relief indicated on topographic maps?

29. How are topographic maps useful in studying landforms?

30. Can you think of any situation in which contour lines would cross? Explain.

SKILL REVIEW

1. Determine the actual area on Earth's surface in kilometers of area C on Figure 3–19.
2. What might have occurred had you not accurately measured the height of the water in Investigation 3–1?
3. How could you construct a model of your house or apartment?
4. Deb's science grades for the first 5 months of school were 63, 85, 95, 82 and 97. Make a line graph of these scores. What can you say about Deb's grades?
5. Make a scale diagram of your classroom showing location of desks, chairs, tables, doors, windows, sinks, and shelves or book cases. Compare your diagram with those of others in the class.

PROJECTS

1. Use Figure 3–8 to construct a scale model of the different landscape areas of North America. Find photographs of landforms that provide examples of each area and show how the surface features of each area differ from one another.
2. Construct three-dimensional models that show how the eroding action of streams causes contour lines on a map to appear to loop upstream.

READINGS

1. Campbell, John. *Introductory Cartography*. Englewood Cliffs, NJ: Prentice-Hall, 1984.
2. Madden, James F. *Wonderful World of Maps*. Maplewood, NJ: Hammond, 1986.
3. Simon, Seymour. *Earth, Our Planet in Space*. New York: Four Winds Press, 1984.

SWISS BANK CORPORATION
1 Kilo
PURE PLATINUM
MADE IN ENGLAND

Matter and Its Changes

At first glance, sugar and table salt appear to be quite similar. Both are made of small, white crystals. Both dissolve easily in water. They are quite different, however. They have different properties. Like sugar and salt, every material on Earth is made up of very small particles called atoms. Atoms, in turn, are made up of even smaller particles. How these particles are arranged determines the material's properties; that is, how it appears and reacts to other objects or materials.

ATOMS

All Earth materials are made up of atoms. You will learn how atoms are structured and how they combine with each other. This will help you better understand Earth as a planet. The answers to questions about Earth materials and how they form and react are the keys to answering bigger questions about the universe itself.

GOALS

1. You will learn about the structure of matter.
2. You will be able to classify different forms of matter.
3. You will learn about the chemical properties of matter.

4:1 Building Blocks of Matter

Matter is anything that has mass and occupies space. Every material we know is made of matter. Stars, planets, rocks, water, and air are all matter. All living organisms are matter as well. Matter consists of elements. **Elements** are basic substances that cannot be broken down or changed into a simpler form by either chemical or physical processes. Thus, elements are the basic materials of our world.

Ninety naturally occurring elements have been found on Earth. Other elements have been produced in the laboratory. Presently, 109 elements are known. Each element is different from all others, and most can combine in thousands of ways. Your body is made of some of the same elements that make up air, water, and rocks. In

How many naturally occurring elements have been found?

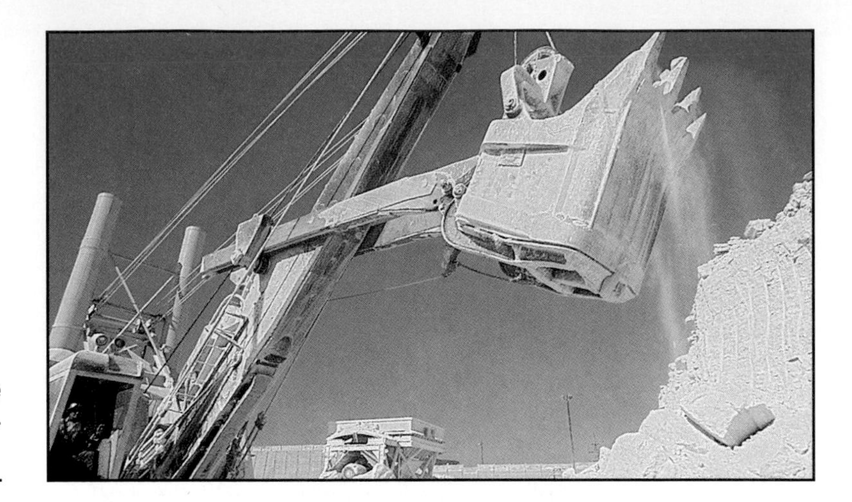

FIGURE 4–1. Sulfur, a native element, is often found in an uncombined form.

turn, the combinations of elements can be separated into the original elements. Of the 90 elements found in nature, only a few are found uncombined. These uncombined elements, such as gold and silver, are **native elements.**

The smallest unit of an element that still retains the properties of the element is an **atom**. Each atom of an element is chemically the same as all the other atoms of that element. Atoms are too small to be seen with an ordinary microscope. Only through the use of X rays can we study the size and shape of an atom.

4:2 Structure of Atoms

To understand how matter changes, we must first understand how atoms behave. Because atoms are too small to be observed easily, we must construct mental models that explain how atoms react with one another.

What three particles compose an atom?

You may already know three of the particles that compose an atom. These particles are the proton, electron, and neutron. A **proton** has one positive electrical charge. An **electron** has one negative electrical charge. A **neutron** has no electrical charge.

According to the modern model, atoms consist of a central positively charged core. The core is surrounded by a cloud of negatively charged particles. The core is called the **nucleus**, and it is made up of positive protons and neutral neutrons. The mass of an atom depends on the number of protons and neutrons present in the nucleus. The **mass number** is the sum of the protons and neutrons. Thus, elements increase in mass as the number of protons and neutrons increases. Electrons move around the nucleus and form a cloud of negative charge. Electrons have different energies. Electrons with low energies are more likely to be found near the nucleus. Those with higher energies are more likely to be found farther away from the nucleus. All the electrons of an atom make up the **electron cloud**. Electrons do not occupy specific places within the cloud. Instead, they should be thought of as being anywhere in the cloud at a given time. Electrons are not considered in the mass of an atom, because they account for only a tiny fraction of the total mass.

The average distance between a nucleus and its outermost electrons determines the size of the atom. Although atoms are very small, there is a large amount of space between the electrons and the nucleus. Even though an atom is mostly empty space, it acts as a unit.

Atoms are electrically neutral; that is, the number of protons is equal to the number of electrons. Each element is made up of atoms. The number of protons in the atoms of an element is unique to that element. For example, all atoms with only one proton are hydrogen atoms. The number of protons in an atom is called the **atomic number**. Sometimes atoms of certain elements lose or gain protons. When a change in proton number occurs, the atom becomes a different element.

Remember that all atoms of hydrogen have only one proton. However, hydrogen atoms can have different numbers of neutrons. The most common form of hydrogen has one proton and one electron. Another form of hydrogen atom has one proton, one electron, and one neutron. A third form of hydrogen has one proton, one electron, and two neutrons. Atoms of the same element that differ

BIOGRAPHY

Maria Goeppert-Mayer

1906-1972

Maria Goeppert-Mayer received the Nobel Prize in physics in 1963 for her work on the electron shells that surround the nucleus of each atom. She worked with Hans Jensen to explain the "magic numbers" of electrons found in each of the shells. By devising the pairing of electrons—called "spin-orbit coupling"—they were able to explain the numbers that had baffled physicists for years.

What determines the size of an atom?

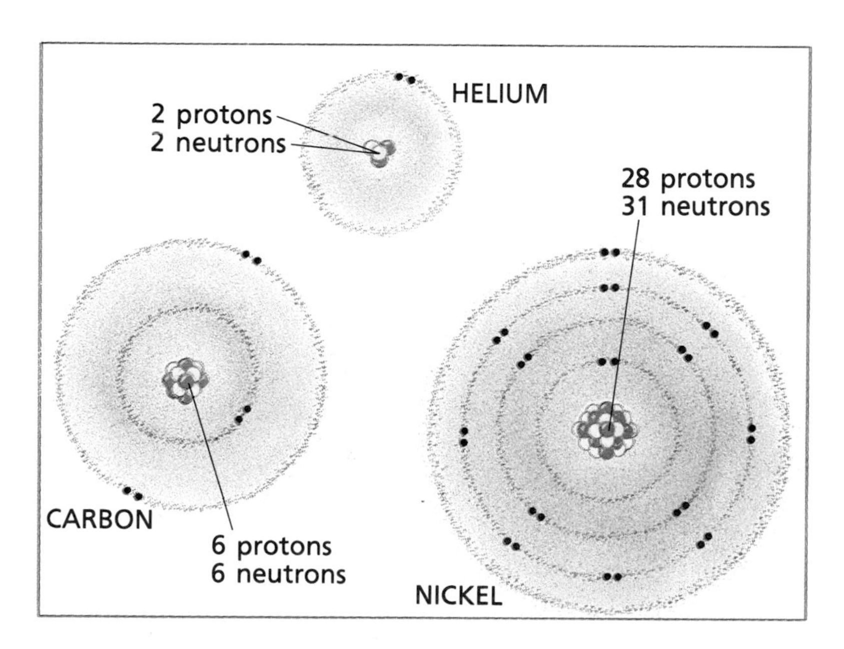

FIGURE 4–2. In an atom, the number of protons equals the number of electrons. Electrons have different energies.

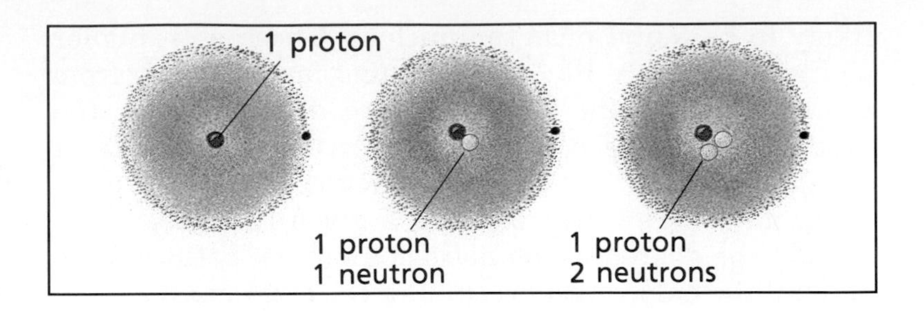

FIGURE 4–3. Isotopes of hydrogen differ in the number of neutrons present in the nucleus. What are their mass numbers?

in neutron number are called **isotopes**. Isotopes of an element all have the same number of protons, but different numbers of neutrons. Most elements have at least two isotopes.

4:3 How Atoms Combine

What determines how atoms behave?

A **chemical property** is a characteristic that depends on the reaction of a substance to form other substances. The number of electrons in the outer part of the electron cloud determines how an atom behaves. Sometimes atoms undergo a chemical change and join to form compounds.

What is the atomic number of gold?

Table 4–1

Selected Elements			
Element	**Symbol**	**Atomic number**	**Mass number***
aluminum	Al	13	27
calcium	Ca	20	40
chlorine	Cl	17	35
copper	Cu	29	64
gold	Au	79	197
helium	He	2	4
hydrogen	H	1	1
iron	Fe	26	56
magnesium	Mg	12	24
nitrogen	N	7	14
oxygen	O	8	16
plutonium	Pu	94	244
silicon	Si	14	28
silver	Ag	47	108
sodium	Na	11	23
sulfur	S	16	32
uranium	U	92	238

*Mass number of the most common isotope.

TECHNOLOGY: ADVANCES

Search for Proton Decay

Most scientists have thought that matter could be neither created nor destroyed. Some scientists, however, believe that all matter has a limited life span, and that in billions and billions of years, all protons that compose matter will have decayed into positrons, positively charged electrons, and pions, subatomic particles. These physicists have built tanks in mines and tunnels to try to detect the decay of a proton in purified water. The tanks are surrounded by light meters that would record the tiny flash of light produced as positrons and pions are emitted from the proton. Using computers, the physicists could then reconstruct the decay event.

Light flashes have been detected in several of these tanks. However, scientists believe that they are seeing the effects of cosmic ray bombardment, or the tiny explosions caused when neutrinos, which move at the speed of light and can pass through Earth, collide with protons. On rare occasions, these explosions look exactly like scientists expect a decaying proton to look like. When the number of flashes that can be expected due to cosmic ray or neutrino bombardment are accounted for, a few recorded flashes remain unexplained. Scientists are continuing to search for definite proof that what they have seen is indeed proton decay.

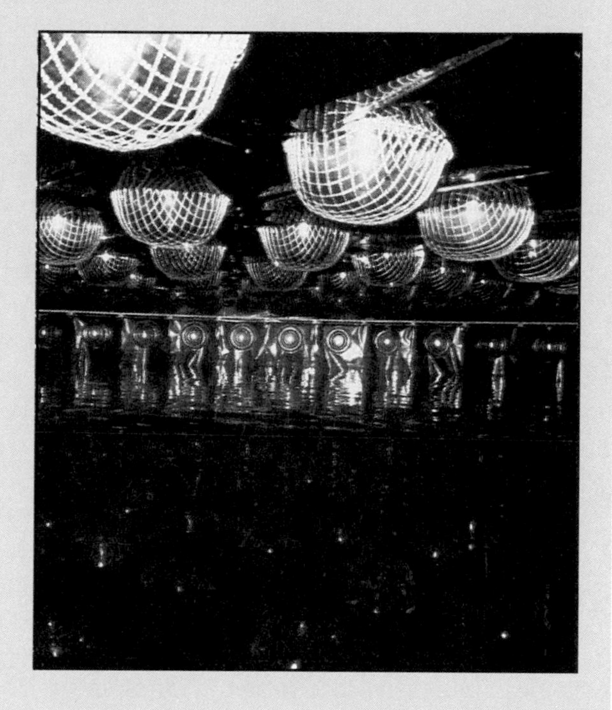

Compounds have properties different from the elements that compose them. Also, the elements in compounds cannot be separated by physical means. We will discuss two types of chemical compounds.

Atoms may chemically combine by sharing the electrons in the outermost part of their electron clouds. A **molecule** is the smallest unit of this type of compound. Each molecule of the compound methane (CH_4) contains one atom of carbon (C) and four atoms of hydrogen (H). Each molecule of water (H_2O) contains one atom of oxygen and two atoms of hydrogen. Some elements naturally occur as molecules in which two or more atoms of the same elements are chemically joined. Atmospheric nitrogen (N_2)

FIGURE 4–4. Nitrogen, water, and methane exist as molecules.

has two nitrogen atoms joined as a molecule. Ozone (O_3) has three oxygen atoms joined.

How does an ion become positively charged?

Atoms are electrically neutral. Under some conditions, atoms gain or lose electrons and become ions. **Ions** are electrically charged particles. Ions are positively charged when they have lost electrons. That is, they have more positive charges in the nucleus than they have negative electrons in the electron cloud. Ions are negatively charged when they gain electrons. Then the nucleus has too few protons to balance the electrons that have been gained. Ions combine chemically to form electrically neutral ionic compounds. In ionic compounds, negative ions join enough positive ions to form a new, neutral substance.

F.Y.I. Mixtures can be classified as homogeneous or heterogeneous. Bottled salad dressing is a heterogeneous mixture. Vinegar is a homogeneous mixture.

A **mixture** is a physical combination of different substances. Mixtures may be combinations of elements or compounds in any proportion or amount. Soil is an example of a mixture. Soil may contain rock fragments and decayed plant or animal matter. These materials can be separated by physical means. The rock fragments may be large enough to be picked out by hand. A sharp needle may be used to separate the particles. Each part that makes up a mixture keeps its own properties.

FIGURE 4–5. Negative and positive ions may join to form ionic compounds (a). Sodium ions and chlorine ions combine to form salt (NaCl) (b).

a

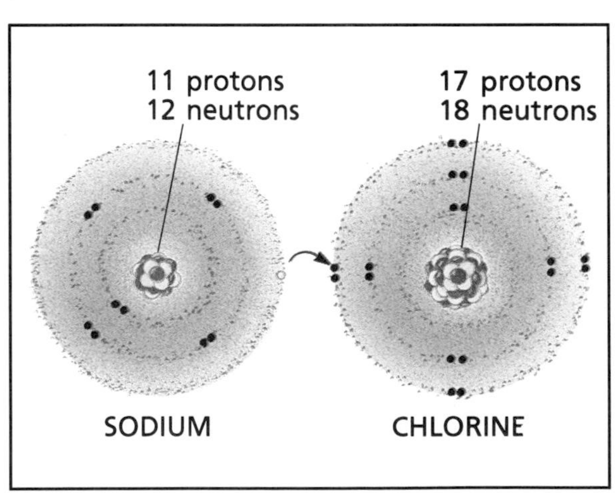

b

Mixtures may be in the form of a solution. A **solution** is a mixture containing two or more substances, one of which is dissolved in the other. Salt dissolved in water is a solution. The salt and water may be separated by physical means.

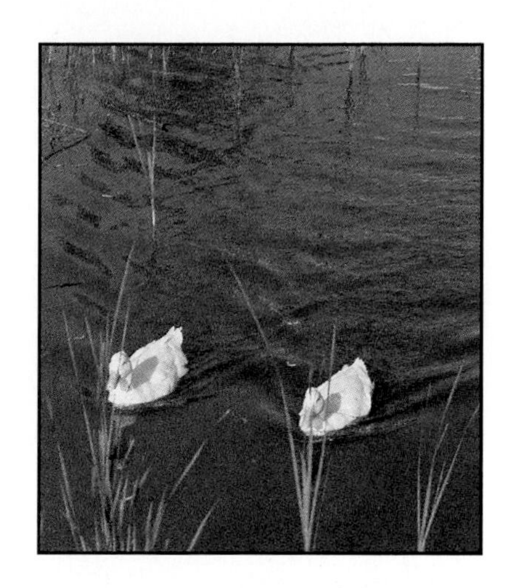

FIGURE 4–6. Pond water is a mixture of water, sand, clay, and other substances.

Review

1. Define matter in terms of elements.
2. Define elements in terms of atoms.
3. Describe the three particles that make up an atom.
4. How do atomic number and mass number differ?
5. ★ Determine the atomic number and mass number of uranium from Table 4–1. How many protons and electrons are present in this isotope of uranium? How many neutrons are present in this isotope of uranium?

PROBLEM SOLVING

Filled to the Rim???

Janice went to her favorite delicatessen on a bitter cold day in January. After carefully studying the menu, she ordered a corned beef sandwich on rye bread and a cup of hot tea. Her waiter brought the food to her table, but much to Janice's dismay, he had filled her teacup to the rim. Janice wondered how she was going to put her usual two teaspoons of sugar into the tea. Rather than risk burning her fingers by trying to pour some of the tea onto the saucer, Janice put a heaping spoonful of sugar into the tea. It did not overflow. Janice was surprised. She then added another spoonful of sugar. Intrigued, she added still another spoonful of sugar. As she ate her sandwich, Janice thought about the seemingly full cup of tea. Why had the tea not spilled from the cup when the sugar was added?

INVESTIGATION 4–1

Forms of Matter

Problem: What are some examples of different forms of matter?

Materials

paper
pencil

Procedure

1. Make a chart like the one shown below. Each column should be headed "Mixtures," "Compounds," and "Elements," respectively.
2. Classify each of the items listed below as an element, mixture, or compound, and list each in the proper column on your paper. If you are working in a group, be sure that everyone agrees on how to classify each item before moving to the next one on the list.

coal	muddy water
clouds	carbon dioxide
diamond	salt
clean air	sand
dusty air	silver
gold	limestone
ice	sugar
hydrogen	uranium
oxygen	water
iron ore	steam

Analysis

1. Classify the substances listed above using the terms shown in data table.

Conclusions and Applications

2. How is a compound formed?
3. Can an element be separated from a compound by physical means? Explain your answer.
4. What are some examples of different forms of matter?

Data and Observations

Mixtures	Compounds	Elements
1.	1.	1.
2.	2.	2.
3.	3.	3.
4.	4.	4.
5.	5.	5.
6.	6.	6.
7.	7.	7.
8.	8.	8.
9.	9.	9.

MATTER

Changes that occur in atoms and molecules of a substance are chemical in nature. These changes always affect the chemical properties and behaviors of that substance. Often, however, objects or materials may undergo considerable change without any alteration of their atomic makeup. This is known as physical change. Physical changes can be more rapid and noticeable than changes in chemical properties of a substance.

GOALS

1. You will learn about the physical properties of matter.
2. You will learn about the different states of matter and how temperature and pressure affect them.

4:4 Physical Properties of Matter

Physical properties of a substance are those that can be observed and measured without changing the chemical composition of the substance under study. Physical properties include mass, hardness, color, taste, odor, and how a material breaks. Physical properties depend on the kinds of atoms and their arrangement in an element or chemical compound. Width, length, and volume are properties that depend on the amount of matter present. Properties that depend on matter itself include density, malleability, conductivity, appearance, and specific gravity.

What are physical properties?

Density is expressed as the measure of mass divided by volume. This ratio is usually given in grams per cubic centimeter (g/cm^3). The density of water at 4°C is 1.0 g/cm^3. The density of Earth is calculated by dividing its mass (5.96×10^{27} g) by its volume (1.08×10^{27} cm^3). Earth's average density is 5.52 g/cm^3. **Specific gravity** is the ratio between the mass of a given substance and the mass of an equal volume of water. This ratio is not expressed in units because the units cancel. Earth's specific gravity is 5.52, or 5.52 times as dense as water.

What is density?

F.Y.I. 1 cm^3 = 1 mL

FIGURE 4–7. Physical properties include the ability to be pounded without losing strength (a), the ability to transfer electric current (b), and appearance (c).

a

b

c

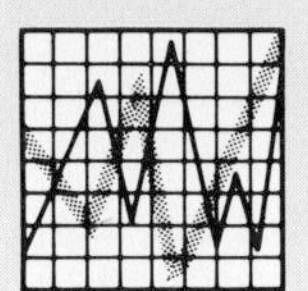

SKILL

Using Laboratory Equipment

Problem: How can you use simple laboratory equipment to make accurate observations?

Materials 

balance (beam)
graduated cylinder (100 mL or larger)
meter stick
thermometer (3)
rock sample
stick or dowel
string
globe
water

Procedure

1. Each of these tasks requires the use of a piece of laboratory equipment to make accurate measurements.
   A. Use a balance to determine the mass, in g, of a rock sample. Be accurate to the nearest 0.1 g.
   B. Use a graduated cylinder to determine the volume, in mL, of the water. Be accurate to the nearest 0.5 mL.
   C. Use 3 thermometers to measure the average temperature, in °C, at a certain location in the room. Be accurate to the nearest 0.5°.
   D. Use a meter stick to measure the length, in cm, of a stick or dowel. Be accurate to the nearest 0.1 cm.
   E. Use a meter stick and string to measure the circumference of a globe. Be accurate to the nearest 0.1 cm.
2. Begin at any station and determine the measurement requested. Record the data and list sources of error that might make your data incorrect.
3. Proceed to the other four stations as directed by your teacher. Complete the procedure, as in Step 2, at each station.
4. Compare your measurements with those who used the same objects. Review the values provided by your teacher. How do the values you obtained compare to those provided by your teacher and those obtained by other students in your class for the same samples?
5. Determine your percentage of error in each case. Use the formula

$$\frac{\text{your value} - \text{teacher's value}}{\text{teacher's value}} \times 100 =$$

% of error.

6. Decide what percentage error will be acceptable. Generally, being within 5 to 7% of the correct value is considered good. If your values exceed 10% error, try the measurement again to see where the error occurred or if the difference in values can be explained.
7. Discuss the kinds of errors that were made in obtaining the values. Determine what can be done to obtain the most accurate results.

Data and Observations

Part	Sample #	Value of Measurement	Causes of Error
A	______	mass = ______ g	
B	______	volume = ______ mL	
C	______ (location)	average temp. = ______ °C	
D	______	length = ______ cm	
E	______ (globe)	circumference = ______ cm	

INVESTIGATION 4–2

Density and Specific Gravity

Problem: How are some physical properties determined?

Materials

Part A
pan balance
graduated cylinder (100 mL)
beaker (100 mL)
small wood block
small metal block
water (10 mL)
piece of clay

Part B
pan balance
graduated cylinder (100 mL)
quartz sample
water (50 mL)

Procedure

Part A: Density

1. With the pan balance, measure the mass of the wood block, metal block, clay, and 10 mL of water. Record your values.
2. Use the graduated cylinder to determine the volume of each sample. Record these values.
3. Calculate the density of each sample by using this equation.

$$\text{density} = \frac{\text{mass}}{\text{volume}}$$

Part B: Specific Gravity

1. Measure and record the mass of the quartz sample.
2. Fill the graduated cylinder with 50 mL of water.
3. Submerge the quartz in the cylinder. Be careful not to lose any water. Record the volume to which the water level has risen.
4. Subtract the original volume of the water from the new volume. This change in the volume of water is the volume of the quartz.
5. Use the formula for density to determine the density of quartz.
6. Determine the specific gravity of quartz using the following equation.

$$\text{specific gravity} = \frac{\text{density quartz}}{\text{density water at } 4^\circ\text{C}}$$

Note: The density of water at 4°C is 1.0 g/mL.

Data and Observations

Object	Mass	Volume	Density
wood			
metal			
clay			
water			
quartz			

Analysis

1. Split the piece of clay into two equal pieces. What is the density of each piece?
2. What is the density of 30 mL of water at 4°C?
3. What is the specific gravity of 30 mL of water?

Conclusions and Applications

4. Does size determine density? Explain.
5. Does size determine specific gravity? Explain.
6. What does determine density and specific gravity?
7. How would you determine the density of a rock sample?
8. How do you determine specific gravity?

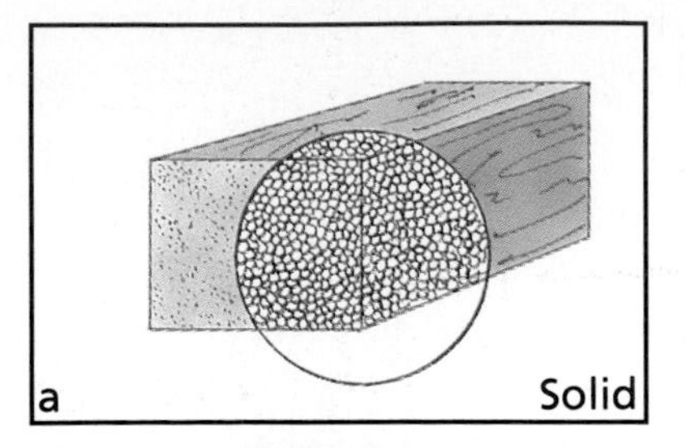

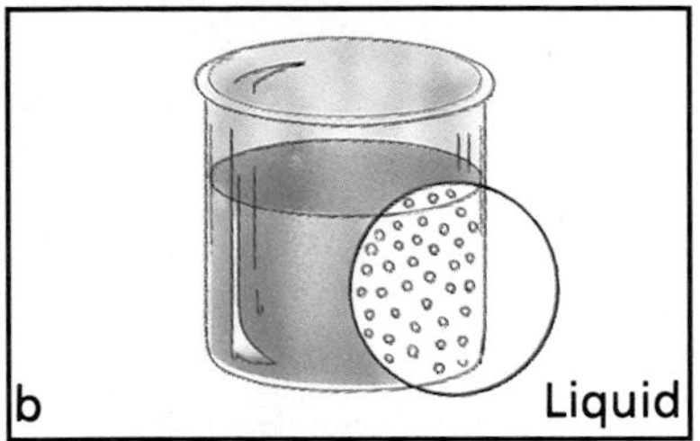

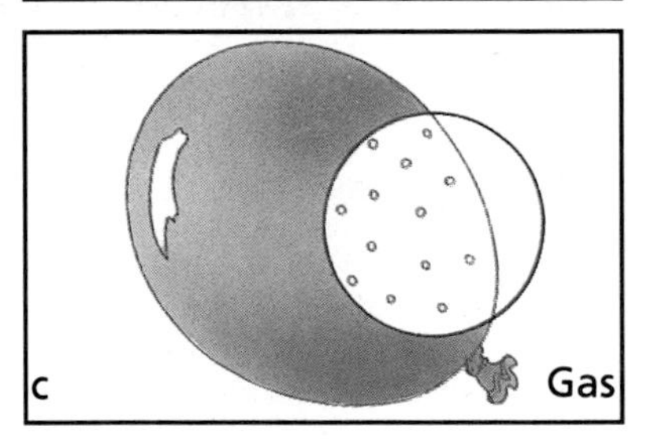

FIGURE 4–8. In a solid (a), each particle has a fixed position. Particles in a liquid (b) remain close to each other as they move. Particles in a gas (c) move freely.

4:5 Physical States of Matter

On Earth, matter occurs in one of three physical states: solid, liquid, or gas. In a solid, each atom or molecule has a fixed position in relation to surrounding atoms. The atom itself may vibrate around a fixed point, but the atom does not move away from the other atoms around it. Solids resist changes in shape and in volume. A few familiar solids include ice, wood, and rocks.

In liquids, atoms or molecules are free to move. The atoms remain close to one another, however, as they change positions. Like solids, liquids resist changes in volume, but a liquid takes the shape of its container.

Atoms or molecules of a gas move freely and independently. Gases fill all the available space in a container, regardless of the container's size or shape. Gases offer no resistance to change in shape. Gases are much less resistant to change in volume than either liquids or solids. Gas pressure occurs when gas molecules bump against the walls of the container.

Many scientists recognize the existence of a fourth physical state of matter, that of plasma. A plasma is a state of matter in which particles are moving freely as in a gas. However, plasma has a higher energy than a gas. All particles in plasma are in the form of ions and free electrons. Plasmas only exist at very high temperatures, such as those present in the sun and other stars.

Differences among the three physical states depend on the distance between atoms or molecules, and the rate of movement of the atoms or molecules. These factors, in

FIGURE 4–9. Matter in the plasma state is generally located in areas of extremely high energy, such as stars.

FIGURE 4–10. Differences among the physical states of water depend on the distance between the molecules and the rate of movement of the molecules.

turn, depend on temperature. In gases, pressure is also a factor. An increase or decrease in temperature causes a similar increase or decrease in the speed of atoms or molecules in solids and liquids. In a gas, a change in pressure causes a change in particle speed.

The addition of energy to atoms or molecules usually will increase the distance between them and the rate of their movement. This will result in an increase of temperature and pressure. For a gas, this is true only if the volume changes. Generally, there is no upper limit to temperature and pressure. However, because temperature is a measure of molecular motion, there is a point at which all motion nearly stops. This point is called **absolute zero.** At absolute zero all energy has been removed. The temperature is as low as it can possibly be. Absolute zero is −273°C, or 0 Kelvin. There is no temperature colder than absolute zero.

F.Y.I. Air is usually thought of as a gas, but at about −190°C, it is a bluish liquid.

Changes in physical state occur at the boiling point, melting point, and freezing point. These temperatures are different for different substances. Liquids change to solids at their freezing points. Solids change to liquids at their melting points. Melting, freezing, and boiling points vary with pressure for each different substance. For example, a decrease in pressure lowers the boiling point. Water at sea level boils at 100°C. But on top of Pike's Peak, about 4300 meters above sea level, water boils at 85°C.

At what temperatures do changes in physical state occur?

The naturally occurring temperature range possible at Earth's surface is limited. Within this range, some substances exist only as solids. Others exist as gases; still others as liquids. Water is unique. It is the only common substance that occurs naturally on Earth as a solid, liquid, and gas.

Review

6. Explain the difference between density and specific gravity.
7. A wood block has a mass of 282 g and a volume of 430 cm^3. What is the density of the wood?
8. What causes a gas to exert pressure?
9. How does the temperature of a gas affect its pressure?
★ 10. Why does pressure affect the boiling point of a substance?

CAREER

Chemist

Joanna Steiner was always curious about why her parents used different products for different household chores. She wondered why they could not use dishwashing soap in the clothes washer.

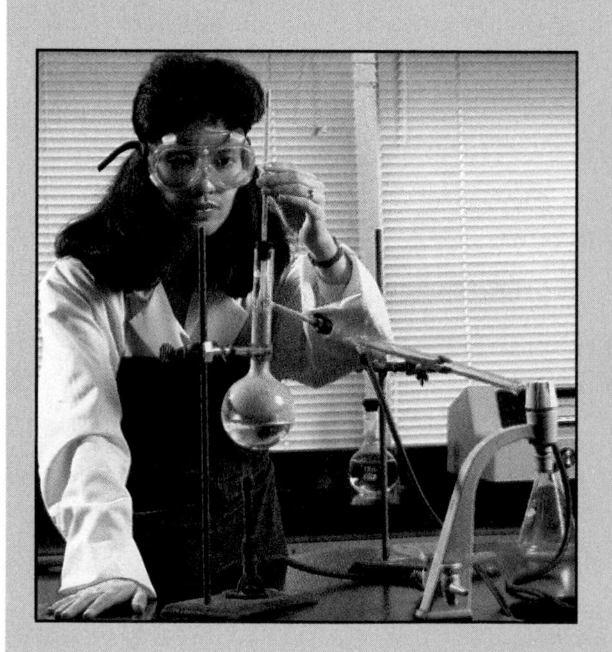

What made these products different? Why was one brand better than another?

It was not until she took chemistry that she began to understand why and how window cleaners cleaned windows and furniture cleaners cleaned furniture. This curiosity lead her to a job as a chemist in a research facility. Today, Ms. Steiner does basic research with materials obtained from plants to design new paints, adhesives, and drugs.

Ms. Steiner uses her knowledge of chemistry and mathematics to develop new products or improve already existing products. She finds that her job takes lots of perseverance, curiosity, experimentation, and the ability to concentrate on detail. Ms. Steiner says there is never a boring moment in her work.

For career information, write:
The American Chemical Society
Career Services
1155 16th St. NW
Washington, D.C. 20036

Chapter 4 Review

SUMMARY

1. Elements are basic substances that cannot be broken down or changed into a simpler form by either chemical or physical processes. 4:1
2. Atoms are composed of protons, electrons, and neutrons. 4:2
3. An isotope is an atom of a given element that differs in the number of neutrons contained in the nucleus. 4:2
4. Atoms that have gained or lost electrons are called ions. 4:3
5. Density is the ratio of the mass of an object divided by its volume. Specific gravity is the ratio of the mass of an object to an equal volume of water. 4:4
6. Matter occurs in one of three physical states on Earth. These states are solid, liquid, and gas. A fourth physical state, plasma, is also recognized by many scientists. 4:5
7. The physical state of matter depends on the distance between the atoms or molecules and the rate of atomic or molecular motion. 4:5
8. Changes in physical state occur at the boiling, melting, and freezing points. 4:5

VOCABULARY

a. absolute zero
b. atom
c. atomic number
d. chemical property
e. compounds
f. density
g. electron
h. electron cloud
i. elements
j. ions
k. isotopes
l. mass number
m. matter
n. mixture
o. molecule
p. native elements
q. neutron
r. nucleus
s. physical properties
t. proton
u. solution
v. specific gravity

Matching

Match each description with the correct vocabulary word from the list above. Some words will not be used.

1. atoms of the same element that differ in neutron number
2. a physical combination of different substances
3. the smallest structural unit of a compound
4. the comparison of the volume of an object to the amount of matter in it
5. anything that has mass and occupies space
6. the sum of the protons and neutrons in the nucleus of an atom
7. pure materials that do not chemically combine with other materials in nature
8. depends on the reaction of a substance to form other substances
9. the core of an atom
10. characteristics that can be observed and measured without changing the chemical makeup of the substance

MAIN IDEAS

A. Reviewing Concepts

Choose the word or phrase that correctly completes each of the following sentences.

1. A substance composed of only one type of atom is called a(n) *(compound, mixture, element).*
2. A(n) *(element, mixture, compound)* can be separated by physical means.
3. Atoms that gain or lose electrons become *(compounds, ions, isotopes).*
4. Atoms of a given element with different mass numbers are *(compounds, ions, isotopes).*
5. The number of *(electrons, protons, neutrons)* determines an element's atomic number.
6. A physical property that cannot be seen is *(color, volume, density).*
7. Solutions are *(mixtures, compounds, elements).*
8. The smallest unit of an element is a(n) *(atom, molecule, compound).*
9. A solid changes to a liquid at the *(boiling, melting, freezing)* point.
10. The way in which atoms combine is dependent on their *(chemical properties, physical properties, physical state).*
11. The physical state of matter where atoms and molecules are closest together is *(gas, liquid, solid).*
12. A uniform solid object is broken into two pieces. The density of either piece will be *(less than, greater than, equal to)* the density of the original piece.
13. Electrons have a *(negative, positive, neutral)* charge.
14. The physical state of a substance that has a higher energy than a gas is a *(liquid, solid, plasma).*
15. The ratio between the mass of a substance and the mass of an equal volume of water is *(density, specific gravity, electron cloud).*

B. Understanding Concepts

Answer the following questions using complete sentences.

16. A brick with a density of 2.7 g/cm^3 is cut exactly in half. Calculate the density of one half. Explain.
17. A block of metal is 6.0 cm in length, 4.0 cm in width and 2.0 cm thick. If it has a mass of 158.4 grams, what is the density of the metal?
18. The density of 6.0 mL of water is 1.0 g/mL. What will the specific gravity of the water be if the volume of water is increased to 12 mL?
19. Where are electrons found in an atom?
20. Describe the relationship between the mass of an element and the number of protons and neutrons in the nuclei of atoms of that element.
21. Identify the effect or result of a change in the number of protons, neutrons, or electrons in the atoms of an element.
22. Describe how the distance and speed between atoms and molecules of a solid or liquid is dependent upon the temperature.
23. Why is water considered to be one of the most unique compounds on Earth even though it is so common?
24. When do changes in physical states of matter occur?
25. How is the size of an atom determined?

C. Applying Concepts

Answer the following questions using complete sentences.

26. How does a decrease in pressure affect the boiling point of water?
27. Why are ions chemically unstable?
28. How are elements and compounds different?
29. Why do liquids and gases take the shape of a container and solids do not?

30. Why are native elements uncommon?

SKILL REVIEW

1. If the carbon atom shown in Figure 4–2 were actually that size, what would the accepted value (to the nearest 0.1 cm) be for the size of an atom of carbon?
2. Use the information provided in Table 4–1 and draw a diagram of an oxygen atom. Assume that the carbon atom in Figure 4–2 is actually that size and use it as a basis for comparison.
3. Suppose you are helping your mother bake a cake. From across the room you see her add two cups of a white powdery substance. You think to yourself, the sugar will make the cake sweet. Is this an observation or an inference?
4. Make a line graph of the atomic and mass numbers for the elements in Table 4–1. *(Hint: Study the data carefully before choosing a scale.)*
5. You are interested in determining the effect of temperature on water's ability to dissolve table salt. Describe the variables and controls of your proposed experiment.

PROJECTS

1. Everyone recognizes that atoms and molecules are very small, but few people realize the tremendous range of sizes that exist between small and large ones. Construct scale models of some of the atoms of the elements shown in Table 4–1 to demonstrate this difference. Using the chemical composition of some common compounds, construct scale models of their molecules. Make a display that will demonstrate these size differences.
2. Some solids show evidence of bending or flowing like liquids. Some gases exhibit the density and flow of liquids. Some liquids are so volatile that they react almost like gases. Find out about these materials that are almost, but not quite, solids, liquids, or gases. Develop a display that gives examples of these materials having the properties of more than one physical state. Allow viewers of your display to try to decide into which physical state the material should be classified.

READINGS

1. Berger, Melvin. *Atoms, Molecules & Quarks*. New York: Putnam, 1986.
2. Palmer, Robert. "What's a Quark?" *Science 85,* November, 1985, pp. 66-70.
3. Taubes, Gary. "Everything's Now Tied to Strings." *Discover,* November, 1986, pp. 34-56.

SCIENCE AND SOCIETY

RADON CONTAMINATION

Radon is now considered one of the foremost environmental health problems in the United States. This odorless, colorless gas is produced by the radioactive decay of one element, uranium, into another element, radium. Radium then decays into two isotopes of polonium—polonium-218 and polonium-214. These in turn emit alpha particles that cling to dust or other materials in the air. When these are inhaled, they may cause lung cancer. Many experts feel that 20 000 cases of lung cancer per year may be caused from the inhalation of radon. This makes radon the second largest cause of lung cancer in the United States. The leading cause of lung cancer is still cigarette smoking.

Background

Radon was first noticed in a residential area of Colorado in 1966 where houses had been built on tailings from uranium mines. Next, it was found in houses built on the sites of old phosphate mines. Several states discovered houses built with concrete blocks made from radioactive phosphate slag. Radon also occurs in high concentrations in many geological formations. Radon in a granite formation was discovered by accident in December of 1984 when Stanley Watras, an engineer in a nuclear plant in Pennsylvania, continually set off the radiation alarms as he left the plant. One day he decided to go through the stations as he entered the plant. The alarms went off. Since he had not been exposed to any on-site radiation yet that day, he realized that the radiation he was carrying must have been brought from his home. When his home was tested, it showed levels of radon contamination 800 times higher than the EPA's safety level of four picocuries for each liter of air. A picocurie is a trillionth of a curie, the common measure of radiation, named after the Curie family, the discoverers of radiation.

There is another unit of measurement used in the measurement of radon levels, the "Working Level," or "WL." One WL is the same as 200 picocuries. The EPA's standard of four picocuries is equal to 0.02 WL. The average house contains one picocurie of radon per liter of air.

Radon seeps into a house from the soil. It can enter through cracks in basement walls, unsealed areas around pipes and

FIGURE 1. Because many people are becoming aware of radon contamination, they are using detection kits.

drains, and so on. How much seeps into a home depends on the bedrock of the area, the type of soil, and the type of home construction. One of the more important features of house design in areas of high radon emissions is the size of the basement. The greater the basement area, the greater the chance of contamination.

Early studies seemed to blame people who had tried to conserve energy. Supposedly, by insulating their homes, they had trapped radon inside the house. The hypothesis was that the radon could not escape through drafts as it could in uninsulated houses. However, it has been proven that the amount of insulation is not a factor in radon contamination.

Case Studies

1. There are varieties of measures that can be taken to try to reduce the flow of radon into a home. These measures include sealing the basement to prevent direct entry and the installation of fans and pumps under the house to reduce the gas pressure and to divert any flow to the outside.

2. The Watras' house required elaborate measures. Uranium ore was found under the family room and pure radium and uranium were found in the basement walls. These elements were probably deposited by the seepage of groundwater carrying the materials. The dirt removed from the Watras' house was so radioactive that dumps in the area would not accept it. It had to be buried on their property. The entire foundation was then encased in plastic to prevent more seepage. A system of drains and vent pipes was installed. After these alterations were made, the house was habitable.

3. The disposal of contaminated soil has become a major problem. The town of Montclair, New Jersey, has been trying to dispose of thousands of drums of radon-contaminated soil since 1985. No one wants to accept it.

Developing a Viewpoint

1. Banning smoking in houses contaminated with radon is one of the first measures recommended by the EPA to treat the problem. Why is this considered important?

2. The EPA has set the standard for safety at four picocuries of radon per liter of air. Some scientists say that figure is "alarmist" and that the safe standard should be between eight and ten picocuries per liter. If you found the level in your basement to be six picocuries, what would you do?

3. Should the law require homeowners to disclose any history of radon contamination in a house to a prospective buyer before a deal is finalized? Explain.

4. Since radon is a natural substance and not a pollutant that can be traced to a given company, who should pay for the cost of cleaning up the contamination?

5. Several bills have been introduced in the United States Congress to fund research and to amend tax laws to allow homeowners to deduct the costs of radon cleanup from their federal income tax. Do you believe the government should become involved in this area? Explain your answer.

Suggested Readings

"A Citizen's Guide to Radon: What It Is and What to Do About It" and "Radon Reduction Methods: A Homeowner's Guide." EPA Regional Offices.

"The Indoor Radon Story." *Technology Review*. January, 1986.

"Measures Offered in Congress on Detection and Cleanup of Radon." *New York Times*. March 13, 1987.

UNIT 2

Telescopes housed in the dome-shaped buildings at Kitt Peak Observatory explore the world beyond Earth. Scientists study stars and other planets to enable them to better understand our planet, Earth, and its star, the sun. What lies beyond Earth? What other tools do scientists use to study the universe?

~10 B.Y.A.* The universe forms as a result of the Big Bang.

1608 Dutch invent the first telescope.

1679 Hevelius measures positions of celestial objects.

EARTH IN SPACE

1781
Frederick Hershel discovers Uranus.

1816
Kaleidoscope is invented.

1868
Helium is discovered in the sun's spectrum.

1938
A hoax of a Martian invasion is broadcast.

1984
Kathy Sullivan walks in space.

EARTH MOTIONS EARTH'S MOON

Chapter 5

Earth-Moon System

The moon is Earth's natural satellite. Earth and moon together form a system that revolves about the sun. For centuries, people wondered what the moon was like. During the twentieth century, humans walked on the moon. Many photographs of the moon have been taken. How do the moon and Earth affect one another? What motions are common to both objects? How did the moon form?

EARTH MOTIONS

GOALS

1. You will learn about Earth's location in space.
2. You will learn about the motions of Earth that cause the seasons.

For thousands of years, humans have carefully observed the day and night skies and used their observations to mark the passage of time. The early astronomers became so exact that they could predict events such as the seasons and eclipses with amazing accuracy. Yet, it was less than 500 years ago before anyone took seriously the idea that Earth itself was moving, as well as those other objects such as the moon, sun, and stars. It was not that no one thought of it. It was just that no one could provide a believable model or explanation for these observations.

5:1 Our Place in Space

When observers first began studying the sky, it appeared that Earth was the center of the universe. This model is called the **geocentric universe**. *Geo* means Earth and *centric* means center. This model was based on observations of the apparent motions of the sun, moon, planets, and stars. When astronomers were able to distinguish between the real motions and the apparent motions of space objects, they realized that the geocentric model needed to be changed. **Real motion** refers to the actual movement of any object in space. **Apparent motion** is the motion of an object as seen by an observer on Earth. Often the real motion can be determined only after careful observation of an object's apparent motion.

How do real and apparent motions differ?

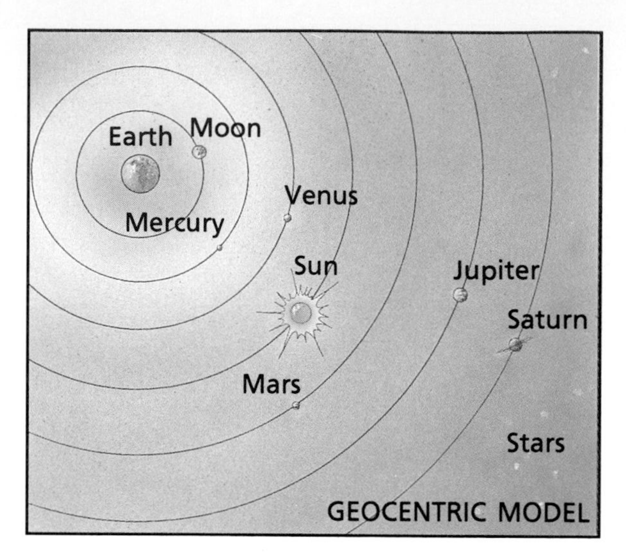

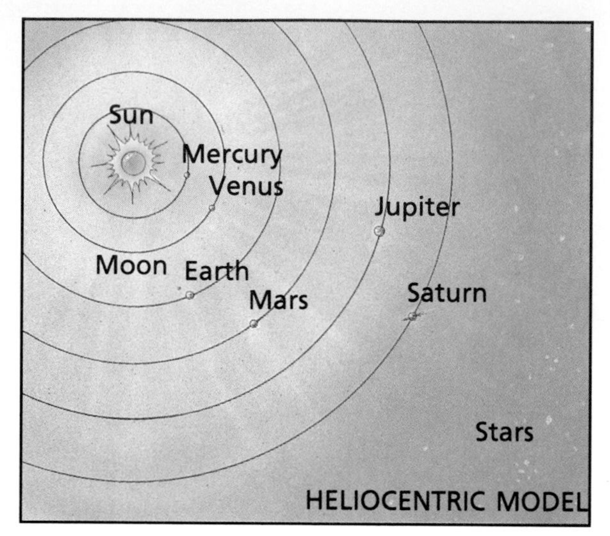

FIGURE 5–1. The geocentric (a) and heliocentric (b) models of the universe were the result of observing apparent motions of space objects.

F.Y.I. Born in Poland in 1473, Mikolaj Kopernik (Nicolaus Copernicus) spent over 30 years observing and studying the planets and stars. His book, *On the Revolutions of the Heavenly Spheres,* was written in 1533, but hidden for 10 years in his home before being published in 1543. Although credited with developing the heliocentric model, he actually discovered the idea by reading some of the manuscripts of a Greek astronomer who lived from 310 B.C. to 230 B.C.

Almost 500 years ago, Nicolaus Copernicus realized that the real motions of Earth, sun, moon, and planets were different from what they appeared to be. He developed a model of the universe known as the **heliocentric universe.** *Helio* means sun. In the heliocentric model, the planets revolved about the sun, which was placed at the center of the universe. **Revolution** is the circling of one object about another.

The heliocentric model was only a single step toward developing a more accurate model of Earth's place in space. After it was determined that neither Earth nor the sun was the center of the universe, astronomers still used a model in which the moon revolved about Earth. Over time, however, it has been shown that the moon's center and Earth's center appear to revolve about a single imaginary point. This point, which is called the **barycenter**, is 1700 kilometers beneath Earth's surface. Present day astronomers use this model and refer to the moon and Earth as a single system revolving about the sun. Each object has its own real motion, and is affected by the presence and real motion of the other.

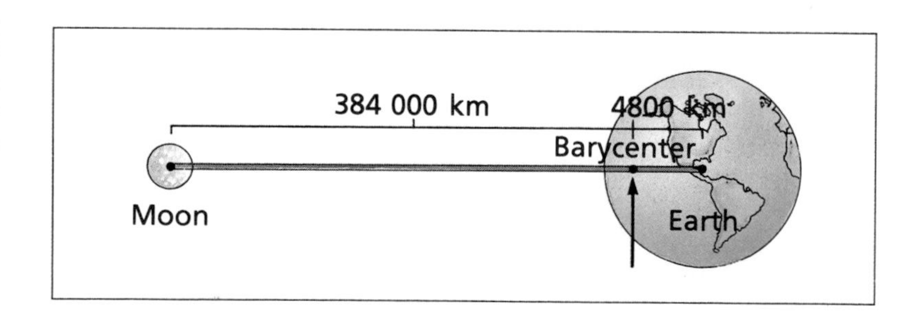

FIGURE 5–2. If a pole connected the centers of Earth and moon, they would balance at their common center of gravity, the barycenter.

5:2 Measurements of Earth

In 1680, Isaac Newton explained why objects move the way they do. He called his explanations the laws of motion. One of Newton's laws states that objects at rest tend to stay at rest. However, once an object is set in motion by an outside force, movement continues in a straight line at a constant speed. This movement continues unless another force acts on the object. This property of matter is **inertia**. Force is needed to make an object move, change direction, or stop moving. The amount of force needed for a given change in motion depends on the mass of the object. The more massive an object, the more force that is required to cause a given change in its state of motion.

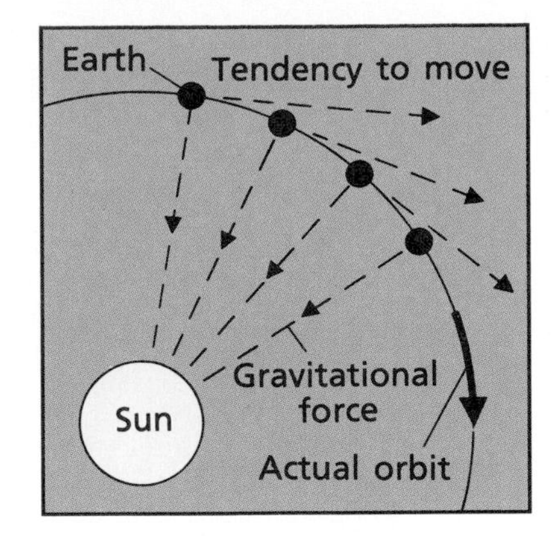

FIGURE 5–3. Gravitational forces and inertia work together to cause Earth to follow a curved path around the sun.

Newton also explained that movements of objects may be related to gravitational forces. **Gravitational forces** depend on the mass of the objects and the distance between their centers. The more mass an object has, the greater the force of gravity it exerts. The sun's gravitational force causes each planet to curve toward it. As a result, planets, including Earth, follow elliptical paths. The actual path followed by a planet is its **orbit**.

On what do gravitational forces depend?

Two real motions of Earth are rotation about its axis and revolution about the sun. **Rotation** is the turning motion of an object. Earth makes one complete rotation on its axis with respect to the sun once every 24 hours. This period is a **solar day**. The time needed for Earth to make one revolution about the sun is the basis for another measurement of time, the **year**. One complete revolution about the sun requires 365¼ days, or one year. As Earth revolves about the sun, its average distance from the sun is 149 600 000 kilometers. Earth is nearest the sun in January, when the sun is about 147 million kilometers away. In July, Earth and the sun are about 152 million kilometers apart.

What are two real motions of Earth?

FIGURE 5–4. Day and night are caused by Earth's rotation on its axis.

Earth's size makes its measurement difficult, but there is much evidence that Earth is nearly spherical. As early as 500 B.C., Pythagoras, a Greek philosopher, concluded that Earth is a sphere. He based his conclusion on observations of Earth's curved shadow falling across the moon.

About 200 B.C., Eratosthenes, a Greek mathematician and astronomer, computed Earth's size. His method was based on two assumptions: (1) Earth is a sphere; and (2) The sun's rays are parallel.

Eratosthenes knew that at noon on June 21, a vertical pole cast no shadow in the city of Syene. Directly north of Syene, in Alexandria, Egypt, a vertical pole did cast a shadow. Eratosthenes measured the angle between an obelisk and its shadow to determine the angle at which light arrived from the sun on June 21. He found the angle to be 7°12′, which is 1/50 of a circle's circumference. He then measured the distance between Alexandria and Syene. This distance, multiplied by 50, gave the circumference of Earth. His measurement was close to the 40 000 km that we use today.

Earth is not a true sphere. It is an oblate spheroid (SFIHR oyd), that is, it is a sphere that bulges at the equator and is flattened at the poles. There are two reasons for this. The first is Earth's rotation on its axis. This causes greater centrifugal force at the equator. The second reason is due to the fact that Earth is made of materials that have different densities.

FIGURE 5–5. Eratosthenes measured the angle of the sun's rays to calculate Earth's circumference (a). What is Earth's apparent shape (b)?

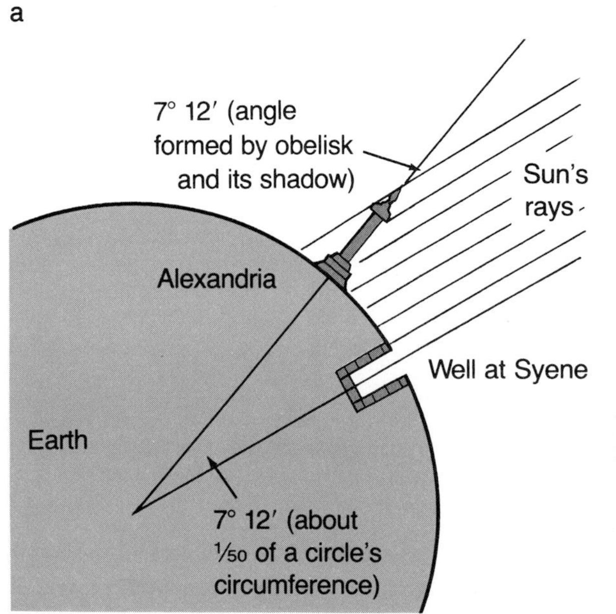

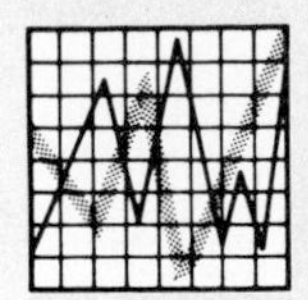

SKILL

Using a Globe

Problem: How do you determine the "roundness" of Earth?

Materials

string
meter stick or metric ruler
globes (2—different sizes)
basketball, volleyball, or soccerball

Procedure

1. Using the string, measure the circumference of the first globe at the equator. Record this measurement in the data table.
2. Measure the circumference of the globe along the prime meridian and International Date Line (180° meridian). Record these measurements in the data table.
3. Determine the "roundness ratio" of the globe by dividing its larger circumference by its smaller circumference. Continue all divisions to four decimal places.
4. Repeat this procedure with a basketball, volleyball, or soccerball. Use any two circumferences that are at right angles (90°) to each other. Record these data.
5. Repeat these measurements with the second globe. Record your data.
6. Use these data and the data of Earth's circumference provided in the Data and Observations table to answer the questions.

Questions

1. What is the roundness ratio of a perfect sphere? Will it always be the same for spheres of any size? Explain.
2. Compare Earth's roundness ratio to the globes and ball. Which is more round? Explain how your data support your answer.
3. How accurate a model of Earth are the globes you used? Explain.
4. How round is Earth and what is the best model to represent this "roundness"?

Data and Observations

Object	equatorial circumference	polar circumference	roundness ratio
Globe #1	cm	cm	
Ball	cm	cm	
Globe #2	cm	cm	
Earth	40 054 km	39 922 km	

TECHNOLOGY: APPLICATIONS

Making a Globe

Globes are uniquely different from maps because globes represent all parts of Earth's surface to scale. Distances, areas, and directions can be observed without the distortion that is caused by map projections.

Several types of globes are made today. One type of globe dates back to 1810. Printed and cut strips of cigar-shaped paper called gores are hand mounted on a sphere made of plastic. In the past, spheres were made of various materials, including plastic, spun aluminum, and wood. These are called hand-covered globes.

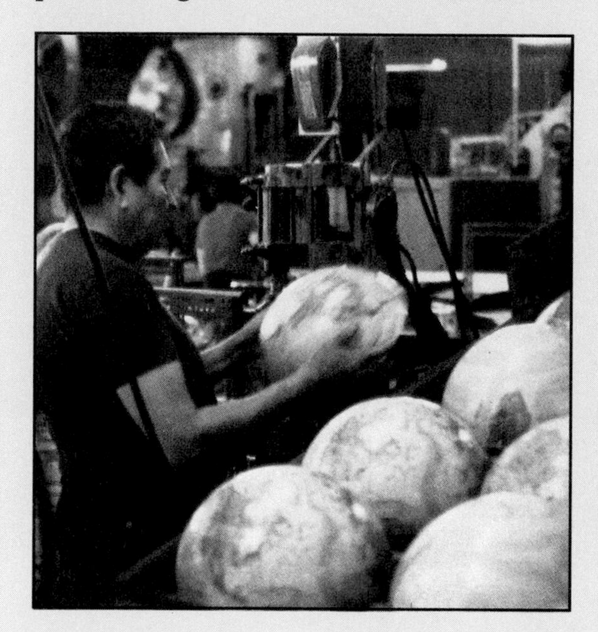

A press-craft globe is another type of globe. This type has been made for 50 years and involves 12 half gores, which stop at the equator, radiating outward from a polar cross. This map sheet of gores is mounted to special composition board, which is also cut into the same "pressed flower" shape and covered with glue. This is then placed into a mold and hydraulically pressed into a hemispheric shape. Each hemisphere is then trimmed at the equator and joined using glue and a fiber reinforcement inside the equator. An equator tape is added at the center and the globe receives a coat of protective lacquer.

Globes are now made by automated machinery that mounts the map to the board, cuts out blank composition board, punches center holes, and performs other tasks that used to be done by hand. Globes are available in various diameters. Currently, seven hundred globe models are made in ten languages.

5:3 Earth Motions and the Seasons

Careful measurements show that the amount of solar energy emitted by the sun changes very little. However, the amount of solar energy received at a particular latitude on Earth varies. As it revolves about the sun, Earth rotates on its axis, which is tilted 23½° from the perpendicular to Earth's orbital plane. Seasons occur because the tilt of Earth on its axis causes the length of daylight to vary. Also, the angle at which the sun's energy strikes a given location changes through the year.

What is the angle of tilt of Earth's axis?

Look at Figure 5–6, which shows Earth in its orbit during one year. When the North Pole is tilted toward

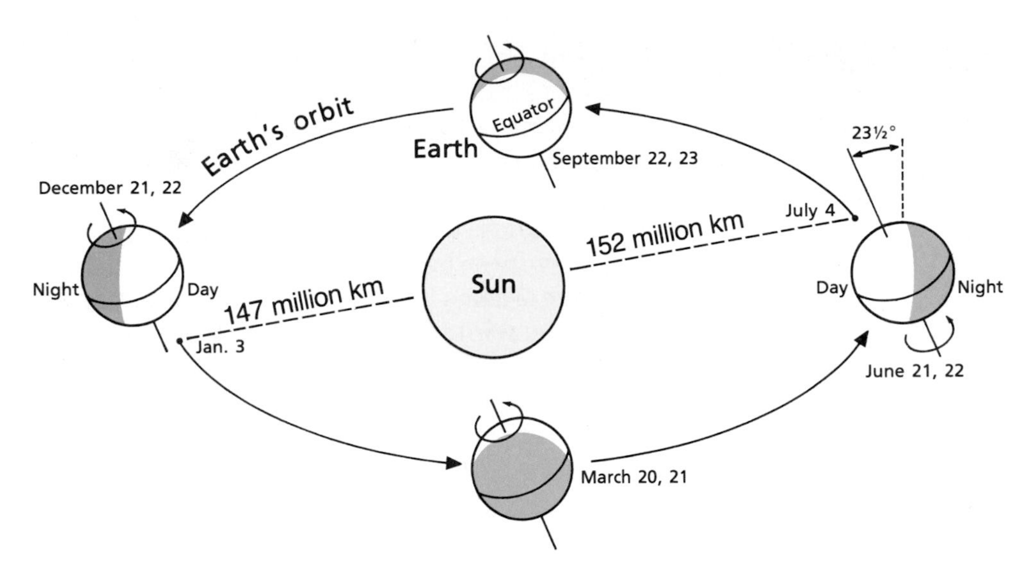

FIGURE 5–6. The tilt of Earth's axis causes the amount and angle of sunlight at a particular latitude to vary throughout the year.

the sun, the Northern Hemisphere has summer. The North Pole is tilted most directly toward the sun on June 21 or 22, the summer **solstice.** On this day, locations north of the Arctic circle have 24 hours of daylight. In the Southern Hemisphere, daylight hours are less and the South Pole has 24 hours of darkness.

After the summer solstice daylight hours begin to decrease in the northern latitudes. On September 22 or 23, the fall **equinox**, Earth's tilt is sideways with respect to the sun. Hours of daylight and darkness are the same in both the Northern and Southern Hemispheres. On this day, fall begins in the north, and spring begins in the south.

FIGURE 5–7. On December 21 or 22, the South Pole is pointed most directly toward the sun, and the winter solstice occurs (a). On June 21 or 22, the North Pole is pointed most directly toward the sun, and the summer solstice occurs (b).

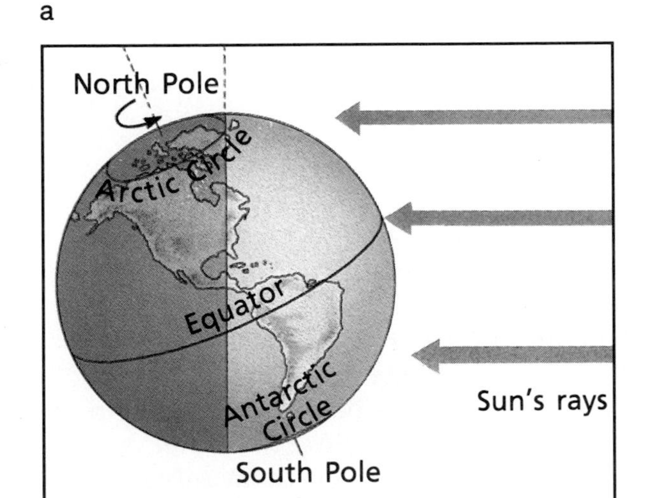

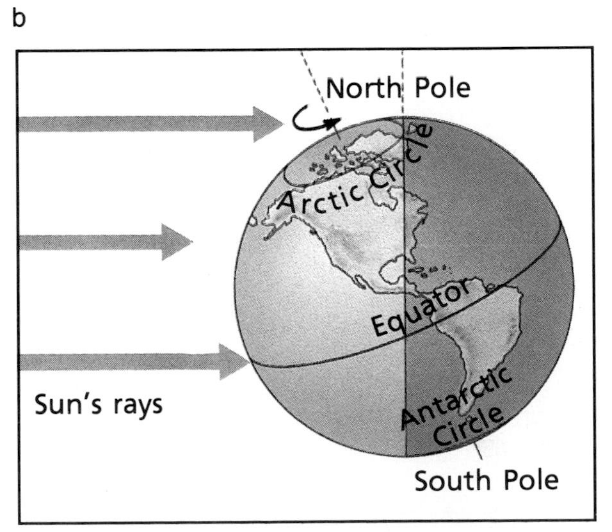

What occurs on December 21 or 22?

Daylight hours continue to decrease in the Northern Hemisphere and increase in the Southern Hemisphere until December 21 or 22, the winter solstice (Figure 5–7). The Northern Hemisphere receives the smallest amount of solar energy because it is tilted away from the sun. On this day, the Northern Hemisphere has its longest night and winter begins. Locations south of the Antarctic circle have 24 hours of daylight and summer begins.

After the winter solstice, daylight hours once again increase in the north. On March 20 or 21, Earth is again sideways with respect to the sun. On this day, called the spring or vernal equinox, daylight and night hours again are equal for both hemispheres. March 21 marks the first day of spring in the Northern Hemisphere, and the first day of fall in the Southern Hemisphere. On June 21 or 22, the cycle of the seasons repeats. At the equator, daylight and darkness are always about equal length.

Review

1. Compare a heliocentric with a geocentric model of the universe. Is either correct? Explain.
2. Explain how Earth and the moon might be thought of as a two-planet system orbiting the sun.
3. Identify two real motions of Earth.
4. How do revolution and rotation differ?
5. ★ Why do equinoxes and solstices not occur on exactly the same date each year?

PROBLEM SOLVING

A Solar Secret

Jennifer and Edmond live in North America and were comparing dates of the year with information they knew about the sun. Jennifer knew very well that summer begins in June and winter in December. Edmond, on the other hand, had read that Earth is closest to the sun in late December and January, and farthest away in late June and July. How then can summers be warmer than winters? Should it not be the other way around? Can you help them solve this puzzle?

CAREER

Astronaut

After medical school, Dr. Margaret Seddon completed an internship and residency with an interest in surgical nutrition. She was later selected as an astronaut candidate by NASA. After completing a one-year training and evaluation period, Dr. Seddon became eligible for assignment as a mission specialist on space shuttle flight crews. Her work at NASA includes working with computer software and acting as a launch and landing rescue helicopter physician. Dr. Seddon made her first flight as a mission specialist in 1985. She currently works in the Astronaut Office.

For career information, write:
National Aeronautics and Space Administration (NASA)
Lyndon B. Johnson Space Center
Houston, TX 77058

EARTH'S MOON

GOALS

1. You will learn about the motions of the moon and the phases that result from these motions.
2. You will gain an understanding of lunar and solar eclipses.
3. You will learn about the history of the moon.

Is the moon visible to an observer on Earth every clear, cloudless night? Is the moon visible only at night? Does the moon appear to rise and set at about the same time each night or day? Surprisingly to some, the answer to all of these questions is no. Careful observations of the moon for a few nights will show the moon slightly farther east in the sky at the same time each evening. During a single evening, however, the moon appears to move rapidly westward. Which of these motions are real? Which are only apparent?

5:4 Measurements of the Moon

The moon rotates on its axis and makes one revolution around Earth with respect to the stars about every 27⅓ days. Because the rotation and revolution of the moon take about the same amount of time, a casual observer on Earth always sees the same side of the moon. However,

Why does a casual observer always see the same side of the moon?

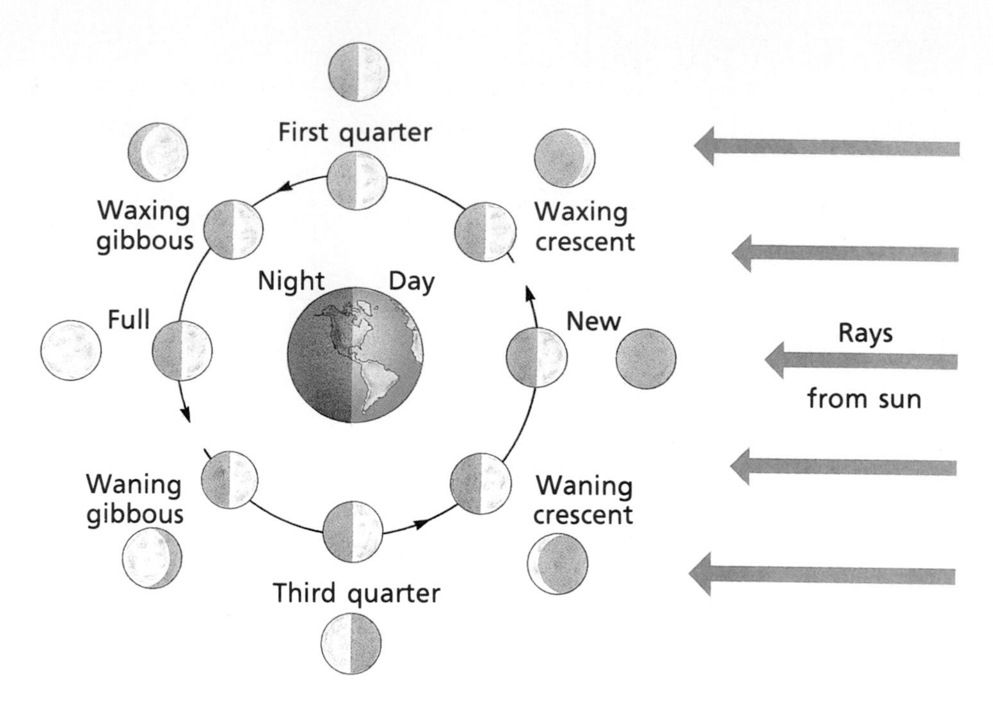

FIGURE 5–8. The outer circle shows the phases of the moon as seen from Earth. The inner circle shows the appearance of the moon to an observer in space. Why do we only see one side of the moon?

F.Y.I. A waxing quarter moon is visible from Earth only during the evening. A waning quarter moon is visible from Earth only during the morning.

When is the moon waning?

because of the 6½° tilt of the moon's axis, a careful observer can see beyond the moon's poles at certain times.

The phases, or changing appearance of the moon as seen from Earth, depend on the moon's position relative to the sun. When the moon lies between the sun and Earth, the side of the moon facing Earth is dark. See Figure 5–8. This is the **new moon** phase. As the moon moves eastward in its orbit, more of its sunlit side becomes visible, and the moon is said to be waxing. When the sun, Earth, and moon form a 90° angle, the moon appears half bright and half dark. This is the **first quarter** phase. When the sun, Earth, and moon align with Earth in the middle, the entire side of the moon facing Earth is bright. This is the **full moon** phase. As the moon moves farther around its orbit, the bright portion becomes smaller, and the moon is said to be waning. At the **last quarter** phase, the side facing Earth is again half bright and half dark. The visible part becomes smaller until the moon cannot be seen. The cycle is complete at new moon phase. If we can see less than a quarter of the moon, it is called a crescent moon. If we can see more than a quarter, it is called a gibbous moon. It takes about 29½ days for the moon to complete one cycle of phases.

The moon's diameter is 3476 kilometers, which is small compared to many members of the solar system. The moon's density is 3.3 g/cm^3. Compare this measurement to the average density of Earth given in Chapter 4.

INVESTIGATION 5–1

Moon Observations

Problem: What moon data can be learned by direct observations?

Materials

paper pencil almanac

Procedure

1. Make a chart similar to the one shown.
2. From an almanac, find the date of the next full moon and determine the time of moonrise on that date. This date will be the first day of your observations.
3. On the date you found in Step 2, record the exact time you saw the moon rise. Give the official time of moonrise (from the almanac) in parentheses. Record the phase of the moon in your chart.
4. In the morning, record the exact time you saw the moon set. If the official moonset time is given, record it also.
5. Record data for seven days. Each day record the phase of the moon.
6. After five days, the moon may rise too late for you to observe it. After the first two or three days, the sun may be too bright for you to see moonset. Use only the official times in your chart.

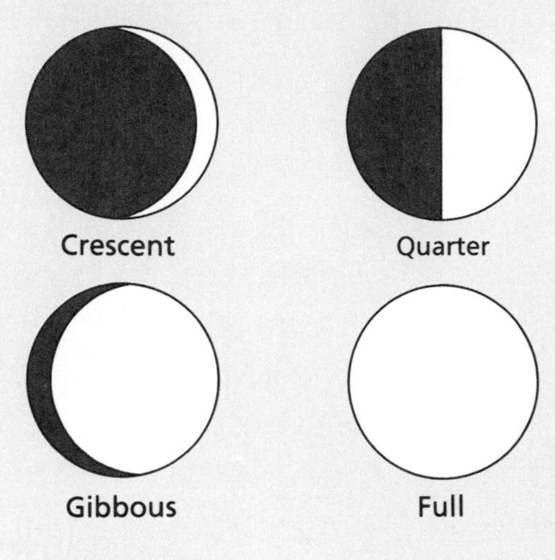

FIGURE 5–9.

7. Calculate the time between moonrise on each succeeding day.
8. Calculate the amount of time the moon is visible each night.

Data and Observations

Date		
Moonrise		
Moonset		
Time visible		
Difference in moonrise time		
Phase of moon		

Analysis

1. How much later does the moon rise each day?
2. What was the moon's shape at the end of seven days? How long before the moon is in new moon phase? In full moon phase?
3. How long does it take for the moon to go from full moon to full moon?

Conclusions and Applications

4. When viewed on any single night, the moon appears to move westward across the sky. When viewed on successive nights, the moon appears to move about 13° eastward through the patterns of stars. Classify each motion as real or apparent, and explain the cause of these motions.
5. How can you determine how much later the moon rises each night?
6. How does the moon's shape appear to change?

5:5 Eclipses

The moon and Earth have been casting shadows on one another throughout their existence. Because the moon's orbital plane is tilted 5° with respect to Earth's orbital plane most of the time, shadows fall either above or below Earth or the moon. However, if Earth, moon, and sun form a straight line, an eclipse is possible. An **eclipse** is the passing of one object into the shadow of another object. The alignment must occur near the point at which the moon's orbit crosses the plane of Earth's orbit. A lunar eclipse occurs when the full moon is in Earth's shadow. A solar eclipse occurs when Earth is in the shadow of the new moon.

When does a lunar eclipse occur?

Although partial solar eclipses occur more frequently than lunar eclipses, fewer people on Earth observe them. During a lunar eclipse, sunlight to the moon's surface is blocked. Everyone on the "night" side of Earth is able to see this type of eclipse. During a solar eclipse, only the observers within the path of the moon's shadow will be able to see the eclipse. Thus, although partial solar eclipses occur more often, lunar eclipses are more often observed.

Why do fewer people observe solar eclipses?

Two shadows form during an eclipse. An inner complete shadow, the **umbra**, is surrounded by a zone of partial shadow, the **penumbra**. The moon remains within the umbra for almost two hours during a total lunar eclipse. During a total solar eclipse, the umbra forms a narrow tract, at most about 270 kilometers wide, which takes less than 7½ minutes to travel across Earth's surface.

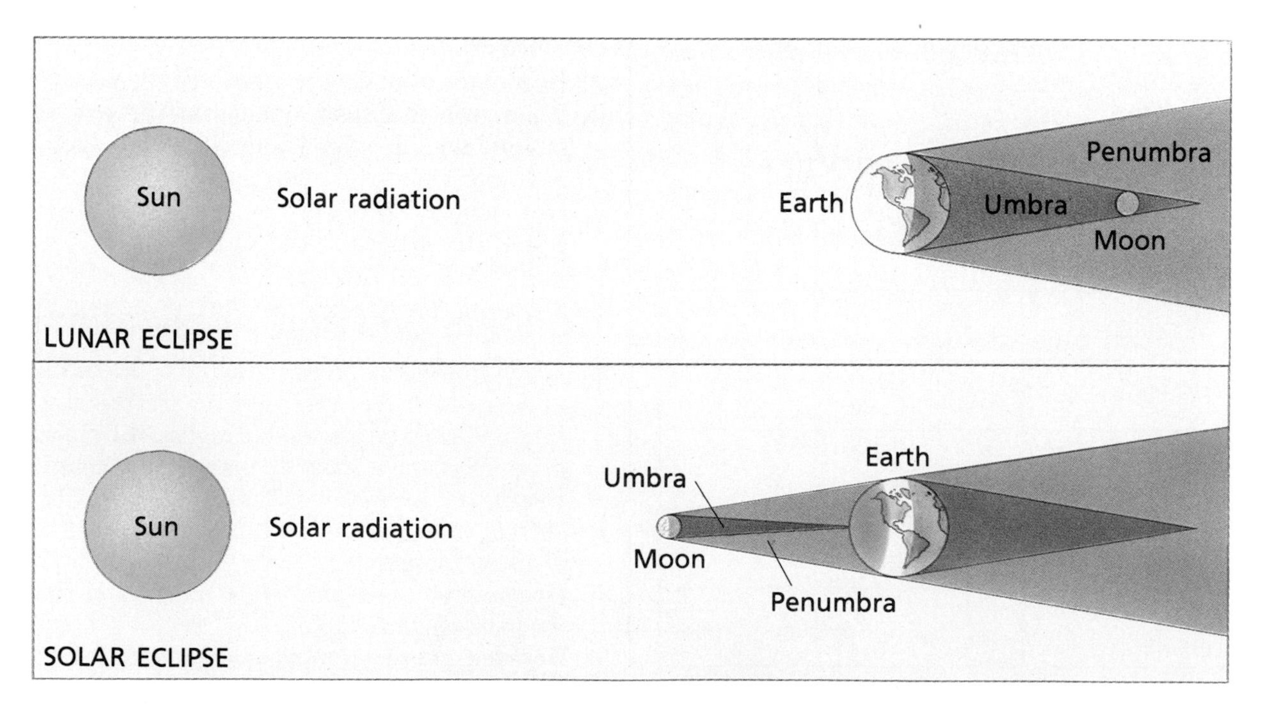

FIGURE 5–10. An eclipse occurs when one object passes into the shadow of another.

INVESTIGATION 5–2

Eclipses

Problem: How do the motions and size of Earth and moon cause eclipses?

Materials

light source (unshaded)
polystyrene ball on pencil
globe
Figures 5–8, 5–10

Procedure

1. Study the positions of Earth, moon, and sun shown in Figures 5–8 and 5–10.
2. Use Figure 5–11 as a guide when moving the moon around the globe to duplicate the exact positions that would have to occur in order for both a lunar and a solar eclipse to take place. Keep the distance between the model moon and the globe reasonably constant.
3. Place the model moon at each of the following phases: first quarter, full moon, last quarter, new moon. Identify which, if any, eclipse could occur during each phase. Record your data.
4. Once the phases when eclipses can occur have been identified, place the model moon at each of those positions. Move it slightly toward and away from Earth. Note the amount of change, if any, in the size of the shadow causing the eclipse. Record this information.

Data and Observations

Phase of Moon	Observations
first quarter	
full	
last quarter	
new	

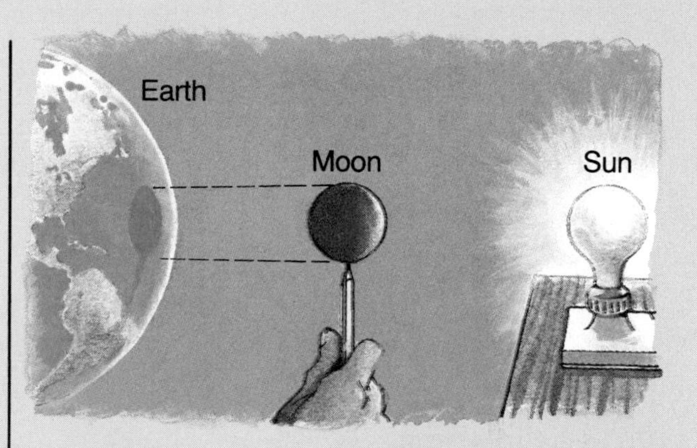

FIGURE 5–11.

5. On a separate sheet of paper, sketch diagrams that show the correct relationships between the positions of Earth, moon, and sun for Steps 3 and 4.

Analysis

1. During which phase(s) of the moon is it possible for an eclipse of the sun and an eclipse of the moon to occur?
2. Describe the effect that a small change in the distance between Earth and the moon has on the size of the shadow causing the eclipse.
3. As seen from Earth, how does the apparent size of the moon compare to the apparent size of the sun? How can an eclipse be used to confirm the answer to this question?

Conclusions and Applications

4. Why do an eclipse of the sun and an eclipse of the moon not occur once each month?
5. Explain how it is possible for a total eclipse of the sun to occur at a time other than noon and a total eclipse of the moon to occur other than at midnight.
6. Why have few people seen a total solar eclipse?

5:6 Moon History

The moon's surface, like Earth's, consists of mountains, valleys, and plains. But, because of its low density and small size, the moon has no atmosphere. Thus, the surface processes that formed the moon's surface are much different from those that formed Earth's.

How are most craters formed on the moon's surface?

Many of the craters on the moon's surface are due to the impact of meteors, space objects that strike the surface. When these objects hit the moon, they explode, breaking the rock into tiny fragments, forming regolith. Regolith makes up the outermost layer of the moon. Regolith is made of dust, rock fragments, and boulders.

A few of the larger craters have bright streaks called **rays** that extend outward from them. The presence of rays indicates that the craters are relatively recent in age. Lunar rocks, unlike Earth rocks, have no traces of water in them. Lunar rocks do have tiny crystals of pure iron, which do not exist on Earth.

BIOGRAPHY

Johannes Kepler
1571-1630

Kepler discovered the laws of planetary motion, laws that govern the movement of planets and their moons. In fact, Kepler's three laws describe the motion of any orbiting body. Spending most of his life in Germany, he resided for a while on the Danish island of Uraniborg with the famous astronomer, Tycho Brahe. Using Tycho's observations Kepler was able to show that his laws are obeyed in nature.

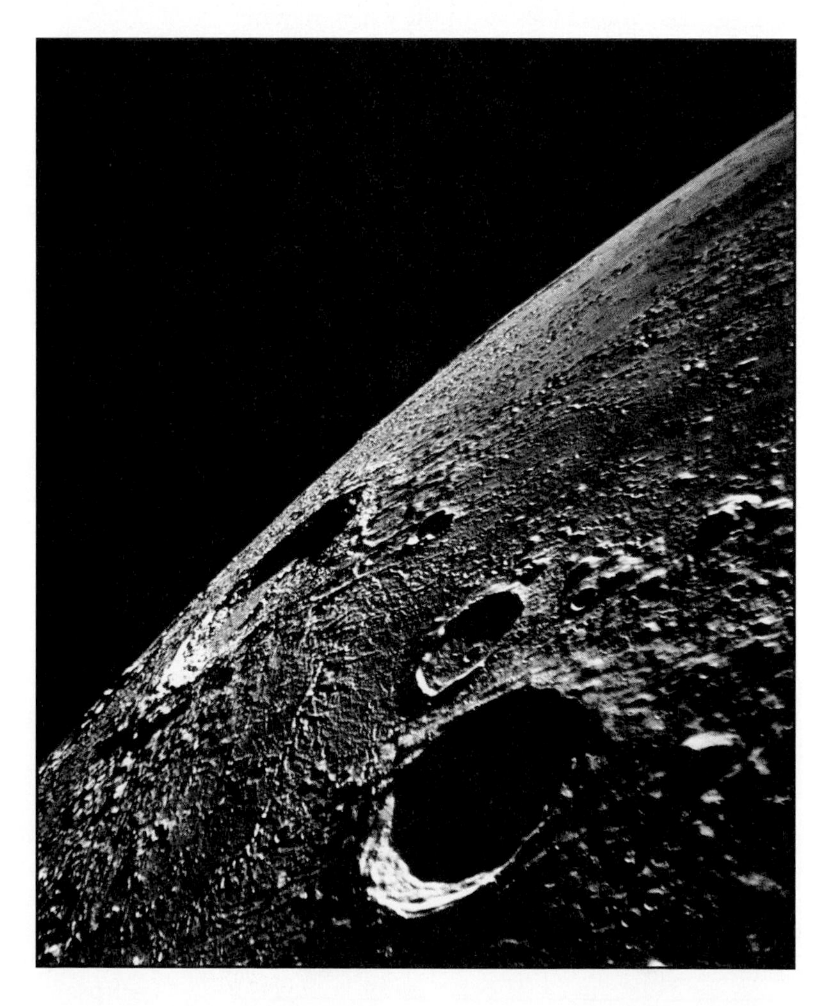

FIGURE 5–12. The large crater in the foreground is nearly 65 kilometers from rim to rim.

FIGURE 5–13. Maria are plains of dark-colored lava on the moon's surface.

F.Y.I. Galileo made the first map of the moon in 1610. In 1647, Hevelius published a second map in which he named 10 mountain ranges including the Apennines and Alps.

Almost as soon as it formed, the moon's surface may have melted to a depth of several hundred kilometers. Between about 4 and 4.3 billion years ago, the surface cooled and hardened, forming mountain ranges. For the next half billion years, the moon's interior apparently heated. Great floods of dark-colored lava covered the low areas on the surface. Galileo named these lunar plains **maria** because of their sealike appearance through a telescope. *Maria* is the Latin word for seas. It is now known that these low-lying regions contain no water. The rocks of the maria are about 3.6 to 3.7 billion years old, and are the youngest rocks on the moon.

What are maria?

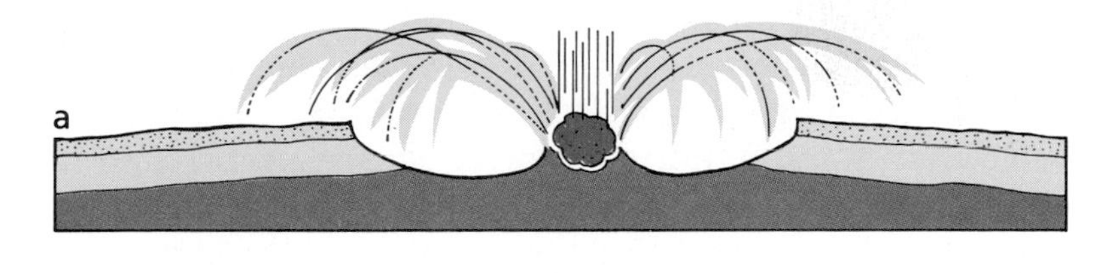

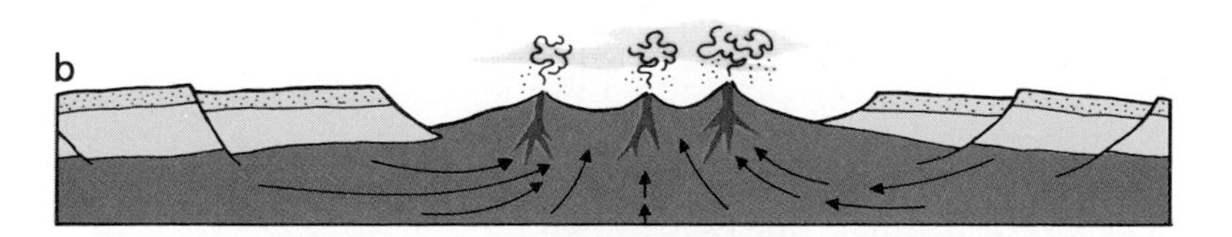

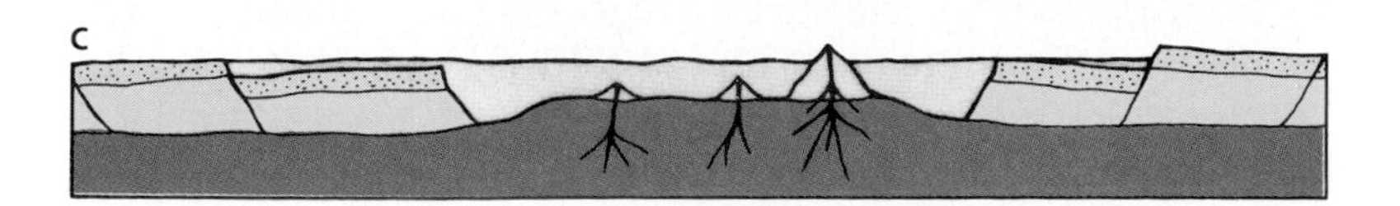

FIGURE 5–14. Maria may have formed when the impact of large objects (a) caused cracks in the moon's surface through which molten material could flow (b). The molten material solidified to form the smooth maria (c).

FIGURE 5–15. Moonquakes are recorded by instruments left on the moon by astronauts.

Instruments left on the moon by astronauts have recorded thousands of moonquakes. Moonquakes occur between 600 and 800 kilometers below the surface. Moonquakes are more frequent when Earth and moon are close together. Some characteristics of the moon's interior are suggested by moonquake records. Scientists have determined that the moon consists of several layers. The outer crust is about 60 kilometers thick on the side facing Earth. It appears to be about 100 kilometers thick on the far side of the moon. Beneath the crust is a rigid layer extending to about 1000 kilometers. Material may be molten below 1000 kilometers. A small iron core with a radius of about 500 kilometers is present.

Questions still exist about the moon's origin. One hypothesis suggests that just under 4.5 billion years ago, when Earth was about 100 million years old, a huge chunk of space debris struck Earth. This object broke through the crust and plunged into the interior. As a result, a huge plume of hot gases and debris was shot into space and formed a cloud around Earth. As the cloud cooled, molecular and gravitational forces caused particles in it to stick together. Within about 1000 years of the collision, the cloud of gas and debris had formed the object we call the moon.

Review

6. Define umbra and penumbra.
7. Explain why lunar eclipses do not occur each month.
8. Identify some of the surface features of the moon. Which of these features are related to processes occurring in the moon's interior?
★ **9.** Identify examples of both real and apparent motions of the moon. Explain the occurrence of each.

Chapter 5 Review

SUMMARY

1. All objects in space exhibit real and apparent motion. 5:1
2. Two early models that helped explain the position of objects in the universe were the geocentric and heliocentric models. 5:1
3. The barycenter of the Earth-moon system is the common center of gravity of the two objects that orbit the sun. 5:1
4. Two real motions of Earth are rotation on its axis and revolution about the sun. 5:1, 5:2
5. Seasons are caused by the tilt of Earth's axis and its revolution around the sun. 5:3
6. The moon rotates on its axis and revolves around Earth once about every 27⅓ days. 5:4
7. Phases of the moon depend on its position relative to the sun. 5:4
8. A lunar eclipse occurs if the moon passes through Earth's shadow. A solar eclipse occurs if the moon's shadow falls on Earth. 5:5
9. Regolith is composed of dust, rock fragments, and boulders. Maria are lunar plains composed of lava. 5:6
10. The moon appears to be about 4.5 billion years old. 5:6

VOCABULARY

a. apparent motion
b. barycenter
c. eclipse
d. equinox
e. first quarter
f. full moon
g. geocentric universe
h. gravitational forces
i. heliocentric universe
j. inertia
k. last quarter
l. maria
m. new moon
n. orbit
o. penumbra
p. rays
q. real motion
r. revolution
s. rotation
t. solar day
u. solstice
v. umbra
w. year

Matching

Match each description with the correct vocabulary word from the list above. Some words will not be used.

1. phase of the moon observed when it is waxing; sun, Earth, and moon at 90° angle
2. bright streaks extending from lunar craters
3. movement of an object as seen by an observer on Earth
4. one Earth revolution around the sun
5. the imaginary common center of gravity for Earth and moon
6. the movement of an object in a straight line unless acted upon by another force
7. one complete rotation of Earth
8. the 21st or 22nd of June or December
9. the passing of one object into the shadow of another
10. the actual path followed by any object around another object

Chapter 5 Review

MAIN IDEAS

A. Reviewing Concepts

Choose the word or phrase that correctly completes each of the following sentences.

1. The actual shape of Earth, as viewed from space, most resembles a *(bowling ball, football, pear)*.
2. When the North Pole is tilted most directly toward the sun in June, a(n) *(equinox, solstice, eclipse)* occurs.
3. The average distance between Earth and the sun is *(140 600 000 km, 149 600 000 km, 151 000 000 km)*.
4. The moon has low density and small size, thus, no *(magnetic force, thermal force, atmosphere)* is found on the moon.
5. Maria are *(highlands, mountains, lava-filled basins)*.
6. When Earth lies in the moon's shadow, a(n) *(full moon, eclipse, crescent moon)* occurs.
7. The rising and setting of the moon are examples of *(heliocentric, real, apparent)* motion.
8. The spinning of an object on its axis is *(libration, rotation, revolution)*.
9. A solar eclipse occurs *(less often than, more often than, as often as)* a lunar eclipse.
10. The moon's average density is *(5.52 g/cm^3, 3.3 g/cm^3, 2.8 g/cm^3)*.
11. Lunar rocks are unlike Earth's rocks because they contain no *(iron, copper, water)*.
12. Copernicus is credited for developing the *(barycenter, heliocentric, geocentric)* model of the universe.
13. The revolution of Earth around the sun determines a period of time called a(n) *(orbit, year, solar day)*.
14. A day on which the length of daylight and darkness is approximately the same for all points on Earth is the *(penumbra, solstice, equinox)*.
15. The phases of the moon occur because of its *(rotation, revolution, position)* in relation to the sun.

B. Understanding Concepts

Answer the following questions using complete sentences.

16. What factors affect the gravitational force existing between any two objects in the universe?
17. Compare the density of Earth to that of the moon. Determine the difference between the average densities.
18. Describe the difference between rotation and revolution.
19. Why have observers on Earth never seen the craters on the back side of the moon?
20. After which phase of the moon is it said to be waxing? Waning?
21. How do real and apparent motions differ? Give examples of each.
22. Identify the dates of equinoxes and solstices.
23. Name and describe the two shadows formed during a total lunar eclipse.
24. Describe the origin of craters on the moon's surface.
25. How did Eratosthenes measure Earth's size? When did he do it?

C. Applying Concepts

Answer the following questions using complete sentences.

26. Why do eclipses not occur every month?
27. How is a satellite held in orbit around Earth?
28. Why would you weigh less on the moon than on Earth?
29. Explain the phenomenon that causes Earth's seasons to occur.
30. How would the moon appear to an observer in space during its revolution

around Earth? Would the moon go through phases? Explain.

SKILL REVIEW

1. Construct a model of the Earth-moon system showing both size and distance to scale.
2. Was Eratosthenes' measurement of Earth based on observations or inferences?
3. The moon's mass is 1/81 that of Earth's. Earth's mass is 5.96×10^{24} kg. The moon's density is 3.3 g/cm^3. Calculate the volume of the moon.
4. Use a globe to determine the latitude and longitude of your city.
5. Figure 5–8 on page 106 is not drawn to scale. Using the dimensions of Earth and moon as a guide, draw Figure 5–8 to scale.

PROJECTS

1. Construct a working model that will effectively demonstrate the lunar phases and eclipses. It should accurately represent the positions of Earth, the sun, and moon.
2. Develop a model that will show how the centers of Earth and the moon appear to move around their common gravitational center of mass, the barycenter.

READINGS

1. Barrett, Norman. *The Moon*. New York: Franklin Watts, 1985.
2. Fisher, Arthur. "Birth of the Moon." *Popular Science*. January, 1987, pp. 60-64, 91-92.
3. Green, Suzanne. *Seasons*. Garden City, NY: Doubleday, 1987.

TOOLS OF ASTRONOMY ROCKETS AND SATELLITES

Exploring Space

Humans were first launched into space in 1961, when the U.S.S.R. put Yuri Gagarin into orbit. He orbited Earth in Vostok I for 108 minutes at an altitude of 327 kilometers. In 1969, astronauts from the United States reached the moon.

TOOLS OF ASTRONOMY

GOALS

1. You will learn about the tools of astronomy.
2. You will learn how distances between objects in space can be expressed.

A tool can be defined as "anything necessary to carry out one's profession or occupation." To the astronomer, instruments such as telescopes and spectrographs are necessary tools used to enhance vision in order to understand the realms of space.

6:1 Telescopes

To observe objects in space, astronomers rely on a valuable tool, the telescope. Optical telescopes include the refracting telescope and the reflecting telescope. The first telescopes were **refracting telescopes,** which use an objective lens to refract or bend light toward the focal plane where the image is located. An eyepiece is placed behind the focus of the objective. The eyepiece enlarges the image. The size of a refracting telescope is limited. As the lens gets larger, the grinding, polishing, and obtaining of perfect glass without bubbles becomes difficult. Also, the weight of a large lens is difficult to support. The largest refractor in the world is the 1-meter telescope at Yerkes Observatory. The objective weighs over 200 kg.

Why is the size of a refracting telescope limited?

Isaac Newton invented the reflecting telescope, which uses concave mirrors to form the image. These mirrors are easier to make than lenses. In a **reflecting telescope,** light is collected by a concave mirror, which produces a small image. Usually a second mirror reflects the image to the eyepiece, where it is magnified. The largest reflecting telescope in the United States is on Mount Palomar, California.

F.Y.I. In a reflecting telescope, the real image is always inverted and smaller than the object.

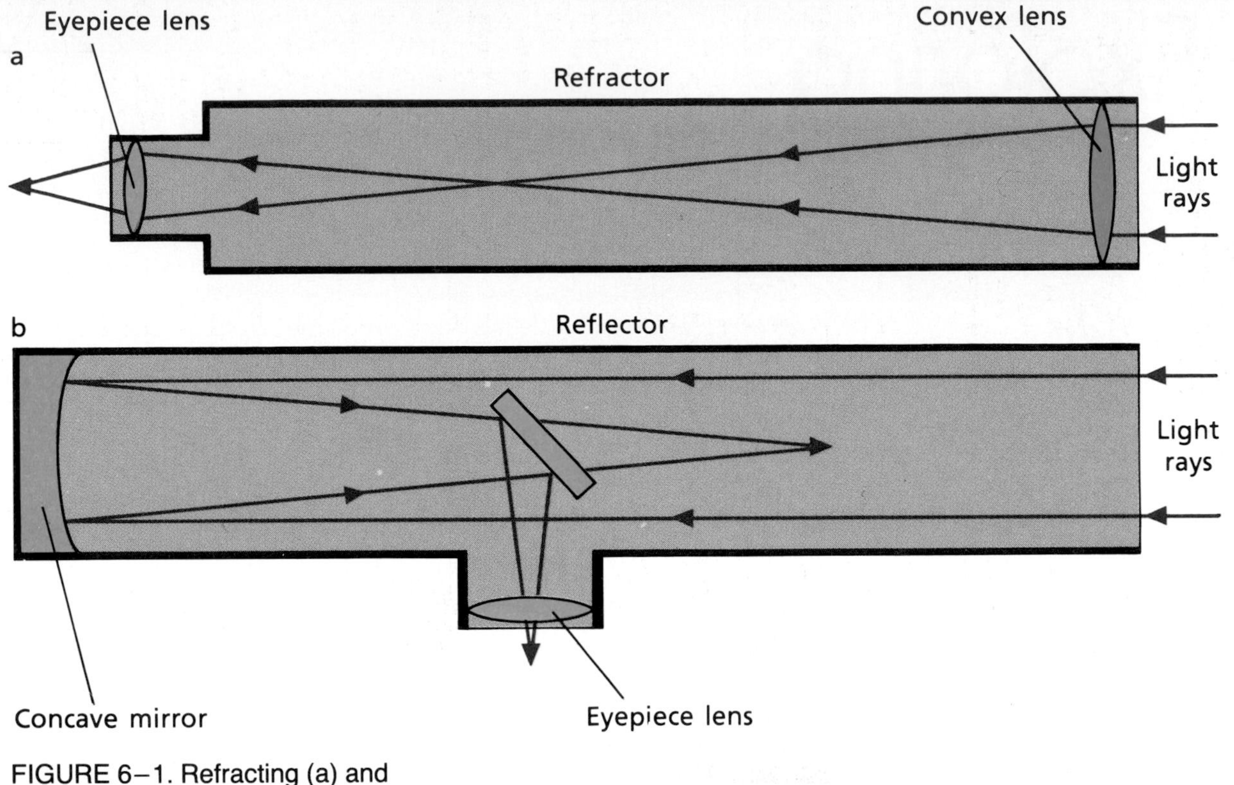

FIGURE 6–1. Refracting (a) and reflecting (b) telescopes use different objects to focus light.

CAREER

Rocket Repair Person

Paul Jones is a rocket repair person. He knows how rockets are put together and how they function. Mr. Jones specializes in repairing the delicate electrical circuits that control the direction and speed of flight. Other people work on various parts of rockets or the different kinds of power systems. It is the job of the repair person to know exactly what makes the rocket engine function.

Mr. Jones learned his skill as a member of the armed forces. Some technical schools also have programs for rocket repair. Since rockets are constantly changing due to new technology, a repair person must be able to adapt the knowledge of the basics to new rockets.

For career information, write:
International Association of Machinists and Aerospace Workers
1300 Connecticut Avenue
Washington, DC 20036

INVESTIGATION 6–1

Telescopes

Problem: How does the path of light differ in reflecting and refracting telescopes?

Materials  

candle
cardboard, white, 50 cm × 60 cm
flashlight
magnifying glass
glass of water
aluminum or silver spoon
plane mirror
convex mirror
concave mirror
white paper

Procedure

1. Observe your reflection in a plane, convex, and concave mirror. Note differences in your image in the three mirrors.
2. Hold an object in front of each of the mirrors. Compare the images as to relative size and position.
3. Darken the room and hold the convex mirror at a 45° angle, slanting downward from your body. Direct the flashlight beam toward the mirror from right angles to your body. Note the size and position of the reflected light.
4. Repeat Step 3 using a plane mirror. Draw a diagram to show what happens to the beam of light.
5. Place the spoon in a glass of water. Diagram the shape of the spoon at the water line.
6. Wrap the white paper around the end of the flashlight so that its beam is very small. Shine the light into a large glass of water, first directly from above, then from a 45° angle to the water surface. Compare the direction of the light rays when viewed from the side of the glass.
7. Light a candle and set it up some distance from the vertically held cardboard screen. **CAUTION:** *Keep hair and clothing away from the flame.* Using the magnifying glass as a convex lens, hold it between the candle and the screen until you have the best possible image.
8. Move the glass closer to the candle. Note what happens to the size of the image. Move the cardboard until the image is in focus.

Data and Observations

Object	Observations

Analysis

1. What is the purpose of the concave mirror in a reflecting telescope?
2. How did you determine the position of the focus of the magnifying glass in Step 7? What does this tell you about the position of all the light rays?
3. In one type of reflecting telescope, a plane mirror is in the tube near the eyepiece. What is the purpose of this mirror?
4. The eyepiece of a telescope is convex. What is its purpose?
5. What is the effect of the concave mirror on your reflection? Of the convex mirror? Of the plane mirror?
6. What effect did the convex mirror have on the beam of light in Step 3?

Conclusions and Applications

7. Discuss your observations of the relationship of the distance between the object and lens and the clearest and largest image you could obtain in Steps 7 and 8.
8. How does the path of light differ in refracting and reflecting telescopes?

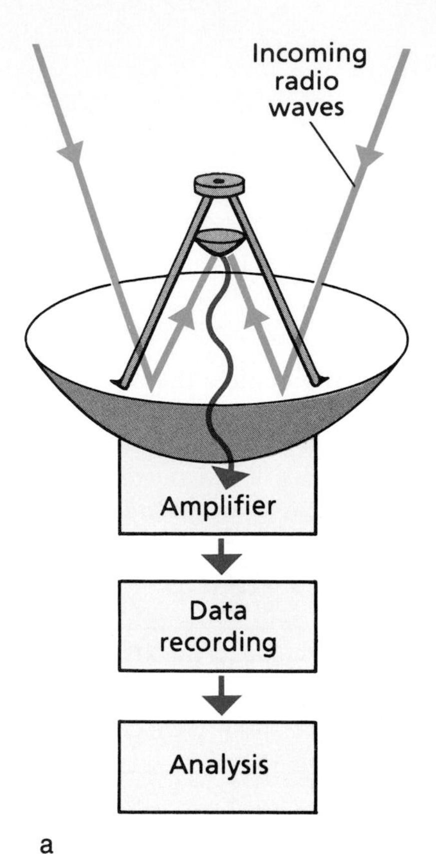

a

b

FIGURE 6–2. A radio telescope (a) is composed of a reflector or dish that collects and focuses radio waves for the receiver. Radio waves from space must be amplified so they can be studied (b).

F.Y.I. Because of Earth's gravity, optical telescopes are limited in size. The glass in lenses of refracting telescopes and in the mirrors of reflecting telescopes begins to distort and slowly flow when constructed and used on Earth's surface. Radio telescopes do not have this disadvantage.

A **radio telescope** consists of a reflector or dish and a receiver. The reflector collects radio waves in the same way an optical telescope collects light waves, and focuses them for the receiver. Radio telescopes measure the intensity of radio waves coming from space. Their resolution is less than that of optical telescopes due to the greater wavelengths of radio waves. Because most radio waves move through clouds, observations can be carried out in all kinds of weather.

6:2 The Spectrograph

The **sun** is a star that produces large amounts of radiant energy each second. This energy is emitted into space in all directions. **Radiant energy** is energy that is transferred from one place to another by means of waves. Familiar forms of radiant energy are infrared, light, and radio waves. Other forms include ultraviolet waves, X rays, and gamma rays. These waves can be arranged in the order of their wavelengths to form an **electromagnetic spectrum.** Wavelength is the distance between two wave crests. Frequency is the number of waves that pass a given point per second. Radiant energy travels in waves at the speed of light, which is 300 000 kilometers per second.

Radiant energy produced by the sun takes about eight minutes to reach Earth. The distance between Earth and the sun is about 150×10^6 km. This distance is known as an **astronomical unit (AU).** An object located twice as far from the sun as Earth receives radiant energy about 16 minutes after it leaves the sun. This object's distance from the sun would be 2 AUs, or 300×10^6 km.

What is an AU?

The AU is a convenient unit for measuring distances within the solar system. However, distances between stars are so vast that measurements and calculations in AUs are difficult. Astronomers often use the distance light travels in one year, a **light-year** (9.5×10^{12} km), for measuring distances in space. These large distances also are measured in parsecs. One parsec equals 3.26 light-years.

A **spectroscope** is an instrument that separates visible light into its various wavelengths. This instrument is called a **spectrograph** when the spectrum is photographed. Each wavelength of light has a characteristic color. From longest to shortest, these colors always are red, orange, yellow, green, blue, indigo, and violet. Red has the longest wavelength; violet has the shortest. All elements emit energy with wavelengths that are unique, much like a person's fingerprints. Using spectrographs, scientists can recognize the elements in the sun by studying the sun's spectrum and comparing it with the spectra of known elements.

What is a spectroscope?

Scientists study an object's spectrum to learn something about its movement. An object's spectral lines are shifted

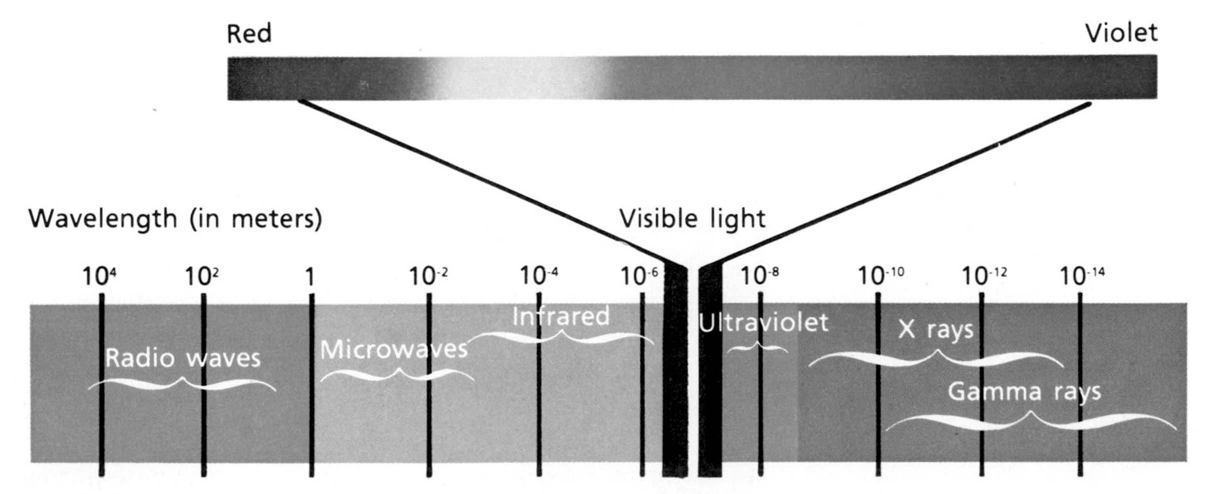

FIGURE 6–3. The electromagnetic spectrum ranges from very long radio waves to very short gamma rays. Visible light covers a very small portion of the electromagnetic spectrum.

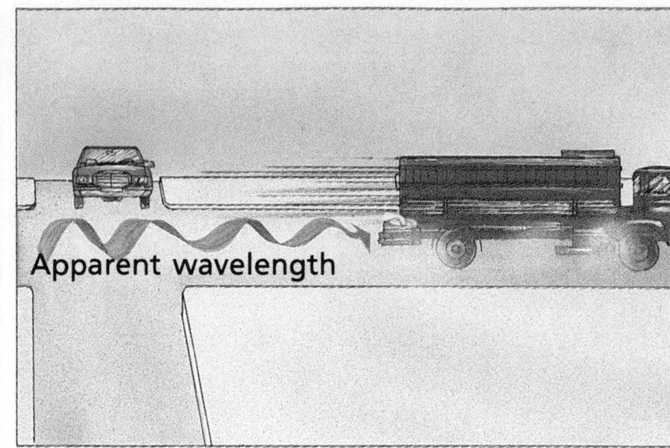

FIGURE 6–4. As a fire truck approaches, the pitch of its siren increases. As the truck recedes, the pitch of its siren decreases. Light waves emitted by approaching or receding objects behave in the same way, causing violet or red shifts in their spectra.

from their normal positions as a result of the motion between the source of the lines and an observer. This is known as the Doppler shift.

Spectrographs can be used to determine whether stars or galaxies are moving toward or away from Earth. As a distant object in space approaches Earth, its wavelengths shorten. This causes a shift of the wavelengths toward the violet end of the object's spectrum. If an object is moving away from Earth, its wavelengths lengthen. This causes a shift toward the red end of the object's spectrum.

Review

1. Describe the differences between reflecting and refracting telescopes.
★ 2. How many AUs are equal to one light-year?

FIGURE 6–5. By studying a star's spectrum (a), an astronomer can determine the direction the star is moving. A spectral shift to the violet (b) indicates the star is moving toward Earth. A spectral shift to the red (c) indicates a star is moving away from Earth. Describe the motion of a star with a spectral shift to blue.

INVESTIGATION 6–2

Spectroscopes

Problem: How are spectroscopes used to analyze visible light?

Materials 

diffraction grating
cardboard or paper tube
light source
colored pencils or crayons
paper
tape
scissors

Procedure

Part A

1. Cover both ends of a cardboard or paper tube with paper.
2. Using scissors, make a thin slit in one end so that only a narrow shaft of light enters. **CAUTION:** *Always use scissors with care.*
3. Make a small hole (0.5 to 1.0 cm in diameter) in the paper covering the other end of the tube.
4. Cover this hole with a piece of diffraction grating. The instrument you have constructed is a simple spectroscope.
5. Look through the diffraction grating, and aim the narrow slit at a light source. **CAUTION:** *Do not use the sun as a light source.* Move the tube slightly to the left or right.
6. On a sheet of paper, sketch what you observe through the diffraction grating.

Part B

Use the spectroscope constructed in Part A and Steps 5 and 6 to sketch the light patterns produced by any additional light sources supplied to you by the teacher. In each case, sketch your observations exactly as the light patterns appear through your spectroscope.

Analysis

Part A

1. Describe your observations as you looked through the spectroscope and moved it.
2. The light source you observed produced many wavelengths of energy at once. What would you expect to observe if the light source produced only a few different wavelengths of energy?

Part B

3. Compare and contrast the spectra of the various unknown substances.
4. Does the light energy being given off from these different substances appear to be exactly the same? Explain your answer.

Conclusions

5. What happens to light that is viewed through a spectroscope?
6. How can a spectrograph help scientists identify elements present in the sun?

Data and Observations

Light Source		Spectrograph Pattern
Part A	white light	
Part B	unknown A	
	unknown B	
	unknown C	

GOALS

1. You will learn how rockets and satellites serve as tools in understanding more about Earth.
2. You will learn some of the history of space exploration.

ROCKETS AND SATELLITES

Bound to Earth's surface by gravity, people have looked outward into space for centuries, wondering and asking about what they observed. They made careful measurements of what they saw. A series of discoveries and technologies, such as gunpowder, rockets, satellites, and photography, allowed people to leave Earth in the 20th century. People were able to reach and walk on Earth's nearest neighbor in space, the moon.

6:3 A First Step into Space

"That's one small step for a man, one giant leap for mankind." These historic words were spoken by Neil Armstrong as he first stepped onto the moon. This voyage and other space voyages are made possible by the combined use of rockets, radios, and computers.

Rockets, called "fire-arrows" by the Chinese, were used as weapons as early as 1232. A fast burning mixture similar to gunpowder was used as fuel. **Rockets** are action-reaction engines based on Isaac Newton's third law of motion. This law states that *for every action force, there is an equal and opposite reaction force.* The fuel container and burning chamber of a rocket is a tube that has a nozzle at one end. As the fuel burns, the expanding hot gases exert an action force on all surfaces of the tube. The tube pushes back with a reaction force on all surfaces except at the nozzle, where hot gases escape. A thrust force builds up at the end of the chamber opposite the nozzle. **Thrust** is the force that propels the rocket forward. At Earth's surface, a large amount of thrust is needed to overcome gravity and air friction. In space, much less thrust is needed to move a rocket forward.

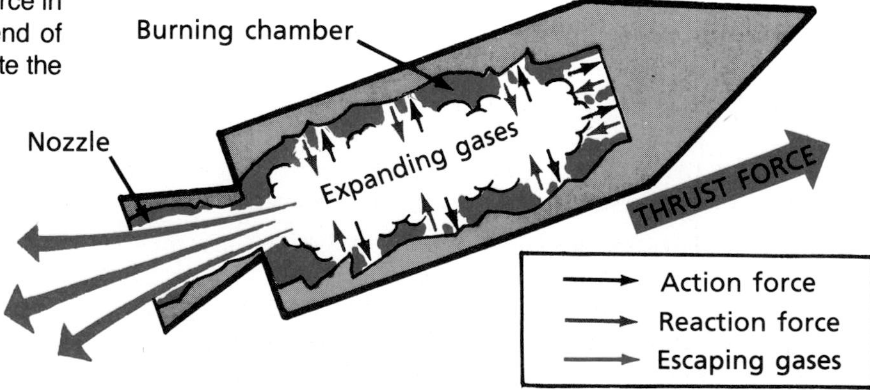

FIGURE 6–6. The thrust force in a rocket builds up at the end of the burning chamber opposite the nozzle.

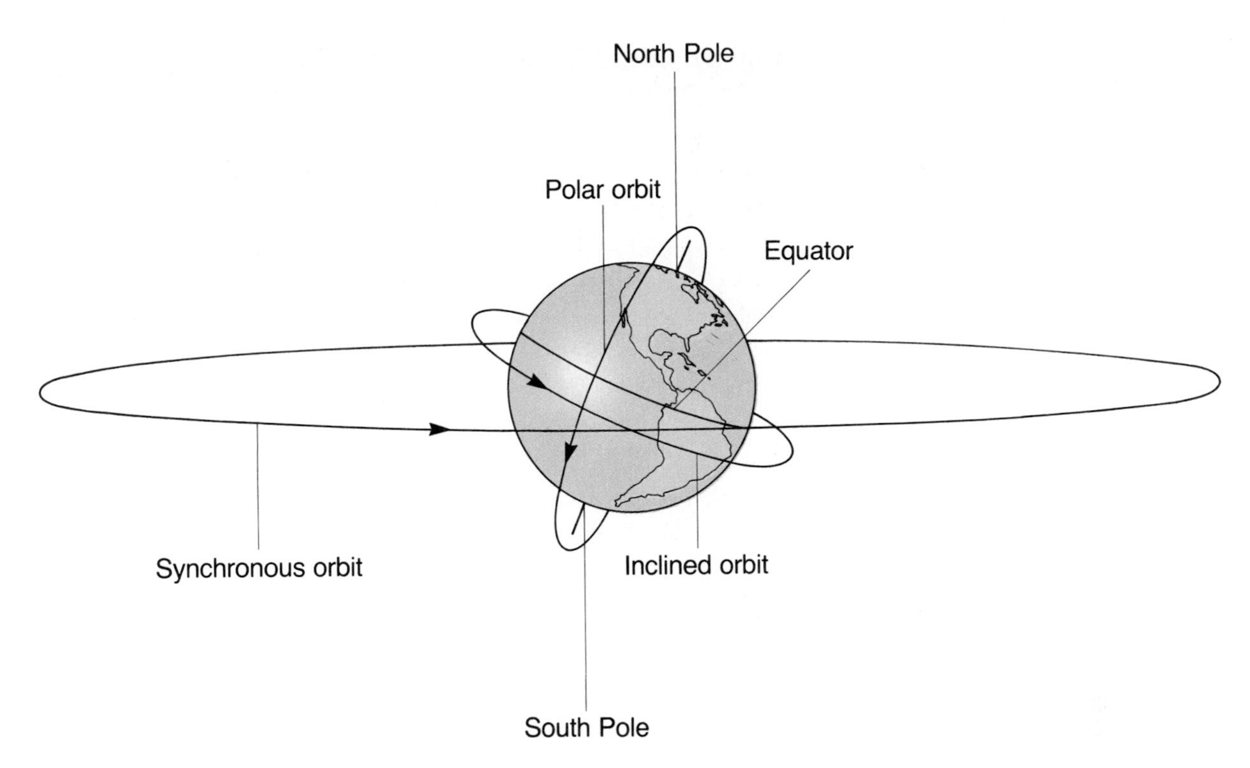

FIGURE 6–7. Satellites can have any one of three kinds of orbits around Earth.

The Space Age began in 1957, when the U.S.S.R. used a rocket to launch Sputnik 1 into Earth orbit. Signals from two radio transmitters on Sputnik 1 were heard around the world. Sputnik 1 was an artificial satellite. A **satellite** is an object that revolves around a larger, primary body. The moon is a natural satellite of Earth. The path, or orbit, may be circular or elliptical. The plane of the orbit of a satellite can form any angle with Earth's axis, but it must always pass through Earth's center. If a satellite's orbit stays over the equator, it is in an equatorial orbit. All other orbits are inclined orbits. If a satellite orbits Earth directly over the poles, it is in a polar orbit. Satellites in polar orbits scan the entire Earth as it rotates below. Satellites in circular orbits at 35 800 kilometers above Earth revolve at the same rate as Earth rotates. These satellites are in a geosynchronous orbit, because the satellites are always over the same geographical sites.

What is a polar orbit?

Satellites are used for communications and to monitor weather. The first reflector communications satellite was a plastic balloon called Echo. On its surface, Echo had a thin aluminum coating that acted like a mirror to reflect light and radio waves. Echo I was used to relay telegrams, telephone calls, and pictures across continents and oceans. Later, television pictures were relayed this way.

FIGURE 6–8. This satellite, known as Solar Max, orbits Earth at an altitude of 570 kilometers. It has seven different instruments under the control of an on-board computer. Since 1980, it has obtained and relayed information about the surface of the sun to scientists on Earth's surface.

What are two current uses of satellites?

The first communications satellite in synchronous orbit, Syncom II, was launched in 1963. This satellite could receive, amplify, and relay signals and thus was called a repeating satellite. Similar satellites now are used by many nations. The largest system, called Intelsat, has 102 member nations with 250 ground stations. This system began in 1964 with the Early Bird satellite. It provides a 240-channel link between Europe and the United States, and is managed by an international group. The American member of this group, Comsat, relayed the moon landing of Apollo 11 to the world. Other uses include transmission of telephone, educational, medical, and other types of communications.

During the 1980s, hundreds of satellites have been placed in orbit. These have standardized components for power supply, data handling, and communications. These parts are all mounted on a standard frame onto which an instrument package can be attached.

6:4 Race to the Moon

The National Aeronautics and Space Administration, or **NASA,** is the agency that oversees the space program in the United States. Plans for the moon project began with the **Mercury** and **Gemini** projects. Mercury flights provided data and experience in the basics of spaceflight. The problems of return to Earth's surface, reactions to weightlessness, and capacity to work in space were solved. Project Gemini was more advanced. Two astronauts took part in each flight. They practiced controlling the spacecraft and working in space. Two Gemini vehicles docked in space. Spacewalking while connected to the spacecraft by a hose that supplied air and communication wires was accomplished.

Three series of space probes, Ranger, Lunar Orbiter, and Surveyor, followed Mercury and Gemini. These probes took pictures of the moon that aided scientists in choosing a spot for the moon landing. Surveyor spacecraft actually landed on the moon.

The final step for placing an astronaut on the moon was a series of spaceflights known as the **Apollo** program. The early Apollo missions provided data and practice for the complex processes involved in a lunar landing. Apollo 11 was launched on July 16, 1969, and returned to Earth on July 24. Neil Armstrong and Edwin Aldrin landed on the moon and spent two hours exploring its surface. Apollo 11 was followed by five more lunar landings. They returned over 2000 samples of moon rock for study. Cameras

F.Y.I. Apollo spacecraft were designed to carry the crew members: one to remain on board while the other two descended to the moon's surface.

a

b

c

FIGURE 6–9. The Juno I rocket has launched American satellites (a). The Redstone rocket was used to launch the first American astronaut into space (b). The Saturn V rocket was used in the Apollo lunar program (c).

a

b

FIGURE 6–10. Astronauts first landed on the moon in 1969 (a). The lunar rover was first used by Apollo 15 astronauts (b).

mapped much of the moon's surface. The Apollo program ended in 1972. However, studies of the lunar samples and of the data from instruments left on the moon continue to give us new clues to the histories of Earth and its satellite, the moon.

6:5 The Space Shuttle

The **space shuttle** is a reusable craft designed to transport astronauts, materials, and satellites to and from space. The shuttle system has four major elements: the orbiter,

PROBLEM SOLVING

Design a Spacesuit

You are to design a spacesuit for an astronaut traveling to the planet Mars. Mars' atmosphere is different from Earth's. An astronaut must be able to survive on the surface of Mars for several hours. List six items that you will include in your design that will insure the astronaut's survival. Place these six items in order of their importance.

Compare your list to the lists of others doing the same task. If differences in the order of importance exist, try to decide which order is more correct. Are six items enough or must there be more before the astronaut can survive on Mars? What items would you add?

an external liquid fuel tank, and two solid fuel rocket booster engines. At launch, all four elements are joined. The engines then are fired, sending the shuttle to an altitude of 40 kilometers in about two minutes. Then the solid fuel rockets drop off and parachute back to Earth. They are recovered for future use. The external liquid fuel tank also drops off, but it is not recovered.

What are the four major elements of the shuttle system?

Once in orbit, the shuttle becomes a controllable satellite staffed by astronauts and scientists. Payloads, such as communication satellites, telescopes, and scientific experiments and instruments, are placed into orbit by the shuttle crew. These payloads can be retrieved later with a mechanical arm and returned to Earth in the shuttle's cargo bay. In September, 1985, two crew members from a shuttle orbiter completed the first successful repair of a malfunctioning satellite while in orbit. Experiments with certain chemicals useful for the treatment of diseases have been successfully accomplished in the environment of the shuttle.

a

b

c

FIGURE 6–11. Crew members on the space shuttle (a) conduct many scientific experiments in space (b and c).

BIOGRAPHY

Franklin Chang

1950–

Franklin Chang dreamed of traveling in space. Chang's dream came true when he flew on the space shuttle Columbia in January, 1986. He broadcast a commentary from the Columbia in Spanish—a first for space shuttle flights. Chang has also worked on the development of the plasma rocket designed for trips to distant stars.

Spacelab is a fully equipped workshop and laboratory designed to fly in the cargo bay of the space shuttle. One part of the spacelab provides a laboratory for scientists. In it, as many as four scientists can carry out experiments with up to three dozen scientific instruments. Another module contains equipment and materials exposed directly to space.

What happens at the completion of each shuttle mission?

At the completion of each mission, the orbiter of the space shuttle glides back to Earth and lands like an airplane. Its landing speed of about 335 kilometers per hour requires a very large landing field.

The tragic loss of the seven members of the space shuttle Challenger in January, 1986, was a shocking reminder that each moment spent off Earth's surface has a potential for disaster when unforeseen events take place. The most dangerous time of any space shuttle flight is between the launch and orbit, while the shuttle is still attached to the external tank and rocket boosters.

6:6 Space Probes

Space probes are rocket-launched vehicles that carry instruments, cameras, and other data-gathering equip-

FIGURE 6–12. The orbiter of the space shuttle returns to Earth when a mission is completed.

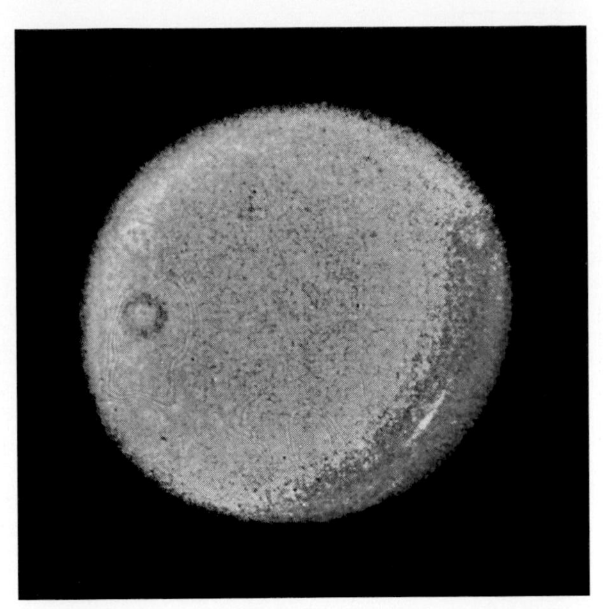

FIGURE 6–13. The Voyager 2 space probe collected new data about Uranus during its encounter in 1986.

ment above Earth's atmosphere. Space probes have a power supply, which is either solar or nuclear, and a radio system that is able to relay pictures and measurements back to Earth.

What kinds of data are collected by space probes?

The mission of a probe involves interplanetary or deep space measurements. Data collected include sun-related activities, temperatures, and the presence and strength of gravity fields of space objects. The approach of a probe to an object and the gathering of data is called an encounter. The encounter may be a flyby, in which case the probe passes by but does not land on the space object. Probes may orbit about an object in space, or they may make a soft landing on it.

The U.S.S.R. made the first attempt at an encounter with Venus in 1961. Deep-space exploration by the United States began in 1962 with the Mariner 2 probe. Other probes, including the Pioneer and the Viking series, sent back pictures and data not available from Earth. Viking probes landed on Mars and conducted tests. Voyagers 1 and 2 have visited Jupiter and Saturn. Voyager 2 has visited Uranus and is continuing on to Neptune. Pioneers 10 and 11 carry a message from Earth to any other life in the universe. If this message is understood by other life forms, they will learn the location of Earth and our sun. Data from probes refine our distance measurements and increase our knowledge of atmospheres, magnetic fields, and gravity fields in space.

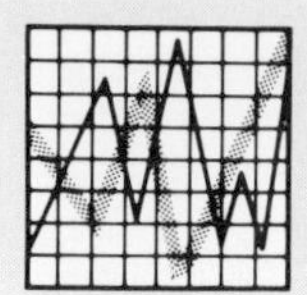

SKILL

Compare and Contrast

Problem: How can you compare and contrast two sides of an issue or question?

Materials

pencil and paper

Situation

You and your classmates will be acting as the Board of Directors for the National Aeronautics and Space Administration. You have been asked to prepare a report. The report is to recommend to the President (or your teacher) whether NASA should emphasize a spacecraft program with people aboard or should concentrate on increasing its efforts in the crewless space probe program. Regardless of how you personally feel about the issue, you are asked to compare and contrast these two space programs. You are to reach an agreement with your colleagues on the Board (classmates) as to how NASA should proceed. Remember, in this situation you must choose one or the other program, but not both.

Procedure

1. Divide into small discussion groups of no more than five students in each group.
2. Within each group, develop a list of similarities and differences and advantages and disadvantages of both space flight programs. Review Sections 6:3 through 6:6.
3. On the basis of your list and the discussion, decide as a group which program should be chosen. Then list what will be sacrificed if that path is chosen.
4. Follow the directions of your teacher in arriving at a full-class decision from the combined groups.
5. Construct tables like the ones shown to list the strengths and weaknesses of each of these programs to support your final decision.

Questions

1. Compare the two types of space programs presented here. In what ways are they similar? In what ways do they accomplish the same goals?
2. Contrast these two programs. In what ways do they differ? In what ways do they accomplish different goals?
3. Which program did your group decide to promote? Identify the main reason for this selection.
4. If you agreed with your group's decision in question 3, what were the best arguments for it? If you did not agree with your group's decision, what do you think will be the biggest problem that NASA will face in the future?

Table 6–1

Spacecraft with a Crew	
advantages	disadvantages

Table 6–2

Space Probes with No Crew	
advantages	disadvantages

6:7 Space Stations

A **space station** has living quarters, work space, and all the equipment and support systems needed for people to live and work in space. Space stations carry telescopes, cameras, computers, and whatever else is needed for research projects.

What types of facilities does a space station contain?

In 1973, the United States launched the space station Skylab. During 1973 and 1974, one crew of three astronauts stayed in orbit for 84 days. Skylab obtained valuable data about space and the reactions of astronauts living in space. Skylab broke up and fell to Earth in 1979.

An international space station was launched in July, 1975. An American Apollo spacecraft and a Soviet Soyuz spacecraft docked while in orbit. The two crews visited each other and performed experiments in astronomy, geoscience, biology, and zero gravity. Following this flight, several Soviet crews made record stays in their own space stations.

FIGURE 6–14. A space station has all the needed equipment to live and work in space.

Future space stations could be laboratories for making products that are eventually returned to Earth. These stations could be power stations that generate electricity from energy from the sun. Shuttle orbiters will be used to reach and service these space stations. In time, space stations might orbit the moon or other planets. Plans for the development of space stations such as the one shown in Figure 6–14 are now being considered as the next major step into space.

Review

3. Describe what happens inside the fuel chamber of a rocket.
4. Draw a diagram illustrating the three types of satellite orbits around Earth.
5. What problems of space exploration were solved by the Mercury flights?
6. What types of activities might be carried out on space stations in the future?
★ 7. The Apollo 11 mission took just over two days to reach the moon from Earth. If that average speed were maintained, how long would it take for an Apollo spacecraft to travel one AU through space?

TECHNOLOGY: APPLICATIONS

Space Program Spin-offs

How does the space program affect your everyday life? Whenever you use a product with an integrated circuit, a battery powered hand tool, or a quartz watch, you are using space program spinoffs. Technology information from spin-offs was transferred to uses different from the original application.

It is difficult to find an area of life upon which spin-offs have not had an impact. Spin-offs in the health services field include rechargeable heart pacemakers, cardiac care systems for emergency rescue vehicles, and a device operated by eye movement that allows a paralyzed person to control a TV or light switch. Spin-offs adapted to aid persons visually impaired include a device that converts regular inkprint into a vibrating form that enables a person to read any printed material, and a paper-money identifier that reacts to the colored fibers in money and generates audible signals that identify denominations. In recreation equipment, NASA technology has been incorporated into bicycle seats, ski goggles, and electric footwarmers. These are just a few of the benefits of NASA-developed technology. Aerospace technology transfers are so much a part of our lives, it is difficult to measure the benefits of the space program.

Chapter 6 Review

SUMMARY

1. Reflecting and refracting optical telescopes differ in the way in which light energy is focused for the image. 6:1
2. A spectroscope separates visible light into its wavelengths. 6:2
3. Three basic units of distance in astronomy are the astronomical unit (150×10^6 km), the light-year (9.5×10^{12} km), and the parsec (3.26 light-years). 6:2
4. Rockets are based on Newton's third law of motion. 6:3
5. Satellites revolve around a larger, primary body. 6:3
6. Artificial Earth satellites are used for many purposes, such as weather analysis and communications. 6:3
7. Space flights to the moon were the result of a long series of space investigations by the Mercury, Gemini, and Apollo programs. 6:4
8. The space shuttle is used to carry people and gear to and from space. 6:5
9. Space probes are rocket-launched vehicles used to gather data about objects outside the Earth-moon system. 6:6
10. Space stations are orbiting vehicles with all the equipment needed for living and working in space. 6:7

VOCABULARY

a. Apollo
b. astronomical unit (AU)
c. electromagnetic spectrum
d. Gemini
e. light-year
f. Mercury
g. NASA
h. radiant energy
i. radio telescope
j. reflecting telescope
k. refracting telescopes
l. rockets
m. satellite
n. spacelab
o. space probes
p. space shuttle
q. space station
r. spectrograph
s. spectroscope
t. sun
u. thrust

Matching

Match each description with the correct vocabulary word from the list above. Some words will be not be used.

1. one Earth-sun distance
2. an object revolving around a larger primary body
3. an instrument that photographs a spectrum
4. uses concave mirrors to collect and focus light
5. distance light travels in one year
6. the force that propels a rocket forward
7. the closest star to Earth
8. this craft orbits Earth as a support system containing instruments and factories
9. an instrument that consists of a reflector and a receiver
10. energy transferred by means of waves

Chapter 6 Review

MAIN IDEAS

A. Reviewing Concepts

Choose the word or phrase that correctly completes each of the following sentences.

1. The average distance between Earth and the sun is *(140 600 000 km, 300 000 000 km, 150 000 000 km).*
2. The program that actually placed an astronaut on the moon was called *(Mercury, Gemini, Apollo).*
3. A reusable craft for transporting astronauts, materials, and satellites to and from space is a *(rocket, space shuttle, space station).*
4. Action-reaction engines using Newton's third law of motion are called *(rockets, satellites, spectroscopes).*
5. An instrument that can detect energy waves from space beyond the visible spectrum is a *(refracting, reflecting, radio)* telescope.
6. Forms of radiant energy arranged in order of their wavelengths form a(n) *(electromagnetic spectrum, infrared wave, radio wave).*
7. The first astronauts who flew alone in space to solve problems of weightlessness and working in space participated in Project *(Mercury, Gemini, Apollo).*
8. The first astronauts to land on the moon's surface were crew members of the *(Gemini, Apollo 11, Apollo 13)* mission in July of 1969.
9. Which element of a space shuttle system is not recovered for reuse? *(orbiter, external fuel tank, solid fuel rocket boosters).*
10. A fully equipped laboratory designed to fly in the cargo bay of the space shuttle is *(Skylab, space station, spacelab).*
11. In a *(radio, refracting, reflecting)* telescope, light is collected by a concave mirror, producing a small image.
12. The basic unit used to measure the distances between stars is the *(parsec, light-year, astronomical unit).*
13. The radiant energy produced by the sun takes about *(30 seconds, 8 minutes, 30 minutes)* to reach Earth.
14. An instrument that is used to determine elements present in the sun and stars is the *(spectroscope, parsec, optical telescope).*
15. The extent to which the spectrum of an object is shifted toward longer or shorter wavelengths is a measure of its *(direction of movement, distance from Earth, temperature).*

B. Understanding Concepts

Answer the following questions using complete sentences.

16. Distinguish between spectroscopes and spectrographs.
17. The moon is approximately 384 000 km from Earth. Express this distance in astronomical units.
18. Describe an advantage of radio telescopes over optical telescopes.
19. What is radiant energy, how is it transferred, and where does the major source of energy come from?
20. Compare and contrast refracting and reflecting optical telescopes.
21. The farthest planet from the sun is at a distance of approximately 30 AUs. About how long does it take radiant energy from the sun to reach that planet?
22. What is Isaac Newton's third law of motion?
23. Why are rockets more useful and efficient in space than near Earth?
24. During what years did astronauts land on and investigate the moon's surface?
25. What are satellites and for what are they used?

C. Applying Concepts

Answer the following questions using complete sentences.

26. Suppose NASA had to choose between continuing the spaceflight program with people aboard or the crewless space probes. Which do you feel is the more valuable program? Explain your answer.

27. Explain the advantages of placing a telescope into orbit around Earth.

28. Describe how it would be possible for a scientist who is color blind to successfully operate a spectroscope.

29. Write a single paragraph history of the NASA program using the following terms: Mercury, Gemini, Apollo program, Apollo 11, Neil Armstrong, space shuttle, Challenger, spacelab, and space station.

30. How are large optical telescopes used?

SKILL REVIEW

1. Draw a straight line on a sheet of paper and measure its length to the nearest 0.1 cm. If the length of the line represents one AU, calculate the actual length of a line that would equal one light-year.

The Skill in Chapter 4 required the use of four different pieces of laboratory equipment: a balance, graduated cylinder, meter stick, and Celsius thermometer. For questions 2 through 5, identify which of these could be used by astronauts in orbit in either the shuttle or spacelab to duplicate the same activity. Explain your answers.

Equipment	*Earth's Surface*	*In Orbit*
2. balance	yes	______
3. graduated cylinder	yes	______
4. meter stick	yes	______
5. thermometer (°C)	yes	______

PROJECTS

1. Design and build a three-dimensional model of a space station. Show how astronauts and materials will be transported to and from Earth's surface.

2. Construct and demonstrate the use of an optical telescope. Show how the light is brought to a focus in this telescope and contrast this to the other type of optical telescope.

READINGS

1. Billings, Charlene W. *Space Station: Bold New Step Beyond the Earth.* New York: Dodd, 1986.

2. Davies, Owen. "Magic Mirrors." *Omni.* June, 1986, p. 32.

3. Toner, M. "Space: Rescues and Rehabs." *Science Digest.* January, 1986, pp. 44–45.

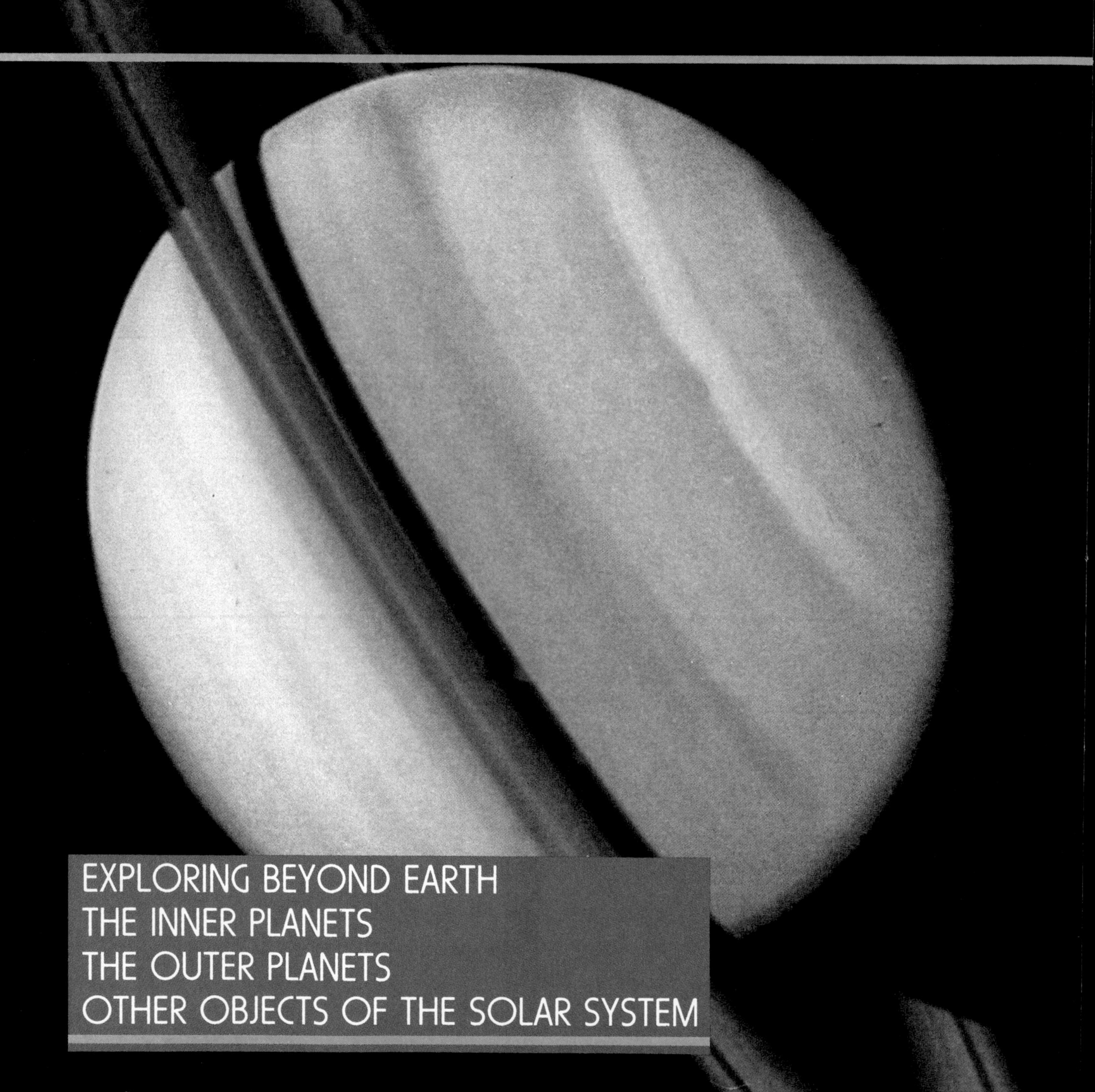

EXPLORING BEYOND EARTH
THE INNER PLANETS
THE OUTER PLANETS
OTHER OBJECTS OF THE SOLAR SYSTEM

Chapter 7

The Solar System

A planet is an object in space that reflects light from a nearby star around which it revolves. The term *planet* means "to wander." To the ancient astronomers, who believed in the geocentric model of the universe, there were seven planets. They were Mercury, Venus, the moon, the sun, Mars, Jupiter, and Saturn. Copernicus had to convince early astronomers that the moon and sun were not planets; but Earth was. Modern astronomers now know of nine planets revolving about a single star, the sun.

F.Y.I. The seven ancient "planets" of the geocentric astronomers apparently are the basis for the seven days of a week. Had stronger telescopes been available before the days of the week were named, a week might consist of nine days rather than seven.

EXPLORING BEYOND EARTH

GOALS

1. You will learn about the possible origin of our solar system.
2. You will learn about the shapes of the orbits of the objects revolving around the sun.

For thousands of years, humans could only look out into the night skies and wonder. Over the past two decades, with the use of a series of space probes, astronomers have extended their investigations outward from Earth to explore its neighbors in the solar system.

7:1 Origin of the Solar System

About when did the sun form?

Scientists believe that the sun formed about 4.6 billion years ago. This occurred when a cold, slowly rotating cloud of gas and dust began to contract due to gravitational forces. The cloud began to rotate faster and faster until it flattened into a disk-shaped mass. Most of the material concentrated in a rapidly spinning core. Compression of the matter in the core caused heating and the sun was formed. The remaining material cooled, contracted due to gravitational forces, and formed the planets.

How were the planets formed?

7:2 Exploring Earth's Neighborhood

Let us look at Earth's neighborhood. Moving outward through space, millions of kilometers past the moon, an explorer would find, not an empty vacuum, but atoms,

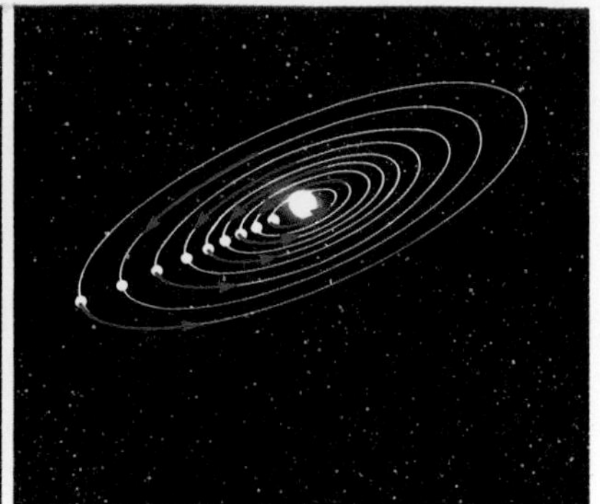

FIGURE 7–1. The solar system may have formed from a rotating cloud of gas and dust.

ions, and isolated molecules. This matter forms a group of objects that extends billions of kilometers in all directions from the sun. This system is known as the **solar system.**

The solar system includes a vast territory. If all the matter in the solar system, excluding the sun, were combined, it would make up less than one percent of the sun's total mass. The sun contains 99.86 percent of the mass of the solar system. Because of the sun's gravitational pull, it is the central object around which other objects of the solar system revolve.

The planets orbit the sun. Early astronomers thought that these orbits were circular and made calculations on that basis. However, they found that their predictions of planet locations were always slightly off. In the early 1600s, a mathematician named Johannes Kepler discovered that the orbit of Mars was elliptical, not circular. An **ellipse** is a closed curve that is elongated. Kepler inferred that if the orbit of one planet was elliptical, then other objects in motion around the sun, including Earth, must also have elliptical orbits.

FIGURE 7–2. Planets have elliptical orbits.

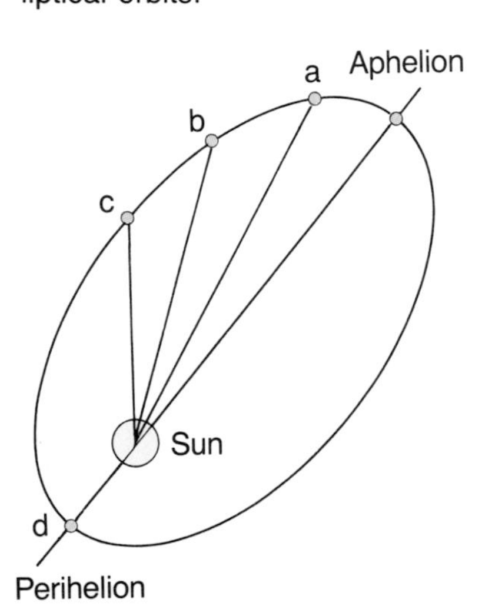

For any object in motion about the sun in an elliptical orbit, there will be some point in the orbit when that object is closest to the sun. Likewise, at another point it is farthest from the sun. The point in an orbit when an object is closest to the sun is the **perihelion.** The farthest point in its orbit from the sun is the **aphelion.** All objects in orbit have perihelions and aphelions.

Review

1. What is an ellipse?
2. Describe how the sun and planets may have formed.
★ 3. What is the difference between the perihelion and the aphelion in the orbit of an object around the sun?

INVESTIGATION 7–1

Planetary Orbits

Problem: What is the most accurate model of the shape of a planet's orbit around the sun?

Materials 

thumbtacks or pins
string
cardboard (21.5 cm × 28 cm)
paper
pen or pencil
metric ruler

Procedure

Part A

1. Place a blank sheet of paper on top of the cardboard and place two thumbtacks or pins about 3 cm apart.
2. Tie the string into a circle with a circumference of 15 to 20 cm. Loop the string around the thumbtacks. With someone holding the tacks or pins, place your pen or pencil inside the loop and pull it taut.

FIGURE 7–3.

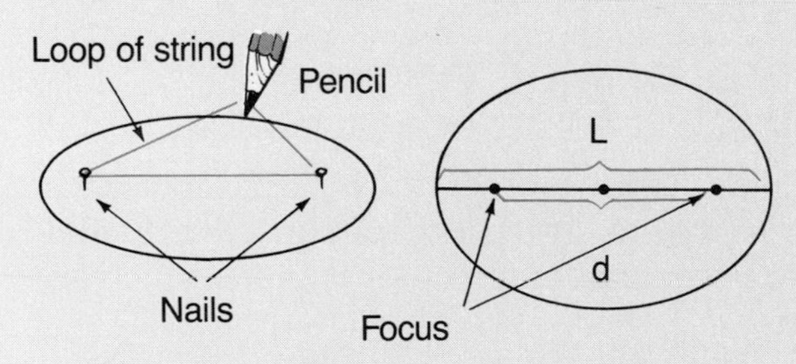

3. Move the pen or pencil around the tacks, keeping the string taut, until you have completed a smooth, closed curve or an ellipse.
4. Repeat Steps 1 through 3 several times. First vary the distance between the tacks and then vary the circumference of the string. However, change only one of these each time. Note the effect on the size and shape of the ellipse with each of these changes.
5. Orbits are usually described in terms of eccentricity (e). The eccentricity of any ellipse is determined by dividing the distance (d) between the foci or tacks by the length of the major axis (L). See Figure 7–3.
6. Calculate and record the eccentricity of the ellipses that you constructed.

Part B

7. Refer to Appendix F to determine the eccentricities of planetary orbits.
8. Construct an ellipse with the same eccentricity as Earth's orbit.
9. Repeat Step 8 with the orbit of either Pluto or Mercury.

Data and Observations

Constructed Ellipse	d (cm)	L (cm)	e (d/L)
#1			
#2			
#3			
Earth's orbit			.017
Mercury's orbit			
Pluto's orbit			

Analysis

1. What effect does a change in the length of the string or the distance between the tacks have on the shape of the ellipse?
2. What must be done to the string or placement of tacks to decrease the eccentricity of a constructed ellipse?

Conclusions and Applications

3. Describe the shape of Earth's orbit. Where is the sun located within the orbit?
4. Name the planets that have the most eccentric orbits.

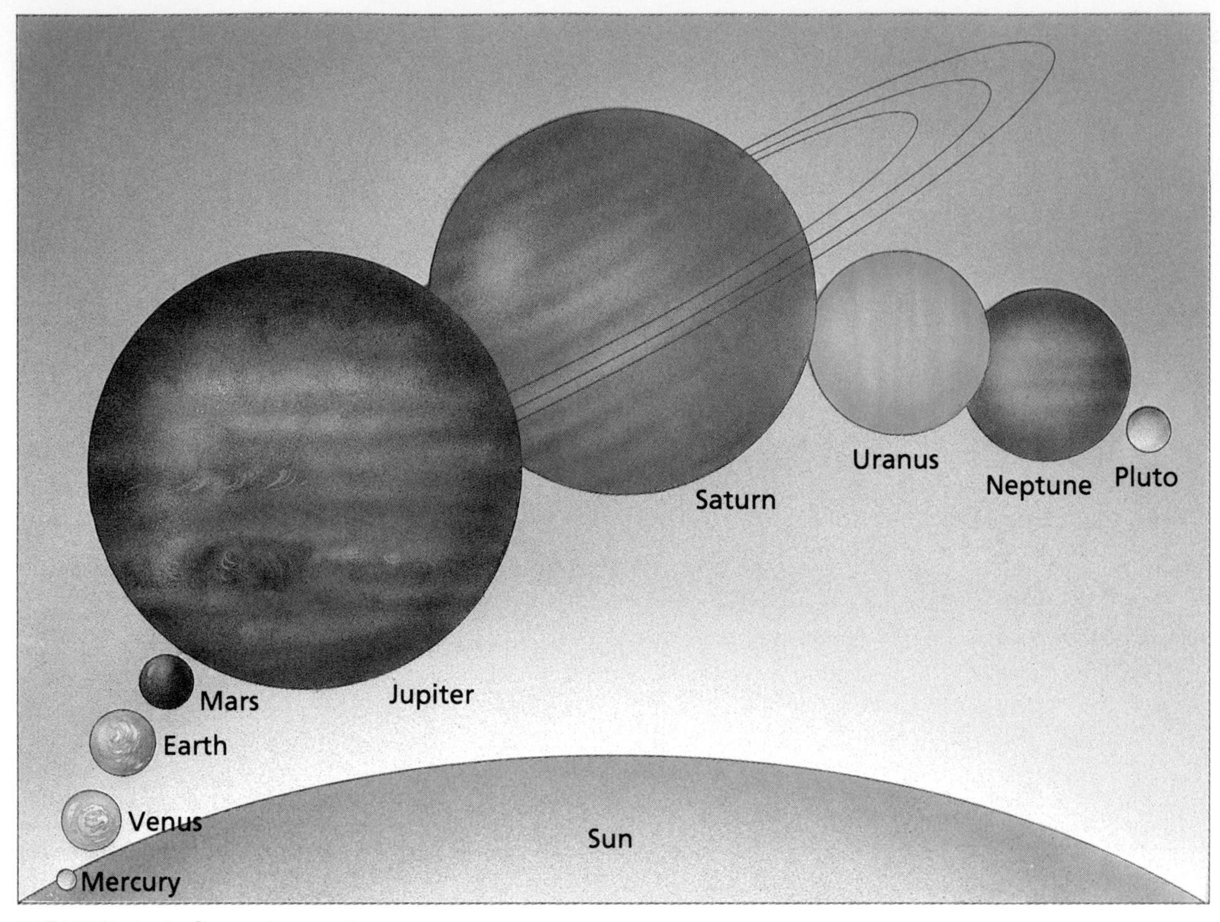

FIGURE 7–4. Our solar system

GOALS

1. You will learn about Mercury, Venus, Earth, and Mars.
2. You will learn how inner and outer planets differ.

FIGURE 7–5. Mercury's orbit

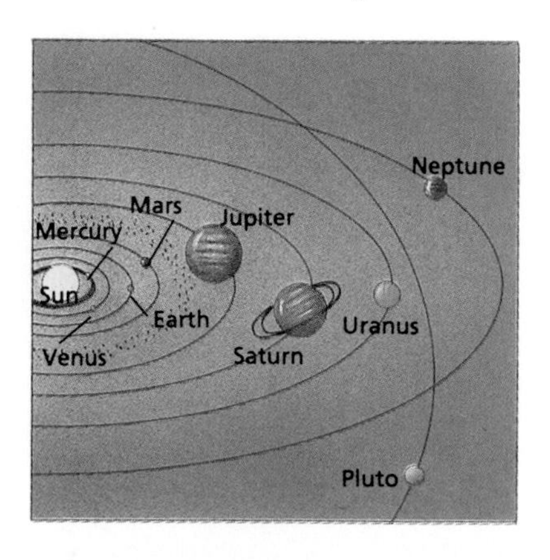

THE INNER PLANETS

Mercury, Venus, Earth, and Mars are the inner planets. These planets are relatively close to the sun. All are solid, rocklike bodies in contrast to the outer planets, which are mostly gaseous.

7:3 Mercury

Mercury is the second smallest planet in the solar system. Usually Mercury cannot be viewed because it is too close to the sun. When Mercury can be seen, it is usually close to the horizon in the early morning or late evening sky. Mercury has a heavily cratered surface like that of the moon. The terrain, however, is more rugged than that of our moon. Some of the cliffs on Mercury are nearly three kilometers high. Mercury's density suggests an iron-rich core, but its magnetic field is extremely weak. Its atmosphere, which is mostly sodium, is very thin. Because Mercury is a relatively dark object, it does not

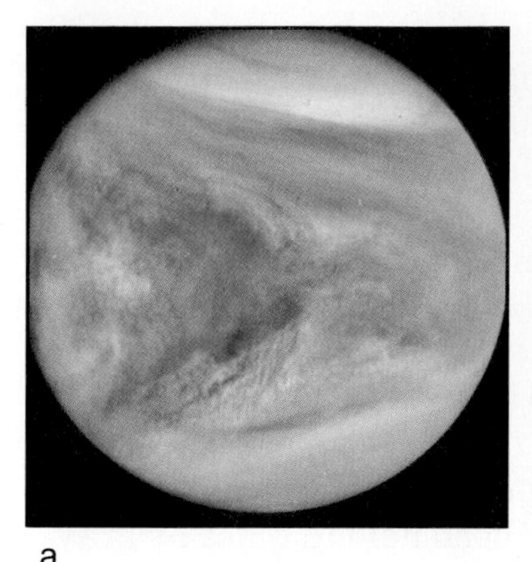

a

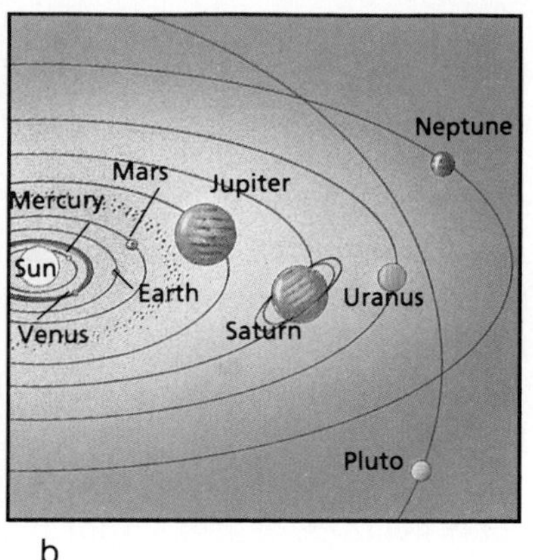

b

FIGURE 7–6. The surface of Venus is not visible because of the dense layers of clouds that cover the planet (a). The orbit of Venus (b). How do the dense clouds affect the temperature on Venus?

reflect much of the sunlight falling on it. Its temperatures range from over 400°C by day to about −150°C by night.

7:4 Venus and Earth

Venus and Earth are similar to each other in size, mass, and shape. Because it is closer to the sun, Venus receives twice as much sunlight as Earth. Nearly three-fourths of this sunlight is reflected into space by clouds in Venus' dense atmosphere and only 2 percent reaches the surface. This small amount is trapped by the cloud cover. Temperatures near the surface can reach over 400°C, due to the greenhouse effect. The atmosphere of Venus is 97 percent carbon dioxide. Droplets of sulfuric acid give the clouds of Venus a yellow color.

F.Y.I. The inner planets are also referred to by astronomers as the terrestrial or Earthlike planets.

What makes Venus' clouds yellow?

Venus appears to go through phases similar to those of our moon. These phases occur because the orbit of Venus around the sun is inside the orbit of Earth. A complete cycle for these phases takes just under two years. Recall that a cycle of Earth's moon phases is completed every 29½ days. Venus has a **retrograde** or opposite rotation from that of most other planets. Rotation is extremely slow, 243 Earth-days.

In the early 1990s, NASA plans to launch a special spacecraft from the space shuttle to map more than 90%

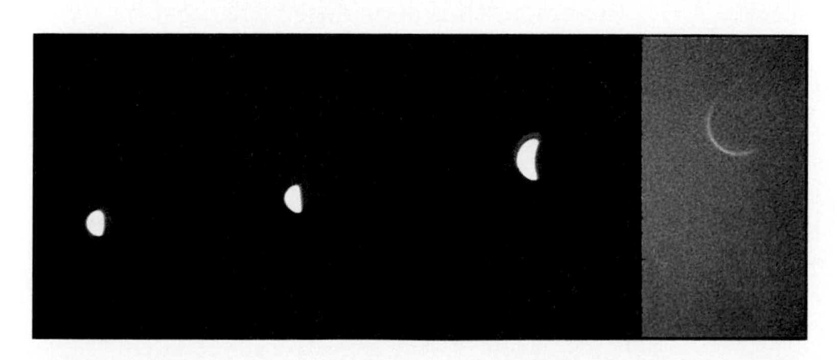

FIGURE 7–7. The first observations of Venus through a telescope were made by Galileo. He discovered that Venus exhibits phases like the moon. The phases of Venus shown here occur over a period of almost two years. Why does Venus go through phases?

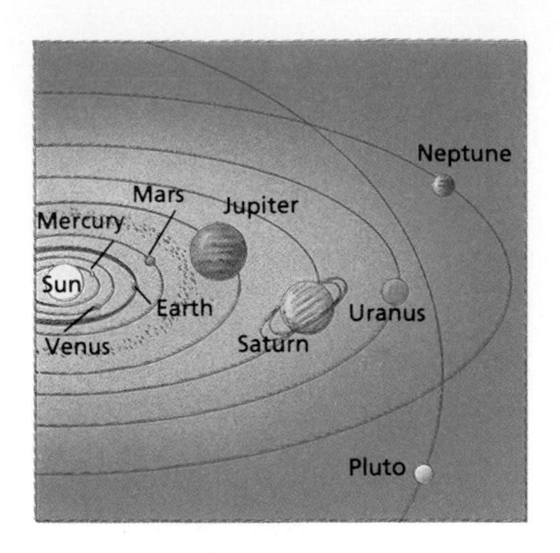

FIGURE 7–8. Earth's orbit

of the surface of Venus. This space probe will use rapid radar pulses to analyze the planet's surface and distinguish features as small as 250 meters.

Earth is the third planet from the sun. It orbits the sun at a mean distance of 150 000 000 km, or one AU. Earth rotates on its axis once in about 24 hours. It revolves around the sun once in about 365 days. Our measurements of time are based on these motions. Within the solar system, Earth days and years are used to describe the motions of other planets.

Earth's atmosphere is unique among the planets in our solar system. Water vapor has been held in place by gravity. Earth's atmosphere moderates temperature, allowing water to exist as a solid, a liquid, and a gas. Oxygen has been added to the atmosphere over time. Because of these factors, Earth may be the only planet in our solar system where life as we know it exists.

TECHNOLOGY: ADVANCES

Big Ear—Are We Alone?

Any object at a temperature above absolute zero gives off radio waves. The sun, galaxies, clouds of complex molecules, and even people, emit radio waves. Big Ear, the third largest single radio telescope in the United States, is constantly monitoring the sky in search of radio signals from space.

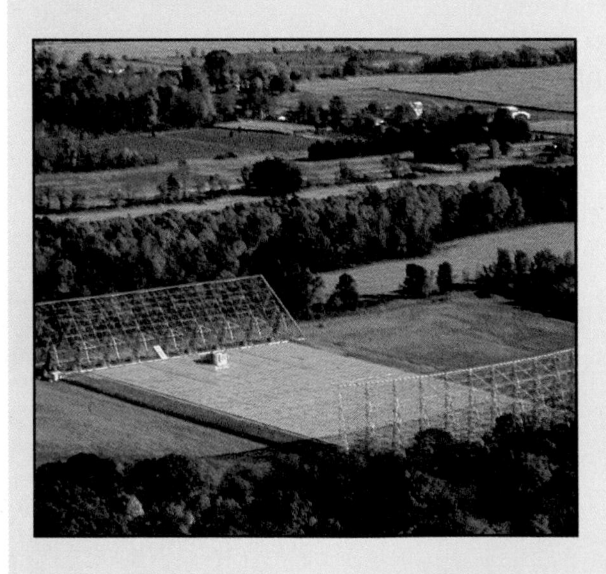

Since 1973, Big Ear has been searching for intelligent signals, which are those emitted by objects other than stars, galaxies, and planets. Some scientists believe that because the sun is a typical star, and Earth is a satellite of the sun, intelligent life may exist on similar planets around similar stars among the 100 billion galaxies in the universe.

The challenge of monitoring deep space presents some difficult problems. Scientists have developed several methods designed to maximize their chances of receiving intelligent signals. Atomic hydrogen gas, which is both abundant and active in space, emits signals at a wavelength of 21 centimeters. Thus, this wavelength is considered the most likely channel on which to find intelligent signals. In addition, radio telescopes must be pointed in the "most likely" direction at the "right" moment. Big Ear is presently "listening" to the solar neighborhood in an attempt to detect intelligent radio signals.

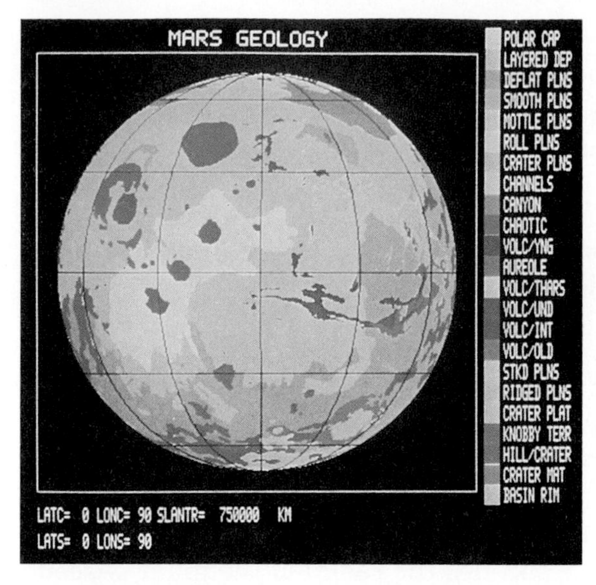

a

b

FIGURE 7–9. Features of Mars' geology are visible on this computer-enhanced image (a). Information about the surface of Mars (b) has been gathered by space probes.

7:5 Mars

Mars is the fourth planet from the sun. Space probes have added much to our knowledge of Mars. We have known about its polar caps for a long time, but many crustal features have since been discovered by probes. The Martian terrain consists of ridges and valleys. Rift zones, areas of fractures or cracks in the crust, extend for over four thousand kilometers across Mars. Olympus Mons, though extinct, is the largest known volcano in the solar system. It rises 25 kilometers above the Martian surface. Martian craters are numerous. Ejected material seems to have flowed away from the craters.

Mars' atmosphere includes clouds and fog. Water vapor condenses at night and evaporates when the sun rises. Most of the atmosphere is carbon dioxide, but small amounts of nitrogen, argon, and oxygen are present. Dust storms are common on Mars. They may be local, or they may spread completely around the planet. The red dust found on Mars is an iron oxide similar to rust that gives the sky a pink hue.

What gases are present in the atmosphere of Mars?

Mars has two satellites. Both are irregular in shape. Phobos is about 29 kilometers in diameter. Deimos is about 21 kilometers in diameter. Surfaces of both satellites are cratered much like Earth's moon. Phobos revolves around Mars about three times during a Martian day. Thus, when viewed from Mars, Phobos rises in the west and sets in the east. Phobos is the only solar system satellite to do so.

FIGURE 7–10. The orbit of Mars

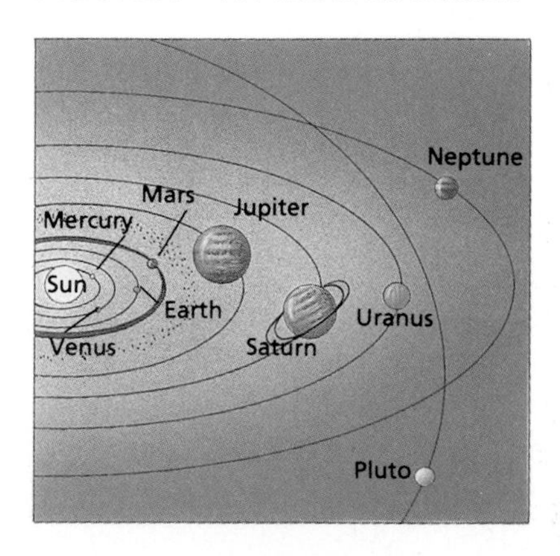

Review

4. Describe Mercury's surface.
5. Why does an observer on Earth see Venus go through phases?

★ 6. How many natural satellites are there among the inner planets? Identify them by name and the planet around which they orbit.

GOALS

1. You will learn about the most distant planets from the sun.
2. You will learn how these planets differ from inner planets.

F.Y.I. The outer planets are also referred to by astronomers as the Jovian or Jupiterlike planets. The exception is Pluto, which is thought to be unlike either the terrestrial or Jovian planets.

What is the largest satellite in the solar system?

FIGURE 7–11. Jupiter's orbit

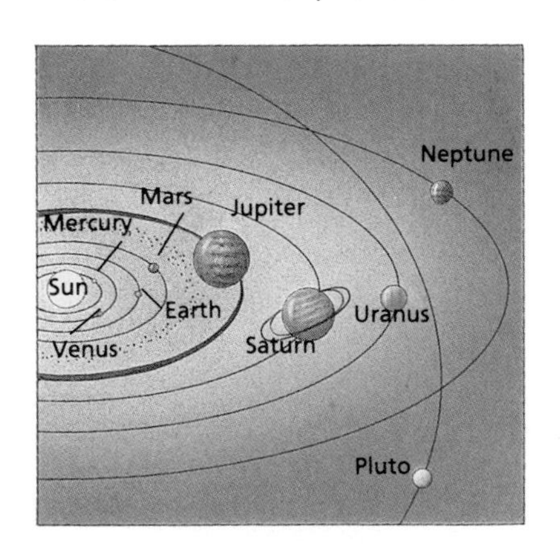

THE OUTER PLANETS

The first four planets from the sun are rocky objects whose mean density is about 4.7 g/cm^3. Between 5 and 30 AUs from the sun lies the realm of the giant gaseous planets. These outer planets are much more massive than the terrestrial planets. The mean densities of the Jovian planets, however, are much lower.

7:6 Jupiter

Jupiter is the largest planet in the solar system. Jupiter and its 16 moons resemble a miniature solar system. Four of these moons are very large. Io, the moon closest to Jupiter, is mostly solid rock. Io has its own very thin atmosphere of sulfur and sodium and a number of erupting volcanoes. Europa is mostly rock with a thick coating of ice. Ganymede, the largest of all the solar system satellites, is half rock, half ice. Callisto's composition is similar to Ganymede's. Some of Jupiter's other moons may be captured space objects.

Scientists believe that Jupiter is mostly liquid and gaseous hydrogen with some helium, much like the sun. Jupiter's core is 30 000°C, six times hotter than the sun's surface. The planet radiates twice as much heat into space as it receives from the sun, but its cloud surface is a cold −140°C.

White to reddish-brown cloud bands alternate around Jupiter. Light bands are due to rising columns of gas. Dark bands are descending gas. These alternating bands are generated by heat from Jupiter's core. Areas of turbulence are similar to Earth's hurricanes, except that Jupiter's clouds probably are composed of ammonia. Lightning has been seen in Jupiter's clouds. The **Great Red Spot** on Jupiter is its most spectacular feature. The hurricane-like Red Spot was first observed about 150 years

a

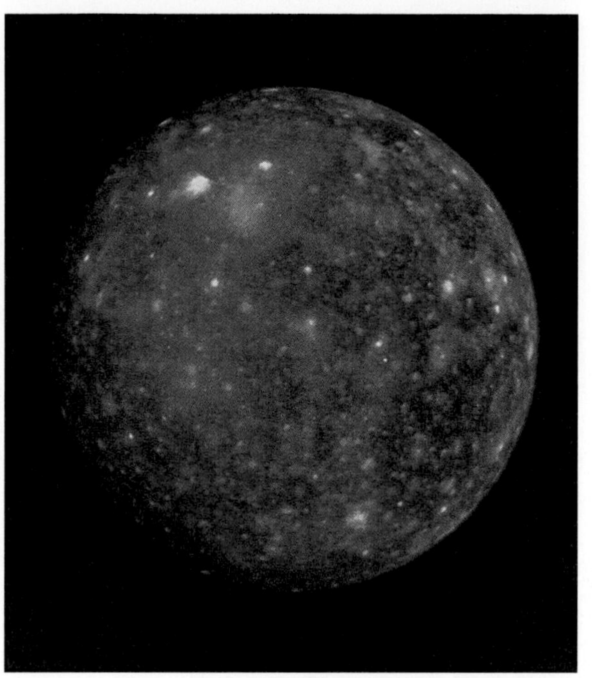

b

FIGURE 7–12. Heat from Jupiter's core produces the light- and dark-colored bands of clouds that encircle the planet (a). A moon of Jupiter, Callisto, is thought to consist of ice and rock (b).

ago. It is a gaseous mass 30 000 to 50 000 kilometers long and 12 000 kilometers wide. It towers upward like a thunderhead, eight to ten kilometers above surrounding clouds.

Voyager 1 photographed a ring around Jupiter that is probably made of fine dust. Jupiter has a strong magnetic field, ten times that of Earth. Electrons trapped in the magnetic field make Jupiter a fairly strong source of radio waves.

7:7 Saturn

Describe Saturn's atmosphere.

Saturn is another gaseous planet like Jupiter. It is 95 times more massive than Earth. Although it is the second largest planet, Saturn has the lowest density. Saturn would float on water! Much data have been gathered about Saturn by Voyagers 1 and 2. Over one thousand rings have been discovered. Saturn also has more than 20 satellites. The largest of Saturn's satellites is Titan, which has a dense atmosphere. Saturn's atmosphere is about 60 percent hydrogen with more methane and less ammonia than Jupiter's atmosphere. Saturn has an internal heat source. It radiates almost three times more energy into space than it receives from the sun. Like Jupiter, Saturn appears to rotate faster at its equator than at the poles.

FIGURE 7–13. Saturn's orbit

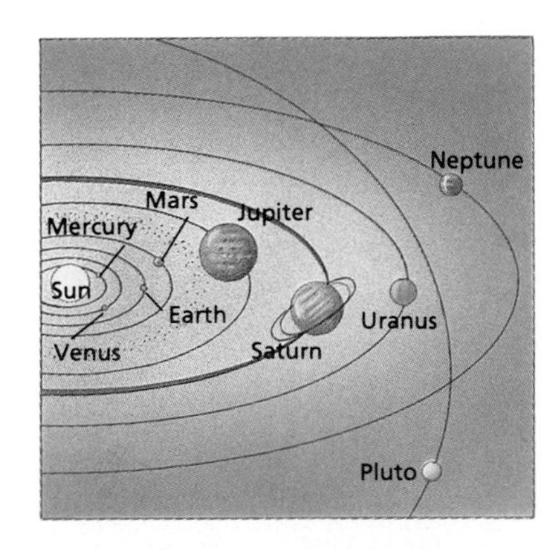

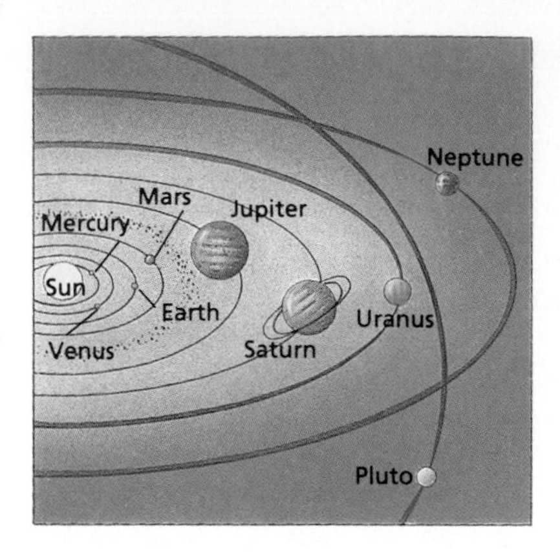

FIGURE 7–14. The orbits of Uranus, Neptune, and Pluto

7:8 Uranus, Neptune, and Pluto

Uranus is a gaseous planet with at least ten dark rings, ten arc-shaped pieces of rings, and 15 satellites. Its rotational axis is inclined to its orbit by 98 degrees, so its rotation is retrograde. Uranus' satellites have retrograde revolution. Uranus is thought to be made of hydrogen and methane gases.

Neptune is also a gaseous planet similar in size and composition to Uranus. Neptune appears to be surrounded by at least three partial ring arcs less than 20 kilometers wide. Neptune has two satellites, one of which, Triton, is larger than Pluto.

Most of the time, **Pluto** is the outermost planet. It is probably composed of frozen water, methane, and ammonia. Pluto's orbit is so eccentric that it moves inside Neptune's orbit some of the time, making Neptune the outermost planet. Pluto is currently inside Neptune's orbit. The diameter of Pluto is about 2400 km. Its mass is about 1/500 that of Earth. Pluto's surface temperature is believed to be about 40 K. Pluto revolves about the sun every 250 years and has at least one satellite, Charon.

BIOGRAPHY

Ursula Marvin
1921–

Ursula Marvin is a meteoritist who was among the first geologists to work with rocks brought back from the moon on Apollo 11. Marvin is now trying to explain the unusual concentration of meteorites on Antarctica and is looking for some sign of life in the frozen rock specimens.

Review

7. Identify the four largest moons of Jupiter.
8. Describe Uranus' rotational direction.
★ **9.** Identify the farthest known planet from the sun.

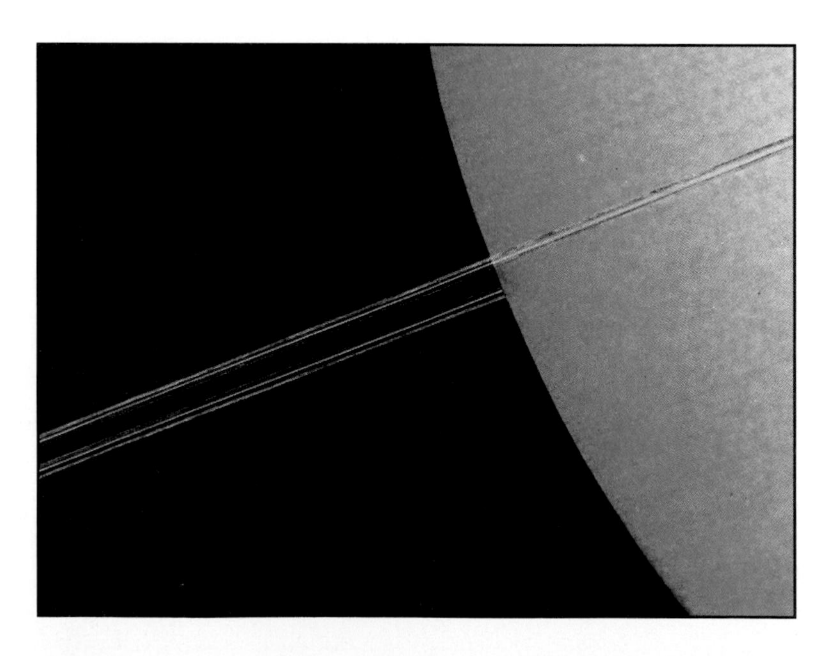

FIGURE 7–15. Uranus is a gaseous planet with a series of rings.

INVESTIGATION 7–2

Solar System Distance Model

Problem: How can you construct a scale model of the distance between the sun and planets in the solar system?

Materials 

adding machine tape
meter stick
scissors
pencil

Procedure

1. Use Appendix F to obtain the mean distance from the sun in AUs for each planet. Record these data in the chart.
2. Using ten centimeters as the distance between Earth and the sun (10 cm = 1 AU), determine the length of adding machine tape you will need to do this investigation.
3. Calculate the scale distance that each planet would be from the sun on the adding machine tape. Record this information.
4. Cut the tape to the proper length.
5. Mark one end of the tape to represent the position of the sun.
6. Put a label at the proper location on the tape where each planet would be if the planets were in a straight line outward from the sun.
7. Complete the chart by calculating the scale distance of each planet from the sun if 1 AU equals 2 meters on a model.

Analysis

1. Explain how the scale distance is determined.
2. How much adding machine tape would be required to construct a model with a scale distance 1 AU = 2 m?

Conclusions and Applications

3. In addition to scale distances, what other information do you need before you can construct an exact scale model of the solar system?
4. Proxima Centauri, the next closest star to our sun, is 4.3 light-years from the sun. Using the scale of 10 cm = 1 AU, how long a piece of adding machine tape would you need to include this star on your scale model?

Data and Observations

Planet	Distance to sun (km)	Distance to sun (AU)	Scale distance (1 AU = 10 cm)	Scale distance (1 AU = 2 m)
Mercury	58×10^6			
Venus	108×10^6			
Earth	150×10^6			
Mars	228×10^6			
Jupiter	780×10^6			
Saturn	143×10^7			
Uranus	288×10^7			
Neptune	451×10^7			
Pluto	592×10^7			

OTHER OBJECTS OF THE SOLAR SYSTEM

GOALS

1. You will learn about comets, meteoroids, meteors, and meteorites.
2. You will learn what asteroids are and where they are found.

Although the planets and their satellites are the most noticeable members of the sun's family, there are other objects that orbit the sun. Comets, meteors, and asteroids are also members of the solar system.

7:9 Comets and Meteors

A **comet** is a mass of frozen gases, cosmic dust, and small rocky particles that orbits the sun. Some comets orbit at regular intervals. Others appear one time and are never seen again. The paths of comets are usually very elliptical.

What happens when a comet approaches the sun?

The solid portion of a comet is called the nucleus. When a comet approaches the sun, a cloud of dust and gas forms a halo called a coma around the nucleus. This coma may have a diameter greater than Jupiter's diameter (142 800 km). Since solar energy causes the gases in a comet to vaporize, its brightness increases as it nears the sun. The tail of a comet always points away from the sun.

Comets are named for their discoverers. Halley's Comet was first recorded in 240 B.C. Halley predicted the comet's return in 1758 based on its previous appearances. It has returned every 75 to 76 years.

FIGURE 7–16. Most comets are thought to originate in a dense comet cloud beyond Pluto (a). A comet (b) consists of a nucleus, a coma, and a tail (c).

a

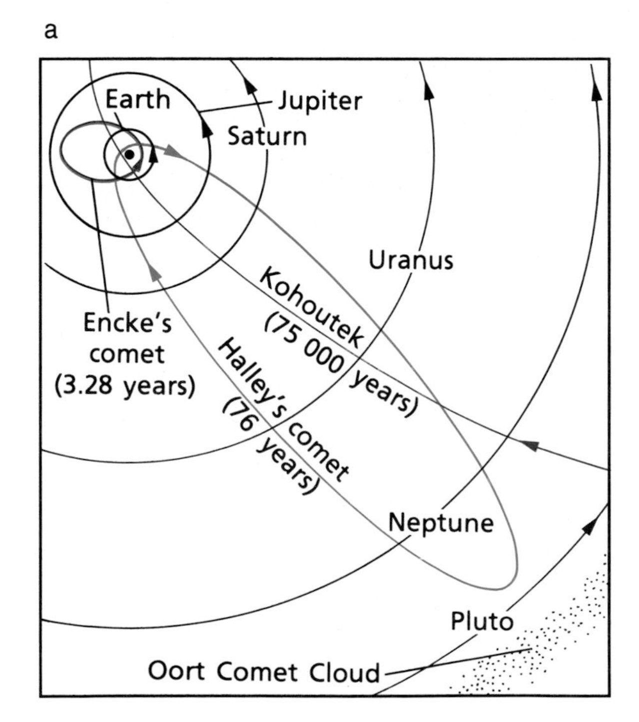

b

c

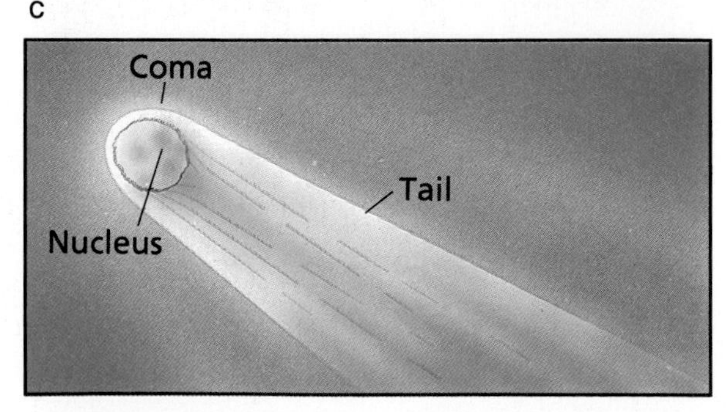

FIGURE 7–17. Halley's Comet

CAREER

Aeronautical Engineer

Susan Althoff is a United States Army civilian engineer. As a researcher at NASA's Langley Research Center, Ms. Althoff works with advanced helicopter models before they undergo laser-assisted wind tunnel tests. The velocimeter shown helps Ms. Althoff predict helicopter rotor performances by measuring complex air flows.

Ms. Althoff's skills in mathematics, physics, and mechanical drawing are essential for her career as an aeronautical engineer. Ms. Althoff uses models to test new designs in huge wind tunnels, and works with other engineers who specialize in other areas.

For career information, write:
American Institute of Aeronautics
and Astronautics
1290 Avenue of the Americas
New York, NY 10019

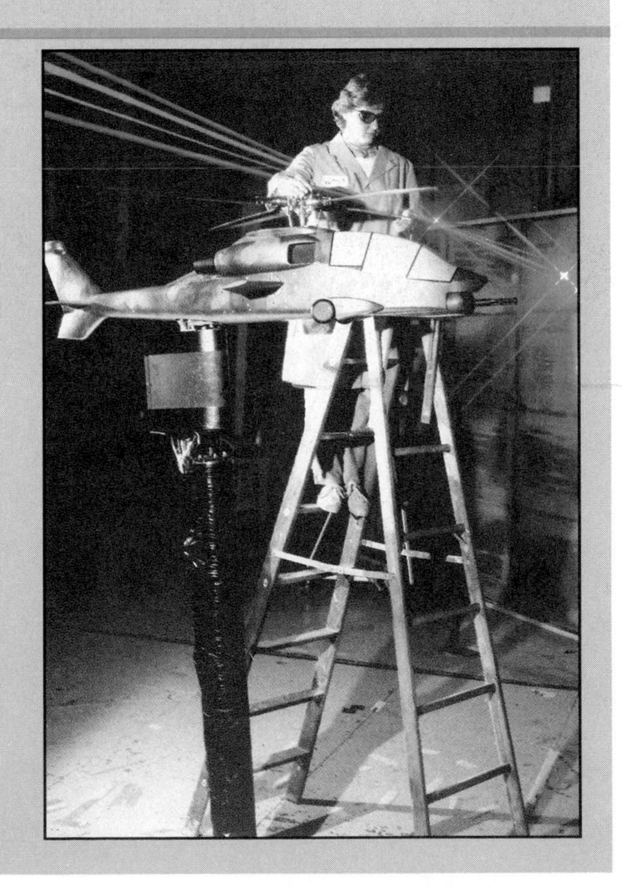

FIGURE 7–18. A meteor shower occurs when Earth passes through a swarm of meteors.

Meteoroids (MEET ee uh roydz) are small fragments of matter moving in space. Meteoroids are important clues to the composition of matter in space. Meteoroids enter the atmosphere with speeds between 12 and 72 km per second. Most meteoroids have masses smaller than one gram and will completely vaporize upon penetrating the atmosphere. **Meteors** are meteoroids that reach Earth's atmosphere. About 25 million meteors visible to the unaided eye occur every 24 hours over the entire planet. When Earth passes through a swarm of these fragments they burn up, forming a meteor shower. **Meteorites** are meteors that strike Earth. Meteor Crater in Arizona was formed by a large meteorite.

FIGURE 7–19. Some meteorites are composed of iron and nickel (a). Meteor Crater (b) in Arizona was formed by a meteorite.

a

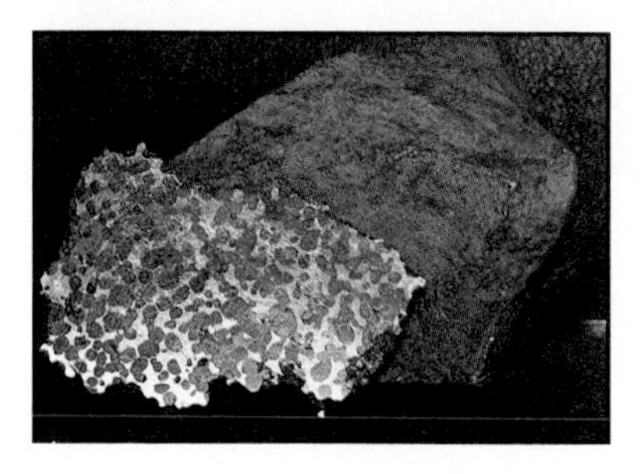

b

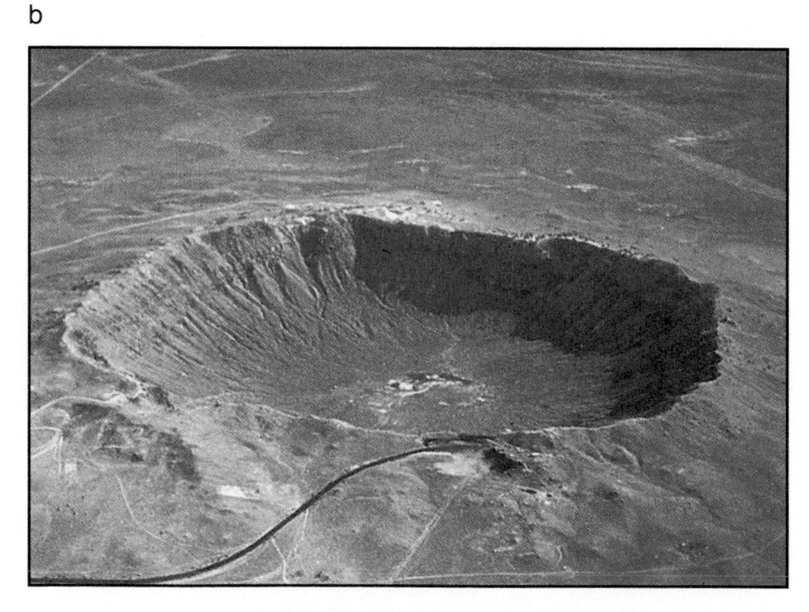

7:10 Asteroids

Asteroids are fragments of matter similar to that from which planets were formed. A region of asteroids is found between the inner rocky planets and the outer gaseous planets. Asteroids range in size from tiny particles to objects over 1000 kilometers in diameter. Several thousand asteroids are large enough to have been named and measured. The largest asteroid, Ceres, was the first to be discovered in 1801. It measures just over 1000 km at its longest dimension. Most asteroids are located in the region between Mars and Jupiter. A few of the moons of Jupiter and of Mars may be asteroids.

The asteriod belt may have formed because that part of the solar system cloud did not have enough mass to condense to form a planet. Another hypothesis states that Jupiter's large gravitational pull may have disrupted any larger body that was forming.

F.Y.I. Although the asteroids are thought of as existing in a "belt," they certainly do not fill the space. Pioneer and Voyager spacecraft passed through the belt without detecting a single asteroid.

Review

10. Identify the largest asteroid.

★ **11.** Compare and contrast meteoroids, asteroids, and comets.

PROBLEM SOLVING

The Hotter It Gets???

Carol and Salvatore were pretending to be a part of an astronaut team whose next mission was to land on the surface of one of the inner planets. In researching information about the surface conditions that they might expect to find on the inner planets, Carol and Sal were surprised to note that Mercury, which is the closest planet to the sun, does not have the highest average surface temperature. Venus, which is almost twice as far from the sun, averages higher surface temperatures. How can you explain this problem?

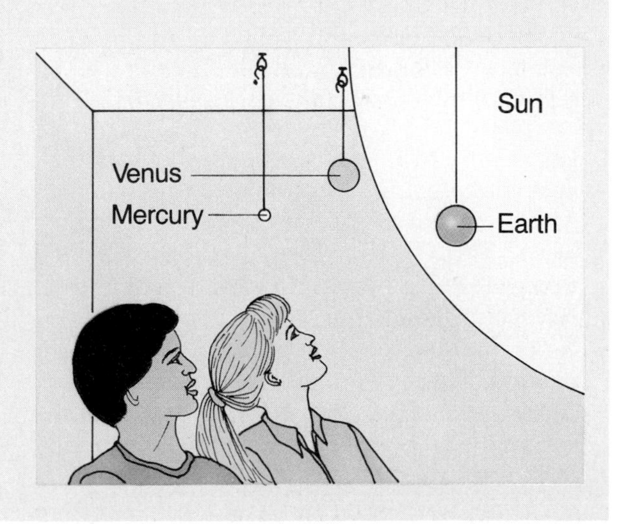

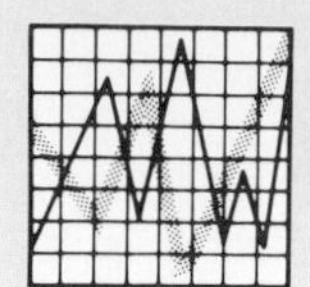

SKILL

Making Scale Drawings

Problem: How do you construct a scale drawing of solar system objects?

Materials 

metric ruler
drawing compass
pencil
paper

Procedure

1. Review the information on the sizes of the planets provided in Appendix F.
2. Using the given data in the Data and Observations chart, select a scale diameter for Earth.
3. Draw a circle with this diameter on your paper.
4. Calculate the scale diameters of the other objects listed, using Earth's diameter as 1.0 km. Record these in the Data and Observations chart.
5. Construct a scale drawing of each of these objects on your paper.

Questions

1. How are the scale sizes for the drawings of solar system objects determined from the scale diameter of Earth?
2. Using the scale size of the diameter of Earth in your diagram, calculate the scale distance for 1 AU.
3. If you were to use your scale diagrams of the sun, planets, and other objects to construct a scale model of the solar system, how far would Earth have to be located, in scale distance, from the sun? How far would Pluto have to be placed?
4. Refer to Investigation 7–2, in which you made a solar system distance model. Using the scale 1 AU = 2 m, how large would the sun and Earth models have to be to remain in scale?

Data and Observations

Solar system object	Actual diameter (km)	Multiple of Earth size	Scale diameter
Sun	1 400 000		
Mercury	4880		
Venus	12 100		
Earth	12 750		
Mars	6800		
Jupiter	142 800		
Saturn	120 660		
Uranus	51 800		
Neptune	49 500		
Pluto	~2400		
Ceres (asteroid)	1000		
Halley's Comet (max. tail length)	40 000 000		

Chapter 7 Review

SUMMARY

1. Scientists believe that our star, the sun, formed about 4.6 billion years ago. 7:1.
2. The solar system could have formed when a cold, slowly rotating cloud of gas and dust began to contract. 7:1
3. Planets revolve about the sun in elliptical orbits. 7:2
4. Perihelion is the point at which an orbiting object is closest to the sun. Aphelion is the farthest point in the orbit. 7:2
5. The inner planets are solid objects that orbit close to the sun. They include Mercury, Venus, Earth, and Mars. 7:3, 7:4, 7:5
6. The outer planets include Jupiter, Saturn, Uranus, Neptune, and Pluto. 7:6, 7:7, 7:8
7. Although Pluto has a greater average distance to the sun, it is presently closer to the sun than Neptune. 7:8
8. Comets, asteroids, and meteoroids are other objects found in our solar system. 7:9, 7:10
9. A region densely populated with asteroids exists between the inner rocky planets and the outer gaseous planets. 7:10
10. A few of the moons of Jupiter and Mars may be asteroids. 7:10

VOCABULARY

a. aphelion
b. asteroids
c. comet
d. Earth
e. ellipse
f. Great Red Spot
g. Jupiter
h. Mars
i. Mercury
j. meteorites
k. meteoroids
l. meteors
m. Neptune
n. perihelion
o. Pluto
p. retrograde
q. Saturn
r. solar system
s. Uranus
t. Venus

Matching

Match each description with the correct vocabulary word from the list above. Some words will not be used.

1. an elongated closed curve
2. the third planet from the sun
3. pieces of rock from space that strike Earth's surface
4. the closest point to the sun in a planet's orbit
5. the planet with the largest known volcano in the solar system
6. currently, the outermost planet
7. an area of atmospheric turbulence; similar to a hurricane on Earth
8. a mass of frozen gases, cosmic dust, and small rocky particles
9. planet most similar to Earth in size, mass, and shape
10. a rotation or revolution in the opposite direction of most planets or satellites

Chapter 7 Review

MAIN IDEAS

A. Reviewing Concepts

Choose the word or phrase that correctly completes each of the following sentences.

1. The largest planet in the solar system is *(Saturn, Mercury, Jupiter)*.
2. Titan is a satellite of *(Saturn, Mars, Jupiter)*.
3. Meteoroids are fragments of matter that are called *(meteors, asteroids, comets)* if they enter Earth's atmosphere.
4. Halley is the name of a(n) *(comet, meteorite, asteroid)* that can be seen from Earth every 75 years.
5. Most of the outer planets are *(rocklike, gaseous, elliptical)* bodies.
6. Venus' atmosphere is mostly *(helium, hydrogen, carbon dioxide)*.
7. Olympus Mons is the name of the largest volcano of the solar system, which is found on *(Earth, Mars, the moon)*.
8. Over one thousand rings surround *(Saturn, Venus, Pluto)*.
9. One planet that goes through phases is *(Venus, Mars, Saturn)*.
10. The Great Red Spot probably is a hurricane-type cloud on *(Venus, Jupiter, Mars)*.
11. When an object in orbit is at its farthest point from the sun it is at *(perihelion, aphelion, retrograde)*.
12. Kepler first discovered *(elliptical orbits, Jupiter, planets)*.
13. An inner planet is *(Neptune, Jupiter, Mars)*.
14. A planet with a retrograde rotation is *(Saturn, Uranus, Pluto)*.
15. The planet that revolves about the sun every 250 years is *(Earth, Mars, Pluto)*.

B. Understanding Concepts

Answer the following questions using complete sentences.

16. Which two planets in the solar system rotate in different directions from the other seven?
17. What is one possible explanation of how the solar system began?
18. Describe the shape of the planetary orbits around the sun.
19. Identify the farthest planet from the sun. Explain your answer.
20. Compare and contrast Venus and Earth.
21. Describe the compositions and appearances of the inner and outer planets.
22. Describe the motion of a comet's tail as the comet orbits the sun.
23. If you constructed a model of the solar system where one AU equaled two meters in scale, how far away from the sun would the scale model of Pluto have to be? How large would the sun be in this scale model?
24. Why is Earth's atmosphere unique among the planets in our solar system?
25. Describe the orbit of a typical comet.

C. Applying Concepts

Answer the following questions using complete sentences.

26. Why is the surface temperature on Venus so much higher than on Earth?
27. Write a short essay discussing the occurrence of water on Mars. Find all the evidence you can for the presence of water in the past and its location on this planet.
28. Why are surface probes or landings on Jupiter or Saturn unlikely events?
29. Describe the relationship between the mass of a planet and the number of satellites orbiting that planet.
30. Identify when Earth is near or at perihelion and aphelion in its solar orbit. Describe the effects on Earth's seasons of the changing distance between Earth and the sun.

SKILL REVIEW

1. The diameter of the sun is about 1 400 000 kilometers. Earth's orbit has a radius of approximately 150 000 000 km. Draw a scale diagram that accurately shows Earth's orbit around the sun.
2. The eccentricity of the orbit of the largest asteroid, Ceres, is 0.0765. Describe a procedure that you would follow to show how that orbit would appear if drawn to scale.
3. Compare and contrast characteristics of Neptune and Uranus.
4. The density of Earth is 5.5 g/cm^3. The density of Jupiter is 1.3 g/cm^3. Use the data in Appendix F to calculate the volume of Jupiter.
5. Compare and contrast the inner planets with the outer planets.

PROJECTS

1. Design and build a three-dimensional scale model of the solar system.
2. Mercury, Venus, Mars, Jupiter, and Saturn can be observed with the unaided eye. Research and construct a display that will show when and where these planets can be observed during the next five years.

READINGS

1. Gold, Michael. "Voyager to the Seventh Planet." *Science 86*. May, 1986, pp. 32-39.
2. Lambert, David. *The Solar System*. New York: Franklin Watts, 1984.
3. Pasachoff, J. M. "Volcanoes in the Solar System." *Science Digest*. February, 1986, pp. 28-33.

STELLAR ASTRONOMY
STARS AS OTHER SUNS
DEEP SPACE ASTRONOMY

Stars and Galaxies

The distances between Earth and the sun or between the sun and the outer planets seem quite large. Yet, compared to the size of galaxies or the universe itself, our solar system is very small. Except for the sun, stars are so far from Earth that even the largest telescopes show them only as tiny dots of light.

STELLAR ASTRONOMY

GOALS

1. You will learn a brief history of astronomy.
2. You will learn about the sun and the processes that operate within it.
3. You will learn about the brightness of a star.

The sun is the star nearest Earth. Its mass is 2.2×10^{27} metric tons, 330 000 times more massive than Earth! We depend on the sun, because life as we know it could not exist without the sun.

8:1 Early Astronomers

In 1543, Nicholas Copernicus wrote a book in which he suggested that the sun, not Earth, was the center of the solar system. Copernicus knew of only six planets, but he placed them in their correct order with respect to the sun. He also calculated the correct scale of distances between Earth and the known planets.

List five early astronomers.

Tycho Brahe, a Danish mathematician, contributed 20 years of observations to astronomy. He was concerned with accurately recording the positions of stars and planets. His data were used by his assistant Johannes Kepler to plot the correct paths of the planets. Recall from Chapter 7 that Kepler proved that the orbits were ellipses.

Galileo and Newton perfected telescopes that enabled astronomers to discover previously unknown objects. With these new tools, it was possible to observe motions and relationships among objects in space.

8:2 The Sun

Typical of other stars, the sun is a nuclear furnace. Most of the sun's mass consists of the gases hydrogen and

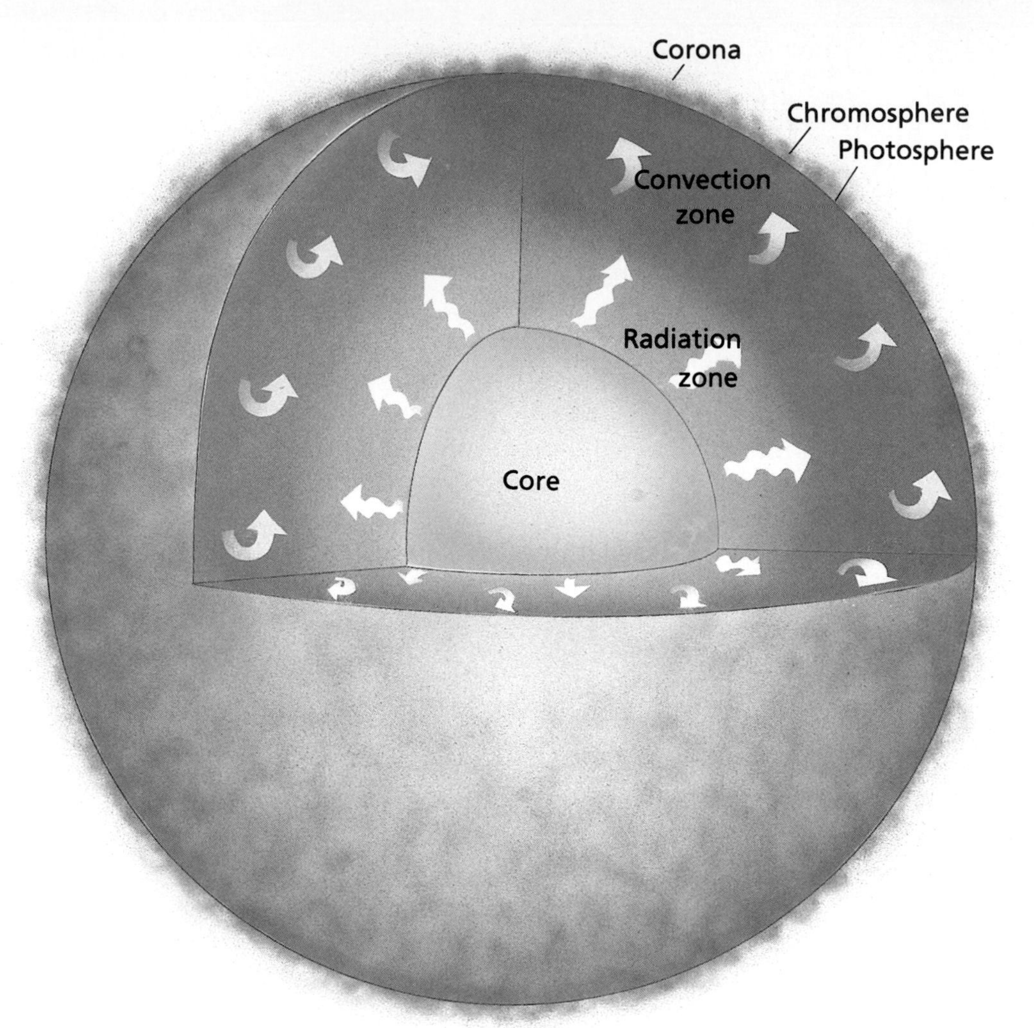

FIGURE 8–1. Energy is produced by fusion in the sun's core. Energy travels through the sun by radiation and convection.

F.Y.I. The light energy that reaches Earth each second is only one two-billionth of the total energy produced by the sun during that time interval.

helium. The rotation rate is not the same for all parts of the surface of the sun. Gases at the sun's equator move faster than those at the poles. The sun's rotation rate ranges from 25 days at the equator to 34 to 35 days at the poles.

The sun has a gaseous core surrounded by several other layers. The core is the inner part of the sun where fusion occurs. Fusion is the nuclear reaction in which hydrogen gas is changed into helium gas. At temperatures near 15 million °C, protons, which normally repel each other, are forced to combine. Energy is released during the reaction and forms a stable helium nucleus of two protons and two neutrons. The nuclear reaction maintains great pressure in the sun's interior, which overcomes the weight of overlying gases. Radiation flows outward from the sun's core to the surface. From the surface, radiation escapes into space.

a

b

FIGURE 8–2. The sun's corona extends millions of kilometers above the photosphere (a). Solar prominences release large amounts of energy (b).

The **photosphere** is the surface of the sun, which emits the visible radiation that we see. Outward from the photosphere is the chromosphere. This layer of hot gas can be seen only during an eclipse or with a special telescope. The **chromosphere** is bright red, extending about 6000 kilometers above the photosphere. The red color of this layer is due to the hydrogen gas present. **Prominences** are gases trapped at the edges of the sun that appear to shoot outward from the chromosphere. Beyond the chromosphere is the transparent zone of the **corona.** This zone of very hot gas is visible only during a total eclipse of the sun or with special instruments.

When is the corona visible?

Radiation from the sun is so intense that you should never look directly toward it. Special telescopes project the sun's image onto a screen that is shaded from direct sunlight. Such a solar telescope is located at Kitt Peak. Its light shaft is 153 meters long. Most of the telescope is beneath Earth's surface, where even temperatures are maintained and air currents are avoided.

Sunspots are relatively cool, dark areas on the sun's surface. Sunspots first appear as small dark areas on the photosphere. Their diameters may be over 1500 kilometers. Temperatures in sunspots are about 4000°C. Surrounding matter is about 5500°C. In a few days, sunspots may expand to thousands of kilometers. Sunspots may last for a day or for several months. Sunspot activity follows an 11-year cycle.

FIGURE 8–3. Two photos of the sun's photosphere taken several days apart show several sunspot groups on the solar surface. Sunspots are usually about 1500°C cooler than the photosphere.

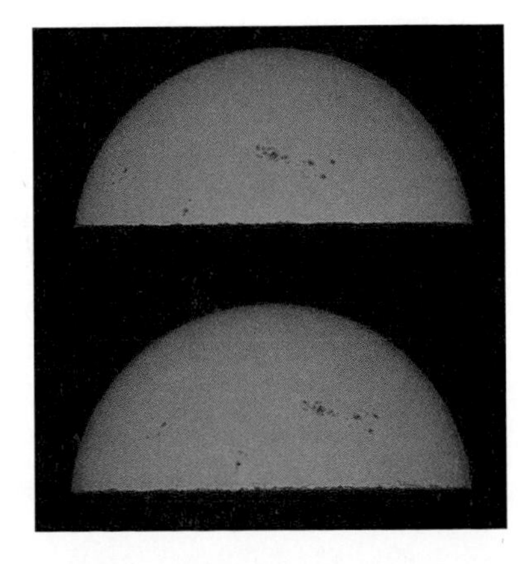

Solar flares are sudden increases in brightness of the chromosphere, often near sunspot groups. Protons and electrons stream outward from these areas at speeds up to 1500 kilometers per second. Many of these protons and electrons reach Earth. Solar flares disturb radio reception and affect Earth's magnetic field. Sometimes gases in

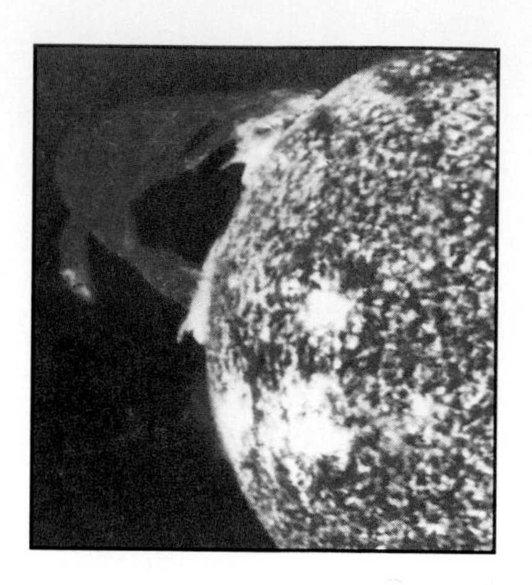

FIGURE 8–4. Solar flares are sudden increases in the brightness of the sun's chromosphere.

Earth's upper atmosphere are excited by solar flares. These gases radiate lights known as the aurora borealis or northern lights. In the Southern Hemisphere, these lights are called the aurora australis.

At a certain distance from its surface, the sun's atmosphere boils away. This **solar wind** is made up of ions and electrons that move outward from the sun. The velocity of the solar wind ranges from 250 to 800 kilometers per second and is very strong after solar flare activity.

8:3 Stars

Stars are hot, bright spheres of gas. Some stars appear brighter than others. The brightness of a star as viewed from Earth is its **apparent magnitude.** A star may appear bright because it is close to Earth, or because it emits more radiant energy than other stars. **Absolute magnitude** is a measure of a star's actual brightness. To compare absolute magnitudes, astronomers determine what brightness a star would have if it were 10 parsecs from Earth. The higher the magnitude number, the dimmer the star. A first magnitude star is about 2.5 times brighter than a second magnitude star, and about 6.25 times brighter than a third magnitude star.

Review

1. Identify activities that occur in each area of the sun.
2. How does a solar flare differ from a solar prominence?
3. ★ How do absolute and apparent magnitude differ?

PROBLEM SOLVING

Star Light, Star Bright

Mary conducted an activity to determine the relationship between distance and the brightness of stars. She used a meter stick, a light meter, and a light bulb. The bulb was mounted at the zero end of the meter stick. Mary placed the light meter at the 20-cm mark on the meter stick and recorded the distance and the light meter reading in the data table. Readings are in luxes, which are units for measuring light intensity. Mary doubled and tripled the distances and took more readings. What is the relationship between light intensity and distance? What would it be at 100 cm?

Distance (cm)	Meter Reading (luxes)
20 cm	4150
40 cm	1050
60 cm	460
80 cm	262.5

INVESTIGATION 8–1

Sunspots

Problem: How can you trace the movement of sunspots?

Materials 

several books
cardboard
clipboard
drawing paper
small refracting telescope
small tripod
scissors

Procedure

1. Find a location where the sun may be viewed at the same time of day for a minimum of five days. **CAUTION:** *Do not look directly at the sun. Do not look through the telescope at the sun. You could damage your eyes.*
2. Set up the telescope with the eyepiece facing away from the sun as in Figure 8–5.
3. Set up the clipboard with the drawing paper attached.
4. Use the books to prop the clipboard upright. Point the eyepiece at the drawing paper.
5. Arrange a shield of heavy cardboard with the center cut out. See Figure 8–5.
6. Move the clipboard back and forth until you have the largest possible image of the sun on the paper. Adjust the telescope to form a clear image.
7. Trace the outline of the sun on paper.
8. Trace any sunspots that appear as dark areas on the sun's image. At the same time each day for a week, check the sun's image and trace the position of the sunspots.
9. Using the sun's diameter as approximately 1 400 000 km, estimate the size of the largest sunspots that are observed.
10. Calculate how many kilometers any observed sunspots appear to move each day.

FIGURE 8–5.

11. At the rate determined in Step 10, predict how many days it will take for the same group of sunspots to return to about the same position as you first observed them.

Analysis

1. Which part of the sun showed up in your image?
2. What are solar flares, and how can they be related to sunspots?
3. What was the average number of sunspots observed each day during this investigation?

Conclusions and Applications

4. How can the movement of sunspots be traced?
5. How can sunspots be used to determine that the sun's surface is not solid like Earth's?

Data and Observations

Date of Observation	Number of sunspot groups (approx)	Estimated average sunspot diam. (km)	Approximate actual movement (km)	Predicted return time (Earth days)

STARS AS OTHER SUNS

GOALS

1. You will learn about the formation and life cycles of stars.
2. You will learn how to use a star chart to locate and identify stars and constellations.

Stars form as a result of the force of gravity acting on particles of gas and dust in an area of space. The amount of material in an area is the single most important factor for how big the star will be, how hot its surface will become, how long it will be in existence, and how it will "die."

8:4 Stellar Evolution

Where do stars form?

A star is born in a low-density cloud of gas and dust. This cloud is called a **nebula.** As the particles in the cloud begin to attract each other and form larger masses, the temperature of the nebula rises. The star begins to glow. When the temperature reaches about 10 million °C, matter changes to energy. Fusion, a nuclear reaction in which energy is produced, occurs in the star's core, releasing radiant energy into space at the star's surface.

When the hydrogen in the core of a small- or average-size star is exhausted, the core collapses, causing its temperature to rise. The heat from the core causes the gases nearby to heat, contract, undergo fusion, and produce helium. Stars in this stage are called **red giants,** or supergiants, and are nearing the end of their existence. A **white dwarf** is the dying core of a giant. It radiates its heat into space as light waves.

A **nova** is an ordinary star that experiences a sudden increase in brightness. The increase can be on the order of tens of thousands of times brighter. The nova then fades slowly back to its original brightness. A supernova is an

FIGURE 8–6. A nebula is a dense patch of gas and dust in space. Many stars, including our sun, are thought to have formed within such clouds.

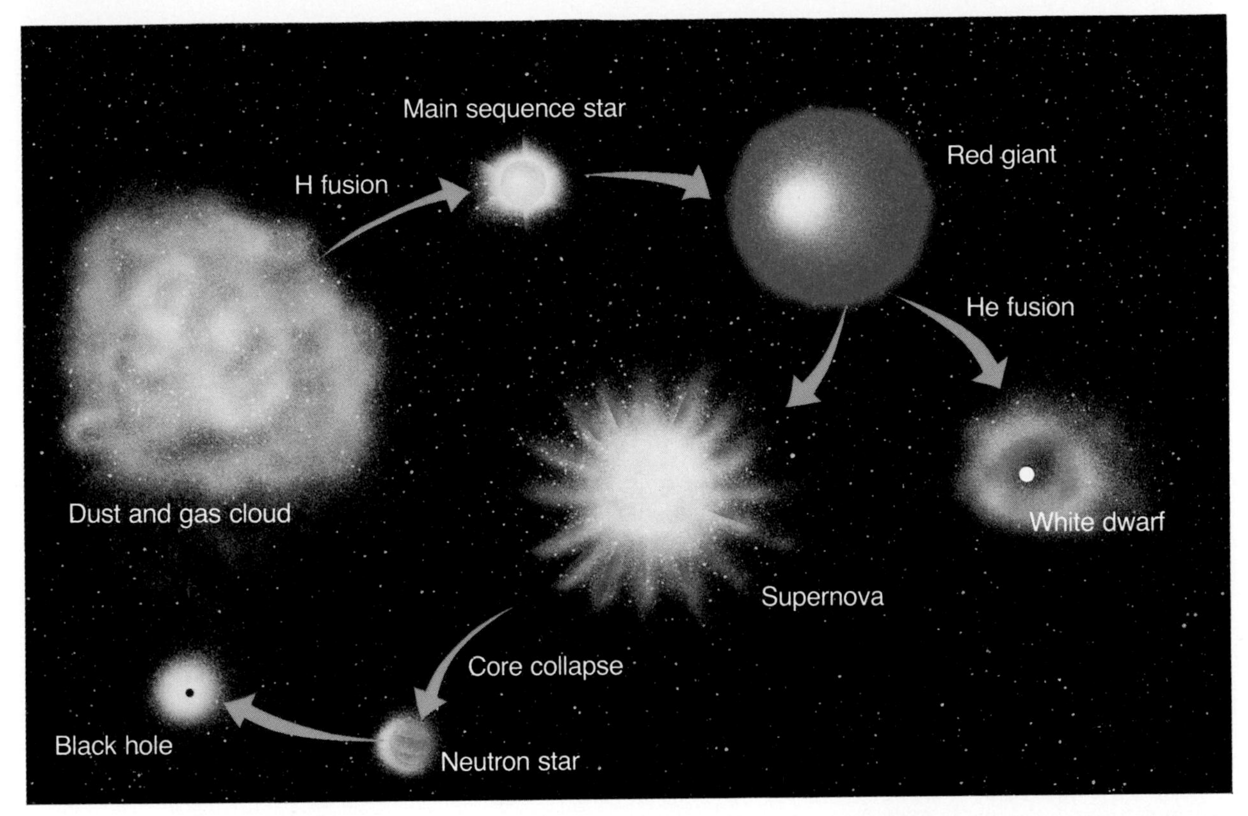

FIGURE 8–7. A star goes through various stages in its life cycle.

even greater outburst of energy. It can outshine billions of suns! A supernova was observed in 1987. Astronomers estimated that light from the supernova had been traveling for about 163 000 years before reaching Earth.

The star left behind from a supernova is a **neutron star.** The collapsing material is forced into such a small space that the protons and electrons join to form neutrons. The matter in a neutron star is so dense that one spoonful of it would have a mass of 1 000 000 000 tons! Pulsars also may be formed when a supernova occurs. A **pulsar** is a rapidly spinning neutron star. Pulsars give off radio waves that enable astronomers, using radio telescopes, to determine the rate of spin of the star.

A **black hole** is a star in which matter is condensed and the gravity field so strong that light cannot escape. Even though no light is observable, astronomers have located X-ray sources that may be black holes.

F.Y.I. In 1987, a supernova occurred in the Large Magellanic Cloud, a small neighboring galaxy. Because of its location, the supernova explosion was visible only to observers in Earth's Southern Hemisphere. It was visible even without the aid of binoculars or telescopes.

What is a pulsar?

8:5 Classification of Stars

Two astronomers in the early 1900s discovered a strong correlation between star type and absolute magnitude.

Where is the main sequence located on the H–R diagram?

Henry Russell and Ejnar Hertzsprung found that stars can be classified by surface temperature and absolute magnitude. A graph of these measurements is called the **Hertzsprung-Russell (H–R) diagram.** Most stars lie along a diagonal line from upper left to lower right on the diagram. These stars are **main sequence stars.** Main sequence stars use up their hydrogen fuel at a steady rate. They are in a state of equilibrium. Blue stars have surface temperatures of about 60 000 K. These are the most massive and hottest stars. They are located in the upper left position in the diagram. White stars have surface temperatures of about 9000 K. Yellow stars are about 6000 K. Red stars range from 2500 K to 4000 K and are the smallest and coolest of the main sequence stars. Stars that lie on either side of the diagonal do not shine by fusion reactions alone. White dwarfs, giants, and supergiants lie outside the main sequence.

8:6 Variable Stars

Variable stars are stars that change brightness. These changes can be actual or apparent. Apparent changes in

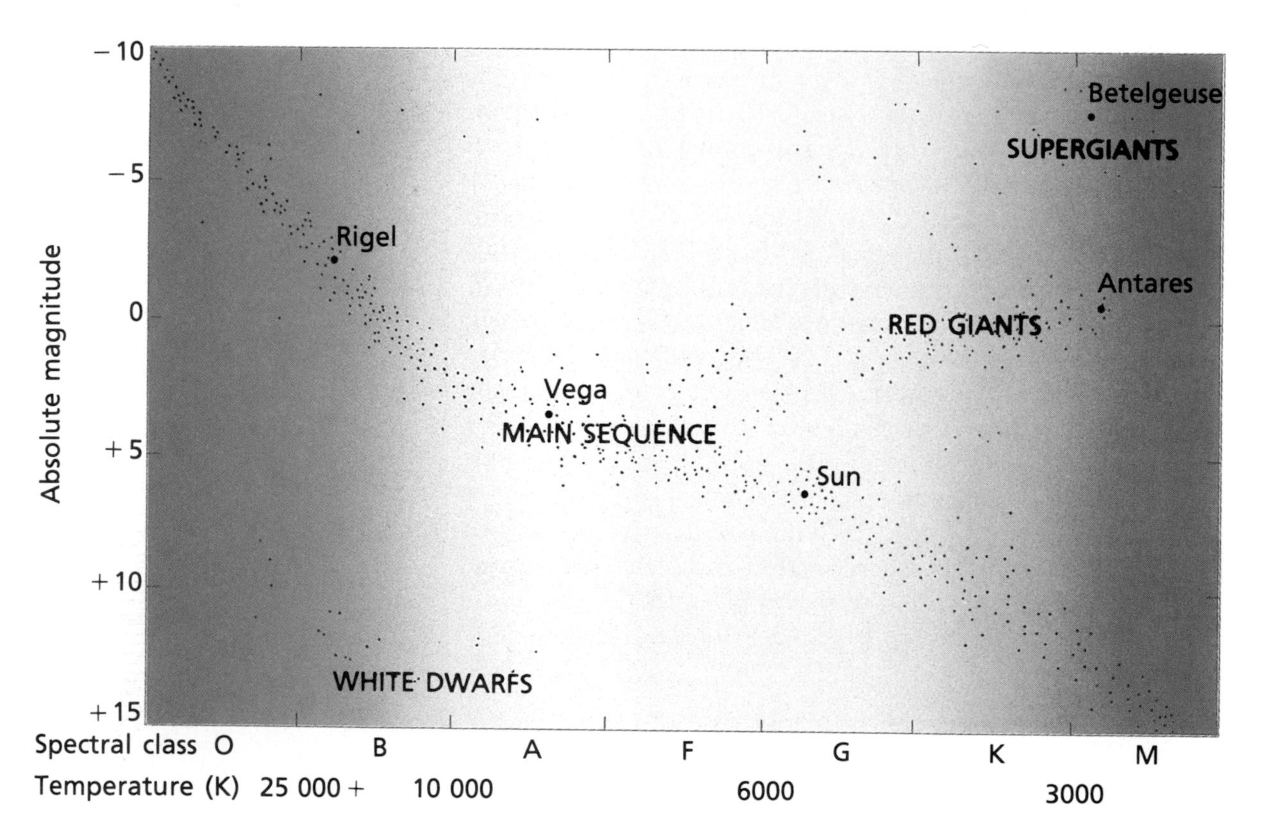

FIGURE 8–8. The H–R diagram is used to classify stars according to their absolute magnitudes and temperatures.

INVESTIGATION 8–2

Stars

Problem: Is the sun an "average" star?

Materials

graph paper
colored pencils

Procedure

1. Copy the graph shown in Figure 8–9.
2. Using a red pencil, plot the temperature and brightness data for the 15 nearest stars shown in the table below.
3. On the same graph, use a blue pencil to plot the data for the 15 brightest stars provided in Table 8–1 on page 170.

Questions

1. How do the brightness and temperature of the sun compare to the brightest stars in the Northern Hemisphere?
2. How does the sun compare to the stars closest to it?

FIGURE 8–9.

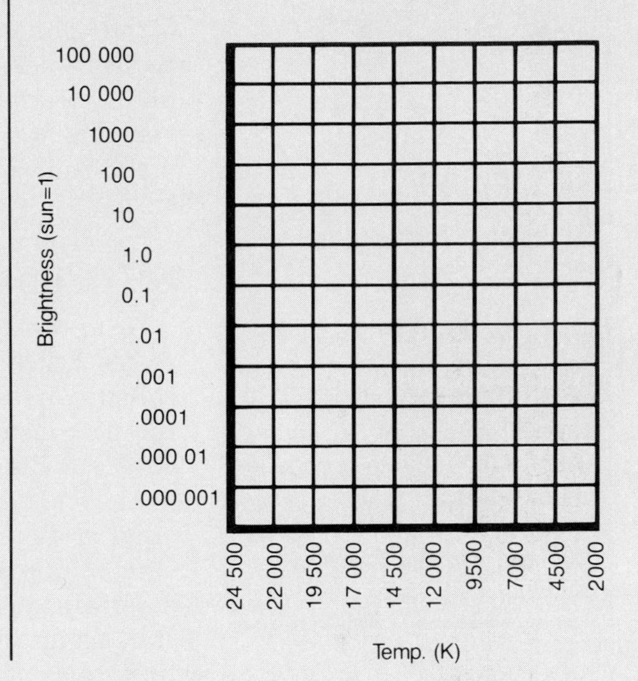

Data Table

The 15 Stars Nearest Earth							
Star	**Temp. (K)**	**Brightness (Sun = 1)**	**Distance (L.Y.)***	**Star**	**Temp. (K)**	**Brightness (Sun = 1)**	**Distance (L.Y.)***
Sun	5 600	1	0.00002	Sirius B	10 700	0.002	8.7
Proxima Centauri	5 800	0.00006	4.3	Luyten 726-8A	2 700	0.00006	8.9
Alpha Centauri A	4 200	1.5	4.3	Luyten 726-8B	2 700	0.00004	8.9
Alpha Centauri B	2 800	0.43	4.3	Ross 154	2 800	0.0004	9.4
Barnard's	2 800	0.00044	5.9	Ross 248	2 700	0.00011	10.3
Wolf 359	2 700	0.00002	7.6	Epsilon Eridani	4 500	0.30	10.7
Lalande 21185	3 200	0.0052	8.1	Ross 128	2 800	0.00033	10.8
Sirius A	10 400	24.0	8.8				

*Light-years from Earth

brightness occur as two or more stars orbit a common point and eclipse one another. These eclipsing stars are one type of **binary star** or binary system. Astronomers are discovering that binary stars are very common. Over half of the stars that are visible from Earth are binary systems.

Actual changes in the brightness of a star can occur as stars expand and contract. The expansion and contraction of the outer gases of a star also can change the intensity of the light emitted from it. Polaris, the North Star, varies in brightness slightly once every four days as its atmosphere expands and contracts. Betelgeuse is an irregular variable star that cycles through a change of over one magnitude as its surface area changes over 40 percent in size.

BIOGRAPHY

S. Jocelyn Bell Burnell
1943–

Jocelyn Bell Burnell was a graduate student when she discovered pulsating radio stars or pulsars. In 1967, she noticed unusual marks on the chart of the radio telescope in Cambridge, England. Now known to be neutron stars, pulsars are the last stage in the evolution of a star.

8:7 Constellations

At first glance, trying to locate any one star in the sky seems difficult. Stargazers use groups or patterns of stars called **constellations** to organize the night sky. Most of the 88 constellations recognized today are named from Greek or Roman mythology.

Astronomers use constellations as landmarks to locate other stars and other space objects in the sky. An object, such as one of the planets or a galaxy, is said to be "in" a certain constellation when it appears to be near that star group. In fact, the stars in each constellation are thousands of times farther away than any object in our solar system, and the individual stars may not be near the object said to be "in" the star group. The constellation simply serves as a reference point in the sky.

FIGURE 8–10. The brightness of a binary system (a) appears to vary. The Crab Nebula (b) formed from a supernova.

a

1
2
3
4
Primary eclipse
Secondary eclipse
Primary eclipse
2 Between eclipses
3
4 Between eclipses
Brightness
1
Binary period
1

b

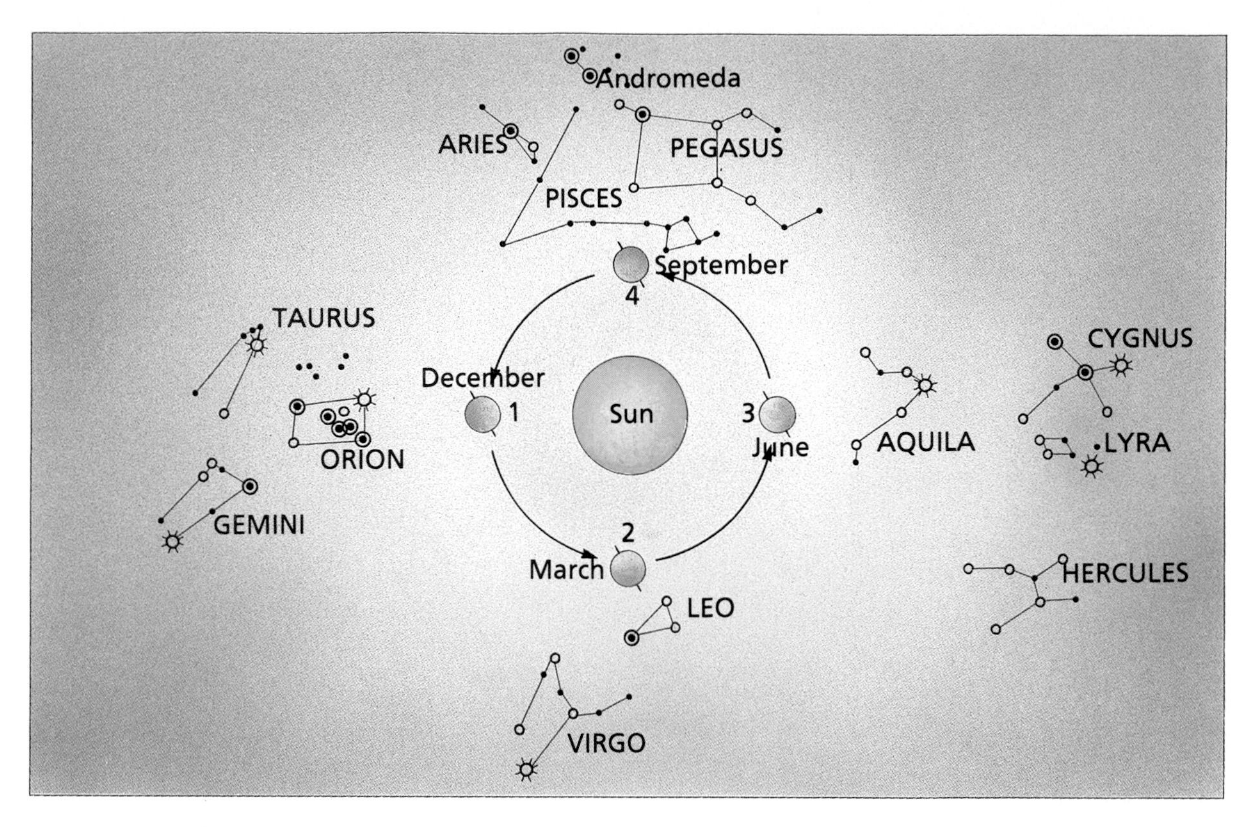

FIGURE 8–11. As Earth revolves around the sun, different constellations come into view.

Because Earth rotates, stars appear to move across the night sky. Most stars appear to travel on or along east-west paths similar to the sun's. The revolution of Earth around the sun causes different constellations to be seen at different times of the year. Look at Figure 8–11. When Earth is in position 1, a person looking up and south at midnight would see the winter star groups of Orion, Taurus, and Gemini. In spring (position 2), a person could observe Leo and Virgo. As Earth revolves, different constellations appear.

Why are different constellations visible at different times of the year?

Stars located above the North Pole appear to move in circles completely above the horizon each night as Earth rotates. Stars above the South Pole stay below the horizon. The revolution of Earth around the sun causes these stars to appear in a different position over the poles as the seasons change. Stars over the North Pole remain visible throughout the year. Stars over the South Pole remain invisible. A person looking northward in the Northern Hemisphere at night can observe the Big Dipper in Ursa Major, Cassiopeia, and Ursa Minor with Polaris. These stars are shown in Figure 8–12.

F.Y.I. Constellations that remain visible throughout the year are referred to as circumpolar constellations.

The 21 brightest stars that can be seen from Earth are called first magnitude stars, and are the first stars noticed

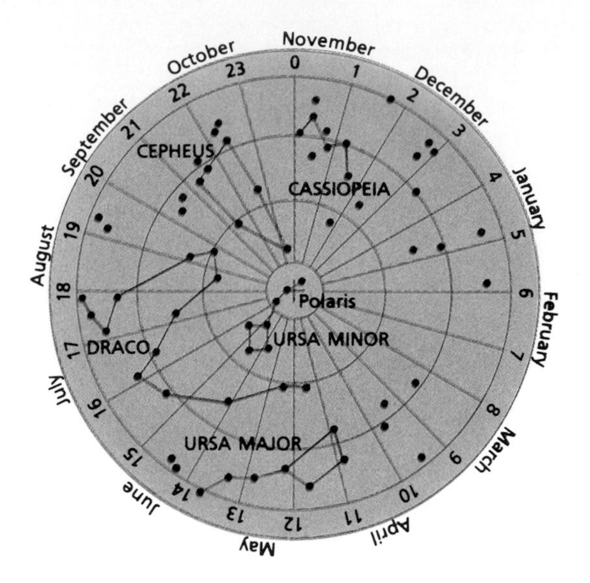

FIGURE 8–12. The positions of the stars above the North Pole change throughout the year, but the stars are always visible.

at night. The 15 commonly observed first magnitude stars in the Northern Hemisphere are listed in Table 8–1. Notice that Polaris is not among them.

Review

4. Identify the stages of evolution for a typical star.
★ 5. Distinguish between a nova and a supernova.

Table 8–1

The 15 Brightest Stars Seen from the Northern Hemisphere

Star	Constellation	Brightness (Sun = 1)	Distance (light-years)	Temperature K	Months observed
Aldebaran	Taurus	180	68	4 200	Dec/Jan/Feb
Altair	Aquila	11	16	8 000	July/Aug/Sept
Antares	Scorpius	6 550	424	3 400	May/June/July
Arcturus	Bootes	104	36	4 500	Apr/May/June
Betelgeuse	Orion	19 700	587	3 200	Jan/Feb/Mar
Capella	Auriga	150	42	5 900	Dec/Jan/Feb
Deneb	Cygnus	65 500	1 630	9 900	July/Aug/Sept
Fomalhaut	Pisces Austrinus	15	23	9 500	Oct/Nov/Dec
Pollux	Gemini	34 000	35	4 900	Feb/Mar/Apr
Procyon	Canis Minor	7.2	11	6 500	Feb/Mar/Apr
Regulus	Leo	92	84	20 000	Mar/Apr/May
Rigel	Orion	59 700	800	11 880	Jan/Feb/Mar
Sirius	Canis Major	24	8.8	10 000	Feb/Mar/Apr
Spica	Virgo	15 000	260	21 000	Apr/May/June
Vega	Lyra	55	26	10 700	July/Aug/Sept

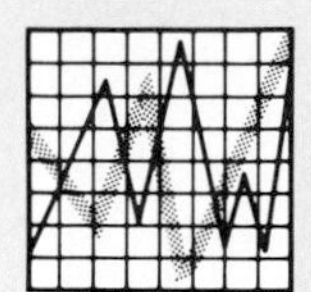

SKILL

Using Star Charts

Problem: How are star charts used to locate and identify stars and constellations?

Materials

star chart for appropriate season from the Appendix
flashlight
red cellophane
tape

Procedure

1. Cover the bulb end of your flashlight with red cellophane. The red light produced by the flashlight will not dim your night vision.
2. Examine the star chart for the appropriate season. This chart indicates the constellations and brightest stars as they appear on the dates and at the times listed. The line labeled *ecliptic* shows the apparent path of the sun through the sky.
3. The horizontal and vertical lines on the chart roughly correspond to latitude and longitude lines on Earth. The horizontal lines are called declination lines, and the vertical lines are called hour circles. Determine your latitude and find the declination line that corresponds to it.
4. To use the chart, face south and hold the book overhead with the top of the chart pointing toward the north and the right side of the chart pointing west.
5. The stars located along the declination line you located above will be overhead as you look at the night sky. The brightest stars are indicated by the symbol ☼, and the star names are indicated. Names of constellations are printed in all capital letters. The constellations north of right ascension 60° are shown in Figure 8–12.
6. Record your observations in a chart like the one below.

Questions

1. Why do you need several star charts in order to locate stars and constellations throughout the year?
2. Why are most stars and constellations above 60° north declination not indicated on seasonal star charts?
3. How are star charts used to locate constellations in the night sky?
4. Name the first magnitude stars that should be visible on the next clear evening.
5. List three or four constellations that should be visible on the next clear evening that cannot be seen on every night of the year.
6. Name three constellations that can be observed on every clear night of the year.

Data and Observations

Date	Season	Constellation	Stars Observed

GOALS

1. You will learn about the properties and distribution of galaxies.
2. You will learn about relative motion in the universe and a theory of how the universe may have begun.

DEEP SPACE ASTRONOMY

What lies beyond Pluto? Are there more planets? If so, are any like Earth? Astronomers have discovered that clusters of stars lie beyond our solar system. By studying these other stars, they hope to discover more about Earth and the origin of the universe.

8:8 Galaxies

How are galaxies classified?

Galaxies are large groups of stars. Most galaxies contain gas and billions of stars. Galaxies are classified according to their appearance. Some are irregular, with many stars and great clouds of gas and dust. Most of the stars in irregular galaxies are in early stages of their life cycles. Elliptical galaxies are smooth ellipsoids. They contain little dust and gas, but may have millions to trillions of stars. Spiral galaxies are disk-shaped and have arms that rotate around a dense center. The arms resemble pinwheels of light and seem to be the places where most new stars form. Some galaxies have a small nucleus from which great quantities of radio waves are emitted. Most large galaxies are about two to three million light-years apart. No two are exactly alike. Some clusters of galaxies may have up to 10 000 galaxies.

F.Y.I. Quasars are starlike objects that emit radio waves. Quasars are believed to be the centers of galaxies formed millions of years ago, or may be matter concentrated around the edges of black holes.

Earth's galaxy, the Milky Way, is seen as a bright band of light that stretches from horizon to horizon. The Milky

FIGURE 8–13. An irregular galaxy in Ursa Major has no structured appearance (a). In cross section, elliptical galaxies are smooth ellipses (b).

a

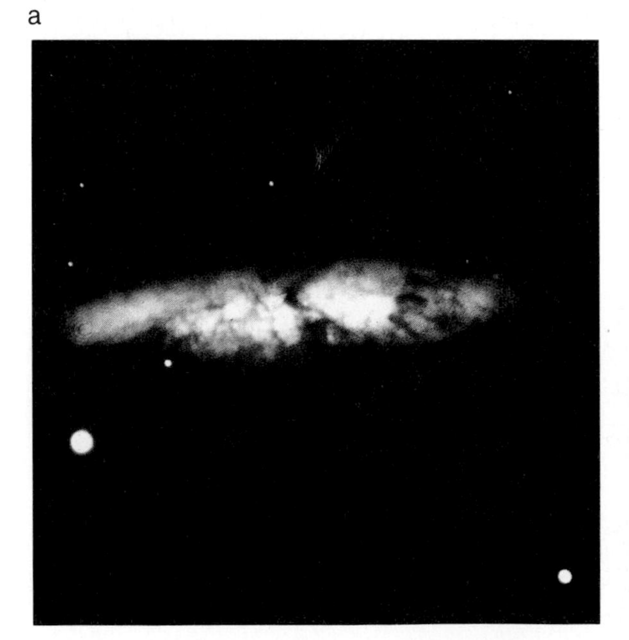

b

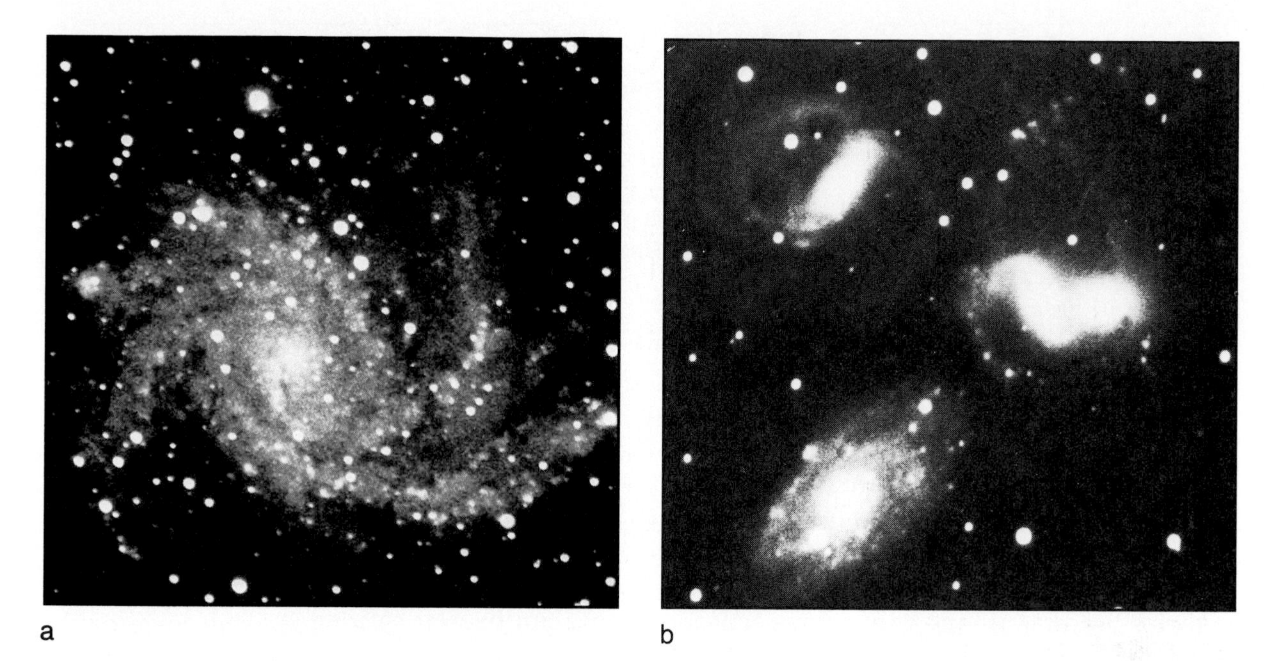

FIGURE 8–14. A spiral galaxy (a) is similar to the Milky Way. Galaxies are found in clusters (b).

Way is a spiral galaxy with a diameter of about 100 000 light-years. It contains between 100 billion to 200 billion stars. In our galaxy, there are over 1000 clusters of stars that have been observed. There are probably many more. These clusters contain thousands of stars. The center of the galaxy is hidden by dark clouds of dust. The sun is near a spiral arm about 30 000 light-years from the center of the Milky Way. The sun carries the whole solar system with it as it orbits the center of the galaxy at about 200 to 300 kilometers per second. One revolution takes about 250 million years to complete.

FIGURE 8–15. The sun is in a spiral arm of the Milky Way, about 10 000 parsecs (30 000 light-years) from the center of the galaxy.

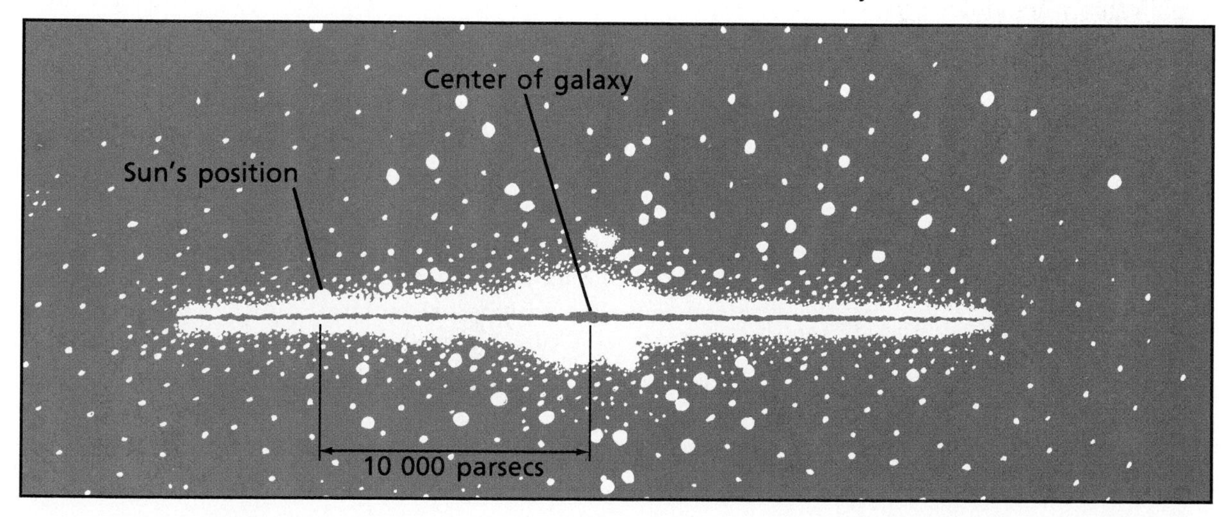

TECHNOLOGY: ADVANCES

The IUE

The International Ultraviolet Explorer, or IUE, is a satellite jointly operated by NASA, the European Space Agency, and the United Kingdom's Science and Engineering Research Council. A space telescope aboard the IUE is designed to observe spectra of celestial objects in ultraviolet light, which is impossible on Earth because of the shielding effect of the atmosphere. The 45-centimeter diameter telescope is attached to two spectrographs.

IUE is unlike most satellite observatories because it is in continuous contact with ground stations. The instruments respond to commands almost immediately, and cameras send spectral images to Earth. An observer can view the data and make decisions as events occur.

Since it was launched, scientists have obtained about 40 000 spectra of stars, planets, galaxies, quasars, and comets. Astrophysicists have also used the IUE to probe nebulae, which are mixtures of hot stars, ionized gases, and opaque dust. By mapping ultraviolet emission, the structure of nebulae, as well as the density and temperatures of the gases present, can be determined.

8:9 Relative Motion and the Universe

As far as can be determined, even the most distant galaxies are composed of elements just like those found on Earth. It also appears that the same forces and relationships that exist on Earth and in our solar system exist throughout all of the known galaxies. By knowing more about forces and motions on Earth, in the solar system, and in this galaxy, scientists can begin to piece together clues about the origin of the universe.

Newton summed up his understanding of motion in three laws. The first law states that an object continues at rest, or in uniform motion, until acted upon by an outside force. The second states that the force exerted on a moving object is equal to the mass multiplied by the acceleration of the object. The third law states that for every action force there is an equal and opposite reaction force.

Albert Einstein determined that everything in the universe is in motion. For example, both Earth and the moon are moving. We can measure their motion with respect to one another. We can measure their relative motions but not absolute motion. **Relative motion** compares the motion of one object to the motion of another object. Only the speed of light is considered to be an absolute measurement.

FIGURE 8–16. The people in car B observe a backward relative motion as car A moves forward. A stationary observer on the sidewalk observes the actual motion of both cars. Are motions of space objects observed from Earth relative or actual?

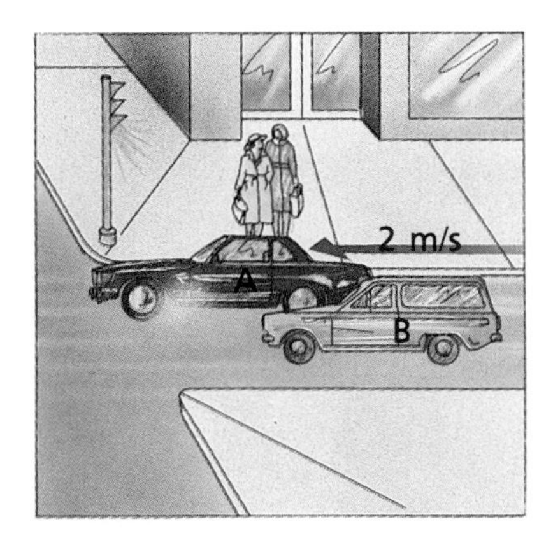

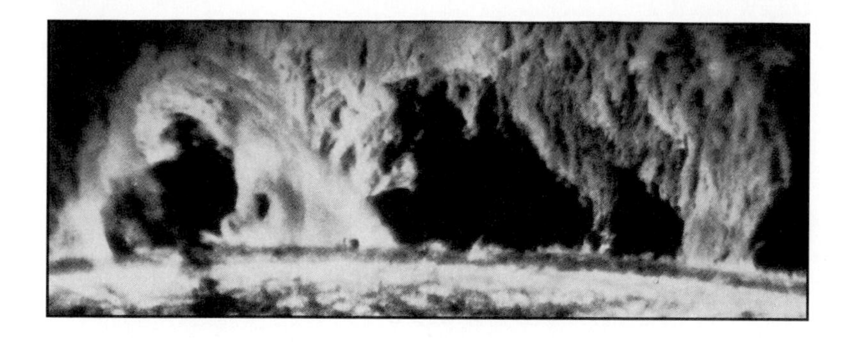

FIGURE 8–17. A massive object like the sun causes space to curve around it and time to slow down.

Einstein also showed that mass is never lost, but it may change to energy. Energy also may be changed to mass. Mathematically stated, this law is $E = mc^2$, where E is energy, m is mass, and c is the speed of light. Einstein further stated that massive objects affect space and time. Space is curved in the presence of massive objects and time slows down. The Viking space probe proved these concepts on a mission to Mars when it found that space curves around the sun and time slows down.

In 1924, Edwin Hubble proved that other galaxies lie beyond the Milky Way. Since then, astronomers have concluded that each galaxy outside the Local Group exhibits some degree of red shift in its spectrum. The Local Group includes the Milky Way and 28 other galaxies. The red shift indicates that the distant galaxies outside the Local Group are moving away from one another.

What is the Local Group?

FIGURE 8–18. Galaxies outside the Local Group appear to be moving. In which direction are they moving?

A number of models have been developed that use the idea of distant galaxies in motion away from each other to explain how the universe may have begun. The most commonly accepted idea for the origin of the universe at the present time is the **Big Bang.** According to this idea, a fireball exploded 10 to 20 billion years ago and matter and energy spread apart in all directions. As the material cooled, hydrogen gas formed and collected into clouds. These clouds formed galaxies during the next half-billion years.

Describe the Big Bang.

At present, within the galaxies, stars form and die while expansion of the universe continues. Most scientists believe that the matter may someday stop expanding and begin falling back toward the point of the original explosion. The matter may then form a small mass that again may explode and expand.

Review

6. In what galaxy is Earth, the sun, and the rest of the solar system located?

7. How are galaxies classified?

★ **8.** What do the spectra of all galaxies outside the Local Group have in common? What does this indicate?

CAREER

Observatory Technician

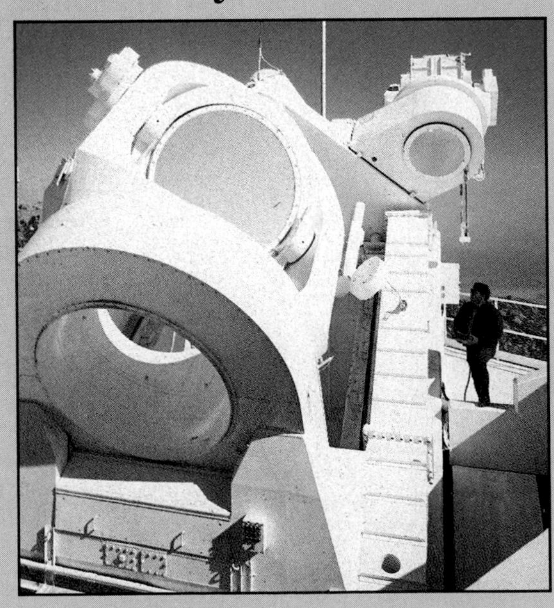

Joshua Jenson is an observatory technician who works at a government-owned observatory at a local university. As an observatory technician, he spends most of his time collecting and analyzing data that comes from both optical and radio telescopes. The radio telescope detects electromagnetic radiation from distant sources. The information goes into a computer. Mr. Jenson makes sure all equipment is running smoothly and correctly, which is quite a job since they are in operation 24 hours a day!

For career information, write:
American Astronomical Society
University of Virginia
Box 3818 University Station
Charlottesville, VA 22903

Chapter 8 Review

SUMMARY

1. The inventions and observations of early astronomers provided the basic understandings that underlie modern astronomy. 8:1
2. Some of the sun's layers include the core, photosphere, chromosphere, and corona. In the sun's core, hydrogen gas is changed to helium gas by fusion. 8:2
3. Stars are hot, bright spheres of gas. 8:3
4. Stars pass through a series of stages in their life cycles. 8:4
5. Stars are classified by temperature and absolute magnitude on the Hertzsprung-Russell diagram. 8:5
6. Variable stars are stars that change brightness. 8:6
7. Constellations are patterns of stars within certain areas of the sky visible during different months of the year. 8:7
8. Galaxies are classified according to appearance. 8:8
9. Scientists believe that red shifts in the spectra of some galaxies indicate that all matter in the universe at one time existed as a dense mass that exploded 10 to 20 billion years ago. 8:9

VOCABULARY

a. absolute magnitude
b. apparent magnitude
c. Big Bang
d. binary star
e. black hole
f. chromosphere
g. constellations
h. corona
i. galaxies
j. Hertzsprung-Russell diagram
k. main sequence stars
l. nebula
m. neutron star
n. nova
o. photosphere
p. prominences
q. pulsar
r. red giants
s. relative motion
t. solar flares
u. solar wind
v. sunspots
w. variable stars
x. white dwarf

Matching

Match each description with the correct vocabulary word from the list above. Some words will not be used.

1. large groups of stars classified by appearance
2. a cloud of gas and dust in which a star is born
3. the actual brightness of a star
4. the dying core of a giant star
5. a rapidly spinning neutron star
6. stars that lie along a diagonal on the H–R diagram
7. the visible surface of the sun
8. a system of two or more stars eclipsing one another
9. small groups of stars that appear to form a pattern in the night sky
10. the brightness of a star as viewed from Earth

Chapter 8 Review

MAIN IDEAS

A. Reviewing Concepts

Choose the word or phrase that correctly completes each of the following sentences.

1. A star in which matter is condensed and its gravity field so strong that light cannot escape is a *(neutron star, black hole, pulsar)*.
2. Binaries are one type of *(variable star, galaxy, nova)*.
3. That mass is never lost was shown by *(Galileo, Einstein, Copernicus)*.
4. In the H–R diagram, stars are classified on the basis of surface temperature and *(red shift, absolute magnitude, apparent magnitude)*.
5. A star that experiences a sudden increase in brightness is a *(nova, sunspot, nebula)*.
6. The inner part of the sun where fusion takes place is the *(corona, chromosphere, core)*.
7. Gases trapped at the edges of the sun are *(prominences, solar flares, sunspots)*.
8. Pulsars are rapidly spinning *(neutron stars, novas, binary stars)*.
9. A white dwarf is in a(n) *(early stage of its life cycle, late stage of its life cycle, main sequence position on the H–R diagram)*.
10. Earth's galaxy is shaped like a(n) *(ellipse, spiral, irregular mass)*.
11. Evidence that most galaxies are moving outward relative to Earth in all directions is the *(theory of relativity, red shift of spectra, Big Bang)*.
12. A star that changes brightness due to expansion and contraction is a *(main sequence star, variable star, constellation)*.
13. A low-density cloud of gas and dust in space is a *(nebula, main sequence star, solar wind)*.
14. The term *magnitude* refers to a star's *(mass, temperature, brightness)*.
15. *(Sunspots, Red giants, White dwarfs)* are relatively cool, dark areas on the sun's surface.

B. Understanding Concepts

Answer the following questions using complete sentences.

16. What are the variables that determine a star's position on an H–R diagram?
17. List the various parts or areas of the sun from its center outward.
18. What physical property of a star may be determined by analyzing the star's color?
19. What is a main sequence star?
20. What is the difference between a nova and a supernova?
21. Distinguish between the apparent and absolute magnitude of a star.
22. What is the single most important determining factor in whether a star will end as a white dwarf, neutron star, or black hole?
23. Distinguish between actual and apparent variable stars.
24. What term is applied to the 21 brightest stars in the night sky as seen from Earth? How many of these can be observed from the Northern Hemisphere?
25. When a star is near the end of its existence in the red giant stage, what happens to the core and the gases nearby?

C. Applying Concepts

Answer the following questions using complete sentences.

26. Why have primitive galaxies (the first galaxies to form) not yet been seen?
27. Why do we observe different stars and constellations throughout the year?

28. Why is Polaris the only star that does not seem to change position during the night in the Northern Hemisphere?
29. How can a black hole be located if it does not emit any light?
30. What type of reactions produce the energy emitted by the stars?

SKILL REVIEW

1. You plan to construct a model of the Milky Way galaxy that is 20 cm wide. If the Milky Way is 100 000 light-years across, what scale will you use?
2. Use the star chart in Appendix J to answer this question. For an observer in North America at 8 P.M. on November 21, what constellation is almost directly overhead?
3. Use the summer star chart in Appendix I to answer this question. For an observer in North America at 10 P.M. on July 21, what first magnitude star is almost directly overhead?
4. What are the two variables used to construct an H–R diagram of stars?
5. Compare and contrast a nova and a supernova.

PROJECTS

1. Design and construct a scale model of the sun, showing the various layers from the core outward to the corona. Point out which processes occur in each layer.
2. Design and construct a scale model of the Milky Way, showing the approximate position of the solar system and the motion of stars about the galactic center.

READINGS

1. Bartusiak, M. "The Bubbling Universe." *Science Digest.* February, 1986, pp. 64-65.
2. Sabin, Louis. *Stars.* Mahawah, NJ: Troll Associates, 1985.
3. Shaham, Jacob. "The Oldest Pulsars in the Universe." *Scientific American.* February, 1987, pp. 50-56.

SCIENCE AND SOCIETY

THE UNITED STATES SPACE PROGRAM—WHERE NEXT?

People have been launched into orbit around Earth and have been sent to the moon. Spacecraft have landed on Mars to collect Martian soil for analysis. Two Pioneer spacecraft are currently farther from the sun than any of the planets in our solar system. Voyager 2 has sent back detailed photographs of most of the outer planets. Instruments on some of the satellites placed in orbit have been aimed at deep space. These instruments have received X rays and infrared signals that otherwise would have been absorbed by Earth's atmosphere.

FIGURE 1. Crewed space shuttle missions have become an important part of the United States space program.

The signals have enabled scientists to identify new objects in space and have provided new information about previously known space objects. For example, rings of matter have been detected around some stars. Some astronomers believe that these rings are the first stages in the formation of a system of planets.

In addition to space exploration, satellites also have been used to study Earth. Some are used to survey Earth's natural resources, while others provide data about weather patterns. Other satellites take numerous photographs of Earth. Communication satellites reflect or transmit electromagnetic waves from one point on Earth to another.

Background

There are many feats possible in space that are impossible on Earth. Scientific experiments could be performed in the weightless environment of space. Satellite technology could provide increased conveniences in our daily lives. However, such endeavors cost money and funding is often not available for every project. Therefore, difficult decisions must be made to determine which projects will be funded.

Recall the space shuttle Challenger disaster in January, 1986. All American satellites were intended to be launched from the space shuttle rather than directly from Earth's surface. The Challenger accident brought the entire United States space program to a temporary halt. Many scientists have criticized the National Aeronautics and Space Administration (NASA) for relying on one program for all its launch needs. These critics would use expendable launch vehicles (ELV) for programs that do not require putting people in space.

Some scientists say that the program to put people into space has taken too much attention away from the more purely scientific research projects. Among these projects are the Ulysses program, which is to investigate the sun, and the Galileo project, which is intended to investigate Jupiter. Even before the Challenger accident, the priority given to the space shuttle was considered by some to be delaying projects like Ulysses and Galileo. The same criticisms have been voiced about the space station, a program that is to follow the space shuttle program.

Debate

The decision to spend money on United States space programs lies in the hands of Congress. In order to see the many sides of this complex issue, your class can hold a mock Congressional budget hearing about space program funding. Suppose four programs were up for consideration: more intensive study of Earth; further uncrewed missions within the solar system; a scientific base on the moon; or a crewed mission to Mars. Choose five students to represent the following people.

1. The Secretary of Defense wants to make sure the space program continues to meet the nation's military needs. These needs include the ability to photograph areas on Earth to which there is no direct access. Military needs also include reliable propulsion systems to launch missiles.

2. An astrophysicist wants to promote continued scientific exploration with uncrewed spacecraft launched independently from the ground and feels there is still much to be learned without endangering people's lives in space.

3. The mayor of a city that is home to a facility for crewed space missions wants to make sure the space program continues to include crewed missions. Shutting down this facility would mean severe economic hardship for the city.

4. Spokesperson for the L-5 Society wants to establish space colonies that orbit Earth equidistant from Earth and the moon. The L-5 Society plans to finance these colonies with profits from solar power satellites and obtain most of their construction materials from the moon.

5. A NASA administrator would like to see increased funding for all types of space programs. Less funding means fewer people employed in space-related industries.

Choose five other students in your class to represent members of Congress. Others could represent reporters or interested citizens. The students representing members of Congress will need to keep the costs as well as the benefits in mind as they debate. They should ask which of these programs will generate income that could pay for the costs in part or in full. At the close of the hearing, the members of Congress must announce which programs they will fund and give reasons for their choices.

Developing a Viewpoint

1. After hearing the space program issues debated, which of the views expressed most closely matches your views? Why?

2. Do you believe the United States should continue to fund the major portion of space exploration or should private companies be encouraged to fund them? Why?

3. Do you feel permanent bases should be established on the moon or other planets? Explain your viewpoint.

Suggested Readings

"Business in Space Gets a Boost." *U.S. News and World Report,* 101. September 1, 1986, pp. 60-61.

"Space Program Said to Lack Direction," *Science,* 237. August 28, 1987, p. 965.

"Leadership and America's Future in Space." *Astronomy,* Vol. 16, No. 1. January, 1988, pp. 8-17.

UNIT 3

Air and water interact to form Earth's atmosphere. The sun warms the atmosphere causing winds. All of Earth's weather occurs in the lowest part of the atmosphere and is a result of the interactions of air, water, and the sun. How is the principle used in these balloons like the interactions in the atmosphere?

~225 M.Y.A
Atlantic Ocean begins to form.

1483
Leonardo da Vinci designs the helicopter.

1513
Portuguese discover the Pacific Ocean.

EARTH'S AIR AND WATER

1714
Gabriel Farenheit constructs mercury thermometer.

1806
Francis Beaufort designs a wind strength scale.

1903
Wilbur Wright flies.

1976
Concorde begins passenger service.

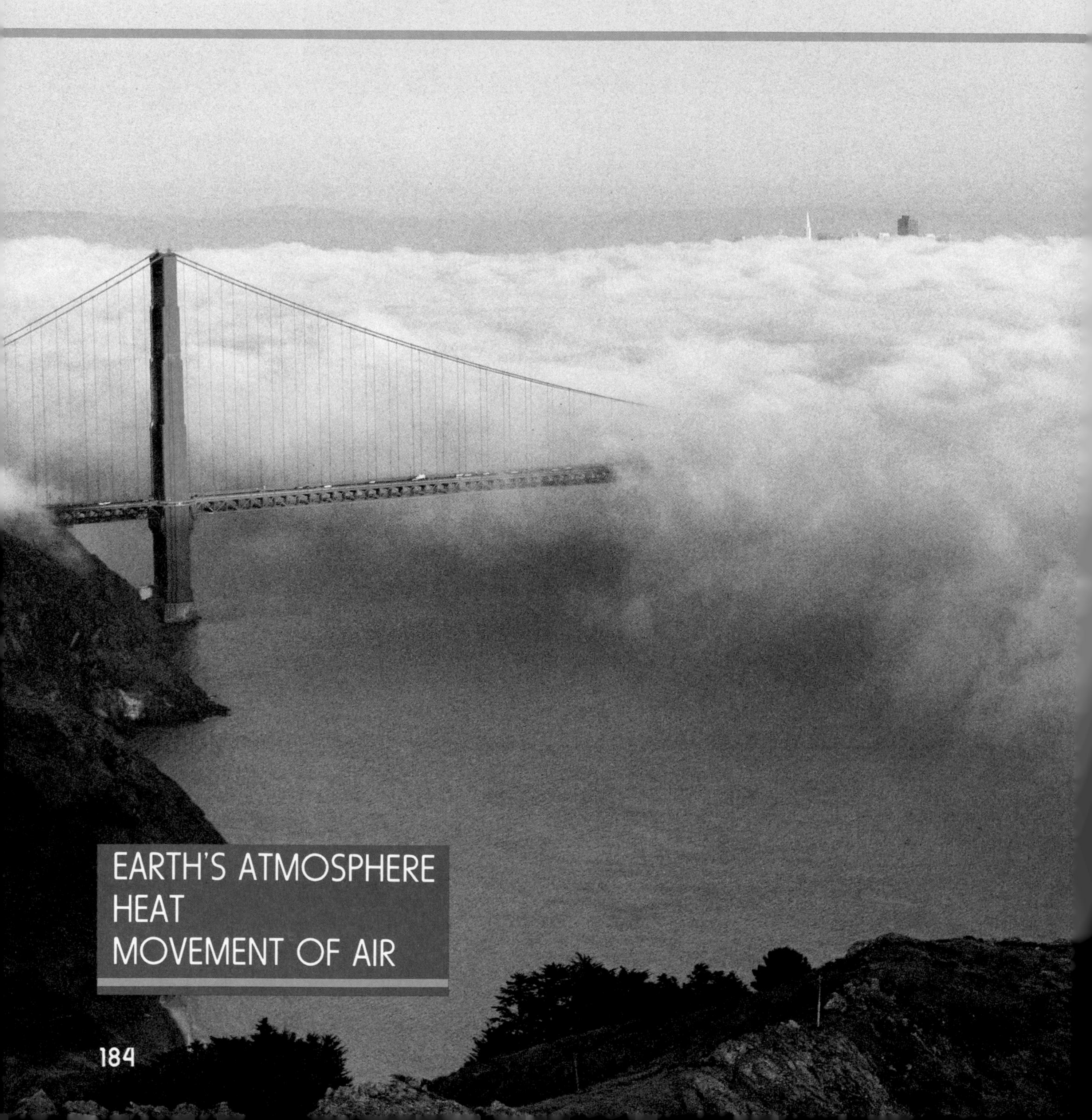

EARTH'S ATMOSPHERE
HEAT
MOVEMENT OF AIR

Air

Just as fish live in the fluid water, you live in the fluid atmosphere. The atmosphere is a mixture of many gases. Water vapor makes up only zero to four percent of atmospheric gases, but it has a large effect on your life as weather.

EARTH'S ATMOSPHERE

Earth is protected by a blanket of air called the atmosphere. Earth's atmosphere is the mixture of gases, solids, and liquids that surrounds the planet.

GOALS

1. You will gain an understanding of the composition and the structure of the atmosphere.
2. You will learn how barometers work.

9:1 Composition of the Atmosphere

Dry air is 78 percent nitrogen, 21 percent oxygen, and one percent other gases, which include carbon dioxide and argon. The amount of water vapor in air varies from zero to four percent. Dust, smoke, and waste gases from industry and transportation are substances in air not included among the gases listed in Table 9–1.

What are the two most abundant gases in the atmosphere?

Nitrogen compounds are plant nutrients. Few plants, though, can take nitrogen directly from the air. Some plants, such as alfalfa, have nitrogen-fixing bacteria in their roots. These bacteria change nitrogen in the atmosphere to a form that can be used by plants. Animals eat plants and return nitrogen to the soil in their body wastes. Nitrogen also is returned to the soil as plants decay. The nitrogen cycle is shown in Figure 9–1.

How is nitrogen returned to the soil?

Oxygen is needed by most life forms. Animals use this gas directly from the air or dissolved in water. Oxygen is made by green plants during photosynthesis. Also, oxygen in the air combines slowly with some elements in rocks. This causes the rocks to break down. During combustion, oxygen combines with substances such as wood, paper, and gasoline.

Carbon dioxide makes up a small percentage of the air, but it is an important gas. Without carbon dioxide, plants could not produce oxygen. The amount of water vapor in the air varies from place to place and from time to time. Water and carbon dioxide are both essential for life processes. Water vapor, like carbon dioxide, helps prevent heat loss from Earth.

F.Y.I. Meteorologists have recently discovered that during certain months, holes develop in the ozone layer over both the North and South Poles. The hole over Antarctica has been as large as the United States.

Another gas that is important to all life is ozone. **Ozone** is a form of oxygen. Ozone absorbs most of the ultraviolet radiation that enters the atmosphere. Ultraviolet radiation can cause skin cancer, genetic mutations, and cataracts. Emissions from some jets, fallout from nuclear explosions, some fertilizers, and fluorocarbons can destroy ozone molecules. This loss of ozone causes an increase in the amount of ultraviolet radiation that reaches Earth's surface. Fluorocarbons that have been used in refrigerants and in propellants for aerosol sprays are especially dangerous. Although some countries have banned the use of these chemicals in products such as aerosol sprays, scientists fear that much damage has been done to the atmosphere. Thus, ozone levels are carefully monitored.

Table 9–1

Gases of Earth's Atmosphere

Gas	Chemical symbol	Percent by volume	Some uses
Nitrogen	N_2	78.09	fertilizers, amino acids, nitroglycerin
Oxygen	O_2	20.95	animal respiration, rocket fuel
Argon	Ar	0.93	electric light bulbs, welding
Carbon dioxide	CO_2	0.03	photosynthesis
Water vapor	H_2O	0 to 4.0	component of all life, absorbs Earth's heat
Neon*	Ne	trace	advertizing signs
Helium*	He	trace	aqualungs, welding, lighter-than-air craft
Methane*	CH_4	trace	home heating and cooking
Krypton*	Kr	trace	wavelength used to define SI units of length
Xenon*	Xe	trace	electronic flash bulbs
Hydrogen*	H_2	trace	welding torch fuel, production of ammonia
Ozone*	O_3	trace	bleach, disinfectant, provides protection against ultraviolet radiation

*Neon through ozone constitute 0.0001 percent of the atmosphere.

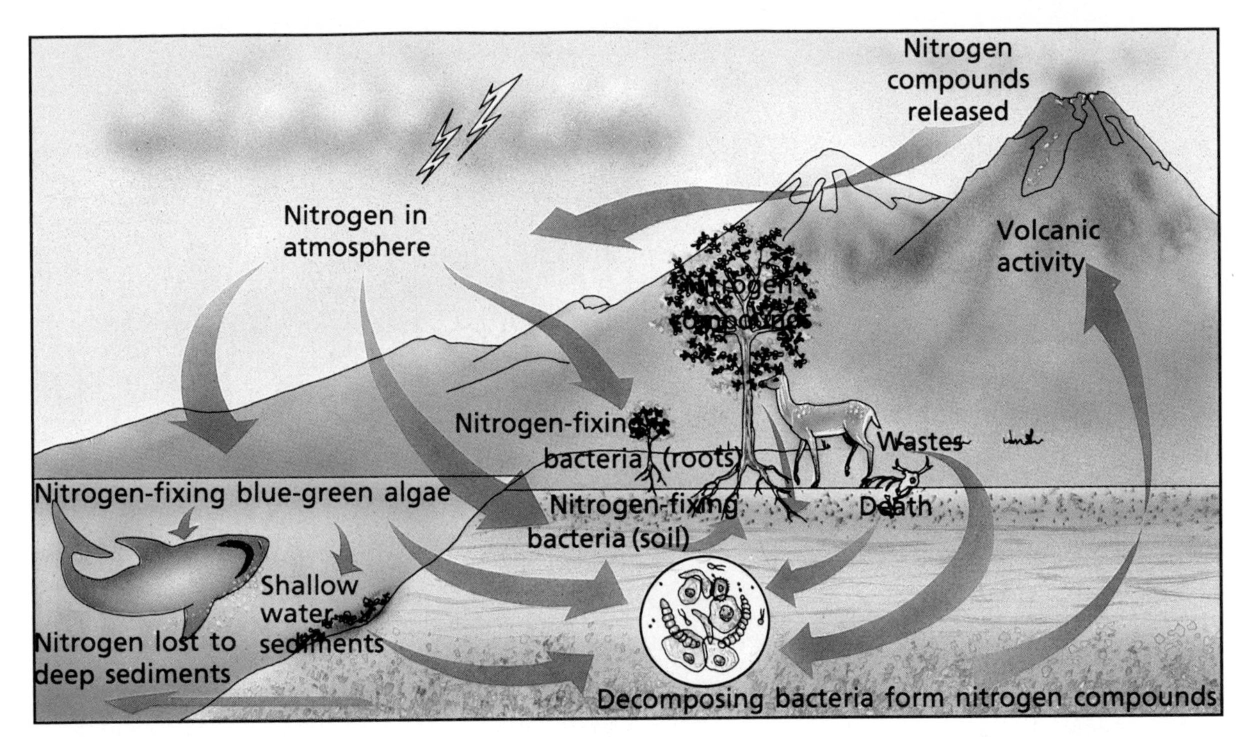

FIGURE 9–1. Nitrogen is present in many forms throughout the nitrogen cycle. What would happen if all nitrogen-fixing bacteria died?

9:2 Structure of the Atmosphere

The atmosphere is divided into four layers based on temperature. The **troposphere** (TROHP uh sfihr) is the layer nearest Earth. It contains 75 percent of the gases of the atmosphere, as well as dust and water vapor. This layer is the zone where weather and clouds occur. Temperatures decrease with increasing height in the troposphere. Near the top of this layer, between 8 and 20 kilometers from Earth, a boundary called the **tropopause** acts as a ceiling to the weather zone. Just below the tropopause are strong winds called the jet streams, which will be discussed later in the chapter.

F.Y.I. The word *troposphere* comes from the Greek word *tropos,* which means to turn or mix.

Above the tropopause lies the **stratosphere** (STRAT uh sfihr), which extends upward to about 50 kilometers from Earth. In the lower part of this zone, temperatures are a constant −50°C. However, at about 50 kilometers, temperatures rise to about 0°C. The stratosphere contains the ozone layer.

Where is Earth's ozone layer?

The **mesosphere** (MEZ uh sfihr) extends upward from about 50 kilometers to between 80 and 85 kilometers. This layer is the coldest zone of the atmosphere. Temperatures decrease to near −100°C at the top of the mesosphere.

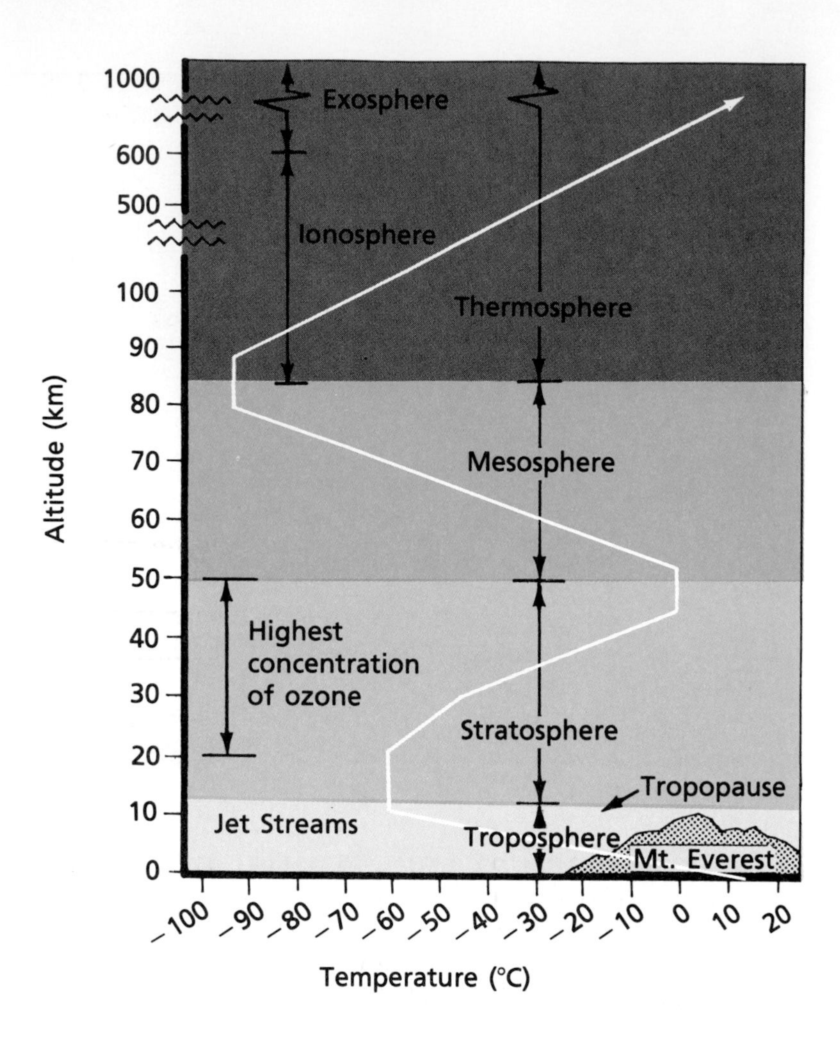

FIGURE 9–2. The atmosphere is divided into four layers based on temperature variations. What are the two parts of the thermosphere?

The **thermosphere** (THUR muh sfihr) extends from about 80 kilometers upward into space. Temperatures in the thermosphere increase quickly because of absorption of energy from the sun. The thermosphere is divided into two parts. The **ionosphere** (i AHN uh sfihr) is a layer of electrically charged particles that begins between 80 and 85 kilometers above Earth's surface. Here, particles of the atmosphere are bombarded by energy from space and form ions and free electrons. These charged particles are useful for communications because at night they reflect radio waves. During the day, however, radio reception is terrible over long distances because the waves are absorbed. When streams of particles from solar flares come in contact with the ionosphere, the ions glow different colors. The resulting display is an aurora. Auroras are brightest near the poles due to the deflection of the particles by Earth's magnetic field. Particles from solar flares also disturb radio, TV, and telephone transmissions.

F.Y.I. Refer to Section 8:2 for more information on solar flares, auroras, and solar wind.

The **exosphere** begins at an altitude of about 500 to 700 kilometers above Earth and extends out to interplanetary space. At these altitudes, atoms and ions are very far apart. Some gases in the exosphere actually escape into space. Particles of the solar wind are concentrated into radiation layers at about 3000 and 16 000 kilometers above Earth's surface. These layers of the exosphere are the Van Allen Belts and are held in place by Earth's magnetic field.

9:3 Atmospheric Pressure

Gases of the atmosphere, like all matter, have both volume and mass. Close to Earth, the atoms and molecules are pushed together due to the pressure of the mass of air above them. As the atoms and molecules become more closely packed, they exert more force. Atmospheric pressure is greatest at sea level and decreases outward from Earth. Air presses down on Earth's surface with a force of 10.1 newtons per square centimeter at sea level.

Air pressure varies at different places at the same elevation. This is caused by differences in air's density. There are more molecules of air above a point having high pressure than above one where low pressure is observed. Warm air is less dense than cool air, because the air molecules are farther apart. Because temperature and altitude determine air pressure, air pressure readings are useful in predicting weather and wind changes, and in estimating altitudes.

The force of air, or air pressure, is measured with a **barometer**. The two kinds of barometers are the aneroid (AN uh royd) and the mercury. The aneroid barometer is

BIOGRAPHY

Evangelista Torricelli
1608–1647

As a professor of mathematics in Florence, now a part of Italy, Torricelli argued that since the atmosphere has weight, it must also have pressure. Torricelli determined that atmospheric pressure determines the height to which fluid will rise in a tube that is inverted into a saucer of the same liquid. By using this procedure, he constructed the first mercury barometer in 1644.

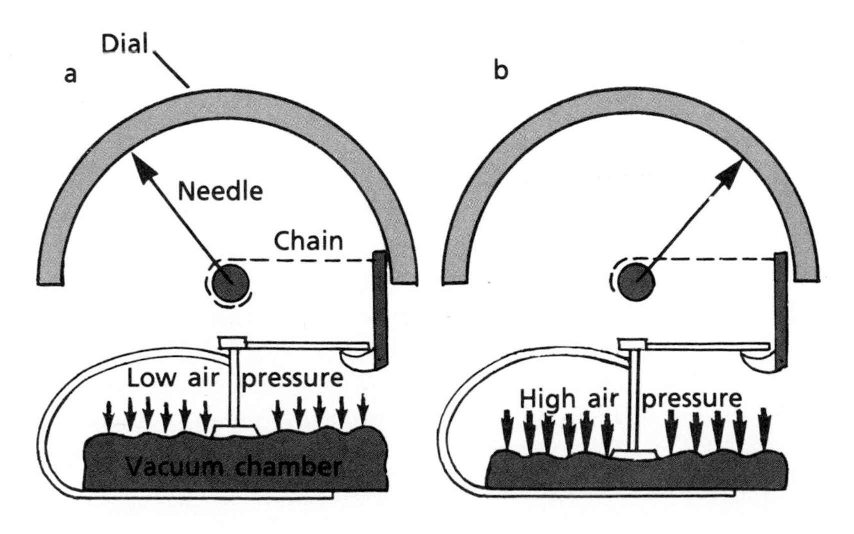

FIGURE 9–3. An aneroid barometer measures air pressure with a vacuum chamber. Low pressure (a) causes the chamber to compress less than high pressure (b).

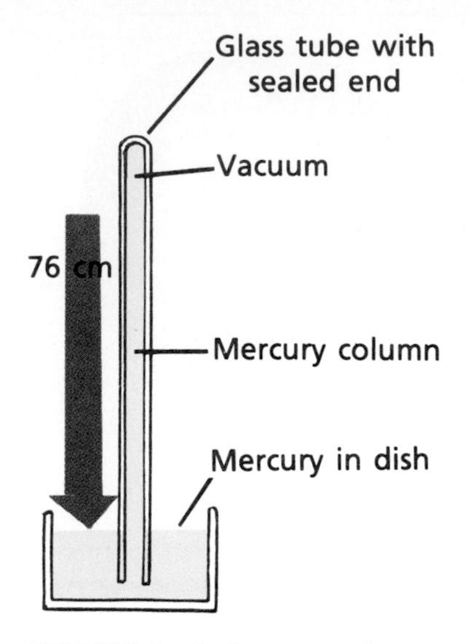

FIGURE 9–4. A mercury barometer measures air pressure using a glass tube.

a metal box from which most of the air is removed. The top of the box is a thin metal disk that bends under pressure. As air presses down on the box, the disk bends inward. The pressure causes a spring to move within the box. The spring is attached to a needle that shows the amount of pressure exerted by the air.

A mercury barometer is more accurate than an aneroid, but a mercury barometer is too large to be practical for many uses. A mercury barometer is made of a long tube closed at one end. The tube is filled with mercury and the open end is placed upright in a dish of mercury. When the open end of the tube is placed in the dish, a vacuum forms between the closed end of the tube and the top of the mercury column. The height of the mercury column is about 76 centimeters at sea level. Mercury moves up and down in the tube depending on the air pressure outside. High pressure forces the mercury in the tube to rise. Low pressure makes it drop. When the pressure inside and outside is equal, the mercury stops moving.

Review

1. What is one use for helium?
2. Why is ozone important to Earth's organisms?
★ 3. Farmers often rotate crops so that the soil will not be depleted of the nutrients that plants need. Why would they sometimes plant alfalfa in these fields?

PROBLEM SOLVING

A Tiring Ride

One summer morning before starting on a bike ride, Ricardo noticed that his front tire was slightly low. Since he was in a hurry and would be riding on a smooth, blacktop road, he decided to wait until noon, when he reached his destination, to add air.

When he finally took his bike to a filling station, he discovered that his tire was firmer than it had been early in the morning. After Ricardo thought about his trip, he understood what had happened to his tire. What was Ricardo's explanation?

INVESTIGATION 9–1

Air Pressure

Problem: How does a barometer work?

Materials 

small coffee can
drinking straw, plastic
rubber balloon
rubber band
heavy paper, one sheet, 28 cm × 21.5 cm
transparent tape
scissors
pen or pencil
metric ruler

Procedure

1. Using Figure 9–5 as a guide, draw a line 8 cm from the right edge of the paper. Draw a second line lengthwise through the center of the paper. The second line should extend 20 cm from the left edge to the 8-cm line.
2. Cut the paper along the 20-cm line. Cut away the section shown in blue.
3. Fold the paper along the 8-cm line.
4. Wrap the section shown in dark blue in Figure 9–5 around the can and fasten with tape. The long edge of the paper should stick up above the can to form a gauge.

FIGURE 9–5.

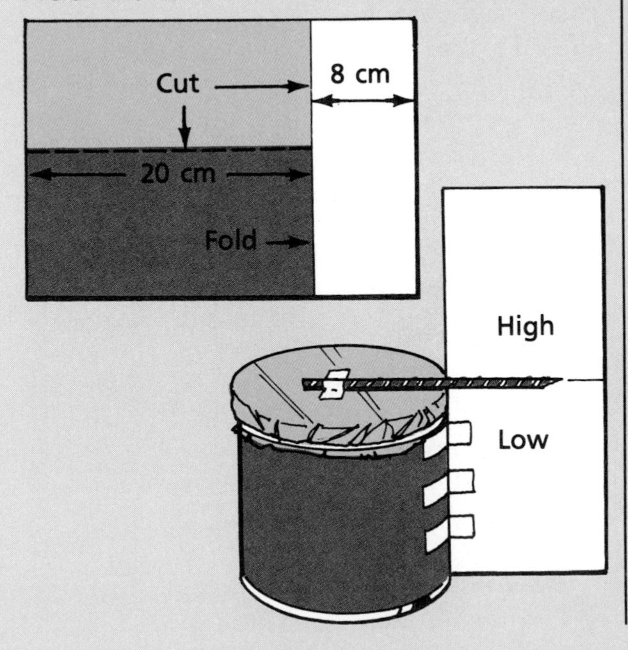

5. Cover the top of the coffee can with the rubber balloon. The balloon must be stretched tightly over the top of the can in order for the barometer to function correctly. Secure the balloon with a rubber band.
6. Trim one tip of the straw into a point. Position the straw so that the pointed end is alongside the gauge. See Figure 9–5. Tape the other end of the straw to the balloon. DO NOT tape the straw to the gauge.
7. Make a horizontal mark on the gauge showing the position of the straw. Write "high" above this mark, and write "low" below this mark.
8. Keep track of the movement of the straw over a period of a week. Also record the weather conditions each day. Record your observations in a data table similar to the one shown.

Data and Observations

Date	Barometric readings (high or low)	Weather conditions

Analysis

1. Explain how your barometer works.

Conclusions and Applications

2. Is your barometer as sensitive as an aneroid barometer? Explain.
3. What type of weather is usually associated with high pressure? With low pressure?
4. How can a weather forecaster use the barometric reading to help formulate a forecast?

GOALS

1. You will learn how energy is absorbed and reflected.
2. You will learn about heat transfer.

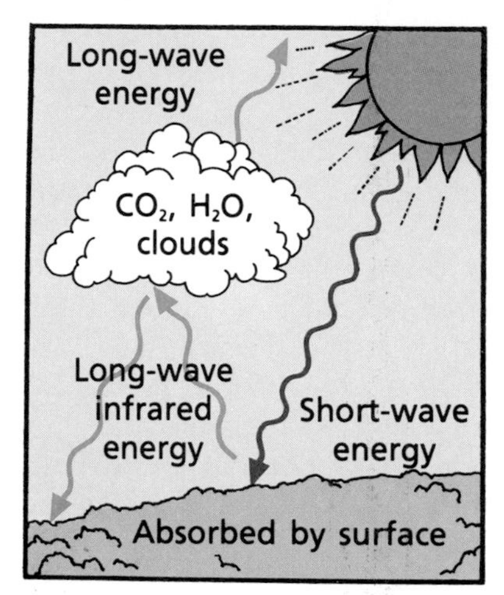

FIGURE 9–6. The greenhouse effect warms air near the surface.

HEAT

The sun is the source of most of Earth's surface heat. Heat is energy transferred as a result of a difference in temperature. Heat received from the sun is transferred to different parts of our planet.

9:4 Energy from the Sun

The sun produces heat and light waves, which are just a small part of the electromagnetic spectrum. The spectrum includes all energy forms, such as long radio waves, shorter X rays, and ultraviolet waves. Some energy waves that reach Earth are reflected back into space by the atmosphere. Some of the energy is absorbed by the atmosphere. Energy that reaches Earth's surface is absorbed by land and water surfaces.

For example, polar ice reflects the sun's energy and absorbs little of it. Water also reflects solar energy, especially when the sun is near the horizon. Sand and snow are other materials that tend to reflect solar energy. Vegetation, dark rocks, and black surfaces such as asphalt

CAREER

Professor of Atmospheric Science

Dr. Lisa Young is a professor of atmospheric science. She conducts college courses in atmospheric science for both undergraduate and graduate students. Dr. Young prepares and delivers lectures, compiles bibliographies for students to read, and stimulates classroom discussions. She also directs the research of students working for advanced degrees, and conducts research, the findings of which may be submitted for publication in professional journals. Dr. Young also advises students on academic curricula and serves on faculty committees. Dr. Young holds a bachelor's and master's degree in math and physics and also a doctorate degree in atmospheric science.

For career information, write:
The American Association of
University Professors
Suite 500
1012 14th Street
Washington, D.C. 20005

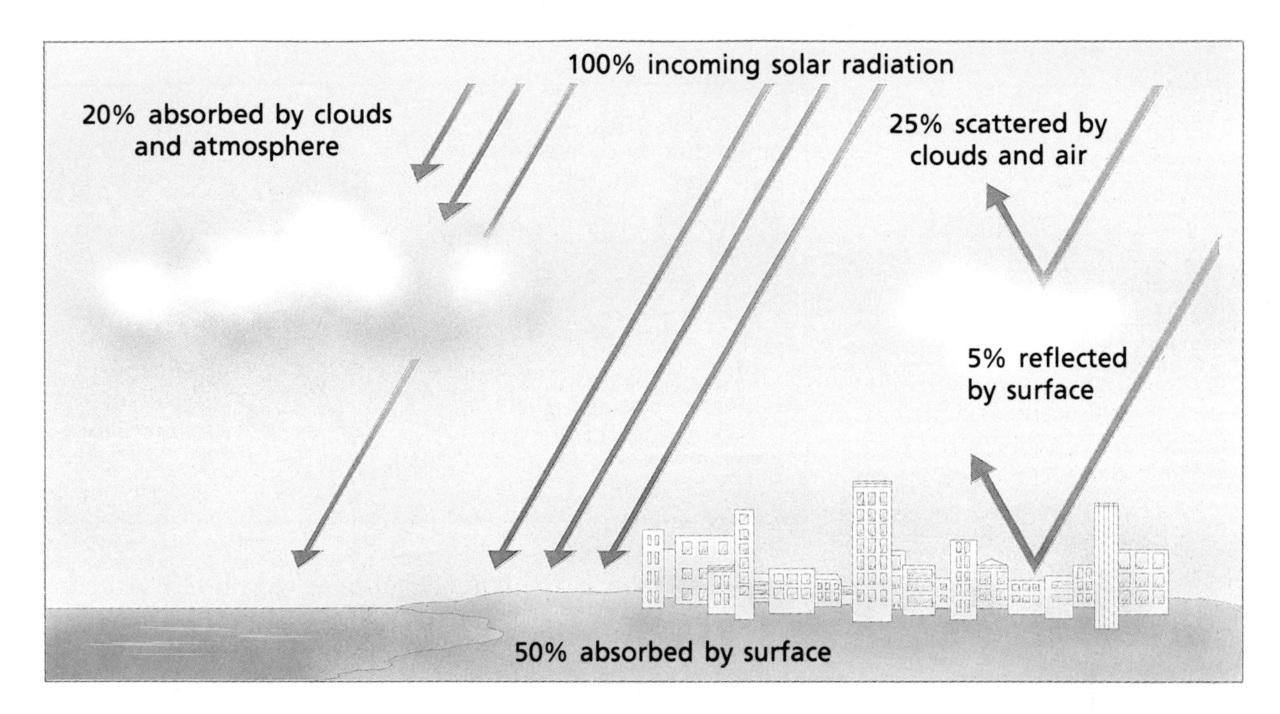

FIGURE 9–7. Half of the solar energy that falls on Earth does not reach the surface. Why does not all solar energy directed at Earth reach the surface?

absorb large amounts of solar energy. The energy is then returned to the atmosphere as infrared radiation and heat. Water vapor and carbon dioxide in the air absorb some of these infrared waves, and some of them are re-emitted to Earth. This absorbed infrared energy keeps heat in the troposphere. The warming effect has been called the **greenhouse effect.**

Earth's surface and the atmosphere lose heat when they are not facing the sun. Land areas lose heat quickly. Water loses heat more slowly. Earth would have great differences in day and night temperatures if it were not for the atmosphere, which traps heat and constantly moves it from one place to another.

9:5 Heat Transfer

Energy from the sun moves through space by radiation. **Radiation** is the transfer of energy by means of electromagnetic waves. A campfire gives off radiant energy in the form of heat and infrared radiation. The side of your body facing the fire becomes warm as heat and infrared waves reach you. The side away from the fire does not receive heat directly and remains relatively cool. Radiation is an important process for transferring infrared radiation absorbed by Earth's surface back to the atmosphere.

What is radiation?

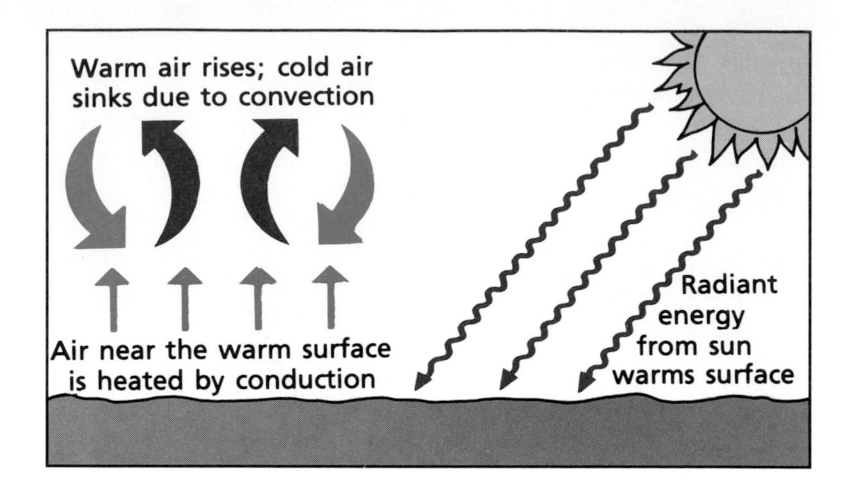

FIGURE 9–8. Solar energy reaches Earth by radiation. Air in contact with Earth's surface is heated by conduction. Heat transfer within the atmosphere occurs by convection.

What is conduction?

Conduction is the transfer of heat through matter by the actual contact of molecules. Molecules are always in motion. Heated molecules move more rapidly than cooler molecules. Heat is transferred from the fast-moving molecules to slow-moving molecules until all molecules are moving at the same rate. A stove heats a pan by conduction. Conduction also occurs at Earth's surface as heated rocks or sandy beaches transfer heat to the surrounding air.

What is convection?

Heat gained by the atmosphere from radiation or conduction is usually transferred by convection. As air absorbs energy, its molecules move faster and farther apart. This movement decreases the density of the air. Cold, dense air sinks, forcing warm, less dense air upward. This transfer of heat due to density differences is called **convection**. Convection currents cause a constant exchange of cold, dense air for less dense warm air. Today, many modern home furnaces use the convection principle for heating.

Review

4. How are water vapor and carbon dioxide involved in the heating of the atmosphere?
5. How does the amount of cloud cover affect the amount of solar energy that reaches Earth?
6. On the average, what percent of the solar energy striking Earth's atmosphere reaches the surface?
7. What is conduction?
8. ★ Explain why you think solar collecting panels are usually painted black.

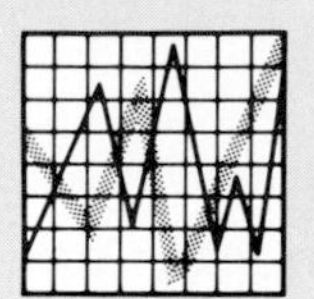

SKILL

Drawing Conclusions

Problem: How can you draw conclusions from scientific data?

Materials  

1000 mL beaker
thermometer
small metal pan
hot plate
metal spoon
water

Procedure

1. Fill the beaker with 200 mL of water.
2. Record the temperature of the water in the beaker.
3. Place the beaker of water near the window in direct sunlight.
4. After one class period, record the temperature of the water.
5. Half fill the metal pan with water and place it on the hot plate.
6. Place the spoon in the water.
7. After heating for 10 minutes, gently touch the spoon. **CAUTION:** *The spoon may be hot.* Record your observations.
8. Place the thermometer on the floor. After waiting for two or three minutes, record the temperature.
9. Place the thermometer on your desk and at the highest point in the room. After two or three minutes, record the thermometer readings at both locations.

Analysis

1. Describe what happened to the temperature of the water in the beaker after exposing it to sunlight.
2. Identify the type of heat transfer that occurred in the beaker.
3. Identify the type of heat transfer that occurred in the pan.
4. Identify the type of heat transfer that is occurring in the room.
5. Describe where the temperature in the room was the highest.

Conclusions and Applications

6. Explain what happened to the temperature of the spoon when the water in the pan was heated.
7. After analyzing all of your observations and conclusions, explain how heat is transferred.

MOVEMENT OF AIR

GOALS

1. You will learn how convection currents and the Coriolis effect influence air circulation.
2. You will learn about the movement of air currents within the atmosphere.

Energy from the sun warms the atmosphere, causing currents of ascending and descending air, or winds. Winds distribute energy around Earth's atmosphere.

9:6 Major Air Circulation

Convection currents in the atmosphere cause cold, dense air from areas near the poles to sink, move along the surface, and force warmer air aloft. This movement of air creates a system of wind and ocean currents around Earth. Generally, the movement of air is away from the equator in the upper atmosphere, and toward the equator at Earth's surface.

If Earth did not rotate, heated air would rise over the equator and move north and south high above Earth's surface. The air would cool and sink at the poles. The cool, sinking air would push air at Earth's surface toward the equator. Resulting winds would move directly toward the equator with equal speed.

Recall that Earth rotates on its axis once every 24 hours. Each point at Earth's equator moves eastward at about 1666 kilometers per hour. Due to Earth's shape, however, each point north or south of the equator moves slower than points along the equator. These differences in Earth's speed of rotation at different latitudes affect air circulation.

What causes the Coriolis effect?

The **Coriolis** (kohr ee OH lus) **effect** is an apparent force that results from Earth's eastward rotation. This apparent force causes moving objects near Earth's surface

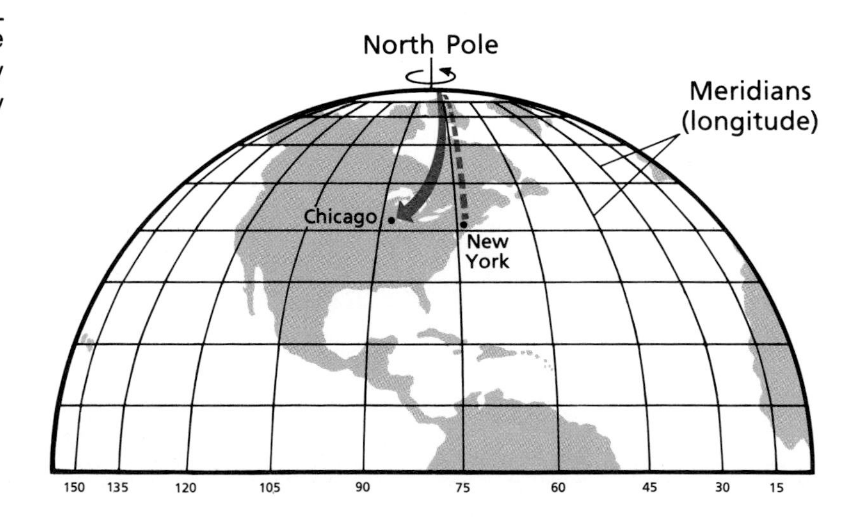

FIGURE 9–9. Air moving from the North Pole toward New York City is deflected toward Chicago by the Coriolis effect.

in the Northern Hemisphere to be turned westward from their original paths. Imagine a mass of air moving from the North Pole toward the equator. To someone standing at the equator, the southbound air mass appears to veer to the west. However, the path of the air mass would appear to be nearly straight to an observer in space. Viewed from the North Pole, masses of air are deflected to the right in the Northern Hemisphere. From the South Pole, air masses are deflected to the left. Seen from some point out in space, overall air movement appears to be from northeast to southwest in the Northern Hemisphere. Most air flow in the Southern Hemisphere is from southeast to northwest.

Differences in the heating and cooling rates of land and water also affect air circulation. Land and water temperatures rise and fall at different rates because land absorbs and loses heat faster than water does. During the day, dense, cool air from over the water flows inland and forces warm air over land aloft. This small-scale circulation is called a **sea breeze**. A sea breeze usually starts to form three or four hours after sunrise. By early afternoon, it has reached its peak intensity. The circulation cell extends both inland and seaward about 20 to 25 kilometers. The cell is usually only one kilometer deep. When the leading edge of a sea breeze reaches land, the temperature decreases by about 5°C or more in less than one hour.

What is a sea breeze?

At night, the land is cooler than the water because the land has given up its heat to the atmosphere. The cool air flows over the warmer water and forces warm air aloft. This circulation is called a **land breeze**. A land breeze normally starts to form in the late evening. It reaches its peak intensity near sunrise.

FIGURE 9–10. Cool air forces warm air to rise. Thus, convection causes a sea breeze (a) during the day and a land breeze (b) at night.

a

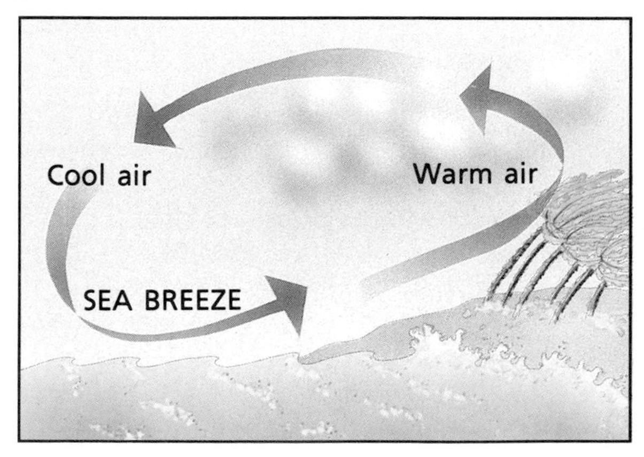

b

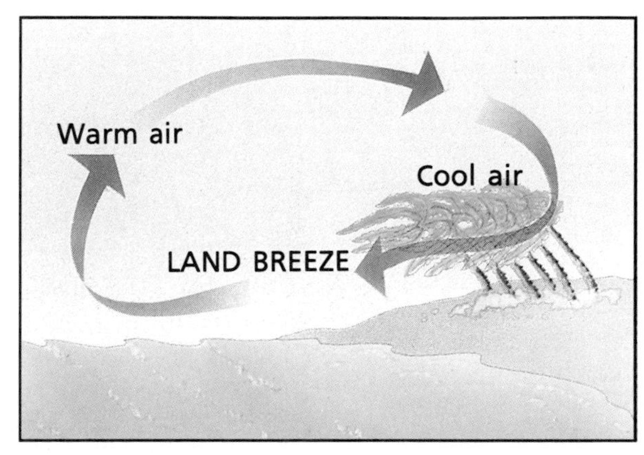

INVESTIGATION 9–2

Heating Differences

Problem: How do soil and water compare in their abilities to absorb and release heat?

Materials 

ring stand
thermometers (4)
soil
water
metric ruler
masking tape
graph paper
colored pencils (4)
clear plastic boxes of identical size (2)
overhead light source with reflector

Procedure

1. Fill one box two-thirds full of soil.
2. Place one thermometer in the soil with the bulb barely covered. Use masking tape to fasten the thermometer to the side of the box as shown in Figure 9–11.
3. Position a second thermometer in the box with its bulb about 1 cm above the soil. Tape the thermometer in place.
4. Fill the second box two-thirds full of water.
5. Position the remaining two thermometers the same way as you did in the soil box.
6. Attach the light source to the ring stand. Place the two boxes about 2 cm apart below the light. The light should be about 25 cm from the tops of the boxes.
7. Record the temperatures of all four thermometers with the light turned off.

FIGURE 9–11.

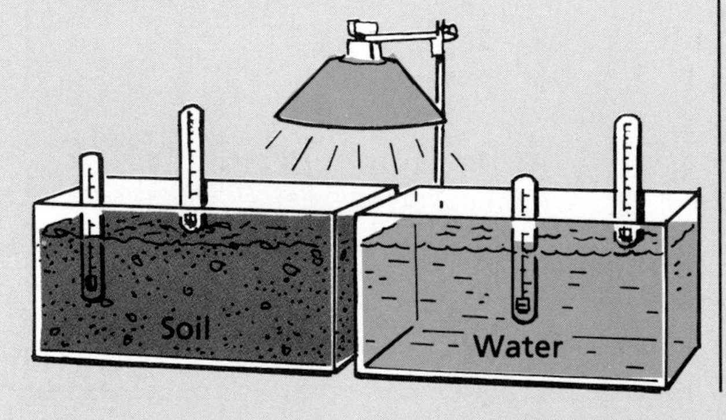

8. Turn on the light. Check to make sure that the bulbs of the thermometers are shielded.
9. Take temperature readings every two minutes for 14 minutes and record.
10. Turn off the light. Take temperature readings every two minutes for 14 minutes and record.
11. Using temperature units on the vertical axis and time in minutes on the horizontal axis, graph your data. Use a different colored pencil to plot the data from each different thermometer.

Data and Observations

Light on

Time in minutes	Thermometer reading			
	1	2	3	4
0				
2				

Light off

Time in minutes	Thermometer reading			
	1	2	3	4
0				
2				

Analysis

1. Did the air heat up faster over water or soil?
2. When the light was turned off, which lost heat faster, water or soil?
3. Compare the temperatures of the air above the water and above the soil after the light was turned off.

Conclusions and Applications

4. Explain how the information you gathered relates to land and sea breezes.

9:7 Wind Systems

Victor Starr, an American meteorologist, designed a model of the major wind systems of Earth. Starr used a shallow pan of water with ice at the center. He heated the outer rim of the pan to set up convection currents. When the convection system was well established, Starr rotated the pan. As the pan rotated, the water broke up into eddies, circular movements that form within the main current. The eddies moved in the belt halfway between the warm outer rim and the cold center. Starr's model suggested that major wind patterns are controlled by a combination of unequal heating and Earth's rotation.

What did Starr's model suggest?

The major wind system of Earth is similar to Starr's model. The **doldrums** are a windless zone at the equator. Here the air seems to be motionless, but actually it is being forced aloft. Vessels often were stranded in this zone during the days of sailing ships. For many days there would be no winds to move them. This zone is like the rim of the pan in Starr's model.

At about 30° latitude, air currents descend to Earth's surface. The **trade winds** blow toward the equator and force the air of the doldrums to rise. In the Northern Hemisphere, these winds blow from northeast to southwest. In the Southern Hemisphere, the trade winds blow from southeast to northwest. Note that winds are named for the direction from which they blow. Trade winds provided a busy route for ships sailing between Europe and the Americas. Today, pilots take advantage of the trade winds when flying west.

How are winds named?

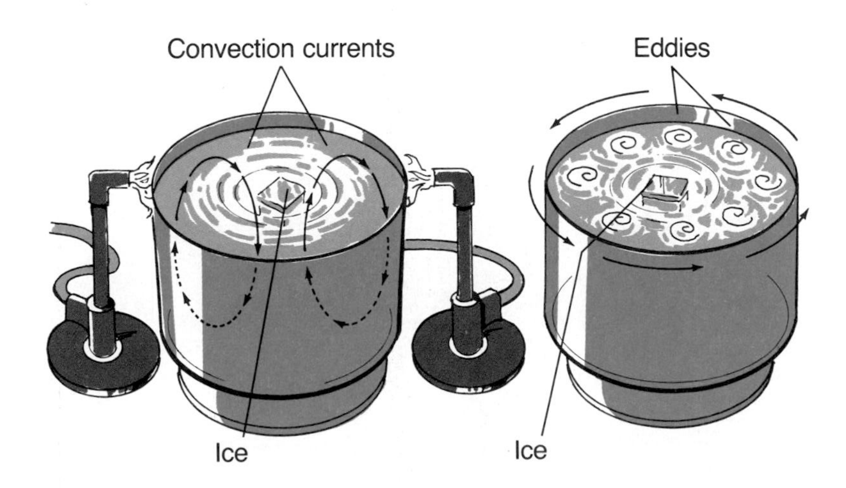

FIGURE 9–12. Starr's model suggested that major wind patterns are controlled by both unequal heating and Earth's rotation.

The **prevailing westerlies** blow from the southwest to the northeast in the zone between 30° and 60° latitude in the Northern Hemisphere. As this warm air moves toward the poles, it encounters cold air from the poles. This zone is similar to the pattern of eddies that formed midway between the warm outer rim and the cold center in Starr's model. The prevailing westerlies are the winds responsible for the movement of weather across the United States and Canada. In the Southern Hemisphere, the prevailing westerlies blow from the northwest to the southeast in this latitude zone.

What are the polar easterlies?

Polar easterlies are cold, dry, dense, horizontal air currents. They move northeast to southwest over the highest latitudes of the Northern Hemisphere. The polar easterlies move southeast to northwest in the Southern Hemisphere. Polar zones are relatively small, but provide large amounts of energy to the zone of the westerlies. The cold polar air displaces the warm air of the prevailing westerlies zone and forces it to rise at about 60° latitude. Near the poles, these currents descend weakly toward Earth.

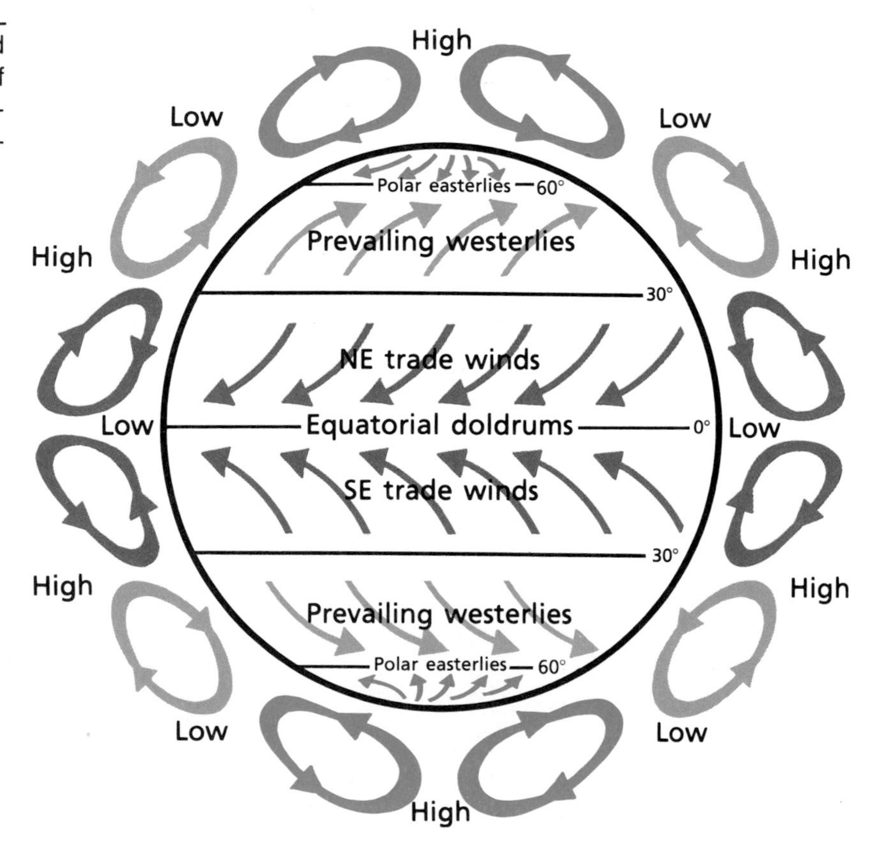

FIGURE 9–13. The major wind systems move from regions of high-density descending currents toward regions of low-density rising currents.

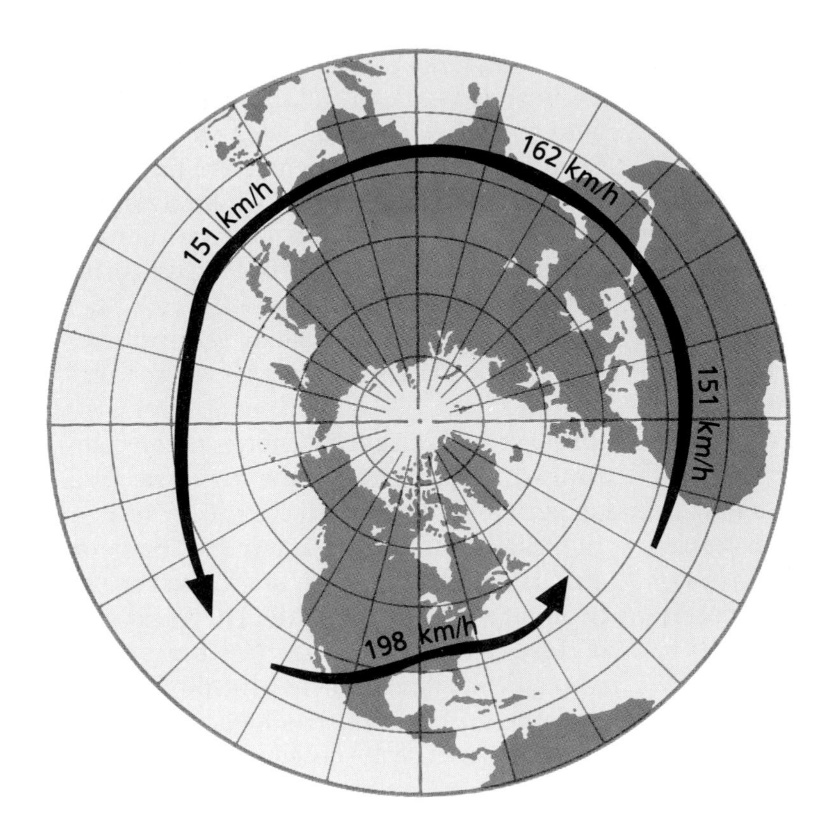

FIGURE 9–14. Jet streams in the Northern Hemisphere flow from west to east in the tropopause. Velocities and paths of these streams are constantly changing. Over which Northern Hemisphere continent are jet stream velocities the greatest?

During the 1940s, American pilots discovered strong belts of wind near the tropopause. These narrow belts of wind, called jet streams, flow from west to east. There are two jet streams in each hemisphere. The jet streams occur near the tropopause over the location where the trade winds meet the westerlies. They also occur over the area where the westerlies meet the polar easterlies. The jet streams resemble very fast-moving, winding rivers. Wind speeds in the jet streams range between 8 and 200 km/h. The jet streams change their positions in latitude and altitude from day to day and season to season.

F.Y.I. During World War II, B-29 superfortresses flying at nearly 322 km/h came to a near standstill when they flew against the flow of jet streams.

Review

9. What is the apparent direction of movement of a mass of air moving from the North Pole to the equator?

10. In which direction do the jet streams flow in the Northern Hemisphere?

★ **11.** The land near a large body of water does not get as cold in the winter, nor as warm in the summer as land farther inland. Why do you think this happens?

TECHNOLOGY: APPLICATIONS

Wind Shear

Wind shear has been blamed for causing more than a dozen commercial airline accidents and killing hundreds of people since 1970. Wind shear is a downward blast of air that occurs when evaporation cools a mass of air, and the air falls because it has become heavier.

There are two types of wind-shear downbursts, the macroburst and the microburst. A macroburst is a large downburst. An intense macroburst can cause widespread, tornadolike damage. Winds, lasting five to twenty minutes, can reach speeds as high as 250 km/h.

However, it is the microburst that is the greatest cause of air carrier death in the United States. Most microbursts are small, only 0.6 to 2.0 kilometers across. Most last from five to fifteen minutes with a peak severity of two to four minutes. When an airplane enters a microburst, it first runs into a headwind that increases the speed of the air that rushes over the wings and gives the plane a lift. Then, after the plane passes through the downdraft in the center of the microburst, it is blown by a tailwind that reduces its lifting ability. Microburst activity is quite frequent at Denver's Stapleton Airport. Researchers estimate that pilots may encounter microbursts at altitudes below 166 km on the average of once every four and a half days during the summer months.

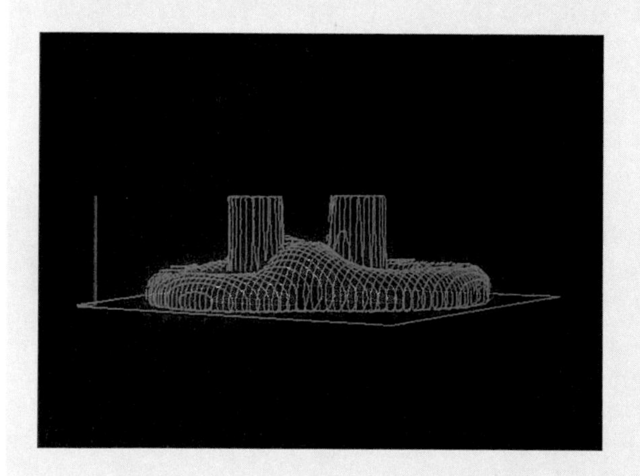

An enhanced low-level wind-shear alert system (LLWAS) has been installed at more than 90 U.S. airports. This system uses a series of anemometers arranged in and around the airports. The airport controller is alerted if there is a significant difference in wind velocity or direction between the anemometer located at the center of the array and any of the other anemometers. However, some microbursts can be so small that they go undetected between the sensors. Other microbursts may be detected, but too late for a warning to be sent.

Therefore, NASA's Langley Research Center is investigating the use of Doppler radar to detect severe turbulence in precipitation that would indicate wind-shear conditions. Such radar could detect hazardous conditions at approach altitudes, thus giving airplane pilots more time to avoid them. Research shows that meteorologists can provide a 2–4 minute warning of microbursts using Doppler radar.

Flight simulators teach pilots how to cope with wind shear. Pilots rely on preflight weather briefings, airborne radar, and radio reports from the tower to avoid severe weather conditions. In the future, commercial planes will be equipped with their own self-monitoring Doppler radar. With better wind-shear detection methods, lives of passengers will be saved, destruction to planes will be prevented, annual fuel savings will reach the billions of dollars, and delays at airports due to severe weather will be reduced.

Chapter 9 Review

SUMMARY

1. Air is composed of gases and contains suspended solids and liquids. 9:1
2. Nitrogen is the most abundant gas in the atmosphere. 9:1
3. Layers of the atmosphere include the troposphere, stratosphere, mesosphere, and thermosphere. The thermosphere includes the ionosphere and the exosphere. 9:2
4. Air has mass and exerts pressure on Earth's surface. 9:3
5. Half of the sun's energy that is directed at Earth's surface is reflected, absorbed, or scattered by the atmosphere. 9:4
6. The greenhouse effect keeps heat close to Earth's surface. 9:4
7. Heat energy is moved from one place to another by radiation, conduction, and convection. 9:5
8. Convection currents resulting from unequal heating affect air circulation. The Coriolis effect influences movement of air masses. 9:6
9. The major wind systems result from the unequal heating of Earth's surface. 9:7

VOCABULARY

a. barometer
b. conduction
c. convection
d. Coriolis effect
e. doldrums
f. exosphere
g. greenhouse effect
h. ionosphere
i. land breeze
j. mesosphere
k. ozone
l. polar easterlies
m. prevailing westerlies
n. radiation
o. sea breeze
p. stratosphere
q. thermosphere
r. trade winds
s. tropopause
t. troposphere

Matching

Match each description with the correct vocabulary word from the list above. Some words will not be used.

1. gas that absorbs ultraviolet radiation
2. layer of the atmosphere that contains 75% of the gases of the atmosphere
3. radio waves are reflected to Earth by this atmospheric layer
4. the apparent force that causes moving bodies to be deflected westward
5. the most important process of transferring heat from place to place in the atmosphere
6. layer of the atmosphere that extends out into interplanetary space
7. winds near the equator
8. occurs at night because the land is cooler than the water
9. instrument that measures air pressure
10. occurs during the day because the water is cooler than the land

Chapter 9 Review

MAIN IDEAS

A. Reviewing Concepts

Choose the word or phrase that correctly completes each of the following sentences.

1. The coldest layer of the atmosphere is the *(stratosphere, thermosphere, mesosphere).*
2. A mass of air traveling from the North Pole to the equator appears to be deflected by the Coriolis effect to the *(east, west, north).*
3. Atmospheric pressure is measured with a *(thermosphere, convection current, barometer).*
4. The most plentiful gas in the air is *(nitrogen, argon, oxygen).*
5. A decrease in the density of a column of air causes the barometric pressure to *(increase, decrease, remain constant).*
6. The most important process of transferring heat from Earth's surface to the atmosphere is *(radiation, convection, conduction).*
7. For a given volume, compared to warm air, cold air is *(more dense, less dense, about the same in density).*
8. The upper boundary of weather is the *(stratosphere, tropopause, ozone layer).*
9. The greenhouse effect is due in part to the presence of *(nitrogen, argon, water vapor)* in the atmosphere.
10. More solar energy is absorbed by *(ice, water, dark soil).*
11. The air of the doldrums is forced upward by the *(trade winds, prevailing westerlies, polar easterlies).*
12. Polar winds in the Northern Hemisphere blow from the *(northeast, southeast, northwest).*
13. Of all of the sun's energy that reaches the atmosphere, *(50%, 75%, 90%)* is absorbed by Earth's surface.
14. As altitude increases, the air pressure *(does not change, increases, decreases).*
15. Without *(ozone, methane, carbon dioxide)*, plants could not produce oxygen.

B. Understanding Concepts

Answer the following questions using complete sentences.

16. Explain the nitrogen cycle.
17. Describe what happens when the ionosphere is bombarded with streams of particles from solar flares.
18. Describe how an aneroid barometer works.
19. Explain what causes the greenhouse effect.
20. Describe how the Coriolis effect influences the way that air circulates.
21. Explain why air rises at the equator and sinks at the poles.
22. Discuss the differences in heat transfer by radiation, conduction, and convection.
23. Explain why some countries have banned the use of fluorocarbons in products such as aerosol sprays.
24. Explain why air pressure is greater closer to Earth's surface than at higher altitudes.
25. Explain why sea breezes occur during the day and land breezes occur at night.

C. Applying Concepts

Answer the following questions using complete sentences.

26. Why are there few or no clouds in the stratosphere?
27. Why is it difficult for scientists to determine the outer limits of Earth's atmosphere?
28. Why do many people wear dark-colored clothing in the winter and light-colored clothing in the summer?

29. Why would sailing ships take one route from Europe to the Americas, and a different route on the return voyage?
30. Why does the greenhouse effect become more of a problem when forests are cut and burned? (*Hint: Consider the process of photosynthesis*).

SKILL REVIEW

1. If you measured the temperature of the air one meter above the ground on a sunny summer afternoon, and measured it again one hour after sunset, you would observe on your thermometer that the temperature dropped at night. What inferences can you make about why the temperature dropped?
2. What evidence is there to support the "conclusion" that many human activities may be contributing to a problem with Earth's ozone layer?
3. In the experiment you did to determine how soil and water compare in their abilities to absorb and release heat, what were the variables and what were the constants?
4. Explain how you could use simple equipment like a metal sinker, a piece of string, water, and a beaker to determine the volume of air contained in a small balloon.
5. Compare and contrast the troposphere and the thermosphere.

PROJECTS

1. Set up an experiment similar to Victor Starr's using the scientific methods discussed in Chapter 1. Report on your results.
2. Investigate how the type of surface affects the absorption and reflection of solar energy by measuring the temperature of the air over asphalt, soil, and water surfaces on a sunny day.

READINGS

1. Asimov, Isaac. *How Did We Find Out About the Atmosphere*. New York: Walker & Company, 1985.
2. Gay, Kathlyn. *The Greenhouse Effect*. New York: Franklin Watts, 1986.
3. Taubes, Gary. "Made in the Shade? No Way." *Discover*. August, 1987, pp. 62–72.

WHAT IS WEATHER?
WEATHER PATTERNS
CLIMATE

Weather and Climate

Weather and climate affect you in nearly everything you do. The cost of food items, type of clothing you wear, and your comfort level are all affected by weather and climate. Pilots, farmers, and construction workers all depend on accurate weather forecasts. How have you been affected by weather today?

WHAT IS WEATHER?

GOALS

1. You will gain an understanding of the factors that control weather.
2. You will learn about the different cloud types and their associations with weather.

Weather is the current state of the atmosphere in terms of air pressure, wind, temperature, and moisture. The interactions of air, water, and solar energy cause weather.

10:1 Factors of Weather

As you learned in Chapter 9, air has mass and volume and exerts pressure. High atmospheric pressure areas may form when air is cooled. In cool air, the molecules move closer together, causing the air to become denser and to sink. Thus, the dense air creates a high-pressure area. Low-pressure areas may form when air is warmed, causing molecules to move farther apart and air to become less dense. Air moves from an area of high pressure to an area of low pressure. This movement of air, caused by density differences, is wind.

How does temperature affect humidity?

Temperature also affects the ability of air to hold moisture. Moisture in the atmosphere is called humidity. Water vapor molecules are held in spaces between other gas molecules. As air is heated, gas molecules move farther apart. Thus, there is more space available for water vapor. On the other hand, when air cools, gas molecules move together. There is less space for water vapor, and it is squeezed out. Warm air can hold more moisture than an equal volume of cold air.

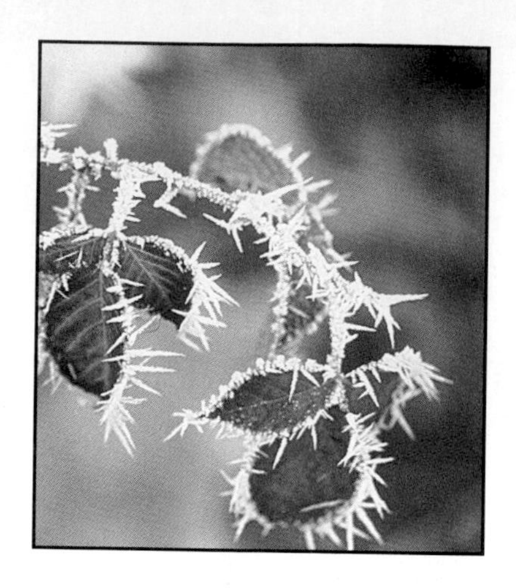

FIGURE 10–1. Frost forms when the dew point is below freezing.

Relative humidity is a measure of the amount of water vapor in a given volume of air, compared to the total amount of moisture that that volume of air could hold at a given temperature. For example, at 20°C, a cubic meter of air can hold a total of twelve grams of water vapor. If only four grams are present, the relative humidity is 33 percent. When air holds all the moisture it can at a given temperature, it is saturated (SACH uh rayt id), and the relative humidity is 100 percent. At 100 percent relative humidity, water vapor may condense, or change from a gas to a liquid. The temperature at which condensation occurs is the **dew point**. On a clear night, a land surface will lose heat rapidly into space. A moist layer of air near the surface will cool and become saturated. Dew then forms at the surface as the water condenses. If the surface air temperature is below freezing, frost forms directly from the saturated air.

Table 10–1

Determining Percent Relative Humidity

Dry bulb Temperature	Dry bulb temperature minus wet bulb temperature °C									
	1	2	3	4	5	6	7	8	9	10
10°C	88	77	66	55	44	34	24	15	6	
11°C	89	78	67	56	46	36	27	18	9	
12°C	89	78	68	58	48	39	29	21	12	
13°C	89	79	69	59	50	41	32	22	15	7
14°C	90	79	70	60	51	42	34	26	18	10
15°C	90	80	71	61	53	44	36	27	20	13
16°C	90	81	71	63	54	46	38	30	23	15
17°C	90	81	72	64	55	47	40	32	25	18
18°C	91	82	73	65	57	49	41	34	27	20
19°C	91	82	74	65	58	50	43	36	29	22
20°C	91	83	74	67	59	53	46	39	32	26
21°C	91	83	75	67	60	53	46	39	32	26
22°C	92	83	76	68	61	54	47	40	34	28
23°C	92	84	76	69	62	55	48	42	36	30
24°C	92	84	77	69	62	56	49	43	37	31
25°C	92	84	77	70	63	57	50	44	39	33
26°C	92	85	78	71	64	58	51	46	40	34
27°C	92	85	78	71	65	58	52	47	41	36
28°C	93	85	78	72	65	59	53	48	42	37
29°C	93	86	79	72	66	60	54	49	43	38
30°C	93	86	79	73	67	61	55	50	44	39

INVESTIGATION 10–1

Relative Humidity

Problem: How is relative humidity determined?

Materials 

identical Celsius thermometers (2)
piece of gauze, 2-cm^2
tape
string
cardboard
beaker of water

Procedure

1. Attach the gauze to the bulb of one thermometer with string as shown in Figure 10–2.
2. Tape both thermometers side by side on the cardboard with the bulbs hanging over the edge of one end. The instrument you have created is a psychrometer.
3. Thoroughly wet the gauze on the thermometer by dipping it into the beaker of water. This thermometer is called a wet bulb thermometer.
4. Wait until the alcohol stops moving in this thermometer and record the temperature.
5. Record the temperature of the dry bulb thermometer.
6. Subtract the wet bulb temperature from the dry bulb temperature.
7. Determine relative humidity using Table 10–1. Find the temperature difference you determined in Step 6 by reading across the top of the table. Keep one finger on this number. Find the dry bulb temperature in the first column of the table. Look across this row until you find the column you marked with your finger. The number at the point where the row and column intersect is the percent relative humidity.
8. Repeat Steps 3-7 at another location inside your school building. Be sure to resoak the wet bulb thermometer at your new test location. Also, wait at least 5 minutes in order to let the thermometers adjust to the new location.
9. Repeat Step 8 at a test site outside of your school building.

FIGURE 10–2.

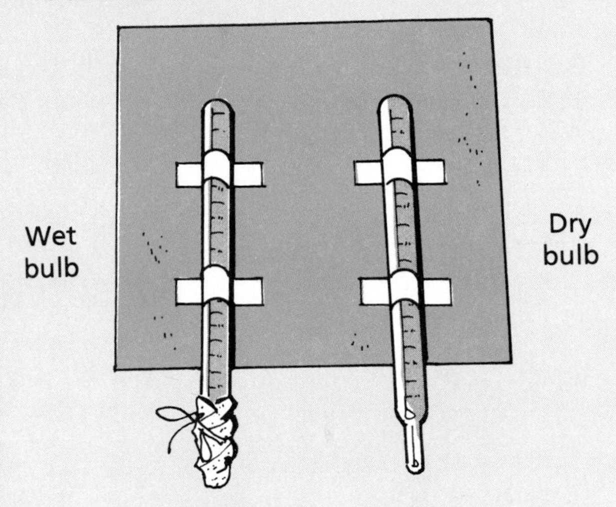

Analysis

1. What was the relative humidity at your three different test sites?

Conclusions and Applications

2. Why did the wet bulb thermometer record a lower temperature than the dry bulb thermometer?
3. What would be the relative humidity if the wet bulb and dry bulb thermometers recorded the same temperatures?
4. How could the relative humidity in your classroom be decreased?
5. Why did the relative humidity vary at your three test sites?
6. How is relative humidity determined?

FIGURE 10–3. The hydrosphere is Earth's water environment. Water moves continuously in the hydrosphere.

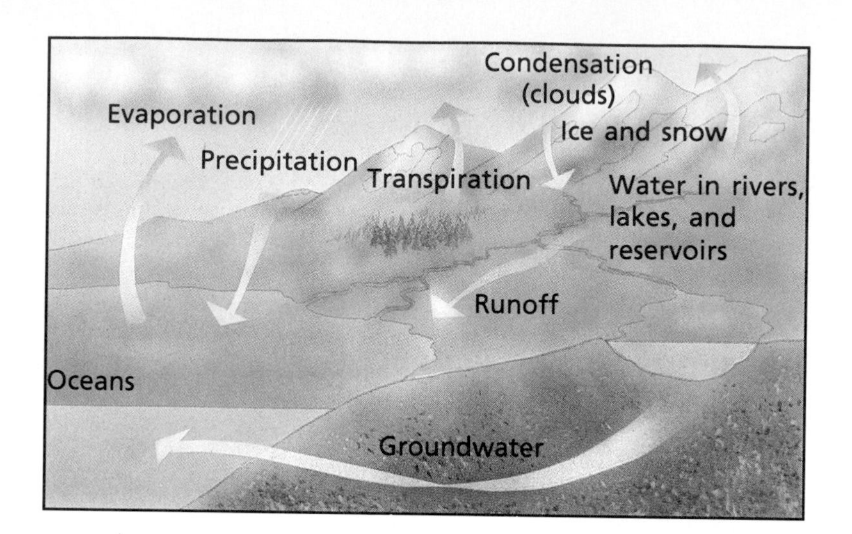

10:2 The Water Cycle and Clouds

Name the parts of the hydrosphere.

Earth is a water planet. In fact, over 70 percent of Earth's surface is covered by water. The **hydrosphere**, Earth's water environment, includes oceans, lakes, rivers, and other bodies of water, groundwater, ice frozen in glaciers, snow, and water vapor in the atmosphere. Water circulates among these different parts of the hydrosphere.

Like the circulation of air, the water cycle is powered by energy from the sun. Evaporation of water and transpiration by plants depend on the sun's energy. Transpiration is the process by which water escapes from the leaves of plants back to the atmosphere. One tree may transpire almost 7000 liters of water into the air during a six-month growing season. The water vapor produced by evaporation and transpiration rises and slowly cools. Condensation occurs if the air is saturated and if dust or salt particles are present. During condensation, water vapor collects around the dust or salt particles, called condensation nuclei, and forms tiny droplets of water.

FIGURE 10–4. Hailstones are layers of ice that form around a small nucleus.

Clouds are collections of tiny droplets of water suspended in the air. Droplets may range from 0.002 to 0.1 millimeters in diameter. In clouds where large amounts of moisture are present, water droplets may join. When these droplets reach 2.0 to 6.5 millimeters in diameter, they may fall as rain. Rain is a form of **precipitation** (prih sihp uh TAY shun). Snow, hail, and sleet are also forms of precipitation.

Hail forms when droplets of water freeze in layers around a small nucleus of ice. Hailstones grow larger as they are tossed up and down by rising and falling air currents that occur during severe thunderstorms. Crop damage from hail amounts to millions of dollars each year.

Sleet forms when raindrops fall through a layer of air below −3°C and freeze. Sleet usually falls during the winter. In summer, sleet melts and falls as rain.

Snow forms when water vapor changes directly to a solid. Snowflakes are usually six-sided crystals occurring in many different patterns. Temperatures in the air must be below freezing for snow to form.

Clouds are classified according to their shapes and altitudes. The four basic types of clouds, according to shape, are cirrus, cumulus, stratus, and nimbus. A **cirrus** cloud is a high, white, feathery cloud usually associated with fair weather. This type of cloud is composed of ice crystals or supercooled water, and may sometimes indicate that bad weather will occur in the near future. **Cumulus** clouds are thick, puffy masses that look like heads of cauliflower with flat bottoms. Cumulus clouds usually develop during the day over land when columns of moist air are forced aloft and cooled to the dew point temperature. **Stratus** clouds occur in layers and often cover the whole sky. Stratus clouds usually are only a few hundred meters thick, but may extend over thousands of square kilometers. Clouds that produce precipitation are **nimbus** clouds. Nimbus clouds are dark gray clouds that have ragged edges. Rain or snow falls continuously from the bottom of these clouds.

Four levels of cloud heights are recognized. High clouds have bases above 6000 meters. Middle clouds generally occur between 2000 and 6000 meters. Low clouds form below 2000 meters. Clouds that have their bases in the low height range but extend upward into the middle or high altitudes are called vertical clouds. Vertical clouds

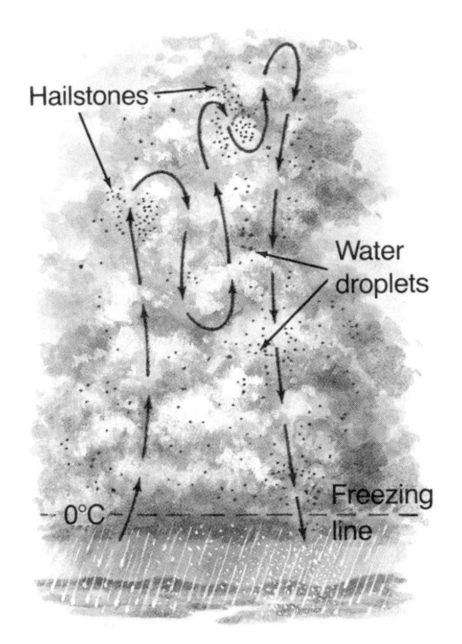

FIGURE 10–5. Hailstones form when many raindrops freeze around crystals of ice as they are tossed about in thunderstorm clouds.

F.Y.I. The word *stratus* comes from the Latin word *stratum*, which means layer or sheet.

FIGURE 10–6. Vertical clouds

VERTICAL CLOUDS (500 to 18 000 m)	Type	Description
	Cumulonimbus	Towering clouds that may spread out on top to form an anvil shape Associated with thunderstorms, heavy rainfall, hail
	Cumulus	Dense, billowy clouds characterized by flat bases; may occur as isolated clouds or as a tightly packed mass Usually associated with fair weather, but may produce precipitation if vertical development is great

FIGURE 10–7. High, middle, and low clouds

HIGH CLOUDS (above 6000 m)	Cirrus	Thin, white, feathery clouds composed of ice crystals or supercooled water Associated with fair weather, but may indicate approaching storms. 
	Cirrocumulus	Thin, white clouds, may look like ripples, waves, or rounded masses Usually associated with fair weather, but may indicate approaching storms Produce "mackerel sky" 
	Cirrostratus	Veillike clouds that may cause halos to appear around the moon or sun Associated with fair weather, but may indicate approaching storms 
MIDDLE CLOUDS (2000 to 6000 m)	Altocumulus	 Light gray clouds in patches or rolls Often precede rain or thunderstorms
	Altostratus	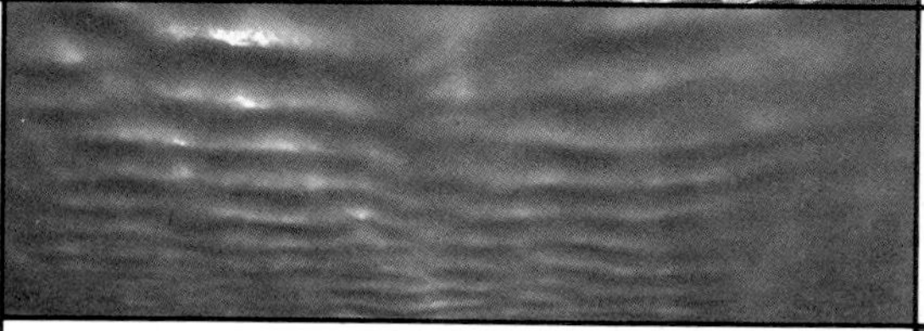 Gray or bluish fibrous clouds; sun or moon appears as a "bright spot" May produce light, continuous precipitation, may indicate approaching warmer weather
LOW CLOUDS (below 2000 m)	Stratocumulus	Soft gray clouds in patches or rolls that may form a continuous layer Occasionally produce light rain or snow 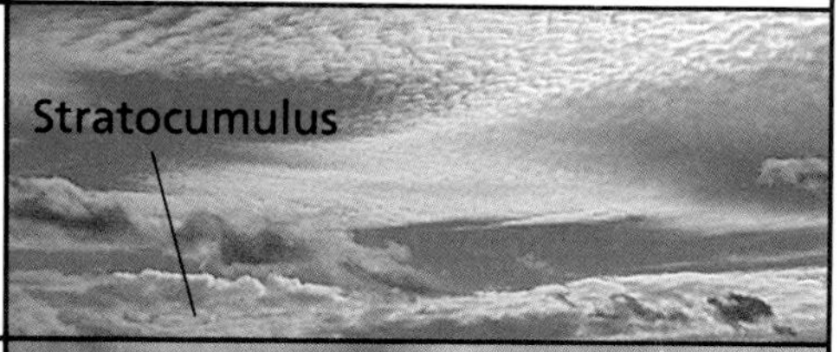
	Stratus	Low layer of gray clouds that may cover the entire sky Associated with light drizzle 
	Nimbostratus	Thick layer of dark clouds that block out the sun Associated with steady, long precipitation 

are often associated with thunderstorms. Figures 10–6 and 10–7 show examples of the different cloud types.

A stratus cloud close to the ground is called fog. Fog is especially common in marine climates. Near the sea, air always contains a large amount of moisture. At night, if the temperature falls below the dew point and the air is still, water vapor condenses in the air and forms fog. Stratus clouds and fog may also occur when warm, very moist layers of air flow across a cold surface.

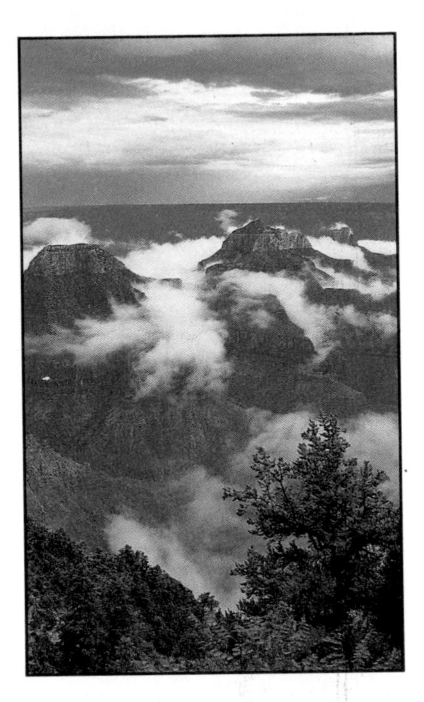

FIGURE 10–8. Fog is a stratus cloud close to the ground.

Review

1. What is the hydrosphere? What are the parts of the hydrosphere?
2. Draw and explain the water cycle.
3. Describe the origin of sleet.
4. ★ Two classrooms have the same humidity. However, the computer room is cooler than the science room. Which room would have the higher relative humidity? Explain your answer.

PROBLEM SOLVING

Where's the water?

Jason and Kim decided that they would help their father fix spaghetti for dinner, so that he could have time to work in the garden. The sauce was already cooking, and it was time to cook the noodles. Jason filled a large pot ¾ full with water, and turned the burner to the highest temperature setting. After a few minutes of watching the pot, Kim and Jason could see that it was still not boiling. So they decided to watch TV while they were waiting.

They got so interested in the program they were watching, they lost track of the time. All at once, Jason remembered the water on the stove, and bolted toward the kitchen. Much to his surprise, the pot was only ½ full of boiling water. On the wall above the stove were droplets of water. After a minute, Jason understood what had happened. He put the noodles in the water, and returned to the TV room to explain to Kim what had happened. What did Jason tell Kim?

GOALS

1. You will gain an understanding of weather associated with warm, cold, stationary, and occluded fronts, and cyclones and anticyclones.
2. You will learn how meteorologists forecast weather.

WEATHER PATTERNS

Weather includes many short-term changes in pressure, temperature, wind direction, and humidity. These changes are generally related to the development and movement of air masses. People can forecast the weather by studying weather patterns.

10:3 Changes in Weather

When a large part of the troposphere stops or moves slowly over a uniform land or water surface, an air mass forms. An **air mass** is a body of air that has the same properties as the source region over which it develops. Air masses extend over thousands of kilometers. They are classified according to their source region. Air masses that develop over continental regions and are relatively dry are identified by "c" for continental. Air masses forming over oceans are moist and are indicated by an "m," meaning maritime. Air masses that develop over high latitudes are cool. These masses are labeled "P" for polar. Low latitude air masses are warm and are labeled "T" for tropical. On a weather map, two symbols are used in combination to describe the temperature and humidity of an air mass. For example, the symbol mT is used for an air mass that is warm and moist.

What type of air masses develop over oceans?

If you could see air, an air mass would appear as a large pile of air over Earth's surface. This pile of air would produce a high-pressure area, called an **anticyclone** (ant ih SI klohn). In an anticyclone, air circulates away

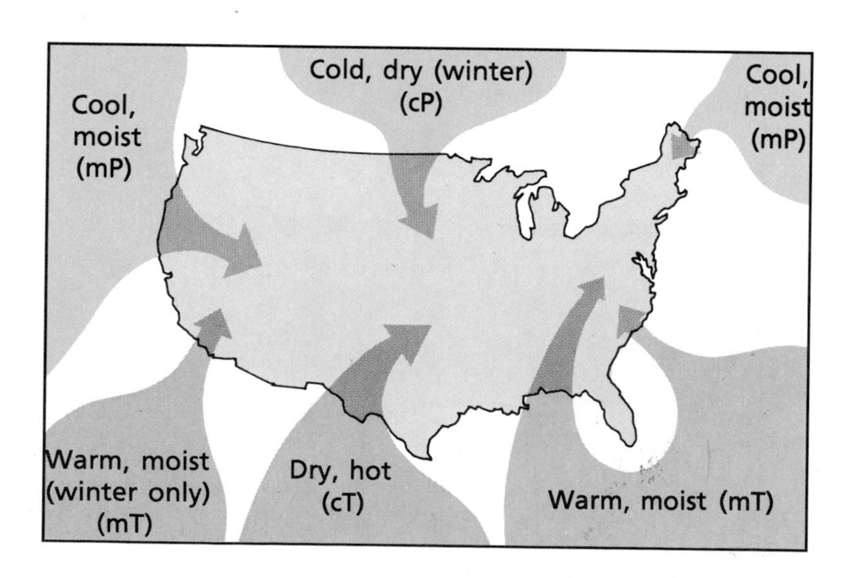

FIGURE 10–9. Six major air masses affect the United States. Which air masses affect your area?

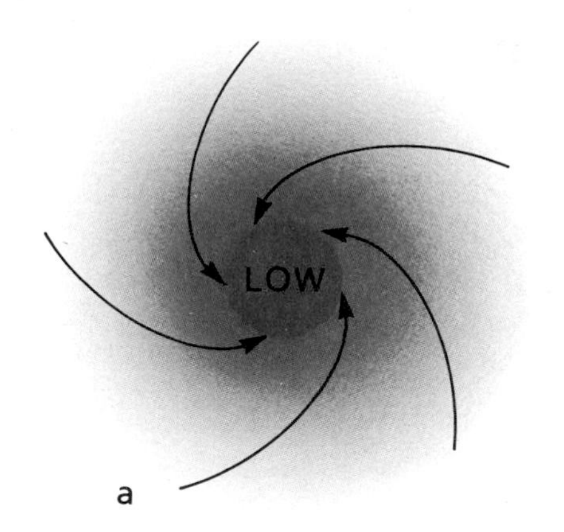

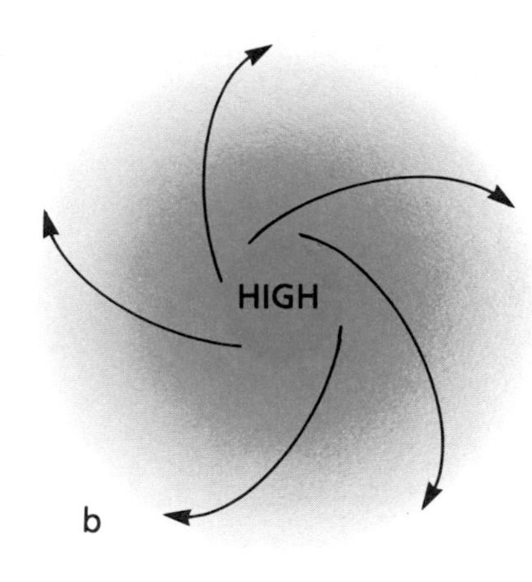

FIGURE 10–10. Winds in the Northern Hemisphere spiral counterclockwise into the center of a cyclone (a). Winds spiral in a clockwise direction as they flow from the high-pressure center of an anticyclone (b).

from the center in a clockwise motion in the Northern Hemisphere. An anticyclone creates areas of fair weather.

What kind of weather is associated with anticyclones?

As the air spirals out from an anticyclone, it enters areas of low pressure, called **cyclones** (SI klohnz). Cyclones appear as depressions or basins in an air mass. The air flows toward the center of a cyclone in a counterclockwise motion. A cyclone may be an area of stormy weather.

Although air masses are modified by the surface over which they move, they tend to keep their unique properties for a long time. When two air masses meet, they tend not to mix. Instead, they form a boundary, or front, separating the air masses. Stormy weather often is associated with fronts.

F.Y.I. Precipitation associated with a warm front will be continuous and may exist for several hours.

A **warm front** develops when a warm air mass meets a cold air mass. The warm air, because it is less dense, slides up over the cold air (Figure 10–11). The first sign of this front is the presence of high cirrus clouds. Later, stratus clouds form as the front continues to move. Nimbostratus clouds may develop and produce rain or snow.

What type of clouds are associated with a warm front?

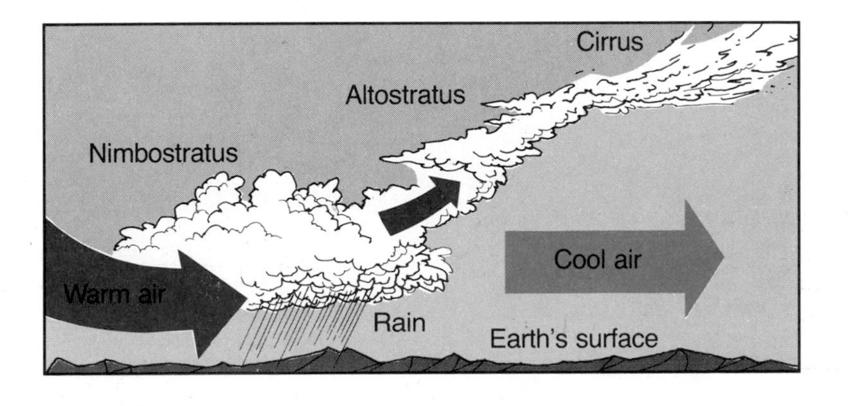

FIGURE 10–11. Warm air slides forward and above cold air, causing precipitation and clouds ahead of a warm front.

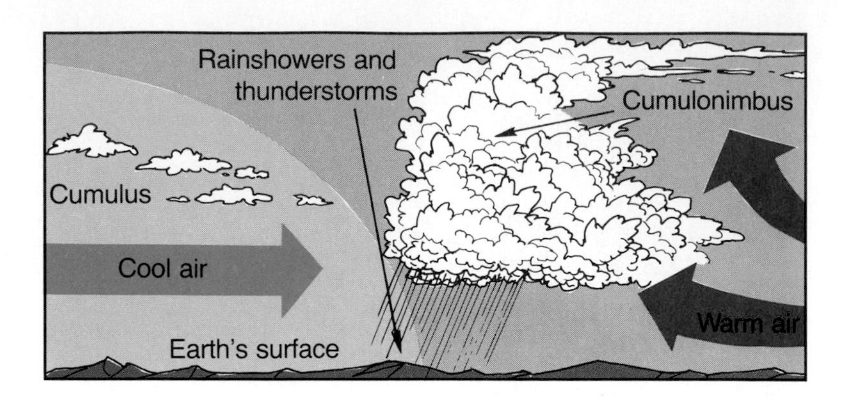

FIGURE 10–12. At a cold front, cold air forces warm air aloft, often producing stormy weather.

F.Y.I. A squall line is an area of intense instability ahead of a fast moving cold front.

A **cold front** forms when a cold air mass invades a warm air mass (Figure 10–12). The cold air forces the warm air rapidly aloft along the steep front. Cumulus and cumulonimbus clouds tend to form along the front, producing rainshowers and thunderstorms. Cooler temperatures follow the passage of a cold front. Cold fronts advance faster in the winter than in the summer. The weather clears shortly after a cold front passes. The passage of a cold front usually brings a shift in winds, a rise in air pressure, a decrease in relative humidity, and a decrease in temperature.

A **stationary front** forms when either a warm front or a cold front stops moving forward. This type of front may remain in the same place for several days. Weather conditions include sluggish winds and precipitation across the entire frontal region.

How does an occluded front form?

An **occluded front** results when two cool air masses merge, forcing the warmer air between them to rise. See Figure 10–13. High winds and heavy precipitation usually are associated with an occluded front.

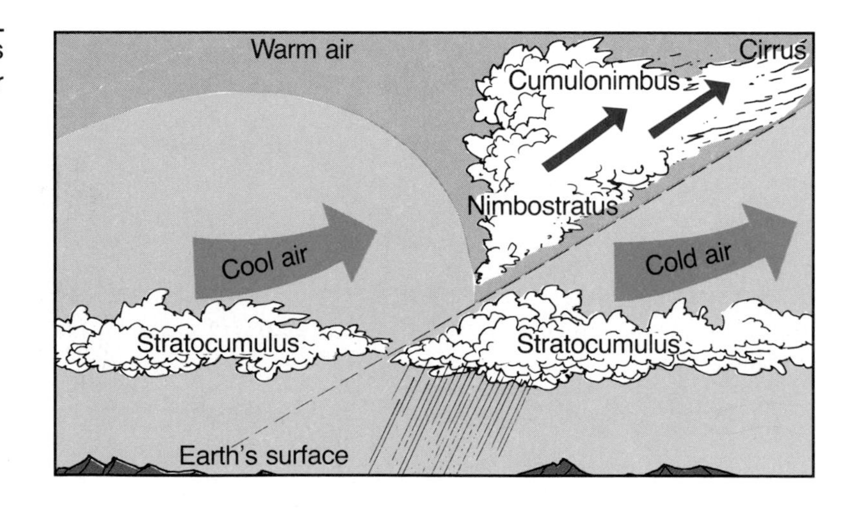

FIGURE 10–13. Warm air is forced above the two cold air masses in an occluded front.

CAREER

Meteorological Technician

Stanley Noonan is a meteorological technician. His duties include working as a weather observer, a forecaster, and as an official in charge of a weather station. Mr. Noonan has worked for industry and local government. He is currently employed by the federal government. The largest employer of meteorological technicians in the United States is the federal government. Technicians like Mr. Noonan may also work for the military services in the Air Weather Service of the U.S. Air Force or the Naval Weather Service of the U.S. Navy. Mr. Noonan's high school preparation for becoming a meteorological technician included courses in physics, chemistry, and mathematics.

For career information, write:
The National Oceanic and
Atmospheric Administration
Department of Commerce
Washington, DC 20230

10:4 Severe Weather

How do thunderstorms form?

Thunderstorms occur when masses of warm, moist air along the surface are forced rapidly upward into colder, dryer layers. As the warm air is forced aloft, it cools and condenses, forming cumulonimbus clouds. These clouds may reach heights of 20 000 meters. After a turbulent period of upward movement of air, rain begins to fali. The rain lasts about 30 minutes. Such storms usually are accompanied by thunder and lightning.

What are some safety measures that you should take during a thunderstorm?

During the violent period of uplift, electric charges build in the clouds. Bolts of lightning leap from cloud to cloud and from the clouds to Earth. Lightning strikes when current flows between regions of opposite electrical charge (Figure 10–14). A single bolt of lightning can discharge millions of volts. When lightning occurs, keep away from open doors and windows and electrical appliances that are connected. Stay out of water and off small boats. A low point in the landscape or a dense grove of trees is the safest shelter if you are outside. Avoid solitary trees because lightning is attracted to the tallest object in an area. Thunder is not dangerous. It is the loud sound caused by the pressure wave that accompanies lightning.

a

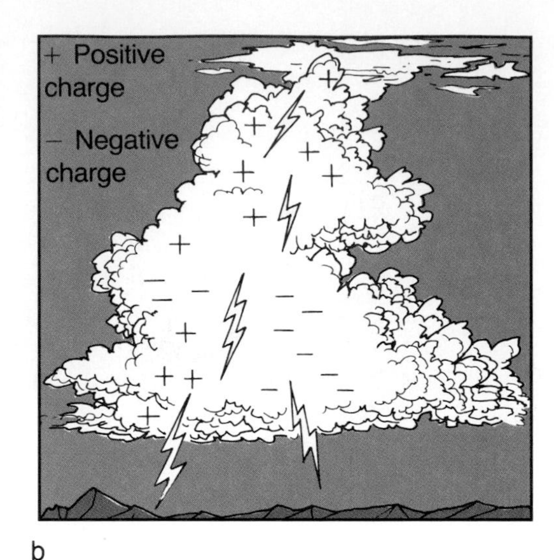

b

FIGURE 10–14. Lightning (a) is caused by the presence of both positive and negative electrical charges within clouds and on the surface (b).

What is a tornado and when do tornadoes most commonly occur in the United States?

A **tornado** is a violent, whirling wind that moves in a narrow path over land. Tornadoes may occur at many different times and places. However, they are most common during the spring in the United States. They are associated with thunderstorms along cold fronts and sometimes with hurricanes. In very severe thunderstorms, warm air may be forced upward with great speed and cause the winds to rotate violently near the surface. Air pressure drops rapidly. A funnel cloud appears at the base of the cumulonimbus cloud (Figure 10–15). As the funnel touches the ground, buildings and trees are ripped apart by destructive winds. Although some tornadoes may last for hours, the average life span of such a storm is 15 minutes.

Blizzards occur during the winter months. A blizzard combines high winds with temperatures of -6°C or lower. Winds often range from 50 to 75 kilometers per hour. Blowing, powdery snow can cause visibility near zero. These severe weather conditions can be dangerous to people and other animals. Thus, exposure should be avoided.

FIGURE 10–15. Tornadoes are associated with severe storms.

Hurricanes are tropical cyclones that form over oceans. They usually form in latitudes between 5° and 20°, and move toward higher latitudes. Air over tropical oceans is very warm and humid. Sometimes centers of very low pressure develop with a rapid inflow of air, forming a nearly circular storm. Air near the center is forced aloft and flows outward at upper levels. Wind speed increases as the storm develops. When wind speeds reach 120 km/h, the storm is a hurricane. An eye forms at the center as air sinks and is warmed by compression. Much of the

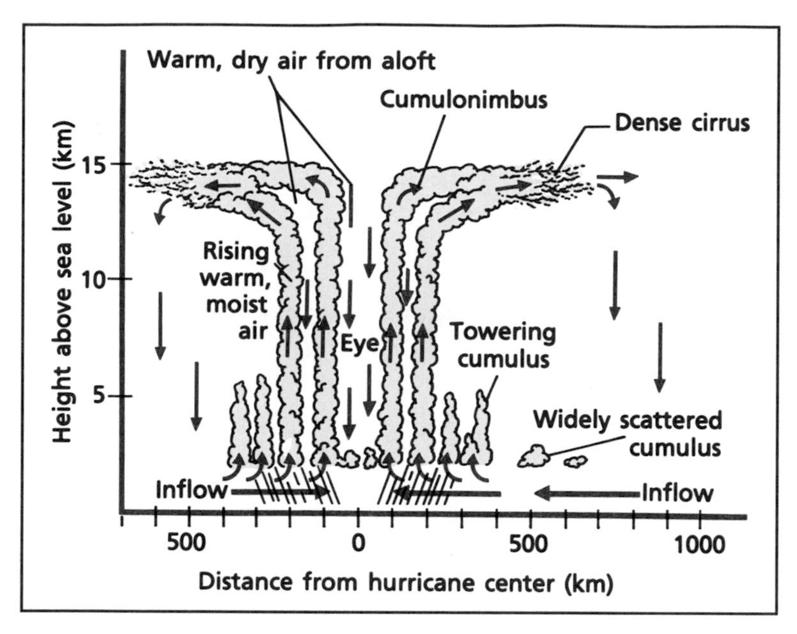

a

b

FIGURE 10–16. Warm, dry air spirals downward in the eye of a hurricane, while warm, moist air rises around the eye (a). A satellite photograph of Hurricane Frederick shows the eye (b).

energy of a hurricane is released through condensation, as it pumps large amounts of moist air into the upper troposphere. Hurricanes that strike land may cause violent winds, heavy rains, and floods. The Gulf and South Atlantic coastal states have experienced many disastrous hurricanes.

The National Weather Service issues advisories when severe weather has been observed, or when the conditions are such that severe weather could occur. When a **watch** is issued, you should prepare for the severe weather. Watches are issued for severe thunderstorms, tornadoes, floods, blizzards, and hurricanes. When forecasters expect severe thunderstorms or tornadoes to occur over the next few hours, they issue watches. Flood watches are issued when heavy rain could cause flooding. A hurricane or blizzard watch is issued when a hurricane or blizzard is a threat to an area.

F.Y.I. Tropical cyclones near North America are called hurricanes. In Eastern Asia they are called typhoons; in India, cyclones; in Australia, willy-willies; and in China, baguios.

During a watch, you should not panic, but you should prepare for what might happen. During all watches, stay tuned to a radio or television station that is reporting weather updates. Be ready to take shelter. Make plans for where you will go if a warning is issued. For example, a hurricane watch is usually issued a few days before the hurricane would reach the area. If you are planning on staying at the same place during the hurricane, use this time to stock up on food. Have a flashlight, batteries, and blankets ready if the emergency should arise. If you are near a beach, do not stay. Leave the area and seek shelter on higher ground.

What is a warning?

When a **warning** is issued, severe weather conditions exist: severe thunderstorms have already been reported in the area; a tornado has been spotted; streams and rivers in the area have reached flood stage. A hurricane or blizzard warning is issued when the storm is actually approaching the area.

You should take immediate action when a warning is issued. If there is a severe thunderstorm warning, you should take shelter inside a protected building in order to avoid lightning and high winds. Do not stand under a tree. During a tornado warning, a siren will be sounded. Go to the central part of a basement, or stand under a stairway, as these are the more supported areas of a building. Stay away from windows. If a flood warning is issued, get to high ground.

When a blizzard warning is issued, stay inside. Have blankets and flashlights ready in case power is interrupted. During a hurricane warning, take shelter on high ground until any threat of flooding is past. Take the same precautions as you would for severe thunderstorms and tornadoes.

Table 10–2

Severe Weather Precautions

Weather Condition	Watch: Be prepared for what might happen.	Warning: Take immediate action.
Hurricane	Stock up on food. Have flashlights, batteries, portable radio, and blankets ready for use. Evacuate beach areas.	Stay on high ground until any threat of flooding is past.
Severe Thunderstorm	Be prepared for possible flooding.	Take shelter inside a protected area. Avoid lightning and high winds. Avoid solitary trees.
Tornado	Be prepared to respond to a possible warning siren.	Go to the central part of the basement or some other well-supported area of the building. Avoid windows.
Flood	Make plans for where you will go if a warning is issued.	Go to high ground.
Blizzard	Stock up on food, flashlights, batteries, portable radio, and blankets.	Stay inside. Have blankets and flashlights on hand.

TECHNOLOGY: ADVANCES

Doppler Radar

A tornado often develops in less than an hour, and destroys lives and property within a matter of minutes. Ground spotters can warn people only two minutes in advance of a tornado touchdown. However, Doppler radar can detect tornadoes 20 minutes before they touch down. In addition to the position and strength of a storm shown by conventional radar, Doppler radar shows wind speed and direction in relation to the radar's antenna. Thus, circular wind patterns indicating tornadoes can be detected.

Doppler radar sends repeated radio pulses and monitors the echoes from storms as far away as 160 kilometers. It measures the shift in frequency of signals reflected from raindrops, and indicates the direction a storm is moving. Radio signals reflected from raindrops moving toward the radar are compressed and shift to a higher frequency. Radio signals reflected from raindrops moving away from the radar are lengthened and shift to a lower frequency. These frequency differences appear on the Doppler radar screen as different colors. Bright green indicates winds coming toward the radar; red indicates winds moving away from the radar. Where red and bright green are close together, rotation is occurring.

Federal agencies have developed a nationwide system of stations, each equipped with Doppler radar called NEXRAD (Next Generation Weather Radar). By using NEXRAD, forecasters can pinpoint tornadoes instead of making general forecasts.

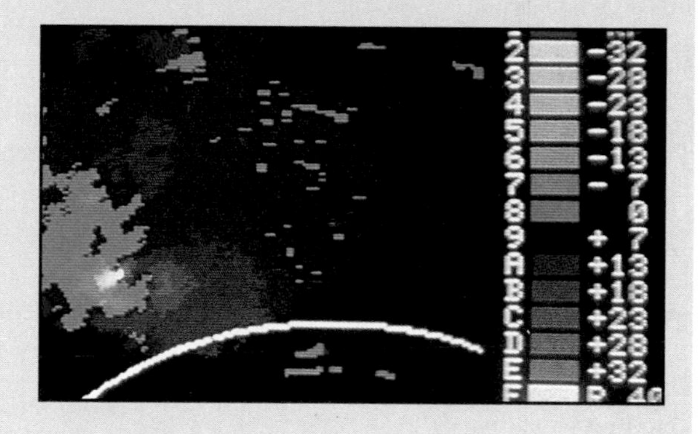

10:5 Forecasting

Weather conditions are important to all of us. A person who studies the weather is a **meteorologist** (meet ee uh RAHL uh just). Predicting the weather requires accurate measurements of temperature, pressure, winds, and humidity. Meteorologists make weather observations by collecting data from instruments and satellites. Instruments used at many weather stations include thermometers and barometers. An anemometer measures wind speed. Rain gauges are used to measure the amount of rain or snow. Computers are used to gather data at many weather stations today. Conditions of the upper atmosphere, including temperature, relative humidity, and atmospheric pressure, are measured with instruments sent aloft by balloons. Weather stations communicate with one another about local weather conditions.

What are some instruments that meteorologists use to predict the weather?

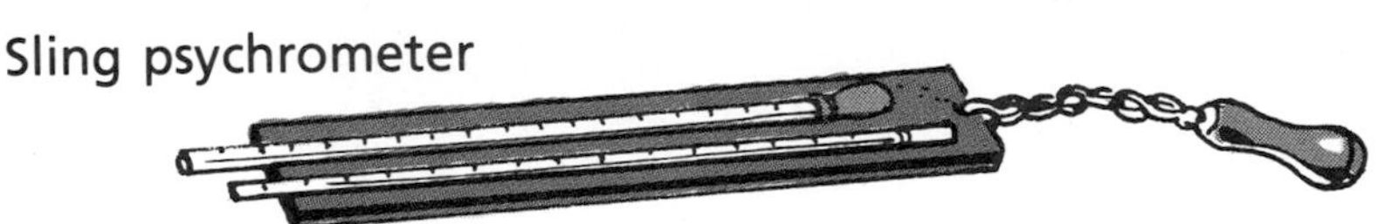

FIGURE 10–17. Wind speed is measured with an anemometer, precipitation with a rain gauge, and relative humidity with a sling psychrometer.

BIOGRAPHY

Leslie Roy Lemon
1947–

Leslie Roy Lemon is the Program Control Manager for NEXRAD Radar Development, UNISYS Corporation. In his research he tries to understand and document how severe storms, including tornadoes, begin and evolve, as well as to improve the use of Doppler radar as a meteorological warning device. Lemon has received the "Special Achievement Award" from the NOAA for the discovery of the tornado "signature" in Doppler radar data.

Since 1960, weather satellites have played an important part in weather forecasting. TIROS I, the first fully equipped weather satellite, carried two television cameras and stored pictures on tape for rebroadcast to Earth. In 1964, the first of a series of NIMBUS satellites was launched. TIROS and NIMBUS satellites are polar-orbiting satellites that circle Earth every 110 minutes. These kinds of satellites orbit Earth directly over the poles. The latest series of weather satellites, GOES, are geostationary satellites, which have an orbital period equal to Earth's rotational period. Thus, they remain in position over the same geographical site. Today, both types of satellites monitor weather conditions. These data are sent to ground stations by microwave transmissions. As a result of satellite photography, scientists who study weather can locate ice floes and icebergs in shipping lanes, issue severe storm warnings, map jet streams, and better forecast the weather.

A weather map is a compilation of weather data from many collecting stations. On a weather map, **station models** describe the local weather of the collecting stations. Each station model shows the wind direction and speed, atmospheric pressure, temperature, dew point, amount and types of clouds, type of precipitation, and other data.

A computer uses the information from these station models to draw isobars (I suh barz) on the map. **Isobars** are lines drawn to connect points of equal air pressure measured in millibars. One thousand millibars equals 76 centimeters of mercury. These lines show the size and position of pressure systems. The spacing between isobars indicates the pressure change occurring over a given dis-

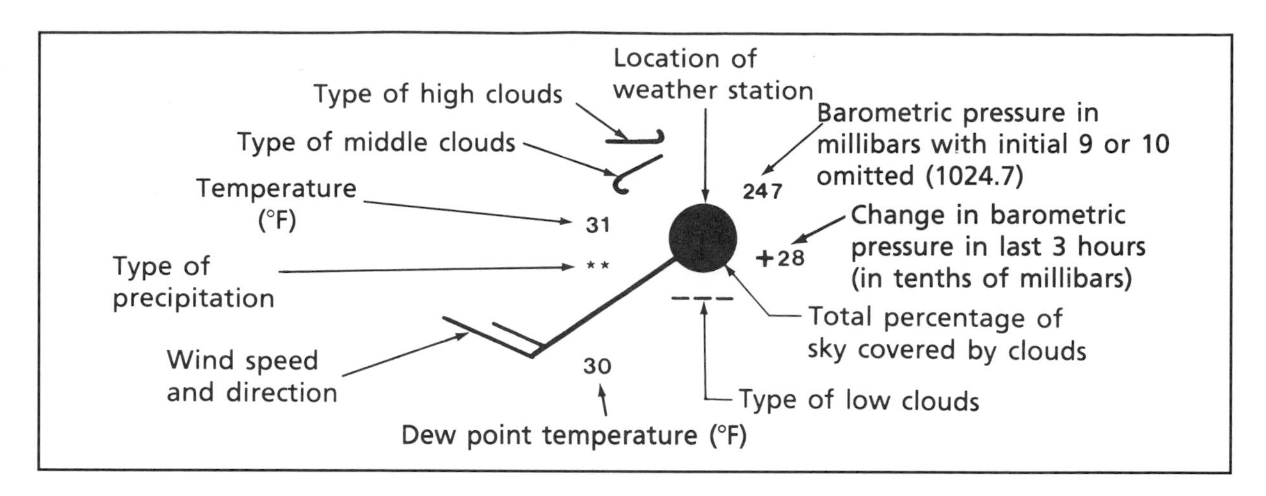

FIGURE 10–18. On a station model, temperature is recorded in degrees Fahrenheit and pressure is recorded in millibars. See Appendix L for a complete listing of station model symbols.

tance, or the pressure gradient. Isobars that are close together indicate a steep pressure gradient and strong winds. Wind usually flows parallel to isobars. **Isotherms** (I suh thurmz) are lines drawn to connect points of equal temperature.

Review

5. Distinguish between a weather watch and a weather warning.
6. Describe weather associated with an anticyclone.
7. What three conditions occur during a blizzard?
8. List three characteristics of the weather that are shown on a station model.
★ 9. Describe the characteristics of an air mass that has been labeled "mT."

FIGURE 10–19. Computers are used to generate many weather maps.

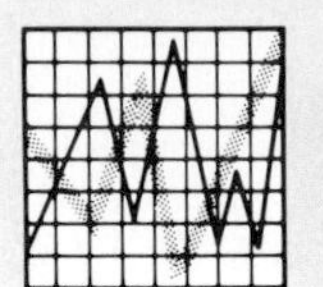

SKILL

Reading a Weather Map

Problem: How do you read a weather map?

Materials

Figure 10–20
hand lens
Appendix L

Procedure

Use the information provided in the questions below, Figure 10–20, and Appendix L to learn how to read a weather map.

Questions

1. Find the station models on the map for Tucson, Arizona, and Albuquerque, New Mexico. Find the dew point, cloud coverage, pressure, and temperature at each location.
2. After reviewing information about the spacing of isobars and wind speed in Section 10:5, determine whether the wind would be stronger at Roswell, New Mexico, or at San Francisco. Record your answer. What is one other way to tell the wind speed at these locations?
3. Determine the type of front near Key West, Florida. Record your answer.
4. The triangles or half circles on the weather front symbol are on the side of the line that indicates the direction the front is going. Determine the direction that the cold front, located over Colorado and Kansas, is going. Record your answer.
5. Locate the pressure system over Winslow, Arizona. After reviewing Section 10:3, describe what would happen to the weather of Wichita, Kansas, if this pressure system should move there.
6. The prevailing westerlies are the winds responsible for the movement of weather across the United States and Canada. Based on this, would you expect Charleston, South Carolina, to continue to have clear skies? Explain your answer.
7. The line on the station model that indicates wind speed shows from which direction the wind is coming, and the wind is named accordingly. What is the name of the wind at Jackson, Mississippi?

FIGURE 10–20.

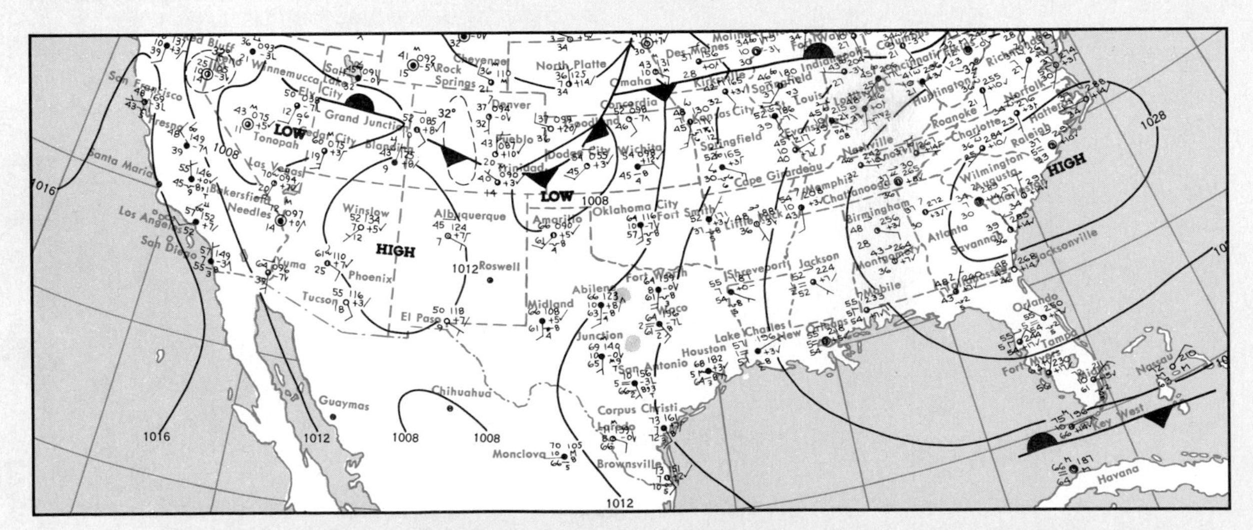

INVESTIGATION 10–2

Weather Observations

Problem: What weather observations can you make yourself?

Materials 

Celsius thermometer
magnetic compass
aneroid barometer

Procedure

1. Copy the chart provided in Data and Observations. You will be making weather observations each day for 7 days by following the directions below. Be sure to make observations at the same time and place each day.
2. Determine the temperature by placing a thermometer in a shaded location. Be sure to observe the temperature in the same location each day.
3. Determine the air pressure using an aneroid barometer.
4. Estimate the amount of sky covered by clouds as clear, overcast, or somewhere in between.
5. Determine the types of clouds using Figures 10–6 and 10–7.
6. Use a compass to determine the direction from which the wind is blowing.
7. Describe the precipitation. Use the terms rain, snow, sleet, hail, fog, or clear.
8. Use the data you collect each day to forecast weather conditions for the following day. Note any trends you see in your observations, such as high cirrus clouds preceding rainy weather.

Analysis

1. Which wind directions were associated with cool days? With warm days?
2. Which wind directions were associated with many clouds? With few clouds?

Conclusions and Applications

3. Was there a relationship between low barometric pressure and the presence of clouds and precipitation? Explain.
4. How accurate were your forecasts for the next day? Give an explanation for any errors that may have occurred in your forecasting.
5. What weather observations can you make yourself?
6. What additional weather observations do meteorologists make?

Data and Observations

Date	Temperature °C	Atmospheric pressure	% Cloud cover	Cloud types	Wind direction	Precipitation	Forecast

CLIMATE

GOALS

1. You will gain an understanding of the various factors that control the climate of an area.
2. You will learn how human activities affect weather and climate.

Climate is the average of all weather conditions of an area over a long period of time. These conditions include average temperatures, air pressures, humidity, and days of sunshine for a period of 30 years. Climate has changed a great deal throughout Earth's history due to natural causes and many human activities.

10:6 Climatology

What is climate?

There are several different ways to classify climates. One way is according to the amount of solar energy an area receives. Using this classification system, there are three distinct zones.

The amount of energy received from the sun depends on the angle at which sunlight strikes Earth. Sunlight strikes Earth at about 90° relative to the horizon at noon in the tropics. These light waves lose the least energy to the atmosphere and transfer the most heat to Earth. The tropics extend from 23½° north latitude to 23½° south latitude. The northern boundary is the Tropic of Cancer and the southern boundary is the Tropic of Capricorn. Year-round temperatures are always hot in the tropics, except at high elevations.

The polar zones extend from the poles to 66½° north and south latitudes. Polar regions are always cold. Solar energy hits these regions at a low angle, and thus loses much energy to the atmosphere. The solar energy that does reach Earth is distributed over a large area, and much is reflected by polar ice. Also, heat is lost from these regions constantly during the winter when these areas experience 24 hours of darkness.

FIGURE 10–21. The sun's rays that strike Earth at a low angle lose more energy to the atmosphere than those that strike at a high angle (a). The amount of solar energy received by different areas of Earth determines climate zones (b). Name the three climate zones.

a

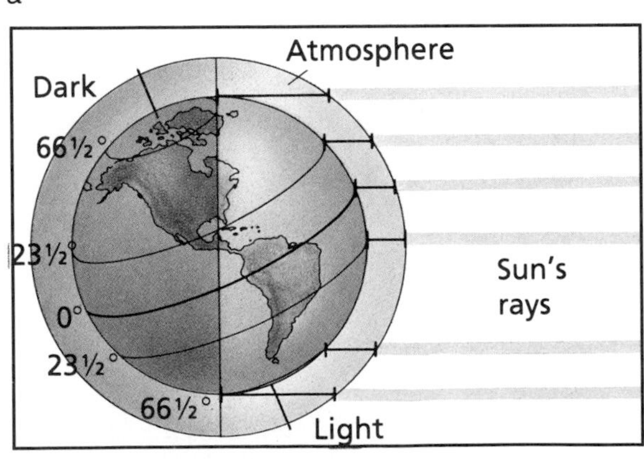

b

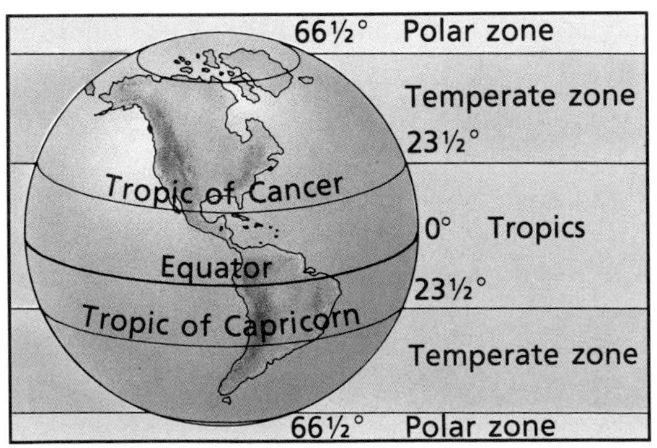

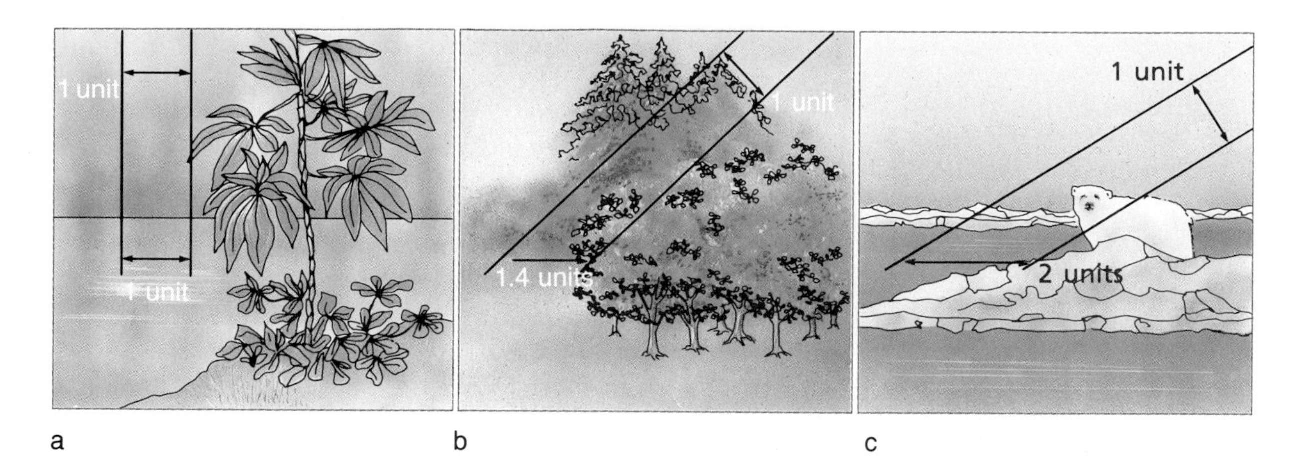

FIGURE 10–22. Solar energy is most concentrated at the equator (a). In the temperate zone (b) and at the poles (c), the same amount of solar energy is distributed over a much larger area due to Earth's tilt.

Between the tropics and the polar zones are the temperate zones. Here, weather generally changes with the seasons. Winters are cold and summers are hot. Spring and fall usually have mild temperatures.

Climates are more complex than the three general divisions of polar, temperate, and tropical. Within each zone, a number of factors affect weather patterns. Large bodies of water influence climate. Coastal areas are warmer in winter and cooler in summer than inland areas. Mountains act as barriers over which winds must flow. As air rises on the windward side of mountains, it cools and drops its moisture. On the leeward side, air heats as it descends and dries out the land. Thus, forests are common on the windward side, and deserts are common on the leeward side of mountains.

FIGURE 10–23. Another way of classifying climate is to use a modified Köppen system. This system determines the climate by averaging temperature, rainfall, and other measurements over a period of thirty years.

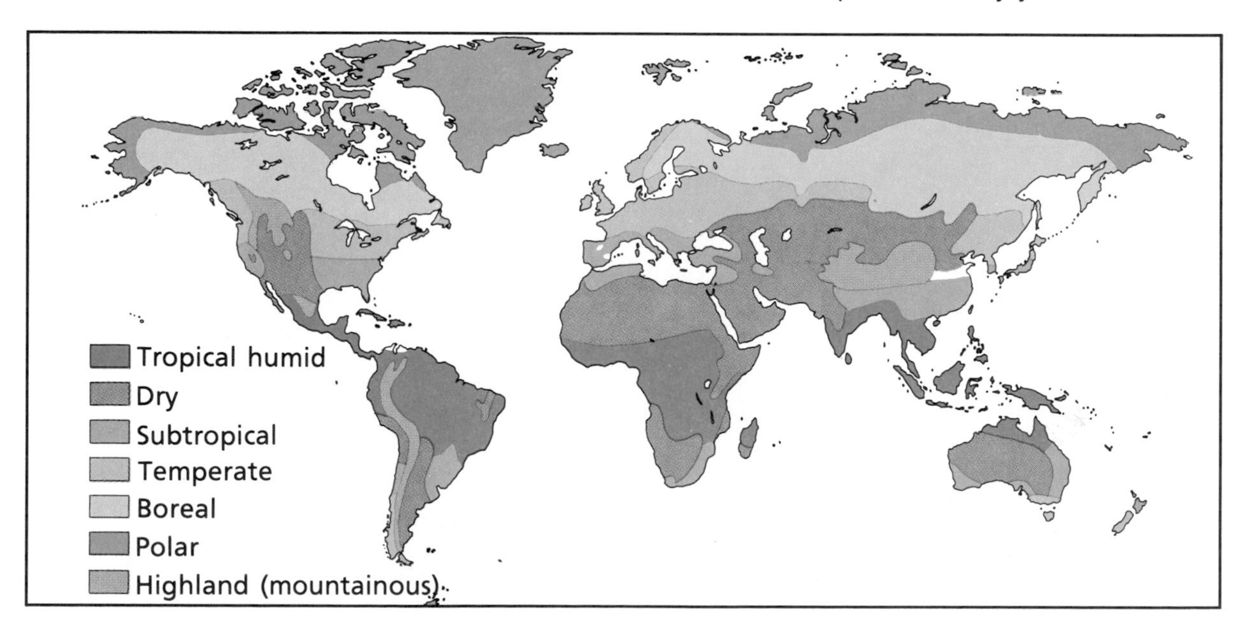

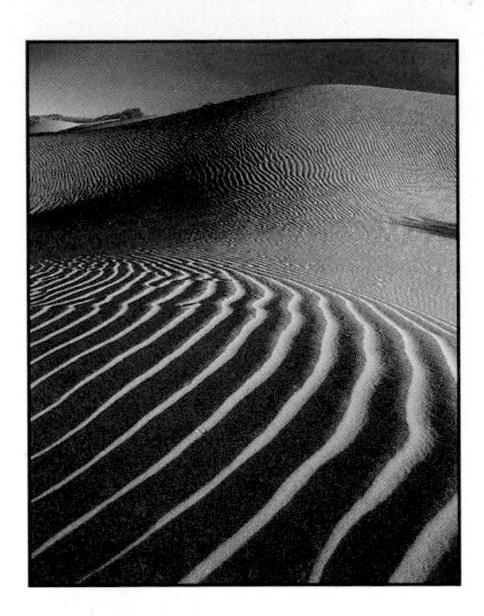

FIGURE 10–24. Deserts receive less than 25 cm of precipitation per year.

Deserts receive less than 25 centimeters of rainfall annually, and there is little cloud cover to screen the sun's radiation. Also, winds that travel for long distances across land lose their moisture and become drying winds. Thus, day and night temperatures may vary as much as 30°C because heat is not retained in the dry air at night. In some deserts, descending currents of air dry the land.

10:7 Cultural Effects of Weather and Climate

There have been attempts to control the weather. Cloud seeding has been used to try to control precipitation. Silver iodide particles are spread in clouds to act as condensation nuclei. This technique has been partially successful in reducing the power of thunderstorms and hurricanes. Cloud seeding also has been used to disperse fog at airports. Rainfall and snowfall have been increased in some areas.

How do cities alter weather and climate?

Other human activities have had unplanned effects on weather and climate. Winter temperatures may be 10°C higher in cities than in a nearby rural area due to large paved areas that absorb the sun's energy. Combustion produces particles that act as condensation nuclei, which produce more rain. This is true especially to the east of big cities.

F.Y.I. Climates can show small-scale variations due to vegetation, buildings, roads, and reservoirs. These variations are microclimates.

These and other human activities may be affecting the climates of Earth. Some scientists believe that increased air pollution and the clearing of forests may lead to a worldwide drop in temperature due to the reflection of solar energy back into space. A solar energy drop of 1.6 to 2 percent could lead to a condition in which oceans would freeze and snows cover areas in the lower latitudes. Other scientists believe that increased amounts of dust and carbon dioxide and the clearing of forests will increase the greenhouse effect. This may raise worldwide temperatures enough to melt polar ice caps. This would cause a rise in sea level.

Review

10. What is the northern boundary of the tropics?
11. Where is solar energy most concentrated?
★ **12.** Explain why some scientists feel that human activities will cause Earth to cool, while others think Earth will get hot.

Chapter 10 Review

SUMMARY

1. Temperature affects pressure, winds, and relative humidity. 10:1
2. Relative humidity is a measure of how saturated air is. 10:1
3. Clouds are masses of condensed moisture and are classified by their general shape and their altitude. 10:2
4. Air masses are classified by their source regions. 10:3
5. Winds blow outward from the center of high pressure areas. Winds blow inward toward the center of low pressure areas. 10:3
6. Violent storms include blizzards, tornadoes, and hurricanes. 10:4
7. The National Weather Service issues watches and warnings. 10:4
8. A station model describes local weather conditions. 10:5
9. Climate is an average of weather conditions over a long period. 10:6
10. Climate zones result from differences in the amount of solar energy areas receive and from the angle at which the sun's energy strikes Earth. 10:6
11. Bodies of water and mountains affect climate. 10:6
12. Cloud seeding has been used to change some types of weather. Scientists are studying the impact of human activities on climate. 10:7

VOCABULARY

a. air mass
b. anticyclone
c. cirrus
d. cold front
e. cumulus
f. cyclones
g. dew point
h. hurricanes
i. hydrosphere
j. isobars
k. isotherms
l. meteorologist
m. nimbus
n. occluded front
o. precipitation
p. relative humidity
q. stationary front
r. station models
s. stratus
t. tornado
u. warm front
v. warning
w. watch

Matching

Match each description with the correct vocabulary word from the list above. Some words will not be used.

1. front that forms when a cold air mass invades a warm air mass
2. temperature at which air is sufficiently cooled to begin condensation
3. lines connecting points of equal air pressure
4. storms that form over tropical water with winds over 120 km/h
5. issued when severe weather conditions exist
6. on a weather map, these describe the weather at collecting stations
7. thick, white, feathery clouds that may indicate approaching storms
8. thick, puffy clouds with flat bottoms
9. also called a high pressure area
10. clouds that form in layers

Chapter 10 Review

MAIN IDEAS

A. Reviewing Concepts

Choose the word or phrase that correctly completes each of the following sentences.

1. The most destructive winter storm in northern latitudes probably is the *(blizzard, anticyclone, tornado)*.
2. The zone that receives the most radiant energy from the sun is the *(polar zone, temperate zone, tropics)*.
3. An air mass developing over the ocean in the tropics would be indicated by the symbol *(cP, mT, mP)* on a weather map.
4. In saturated air, the relative humidity is *(zero, 50%, 100%)*.
5. A *(cirrus, stratus, cumulus)* cloud is a high, white, feathery cloud associated with fair weather.
6. *(Stratus, Vertical, Nimbus)* clouds have their bases in the low height range and extend upward into the middle or high altitudes.
7. If deserts are found near mountains, they are usually on *(east, windward, leeward)* sides of the mountains.
8. In a high pressure area, winds *(blow toward the center, blow away from the center, do not exist)*.
9. Tornadoes are most common in the United States in *(spring, fall, winter)*.
10. A(n) *(cold, stationary, occluded)* front forms when two cold air masses force warm air aloft.
11. During a *(tornado, blizzard, flood)* warning you should go to the central part of your basement.
12. *(Isotherms, Isobars, Station models)* are lines that connect points of equal temperature.
13. A *(cold, warm, stationary)* front forms when a warm air mass meets a cold air mass.
14. Wind speed is measured with a(n) *(thermometer, meter, anemometer)*.
15. Clouds that produce rain are *(nimbus, stratus, cumulus)*.

B. Understanding Concepts

Answer the following questions using complete sentences.

16. Describe the weather conditions of an air mass labeled "mP."
17. What should you do if you hear a tornado siren?
18. What happens when a warm air mass meets a cold air mass?
19. Explain how mountains affect climate.
20. Describe a desert climate.
21. Explain why a person living in a large city will experience higher summer temperatures than a person living in a nearby rural area.
22. Describe how meteorologists use the information that is photographed by weather satellites.
23. What kind of weather is associated with a cold front?
24. Describe how the spacing of isobars can be used to tell whether an area is experiencing high winds.
25. Explain what you can tell by looking at an isotherm on a weather map.

C. Applying Concepts

Answer the following questions using complete sentences.

26. Why is severe weather often associated with an occluded front?
27. How does temperature affect the type of precipitation that will form?
28. Why do desert areas seldom experience fog?
29. How can you forecast the weather using clouds?
30. Why do some scientists believe that human activities will make Earth colder?

SKILL REVIEW

1. Compare and contrast low and high pressure areas.
2. Use Figure 10–20 to determine the wind direction at Albuquerque, New Mexico.
3. Make a station model that shows the following information: temperature 70°F, barometric pressure 224 millibars, clear skies, wind direction south.
4. Why are station model symbols used on weather maps?
5. If you live in the United States, and you observe a weather map that shows a cold front and a low pressure system to the west of where you live, what can you infer about the future weather conditions of your area?

PROJECTS

1. Design and construct several different homemade weather instruments. Your instruments might include an anemometer and a rain gauge. Use them to make weather observations for one week. Compare the accuracy of your instruments with reported measurements from radio or TV weather reports.
2. Use weather maps and local weather information to forecast probable local weather conditions for the next few days. Indicate the scientific basis for your predictions. After several days, evaluate the accuracy of your predictions.

READINGS

1. Alper, Joseph. "Mostly Sunny and Cooler . . . with a chance of Flurries." *Science 86,* January/February 1986, pp. 66-73.
2. Ludlum, David M. *The Weather Factor.* Boston: Houghton Mifflin, 1984.
3. Sabin, Louis. *Weather.* Mahwah, NJ: Troll Associates, 1985.

OCEAN WATER
WATER IN MOTION

Earth's Ocean

Every day your life is influenced by the ocean. Recall from Chapter 10 that the ocean plays a major role in determining Earth's weather and climate. The ocean is used by people for transportation, recreation, food, minerals, dumping grounds for wastes, energy, water, and scientific study.

OCEAN WATER

Earth is the water planet. The ocean is a continuous body of saltwater that covers about 70 percent of Earth's surface. The mass of the ocean is about 1.4×10^{24} g. Oceans contain water that is continuously cycled among the different parts of the hydrosphere.

GOALS

1. You will learn a hypothesis that explains a possible origin of oceans.
2. You will learn about the composition of ocean water.

11:1 Origin of Oceans

About 4 billion years ago, water from deep within Earth was released at Earth's surface through volcanic activity. This water began to accumulate to form Earth's oceans. Some ancient rocks (over 3.5 billion years old) show features that suggest they were deposited in water.

As with many of Earth's features, oceans are always changing. The size and the location of ocean basins have changed throughout geologic time. Sea level has been as much as 200 meters lower in the past when water was held in the ice sheets that once covered large areas of Earth's surface. Sea level has also been higher than it is today. Let us now look more closely at the matter that covers nearly three-fourths of Earth's surface.

F.Y.I. For more information on the location of the ocean basins throughout geologic time, refer to Chapter 20.

11:2 Composition of Ocean Water

Recall from Chapter 4 that mixtures are physical combinations of different substances. Oceans are mixtures. Some substances are dissolved from rocks and carried to

a

b

FIGURE 11–1. Oceanographers collect samples of ocean water in bottles such as these to determine what elements are present (a). Fresh water is produced at desalination plants (b).

the ocean by rivers. Others occur in solution in the water vapor given off by submarine volcanoes. Chlorine is one gas that is added to ocean water by volcanic action. Sodium is a soluble product of the weathering of some rocks and is carried to the ocean by surface waters. Sodium and chlorine combine to form halite, or common salt. Other solids dissolved in seawater include calcium chloride, magnesium chloride, and sodium and potassium compounds. **Salinity** is a measure of the quantity of dissolved solids in ocean water. The average salinity of ocean water is 35 parts dissolved solids per thousand parts water.

What is salinity?

Ocean water is low in both silica and calcium because marine organisms use these substances in their life processes. Many marine animals use calcium to form bones or silica and calcium to form shells. Many marine plants have silica shells. Ocean water has traces of many other dissolved substances, including gases from the atmosphere. Oxygen and carbon dioxide are dissolved in seawater and are used by marine organisms.

The ocean is a storehouse of fresh water. Many areas of the world do not have enough fresh water. Using a technique called **desalination,** salts can be removed from ocean water to produce fresh water. One method of desalination is similar to the water cycle. Ocean water is heated until water vapor forms. The vapor is collected and cooled. The condensed product is fresh water.

What is one method of water desalination?

Review

1. What is salinity?

★ **2.** How do most scientists think Earth's oceans formed?

INVESTIGATION 11–1

Desalination

Problem: How does desalination produce fresh water?

Materials   

pan balance
table salt
water
500 mL beakers (2)
1000 mL flask
1-hole rubber stopper
rubber tubing
hot plate
cardboard
ice
shallow pan
glass tubing bent at right angle
glycerine
towel
scissors
washers
goggles

Procedure

1. Dissolve 18 g of table salt in a beaker containing 500 mL of water. Carefully taste the solution. **CAUTION:** *Be sure the glassware is clean.*
2. Put the solution into the flask. Place the flask on the hot plate. Do not turn on the hot plate.
3. Assemble the stopper, glass tubing, and rubber tubing as shown in Figure 11–2. To do this, rub a small amount of glycerine on both ends of the glass tubing. Hold the tubing with a towel, and gently slide it into the stopper and rubber tubing.
4. Insert the stopper into the flask. Make sure the glass tubing is above the surface of the solution.
5. Use the scissors to cut a small hole into the piece of cardboard. Insert the free end of the rubber tubing through the hole. Be sure to keep the tubing away from the hot plate. See Figure 11–2.
6. Place the cardboard over a clean beaker. Add several washers to the cardboard to hold it in place.
7. Set the beaker in a shallow pan filled with ice.
8. Turn on the hot plate. Bring the solution to a boil. Observe what happens in the flask and in the beaker.

FIGURE 11–2.

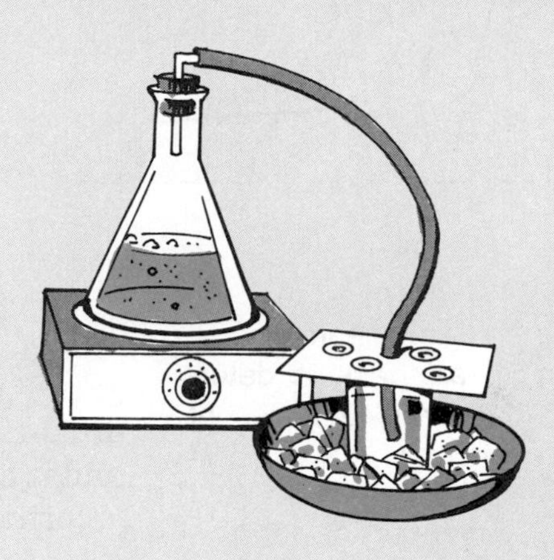

9. Continue boiling until the solution is almost, but not quite, boiled away.
10. Turn off the hot plate and let the water in the beaker cool.

Analysis

1. What happened to the water in the flask as you boiled the solution?
2. What happened inside the beaker? Explain your answer.
3. Taste the water in the beaker. Is it salty?
4. What remains in the flask?
5. Is the combined water in the flask and in the beaker the same volume you placed in the flask at the beginning of the investigation? Explain.
6. Examine the sides of the flask and describe what you see.

Conclusions and Applications

7. How might the desalination process be used to extract minerals from seawater?
8. How does desalination produce fresh water?

CAREER

Marine Maintainability Engineer

Sam Schultz is a marine maintainability engineer. His job is to analyze the engineering designs of proposed projects such as desalination plants, naval vessels, or any type of device for obtaining energy from seawater. He reviews engineering specifications and drawings during the development of these projects. Mr. Schultz also proposes design refinements to make structures more efficient. He designs tests that will be used for testing the quality of the design and the materials that are used.

To become a maintainability engineer for marine-related structures, Mr. Schultz needed to have mechanical ability and understand the complicated relationships of moving water. While working on his bachelors and masters degrees in engineering, he took a variety of courses including physics, chemistry, biology, geology, math, and engineering.

For career information, write:
National Society of Professional Engineers
2029 K. Street N.W.
Washington, DC 20006

GOALS

1. You will learn the difference between surface currents and density currents.
2. You will gain an understanding of the factors influencing currents, waves, and tides.

WATER IN MOTION

Ocean water is in constant motion due to currents, tides, and winds. These movements tend to distribute heat and to equalize temperatures. However, heat added at the equator prevents equilibrium from ever occurring.

11:3 Surface Currents

Some **surface currents** are caused by winds. They have circulation patterns similar to those in the atmosphere. For example, surface currents in the tropics are set in motion by the trade winds that drive the ocean water before them.

In the Atlantic, westward moving surface currents separate when they meet South America. One current flows to the northwest and becomes the Gulf Stream. The other surface current flows to the southwest and becomes the Brazilian Current. These currents carry heat to the ad-

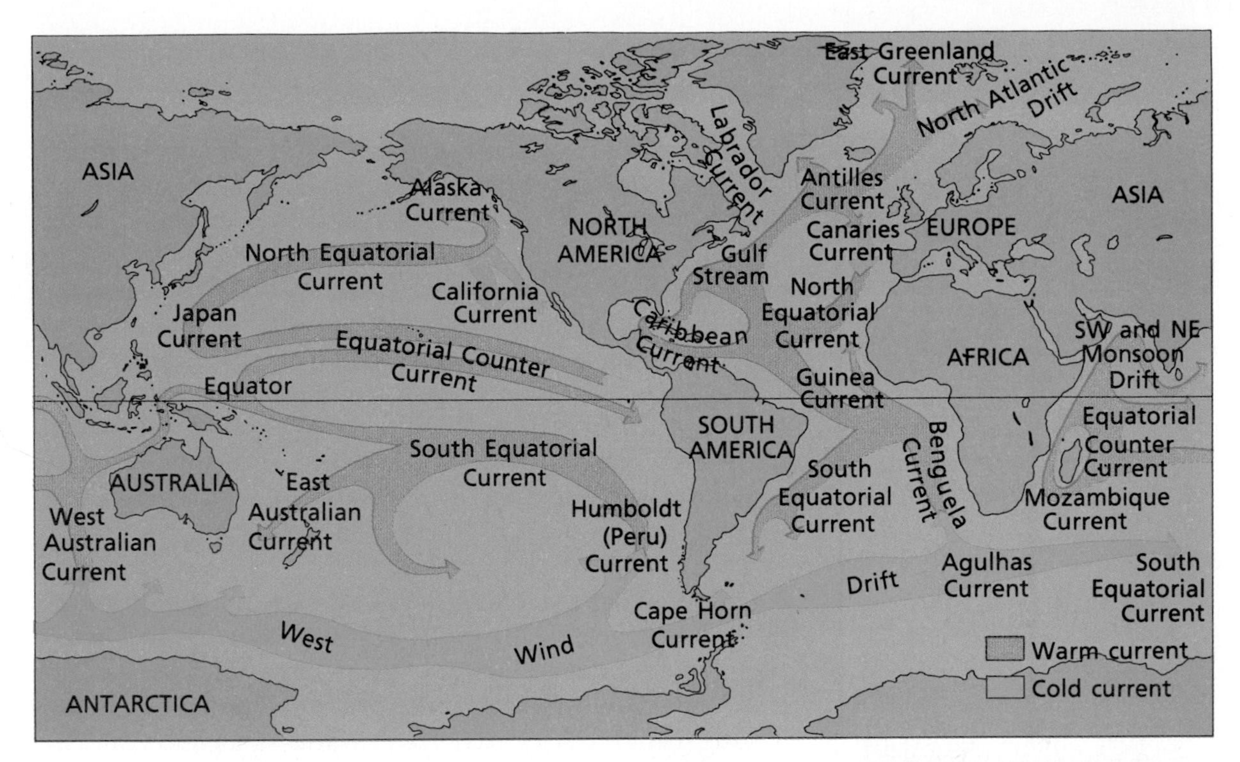

FIGURE 11–3. The patterns of ocean surface currents are determined by the force and direction of winds and by landmasses that act as barriers. Name the major cold current.

jacent land areas and toward the poles. Eastward moving currents are blocked by Europe in the north and Africa in the south, are turned toward the equator, and eventually move westward.

In the Pacific, westward moving currents are turned northward by Asia and southward by Australia. These currents then move eastward until they meet the Americas, where they are turned toward the equator.

In what direction do surface currents move in the Northern Hemisphere?

Most currents north of the equator move in a clockwise pattern. Most of the currents south of the equator move in a counterclockwise pattern. Recall from Chapter 9 that these movements are due to the Coriolis effect.

11:4 Density Currents

A current formed by the movement of more dense seawater toward an area of less dense seawater is a **density current.** Temperature and salinity affect the density of seawater. Thus, density currents are often called **thermohaline currents.** *Thermo* refers to temperature, and *haline* refers to salinity. Density may be less than average if seawater is diluted by melting ice in polar regions, heavy rainfall near the equator, or rivers as they empty fresh water into the ocean. Density may be greater than average when freezing or evaporation of surface water

How can the density of seawater decrease?

BIOGRAPHY

Alexander Von Humboldt
1769–1859

Alexander Von Humboldt is famous for having investigated variations of magnetic intensity, magnetic disturbances, earthquakes, the origin of tropical storms, and animal and plant distribution. One of his more notable discoveries was that of the Humboldt Current, also known as the Peru Current.

causes dissolved salts to be concentrated in a smaller volume of water. Also, cold seawater is denser than warm seawater.

All deep water circulates because of density currents. Deep ocean density currents move cold, dense water from the polar regions to less dense regions near the equator. Cold water at the poles sinks and moves as a mass through the ocean. These slow moving currents may remain away from the surface for thousands of years. Where wind-driven surface currents carry water away from an area, such as along continents, an upwelling of deep ocean water occurs. An **upwelling** of cold water carries high concentrations of nutrients to the surface, creating a highly productive fishing area. The coasts of Oregon and Washington and the coast of Peru are two areas of productive upwelling.

One local density current caused by evaporation occurs in the Mediterranean Sea. A submerged ridge lies across the Straits of Gibraltar. The ridge almost isolates the sea from the Atlantic Ocean. Because few rivers flow into the Mediterranean Sea, little fresh water is added. An enormous quantity of water is evaporated because of the hot, dry climate of the region. Without additions of fresh water, the sea becomes more and more saline. This very salty, dense water sinks and is pushed across the ridge and out into the Atlantic. This deep Atlantic current spreads westward as far as the Bahama Islands. To replace the deep, dense layer of water, less salty surface water from the Atlantic Ocean flows across the submerged ridge into the Mediterranean.

11:5 Waves

Waves are movements in which water alternately rises and falls. Ocean waves get their motion from winds, tides,

a

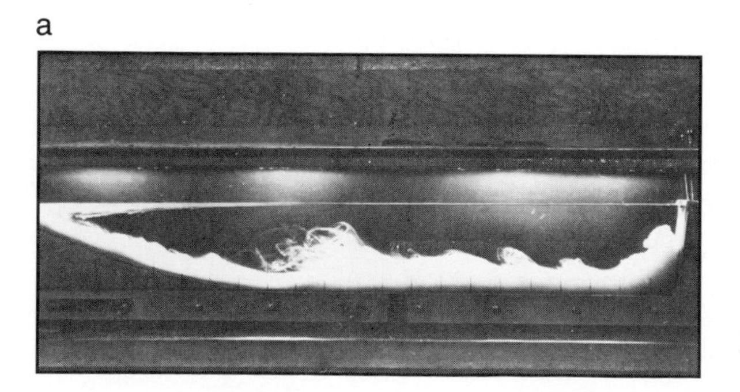

b

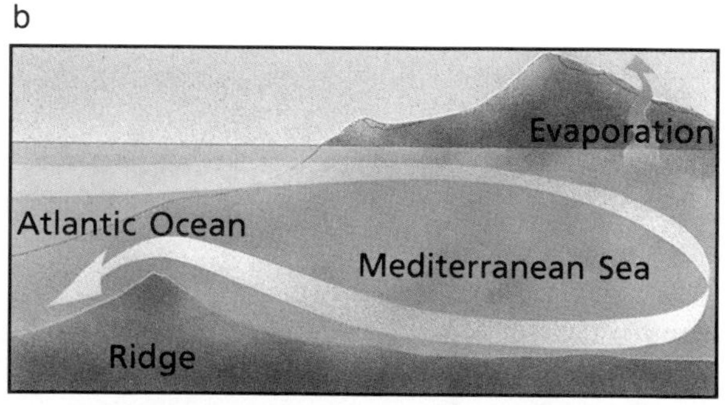

FIGURE 11–4. Density currents form when more dense seawater moves toward less dense seawater (a). A density current in the Mediterranean Sea pushes salty water into the Atlantic (b).

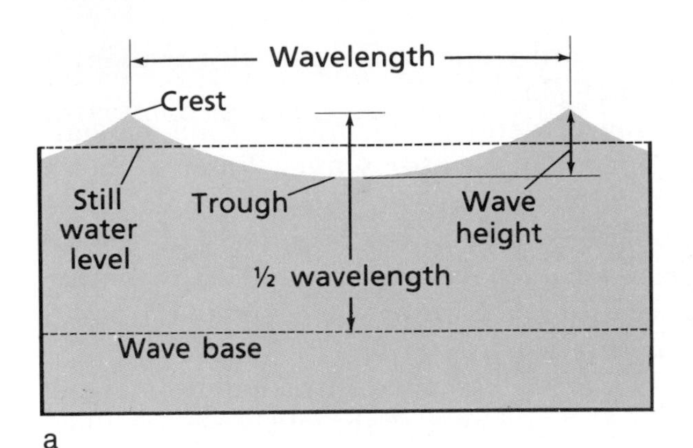

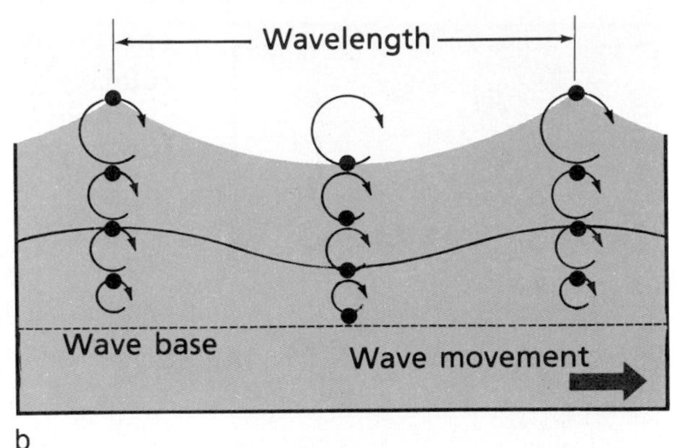

FIGURE 11–5. Characteristics of a wave (a) Particles of water move in a circle in deep water waves. Wave motion decreases with depth (b).

and earthquakes. Lakes also have waves, most of which are caused by wind.

Several terms are used to describe waves. **Crest** is the highest point of the wave. **Trough** is the lowest point of the wave. **Wave height** is the vertical distance between crest and trough. **Wavelength** is the horizontal distance between successive crests or troughs. **Wave period** is the time it takes two successive crests to pass a given point. **Wave base** is the depth of water equal to one-half the wavelength.

Define wavelength.

Although it appears that water is moving forward as a series of waves passes a given point, each particle of water in a wave actually moves around in a circle. The diameter of the circle for a particle of surface water is equal to the wave height. With depth, water particles move in smaller and smaller circles until all motion ceases at the wave base. Wave energy is passed forward and down as water particles move quickly from crest to trough, and then more slowly from trough to crest. Thus, wave

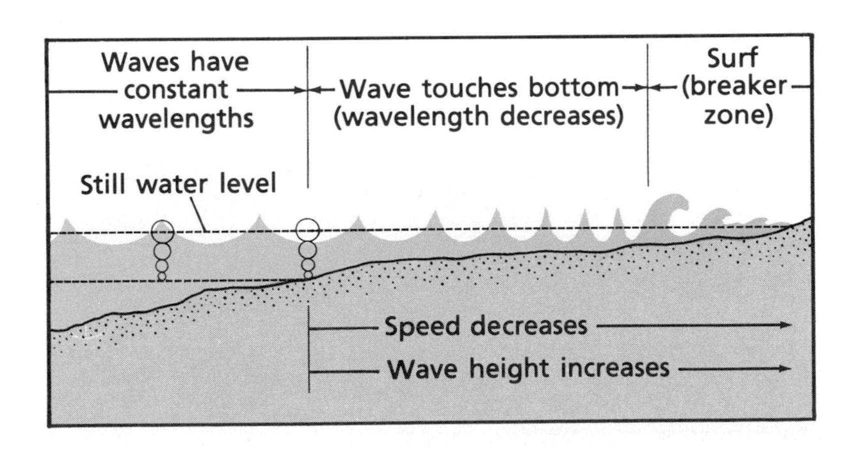

FIGURE 11–6. Wavelength decreases and wave height increases as waves move onto shore.

FIGURE 11–7. Surfers take advantage of breaking waves.

energy moves forward while water particles remain in relatively the same place.

A wave moving in water that is deeper than one-half its wavelength is a **deep water wave.** Water particles in deep water waves are free to complete their orbits without touching bottom. A wave in water that is shallower than one-half its wavelength is a **shallow water wave.** Particles in shallow water waves strike the ocean bottom, causing friction.

What is surf?

When deep water waves move into shallower areas, such as near shore, they change shape and become shallow water waves. The first line of deep water waves to strike bottom slows down. As it slows, its wavelength shortens, and its wave height increases. The same volume of water now is crowded into the shorter distance between crests. Waves continue to grow higher and higher as water depth decreases. The top of a wave moves faster than the bottom. The wave becomes lopsided and collapses or breaks. As the crest falls, all the wave energy is directed toward the bottom. Each wave, as it reaches this position, tumbles and strikes bottom. The zone where waves first fall is called the breaker zone. Between the first breaking wave and the shore, waves form and break many times. Breaking waves are called **surf.** After a wave reaches shore, a backwash toward the sea occurs.

FIGURE 11–8. Seismic sea waves are caused by a sudden shift of the ocean floor due to earthquakes (a). A seismic sea wave caused much damage to Seward, Alaska, in 1964 (b).

a

b

Wind waves form when wind blows across a body of water. The energy of the wind is passed to the surface of the water, forming waves. The waves increase in height as the wind continues to blow, which increases the surface area of the water. The more surface of the water that comes into contact with the wind, the higher the waves. Once formed, waves continue for long distances with or without the wind. But waves stop forming when the wind stops. The height of the waves depends on the velocity of the wind, the distance over which the wind blows, and the length of time the wind blows.

Seismic sea waves, or tsunamis (soo NAHM eez), are shallow water waves caused by a sudden shift of the ocean floor. The shift may be caused by an earthquake or a submarine volcanic eruption. A seismic sea wave has a very long wavelength and a deep wave base. Thus, each wave carries a huge amount of water into the shore zone. When a seismic sea wave approaches the shoreline, water is withdrawn from the shore zone. This low trough is followed by a towering wave crest. The seismic sea wave may be large enough to carry an ocean liner onto shore and leave it stranded when the water retreats.

F.Y.I. *Tsunami* is a Japanese word that means "wave in the harbor."

Why are seismic sea waves so destructive?

PROBLEM SOLVING

Testing the Water

Parul decided that her German shepherd puppy, Skippy, really needed a bath. She got a huge metal washtub and began filling it with water from the garden hose. She filled it 3/4 full. When she tested the temperature with her hand, she thought it was just too cold to put Skippy into. So, she decided to heat some water on the stove, and add that to the wash tub. She did this, and retested the water. To her surprise, the surface of the water was very hot, but the bottom 3/4 of the water was still quite cold. Why was all of the water in the tub not at the same warm temperature?

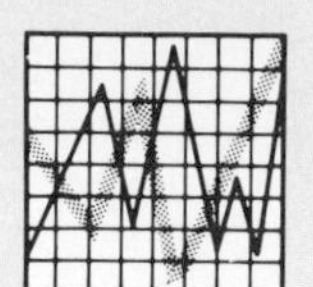

SKILL

Cause and Effect

Problem: How do you determine cause and effect relationships?

Background

One technique often used by scientists to research a topic is to test how one factor changes another. The condition that makes something change is called the cause. The resulting condition is the effect.

Materials  

white typing paper
clear plastic storage box
electric fan, three-speed
overhead light source
clock or watch with second hand
ring stand
water
metric ruler
protractor

Procedure

1. Place the light source on the ring stand. Position the plastic box on top of the paper under the ring stand.
2. Arrange the light source so that it shines directly on the box.
3. Pour water into the box to almost fill it.
4. Place the fan at one end of the box as shown in Figure 11–9. Start it on slow. **CAUTION:** *Do not allow any part of the fan or cord to come in contact with the water.*

FIGURE 11–9.

5. After three minutes, measure the height of the waves caused by the fan. Record your observations in a chart similar to the one shown. Observe the shadows of the waves on the typing paper through the plastic box.
6. After five minutes, measure the waves and record your observations in the chart.
7. Repeat Steps 4 to 6 with the fan on medium, and then with the fan on high.
8. Turn off the fan and observe what happens.

Data and Observations

Fan Speed	Time	Wave height	Observations
Low	3 min.		
	5 min.		
Medium	3 min.		
	5 min.		
High	3 min.		
	5 min.		

Questions

1. From your data sheet, is the wave height affected by the length of time that the wind blows? Explain.
2. Is the height of the waves affected by the force (velocity) of the wind? Explain.
3. How does an increase in fan speed affect the pattern of the shadows of waves on the typing paper?
4. What caused shadows to appear on the paper below the plastic storage box?
5. What was the effect when you turned off the fan?
6. What three factors cause the wave height to vary in the oceans?

11:6 Tides

Tides are shallow water waves caused by the gravitational attraction among Earth, moon, and sun. Due to gravitational and rotational forces, all particles of Earth, moon, and sun are attracted to one another. The moon's gravitational force causes a bulge of ocean water to face the moon. The higher rotational force of the Earth-moon system about the barycenter causes a second bulge to form on the side of Earth opposite the moon. Water to fill the bulges is drawn away from the area of ocean between the bulges. The bulges of water are called high tide, and the area between the bulges is called low tide.

What are tides?

As Earth rotates, different locations on Earth's surface pass through the high and low positions. Many coastal locations, such as the Atlantic Coast of the United States, experience two high tides and two low tides each day, which are termed **semidiurnal.** However, because ocean basins vary in size and shape, some coastal locations, such as many along the Gulf of Mexico, are **diurnal,** with one high and one low tide each day. Due to the motions of the moon and Earth in space, the time of high tide occurs at any given location about one hour later each day.

How do diurnal and semidiurnal tides differ?

The sun also has an effect on tides. Although the sun is much bigger than the moon, its tidal effect is less than the moon's because the sun is farther away from Earth. When the sun, moon, and Earth align, spring tides occur. These tides have nothing to do with the season. During **spring tides,** high tides are highest and low tides are

FIGURE 11–10. Spring tides occur when gravitational forces of the sun and moon are combined (a and b). Neap tides occur when moon, Earth, and sun are at right angles (c and d).

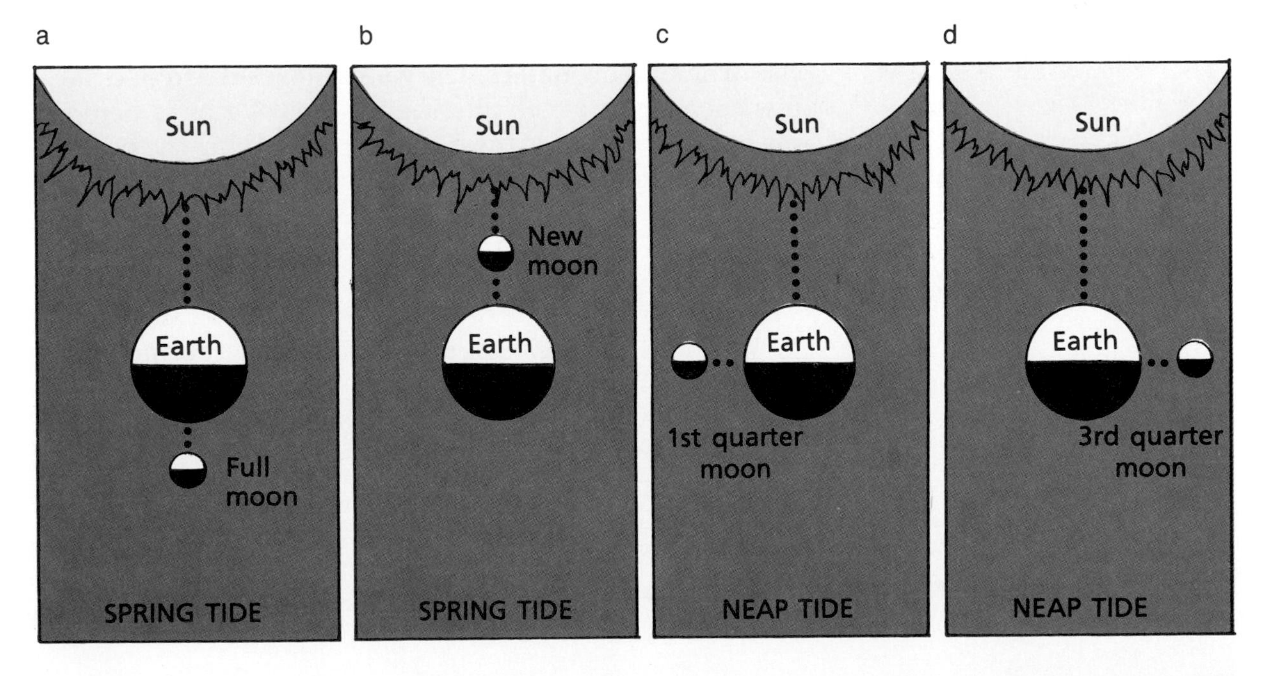

a

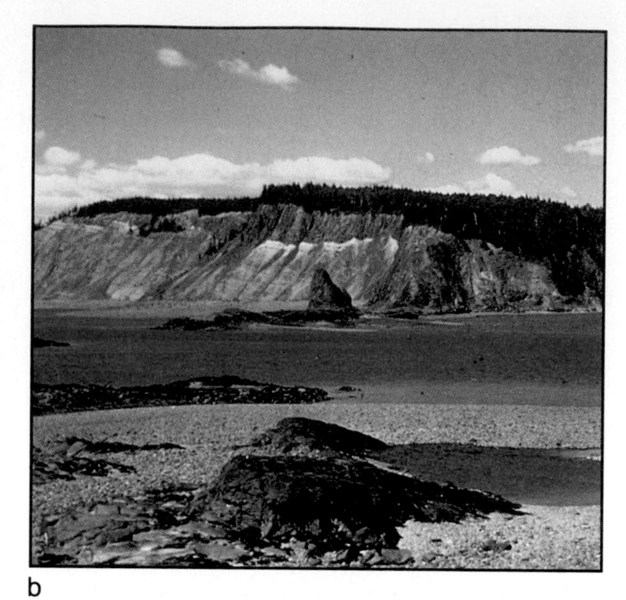

b

FIGURE 11–11. The difference between high tide (a) and low tide (b) in the Bay of Fundy, Nova Scotia, can be easily observed. What is this difference called?

lowest. Therefore, the **tidal range,** or difference between high tide and low tide, is greatest. Spring tides occur twice each month when the moon is in full or new positions.

When do neap tides occur?

Neap tides, or minimum tides, occur when the sun, Earth, and moon form a right angle. The gravitational forces of the moon and sun are then pulling in different directions, and the tidal range is at its lowest. Neap tides occur during the first and third quarter moon positions.

Along some coasts, tides scour deep, narrow channels. Tides speed up through these narrow openings, cutting the channels even deeper. On irregular coastlines, tides may be very high. Tides in the Bay of Fundy, Nova Scotia, rise over sixteen meters. The water moves through a narrow opening into a funnel-shaped bay. Water is held in the bay by the force of the incoming tide. Along the nearby shore, the tide rises only about seven meters.

Review

3. Why are density currents also called thermohaline currents?

4. What happens to wave energy as motion is passed downward in the water?

5. Diagram the positions of Earth, sun, and moon during spring tides and during neap tides.

★ **6.** Compare the circulation of ocean surface currents to the circulation of the atmosphere.

INVESTIGATION 11–2

Tidal Range

Problem: How does the moon affect tidal range?

Materials

graph paper red pencil blue pencil

Procedure

1. Copy the graph onto graph paper.
2. Plot each data point for high tide. Connect these data points with a red pencil.
3. Plot each data point for low tide. Connect these data points with a blue pencil.

Data and Observations

FIGURE 11–12.

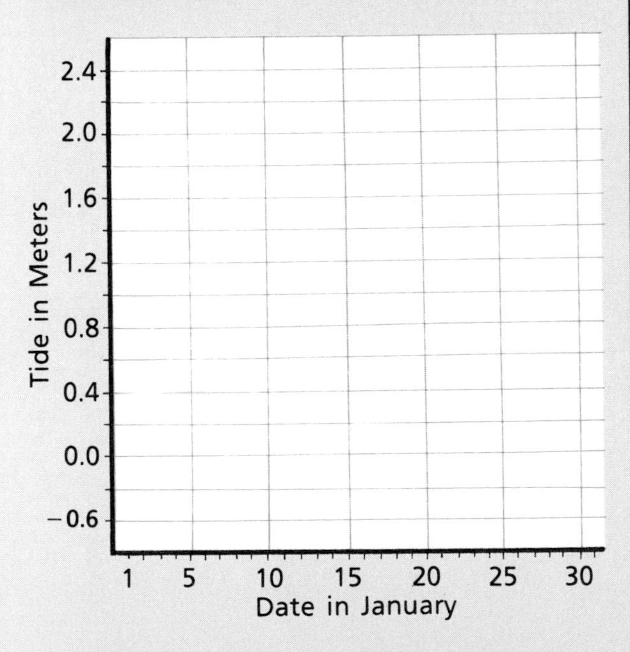

Date	Height of high tide (meters)	Height of low tide (meters)
1	1.4	0.5
2	1.5	0.4
3	1.7	0.2
4	1.8	−0.1
5	2.1	−0.3
6	2.2	−0.5
7	2.3	−0.6
8	2.3	−0.6
9	2.3	−0.6
10	2.1	−0.5
11	1.9	−0.2
12	1.6	−0.1
13	1.6	0.2
14	1.6	0.4
15	1.6	0.4
16	1.6	0.2
17	1.7	0.1
18	1.7	−0.1
19	1.8	−0.2
20	1.9	−0.2
21	1.9	−0.2
22	1.9	−0.2
23	1.9	−0.2
24	1.8	−0.2
25	1.7	−0.1
26	1.6	0.0
27	1.4	0.1
28	1.3	0.3
29	1.5	0.4
30	1.5	0.5
31	1.5	0.3

Data collected at San Diego, California, in January.
0 = average height of sea level without regard to tides.

Questions

1. What day had the lowest tidal range?
2. On what days would you suspect that the moon was new or full? Explain.
3. On what days would you suspect that the moon was in first or third quarter positions? Explain.
4. Did there seem to be any pattern to your graph? Describe any pattern observed.

TECHNOLOGY: APPLICATIONS

Tidal Energy

The tidal range in the Bay of Fundy, Nova Scotia, can be 16 meters, the greatest in the world. The power of the incoming tide there has been calculated to have more force than 8000 freight locomotives or 25 million horses. Engineers have proposed that a giant dam be constructed in order to tap this power. It would contain 128 turbines, and be capable of producing 4560 megawatts of power each year. The dam would cost $6 billion and take 10 years to build.

The power plant would produce energy by admitting the rising tide through open floodgates in the dam. When the water reached its highest level, the gates would close. As the tide reversed, the trapped water would flow through turbines, and electricity would be produced.

Using the tides for energy is not a new idea. Centuries ago people harnessed the tides with tidal mills, large water wheels. In 1966, the first large tidal power plant was built in France. It stretches across the River Rance and generates 500 megawatt-hours of electricity a year. Computers are used continuously to control the pitch of the turbine blades as the water level changes. It is designed to operate with the water flowing in at high tide, and out at low tide.

There are also tidal dams in operation in the Soviet Union and China, but none compare in size to the proposed power plant in the Bay of Fundy. At the Bay of Fundy, scientists are studying the possible ecological disturbances that would occur if a large tidal power plant were built. They fear that harnessing the tide may cause coastlines to flood, the fishing industry to suffer, and bird migrations to be disrupted. They also question what building a dam this large will do to tidal patterns as far south as Boston, Massachusetts, 400 kilometers away. Scientists fear that it would result in a higher tidal range at these southern locations. At high tide, the effects may include flooded farms, contaminated wells, and damage to salt-marshes. At low tide the channels would be impassable, and the piers would be useless.

A small-scale dam is already operating on one arm of the bay. Scientists are studying this to more accurately predict what might happen if a larger dam is built. If this larger dam is constructed, it will mean that electrical power can be produced at reasonable rates, thousands of jobs will be created, and new aquatic habitats will be created for some types of organisms. Also, the dam will provide flood and erosion control in Nova Scotia.

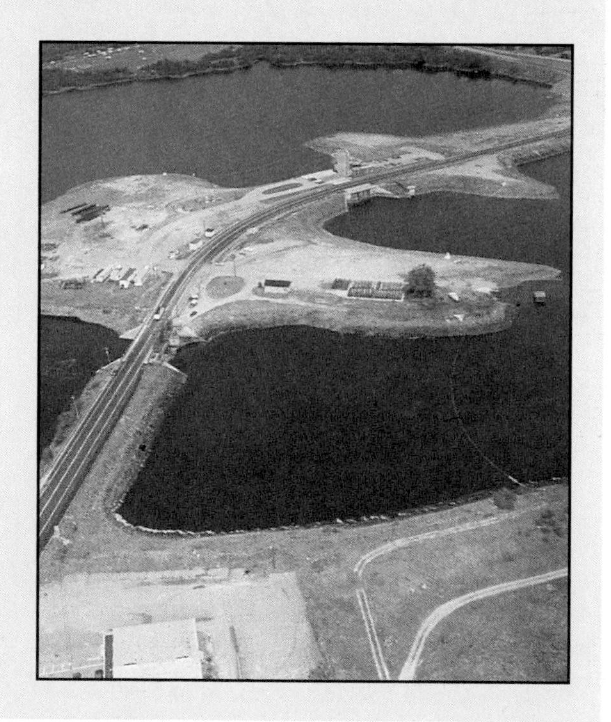

Chapter 11 Review

SUMMARY

1. Most scientists think the oceans were formed due to volcanic activity 4 billion years ago. 11:1
2. Dissolved substances enter the ocean with river water and with water vapor released by submarine volcanoes. 11:2
3. Fresh water is extracted from seawater by desalination. 11:2
4. Some surface currents are caused by winds, and thus have circulation patterns similar to the atmosphere. 11:3
5. Density currents may be caused by dilution, freezing, evaporation, or water temperature differences. 11:4
6. Water particles in a wave move in circular patterns. When waves enter shallow water, the wave height increases. 11:5
7. Tides are shallow water waves caused by the gravitational attraction among Earth, moon, and sun. 11:6
8. Spring tides have maximal tidal range. Neap tides have minimal tidal range. 11:6

VOCABULARY

a. crest
b. deep water wave
c. density current
d. desalination
e. diurnal
f. neap tides
g. salinity
h. seismic sea waves
i. semidiurnal
j. shallow water wave
k. spring tides
l. surf
m. surface currents
n. thermohaline currents
o. tidal range
p. tides
q. trough
r. upwelling
s. wave base
t. wave height
u. wavelength
v. wave period

Matching

Match each description with the correct vocabulary word from the list above. Some words will not be used.

1. waves caused by earthquakes or volcanic eruptions
2. method of extracting fresh water from seawater
3. measure of the amount of dissolved solids in seawater
4. currents caused by the wind
5. occur when the moon is full or new
6. tides with a minimal tidal range
7. the lowest point of a wave
8. depth at which all wave motion stops
9. distance between two crests of a wave
10. cold, nutrient-rich water brought up from the deep by surface currents moving away from a continent

Chapter 11 Review

MAIN IDEAS

A. Reviewing Concepts

Choose the word or phrase that correctly completes each of the following sentences.

1. A(n) *(density current, upwelling, surface current)* carries high concentrations of nutrients to the surface, creating a productive fishing area.
2. Many marine animals form bones and shells from *(potassium, manganese, calcium)*.
3. Tides are *(seismic waves, shallow water waves, deep water waves)*.
4. Tides that form when the sun, moon, and Earth align are *(high, neap, spring)*.
5. Wave height is the vertical distance measured from *(crest to crest, crest to trough, crest to wave base)*.
6. Wavelength is measured from *(crest to crest, crest to trough, crest to wave base)*.
7. Currents that move cold, dense water from the polar regions toward the equator are *(tidal, density, shallow water)* currents.
8. Most currents north of the equator move in a *(southern, clockwise, counterclockwise)* direction.
9. The ocean covers approximately *(90, 70, 50)* percent of Earth's surface.
10. Low *(temperature, salinity, density)* causes water to sink.
11. The density current in the Mediterranean Sea is caused by high *(rainfall, evaporation, amounts of mud and sand)*.
12. With the exception of seismic sea waves and tides, waves are caused by *(gravity, density differences, wind)*.
13. The *(Gulf Stream, California Current, Agulhas Current)* warms the land adjacent to it.
14. The probable source of water that formed the oceans is *(the atmosphere, volcanoes, tides)*.
15. The length of time it takes for two successive wave crests to pass a given point is the *(wavelength, wave base, period)*.

B. Understanding Concepts

Answer the following questions using complete sentences.

16. What is a seismic sea wave, or tsunami, and why and how can it be so destructive?
17. Explain why ocean water is low in silica and calcium.
18. How does evaporation affect circulation of the Mediterranean Sea?
19. Describe what happens to the shape of a deep water wave as it enters shallow water.
20. What area of Earth's surface is covered by oceans?
21. Discuss how salinity affects density currents.
22. Name three gases and seven dissolved substances found in ocean water.
23. Describe the effect of water temperature on water currents.
24. Describe one method of desalination.
25. Explain the difference between a deep water wave and a shallow water wave.

C. Applying Concepts

Answer the following questions using complete sentences.

26. Why is a wave that has a wavelength of 10 meters in water with a depth of 8 meters considered a deep water wave?
27. Why does the greatest tidal range occur during a full or new moon?
28. Why are upwellings economically important?
29. Why does the density of seawater vary?

30. Why do waves break in shallow water? Illustrate your answer in a diagram.

SKILL REVIEW

1. Describe what caused water to condense in the beaker you placed in the pan filled with ice in Investigation 11–1.
2. What is the effect of boiling salt water?
3. What is the effect on the density of seawater if salinity is increased?
4. A wave has a wavelength of 6 meters and is moving in water that is 3.5 meters deep. Can you conclude that the wave is a deep water wave? Explain.
5. You observe that near your motel on Miami Beach, Florida, there is a high tide at 12:01 A.M. and another at 12:51 P.M. Can you infer that Miami Beach's tides are diurnal? Explain.

PROJECTS

1. Design an experiment to test the density of water at different temperatures. Use the scientific method detailed in Chapter 1 as your model for an experimental design.
2. Design a way to desalinate water without using a hot plate.

READINGS

1. Bramwell, Martin. *Oceans*. New York: Franklin Watts, 1984.
2. Duxbury, A.C., and A. Duxbury. *An Introduction to the World's Oceans*. Reading, MA: Addison-Wesley, 1984.
3. Sandok, Cass R. *The World's Oceans*. New York: Franklin Watts, 1986.

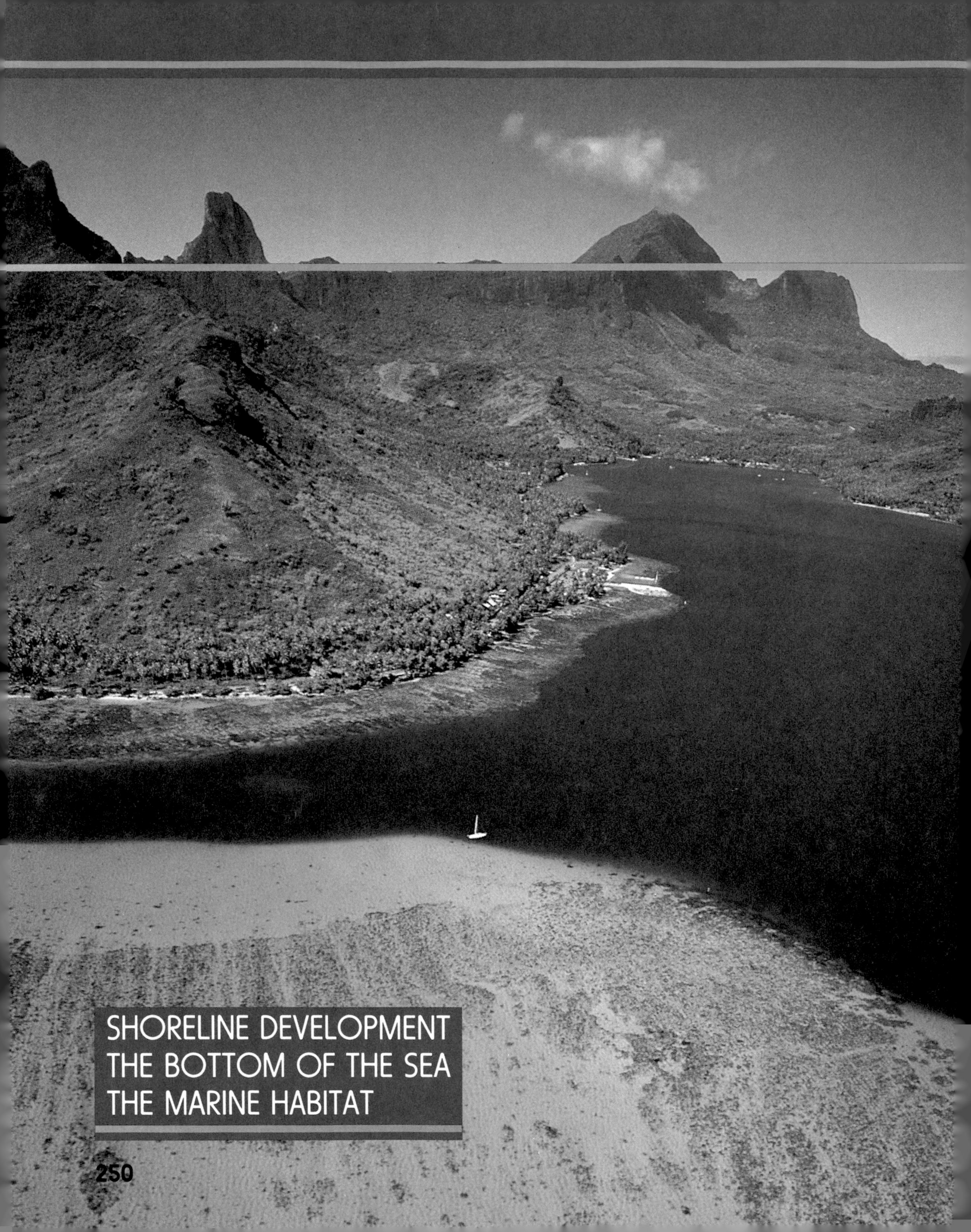

SHORELINE DEVELOPMENT
THE BOTTOM OF THE SEA
THE MARINE HABITAT

Oceanography

Did you know that the bottoms of the oceans are much more mountainous than any other place on Earth, or that seventy percent of the world's oxygen is produced by marine plants? The oceans are certainly unique and important on this planet we call Earth.

SHORELINE DEVELOPMENT

GOALS

1. You will learn how water shapes a shoreline.
2. You will learn about different types of shore deposits.

The shoreline is the boundary where land and sea meet. Shorelines mark the average position of sea level, which is the average height of the sea without consideration of tides and waves. During historical times, sea level has been fairly constant. Thus, sea level is considered to be zero elevation.

12:1 Shore Zones

The **shore zone** includes the area lying between high and low tides. The boundaries of the shore zone change often, sometimes from hour to hour, depending on the wind and tides. Near the shore, materials are in constant motion. These materials, called sediments, are moved back and forth by incoming and outgoing waters. Waves are cutting agents in the shore zone. Wave-produced features along the shore depend on the type of shore zone. Shore zones may be steep and rocky, or they may be sandy and gently sloping.

A steep, rocky shore develops a number of distinctive features. Storm waves scour a hollow or notch at the height of the wave action. The notch is above sea level and is easily seen after the storm subsides. A **stack** is an island of resistant rock left after weaker rock is worn away by waves and currents. Rock fragments broken from cliffs are ground up and become scouring tools. As rock particles roll back and forth, they cut a smooth, flat surface in the exposed rock just below sea level. This bench,

What is a stack?

a

b

FIGURE 12–1. Notches may be widened into sea caves when material is eroded (a). Rock fragments carried by waves may cut a marine terrace (b).

How do longshore currents form?

or marine terrace, may be worn inland until the water is too shallow for the waves to cut any further.

On most shores, waves rearrange the shore sediment. Waves approaching the beach at a slight angle create a current of water that flows parallel to the shore. This **longshore current** carries loose sediment almost like a river of sand. A longshore current that does not contain many sediments will pick up sand and erode beaches. The general smooth flow of the longshore current is often interrupted by rip currents. **Rip currents** are narrow currents that flow seaward at a right angle to the shoreline. They may be dangerous to swimmers who become caught in them.

12:2 Shore Deposits

All material eroded from one shore zone may be transported to and deposited in another shore zone. Beaches

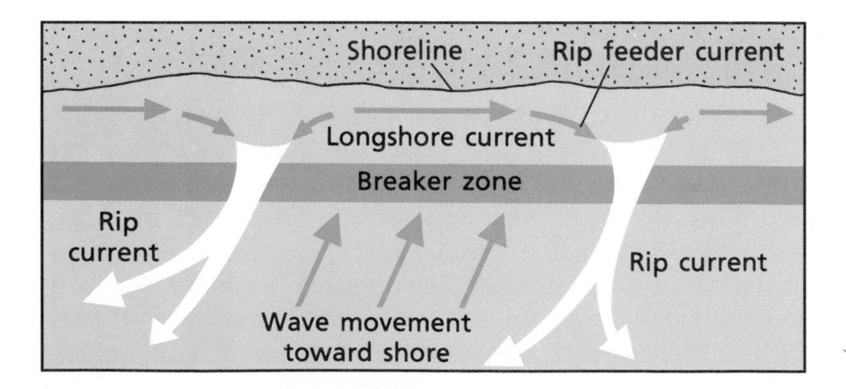

FIGURE 12–2. Rip currents return water moved by longshore currents to deeper water.

Table 12–1

Sand Composition	
Composition	**Characteristics**
quartz	clear and glassy; can be milky; often rounded
feldspar	tan, gray, pink, or salmon; dull luster
hornblende	dark green or black
mica	shiny, soft; breaks into thin sheets
magnetite	dark colored; attracted to magnet
garnet	pink or red; glassy luster
olivine	olive green; glassy luster
calcite	white; pearly luster; breaks into small cubelike pieces
rock fragments	any color; made of more than one substance
shell fragments	can include corals and other marine organisms

are deposits of loose material that parallel the shore. They extend seaward to between nine and forty meters below sea level, and extend landward to the coastline.

What is sand?

Beaches are made of a variety of materials. Some are made of pieces of rock, while others are made of broken shells. These fragments range in size from large cobbles to fine sands. Most beaches are made of fragments that are sand-size. Sand is a textural term that describes sediment with a diameter of 1/16 millimeter to two millimeters. Waves can break rocks and shells into sand-size

FIGURE 12–3. Waves along an irregular coast tend to erode headlands and deposit material in bays (a). Along this shore, waves strike parallel except as they enter the bay. The sandy beach is caused by deposition of sediment carried by waves (b).

a

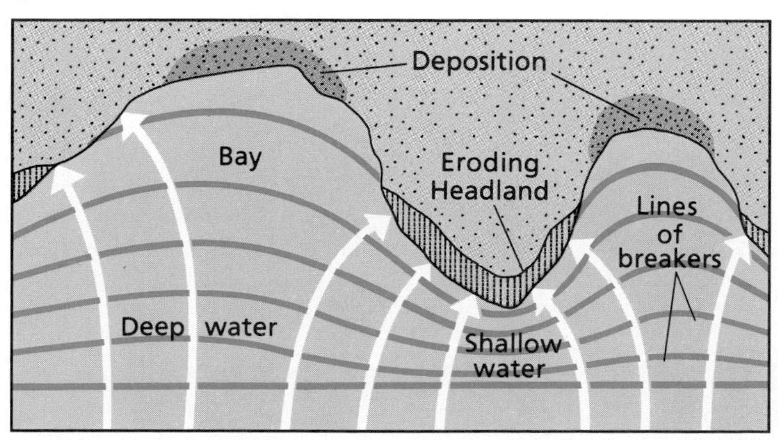

b

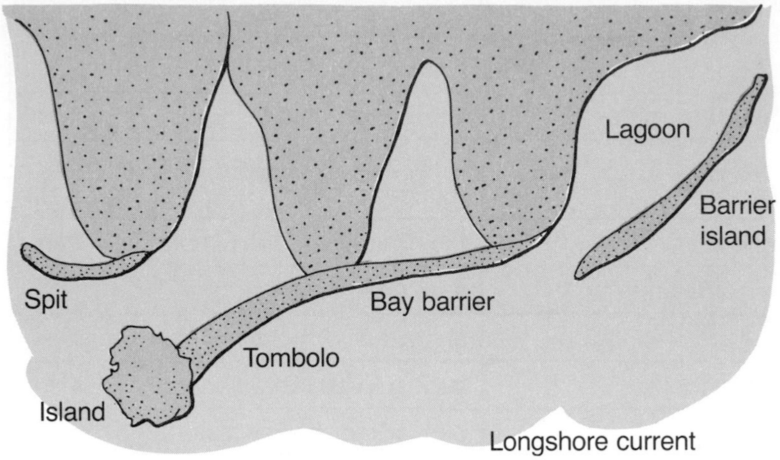

FIGURE 12–4. Beaches, tombolos, and bay barriers are some of the features that develop as sediments are eroded and redeposited along the shore.

Of what types of sand are most beaches composed?

particles, but cannot easily bump the sand grains together hard enough to break them into smaller fragments. Most beach sands are composed of grains of resistant material like quartz and orthoclase, but coral or basalt sands also are found in some locations. The black sands of Hawaii are made of basalt. The white sand beaches of Jamaica are made of mostly coral and shell fragments.

Sand is constantly being carried down the beach by longshore currents. The current may carry loose materials eroded from the headlands into the bay to form a beach along the bay shore. The longshore current may drop its load of sand as soon as it turns into the bay and loses velocity. The resulting deposit is called a **spit.** Eventually, a spit may be built across the mouth of the bay. When a bay is completely cut off, the deposit is known as a bay barrier. The bay then may become a lagoon, which, due to poor circulation, later fills with sediment

FIGURE 12–5. Mobile Bay is a large natural harbor protected by offshore spits and barrier islands. These sandy islands are used as recreation areas.

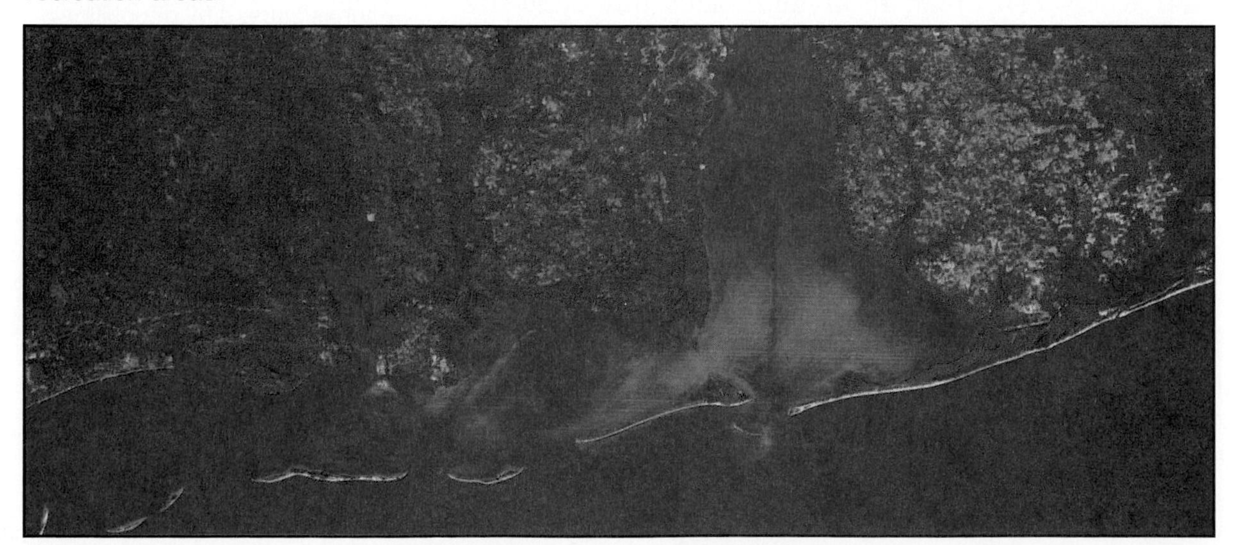

formed from debris and decomposed organic material. If sand is deposited in such a way as to connect an island to the mainland, a tombolo is formed.

Barrier islands are deposits of sand that parallel the shore but are separated from the mainland. The eastern shore of North America and the northern shore of the Gulf of Mexico have many of these islands. Barrier islands probably begin as underwater ridges of sand that are formed by breaking waves. Hurricanes and other storms may raise the ridges to sea level as large amounts of debris are dumped near shore. A decrease in sea level also may bring the deposit above the waves. Once exposed, the wind blows the loose sand into dunes, which keep the island above sea level. Gradually, beaches extend the barrier island shore zone toward the sea.

Review

1. What are the two types of shore zones?
2. Describe two features of steep, rocky shore zones.
3. Describe how a spit is formed.
4. How does a spit become a bay barrier?
★5. Why is sand absent from many of the world's beaches?

CAREER

Scuba Diver/Salvage Diver

Sally Schleder is both a scuba diver and a salvage diver. She performs shallow underwater jobs while wearing SCUBA (self-contained underwater breathing apparatus), mask, flippers, and insulated suit. In deeper water, she wears a more cumbersome pressure-resistant suit made of a magnesium alloy.

Ms. Schleder's duties as scuba diver and salvage diver are varied. She is hired by individuals, government agencies, and private industries to locate and recover submerged automobiles, boats, bodies, and valuable articles. Her duties also include repairing underwater installations such as docks or oil derricks.

When Ms. Schleder became interested in becoming a scuba diver, she took a certified SCUBA course. When she decided to become a salvage diver, she took additional technical courses.

For career information, write:
The Marine Technology Society
1730 M. St., N.W.
Washington, DC 20036

INVESTIGATION 12–1

Beach Sand

Problem: What are some characteristics of beach sand?

Materials

3 samples of different types of beach sand
stereomicroscope
magnet

Procedure

1. Use the stereomicroscope to examine the sand samples. Copy the Data and Observations chart and record your observations of each sample.
2. Describe the color of each sample.
3. Using Figure 12–6, describe the average roundness of the grains in each sample.
4. Place sand grains from one of your samples in the middle of the circle shown in Figure 12–7. Use the upper half of the circle for dark-colored particles, and the bottom half of the circle for light-colored particles. Determine the average size of the grains.
5. Repeat Steps 2–4 for the other two samples.
6. Pour a small amount of sand from one sample into your hand. Describe its texture as "smooth," "rough," or "sharp." Repeat for the other two samples.
7. Describe the luster of the grains as "shiny" or "dull."
8. Determine if a magnet will attract grains in any of the samples.
9. Use Table 12–1 to identify the types of fragments that make up your samples. Record the compositions in the chart.

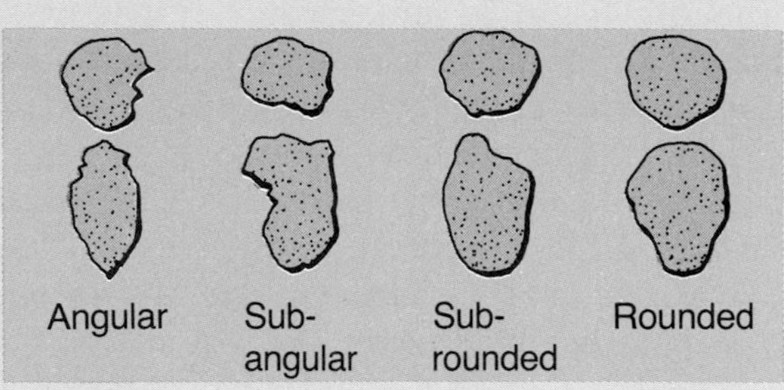

FIGURE 12–6.

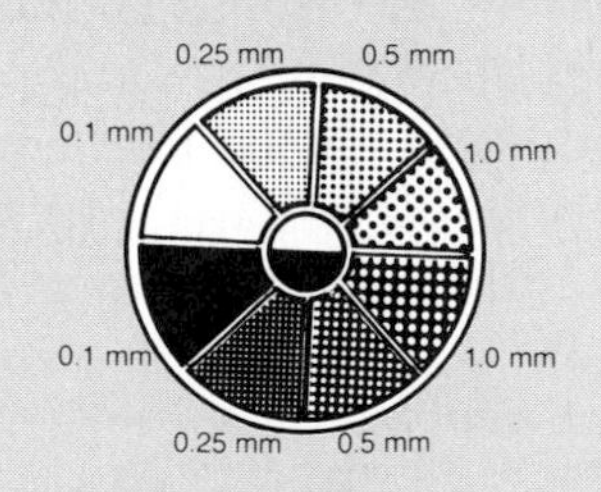

FIGURE 12–7.

Analysis

1. Were the grains of a particular sample generally the same size?
2. Were they generally the same shape?

Conclusions and Applications

3. What are some characteristics of beach sand?
4. Why are there differences in the characteristics of different sand samples?

Data and Observations

Sample	Colors	Roundness	Grain size	Texture	Luster	Composition
1						
2						
3						

THE BOTTOM OF THE SEA

GOALS

1. You will learn about the topography and the deposits of the ocean floor.
2. You will learn how topographic maps and map profiles of the ocean floor are made.

Earth's surface at the bottom of the sea is similar to the topography of the continents. There are mountains, valleys, and plains on the ocean floor. These features are much more pronounced than the equivalent continental landscapes. Sediments cover the floor almost everywhere.

12:3 Deep Sea Deposits

Deep ocean sediments contain shell fragments, volcanic ash, organic matter, and fine sediment. Some sediment carried to the sea by rivers reaches the abyss, the part of the ocean with depths of 2000 to 6000 meters.

What is ooze?

Ooze is a common deposit of the abyss. **Ooze** is sediment that contains at least 30 percent plant and/or animal shell fragments. Oozes containing calcium carbonate, a mineral commonly found in shells, are found only in bottom sediments at depths of 5000 meters or less. Below this depth, the water is so cold that the calcium carbonate dissolves before it can accumulate on the ocean floor. Between depths of 4000 and 6000 meters, ocean floor deposits include clay and oozes composed of silica. Below 6000 meters, deposits are mostly clay.

What are authigenic deposits?

Some sediments of the deep ocean floor form as the minerals dissolved in seawater crystallize. Deposits that form directly from seawater in the place where they are found are called **authigenic deposits.** Manganese nodules are authigenic deposits found over large areas of the ocean floor. These rounded nodules may form as minerals collect around a small nucleus. Manganese nodules are important deep ocean deposits because they contain high concentrations of elements that are important to many industries.

a

b

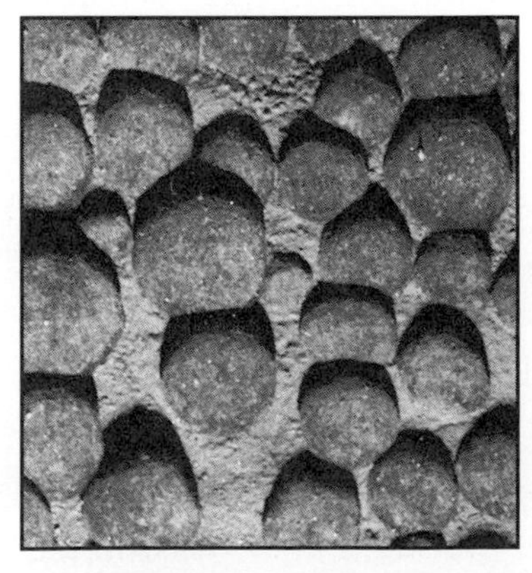

FIGURE 12–8. Ooze (a) is sediment composed of at least 30 percent organic fragments. Manganese nodules (b) are authigenic deposits that may be economically important.

12:4 Topography of the Ocean Floor

The coast is the border of land along the shoreline. The width of the coast may vary from several meters to several kilometers, depending on the relief of the land. The coast belongs to the continent, but it is linked to the sea by contact with ocean waves and currents.

What is the continental shelf?

The **continental shelf** is the relatively flat part of the continent that is covered by seawater. The shelf lies between the coast and the continental slope. The width of the shelf depends on changes in sea level. The shelf is usually narrow where mountains are close to shore. In areas such as the Atlantic and Gulf coastal zones, the shelf is wide and gently sloping.

The **continental slope** is the steeply dipping surface between the outer edge of the continental shelf and the ocean basin proper. As with the shelf and the coast, the slope too belongs to the continent. If you think of the ocean basin as a cake pan, the continental slope is the side of the pan.

F.Y.I. The abyssal plains are the flattest places on Earth. Abyssal plains are more extensive in the Atlantic than in the Pacific.

The **abyssal plain** is the flat, almost level area of the ocean basin. Turbidity currents are currents laden with mud and sand. These currents deposit sediments from continental shelves and slopes into the abyssal plain. Many volcanic peaks rise from the ocean basin plains. These peaks may be single mountains or mountain chains. A **seamount** is a volcano that does not rise above sea level. Volcanoes that do rise above sea level form islands. Some volcanic islands later become submerged due to sinking, or due to a rise in sea level. This type of structure, called a guyot, has a flat top formed by erosion during the time it was at sea level.

A **rift zone** is a system of cracks in Earth's crust through which molten material rises. One rift zone extends south from Iceland through the center of the Atlantic Ocean. Molten material that comes up from Earth's interior through rift zones becomes new crust. Earthquakes and volcanoes are common along such rift zones. Rifting also occurs in eastern Africa in the Great Rift Valley.

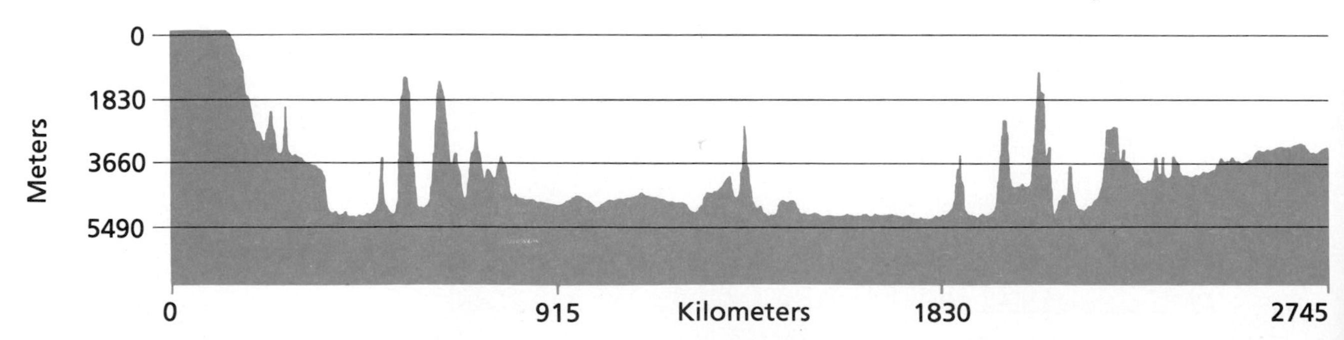

FIGURE 12–9. Using data obtained from echo soundings, oceanographers draw profiles of the ocean floor. This profile shows the floor of the Atlantic Ocean from New York to the northwest coast of Africa. What are the pointed structures below zero meters?

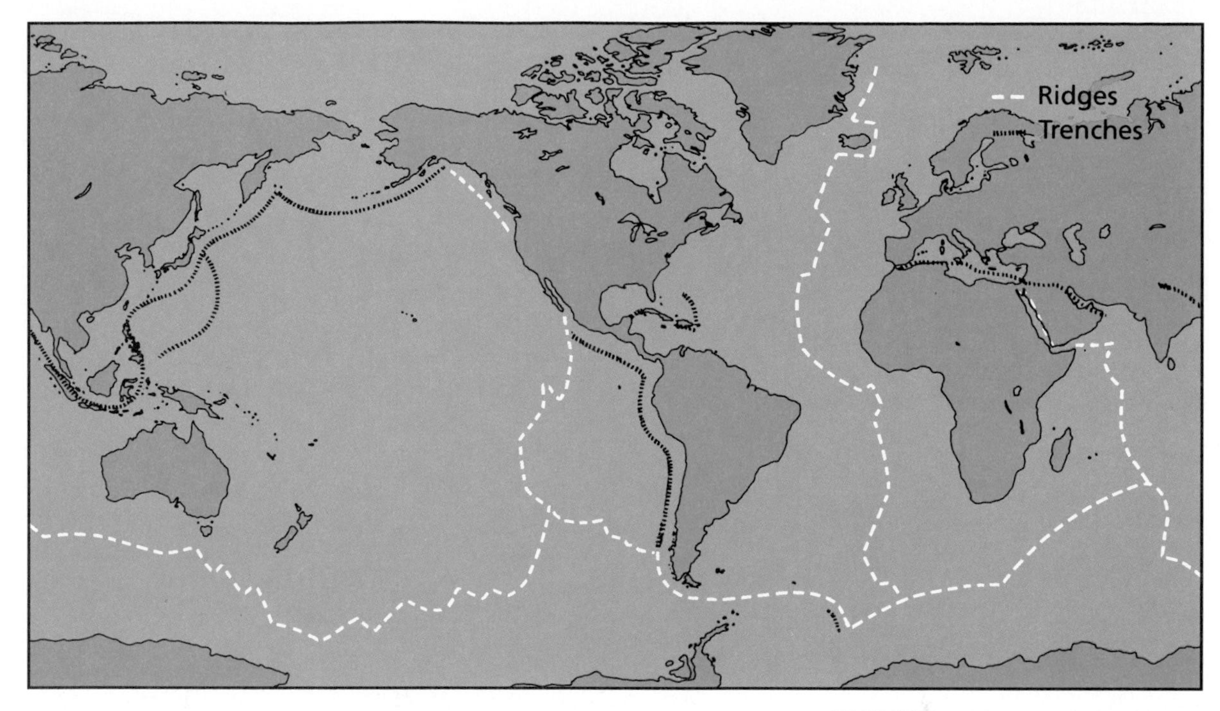

FIGURE 12–10. Mid-ocean ridges extend through the middle of the Atlantic and Indian Oceans. Trenches also are found on the ocean basin floor.

Forces within Earth move the crust apart at the rifts. These forces cause the crust to buckle, forming the **mid-ocean ridge,** an underwater mountain chain that rises from the ocean basins. Some of its peaks form islands, but most of the chain is submerged. The rift zone extending south from Iceland runs through the Mid-Atlantic Ridge. The Mid-Atlantic Ridge is 482 to 4830 kilometers wide, and is about 75 000 kilometers long. Figure 12–10 shows the extent of mid-ocean ridge systems.

What is a mid-ocean ridge?

An **oceanic trench** is a deep trough on the ocean basin floor where oceanic crust is being forced below continental or other oceanic crust. Many ocean trenches are longer and deeper than any on the continents. The Grand Canyon is about 1.6 kilometers deep. The Marianas Trench, near the Mariana Islands in the western Pacific Ocean, is at least 11 kilometers deep.

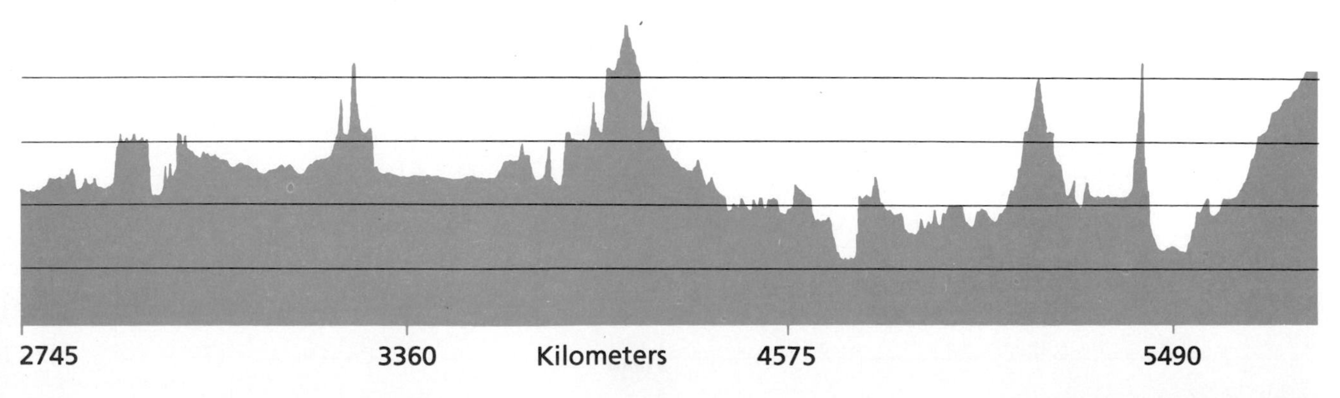

SKILL

Making Map Profiles

Problem: How can you make a map profile?

Background

Map profiles are used to show the side view of topographic features. Figure 12–9 is a map profile. You will learn to make a profile.

Materials

Figure 12–11
graph paper
metric ruler
paper and pencil

Procedure

1. Place the edge of a piece of paper along the X–Y line on the map (Figure 12–11).
2. Make a mark on the paper where each contour line touches the paper. Label each mark with the elevation of that contour line. You may need to refer to Investigation 3–1.
3. On your graph paper, label the vertical axis "Elevation in meters." Show a range of 0 to 70. The horizontal axis, "Distance in kilometers," should be the same length as line X–Y.
4. Place the paper with the labeled contour marks along the horizontal axis. Place a small dot on the graph directly above each mark at the proper elevation. Connect the dots with a smooth line. The line is a profile of the topography along line X–Y.

Questions

1. What is a map profile?
2. Why did your vertical axis have to begin at 0 meters?
3. Why did your horizontal axis have to be equal in length to the X–Y line?
4. How do you construct a profile from a topographic map?

FIGURE 12–11.

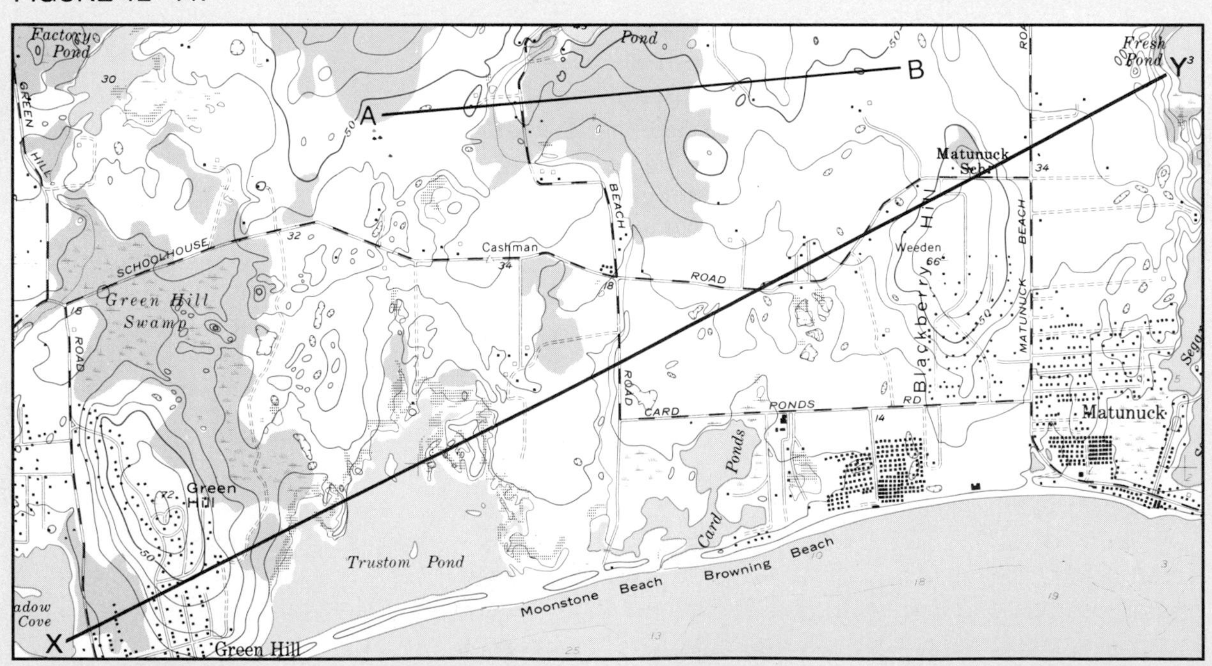

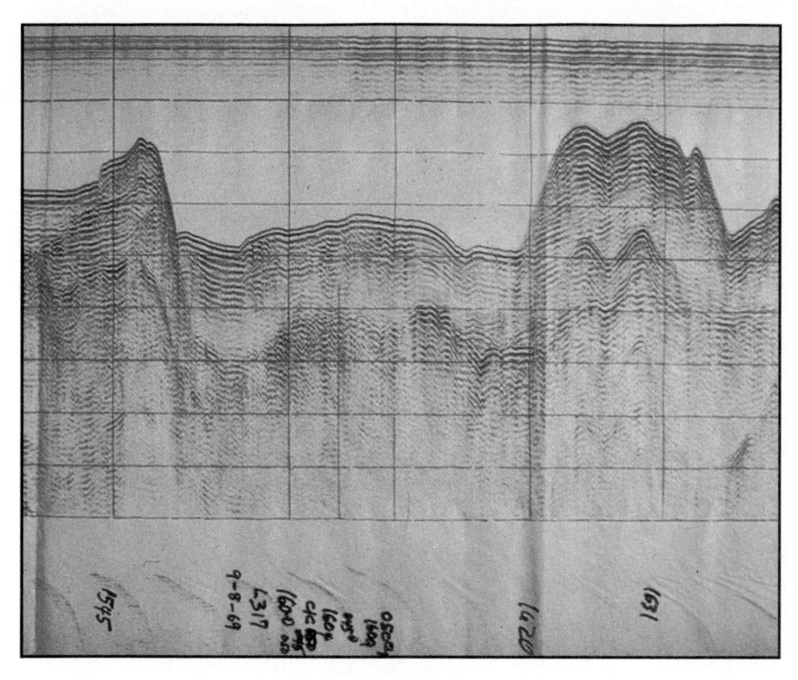

a

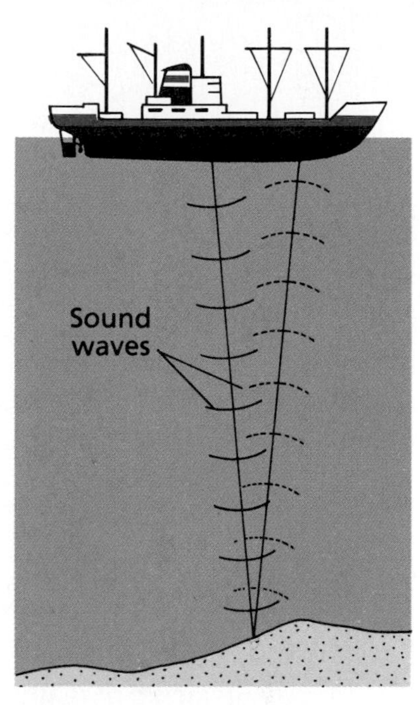

b

FIGURE 12–12. Sea-floor profiles (a) are produced when sound waves from the surface are reflected off the ocean bottom and received at the surface (b).

12:5 Mapping the Ocean Floor

Mapping the ocean has been done using methods such as echo sounding, sonar, radar, and seismic surveys. All of these methods depend upon a similar principle. Waves of some kind, for example, sound, are sent down from the ocean surface. The waves are reflected from the ocean bottom at the same angle that they are received. The time it takes the vibration to return to the surface plus the wave's velocity are recorded at the surface. The distance to the bottom can then be calculated using the equation $D = \frac{1}{2}t \times v$. In this equation, D is the depth of the ocean, t is the time for the vibrations to make a round trip, and v is the velocity of sound in water, 1500 meters per second.

What methods do oceanographers use to map the ocean floor?

Seismic surveys use vibrations similar to sound waves that originate from explosions. Seismic waves pass through rock layers beneath the ocean basins. Each layer of rock has a different density, which causes the waves to travel at different speeds. By matching the velocity of the waves to different rock types, the surveys show the kinds of rock that are present and how many layers there are. These studies have shown that the ocean floor is mainly basalt covered by very thin sediments.

F.Y.I. Magellan made the first attempt at accurately measuring the ocean bottom in 1521 by using a weighted line of known length.

A **gravimeter,** an instrument that can precisely determine variations in gravitational pull caused by topographic features, also is used to map features of the ocean floor. Massive underwater structures have a stronger

What is a gravimeter?

gravitational pull than smaller structures. By measuring the gravitational differences, detailed maps showing sea-floor features can be constructed.

Review

6. How do the continental shelf and slope differ?
7. How is a guyot formed?
8. What occurs at an oceanic trench?
9. What is an authigenic deposit?
★ 10. Calculate the ocean depth if sonar vibrations take 10 seconds to return to a ship.

TECHNOLOGY: ADVANCES

Mapping the Ocean Floor

A geophysicist at Columbia University Lamont-Doherty Geological Observatory developed a technique that uses satellite information to map the ocean floor. This technique, called geotectonic imagery, relies on the fact that gravitational force between ocean water and structures on the ocean floor is greater over more massive structures, causing water to pile up and form bulges.

SEASAT, a satellite launched in 1978, used a radar altimeter to compute the round trip travel time of pulses sent to the ocean surface. The distance to the ocean surface was calculated with an accuracy of five to ten centimeters. Scientists used these data and computers to make highly detailed color topographic maps of the ocean floor.

At the end of this decade, NASA plans to orbit a satellite called TOPEX (topographic experiment) that will be able to sense ocean-surface variations of less than 2.5 centimeters. These data will be used to show features that are as small as 13 kilometers across. As well as being useful for mapping and determining the history of the ocean floor, this information will help submarines navigate safely.

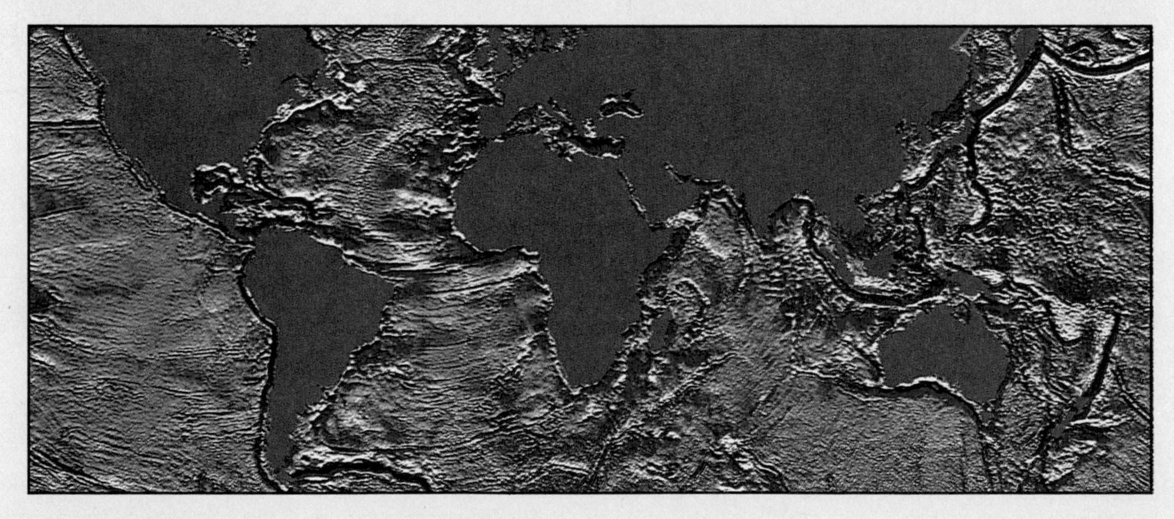

INVESTIGATION 12–2

Ocean-floor Profile

Problem: How can you make a profile of the ocean floor?

Materials

graph paper
colored pencils (blue and brown)

Procedure

1. Set up a graph as shown in Figure 12–13.
2. Examine the data listed in the Data Table. This information was collected at 29 oceanographic locations in the Atlantic Ocean. Each station was along the 39° north latitude line from New Jersey to Portugal.
3. Plot each data point listed in the table. Then connect the points with a smooth line.

Data Table

Station number	Distance from New Jersey (km)	Depth to ocean floor (m)
1	0	0
2	160	165
3	200	1800
4	500	3500
5	800	4600
6	1050	5450
7	1450	5100
8	1800	5300
9	2000	5600
10	2300	4750
11	2400	3500
12	2600	3100
13	3000	4300
14	3200	3900
15	3450	3400
16	3550	2100
17	3600	1330
18	3700	1275
19	3950	1000
20	4000	0
21	4100	1800
22	4350	3650
23	4500	5100
24	5000	5000
25	5300	4200
26	5450	1800
27	5500	920
28	5600	180
29	5650	0

FIGURE 12–13.

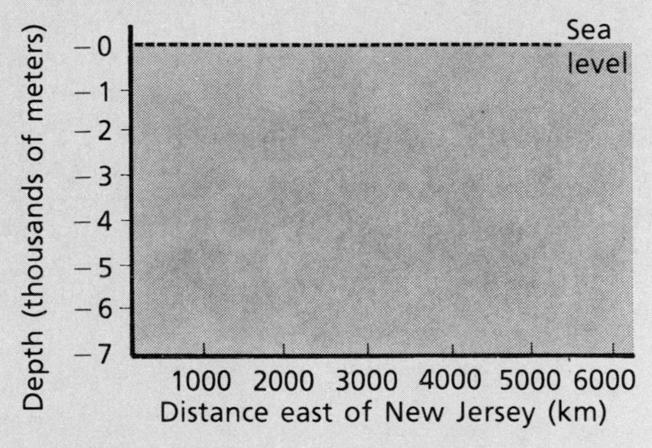

Questions

1. What ocean floor structures occur between 160 and 1050 km from the coast of New Jersey? Between 2000 and 4500 km? Between 5300 and 5600 km?
2. A profile shows a vertical slice through some feature of Earth's surface. In this case, you have constructed a profile of the ocean floor along the 39°N latitude line. If a profile is drawn to represent an accurate scale model of a feature, both the horizontal and vertical scales of the profile will be the same. What is the vertical scale of your profile? What is the horizontal scale?
3. Does the profile you have drawn give an accurate picture of the ocean floor? Explain.

GOALS

1. You will gain an understanding of the life processes in the ocean.
2. You will discover the wide-ranging effects of pollution on the ocean.

THE MARINE HABITAT

Organisms abound in the ocean. Many life forms simply drift with the surface currents while others swim through the water on their own power or crawl over the sediments. Still others are permanently attached to the ocean bottom. Many human activities pollute the oceans and change the habitats of these marine organisms.

12:6 Ocean Life

Drifting plants and animals are called plankton (PLANK tun). For the most part, plankton are microscopic. Objects that are microscopic are too small to be seen with the unaided eye.

All ocean plants live within 200 meters of the surface where there is sunlight. An important group of ocean plants is algae. One-celled, yellow-green algae called diatoms, compose some of the plant portion of plankton, or **phytoplankton.** Each diatom has a cell wall or shell of silica. Diatoms are the main source of food for many sea animals. Diatoms multiply rapidly when storms or upwellings bring large quantities of nutrients to the surface from deep water. Diatoms may cover the surface in these areas, attracting large numbers of marine animals.

Why are diatoms important?

Kelp, a large brown algae, lives in cool water near shore. Kelp may grow up to 0.6 meters per day and reach 60 meters in length. A holdfast organ anchors kelp to the ocean floor and gas-filled bladders keep the plant upright in the water. Kelp has over 300 commercial uses, including textiles, cosmetics, ice cream, and drugs. Kelp also can be made into methane fuel.

Marine animals are classified by their habits and by the depth of water in which they live. The animal portion

FIGURE 12–14. Diatoms provide the main source of food for many ocean organisms (a). Heavy surf can dislodge the holdfast organ of kelp and strand the algae on the beach (b).

a

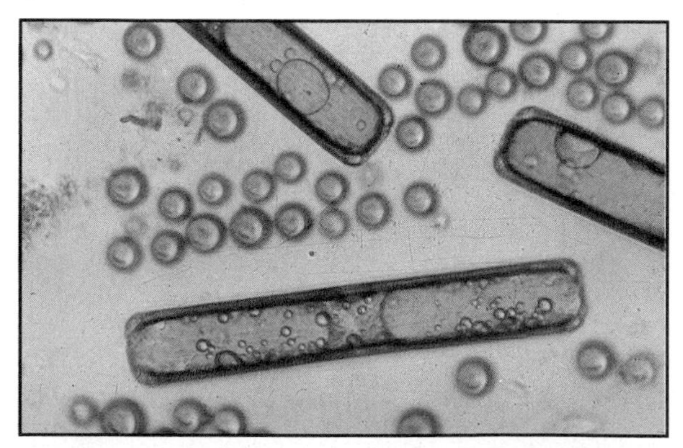

b

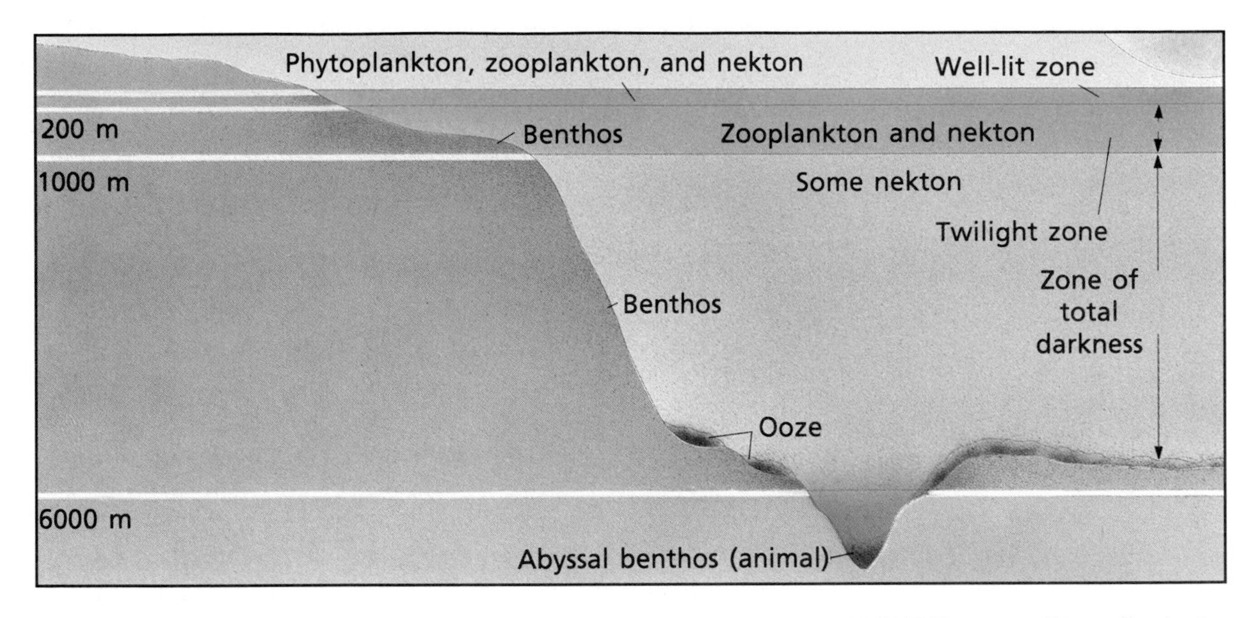

FIGURE 12–15. Plants live in the upper 200 m of the ocean. Few animals survive below 200 m.

of plankton, or **zooplankton,** is composed of eggs and immature stages of fish, lobsters, jellyfish, crabs, and other animals, as well as some adult forms. Most zooplankton depend on surface currents for movement, but some are capable of limited independent movement. These animals migrate to the surface at night to feed on phytoplankton, and then sink to greater depths during the day. Zooplankton can form a layer that is thick enough to reflect sonar. This layer, when present, is called the deep scattering layer.

Nekton (NEK tun) include all swimming forms of fish and other animals from tiny herring to huge whales. Nekton can move from one depth and place to another. Some nekton prefer cold water, others like warm regions. Some of them roam the entire ocean. Many nekton come to the surface only at night to feed on plankton. Others remain in deep parts of the ocean and must rely on special organs to attract other fish for food. The angler fish has a luminous lure that dangles over its forehead and attracts smaller fish. The viper fish has light-producing organs lining the inside of its mouth that guide unwary fish into its digestive tract.

Benthos (BEN thohs) are bottom dwellers. Many live in shallow water near the margins of islands and continents where sunlight reaches the seafloor. Benthic animals include coral, snails, starfish, and clams. Some benthic fish that live in the dark abyss eat partially decomposed matter that sinks to the ocean bottom. These fish have long feelerlike organs with which they detect food. Some benthic animals that are attached to the ocean

BIOGRAPHY

Eugenie Clark
1922–

Eugenie Clark became interested in a career in marine science when, as a child, she spent Saturdays at the old New York Aquarium. Clark's most noted research has been on the reproductive behaviors of fish, the behavior of sharks, and the fish of the Red Sea.

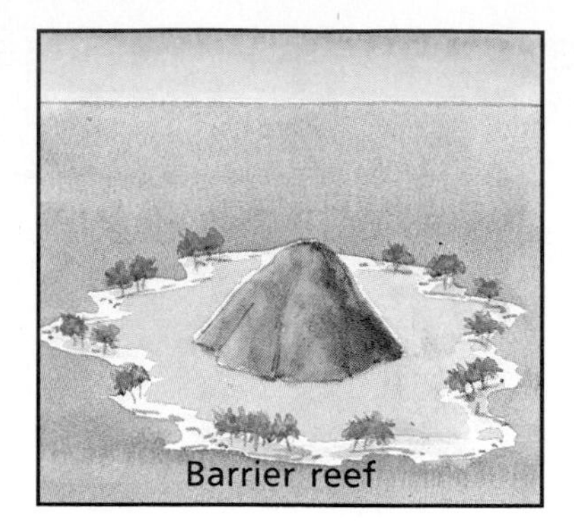

FIGURE 12–16. An atoll forms when the volcanic island within a fringing reef (a), sinks (b), leaving a circular reef surrounding a lagoon (c).

bottom filter food particles from currents of water. Other attached benthos have specialized organs that sting prey that come near them.

What conditions are necessary for a coral reef to form?

Corals are attached benthos that live in warm equatorial waters. Each coral animal builds a small calcite structure that looks something like a cell. Each structure is joined to several other structures to form a colony, or coral reef. Once the reef structure is begun, other organisms thrive near the reef and add to its growth. In order for a reef to develop, water temperature must be between 20° and 26°C. The water must be free of sediment and must circulate freely in order to bring food to the animals.

F.Y.I. Usually only the outermost layer of the reef that is exposed to the surf is actually living coral. The rest is made of skeletons of dead coral.

There are three main types of coral reefs. A **fringing reef** develops close to the shore of a continent or island. Little or no water separates a fringing reef from the shore. Fringing reefs are common around volcanic islands in the South Pacific. Many of these islands slowly sink below sea level. The reef, however, continues to grow upward through the water. A quiet body of shallow water, or lagoon, forms between the sinking island and the coral reef. This type of reef, separated from the landmass by water, is called a **barrier reef.** Sometimes the landmass sinks completely beneath the water, leaving a ring of coral surrounding a lagoon. This type of reef is an **atoll.**

FIGURE 12–17. Circular atolls can be seen on this satellite image.

FIGURE 12–18. There are advantages and disadvantages to life in the sea. Many animal species live in the shallow areas of the continental shelves. Why?

12:7 Life Processes in the Ocean

The ocean environment provides organisms with important gases, nutrients, food, and moisture. It carries away wastes and creates a medium for reproduction. Buoyancy (BOY un see), or upward lift of water, makes movement in seawater very easy. Little protection is needed against heat or cold because temperature change in seawater is small. Despite all the advantages of living in an ocean environment, many organisms do not live long. Most organisms are in constant danger of being eaten. For this reason, many animals produce thousands of eggs, but only a few successfully hatch and survive to a reproductive adult stage.

What do plants need to undergo photosynthesis?

Ocean plants can live only as deep as sunlight penetrates. They are much more common near shore, where nutrients from rivers are concentrated. During photosynthesis, plants use carbon dioxide, dissolved nutrients, and sunlight to produce sugar and starch. Oxygen, a by-product of this reaction, is required by animals to utilize the energy in the food they eat. Animals, in turn, produce water and carbon dioxide, which are then recycled by plants. Energy is transferred from plants to plant-eating animals to animal-eating animals through food chains. Thus, there are many more animal species in the shallow

a

b

FIGURE 12–19. Sulfur compounds escaping from a "black-smoker" deep sea vent (a) provide the energy for chemosynthesis. Tube worms and clams are two types of animals found near these vents (b).

areas of the continental shelves where plant life is more plentiful than in either the open or deep ocean.

Until recently, scientists thought that all food chains on Earth depended on the sun for energy. In 1977, totally new groups of animals were discovered living near hot springs in the deep, sunless water in rift zones on the mid-ocean ridge system. Large tube worms and giant clams were clustered around openings where hot water escaped from volcanic rocks and warmed the water. Dissolved sulfur compounds seeping from these openings were being used by bacteria to produce food and oxygen. This chemical process, called **chemosynthesis,** provided the energy and oxygen necessary for the survival of the community. Since 1977, many other such communities of animals have been found near rift zones.

12:8 Pollution and Its Effects

Pollution is the introduction of substances into an environment that produces a harmful change in the environment. Most pollution caused by humans is concentrated along coastal areas. In some locations, wastes are pumped directly into the ocean. Wastes from industry often contain concentrations of metals such as copper, lead, and mercury that are harmful to organisms. Solid wastes such as foam packing materials collect on beaches. Pesticides and herbicides used in farming often reach the ocean in runoff. They become concentrated in the tissues of marine organisms, resulting in reduced populations.

Why are pesticides a threat to marine organisms?

Power plants and other industries sometimes use water in cooling processes. The warm water that results is

FIGURE 12–20. Solid wastes litter many beaches.

pumped into the ocean, producing thermal pollution. Organisms adapted to cooler water are killed by this dumping of warm water.

Large amounts of sediments introduced into the ocean may also cause harmful changes. Human activities such as agriculture, deforestation, and construction can cause silt to accumulate in some coastal areas. Filter-feeding animals such as oysters and clams cannot survive in silt-filled water. Also, the filling of saltwater marshes for land development destroys many marine habitats.

PROBLEM SOLVING

Washed Ashore

Factory X and Factory Y are both located along Barney Beach. Factory X produces electricity by burning coal. Factory Y generates power using nuclear fission, a process that produces vast amounts of energy. Factory X is located about 20 kilometers north of Hometown. Factory Y is about 10 kilometers south of the city. Recently, many of the fish along the beaches in the area were washed ashore. Many of the organisms in the shallow waters have also died as a result of thermal pollution. What do you think is the source of the pollution? Explain your answer.

FIGURE 12–21. Oil companies sometimes use floating booms to keep oil spills from spreading.

How does oil get into the ocean?

Another pollution concern is oil spills in the ocean. Major spills have resulted from tanker collisions and breakups and from leaks at offshore oil wells. Small amounts of oil frequently are pumped out with a ship's waste water. Oil discarded from cars and industry has been dumped into streams and carried to the ocean. Some oil comes from natural seeps on land or on the ocean floor. One estimate suggests that about five million metric tons of oil are added to the oceans every year.

In 1977, the United States passed The Clean Water Act. This act deals with the problems of city and industrial wastes, and oil and hazardous spills. The Ocean Dumping Act controls dumping of wastes at sea and the establishment of marine sanctuaries. Although ships and industrial plants have been fined heavily when found guilty of polluting, much damage is done. Because all of the oceans are interconnected, some activities that pollute have wide-reaching effects. Therefore, pollution is an international issue of concern.

Review

11. What is zooplankton?
12. How do benthic animals get food?
13. What is photosynthesis?
14. List six human activities that pollute Earth's ocean.
★ **15.** Describe why kelp is not considered part of the phytoplankton.

SUMMARY

1. Wave action on steep, rocky shores produces notches, stacks, and marine terraces. 12:1
2. Shore deposits include beaches, spits, bay barriers, tombolos, and barrier islands. 12:2
3. Ooze contains at least 30 percent organic matter. 12:3
4. The principle topographic features of the ocean include the continental shelf, continental slope, abyssal plains, the mid-ocean ridge, trenches, seamounts, and guyots. 12:4
5. Mapping the seafloor is done with echo sounding, sonar, radar, and seismic surveys. 12:5
6. Marine animals are classified according to their habits and the depth of the water in which they live. 12:6
7. Coral reefs grow in shallow water in warm climates where water circulation provides an adequate food supply. 12:6
8. The marine environment provides organisms with important gases, nutrients, food, moisture, waste disposal, and buoyancy. 12:7
9. Carbon dioxide and oxygen are cycled among marine organisms. 12:7
10. Many human activities cause marine pollution. 12:8

VOCABULARY

a. abyssal plain
b. atoll
c. authigenic deposits
d. barrier islands
e. barrier reef
f. benthos
g. chemosynthesis
h. continental shelf
i. continental slope
j. fringing reef
k. gravimeter
l. longshore current
m. mid-ocean ridge
n. nekton
o. oceanic trench
p. ooze
q. phytoplankton
r. rift zone
s. rip currents
t. seamount
u. shore zone
v. spit
w. stack
x. zooplankton

Matching

Match each description with the correct vocabulary word from the list above. Some words will not be used.

1. microscopic, floating forms of marine plant life
2. manganese nodules are an example of this type of deposit
3. process during which energy and oxygen are produced by bacteria
4. carries loose sediment almost like a river of sand
5. block of resistant rock left along rocky shores
6. makes up the deep scattering layer
7. flat, level part of the ocean basin
8. area where most marine organisms are concentrated
9. underwater volcano
10. a sand deposit that curves into a bay

Chapter 12 Review

MAIN IDEAS

A. Reviewing Concepts

Choose the word or phrase that correctly completes each of the following sentences.

1. Atolls form around submerged *(stacks, volcanoes, spits)*.
2. Dangerous currents that may pull swimmers into deep water are *(longshore, rip, bottom)* currents.
3. The shore deposit that cuts off a river or bay is a *(bay barrier, spit, stack)*.
4. Barrier islands lie *(at right angles, parallel, attached)* to the shore.
5. Black sands of Hawaii are made of *(limestone, quartz, basalt)*.
6. *(Gravimeters, Satellites, Seismic surveys)* use vibrations to determine the topography of the seafloor.
7. The velocity of sound in water is *(1000, 1500, 2500)* meters per second.
8. Stacks, marine terraces, and notches are features of *(rocky shores, the deep ocean floor, sandy shores)*.
9. Coral, clams, and starfish are examples of *(phytoplankton, benthos, nekton)*.
10. Thermal pollution is caused by too much *(oil, solid waste, heat)*.
11. A deep trough on the ocean basin floor where oceanic crust is being forced below continental or other oceanic crust is a(n) *(rift zone, mid-ocean ridge, oceanic trench)*.
12. Sand is a textural term that describes sediments with a diameter of 1/16 millimeter to *(2, 6, 10)* millimeters.
13. Most beaches are composed of *(quartz, coral, basalt)* sand.
14. *(Oozes, Clays, Authigenic deposits)* are composed of at least 30 percent organic fragments.
15. *(Oceanic trenches, Mid-ocean ridges, Abyssal plains)* extend south from Iceland through the middle of the Atlantic Ocean.

B. Understanding Concepts

Answer the following questions using complete sentences.

16. Why is pollution an international problem?
17. Why are there more organisms in shallow water near continents than in other parts of the oceans?
18. Describe an ocean food chain.
19. Explain why kelp is economically important.
20. Describe the gravimeter method of mapping sea-floor topography.
21. Explain why sea-floor deposits below 5000 meters do not contain calcium carbonate.
22. Explain why manganese nodules may be the most economically important deep ocean sediment.
23. Explain why longshore currents are important in shaping a beach.
24. What is a tombolo?
25. Describe a continental slope.

C. Applying Concepts

Answer the following questions using complete sentences.

26. How are scientists currently mapping the ocean floor?
27. Why is the rift zone not filled up if molten material is constantly rising?
28. Why does the shore zone need to be remapped frequently?
29. What happens to the carbon dioxide in solution in ocean water? Why is there more carbon dioxide in deep water than in shallow water near the shore?

30. Would you expect to find coral reefs growing around the volcanoes off the coast of Alaska? Explain your answer.

SKILL REVIEW

1. Compare and contrast the topography of the ocean floor with the topography of Earth's land surface.
2. If you wanted to make a profile map to show the topographic features along line A–B on Figure 12–11, how many centimeters long would you need to make the x-axis of your graph?
3. How do you label the vertical and horizontal axes of a map profile?
4. What is the effect on organisms when saltwater marshes are filled with dirt by land developers?
5. Compare and contrast photosynthesis and chemosynthesis.

PROJECTS

1. Use a stream table and wooden blocks to create and observe the effects of longshore currents on a coastline.
2. Design an experiment that would clean up an oil spill from a beach. Keep in mind that detergents may cause additional harm to the environment.

READINGS

1. Blair, Carvel. *Exploring the Sea: Oceanography Today*. New York: Random, 1986.
2. Parker, Henry S. *Exploring the Oceans: An Introduction for the Traveler and Amateur Naturalist*. Englewood Cliffs, NJ: Prentice-Hall, 1985.
3. Yulsman, Tom. "Mapping the Seafloor." *Science Digest*. May, 1985, p. 32.

SCIENCE AND SOCIETY

OZONE DEPLETION

Although ozone (O_3) is only one of the seven gases that make up 0.0001 percent of Earth's atmosphere, it is very important to Earth. The ozone layer in the stratosphere protects Earth from harmful solar radiation. If this layer is destroyed, it could lead to an increase in skin cancer, a loss of crops, and perhaps an overall global warming. Global warming could lead to a rise in sea level, causing major flooding of coastal cities, among other problems.

Background

In 1974, two chemists in California, Sherwood Rowland and Mario Molina, found through the use of a computer model that gases called chlorofluorocarbons (CFCs) were rising into the stratosphere and causing the unstable ozone molecules to break apart. Although these gases are chemically inactive in the troposphere, once they enter the stratosphere they can be broken down by ultraviolet light in a reaction called photolysis. This reaction releases the chlorine atoms that bond with the unstable ozone making a new product and releasing oxygen. Rowland and Molina estimated that if these gases continued to be released at the 1972 rate, between seven and 13 percent of the ozone would be destroyed in 100 years—roughly double the normal rate.

This was disturbing news. In 1975, the President's Council on Environmental Quality created a task force to look into the matter. The task force decided that unless new information revealed no problem, the manufacture and use of these chemicals would need to be controlled. The CFC industry was furious. CFCs were the main propellant in spray or aerosol cans. CFCs were also important as refrigerants in cooling and heating equipment and as the foaming material in substances such as polyurethane and Styrofoam®. The industry spokespeople took out double-page advertisements in newspapers to say that ozone depletion was only a hypothesis based on a computer model and that there were no measurements to prove anything was wrong at all.

In September of 1976, The National Academy of Sciences released the first of several reports. The report said that The Academy believed that Rowland and Molina's hypothesis was basically correct. The ozone layer was being destroyed and if this

FIGURE 1. This satellite image clearly shows the extent of the hole in Earth's ozone layer.

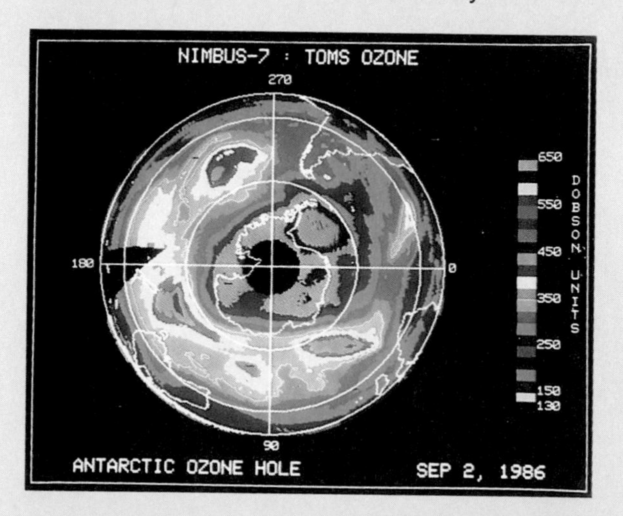

continued, more ultraviolet radiation would reach Earth. This increase could cause an increase in skin cancer and could also cause global warming.

The Environmental Protection Agency (EPA) suggested a ban on the use of CFCs in spray cans by 1979. This action would help the situation, but 50 percent of the chemical was used in the United States in plastic foams or for cooling.

Case Studies

1. By 1978, a ban on the use of CFCs in spray cans was enacted in the United States. In September of 1979, The National Academy of Sciences released another report. It said that a loss of ozone would lead to food shortages as cropland might dry up. The report urged a worldwide ban on aerosol or spray cans because other uses of CFCs had already increased. Something had to be done. This report was one of the first to say that the "wait and see" approach taken by both government and industry was dangerous.

2. At a meeting in Oslo, Norway, in 1980, most countries agreed that something had to be done. The CFC industry started a lobby to talk to people in government. The Alliance for Responsible CFC Policy lobbied successfully when the EPA tried to have Congress impose controls on certain uses of CFCs.

3. An event in 1985 changed everything. British scientists reported the existence of an ozone "hole" appearing every October over Antarctica. The British findings were confirmed by satellite monitoring. Indeed, there does seem to be a hole in the ozone layer. One possible explanation for the hole is that the CFCs have not been exposed to any light during the long Antarctic winter. Once the light reappears, the chemical reactions begin again and the ozone is broken down.

Developing a Viewpoint

1. Worldwide, more and more people are becoming conscious of the need to reduce the levels of CFCs in the atmosphere. Industries, however, are complaining that current proposals to limit their use "go too far too fast." Why are these people concerned? How might jobs be affected?

2. Not all CFCs are dangerous, but the most useful types are. There are some options for producing safe CFCs that are not toxic or flammable. Consumers would need to decide if it is worth paying the increased costs of environmentally safe goods. Suppose you are given a choice between two cars with air conditioning systems. Car A has a system that uses safe CFCs but its cost is several hundred dollars more than car B. Would you be willing to buy car A? Explain your answer.

Suggested Readings

Mark Crawford. "EPA to Cut U.S. CFC Production to Protect Ozone in Stratosphere." *Science*. December 11, 1987.

Richard A. Kerr. "Winds, Pollutants Drive Ozone Hole." *Science,* Vol. 238. October 9, 1987.

Michael D. Lemonick. "Culprits of the Stratosphere." *Time*. September 21, 1987.

Gary Taubes. "Made in the Shade? No Way." *Discover*. August, 1987.

S. Weisburd. "Ozone hole at Southern Pole." *Science News*. March 1, 1986.

UNIT 4

Water, wind, and ice sculpt Earth's surface. These forces work by themselves or in combination to create Earth's widely varying landscapes. These tufa towers are composed of calcium carbonate and are the result of evaporation at Earth's surface. What are some specific natural processes that change Earth's surface?

10 000 Y.A. | 1500 | 1600 | 1700

~10 000 Y.A.
Glaciers cover much of North America.

1535
St. Lawrence River discovered.

1687
Newton develops concept of gravity.

SURFACE PROCESSES

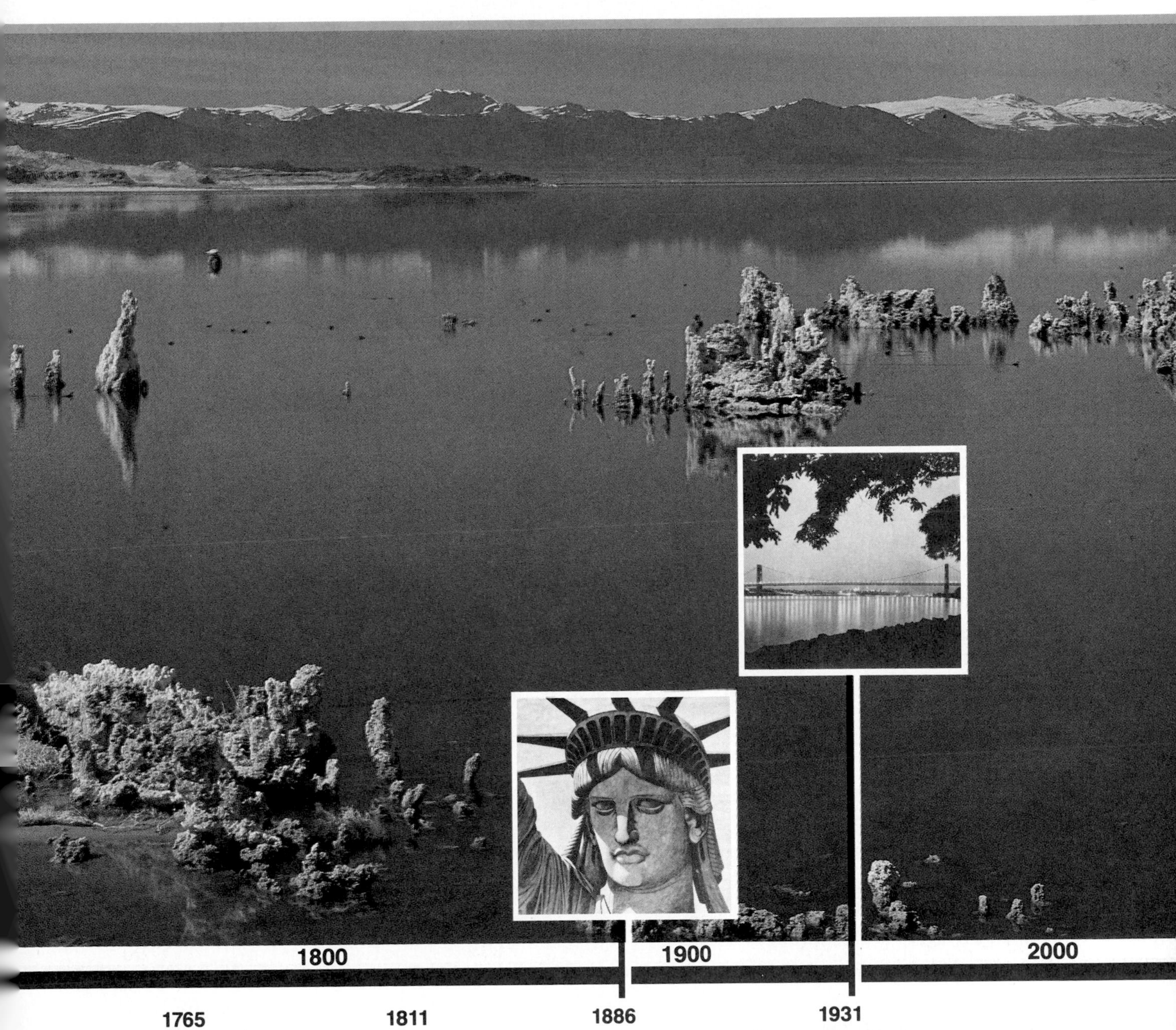

1765
Stamp Act
Congress draws up
rights and liberties.

1811
John Stevens
invents steam-
powered ferryboat.

1886
Statue of Liberty
is dedicated.

1931
George Washington
Bridge opens.

CHANGES AT EARTH'S SURFACE
SOILS
WIND

Weathering and Erosion

Just as the paint on a house is affected by nature's elements, so too is nearly every object on Earth. Weathering and erosion alter the composition or form of many different objects. Even Earth's mountains are changed and broken into smaller fragments by forces and substances.

CHANGES AT EARTH'S SURFACE

Rocks are changed at or near Earth's surface and the resulting fragments are moved to new locations. The speed with which these changes take place depends on environmental conditions.

GOALS

1. You will gain an understanding of weathering and erosion.
2. You will learn about different types of mass movements.

13:1 Weathering

Changes that rocks undergo at or near Earth's surface are called **weathering.** Weathering includes disintegration (dis ihnt uh GRAY shun) and decomposition (dee kahm puh SIHSH un). **Disintegration,** or physical weathering, is the mechanical processes that break large masses of rock into smaller fragments. **Decomposition,** or chemical weathering, is the process that forms new substances from minerals in the rock. Air, water, and substances dissolved in water react with minerals in rocks. In time, disintegration and decomposition together break down even the most resistant rocks. The rate of weathering depends on the composition of the rock, the climate, and the texture of the rock.

Igneous rocks form deep within Earth and are unstable at the surface. The first minerals to crystallize from magma decompose fastest when attacked by weathering. Thus, iron-magnesium minerals and calcium-rich feldspar tend to decompose first. Minerals such as orthoclase or potassium feldspar crystallize at lower temperatures and

F.Y.I. For more information on igneous, metamorphic, and sedimentary rocks, see Chapters 17 and 18.

a

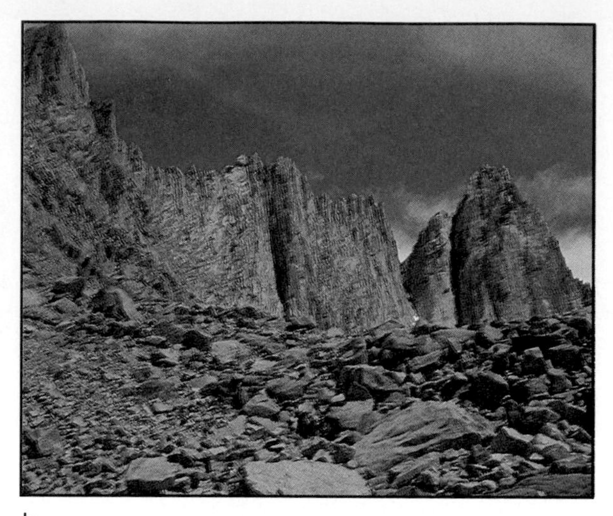

b

FIGURE 13–1. Little chemical weathering occurs in this tundra region in the Yukon Territory because of low temperatures and lack of moisture (a). Water freezing and thawing in cracks in rocks is an important factor in disintegration (b). What is another cause of disintegration?

weather at a slower rate. Quartz, which forms last from magma, is very resistant to weathering. Metamorphic rocks are formed from heat and pressure. Minerals in these rocks generally weather more rapidly than igneous rocks when exposed. Sedimentary rocks form at or near Earth's surface. Shale and limestone, which are sedimentary rocks, weather rapidly in humid climates.

13:2 Effects of Climate on Weathering

Decomposition is most rapid where moisture and warmth are present. Thus, it is more rapid in tropical zones and during the summer in temperate zones. High temperatures, plentiful moisture, and decaying vegetation speed up chemical reactions. Humic acid, an acid added to the soil by vegetation, speeds up decomposition. Carbonic acid formed in the atmosphere and surface waters also causes decomposition. In desert areas and polar regions, the lack of moisture and/or low temperatures keep chemical weathering at a minimum.

What process is important in disintegration?

Disintegration is important in temperate zones any time freezing and thawing alternate. Water gets into cracks and pores in a rock. When water freezes, it expands and exerts pressure on the rock. This pressure widens cracks and loosens mineral grains. Piles of broken rock fragments at the foot of mountain slopes show ice wedging to be an important weathering process. Disintegration also occurs as plants take root in rock cracks. As the roots grow, they put pressure on the rock, widening cracks and loosening rock fragments.

13:3 Products of Weathering

Weathering products include fragments of rock and chemically altered sediments. New sediments formed by decomposition include clay, carbonates, soluble forms of silica, and limonite. Clay, the most abundant of the new sediments, forms from chemical weathering of either iron-magnesium minerals or from feldspars. Carbonates form from decomposition of calcium feldspar. Soluble silica forms during all of these reactions. Limonite forms from the decomposition of iron-magnesium minerals. Quartz grains may be separated from rocks such as gneiss or granite. However, quartz changes very little as rocks crumble and other minerals decompose. Some weathering products stay in place and form soils or loose sediment. Other products are carried away by wind, water, gravity, or moving ice in a process called erosion and are deposited in a new environment.

FIGURE 13–2. Lichens, a combination of a fungus and an alga that grows on bare rock, produce acids that aid in the decomposition of a rock surfaces.

F.Y.I. For more information on minerals, refer to Chapter 16.

13:4 Erosion

Erosion is the process by which weathering products are carried away and redeposited. Sometimes erosion is rapid, as it is when caused by floods. Usually the surface changes are gradual, and it takes a long time before the changes are noticed. Weathering processes break down the surface rocks. The agents of erosion—gravity, wind, water, and glaciers—move products of weathering from their place of origin and deposit them elsewhere.

FIGURE 13–3. Heavy rains or earthquakes can help loosen material on a steep slope causing a landslide (a). A mudflow results when rain loosens the weathered debris on a slope, causing it to slide downhill (b).

a

b

FIGURE 13–4. The movement of rocks and rock fragments downslope results in talus piles of unsorted material.

During their journey, rock fragments carried by an erosional agent are used as tools. They scour the surfaces over which they travel. The fragments themselves may be broken or polished during this journey. Eventually an erosional agent like wind or water may change velocity or a glacier may melt and drop its load of fragments. These fragments form a new deposit. Erosional processes include transporting the fragments and scouring the land surface.

Name two processes included in erosion.

All erosional agents, except gravity, are in motion. Wind, surface water, groundwater, and glaciers are the most active erosional agents. Gravity, although not in motion, causes motion and plays an important part in erosion. Gravity enables materials to move down slopes toward rivers. Gravity also causes rivers to flow toward the ocean and glaciers to move down mountain slopes.

How does gravity cause erosion?

In your study of surface processes, think about these questions: How are materials carried? What happens to the materials being carried? Finally, what happens to the materials when the agent changes its velocity?

13:5 Mass Movements

Probably the first step in wearing away the land is mass movement. Gravity pulls loose material down a slope. Some mass movements are slow, while others are fast. In either case, weathered fragments are moved from higher to lower elevations. Eventually, these materials come to rest at the bottom of the slope. The material that accumulates at the foot of the steep slope or cliff is called **talus** (TAY luhs). Talus is recognized most easily in semiarid climates, but talus can be seen in humid climates as well.

In some areas, gravity works alone to make materials roll, fall, or slide down a slope. In other areas, water helps move the material. Water adds weight and makes the fragments slippery. Any process that makes a slope steeper also aids mass movement. Sometimes slopes are steepened because they are undercut by waves or rivers. People cut slopes to build houses or roads. Some kind of mass movement always occurs on a steepened slope.

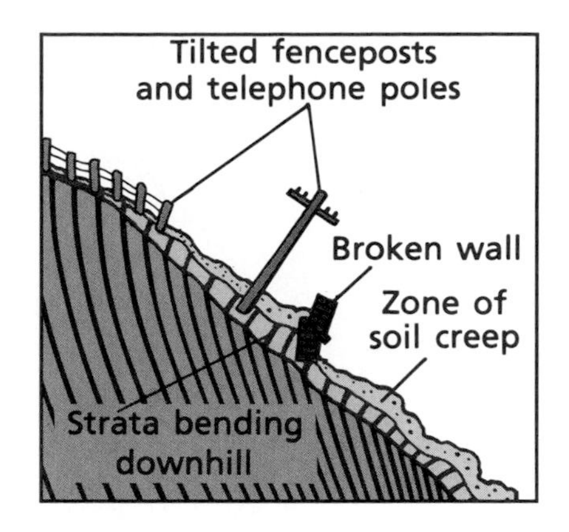

FIGURE 13–5. Tilted telephone poles and fenceposts are indications that creep is occurring.

Many terms are used to describe mass movement. Some terms classify the movements according to the size of the material moved downward. On rides through mountains, you may see the sign, "Beware of Falling Rock." Large masses of fallen rock are called **rockfalls.** Rockfalls occur as freely falling or bounding rock segments move down a steep slope. Any rockfall can be dangerous.

In what two ways are mass movements classified?

Some terms refer to the speed of movement. **Landslides** are rapid movements of large amounts of material. Sometimes large blocks break away from steep mountain slopes. These masses then move downslope as a unit or break up into smaller masses. Landslides often follow long periods of rain, or they may be started by earthquake vibrations. Landslides can carry millions of metric tons of rock to lower elevations.

Mudflows also are rapid movements. These flows occur after heavy rains. Usually, mudflows occur in semiarid

FIGURE 13–6. Slump (a) occurs when weak layers of rock slip down a hillside leaving a curved scar (b).

a

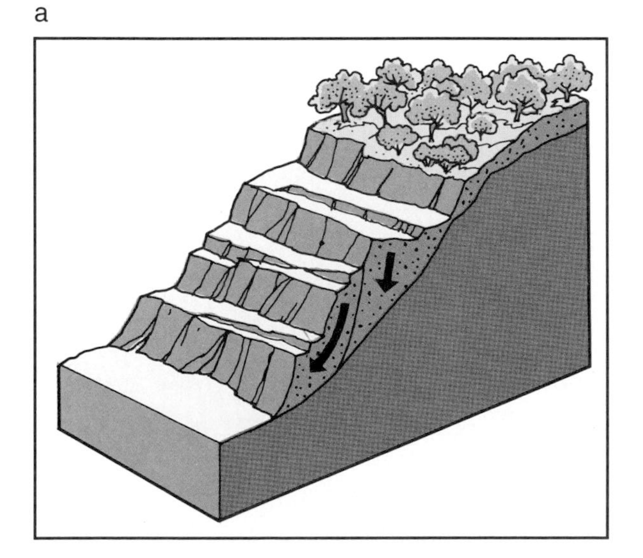

b

BIOGRAPHY

Eve Balfour

1898–

Eve Balfour wanted to be a farmer from the time she was 12 years old. She studied agriculture at Reading University College. She experimented with soils on her farm and discovered the importance of soil bacteria and fungi. Balfour helped form the Soil Association in 1945.

regions. There, long periods of weathering form large amounts of debris that gather into channels. When rains do come, the fine-grained particles wash down the slope. The mudflow follows old channels down to the valley below. There, the thick mud spreads out into a cone-shaped mass. A mudflow can move whatever is in its path, including houses and boulders. Many houses in southern California have been carried away by mudflows caused by heavy rains.

Some downslope movements are extremely slow. It is almost impossible to see what is happening. **Creep** is this kind of mass movement. Creep occurs on slopes primarily in humid regions. Sometimes wavelike bulges of vegetation near the bottom of a slope suggest that creep is occurring. Sometimes the tilt of trees, telephone posts, or fences indicates creep. Creep is common in areas where periods of freezing and thawing alternate. This process causes particles to loosen and move down the slope. Soil moisture helps to lessen the resistance of the fine particles to movement.

Slump is another type of mass movement that occurs on steep slopes. Slump occurs when loose material or rock layers slip downward as a unit. The material does not travel very fast nor very far. Slumping leaves a curved scar where the slumped material originally rested. Weathering and the presence of water contribute to slump.

There is one kind of mass movement that does not require a steep slope. It occurs in the arctic and subarctic regions where the ground stays frozen far below the surface throughout the year. Because the lower layer is frozen, water cannot drain downward, and the surface becomes saturated. The upper layer of broken rock and weathered fragments moves downhill over the underlying permanently frozen layer, even on very gentle slopes.

Review

1. What is the difference between disintegration and decomposition?
2. What are the differences between weathering and erosion?
3. Describe landslides and explain what may cause a landslide to occur.
4. What are the four agents of erosion?
5. ★ What is the relationship between gravity and mass movement?

SOILS

Soil covers most land surfaces. To a farmer, soil is material in which plants will grow. To an engineer, soil is any unconsolidated material. To an earth scientist, soil is a weathered zone of rocks to which organic materials have been added.

GOALS

1. You will learn about the formation and composition of soils.
2. You will learn the differences among soil horizons.

13:6 Soil and Soil Profiles

Soil is a mixture of weathered rock and decayed organic matter. Soils require centuries or several thousand years to form. Soils range in depth from 60 meters in some areas to just a thin layer on top of bedrock in others. Most soils are composed of about 50 percent rocks and minerals, with the rest being air, water, and organic matter.

What is the composition of most soils?

Residual soils are soils that form in place by the gradual weathering of parent rock. As the rocks weather, layers of different colors and textures form. These layers are called horizons. A **soil profile** is a vertical section of all horizons that make up a soil. Figure 13–7 shows a soil profile. The *A* horizon is the topmost layer, sometimes called topsoil. This horizon contains organic matter, roots, and organisms such as worms and insects that break down organic matter into humus and aerate the soil. **Leaching** is a process by which some soil components are dissolved and carried downward by water. Water moving downward through the *A* horizon dissolves some soil components and carries them to deeper layers. Water also carries fine sediment such as clay to lower horizons in the leaching process. The *B* horizon is lighter-colored than the *A* horizon and has fewer organisms. Some roots extend into

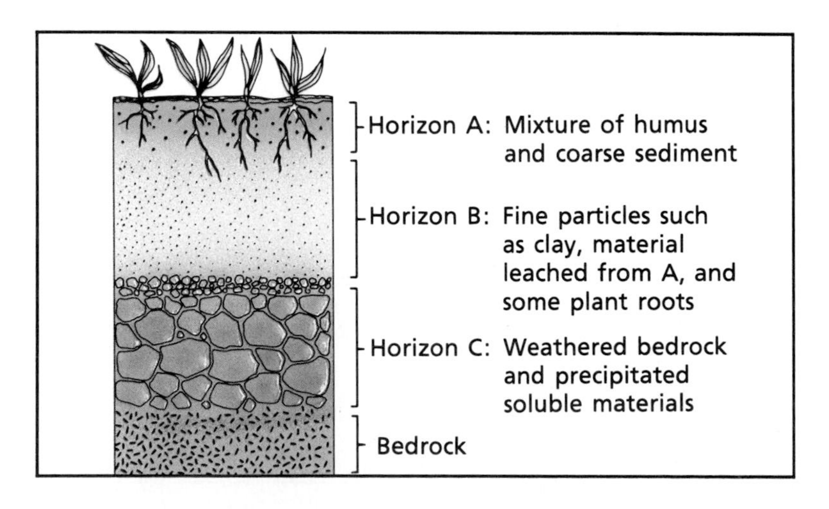

FIGURE 13–7. A soil profile includes three distinct layers.

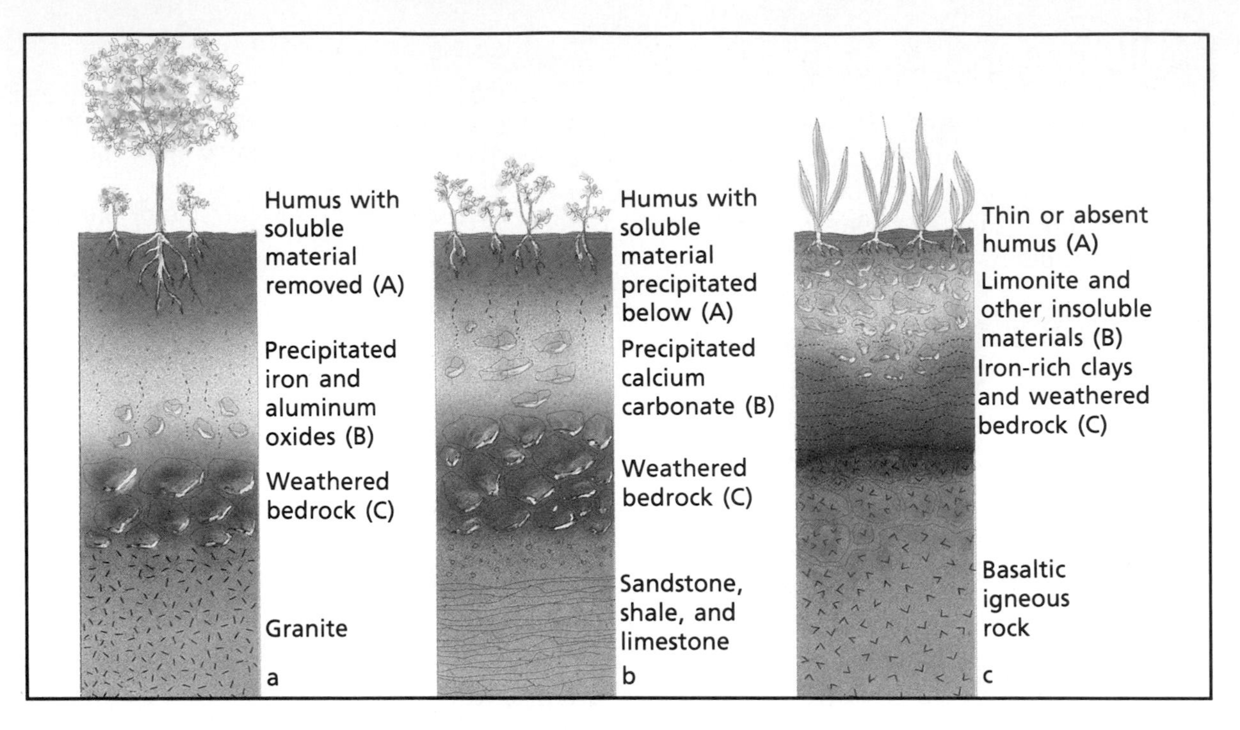

FIGURE 13–8. The soil profile is different for soils developed in humid (a), arid (b), and tropical (c) regions. It depends on the type of bedrock present and amount of rainfall. Compare the B horizons.

this horizon. It also contains iron oxides and other materials that were leached from the *A* horizon. Clay accumulates in this layer. The *C* horizon is rich in materials that were leached from the *B* horizon. The *C* horizon also contains partly weathered bedrock fragments. Below the *C* horizon is the bedrock.

Some soil has no relation to the bedrock below. **Transported soil** is soil that has been removed from one area by erosion and deposited in another location. Horizons are usually poorly defined or absent in transported soils.

13:7 Classifying Soils

The type of soil that forms in a particular region depends on the bedrock, climate, topography, and organisms

Table 13–1

Soil Characteristics			
Color	**Indicates**	**Texture**	**Indicates**
dark brown	well drained	gritty	sandy
red	well drained	sticky	clay
pale color	top soil removed	flourlike	silty
mottled	waterlogged		
gray with yellow or red	poorly drained		
black	organic matter present		

INVESTIGATION 13–1

Soil Characteristics

Problem: What are the characteristics of soil?

Materials

soil sample
sand
clay
gravel
hand lens
water
paper
watch
graduated cylinder
plastic coffee can lids (3)
cheesecloth squares
rubber bands
pencil
250-mL beakers (3)
thumbtack
scissors
large polystyrene cups (3)

Procedure

1. Describe the color of the soil sample.
2. Spread some of the sample on a sheet of paper and examine it with a hand lens. Estimate the percentage of each different kind of particle in your sample.
3. Sketch any organisms that you see.
4. Place a small amount of the soil in your hand, wet it, and rub it between your fingers. Use Table 13–1 to describe the texture of the soil. The best soil for growing plants contains equal parts of sand, silt, and clay mixed with organic matter.
5. Punch the same number of holes in the bottom and around the lower part of each of three polystrene cups with a thumbtack.
6. Cover the holes in each cup with a square of cheesecloth. Secure the cloth with a rubber band. See Figure 13–9.
7. Cut a hole in each plastic lid so that a cup will fit just inside the hole.
8. Place each cup in a lid and place each lid over a beaker.
9. Label the cups A, B, and C.
10. Fill cup A half full of dry sand, cup B half full of clay. Half fill cup C with an equal mixture of sand, gravel, and clay.
11. Use the graduated cylinder to pour 100 mL of water into each cup. Record the time the water was first poured into each cup and when the water first drips from each cup.

FIGURE 13–9.

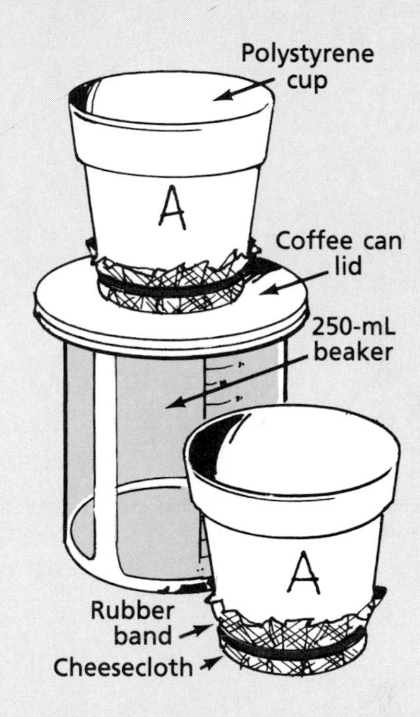

12. Allow the water to drip for 25 minutes, then measure and record the amount of water in each beaker.

Analysis

1. Based on your examination of the soil sample in Steps 1-4, describe your soil sample in as much detail as possible.

Conclusions and Applications

2. Would the soil you tested in Steps 1-4 be good for growing plants? Explain.
3. Permeable soils permit water to move through them quickly. They are well drained. Which of the soils in Steps 5-12 is most permeable? Least permeable?
4. How does the addition of gravel affect the permeability of clay?
5. What are three characteristics of soil?

a

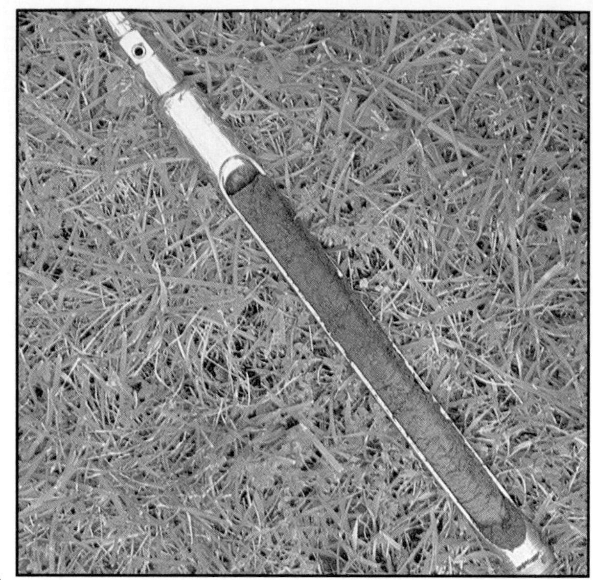

b

FIGURE 13–10. Scientists use special tools to sample the soil (a). A soil sample indicates the thickness of soil layers (b).

present. Soils vary in the type of rock that forms the mineral component, the mix of organic matter, the water content, the texture, and the age. Thus, soil can be classified in many different ways based on its properties.

Soils in humid regions contain bits of leaves, twigs, and animal matter in the *A* horizon. Clays and iron oxide move from the *A* to the *B* and *C* horizons, causing layers to appear stained. Soluble materials such as gypsum, calcite, and salt are removed by surface and ground waters.

On what does the thickness of soil depend?

The thickness of the soil zone depends on climate, slope, and the length of time the soil has been developing. Where plant life is abundant, soils are thick, because decayed vegetation speeds up chemical weathering. Residual soils are especially thick on gentle slopes, where the *A*, *B*, and *C* horizons may total 1.5 meters. Such deposits are found at the foot of mountains like the Appalachian Mountains in the southeastern United States, and in the Pacific Northwest. In the northern United States and Canada, great ice sheets removed much of the original soil, and thick residual soils have not had time to reform.

F.Y.I. There are 15 000 different soil types in the United States, and perhaps hundreds of thousands in the world.

Soils are thin in arid regions where decomposition is slow, in prairie regions where winds erode soils, and in mountainous regions where soils are moved downhill by gravity. Whenever plant life is removed from the soil, such as in a plowed field, the topsoil can be more easily eroded by wind. High winds in the 1930s created dust storms when exposed farmland soil was eroded from the Great Plains area. In recent years, poor agricultural practices have contributed to a loss of six billion tons of topsoil per year in the United States. This erosion, caused by wind and water moving over unprotected soil surfaces, reduces soil fertility and results in higher farm costs.

Review

6. How does residual soil form?
7. What types of material are found in each soil horizon?
8. How do organisms affect the topsoil?
9. What factors influence the thicknesses of soil zones?

 10. Discuss the effect of human activities on soil erosion.

WIND

Wind is an active erosional agent in arid regions like deserts. Along the coast where sand is abundant and vegetation is lacking, windblown fragments are deposited when the wind dies down.

GOALS

1. You will gain an understanding of wind as an agent of both erosion and deposition.
2. You will study sand dune formation.

13:8 Wind Erosion

Winds erode the surface through deflation (dih FLAY shun) and abrasion (uh BRAY shun). **Deflation** is the removal of loose material from the ground surface. Small

FIGURE 13–11. Desert pavement (a) occurs when winds sweep an area clear of loose fragments. Dust storms (b) may occur when winds erode loose material from an area.

a

b

FIGURE 13–12. An oasis may develop where winds erode material to a depth where water is present.

depressions are formed by this process in areas where cementing material has been dissolved from the rocks. Some areas are swept clear of loose fragments by winds. The rock surface left behind, called **desert pavement,** consists only of boulders and pebbles. These materials are too heavy to be moved by the wind.

Where do windblown dust particles originate?

Winds pick up dust particles from deserts, exposed dry riverbeds, and dry glacial lakebeds. In semiarid areas, winds remove huge amounts of topsoil during periods of little rainfall. During the 1930s, the Great Plains became known as the "dust bowl." Because of drought, vegetation died and no longer held the soil in place. In some places, more than one meter of topsoil was carried away by winds. Some of this soil was blown eastward toward the Atlantic Coast. The land, laid bare by removal of the topsoil, was of little value for farming for many years.

How is an oasis formed?

Occasionally, winds erode material down to a depth where water is present. When water is available near the surface, vegetation can grow. Trees, shrubs, and grasses take root and form an **oasis** (oh AY sus), a fertile green area within a desert. The vegetation acts as a barrier against further erosion. Sometimes, however, sand dunes bury an oasis. Oases prove that where water is available, a desert can be fertile.

Most sand grains roll along the surface. Others move by skipping a short distance. As these grains fall, they bump into other grains. These grains then skip a short distance and fall onto still other grains. **Abrasion** is a scouring action of particles carried by the wind. Sand grains act like a sandblasting machine as they strike the

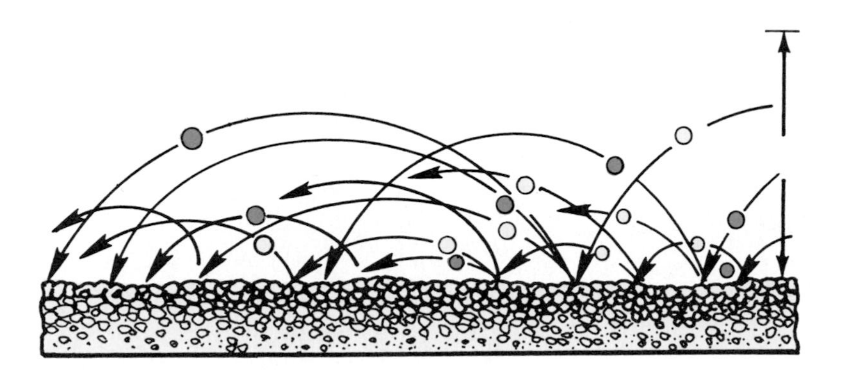

FIGURE 13–13. The abrading action of sand grains decreases above a height of one meter.

surface rocks. Surfaces of both the rocks and sand grains are pitted and polished by this sandblasting. In desert areas, telephone poles and fence posts are sometimes protected by piles of rock to about one meter above the ground. This protects the poles from the sandblasting action.

In time, rocks subjected to sandblasting develop flat surfaces facing the wind. If the rock is blown over, a different face is then flattened. These two faces meet at a sharp angle, which shows they are products of wind erosion, not water. Sand-size grains bump together and round off their corners. Sandblasted grains also may have surfaces that look like frosted glass or may appear pitted. The presence of frosting indicates that the grains were carried by the wind sometime during their history.

What does a wind-eroded rock look like?

FIGURE 13–14. Rocks are used to protect telephone poles from abrasion in some desert areas (a). Rocks eroded by wind develop sharp angles and flat faces (b).

a

b

INVESTIGATION 13–2

Wind Erosion

Problem: Which factors affect wind erosion of different surface materials?

Materials 

goggles
flat pans (5)
1250 mL fine sand
1000 mL clay
250 mL gravel
hair dryer
protractor
sprinkling can
water
cardboard sheet
marker
masking tape
metric ruler

Procedure

1. Label the pans A, B, C, D, and E using the masking tape and marker.
2. Put 500 mL sand into A and B. Put 500 mL clay into C and D. Mix 250 mL sand and 250 mL gravel and put it into pan E.
3. Use the sprinkling can to dampen the material in pans A and C.
4. Hold the hair dryer 10 cm from pan A at an angle of 45°. See Figure 13–15. Tape the cardboard to the other end of the pan. Direct a stream of air onto the pan for one minute. **CAUTION:** *Wear your goggles.* Record in your table every effect of the air that you observe.
5. Repeat Step 4 for pans B, C, D, and E.
6. Smooth out the "soil" in each pan.
7. Change the angle of the hair dryer to 10°. Repeat Step 4 for all pans using this new angle. Record all observations. **CAUTION:** *Wear your goggles.*
8. Smooth out the "soil" in each pan.
9. Repeat Steps 4 through 7 for all pans from a distance of 20 cm. Hold the hair dryer at an angle of 45°. (The distance of the hair dryer to the pan represents the force.) **CAUTION:** *Wear your goggles.* You may need to redampen the "soil" in pans A and C before completing this step. Record your results in the table.

FIGURE 13–15.

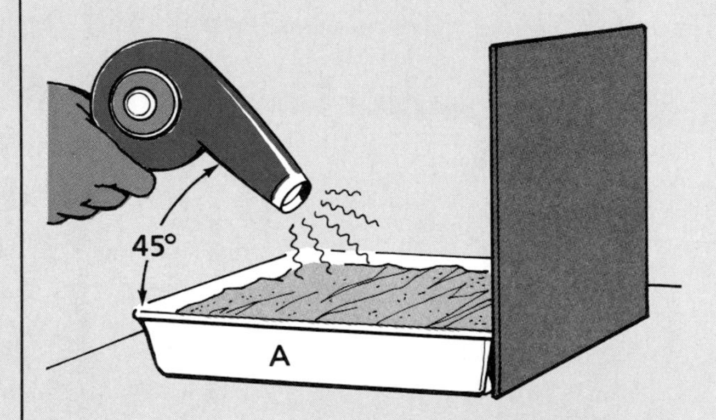

Data and Observations

Pan	10 cm		20 cm	
	45°	10°	45°	10°
A				
B				
C				
D				
E				

Questions

1. Were you able to get a desert pavement in any of the pans? Which one(s)?
2. How do dry sand and clay react to the wind? Explain how moisture changed this reaction.
3. How does the addition of gravel to the sand affect its reaction to the wind?
4. How does the change in force (distance of hair dryer to pan) affect movement of sediment grains? The angle of the wind?
5. Is wind a more effective erosional agent in wet or dry climates? Which pans give evidence to support your answer?

CAREER

Civil Engineer

As a civil engineer, Jane Hubbard is qualified to plan, design, and direct the construction and maintenance of structures such as roads, railroads, airports, bridges, harbors, channels, and dams. Other responsibilities include irrigation projects, pipelines, powerplants, water and sewage systems, and waste disposal units. She may perform technical research and utilize computers to develop solutions to specific engineering problems. She must have knowledge of state and local building codes and ordinances. Also, she must have knowledge of the environment of the area where the structure is to be built, including how the structure will be affected by weathering and mass movements.

To become a civil engineer, Ms. Hubbard graduated with a bachelors degree in engineering, acquired four years of relevant work experience, and passed a state examination. Ms. Hubbard's career requires that she be able to work as part of a team, be creative, have an analytical mind, and have a capacity for detail. Also, she needs to be able to express herself well—both orally and in writing. Ms. Hubbard has continued to take courses throughout her career so she can learn about new technology.

For career information, write:
American Society of Civil Engineers
345 E. 47th Street
New York, NY 10017

13:9 Wind Deposits

The materials carried by wind are well sorted by size because the heavier sand comes to rest first. Sand is sediment of clastic grains ranging in diameter from 1/16 mm to 2.0 mm. The dominant mineral is generally quartz. Silt, being finer, is carried farther than sand and is deposited at a greater distance from its source. Dust, the finest of all the material, gets into the higher air currents. It is suspended in the air and supported by the upward push of air currents. Most dust is deposited only during rain or snowstorms. Dust is important in the atmosphere because it provides condensation nuclei.

What is loess?

Loess (les) is a wind-blown deposit of fine dust particles gathered from deserts, dry riverbeds, or old glacial lakebeds. Loess consists of very thin, angular particles that tend to pack together into a dense mass without layers. In North America, loess deposits are found on hilltops

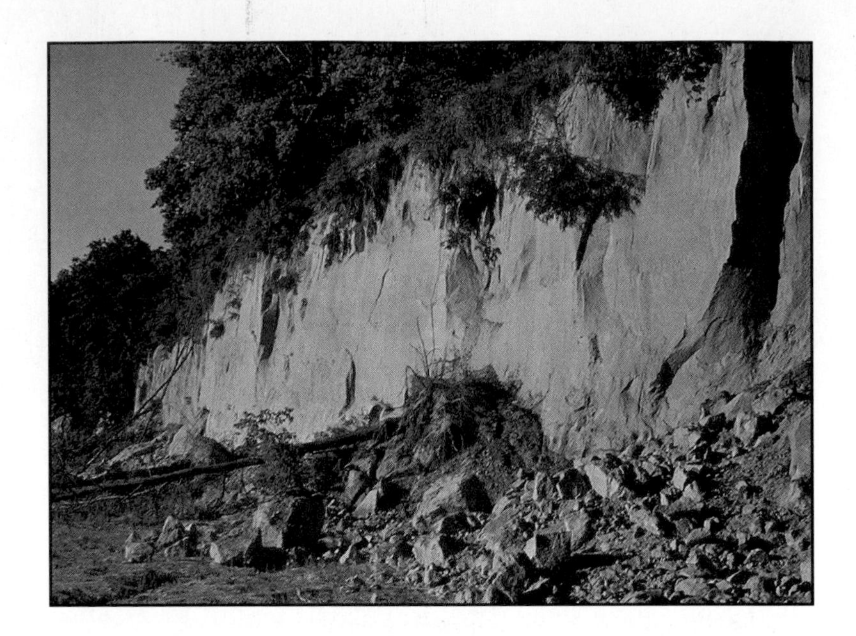

FIGURE 13–16. Deposits of loess lack any visible layers.

and in valleys near the Mississippi River. This material was carried by the strong winds that blew from the ice sheet that once covered the northern United States and Canada. Loess deposits in China are windblown material from the Gobi and Ordos Deserts.

What is the most common type of wind deposit?

Sand dunes are the most common type of wind deposits. A large rock, clump of vegetation, or even a small pile of pebbles is large enough to start a sand dune. Blowing sand grains are dropped by the wind when it blows across an obstacle. These grains begin to collect and a sand drift occurs. As the drift grows, a sand dune forms. Dunes are found in arid and semiarid regions and along shores of seas or large lakes where sand is plentiful. When winds slow down or meet an obstacle, they drop their load of sand. The obstacle need not be large. Even small plants often start the growth of a dune. Along shores, plants may take root in the piles of sand. The loose sand then becomes a fixed dune. Many dunes along the shores of the Great

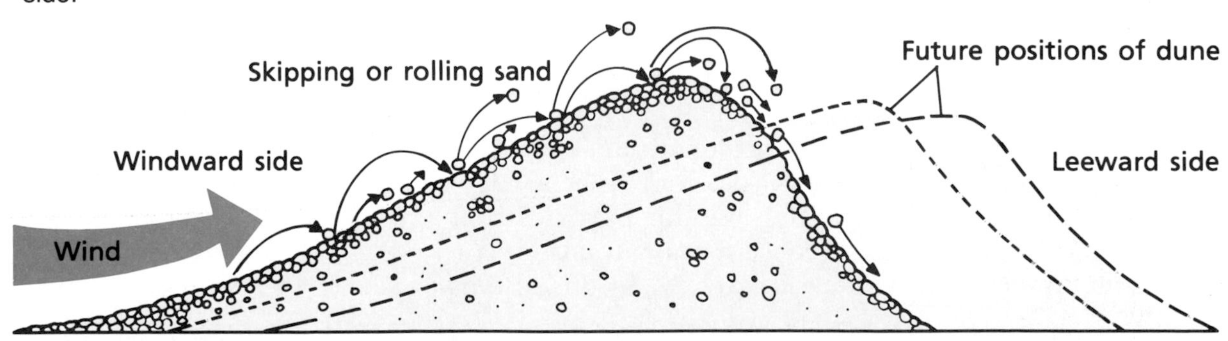

FIGURE 13–17. Sand dunes migrate along the surface as sand is rolled up the windward slope and tumbles down the leeward side.

a

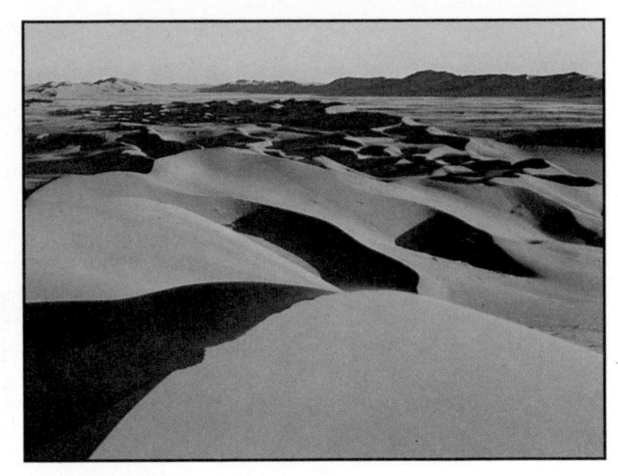

b

FIGURE 13–18. Dunes may become stabilized in areas where vegetation grows on them (a). Where one strong wind prevails, dunes form long, broken lines at right angles to the wind (b).

Lakes, for example, are now tree covered. These dunes are anchored by vegetation. Dunes without vegetation continue to move inland until they become stabilized.

If dunes are not anchored by vegetation, they tend to move forward with the prevailing wind. Each time the wind blows, it carries sand from the windward side (the side facing into the wind) of the dune and drops it on the leeward side (the side away from the wind). The loose sand grains slide down the slope. When the sand is piled too high, a whole sheet of sand slips down the leeward side of the dune. This slipping continues until the slope is able to maintain itself.

What happens if sand dunes are not anchored by vegetation?

Dunes have a gentle slope on the windward side and a steep slope on the leeward side. In dunes that have formed over a long period of time, the layers lie at steep angles to each other. Despite the fact that dunes move with the wind, the layers eventually become cemented. Successive layers can be seen in road cuts through dunes in arid and semiarid areas.

Dunes have many different patterns in ground plan, which is the shape seen when viewed from an airplane.

FIGURE 13–19. As material moves up the windward side of the dune (a), it accumulates at the top of the slipface (b). Eventually, this material forms a series of layers on the leeward side of the dune (c).

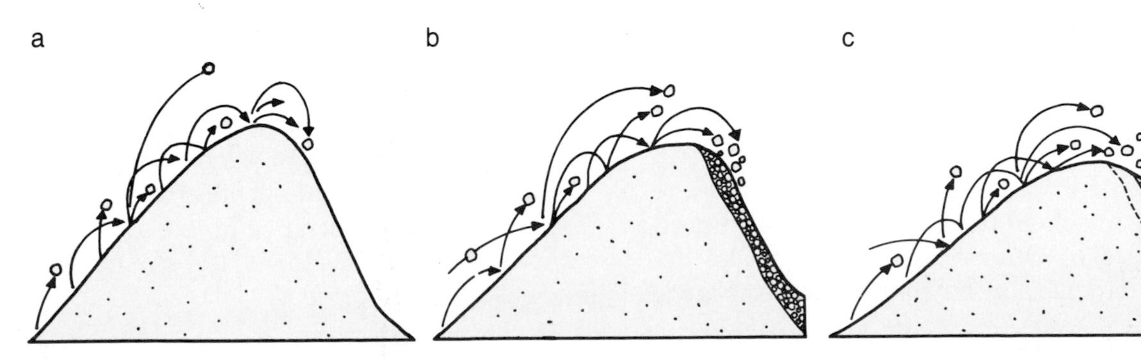

FIGURE 13–20. Barchan dunes are the most common dune shape.

The shape of a dune may change from time to time if there is no strong prevailing wind direction. The shape also may change if the prevailing winds are from two

PROBLEM SOLVING

Darleen's Dilemma

Darleen decided that for her science fair project she would test the permeability of soils from three different areas of her back yard. She decided that she liked the way that she had tested the soils at school using the investigation titled "Soil Characteristics." So, she decided to use similar materials. See Figure 13–9 for the materials she used. She could not find three cups that were the same size but she went ahead and used the cups she did have. She poked some holes into the bottom of two of the cups with a pin. Then she decided that the holes were too small and used a paper clip to poke the holes into the third cup. Since she had poked such small holes into the first two cups, she decided to poke fewer holes into the third cup. Then she filled each cup about half full with soil. Finally, she covered the bottom of each cup with cheesecloth and poured water into each cup until the water just touched the rim. At the end of 25 minutes, Darleen measured how much water was in the beaker below each cup.

Darleen's project did not receive a very high rating at the science fair. Can you explain the reason? How could she have made it a better project?

Table 13–2

Sand Dunes		
Name of dune	**Ideal shape**	**Other characteristics**
barchan		most common form, may reach a height of 30 m and a length of 350 m
transverse		similar in size to the barchan, but not curved, form where sand is plentiful
parabolic		may be 30 m high, form along seacoasts in areas where vegetation holds down the sand
longitudinal		range from 3 to 90 m high and 60 to 100 m in length, form in areas with variable winds and little sand
star		form with a central high point with four or five arms radiating outward, found where the wind direction shifts frequently

alternating directions. This situation is common along shore zones. For example, along the Gulf Coast, winds during summer tend to be from the southeast. During winter, the strong winds come from the north or northwest. Dunes formed in response to summer winds are rearranged by the winter winds. Rounded, symmetrical dunes tend to form in response to these two prevailing wind directions.

FIGURE 13–21. Transverse dunes form where sand is plentiful.

FIGURE 13–22. Migrating dunes may bury forests. These dunes are covering trees on the south shore of Lake Michigan.

What is the most common dune shape?

Where there is one strong prevailing wind direction, dunes usually form first at right angles to the wind. Such dunes may form long, broken lines across a desert. In time, as sand supplies are less abundant, the dune shape may be rearranged into a crescent shape with the points carried toward the leeward direction. These crescent-shaped dunes are called **barchans.** Barchans are the most common dune shape. They have been found on Mars as well as Earth. A similar form that points in the opposite direction is found in areas where vegetation holds the dune in place. Vegetation holds down the ends of these **parabolic dunes,** where the wind moves the center sand to the leeward. Star dunes form in many deserts where wind directions change often, but come from at least three different directions from time to time. **Star dunes** have a central high point with sand extending outward in four or five arms.

Ripples commonly form on the surface of dunes. Ripples are surface features that look like miniature dunes with a gentle face toward the wind and a steep leeward slope. If a dune later becomes cemented to form a rock, ripple marks can help indicate the environment in which the rock formed.

Review

11. What is an active erosional agent along coasts and in arid regions?
12. In what two ways does this agent erode the surface?
13. Describe the appearance of a wind-carried sand grain.
14. What is loess?
★ **15.** Contrast deflation and abrasion.

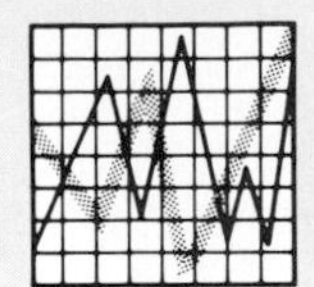

SKILL

Limiting the Number of Variables

Problem: What is the effect of vegetation on the formation of sand dunes?

Background

Recall from Section 1:4 that every experiment has variables. In order to test a hypothesis, the number of variables being tested must be limited so only one factor is affecting the outcome. It is necessary to recognize variables so they can be limited. Read this example and answer the questions.

Place 1 liter of water in each of 3 identical nonelectric coffee pots. Add 1 level coffee scoop of coffee to the first pot, 2 to the second pot, and 3 to the third pot. Turn the burners on high and brew each pot for 10 minutes. Which variables in the example will be exactly the same? Which variable is being tested?

Materials  

identical flat pans (2)
sand (1000 mL)
hair dryer
clock with second hand
small clump of grass
metric ruler
protractor
cardboard
goggles

Procedure

1. Suppose you want to study the effect of vegetation on the formation of sand dunes. You have been given the materials listed above.
2. Read and answer these questions.

Questions

1. Why would the pans need to be identical in size and shape?
2. How much sand would you want to put into each pan? How would you make your measurements precise?
3. What should you do to the sand surface in each pan? Why?
4. What factor are you testing in this experiment? How will the two pans differ?
5. Why will you use the hair dryer? Why should the angle and distance of the hair dryer be the same for both pans? **CAUTION:** *Goggles should be worn to protect the eyes.*
6. How could you be sure that the angle and distance of the hair dryer are identical for each pan?
7. Why should the speed of the hair dryer you use to blow the sand be the same for both pans?
8. How could you make sure that the time that the hair dryer blows on each box is the same?
9. Why should you repeat this experiment several times? Should you ever change the placement of the grass clump in the one pan?
10. Why is measuring an important skill to use when controlling variables?
11. List the variables that are limited in this experiment. Which variable is tested?

Data and Observations

Limited Variables	Tested Variable

TECHNOLOGY: APPLICATIONS

Preventing Desert Expansion

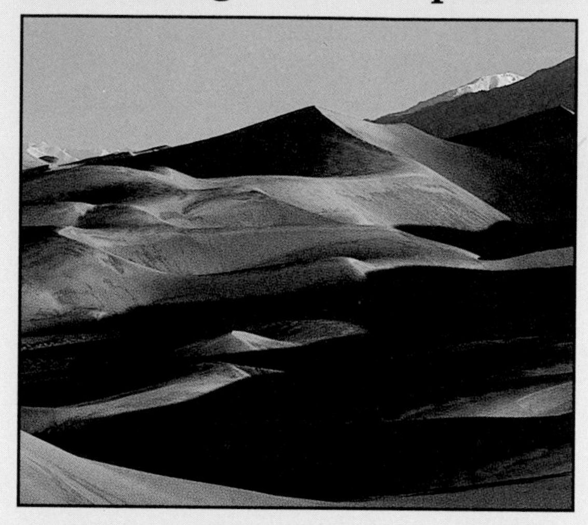

Arid and semiarid regions cover about one-third of Earth's land surface. These areas are expanding in many parts of the world. Only a small portion of this barren land is caused by climatic conditions.

Human activity contributes to most of the desert expansion in the world. Desert expansion, also called desertification, is occurring in Ethiopia, China, the United States, and other places. Desertification claims an amount of land approximately the size of the state of Maine each year.

As Earth's population increases, more land must be used for farming. Overfarming, poor farming practices, overgrazing, and removing trees and vegetation speed up desert expansion. Droughts and the misuse of water resources also can aid desert expansion. Wells drilled for irrigation to halt desert expansion can accelerate the process by attracting more people and herds to the fertile spot. In some areas, farmers have tapped aquifers to such an extent that reservoirs are no longer usable. Heavy irrigation over long periods of time has increased soil salinity in many areas, thus diminishing soil productivity.

Reclamation projects are underway in many countries to halt desert expansion. Experimental farms on the Negev Desert in Israel are growing barley and wheat, as well as olive, fig, and pistachio trees. Although less than ten centimeters of rain falls per year in this area, runoff agriculture has transformed hills into giant funnels that bring rainwater to the terraced valley fields. Also, plastic mulch is used to slow evaporation. In addition, trickle irrigation is sometimes used to water the plants. This method uses a series of thin plastic pipes with holes that deliver the exact amount of needed water to the roots of each plant. An idea for the future is to build big plastic domes over the land. Held up by air pressure, they would provide an ideal climate. Solar collectors on the domes could run refrigeration units during hot weather.

In Iran, desert areas are reclaimed by treating sand with an asphaltlike petroleum residue. The oil stabilizes the sand and allows it to retain moisture. This process enables the planting of seeds that add to the soil's fertility and ability to hold water.

A highly technical irrigation method is being tested in California to make more effective use of water. A computer monitors the amount of sunshine and wind that the plants receive, as well as the air temperature, ground moisture, and humidity. Crops then are given the exact amount of water they need.

Scientists are trying to develop machines similar to those used to filter impurities from human blood to remove salt from seawater. The water left behind could be used for irrigating the deserts. With increases in world population and climatic changes, people will need to continue developing new technologies for the prevention of desertification.

Chapter 13 Review

SUMMARY

1. Distintegration breaks large rocks into fragments. Decomposition is chemical weathering. 13:1
2. Freezing and thawing cause disintegration of rocks. 13:2
3. Weathering products include fragments of rock and chemically altered sediments such as clay, carbonates, silica, and limonite. 13:3
4. Erosion is the wearing away of the land by the actions of wind, water, glaciers, and gravity. 13:4
5. All mass movements occur due to gravity, but may be aided by water. 13:5
6. A soil profile consists of three distinct horizons. Soils form in place by weathering and from transported material. 13:6
7. Types of soil in a particular region depend on the bedrock, climate, topography, and organisms present. 13:7
8. Wind is an effective erosional agent in deserts and along coasts. Wind erodes by deflation and abrasion. 13:8
9. Materials carried by wind are sorted by size. 13:9

VOCABULARY

a. abrasion
b. barchans
c. creep
d. decomposition
e. deflation
f. desert pavement
g. disintegration
h. erosion
i. landslides
j. leaching
k. loess
l. mudflows
m. oasis
n. parabolic dunes
o. residual soils
p. rockfalls
q. slump
r. soil
s. soil profile
t. star dunes
u. talus
v. transported soil
w. weathering

Matching

Match each description with the correct vocabulary word from the list above. Some words will not be used.

1. dunes that form where vegetation holds the points of the crescent-shaped dune in place
2. windblown deposit that packs into a dense mass without layering
3. develops when winds erode to a depth where water is present
4. soils that form by the gradual weathering of parent rock
5. mass movement in which resistant rock or other material is carried downslope by a weak layer
6. material that is too heavy to be moved by the wind
7. dunes formed when the wind blows from at least three directions
8. chemical weathering
9. material that accumulates at the foot of a cliff or steep slope
10. occurs as water moving downward through a soil horizon dissolves soil components and carries them to deeper layers

Chapter 13 Review

MAIN IDEAS

A. Reviewing Concepts

Choose the word or phrase that correctly completes each of the following sentences.

1. Chemical weathering is most rapid in *(temperate, tropical, arid)* climates.
2. Disintegration is *(physical, environmental, chemical)* weathering.
3. Thick soil forms in *(mountain, polar, humid)* regions.
4. In a soil profile, the *(A, B, C)* horizon contains many organisms, humus, and roots.
5. *(Transported soil, Residual soil, Topsoil)* has no relationship to the bedrock below.
6. The loess deposits along the Mississippi River originated mainly from *(riverbeds, deserts, glacial sediments)*.
7. Rocks with angular corners and sharp faces are *(water-abraded, talus, wind-abraded)*.
8. Rounded, symmetrical dunes tend to form in response to *(one, two, three)* prevailing wind directions.
9. *(Sedimentary, Metamorphic, Igneous)* rocks tend to weather quickly.
10. An example of slow mass movement is *(creep, a landslide, a mudflow)*.
11. *(Water, Ice, Wind)* is the most active erosional agent in arid regions.
12. Changes that rocks undergo at or near Earth's surface are called *(erosion, leaching, weathering)*.
13. All erosional agents except *(wind, gravity, water)* are in motion.
14. The scouring action of sand grains is *(abrasion, deflation, slumping)*.
15. The most common type of wind deposit is *(loess, a sand dune, talus)*.

B. Understanding Concepts

Answer the following questions using complete sentences.

16. What causes sand dunes to migrate?
17. How do roots cause disintegration?
18. What is leaching and how does it occur?
19. Explain how talus is formed.
20. How does mass movement occur in the arctic in areas without steep slopes?
21. Describe how residual soils form.
22. What are barchans?
23. What is erosion?
24. Explain why people living in desert areas pile rocks to about one meter around telephone poles and fence posts.
25. Explain why rocks in desert areas have flat surfaces.

C. Applying Concepts

Answer the following questions using complete sentences.

26. How might forest fires speed up the weathering process?
27. Part of the island of Hawaii receives approximately 400 cm of rain per year. Another part of the island receives approximately 12.5 cm of rain per year. How would decomposition rates vary from one part of the island to the other? Would disintegration be an important process on Hawaii? Explain.
28. How does the wind sort materials?
29. How can creep be recognized? Why is creep common only in humid climates?
30. How can dunes be stabilized so that they do not migrate?

SKILL REVIEW

1. Compare and contrast weathering and erosion.
2. Suppose you want to test the effect of a thin layer of oil on the erosion of sand. You decide that you will experiment by using two pans of sand, a control pan and a test pan. Describe how you could limit the number of variables in your experiment.

3. In an experiment to test the effect of oil on the erosion of sand, which pan is the control, the one with oil, or the one without oil?
4. Why was it important to measure exactly 100 mL into each polystyrene cup you used in the investigation on soil characteristics?
5. What is the effect of moisture and warm temperature on decomposition?

PROJECTS

1. Experiment to test the effect of a thin layer of oil on the erosion of sand.
2. Collect additional soil samples and compare their characteristics to the sample you studied in Investigation 13–1, Soil Characteristics.

READINGS

1. Barnhardt, Wilton. "The Death of Ducktown." *Discover*. October, 1987, pp. 34-43.
2. Leutscher, Alfred. *Earth*. New York: Dial Books for Young Readers, 1984.
3. Morgan, R. P. *Soil Erosion and Its Control*. New York: Van Nostrand Reinhold, 1986.

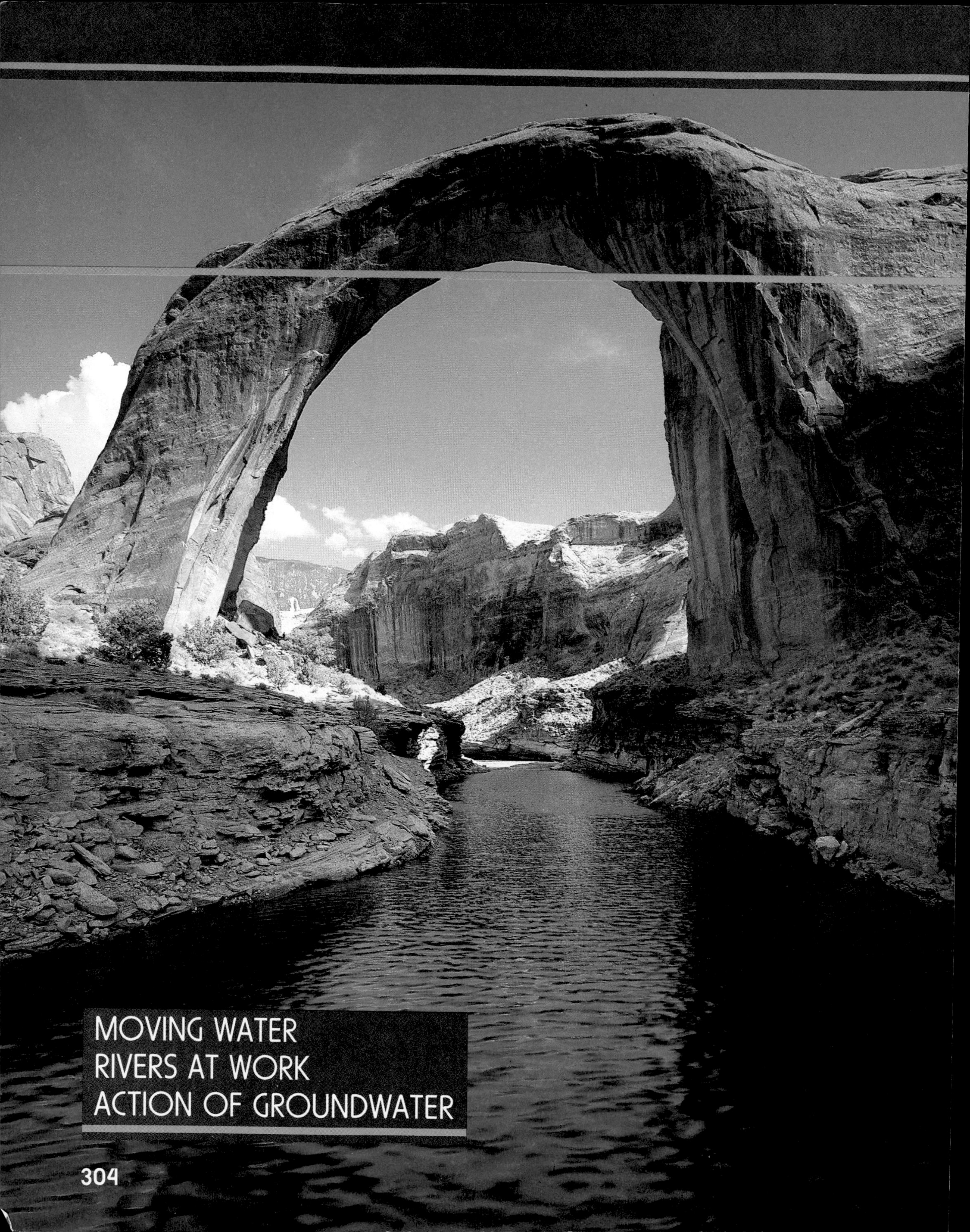

MOVING WATER
RIVERS AT WORK
ACTION OF GROUNDWATER

Water Systems

Running water is Earth's main agent of erosion. Water is responsible for some of the most spectacular landforms on our planet. Rivers change Earth's surface by eroding and depositing sand, silt, and clay. People use river systems for transportation and recreation. How do rivers affect your life?

MOVING WATER

GOALS

1. You will learn how river systems form.
2. You will learn about water erosion in arid and semiarid climates.

Recall from Section 10:2 that water moves through Earth's hydrosphere. When precipitation in a region is greater than evaporation, a river system may develop. In arid and semiarid regions, water is especially important as an agent of erosion.

14:1 Runoff

Water evaporates from the ocean, moves through the atmosphere, and precipitates as rain or snow on the land. Much precipitation either evaporates immediately or is used by plants and animals. Some joins the underground water system. Nearly 40 percent of precipitation flows across the surface and back into the ocean as **runoff.**

What factors determine the amount of precipitation that becomes runoff?

The amount of precipitation that becomes runoff depends on the type of land surface, the slope of the land, and the amount of rainfall. Gentle rains and light snows usually evaporate or sink into the ground. They are also used by organisms. Most heavy, fast downpours become runoff. When the ground is saturated by long periods of rain, additional precipitation becomes runoff.

Steep slopes shed water quickly. Gentle slopes or flat areas hold water in place until it evaporates or sinks into the ground. **Permeable** rocks have spaces between grains. Permeable rocks at the surface allow water to sink into the ground. The particles of some other types of rocks are pressed together so tightly that fluids cannot move through

What is the difference between permeable and impermeable rocks?

F.Y.I. See Section 18:6 for more information about impermeable and permeable rocks.

them. These rocks are **impermeable** (ihm PUR mee uh bul). Impermeable rocks cause rapid runoff.

During warm weather, most water evaporates and runoff is minimal. During cold weather, precipitation may be trapped as snow or ice. If thawing is slow, water sinks into the ground and runoff is minimal. If thawing is fast, runoff is greater. Vegetation reduces runoff by holding water in the soil.

14:2 Development of River Systems

River systems form where runoff follows the same channel after every rainfall. Because of gravity, water continuously flows downhill. If the rock that makes up the land surface is resistant to erosion, the water cuts a path through it and continues to flow in this same channel.

What is a drainage system?

Water flows from high elevations as a sheet that gradually flows into a network of small rills as it moves downhill. Rills join to form creeks, which join to form streams. Streams join to form the main river. This network of channels is a **drainage system.** The area drained by this channel system is the **drainage basin.** Drainage basins are separated from one another by a line of high ground called a **divide.** Divides are constantly being eroded as rivers extend their headwaters to higher elevations. This process is called headward extension.

As water flows through a drainage system, it constantly erodes, transports, and redeposits sediment. In time, a river reaches a balance between these three processes. Then a river has a **profile of equilibrium** (ee kwuh LIHB ree um) along its length. Each tributary within a drainage system develops its own profile and has its own history. Profiles remain constant as long as the water's velocity

a

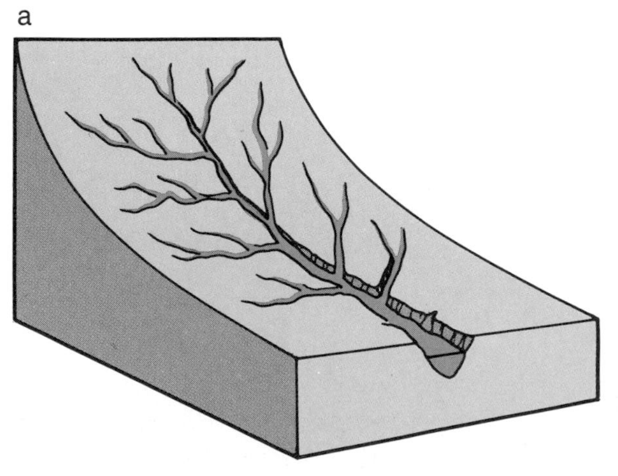

b

FIGURE 14–1. A drainage system is composed of rills, creeks, and streams that join to form a main river (a). Drainage basins are separated from one another by high areas called divides (b).

a

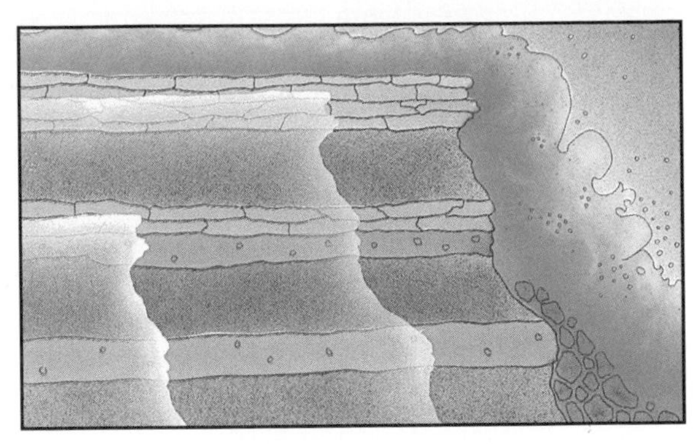

b

FIGURE 14–2. The erosion of waterfalls (a) is an example of headward extension. Water cascading over the falls wears away the soft rock, leaving the upper layer unsupported. This layer then drops to the bottom of the falls. The shading (b) indicates future contours of the falls.

and the level of the land surface remain the same. However, a stream's profile will change if any changes occur in its drainage system, such as: (1) uplift or lowering of the headwater area; (2) rise or fall of sea level; (3) different amount of rainfall that changes the velocity of the water; or (4) wearing down and breaching of the divide.

14:3 Stream Patterns

Drainage systems are made up of an interconnecting network of streams that together form patterns. These patterns provide a clue to the type of rock in a particular area or to the presence of faults and folds.

Which stream pattern is the most common?

Streams flowing over rocks of uniform resistance usually produce a *dendritic* (den DRIHT ik) or treelike stream pattern. A main river like the Mississippi River represents the tree trunk. Tributaries like the Missouri and Ohio Rivers are the large branches. Streams, creeks, and rills are the small branches and twigs. The dendritic pattern is the most common type of stream pattern.

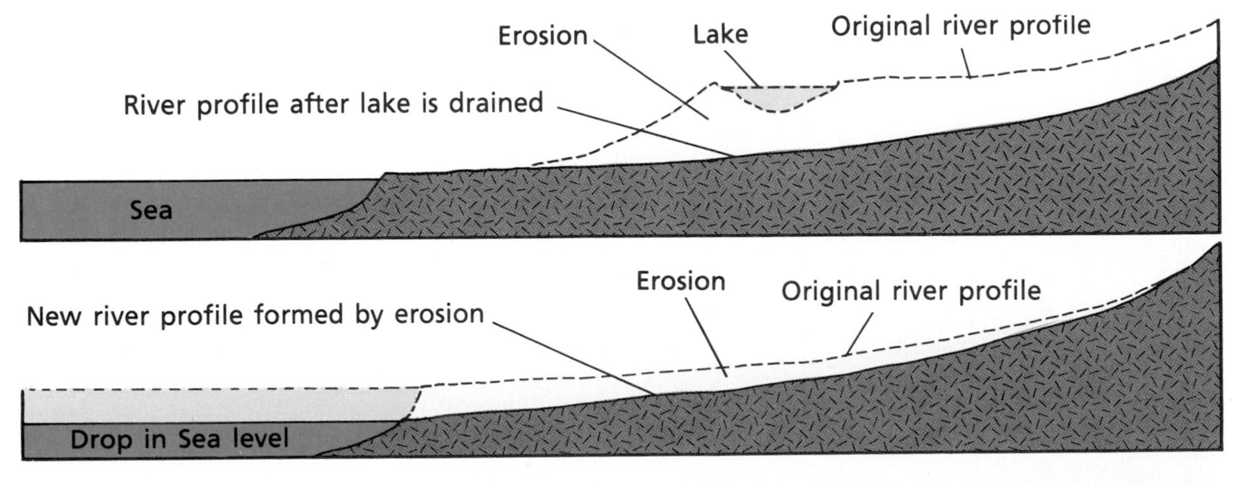

FIGURE 14–3. Streams erode their channels to the lowest point possible. What happens if changes occur along the channel?

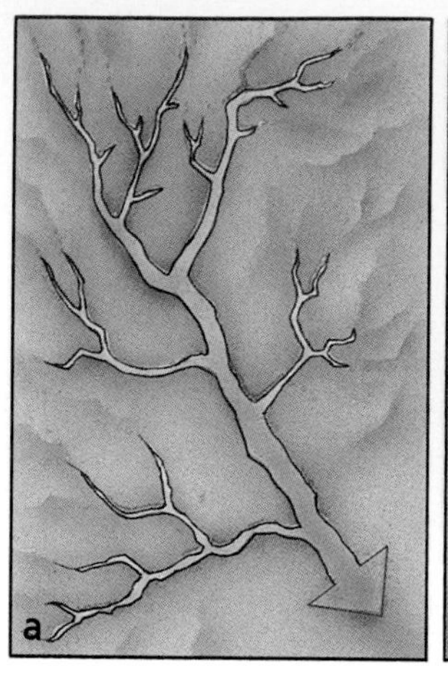

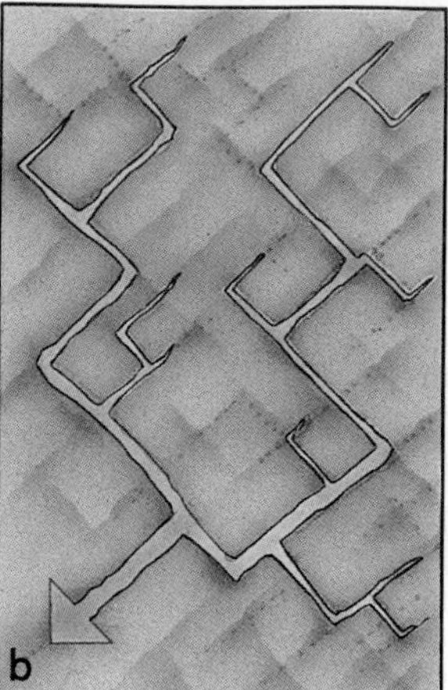

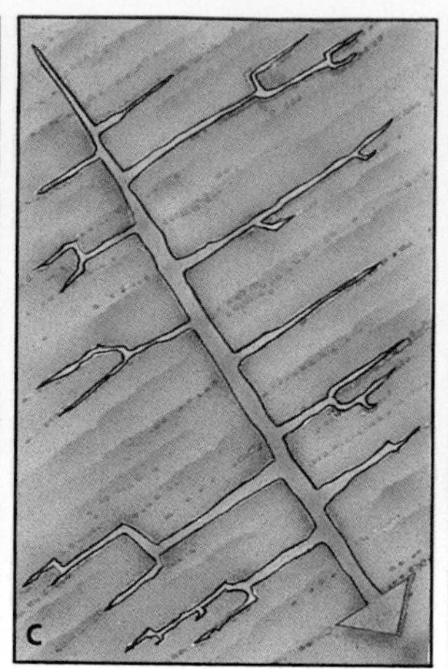

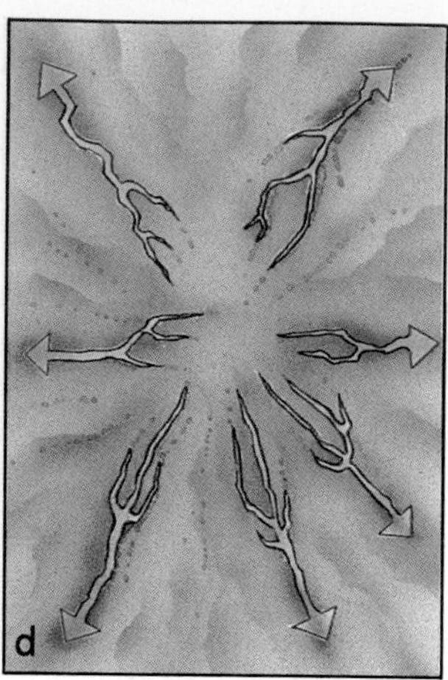

FIGURE 14–4. Four common drainage patterns are dendritic (a), rectangular (b), trellis (c), and radial (d).

Some rocks are jointed or fractured in a rectangular pattern. Thus, drainage that develops on these rocks follows a *rectangular* pattern. The main stream makes right angle turns, and tributaries enter the main stream at right angles. *Trellis* patterns develop in alternating resistant and weak layers of tilted sedimentary rock. Here the main stream is older and well established. It cuts across resistant rock, while the tributaries follow valleys of weak rock. *Radial* patterns are formed by streams that flow outward from a central location. Usually such streams develop on the slopes of volcanoes.

PROBLEM SOLVING

Washed Out

David stared at the cloudy sky. He knew from weather reports that northern areas of the state had received heavy rain for three days. However, his hometown had not received a drop. He hoped the rain would come and go by the weekend, so the big baseball game would not be rained out. His team was in the playoffs and the day of the big game was almost here.

It never did rain that week. Saturday morning arrived and David had hardly slept. He arrived at the field early and could not believe his eyes. The playing field next to the river was underwater. How could the river be flooded? David thought for awhile and answered the question. What did he conclude?

SKILL

Forming Hypotheses

Problem: How do you form a logical hypothesis?

Materials

pencil
paper

Background

A hypothesis is an educated guess about the solution to a problem. Before designing an experiment, you must examine the problem and the variables involved. You must determine all the factors that may be involved. Read this problem and analyze the variables.

What affects a plant's growth rate?

Did you list the type of plant, age of plant, amount of light, angle of light, type of light, amount of water, type of water, presence or absence of plant fertilizers? Did you find other variables? After you have analyzed the variables, you can choose one and form a hypothesis. In doing so, you must record what you expect to test and what you think the response will be.

Procedure

1. Imagine that you are a hydrogeologist and your job is to locate the origin of a particular pollutant. You have located large amounts of mercury in the river at point X. See Figure 14–5. There are 6 factories along the river, and you are certain that the mercury is coming from one of them. You can already hypothesize that the pollutant did not come from factory 6. How do you know this?
2. You do some research and find that factories 1, 2, 3, and 5 all use mercury in some way in their production lines. Pick the best hypothesis from the following: (A) All four factories are dumping mercury. (B) Factory 3 is dumping the mercury. (C) There is insufficient evidence to point the blame at all or one particular factory.
3. Suppose you picked either hypothesis A or C, and you decided to test your hypothesis. You test the water at points Y and Z, and you find no mercury at these locations. You will need to change your hypothesis. Pick the best new hypothesis: (A) The mercury is coming from both factories 2 and 5. (B) The mercury is coming from either factory 2 or 5. (C) The mercury is coming from both factories 1 and 3. (D) The mercury is coming from either factory 2, or 5, or both 2 and 5.

FIGURE 14–5.

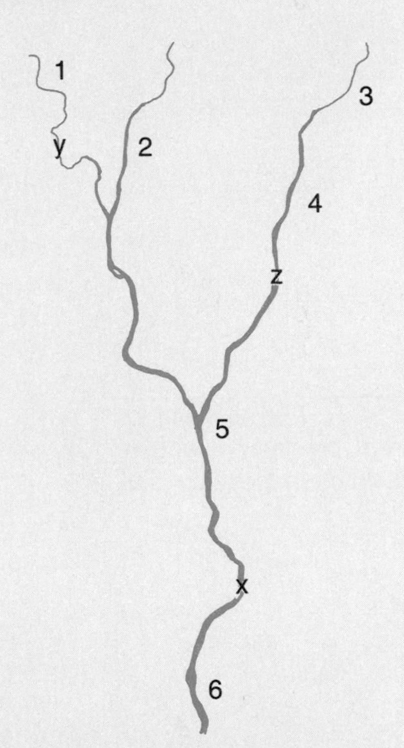

Questions

1. How could you test your hypothesis in Step 3 if you chose hypothesis A, B, or D? How would your results affect your hypothesis?
2. What are the steps to take in forming hypotheses about a particular problem?

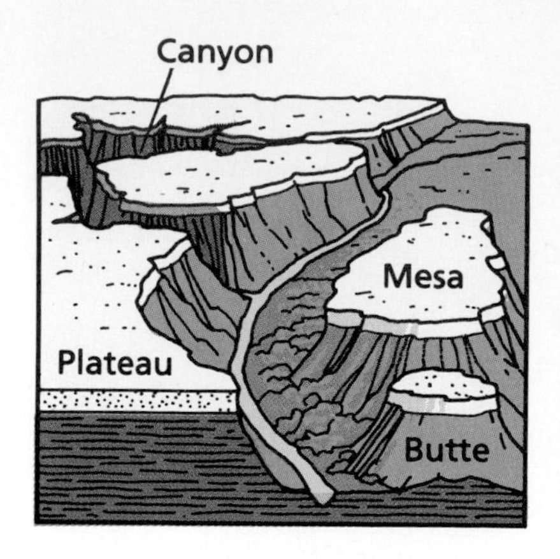

FIGURE 14–6. The erosion of softer material from beneath a resistant rock layer may result in the formation of mesas and buttes.

14:4 Runoff in Arid Regions

In arid and semiarid regions, the occasional rains may cut deep channels down steep cliffs or mountains. Between rains, weathered sediment accumulates in these dry channels. During and after a rain, runoff carries the loose material to the foot of the slope. Here the velocity of the water slows down. Coarse sediments such as sand and gravel are dropped first. Fine sediment, like mud, is carried farther. These sediments form apron-shaped deposits called **alluvial** (uh LEW vee ul) **fans.**

Arid regions receive less than 25 cm of rain per year. Semiarid regions have between 25 cm and 50 cm of rain annually. Most rain evaporates quickly. Little vegetation grows to hold soil in place. Consequently, heavy rains cause erosion. Soft material is removed, and steep resistant rocks rise abruptly from desert plains. **Mesas** (MAY suz) and buttes (BYEWTS) are flat-topped hills covered by a resistant rock layer such as lava, sandstone, or limestone. These hard layers protect soft, less resistant rock, such as shale, beneath them.

In the western United States, between the Rockies and the Sierra Nevadas, the uplift of mountains has left low areas with no drainage to the ocean. Interior drainage is toward the lowest elevations of the area between the mountains. Surface water may collect in this low **playa** (PLI uh) area and form a lake. Runoff carries the fine sediments to the playa, and when the water evaporates, minerals such as calcite, halite, and borax remain.

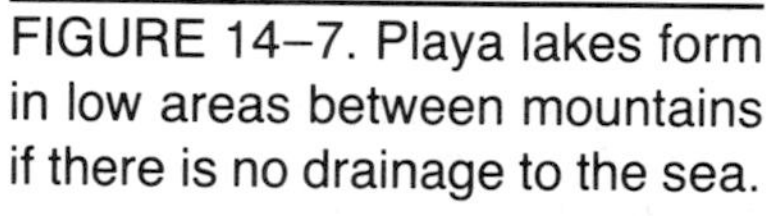

FIGURE 14–7. Playa lakes form in low areas between mountains if there is no drainage to the sea.

Review

1. What are buttes?
2. What minerals can be found in playas?
3. ★ What are three factors that reduce runoff?

RIVERS AT WORK

GOALS

1. You will be able to describe factors that affect the velocity of a river.
2. You will learn how the velocity of a river affects its ability to erode its bed and deposit sediments.

The extent to which a river erodes its bed depends on its velocity and the size of the load it is carrying. Usually, only about one-fourth of the material eroded by runoff reaches the ocean. The rest remains as sediment within the drainage basin.

14:5 Erosion by Rivers

What factors affect a river's ability to erode its bed?

Velocity depends on the slope of the river bed and on the volume of water. During flood periods, the velocity of a river increases because of an increased volume of water. Water erodes the bed and sides of the river channel very rapidly and rivers flow in a turbulent fashion. As water tumbles down, rises, and tumbles down again, it picks up fragments from the bed and sides of the channel. These rock fragments are dashed against the bottom and sides for further scouring action.

Sediment picked up and carried by water becomes the **suspended load** of a river. The **bed load** is material rolled along the river bottom because it is too large and heavy to be carried. Solid materials are abraded into finer particles as they are carried or rolled along the bed. Some material from the bed and sides is dissolved by the river and is carried along in solution.

If a river channel is cut into resistant rock, the channel will be deep and narrow. If the channel walls are soft, the material slumps down into the river and is carried away. In time, mass movements widen the valley far beyond the river channel itself. As a drainage system approaches a profile of equilibrium, the rivers within that system begin to **meander** (mee AN dur). This means they wander from side to side across their floodplains or val-

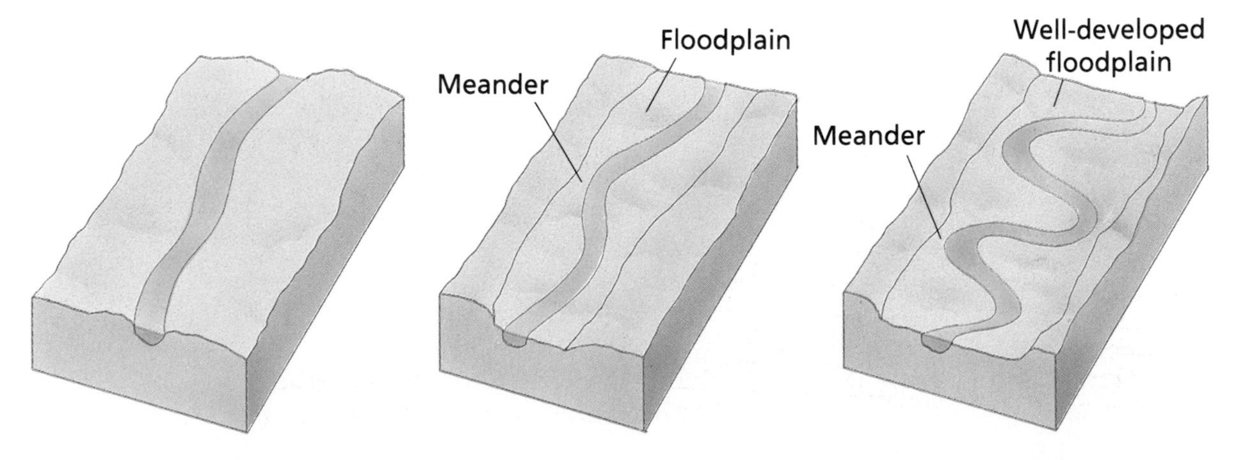

FIGURE 14–8. Meanders develop as a river erodes its floodplain. What is a cut-off meander?

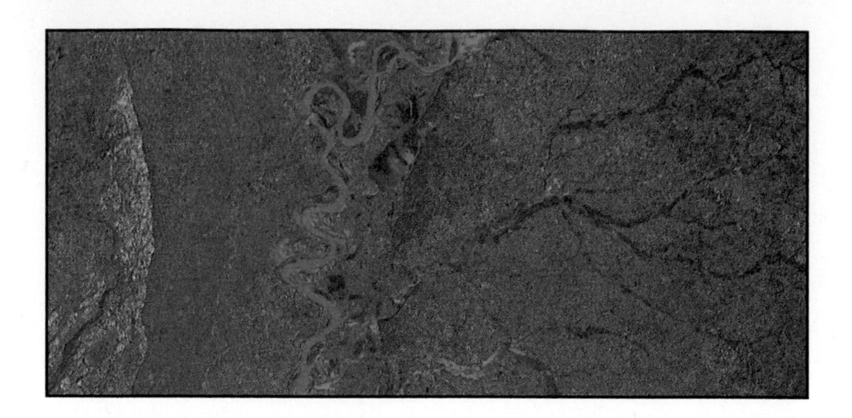

FIGURE 14–9. Meanders are well developed in the lower Mississippi River.

leys. Meandering sometimes begins because of an obstacle or irregularity in the channel. Meandering usually starts where the river profile flattens.

Where is the velocity of a river fastest?

The velocity of water in a river is usually fastest at the center. Friction with sides or the bottom slows water down. As a river meanders, the water velocity increases on the outside of the river bends. It decreases on the inside of the curves. Erosion on the outside of the curve removes about the same amount of sediment that is deposited on the inside of the curve. Meander curves widen and move downstream in time. As floodwater spills from one meander to the next across a narrow neck of land, a meander may disappear. This cut-off meander is called an **oxbow lake.** An oxbow lake dries up and fills with vegetation.

F.Y.I. Because the Rio Grande River meanders, the border between the United States and Mexico—formed by the river—has changed over time. Mexico has had to cede some land to the U.S.

CAREER

Hydrogeologist

Rhonda Hakundy is a hydrogeologist who samples and tests the water from groundwater monitoring wells. On a typical day, Ms. Hakundy goes to the wells, collects water samples, and takes them back to the laboratory to analyze.

Ms. Hakundy has a masters degree in geology, with a strong background in fluid mechanics, hydrology, geochemistry, agricultural hydrology, engineering, climatology and weather.

For career information, write:
American Geophysical Union
2000 Florida Ave. NW
Washington, DC 20009

INVESTIGATION 14–1

Stream Velocity

Problem: How does the slope of a stream affect its velocity and load?

Materials

stream table
plastic pails (2)
rubber tubing
meter stick
wooden blocks (2)
stopwatch
small cork
sand to fill stream table
books (2)
plastic sheet
water

Procedure

1. Arrange the stream table as shown.
2. Make a stream channel down the center of the sand so that it ends at the short length of rubber tubing.
3. Measure and record the length of the stream channel.
4. Fill the pail at the top of the table with water. Set up a siphon using a long piece of rubber tubing.
5. Put one block of wood under the upper end of the stream table.

FIGURE 14–10.

6. Put the cork at the upper end of the stream bed. Start the water into the stream bed.
7. Record the time the cork takes to travel the length of the stream channel.
8. Observe and record whether or not the water carries material other than the cork downstream. Stop the flow of water. Allow excess water to drain from the stream table.
9. Repeat the procedure in Steps 6-8 two more times. Record the average flow time in a table.
10. Stack another block on top of the first at the upper end of the stream table.
11. Repeat Steps 6, 7, 8, and 9 and record all observations in a table.

Data and Observations

Slope	Stream length	Flow time	Observations
one block			
two blocks			

Analysis

1. Calculate the velocity of the stream for a slope of one block and for a slope of two blocks.

$$\text{velocity} = \frac{\text{distance}}{\text{time}}$$

2. Did your stream meander?

Conclusions and Applications

3. How does the increase in slope affect the amount of sediment the stream carries?
4. What was the purpose of the cork?
5. How does the velocity of the stream change when you increase the slope?

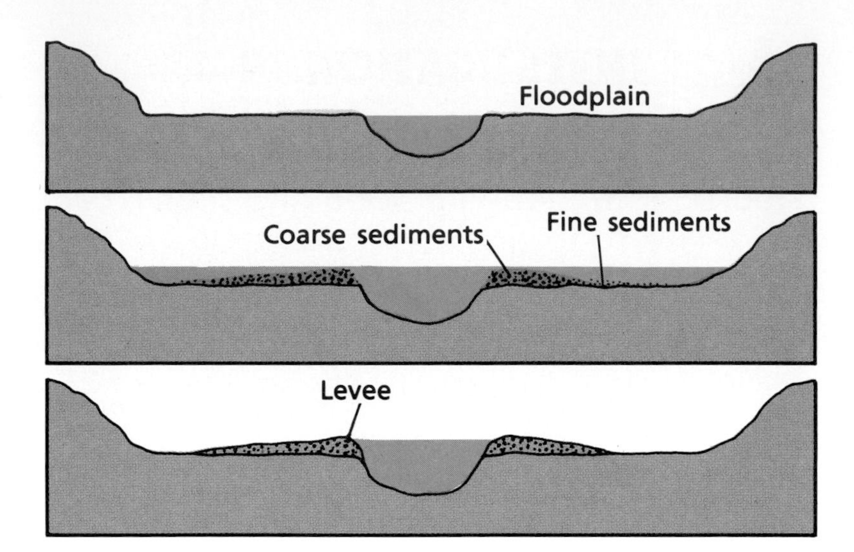

FIGURE 14–11. Levees develop when a river drops sediments along its banks during a flood.

14:6 River Deposits

Some small, quickly-moving tributaries carry very little sediment and seldom deposit their load. At lower levels, rivers may be heavily loaded with sediment. A small decrease in velocity causes a river to deposit sediment along its sides, in the channel, or at its mouth.

A river carries very large amounts of sediment during floods. When the river spills over its banks, coarse sediment is dropped along the river's margin. This sediment forms a low, ridgelike levee (LEV ee) parallel to the channel. Fine sediments are carried farther away from the channel and form a **floodplain** of fertile soil.

Where are some river deposits formed?

Many rivers deposit sediment within their channels during dry seasons. This material is swept out during floods, but more is deposited in the next dry period. Rivers deposit sediments at their mouths if they empty into a quiet body of water such as the ocean or a large lake. The river flow splits into a number of channels that may become filled with sediment as water loses its velocity.

The sediment is often deposited beyond the river mouth in the shape of a fan called a **delta.** Sediment from the

FIGURE 14–12. Deltas form at the mouths of rivers that empty into quiet bodies of water. Over time, sediments may be deposited in a bird's foot shape.

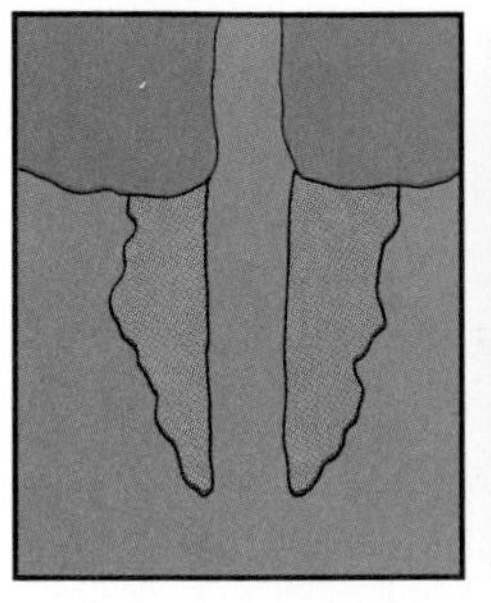
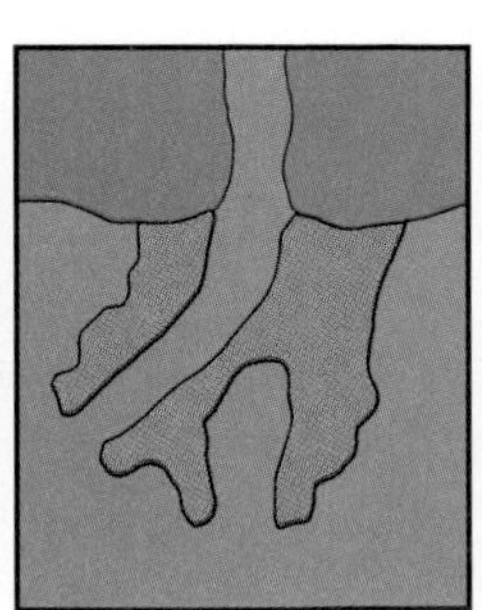
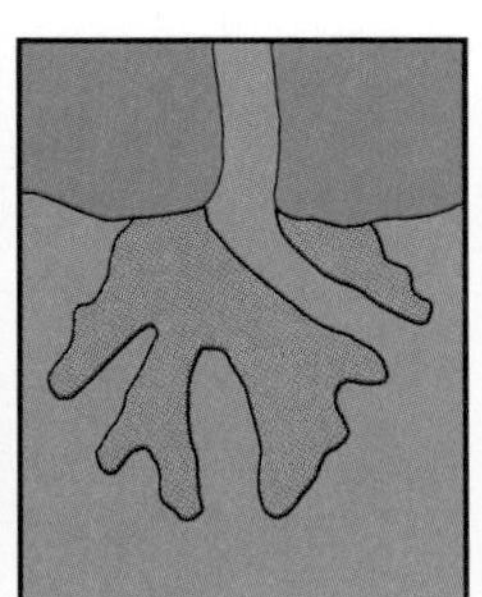
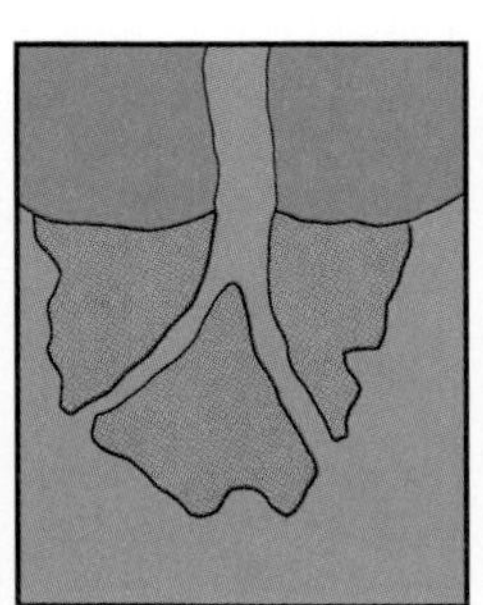
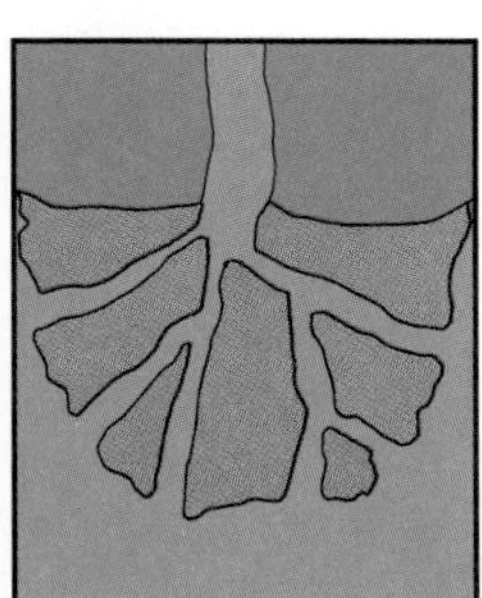

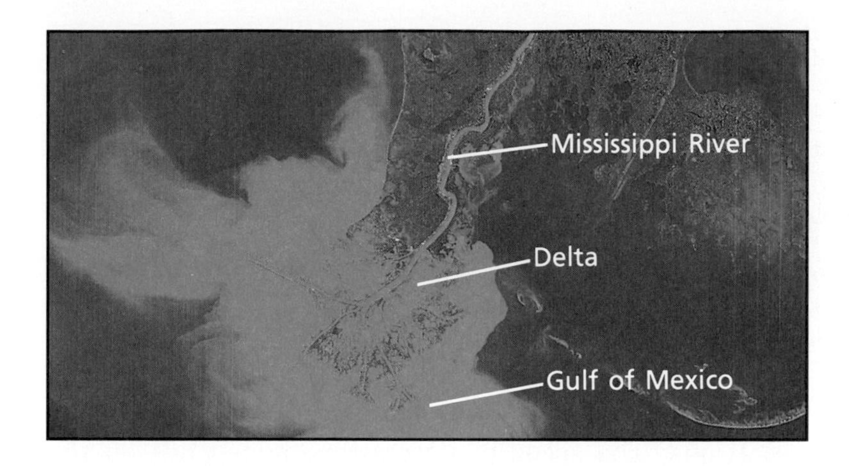

FIGURE 14–13. This satellite image shows that the Mississippi River delta has many small channels that carry water and sediment into the Gulf of Mexico.

Mississippi River has formed five large deltas in the last 5000 years. The river also has supplied sediment for many shore features along the Louisiana and Texas coasts.

Review

4. Upon what does the velocity of a river depend?
5. Where is the velocity of a meandering river fastest?
★ 6. Sketch a meander. Where does it occur?

TECHNOLOGY: APPLICATIONS

Flood Control

At Tsukuba Science City, northeast of Tokyo, Japan, several watershed models have been built. These models have rivers, dikes, dams, spillways, and reservoirs. One 76-meter long "rain room" has a rainmaker located 16 meters above the floor. This apparatus can vary the size of raindrops that fall to duplicate any type of rainstorm.

Experiments that these hydrologists and engineers perform enable them to make better predictions that aid in the evacuation of people from threatened areas during periods of high rainfall. They also enable engineers to design spillways that meet the waterflow characteristics of a particular dam, to minimize reservoir sedimentation, and to control riverbed erosion.

ACTION OF GROUNDWATER

GOALS

1. You will learn how the position of the water table affects the formation of streams, artesian wells, geysers, and hot springs.
2. You will learn how groundwater dissolves some minerals and deposits others.

Earth's upper crust is similar to a porous sponge. Many sediments have spaces between grains called pores. These pores provide places for water to collect and allow water to move from one location to another.

14:7 Groundwater

Why does groundwater move from higher to lower elevations?

When water sinks into the porous parts of the crust, it is called **groundwater.** Gravity causes groundwater to move through interconnected pores from higher to lower elevations. The speed of its movement depends on the permeability of the rocks and the gradient, or slope. Groundwater flows more slowly than surface water because of friction within the pores. There is less friction in large pores than in small pores. Groundwater flows at an average rate of less than 1.5 meters per day.

Groundwater is an important part of the water cycle. About 0.25 cm of the average 76.2 cm of rain that falls each year at a given location sinks into the ground. Some of this water returns to the surface in streams, swamps, and springs. Some returns to the oceans.

What is the water table?

Groundwater moves downward through permeable rock until it reaches an impermeable layer. Above the impermeable layer, pores fill with water to form a **zone of saturation.** The upper surface of this zone is the **water table.** Above the zone of saturation, pores are filled with air between rainstorms. In this **zone of aeration,** water and oxygen react with elements in the rocks to form clay, carbonates, and soil. Some water sinks just below the surface and becomes soil moisture. This layer, usually just a few centimeters thick, supplies water to plants.

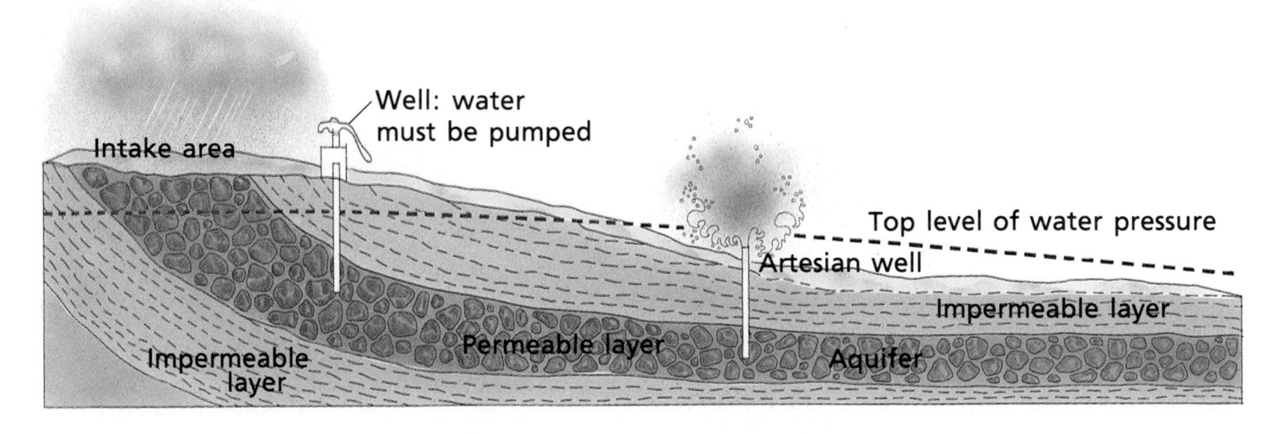

FIGURE 14–14. Artesian wells occur when a sloping aquifer is surrounded by impermeable rock layers.

a

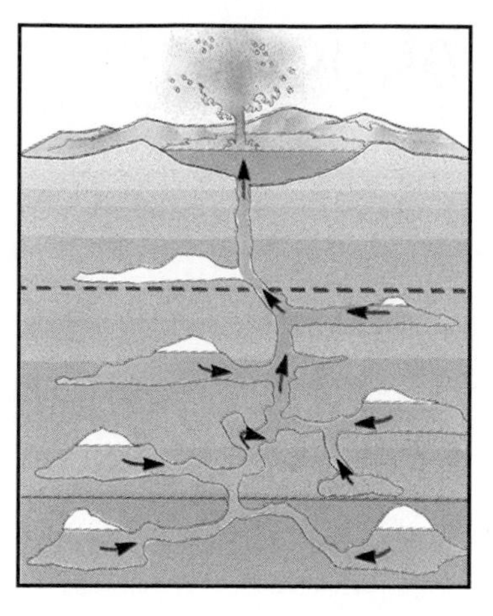

b

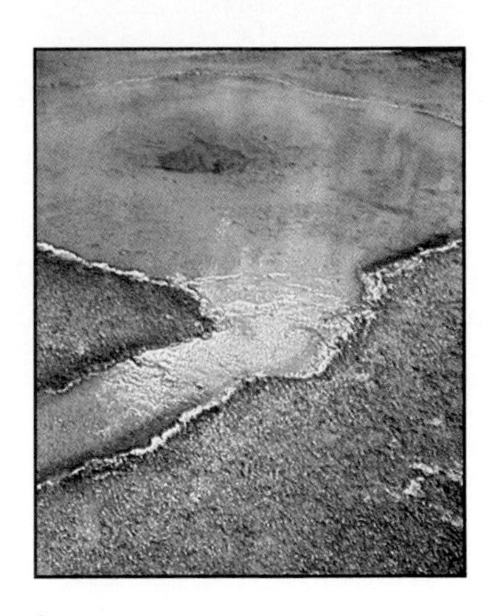

c

FIGURE 14–15. A geyser (a) erupts when water changes to steam (b). A hot spring (c) has a wider opening than a geyser.

Rainwater moves directly down to the zone of saturation if all rocks are permeable. If some layers are impermeable, water may be trapped above them in a perched water table. Water trapped above an impermeable layer flows along the surface of this layer until it reaches an opening of some kind.

Permeable rock layers that are filled with water are called **aquifers** (AK wuh furz). Gravel and sand are the best aquifers because they have the largest continuous openings. Most sands and gravels yield water when wells are drilled into them. This water must be pumped in order to reach the surface. Where the groundwater is removed by wells, it may be used up faster than it can be replaced by rainwater. Large withdrawals of groundwater lower the water table and may cause the land above the aquifer to sink.

When an aquifer is located between two impermeable layers and the upper part of the aquifer reaches the surface, the water moves through the aquifer in a manner that is similar to the flow in a pipe. Water rises toward the surface through any available openings, which may be pores or cracks in the overlying rocks or drilled wells. Water does not rise as high as the intake area because of friction. But if the outlet is lower than the intake, water will flow out at the surface. **Artesian** (ahr TEE zhun) **well** water is under pressure due to the mass of the column of water in the aquifer.

Geysers (GI surs) are hot springs that have a small surface opening. Water is heated at a depth within the crust and is forced upward by steam pressure at fairly regular intervals. Steam is produced by heat from recently intruded igneous rock or magma. Hot springs occur

BIOGRAPHY

Jay H. Lehr
1936-

Jay Lehr is the executive director of the National Water Well Association. He is an authority on groundwater model studies. His models use dyes placed in clear plastic boxes full of porous sands to show how groundwater moves. Lehr is a specialist in groundwater pollution, and in surface and groundwater law.

INVESTIGATION 14–2

Artesian Wells

Problem: How does an artesian well work?

Materials

30-cm rubber tube	funnel
protractor	water
scissors	sink
beaker	tape

Procedure

1. Cut end A at an angle. See Figure 14–16.
2. Use the scissors to cut a hole 2 mm in diameter in the tube 2 cm from end B.
3. Copy the Data and Observations table.
4. Hold the tube over the sink at a 45° angle as shown in Figure 14–16.
5. Place a funnel into end A and use a beaker to pour water in the tube in a steady stream.
6. Be careful not to pour water down the outsides of the tube.
7. Observe what happens at the 2-mm hole. Record your observations in the chart.
8. Reduce the angle that you hold the tube to 20°. Record what happens at the hole.
9. Repeat Steps 5-7 for angles of 30° and 40°. Record your observations.
10. Increase the angle that you hold the tube to 60°. Record what happens at the hole.

FIGURE 14–16.

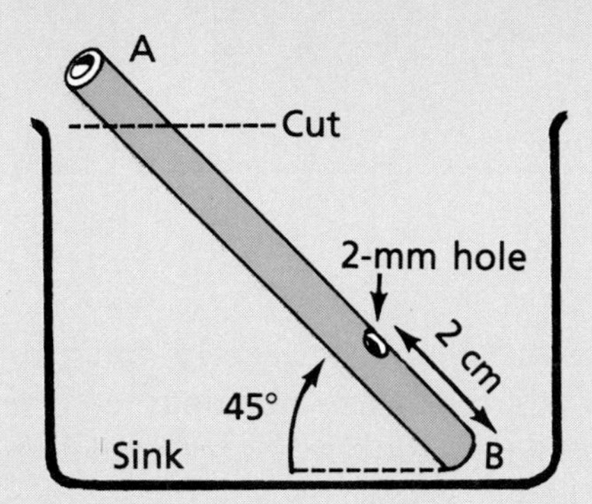

Analysis

1. Which angle allowed the greatest flow of water out of the hole? Which allowed the least?

Conclusions and Applications

2. Compare the inside of the tube to an aquifer. How are they similar?
3. Compare the rubber sides of the tube to impermeable rock. How are they similar?
4. What does the 2-mm hole represent?
5. How does an artesian well work?

Data and Observations

Angle	Observations
45°	
20°	
30°	
40°	
60°	

FIGURE 14–17. A sinkhole may develop when groundwater dissolves underlying limestone.

in the same area as geysers, but openings are not constricted. Calcium sulfate, calcium carbonate, and various sulfur compounds are dissolved in the hot water, giving many hot springs unusual colors, tastes, and odors.

14:8 Solution by Groundwater

Rainwater is a weak acid that will dissolve limestone, which is calcium carbonate. Groundwater circulating in limestone areas is called "hard" water because it contains many carbonates in solution. As water seeps downward in limestone regions, the limestone is dissolved. This leaves openings in the bedrock that may be enlarged into caves. Large caves or caverns may contain underground streams. But eventually, the stream rejoins the surface drainage system. The Green River of Kentucky flows through Mammoth Cave for some distance.

What types of large features are formed by groundwater?

Large features formed by solution include natural bridges, sinkholes, and caves. Some valleys form when the roof of an underground passage collapses. If a section of the roof remains, it forms a natural bridge. **Sinkholes** are funnel-shaped depressions dissolved from limestone along intersecting cracks or joints. The Bottomless Lakes of New Mexico are water-filled sinkholes, the bottoms of which are lower than the water table. Florida also has many lakes formed in sinkholes.

F.Y.I. Sinkholes "swallowed" cars and houses in Winter Park, Florida, in 1981.

14:9 Groundwater Deposits

How does a stalactite form?

Groundwater usually contains large amounts of dissolved minerals, especially if the water flows through limestone. As groundwater drips through the roof of a cave, gases dissolved in the water can escape into the air. A deposit of calcium carbonate is left behind. **Stalactites** are iciclelike structures of calcium carbonate that hang from the roofs of caves. Water also falls to the cave floor and evaporates, depositing calcium carbonate and building a **stalagmite.** A stalactite and stalagmite may merge to form a column.

Valuable metals including copper, silver, lead, and zinc may be deposited by heated groundwater in rock openings called veins. These veins are located in areas of igneous activity where hot groundwater collects metallic ions in solution and later deposits them as metallic ores. Groundwater also deposits cementing material in the pore spaces of some sediments to form sedimentary rocks.

Review

7. How do stalactites and stalagmites form?
8. What valuable metals are deposited by groundwater?
9. List three structures formed by groundwater.
10. Why is the water in limestone regions called "hard" water?
★ **11.** Describe the movement of groundwater using the terms permeable, impermeable, zone of saturation, zone of aeration, water table, and aquifer.

FIGURE 14–18. As water containing calcium carbonate in solution drips from a cave's roof, stalactites and stalagmites may form.

Chapter 14 Review

SUMMARY

1. Runoff results if precipitation over land is not evaporated, used by organisms, or absorbed by the soil. 14:1
2. A river eventually reaches a balance between the processes of erosion, transportation, and deposition. 14:2
3. The dendritic stream pattern is the most common type. 14:3
4. Alluvial fans, mesas, buttes, and playas are features of arid and semiarid regions. 14:4
5. The velocity of a river is usually fastest at its center. 14:5
6. River deposits include deltas, levees, and floodplains. 14:6
7. Gravel and sand make the best aquifers because they have large continuous pores between grains. 14:7
8. Solution by groundwater forms natural bridges, sinkholes, and caves. 14:8
9. Groundwater deposits include stalactites, stalagmites, and veins of copper, silver, lead, and zinc. 14:9

VOCABULARY

a. alluvial fans
b. aquifers
c. artesian well
d. bed load
e. delta
f. divide
g. drainage basin
h. drainage system
i. floodplain
j. geysers
k. groundwater
l. impermeable
m. meander
n. mesas
o. oxbow lake
p. permeable
q. playa
r. profile of equilibrium
s. runoff
t. sinkholes
u. stalactites
v. stalagmite
w. suspended load
x. water table
y. zone of aeration
z. zone of saturation

Matching

Match each description with the correct vocabulary word from the list above. Some words will not be used.

1. deposit found on the floor of a cave
2. the high ground between two drainage systems
3. nearly 40 percent of precipitation becomes this
4. deposits at the foot of a steep slope in arid regions
5. an isolated meander curve
6. funnel-shaped depressions dissolved from limestone
7. a lake between two mountains that has no drainage to oceans
8. sediment deposited beyond the mouth of a river in the shape of a fan
9. large fragments of sediment that are moved along a river bottom
10. the upper boundary of the zone of saturation

Chapter 14 Review

MAIN IDEAS

A. Reviewing Concepts

Choose the word or phrase that correctly completes each of the following sentences.

1. Streams that develop on domes of igneous rock have a pattern that is *(rectangular, radial, dendritic)*.
2. Meanders are associated with *(playa lakes, drainage basins, river floodplains)*.
3. Groundwater does not rise as high in an artesian well as the intake area because of *(deposition, impermeable layers, friction)*.
4. During floods, a river's velocity increases because of an increase in *(slope, dissolved load, water volume)*.
5. The velocity of a river is fastest at the *(center, sides, bottom)* of its channel.
6. Groundwater in *(limestone, sandstone, clay)* regions carries the greatest amounts of dissolved minerals.
7. Groundwater flows *(more slowly than, more quickly than, the same speed as)* surface water.
8. The most common stream pattern is *(radial, rectangular, dendritic)*.
9. *(All, Impermeable, Permeable)* rocks may serve as aquifers.
10. A meandering stream deposits sediment on *(the inside, the outside, both the inside and the outside)* of its meander bends.
11. The large branches of a drainage system are called *(rills, rivers, creeks)*.
12. Rivers sometimes develop a *(dendritic, radial, rectangular)* pattern when they flow over jointed rocks.
13. A(n) *(permeable, impermeable)* rock forms the base of the zone of saturation.
14. The network formed by a main river and its tributaries is a *(zone of saturation, drainage system, sinkhole)*.
15. Sediment picked up from the bed and sides of a river and carried by the water becomes the *(bed load, alluvial fan, suspended load)*.

B. Understanding Concepts

Answer the following questions using complete sentences.

16. Explain how meanders form.
17. What is the process that forms natural bridges?
18. How does temperature affect runoff?
19. Explain how the resistance of the land affects the direction of the path of a stream.
20. What gives hot springs color and odor?
21. Describe how alluvial fans form.
22. What determines the extent to which a river erodes its bed?
23. Describe how a levee forms.
24. Why does groundwater flow more slowly than surface water?
25. Explain why gravels and sands are the best aquifers.

C. Applying Concepts

Answer the following questions using complete sentences.

26. How would the damming of a stream affect its profile of equilibrium?
27. Is there more runoff from a square kilometer of parking lot than from a square kilometer of farmland? Explain.
28. How can aquifers become depleted?
29. Why is water an important erosional agent in desert areas?
30. How does the presence of a bog indicate the depth of the water table?

SKILL REVIEW

1. Compare and contrast a delta and an alluvial fan.

2. Use Figure 14–5. Suppose you have tested the water along this river at points Y and Z, but you have not yet checked X. Y does not appear to be polluted, but toxic substances are found at Z. Pick the best hypothesis from the following and explain your answer. (A) Pollutants will be found in the water at X. (B) Factory 1 is polluting the water. (C) Factories 3 and 4 are polluting the water.
3. What is the effect of heavy rains on arid and semiarid regions?
4. Water leaks into your basement. What can you conclude about the size of the pores in the cement walls and floor? Are they permeable or impermeable?
5. You discover a structure growing from the cement roof of an underground parking garage. It reminds you of a stalactite. What can you hypothesize about the way it was formed?

PROJECTS

1. Devise an experiment to estimate the amount of runoff from a paved parking lot or driveway.
2. Locate a major river in your area. On a map, preferably a topographic map, find the local creeks or rivers that drain into your major river. Trace the drainage from your area to the ocean.

READINGS

1. Bowen, R. *Groundwater*. New York: Elsevier, 1986.
2. Emil, Jane. *All about Rivers*. Mahwah, NJ: Troll Associates, 1984.
3. Mayer, Alfred. "Between Venice & the Deep Blue Sea." *Science 86*. July/August 1986, pp. 55-57.

ICE IN MOTION
ICE AGES

Glaciers

Although you may not live in an area where glaciers are present today, your region likely was changed by glaciers of the past. Today, glaciers are found only at high elevations and in polar regions such as Greenland and Antarctica. Yet, nearly 75% of Earth's total freshwater supply remains frozen in glaciers. Currently, ice covers 10% of all of Earth's land.

ICE IN MOTION

Glaciers are large masses of ice in motion. A glacier can carry an even more massive load of eroded debris than a stream. At various times in the life of a glacier, the glacier will melt, drop its load, and form new deposits.

GOALS

1. You will learn the characteristics of different types of glaciers and how these glaciers develop.
2. You will examine both the erosional and depositional effects of glaciers.

15:1 Development of Glaciers

Glaciers are thick masses of ice that move slowly on land. Glaciers form where more snowfall is accumulating than is melting, and temperatures are cold enough for the snow to remain throughout the year. New snow falls on the surface of the old snow. As the snow accumulates, rounded ice granules form in the upper layers of snow. These granules are like snow that has been on the ground for a few weeks. When 30 to 60 meters of snow and ice granules accumulate, their mass causes bottom granules to become compressed and form glacial ice. The bottom ice becomes pliable, somewhat like putty, and begins to move. If there is a slope, movement will be downhill. When the underlying surface is nearly flat, the glacier spreads outward in all directions, like spilled honey.

Movement of glacial ice is called flow. There are two basic types. Plastic flow involves movement within the ice. When the pressure or load on the ice is greater than 50 meters, the plastic flow causes the layers of ice to slide over one another. The second type of flow is called basal

F.Y.I. Although glaciers usually move only a few centimeters a day, in 1966 a glacier at Mount Steele in the Yukon was observed to be moving 60 centimeters an hour.

BIOGRAPHY

Matthew A. Henson

1866-1955

Beginning in 1891, Matthew A. Henson accompanied Admiral Robert E. Peary on many expeditions to locate the North Pole. Henson was indispensable in helping Peary get to the Pole. It was Henson who actually placed the American flag on the exact spot known as the North Pole on April 7, 1909.

slip. Basal slip occurs when meltwater from the glacier acts as a lubricant to help the ice flow over rocks.

Glaciers are classified as valley, piedmont (PEED mahnt), and continental. **Valley glaciers,** also called alpine glaciers, form at high elevations where snow lasts year after year. Valley glaciers take over the valleys of a river system that predated the colder temperatures. These glaciers begin as snowfields that collect in hollows or valleys. These hollows may be hidden from sun and wind. Thus, snow can remain for years as long as temperatures are low. In humid climates, accumulation of snow may be rapid. In drier regions, snow may be so scarce that glaciers cannot form no matter how cold it is.

Changes in weather conditions cause changes in the elevation of the end position of a glacier. Warmer temperatures cause the glacier to melt, and it appears to retreat. Colder temperatures cause the glacial front to advance. Valley glaciers extend down the slope to the altitude where melting offsets additions of new snow.

The rate of movement of a valley glacier varies. The central part of a glacier moves faster than the edges because of friction along the valley walls. The movement of a glacier can range from a few centimeters to a meter a day. Sudden movements, called surges, may occur after long periods of little movement. Surges may travel at the rate of six kilometers per year and may last two to three years.

When a valley glacier extends onto a plain, it becomes a **piedmont glacier.** At the foot of the mountains, ice forms a continuous sheet as glaciers spread out and join.

FIGURE 15–1. Valley glaciers form at high elevations in old river valleys.

FIGURE 15–2. Most of the continent of Antarctica is covered by a glacier. What type?

In Alaska, the Malaspina Glacier extends for 80 kilometers along the base of St. Elias Mountain. During the last ice age, piedmont glaciers along the Rocky Mountains merged with the continental glacier that moved south from Canada.

Continental glaciers are great masses of ice that are found in high latitudes near Earth's polar regions. Unlike alpine glaciers, continental ice sheets are not confined to valleys or hollows. Presently, they cover nearly all of Greenland and Antarctica. Movement of ice in these glaciers occurs outward from the center. Movement is very slow and usually averages only a few meters a year. Today, continental glaciers range in thickness from 2500 to over 4000 meters. This type of glacier once covered and smoothed all of the mountains in New England. In Canada, they removed the soil and exposed the bedrock. In the past, continental glaciers have covered North America as far south as the Missouri and Ohio Rivers.

F.Y.I. The Barnes Ice Cap of Baffin Island, in Canada, is one of the remaining portions of the continental glacier that covered much of Canada and parts of the northern United States 15 000 to 20 000 years ago.

15:2 Erosion by Glaciers

Glaciers do not cover as much of Earth's surface as stream systems do, nor do they persist over such long periods of time. When these two erosional agents are compared, streams are more effective. However, glaciers, where they are present, completely change Earth's surface. They move over the land, scraping away all loose fragments and piling this debris in massive mounds.

How do glaciers erode bedrock?

Glaciers erode by abrading and plucking. Abrading is the scouring of the bedrock surface over which the ice

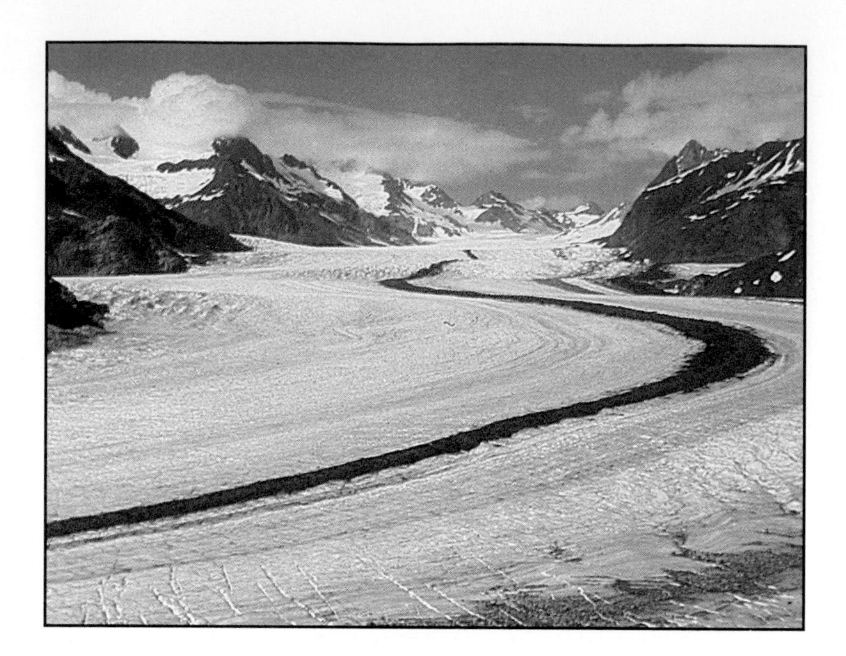

FIGURE 15–3. Material plucked from valley walls may look like streams of dirt flowing through the ice. How does plucking occur?

moves. This process is similar to abrasion by streams and wind. **Meltwater,** water resulting from the melting of glacial snow and ice, flows downward until it penetrates cracks in the bedrock. Meltwater freezes in the cracks. As it freezes, it expands, wedging blocks of rock loose. This process, the lifting out and removal of fragments of rocks, is called **plucking.** This rock becomes part of the moving glacier. In this way, massive loads of boulders, gravel, and sand are added to the bottom of the glacier. As the ice moves forward, this debris abrades, scours, and polishes the underlying bedrock. Gouges and scratches made by rocks frozen in the ice often mar the smooth surface of the bedrock. Long, parallel scratches made by the rocks embedded in the ice are called **striations.** Glaciers erode soft rock more deeply than hard rock, leaving many steplike irregularities in the bedrock. As the glacier advances, the debris frozen into the bottom of the glacier is itself ground up into smaller fragments.

What are striations?

All glaciers erode the rocks over which they move. Valley glaciers, in addition to abrading the sides and floors of a valley, erode headward within the valley. Near the head of the valley, the glacier is in contact with the wall rock. As the rock heats during the day, it causes the ice to melt. Meltwater seeps down from the ice into cracks in the wall rock. This meltwater expands when it freezes. This causes fragments from the wall rock to break off. These rocks are added to the edges of the glacier and carried away by the ice. This type of plucking is called ice wedging. With time, new snow is added to the glacial

What is ice wedging?

ice at the head of the valley. Through this headward extension, the head of the valley is enlarged and a cirque is formed. A **cirque** is a small round basin with one side open to the valley.

When two or more cirques erode a mountain summit from several directions, a sharpened peak is formed. This peak is called a **horn.** Many of the peaks of the Swiss Alps are horns, the most famous of which is the Matterhorn. The Matterhorn was formed as headward extension by valley glaciers reached the mountain summit from several directions.

How was the Matterhorn formed?

Water-eroded stream valleys are usually V-shaped. The tributary stream valleys are at the same elevation as the main river valley. After glaciation, the main valley is U-shaped, deeper, and straighter. The ability of the glacier to deepen a valley depends on the thickness of the ice. Thus, the main valley glacier erodes its valley much deeper than the tributary glaciers erode their valleys. When the ice melts, the outlets of tributaries are high above the main valley floor. These high valleys are called **hanging valleys.** Usually, rivers flow again in these valleys. Waterfalls form where the hanging valleys join the main river.

How have river basins been changed by glaciation?

Erosion by valley glaciers is small compared to the erosion of Earth's surface by continental glaciers. Continental glaciers tend to override the land, removing soil and some bedrock. They also cut deep channels and grooves in the solid bedrock. About 10 000 years ago, the drainage system of North America was changed by continental

FIGURE 15–4. A cirque is the natural hollow in which a glacier developed (a). The Matterhorn in Switzerland was formed as glaciers eroded cirques toward the mountain's summit (b).

a

b

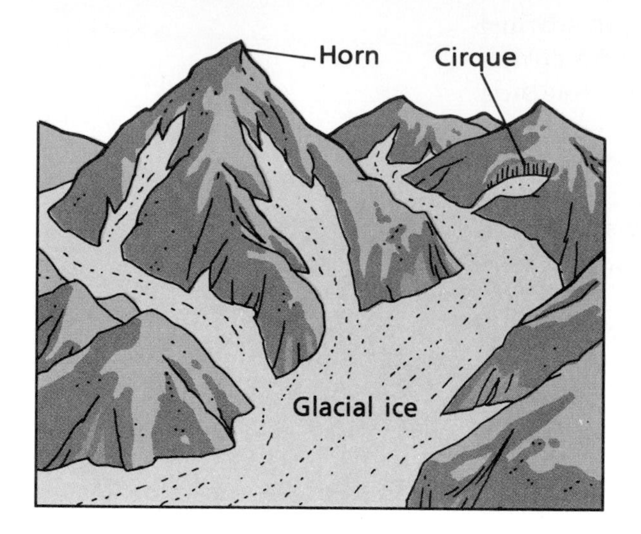

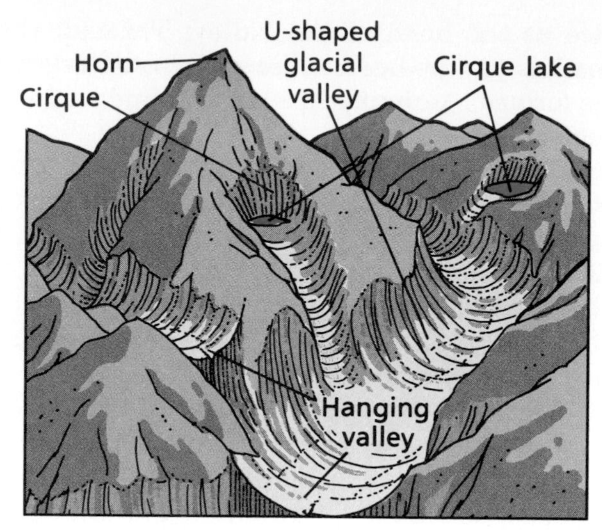

FIGURE 15–5. The presence of horns, cirques, and hanging valleys indicates that an area was once eroded by valley glaciers.

glaciers. Depressions gouged by glacial ice and later filled with water formed lakes for which Canada is famous. Depending on the direction of ice movement, some old river channels were filled with debris. Others were enlarged and deepened by the ice. The Great Lakes located between Canada and the United States were formed in old river valleys. These valleys were deepened as the ice moved south. The bottoms of some of the Great Lakes are below sea level. Only a great sheet of ice could dig a basin to such a depth.

15:3 Glacial Deposits

The first material dropped as a glacier melts is unsorted and unlayered. This material is a jumble of boulders, sand, and clay called **till.** Meltwater flowing away from the margins of the glacier transports some of the debris. Like river deposits, this sediment is sorted and deposited as the velocity of the meltwater decreases. These sorted, layered deposits of glacial debris are **outwash.**

What are terminal moraines?

Ridges of till form during periods when the forward movement of ice is equal to or less than the rate of melting. These ridges, or **terminal moraines** (muh RAYNZ), are deposited at the margins of a glacier. On the Malaspina Glacier in Alaska, the moraine has been in one place long enough for soil to form and trees to grow.

During the most recent ice age, when continental glaciers reached their southernmost positions in the United States, a great loop of terminal moraines was deposited. As the ice melted faster than the glacier advanced, ground moraine was formed. **Ground moraine** is till dropped from the base of an ice sheet when it melts. Continental deposits of ground moraine cover hundreds of square kilometers. Ground moraine tends to fill valleys and cover hills, making the resulting land surface fairly smooth.

Certain areas of ground moraine contain features called drumlins. Drumlins are streamlined hills that form parallel to the direction of the ice flow. The average height of a drumlin is 15 to 50 meters. They range from 900 to 2000 meters in length and 180 to 460 meters in width. Drumlins form beneath the moving ice.

What are drumlins?

Kames are small knobby hills of sand and gravel. These hills are steep-sided and commonly form between lobes of ice that advanced from different directions. Some kames form when large blocks of ice at the end of a glacier become stagnant and melt.

Meltwater drops sand and gravel in wide, gently sloping deposits that fan out from the front of the terminal moraine. These outwash plains are similar to deltas, but the material in them is relatively coarse. Other layered deposits also are formed by glaciers. Meltwater forms rivers within or beneath the ice. These rivers deposit sand and gravel within their channels. When the ice melts, the outwash is deposited in long winding ridges called **eskers.**

What are eskers?

FIGURE 15–6. Continental glaciers deposit both sorted and unsorted sediments as they melt.

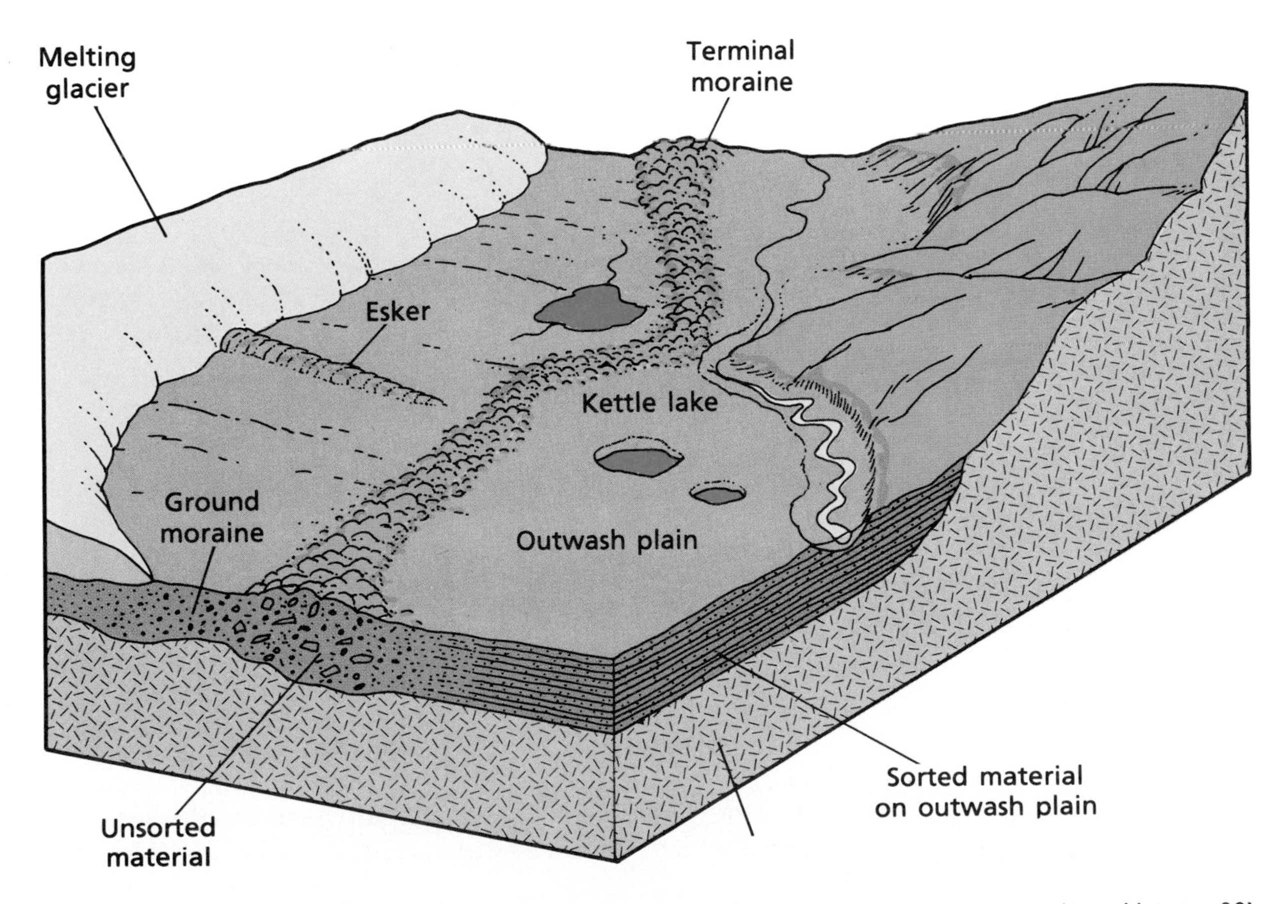

FIGURE 15–7. Kames are deposited by glaciers.

A number of lakes are associated with glaciated regions. **Kettle lakes** are small basins formed when blocks of ice, surrounded by debris, melted very slowly. After the ice melted, water filled the basins and lakes were formed. Most of Minnesota's lakes are kettle lakes.

Review

1. What conditions must exist for a glacier to form?
2. Why is soil extremely thin where continental glaciation has recently occurred?
3. What are four indications that an area has been eroded by valley glaciers?
★ 4. How can you determine if a glacial deposit was formed by meltwater rather than directly by melting ice?

PROBLEM SOLVING

Freda's Frozen Failure

Freda had promised her dad she would mow the lawn and bag the grass before he got home from work. It was a hot afternoon and she knew she would be very thirsty by the time she finished such a big job. She decided to make herself a pitcher of lemonade as a reward. She poured the lemonade into a glass jar, put a lid on it, and placed it in the freezer to make it really cold.

At the end of two hours she finally returned for a glass of lemonade. She opened the freezer to find the glass jar broken. It did not take Freda long to guess what had caused the glass to break. What was Freda's hypothesis?

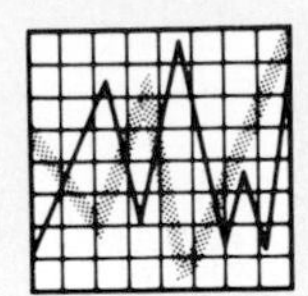

SKILL

Interpreting a Glacial Map

Problem: How are glacial features identified on a topographic map?

Materials

Figure 3–19
metric ruler
paper
pencil

Background

Recall from Chapter 3 that landforms can be identified on topographic maps. Figure 3–19 is a topographic map of a glaciated region. Recall that glaciers erode the land over which they move. Glaciers also change Earth's surface by depositing till and outwash.

Procedure

1. Study Figure 3–19.
2. Compare Figure 3–19 with the features shown in Figure 15–6.
3. Make a table similar to the one shown in Data and Observations in which to record your data.

Questions

1. What is the contour interval of this topographic map?
2. Locate the lake in the northwest corner of the map. What is its elevation?
3. What type of lakes are Stink Lake and Lake Hester?
4. How did these three lakes form?
5. What type of material forms the ridges on both sides of Stink Lake?
6. What are these ridges called?
7. How did these ridges form?
8. What do these ridges tell you about the direction of glacial movement?
9. How long is the ridge to the southwest of Stink Lake?
10. Examine the northeastern corner of the map. Notice the groups of closed contour lines. What are these small hills? How did they form?
11. What type of glaciation formed the features on this topographic map of a part of North Dakota?
12. Which features on the map are erosional? Which features are the result of deposition?
13. Review drainage patterns from Chapter 14. Can a specific drainage pattern be identified in this glaciated area? Explain your answer.
14. Describe how the map would appear if a contour interval of 50 feet were used.

Data and Observations

Area	Type of feature	How was feature formed?
Stink Lake, Lake Hester		
ridges on either side of Stink Lake		
northeast corner of map		

INVESTIGATION 15–1

Glacial Erosion and Meltwater

Problem: How do valley glaciers affect the surface? How does meltwater sort material?

Materials

stream table with sand
pail
ice block containing sand, clay, and gravel
wood block
metric ruler
overhead light source with reflector

Procedure

1. Set up the stream table as shown.
2. Make a river channel. Measure and record its width and depth. Draw a sketch that includes these measurements.
3. Position the light source so that the light shines on the stream bed as shown.
4. Place the ice block in the river channel at the upper end of the stream table.
5. Gently push the "glacier" along the river channel until it is halfway between the top and bottom of the stream table and is positioned directly under the light.
6. Turn on the light and allow the ice to melt. Observe and record what happens.
7. Draw a sketch indicating where the sand, gravel, and clay frozen in the "glacier" were deposited.
8. Measure and record the width and depth of the glacial channel. Draw a sketch of the channel and include these measurements.

Data and Observations

	Width	Depth	Observations
River			
Glacier			

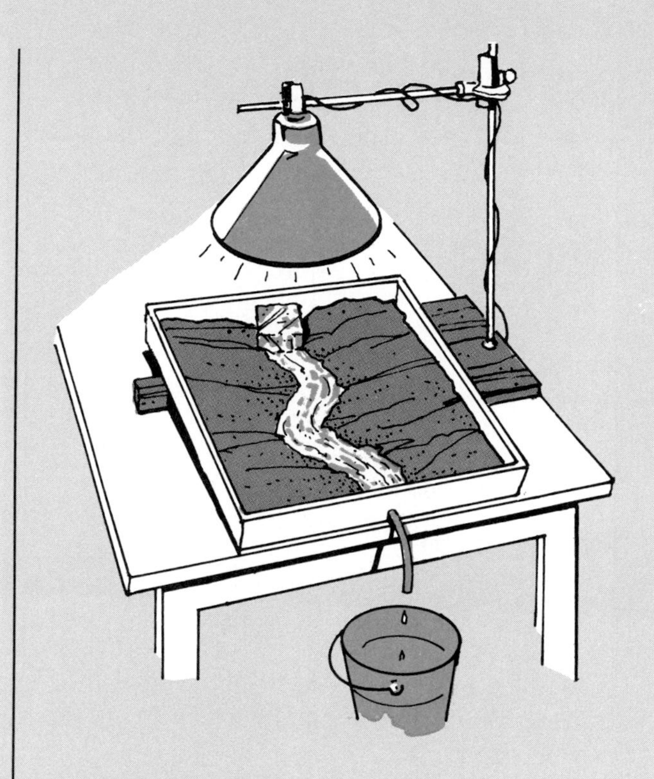

Analysis

1. How does the glacier affect the surface over which it moves?

Conclusions and Applications

2. Explain how you can determine the direction from which a glacier traveled by considering the shape of the channel and deposits.
3. Can you determine the direction of glacial movement from sediments deposited by meltwater? Explain.
4. How would tributary streams flowing into the main channel before glaciation be affected by the changes in the glaciated stream channel?

ICE AGES

Worldwide glaciation has had diverse effects on Earth. Geologists are certain that there have been at least three times in Earth's past when glaciers have covered large areas of the world. Probably there have been other such times, but these three periods of glaciation are well-confirmed by evidence.

GOALS

1. You will examine the effects of glaciation on sea level.
2. You will learn of several hypotheses that explain what may cause ice ages to alternate with warm intervals.

15:4 Effects of Glaciation

In addition to the erosional and depositional features left by glaciers, other evidence of glaciation has been found. Large glacial lakes exist in regions that are presently arid. Large valleys and canyons that were cut by tremendous amounts of water are today occupied by small streams. Glacial moraines found on the edges of central African jungles indicate that these tropical areas were once covered by glaciers.

Many huge lakes formed as the glaciers melted. Some lakes formed between the ice front and the moraines. Others, like Lake Bonneville, formed in the basins between mountains when precipitation was high. Lake Bonneville in what is now Utah covered about 30 000 square kilometers. The Great Salt Lake, Utah Lake, and

F.Y.I. In the U.S.S.R., scientists are studying ways of melting glaciers to relieve the droughts that occur in central Asia.

a

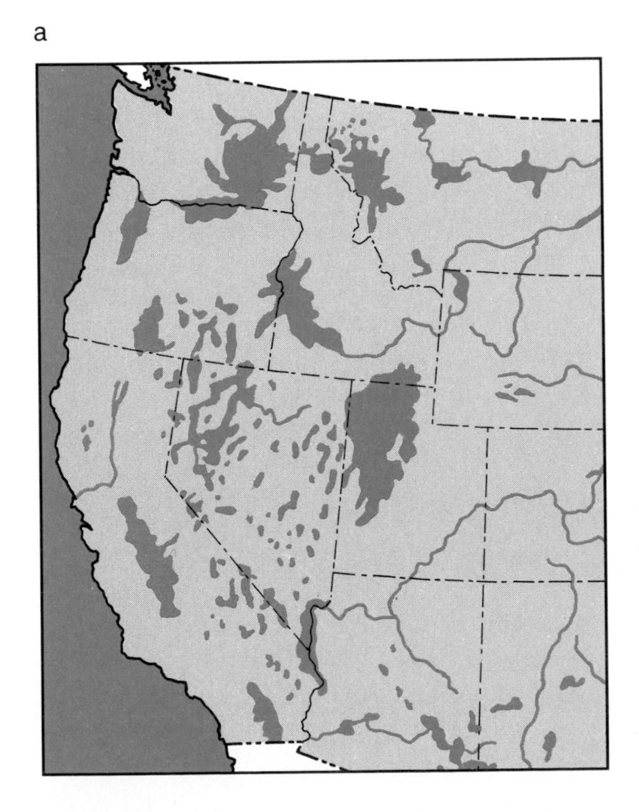

b

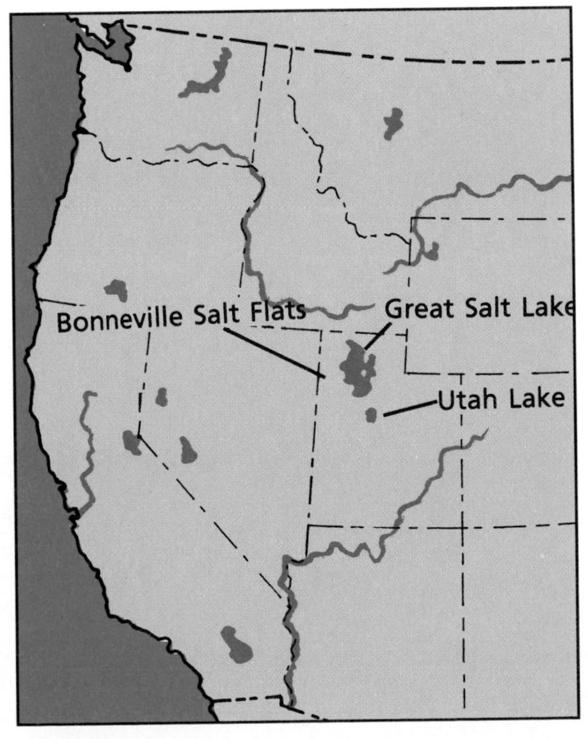

FIGURE 15–8. Many lakes formed in the western United States as continental glaciers melted (a). Only remnants of these lakes now exist (b).

FIGURE 15–9. Terminal moraine deposited by a glacier serves as a dam for this river.

the Bonneville Salt Flats are the remnants of the much larger Lake Bonneville.

Other lakes formed because they were dammed behind terminal moraines deposited by glaciers. Examples are Jenney Lake in Grand Teton National Park in Wyoming and St. Mary Lake in Glacier National Park in Montana. These lake basins were formed by ice sheets. When the glaciers extended into this area, the ice sheets covered stream-eroded lowlands and gouged the bedrock. The Finger Lakes in New York were also formed by this type of erosion by ice sheets.

How does glaciation affect sea level?

Sea level was also affected by glaciation. As the ice melted and returned water to the ocean, sea level rose. During the last glacial period, sea level was about 110 meters lower than it is now. If all the ice present today in Greenland and Antarctica melted, sea level would rise about 60 to 70 meters. Rises in sea level due to glacial melting are accompanied by rises in the land surface. The mass of the ice sheets causes the land to subside. When the glaciers melt, this pressure is removed, and the land begins to slowly rebound. This rebounding of the land continues today in areas once covered by glaciers.

FIGURE 15–10. Sea level was lower during the last ice age.

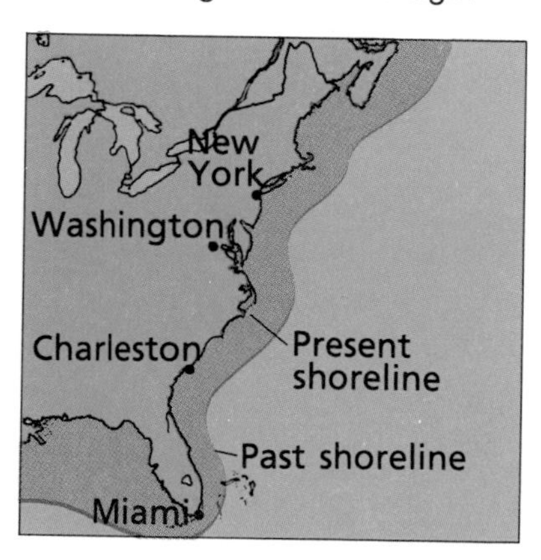

15:5 Glacial Origins

An **ice age** is a period of time when ice sheets and alpine glaciers are far more extensive than they are today. The earliest ice age occurred over 600 million years ago. A second major period of glaciation took place about 200 million years ago. The most recent ice age began about eight million years ago. The last ice sheet finished retreating about 7000 years ago.

The most recent research into the cause of ice ages suggests that the amount of ice on Earth is greatest every 22 000, 42 000, and 100 000 years. Scientists relate these figures to three changes in Earth's movements. Earth's axis is now tilted 23½° to the plane of its orbit. This tilt increases to 25° and then decreases to 22°. When the tilt is at its maximum, 25°, Earth's poles receive more energy from the sun. When the tilt is at its minimum, 22°, the poles receive less energy. This decrease occurs about every 42 000 years.

What do scientists believe cause ice ages to occur?

Another factor affecting glaciation may be the change in the shape of Earth's orbit around the sun. The orbit approaches a circle every 100 000 years. When the orbit is more circular than at present, Earth is farther from the sun and temperatures are colder than normal.

F.Y.I. During the last ice age, more than three-tenths of the world's land surface was under thick ice and snow.

The third factor also involves Earth's axis. Due to differences in gravitational forces on Earth, the axis wobbles every 22 000 years. According to this hypothesis, when a more circular orbit and decreased tilt occur together, an ice age may begin. A more elliptical orbit and greater tilt bring warm interglacial periods.

At least for the past three million years, ice ages have alternated with warm intervals. Some interglacial intervals were warmer than our climate today. Some of the ice ages lasted 60 000 years. The most recent ice age

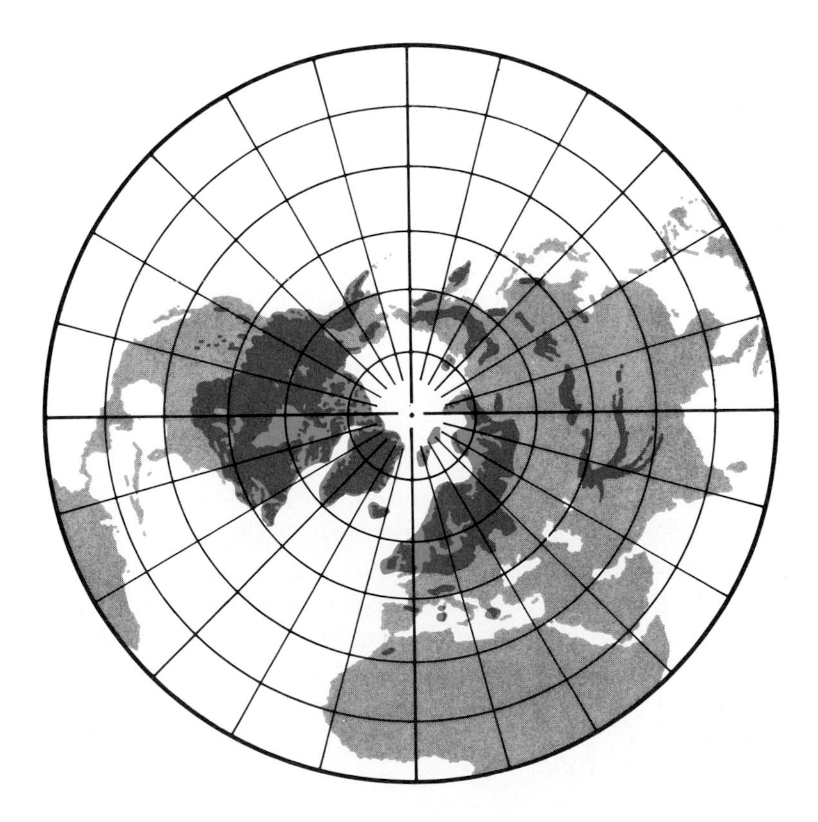

FIGURE 15–11. The extent of glaciation in the Northern Hemisphere during the last ice age is shown on this polar projection.

When might another ice age begin?

reached a peak 18 000 years ago. Then, Earth began to get warmer. The present interglacial period began about 11 500 years ago. Most of the interglacial periods have lasted about 12 000 years. Some scientists predict the beginning of another ice age within 1000 years or so.

These hypotheses explain the variations in the most recent ice ages. They do not explain why glaciers have occurred so rarely over the long span of geologic time. Present hypotheses do not answer all questions about glaciation, but a great deal of research is being done in this field.

Review

5. Compare sea level during the last ice age with sea level today.

6. How does the tilt of Earth's axis affect the amount of solar energy Earth receives?

★ **7.** Why do some scientists believe that we are now in an interglacial period?

CAREER

Glaciologist

Dr. Louis Davis is a glaciologist who studies the movements of glaciers. His team of researchers places red flags in the ice on a glacier. Attached to these flags are radio-equipped collars, similar to those used to track animals. Using microwave equipment in a research helicopter, Dr. Davis and his team measure the movement of the glacier monthly at the five points. Dr. Davis' team uses aerial and satellite photography and time-lapse cameras to further track the glacier's movement. This research group is also concerned with measuring the depth of the water at the glacier's face, and calculating the rate of calving, or the breaking away of giant chunks of ice into the water.

During Dr. Davis' career as a glaciologist, he has investigated the composition, structure, and physical history of glaciers. He has divided his time between field work and laboratory work. In his research, Dr. Davis utilizes the principles of physics, mathematics, and chemistry to compile data for professional papers.

For career information, write:

Society of Exploration Geophysicists
P.O. Box 3098
Tulsa, OK 74101

INVESTIGATION 15–2

Rebounding

Problem: How does a glacier affect the elevation of land over which it is located? What happens to the land after a glacier is gone?

Materials

pan balance
sheet of construction paper
small pans (2)
sand
ice cubes (3)
pencil
paper

Procedure

1. Place the construction paper so that it stretches across the tops of both pans of the balance, but does not hang over the edges of the pans. See Figure 15–12. The construction paper will represent the lower boundary of Earth's crust.
2. Place the small pans on top of the construction paper, each on a different pan of the balance.
3. Pour sand to fill one of the pans.
4. Repeat for the other pan, until they balance. These pans will represent the upper crust of Earth. Observe and draw the side view of the construction paper. Make sure you make your observations at eye level.
5. Place an ice cube on top of pan A. Observe and draw the side view of the construction paper. Add a second ice cube. Observe and draw the side view of the construction paper. Add the third cube to pan A and draw your observations. What does the addition of ice cubes represent in this experiment?
6. Remove one cube at a time from pan A. Observe and draw the side view of the construction paper as you do this. Draw different pictures as you remove each ice cube. What does the removal of the ice cubes represent in this experiment?

FIGURE 15–12.

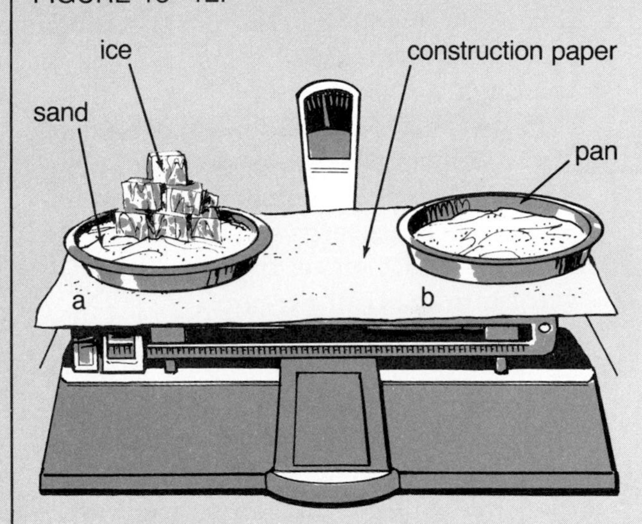

Analysis

1. Describe how the addition of ice cubes to pan A changed the side view of the construction paper and the position of pan A.
2. Describe how the removal of the ice cubes changed the construction paper and the position of pan A.

Conclusions and Applications

3. How does the weight of a glacier affect the land beneath it?
4. What happens to the land when the glacier is gone?
5. Why are some parts of North America rebounding?
6. What happens to the water when a glacier melts?
7. Describe how the land and sea in the Antarctic would change if the glacier melted.

TECHNOLOGY: APPLICATIONS

Irrigation from Icebergs

More than 75% of Earth's fresh water is locked up in glacial ice. People in many countries are studying ways of using some of this ice to irrigate areas where drought is a problem. Some scientists think that by the end of this century, giant tugs will be towing huge icebergs from the South Atlantic to be used to irrigate fields in the Sahara Desert.

In the North Atlantic each year, about 16 000 icebergs break loose from Greenland's glaciers and float away in the ocean. These icebergs slowly float in the currents and melt as they move southward into the Gulf Stream.

Scientists have calculated that it would be possible to tow icebergs as large as 270 square kilometers to areas where the fresh water could be used to water deserts. Such an iceberg would contain more than 7080 million cubic meters of water. This amount of water is equal to the amount of water that flows over Niagara Falls in two months.

A supertanker-size tug could tow an iceberg of this size at a rate of about 32 kilometers per day. A trip from its origin to its final destination could take from 6 to 12 months. Although some of the iceberg would melt during the move, much of it would still reach its destination, especially if the iceberg were insulated with plastic foam or canvas. The transported iceberg could be left near the shore to melt. Since fresh water is less dense than seawater, the melted water would float on top of the seawater. This fresh water could be pumped ashore and into an irrigation pipeline system. Using icebergs for irrigation would be feasible for areas like the Sahara Desert, the Yuma Desert of California, the Atacama Desert of Chile, and for Saudi Arabia and Western Australia because they are all close to ocean or sea ports.

Chapter 15 Review

SUMMARY

1. Glaciers are masses of ice in motion. The three types of glaciers are valley, piedmont, and continental. 15:1
2. All glaciers carry debris frozen in the ice. This debris abrades the surface over which it moves. 15:2
3. Valley glaciers erode headward by ice wedging and plucking. 15:2
4. Glacial deposits of till are unsorted, unlayered, and contain boulders, sand, and clay. 15:3
5. Meltwater deposits, or outwash, are sorted by size and deposited in layers. 15:3
6. Lakes associated with glaciers may be old river valleys that were deepened and then dammed by moraine deposits. 15:3
7. Sea level rose when ancient glaciers melted. 15:4
8. The causes of ice ages are still unknown. Fluctuations in Earth's movements may cause ice ages. 15:5
9. Ice ages have alternated with warm intervals for the past three million years. 15:5

VOCABULARY

a. cirque
b. continental glaciers
c. eskers
d. glaciers
e. ground moraine
f. hanging valleys
g. horn
h. ice age
i. kettle lakes
j. meltwater
k. outwash
l. piedmont glacier
m. plucking
n. striations
o. terminal moraines
p. till
q. valley glaciers

Matching

Match each description with the correct vocabulary word from the list above. Some words will not be used.

1. form at high elevations where snow lasts from year to year
2. form at the margins of glaciers
3. leaves deposits that are sorted and layered
4. formed as debris carried by a glacier scratches and gouges bedrock
5. material dropped from the base of a glacier as it melts
6. formed when valley glaciers erode into the mountain summit from several directions
7. formed when ice blocks melted, leaving holes that later filled with water
8. a glacier at the foot of a mountain
9. sorted debris deposited by rivers flowing within or beneath ice
10. the combination of freezing and pulling that erodes and removes rocks in a glacier's path

Chapter 15 Review

MAIN IDEAS

A. Reviewing Concepts

Choose the word or phrase that correctly completes each of the following sentences.

1. Today a continental glacier is found in *(Canada, Switzerland, Greenland).*
2. Continental glaciers once covered *(all of North America, none of North America, North America south to the Missouri and Ohio Rivers).*
3. Part of what once was Lake Bonneville is *(Lake Superior, Great Salt Lake, the Matterhorn).*
4. Colder climates move valley glacier fronts to *(high elevations, higher latitudes, lower elevations).*
5. The last ice age ended about *(16 000, 60 000, 11 500)* years ago.
6. Glacial valleys are *(U-shaped, V-shaped, W-shaped).*
7. Regions affected by valley glaciers can be recognized by the many waterfalls that spill from *(kettle lakes, horns, hanging valleys).*
8. When a glacier melts, the sea level and the land surface both *(sink, rise, remain the same).*
9. Deposits that commonly run perpendicular to the ice front are *(terminal moraines, cirques, eskers).*
10. When the tilt of Earth on its axis is *(25°, 23½°, 22°)* the poles receive less energy.
11. Some scientists predict the beginning of another ice age within *(10, 100, 1000)* years.
12. *(The Great Lakes, New York's Finger Lakes, Minnesota's thousands of lakes)* are kettle lakes.
13. *(Ground moraine, Terminal moraine, An esker)* tends to fill hollows and cover hills, making the resulting land surface fairly smooth.
14. During the last glacial period, sea level was about *(110 meters higher than, 110 meters lower than, the same as)* it is now.
15. Small knobby hills of sand and gravel deposited by glaciers are *(cirques, deltas, kames).*

B. Understanding Concepts

Answer the following questions using complete sentences.

16. Explain why sea level is lower during an ice age.
17. Describe meltwater deposits.
18. What causes a glacier to form?
19. How does the movement of a glacier affect the underlying bedrock?
20. What is a cirque?
21. Explain how the freezing of water helps glaciers erode bedrock.
22. Describe the two types of glacial flow.
23. How do changes in weather conditions cause changes in the elevation of the end position of a valley glacier?
24. What are the two ways that glaciers erode rocks?
25. How were the Great Lakes formed?

C. Applying Concepts

Answer the following questions using complete sentences.

26. Why are some valleys V-shaped and others U-shaped?
27. How might ice ages affect marine organisms?
28. Why does a continental glacier continue to move even when the ground surface is fairly level?
29. How does the elevation of the land surface change after a continental glacier melts?
30. How do the changes to the landscape made by valley glaciers differ from those made by continental glaciers?

SKILL REVIEW

1. Determine which type of graph would be most effective in showing the surface areas of the five largest kettle lakes in Minnesota.
2. Which type of graph would be most effective in showing Earth's average temperature from 1969 to 1989?
3. Use the map of Voltaire, ND, Figure 3–19 to answer these questions. What are the elevations of Lakes Hester and Stink Lake? What type of terrain is found southwest of the long ridge along Stink Lake? How much area is covered in this map? What is the map scale?
4. Compare and contrast valley and continental glaciers.
5. What are some effects of erosion by continental and valley glaciers?

PROJECTS

1. Examine topographic maps for glacial features. Classify the features you find as either erosional or depositional.
2. Design an experiment to study how the thickness of a glacier affects its ability to erode.

READINGS

1. Bramwell, Martyn. *Glaciers & Ice Caps.* New York: Franklin Watts, 1986.
2. Imbrie, John, and Katherine P. Imbrie. *Ice Ages: Solving the Mystery.* Cambridge, MA: Harvard University Press, 1986.
3. Radlauer, Ruth, and Lisa S. Gitkin. *The Power of Ice.* Chicago: Childrens Press, 1985.

SCIENCE AND SOCIETY

GROUNDWATER: OUR FIRST PRIORITY

People depend on pure, clean, drinkable water. Without it, we cannot survive. In some parts of Earth, clean water is very scarce. In certain areas of developing countries, people spend much of each day hauling water from wells several kilometers from their homes. In the United States, many people find it hard to tolerate even a temporary water shortage.

Background

Prior to the agricultural and industrial revolution in the United States, there was plenty of water for all. Now, however, population and standard of living have threatened to deplete drinking water supplies. Agriculture and industry use approximately 95 percent of that supply. Household uses have expanded greatly. Adding to the problem is the fact that many aquifers have become polluted, thus diminishing the usable supply.

Water is easily polluted. It is known as "the universal solvent." This does not mean that every substance can be dissolved in it, but a large percentage of substances do dissolve in water. Water is a good solvent of organic materials, especially those containing oxygen. These include alcohols, sugars, and organic acids. Water is also good for dissolving some salts. Among these salts are the positively charged ions of metals, which in certain amounts are poisonous to humans. These ions include copper, cadmium, mercury, and lead. Lead pipes were used at one time to transport water supplies.

Water also has the unique property of being able to dissolve, in small amounts, substances not generally considered to be water-soluble. One of these, benzene, is a threat to the water supply in many areas. Some other insoluble products react with water to produce new substances that are soluble. Still other substances, such as petroleum products, simply float on the surface. Some pollutants are not toxic or poisonous, but create a large biological oxygen demand, which has the effect of removing oxygen from the water and making it more difficult for the water to support life and purify itself naturally.

FIGURE 1. Many countries have problems with providing enough moisture for growing their crops.

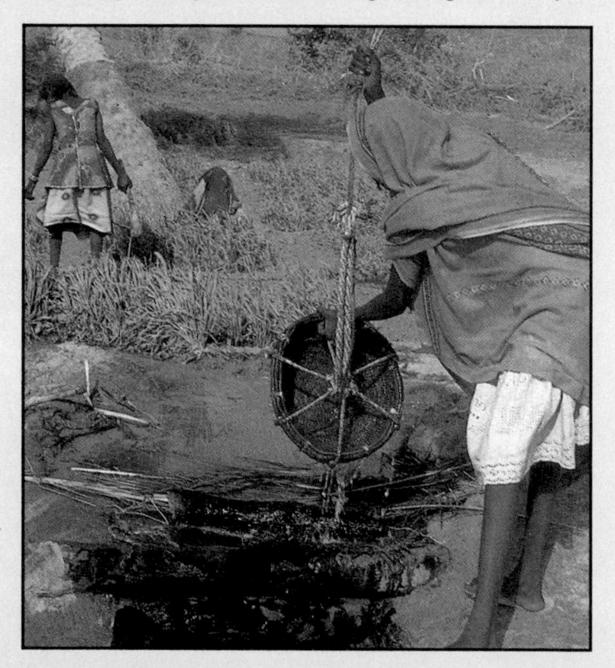

Case Studies

1. Modern agriculture uses large amounts of fertilizers, herbicides, and pesticides. These may run off into nearby waterways. Herbicides and pesticides now considered too dangerous to use are often stored in drums and buried in landfills where they may leak into the soil and eventually leach into the water supply.

2. Industry also produces chemical wastes. Most industries try to comply with regulations, but it is difficult to find appropriate disposal sites. Some companies ignore the law and simply discharge their wastes into rivers and streams or dump them illegally.

3. It seems difficult to believe that the average American household is also a polluter. Think of all the substances around homes and put down drains. It has been estimated by the Environmental Protection Agency that each American family generates approximately one metric ton of waste per year. Much of this waste is hazardous. It includes pesticides, medicines, paints, cleaning products, used batteries, and automotive products. These household wastes are not regulated by the federal laws dealing with the disposal problem.

4. In July of 1976, there was an explosion in a soap and deodorant factory in Italy. A large cloud of dioxin, a deadly poison, was released into the atmosphere. Several hundred nearby residents were evacuated, suffering from burns and internal disorders. Hundreds of animals died and many others were slaughtered because they had become unfit for consumption. The cloud dissolved within one half hour, but the effects were carried far and wide. In addition to the atmospheric pollution from the dioxin, some of it seeped into the ground where it was absorbed by plant roots and where it contaminated the groundwater.

Developing a Viewpoint

1. The XYZ Chemical Company, the leading employer in your town, has been ordered to stop the sale of its leading product, a pesticide found to be extremely toxic to humans. They have been ordered to dispose of all the product in their warehouse within the next six months. They have applied for a permit to bury the containers in a landfill near the plant. If this is not approved, they will be forced to ship the material elsewhere. This would be very expensive and could cause large layoffs at the plant. If you were a member of the elected town council, how would you vote? If you were living near the proposed landfill, how would you vote? What kinds of information do you need in order to make an informed choice?

2. Mr. Brown has raised chickens near your town for many years. One day he announces he is selling his farm to a business that will enlarge the operation many times. The new owners will be in town to meet with concerned town residents in one week. What questions do you think will be raised? Make a list of the pros and cons of such an agricultural expansion.

Suggested Readings

Goldin, Augusta. *Water: Too Much, Too Little, Too Polluted?* New York: Harcourt Brace Jovanovich, 1983.

Gordon, Wendy. *Citizen's Guide to Groundwater Protection.* New York: Natural Resources Defense Council, 1984.

Marjorie Sun. "Groundwater Ills: Many Diagnoses, Few Remedies." *Science.* June 20, 1986.

UNIT 5

Some rocks form at high temperatures and pressures deep within Earth. Others form at or near Earth's surface. The salty waters of the many volcanic springs in the Great Rift Valley of East Africa are evaporating in the desert heat forming these deposits. What other types of rocks form at or near Earth's surface? What kinds of rocks form deep within Earth?

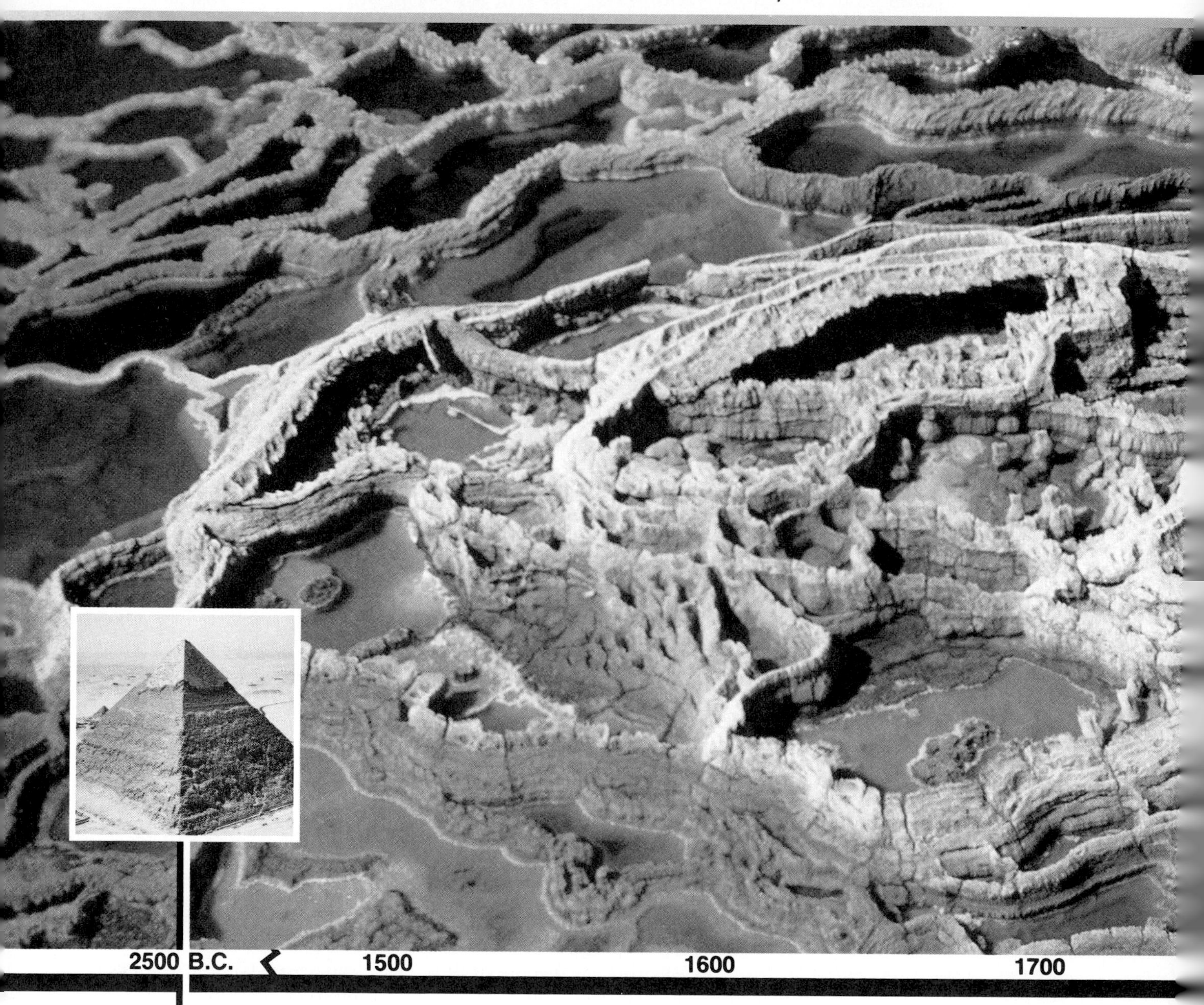

2500 B.C. — 1500 — 1600 — 1700

~2500 B.C.
Egyptian pyramids built.

1565
First pencils manufactured in England.

ROCKS AND MINERALS

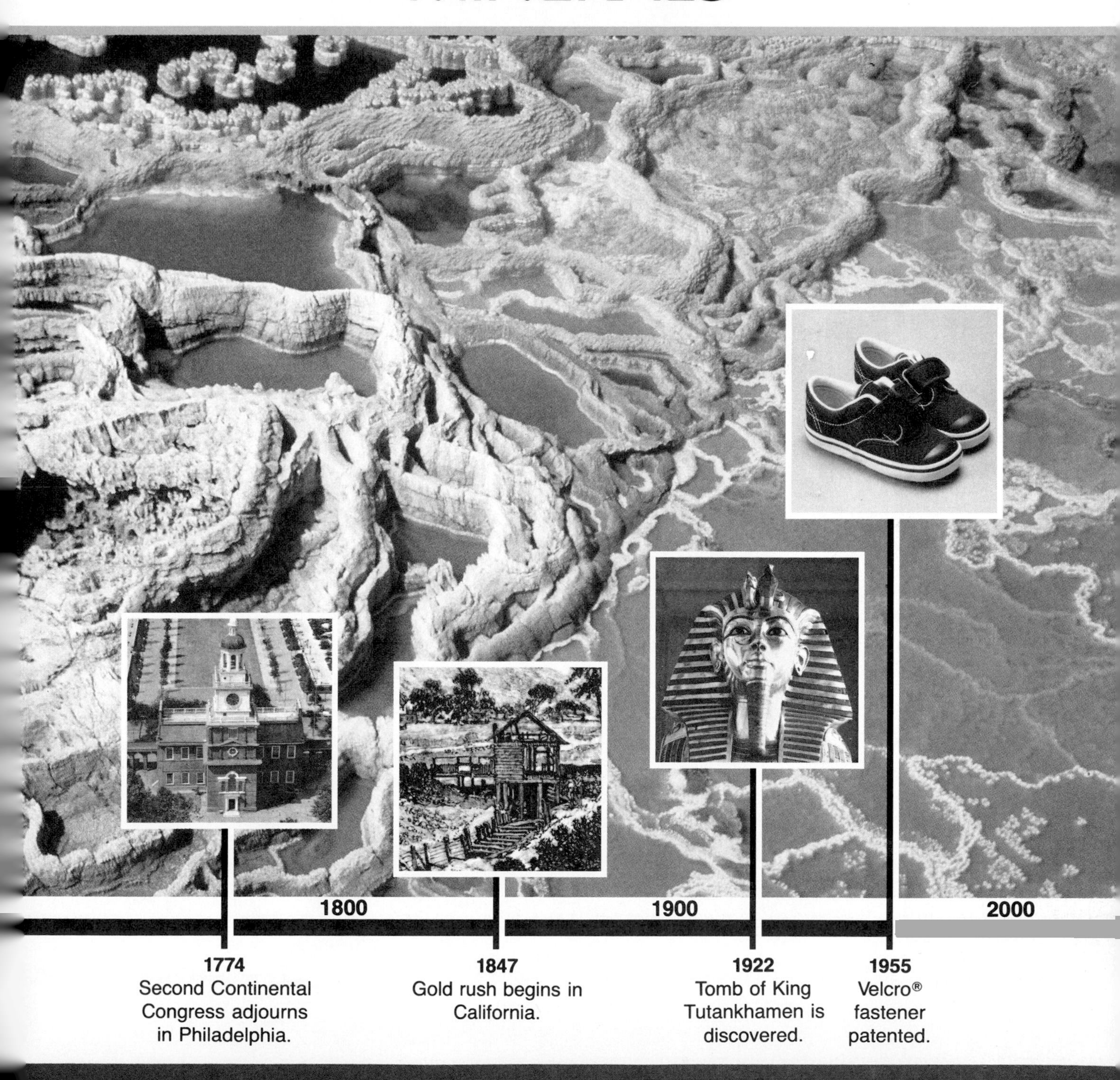

MINERALS
MINERAL IDENTIFICATION

Minerals

You have probably seen a chunk of coal and a diamond ring. They appear quite different. A piece of coal is usually jet black. Most diamonds are colorless. What do coal and diamonds have in common? Both are made of carbon. Diamond is a mineral but coal is not. In this chapter, you will learn about matter that makes up much of Earth.

MINERALS

GOALS

1. You will learn how minerals form.
2. You will learn about the six crystal systems and how minerals differ.

Earth's crust has 90 naturally occurring elements. Some elements are abundant, others are rare. Of all these elements, only eight make up 98 percent of the total mass of Earth's crust. These elements combine to form Earth materials called minerals.

16:1 What Is a Mineral?

Most of the solid portion of Earth's crust consists of minerals. A mineral is a special kind of matter. The atoms of a mineral always have the same internal pattern or crystalline structure. This orderly arrangement of atoms cannot be seen with the unaided eye, but can be determined by use of X rays. The chemical composition of a mineral is always the same within certain limits.

A substance is not a mineral unless it meets five requirements. (1) Minerals are inorganic. (2) They are formed in nature. (3) They are solids. (4) Atoms of a mineral have the same crystalline pattern. (5) The chemical composition of a mineral is the same with only minor variations. Applying these requirements to Earth materials, we can now define a mineral. A **mineral** is a naturally occurring, inorganic, crystalline solid, with a definite chemical composition. Table 16–1 lists the most abundant elements that make up minerals.

What is a mineral?

F.Y.I. The eight most common elements also make up 99 percent of the total volume of Earth's crust. Oxygen alone, however, makes up 94 percent of the crustal volume, leaving only 5 percent for the other seven elements.

Table 16–1

Most Abundant Elements of Earth's Crust

Element	Symbol	Approximate relative abundance by mass
oxygen	O	46%
silicon	Si	28%
aluminum	Al	8%
iron	Fe	6%
magnesium	Mg	4%
sodium	Na	3%
calcium	Ca	2%
potassium	K	2%

Minerals are inorganic. Thus, they are not formed by life processes. Although many marine organisms make shells from mineral matter, the shells are not minerals. Minerals can form from magma. **Magma** is a molten material found beneath Earth's surface. Minerals can also be deposited from solutions or gases.

16:2 Internal Structure of Minerals

What is a crystal?

A **crystal** is a solid bounded by plane surfaces and has a definite shape due to its internal atomic arrangement. The positions of the atoms within the solid produce the visible shape of the crystal.

Few crystals are perfect. When crystals start to form in a magma or a solution, they interfere with one another. Lacking room to grow, minerals harden into masses of tiny grains in which the internal crystal form can be recognized only with a microscope. Sometimes X rays

FIGURE 16–1. Crystals provide clues to the arrangement of atoms within a mineral and serve as one means of mineral identification.

Table 16–2

Crystal Systems				
System name	**Axes of intersection**	**Ideal shape**	**Length of axes**	**Example**
cubic or isometric	90°	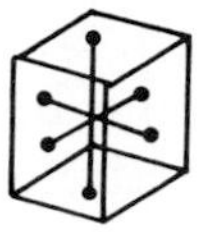	all three axes equal	 Halite
tetragonal	90°	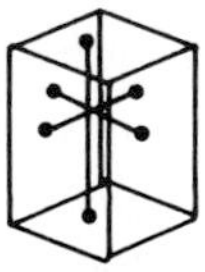	two horizontal axes equal third axis different	 Zircon
hexagonal	60° 90°	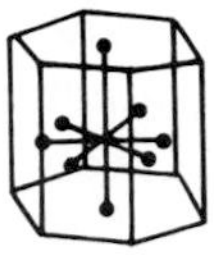	three horizontal axes equal fourth axis different	 Quartz
orthorhombic	90°	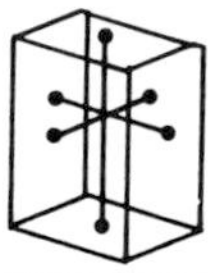	all three axes unequal	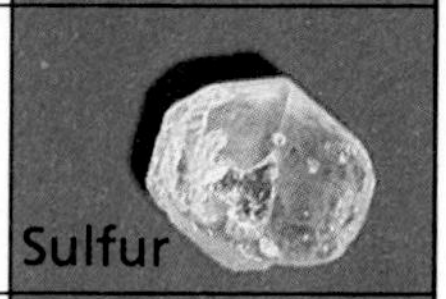 Sulfur
monoclinic	90° third axis different from 90°	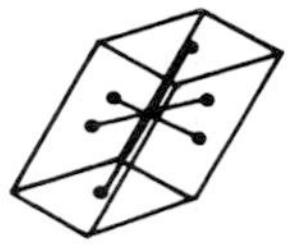	all three axes unequal	 Orthoclase
triclinic	all 3 axes different from 90°	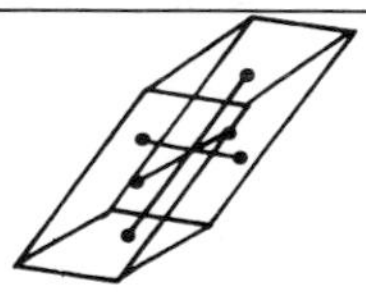	all three axes unequal	 Plagioclase

must be used to determine the internal patterns. Despite the fact that it may be difficult to observe, all minerals have definite internal atomic patterns in one of six possible crystal systems.

How are the six crystal systems defined?

The six crystal systems are defined by three or four imaginary axes that intersect at the center of a perfect crystal. An axis is a straight line around which a crystal is symmetrical. The length of the axes and the angles at which they meet determine a crystal's external shape. Axes are usually drawn at right angles to pairs of faces of the crystal. A crystal face is a smooth surface that has a geometric shape and reflects the internal atomic arrangement of the crystal.

Minerals have exact compositions within certain limits. A mineral always contains the same kinds of atoms. The atoms are arranged in the same way and in the same

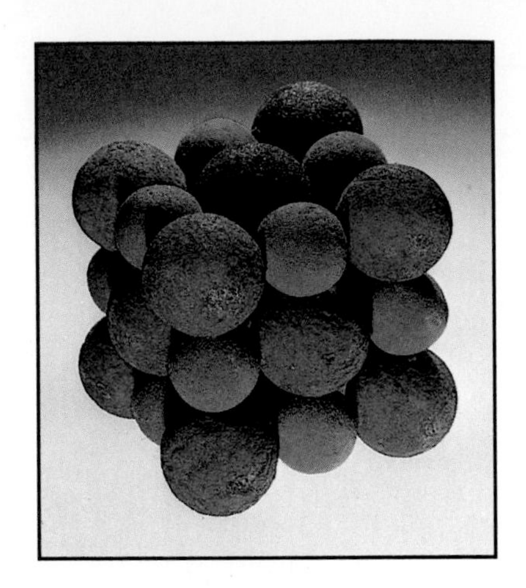

FIGURE 16–2. Halite always has atoms of sodium (red) and chlorine (blue) arranged in the pattern shown here.

proportions. Table 16–1 shows that the two most abundant elements, by mass, in Earth's crust are oxygen and silicon. When these two elements combine with no other elements present, they form the mineral **quartz.** Every specimen of quartz has two oxygen atoms for every atom of silicon. Quartz samples may contain billions of oxygen atoms and billions of silicon atoms, but the ratio of two oxygen atoms to one silicon atom stays the same.

Chemists use a kind of shorthand to record the composition of a substance. Each element has its own symbol. Symbols are combined to represent the formulas of chemical compounds. Minerals are either elements or compounds, and thus can be identified using chemical symbols and formulas. Using this shorthand, the formula for quartz, a compound, is SiO_2.

Some atoms or ions in nature are so similar that they can trade places. Imagine atoms and ions as blue and white building blocks. Blue blocks can be traded for white blocks if they are the same size and have the same charge. If substitution continues, eventually a white building becomes a blue building. Mineralogists decide when enough changes have occurred to rename the mineral. Name changes usually are based on physical properties. For example, the color or density of a mineral might change, or possibly the way the mineral breaks might be different. Plagioclase minerals are a group that substitutes one element for another. Sodium atoms can substitute for calcium atoms, changing anorthite ($CaAl_2Si_2O_8$) to albite ($NaAlSi_3O_8$). These are two different varieties of the mineral plagioclase.

Review

1. What element is most abundant in Earth's crust?
2. What is the formula for quartz?
3. ★ Describe characteristics common to all minerals.

FIGURE 16–3. Anorthite (a) and albite (b) are two different varieties of plagioclase.

a

b

INVESTIGATION 16–1

Crystal Formation

Problem: In what two ways can crystals form?

Materials  

salt solution
sugar solution
large test tube
toothpick
cotton string
hand lens
shallow pan (2)
thermal mitt
test tube rack
cardboard
table salt
granulated sugar
hot plate

Procedure

1. Pour the sugar solution into one of the shallow pans. Use the hot plate to gently heat the solution.
2. Place the test tube in the test tube rack. Using a thermal mitt to protect your hand, pour some of the hot sugar solution into the test tube. **CAUTION:** *The liquid is hot. Do not touch the test tube without protecting your hands.*
3. Tie the thread to one end of the toothpick. Place the thread in the test tube. Be sure that it does not touch the sides or bottom of the tube.
4. Cover the test tube with a piece of cardboard and place the rack containing the test tube in a location where it will not be disturbed.
5. Pour a thin layer of the salt solution into the second shallow pan.
6. Place the pan in a warm area in the room.
7. Leave both the test tube and the shallow pan undisturbed for at least one week.
8. Examine sample grains of table salt and sugar with the hand lens. Note any similarities or differences.

FIGURE 16–4.

9. At the end of one week, examine each solution and see if crystals have formed. Use a hand lens to observe the crystals.

Analysis

1. Describe the crystals that form from the salt and sugar solutions. Include a sketch of each crystal.
2. What happened to the salt water in the shallow pan?
3. Did this same process occur in the test tube? Explain.

Conclusions and Applications

4. What caused the formation of crystals in the test tube? What caused the formation of crystals in the shallow pan?
5. In what two ways can crystals form?
6. Are salt and sugar both minerals? Explain your answer.

MINERAL IDENTIFICATION

GOALS

1. You will learn how physical properties are used to identify some minerals.
2. You will learn about rare gems and common rock-forming minerals.

Look at the mineral on page 348. What is its color? Describe its luster—is it shiny or dull? Physical properties such as color and luster can be used to identify minerals. Other properties include feel, smell, and relative mass. Physical properties are a result of the chemical composition of a mineral.

16:3 Physical Properties of Minerals

How are minerals identified?

Minerals are identified by their physical properties. You can learn to recognize many minerals by appearance. Other properties used to identify minerals are feel, smell, taste, and how they sound when tapped. The most useful properties for quick identification are appearance and feel. After you have seen a number of pieces of the same mineral, you can usually recognize that mineral. But to be sure of its identification, you need to test it. Physical properties that should be tested include luster, hardness, streak, and heft, or relative mass. Form and cleavage also may be useful if they are recognizable.

What are some properties that you should test to be sure of a mineral's identification?

Physical properties of hundreds of minerals have been determined and arranged in charts like Appendices M and N. Answering the following eight questions can help you identify an unknown mineral. The more carefully you observe a mineral, the more easily you will recognize it in the future.

1. Is the mineral shiny? **Luster** refers to the way light is reflected from a mineral's surface. There are two

a

b

FIGURE 16–5. Pyrite has metallic luster (a). Talc's luster is characterized as pearly (b).

CAREER

Goldsmith

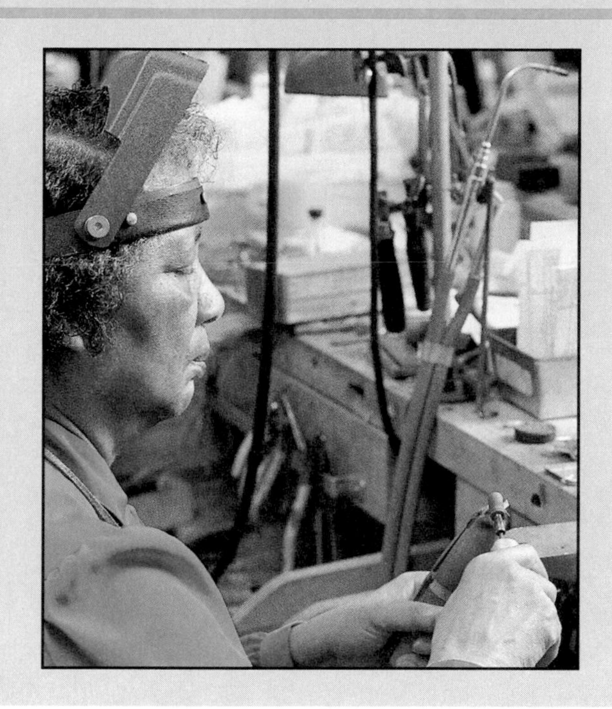

Gilda Robertson is a goldsmith for a local jewelry store, where she designs and makes jewelry for the store's clientele. Ms. Robertson often uses gold leaf in her designs. Gold leaf is made by pounding gold so thin that nearly 250 000 sheets of this gold foil, when piled on top of one another, would be only 1 centimeter thick! Ms. Robertson uses the gold leaf to cover leaves and shells to make attractive broaches and to make gold-plated jewelry. A career as a goldsmith requires artistic ability and creativity.

For career information, write:
Gemological Institute of America
1660 Stewart Street
Santa Monica, CA 90404

types of luster, metallic and nonmetallic. If a mineral looks like a metal, regardless of its color, it has a metallic luster. Pyrite and galena have metallic luster. Nonmetallic luster is described as dull, pearly, silky, glassy, or brilliant. Diamond has a brilliant luster. Minerals that have nonmetallic luster may also be transparent. A transparent mineral allows light to pass through it as if it were made of glass.

2. How hard is the mineral? Hardness is one of the most useful properties used in mineral identification. **Hardness** is a mineral's resistance to being scratched. A harder mineral always scratches a softer one. Friedrich Mohs, a German mineralogist, worked out a scale of hardness used to identify minerals (Table 16–3). In

F.Y.I. Friedrich Mohs devised his scale of relative hardness of minerals in 1812.

Table 16–3

Mohs' Scale of Hardness	
1—talc	6—orthoclase
2—gypsum	7—quartz
3—calcite	8—topaz
4—fluorite	9—corundum
5—apatite	10—diamond

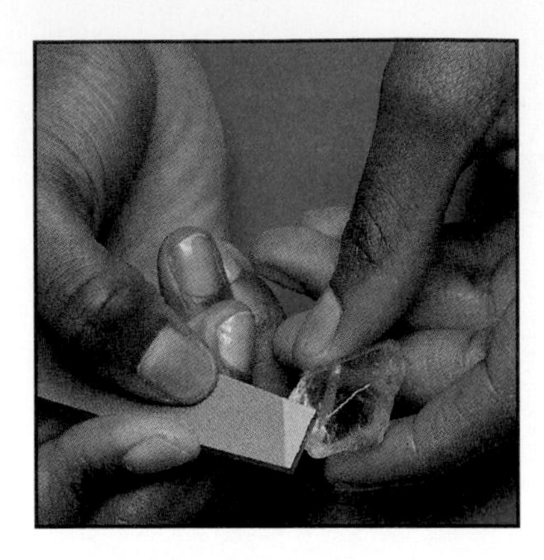

FIGURE 16–6. Fluorite is easily scratched by a steel file.

How is the hardness of an unknown mineral determined?

this scale, ten reference minerals are arranged in order of increasing hardness. Each mineral is given a number between one and ten. Each reference mineral will scratch any mineral with a lower number on Mohs' scale.

The hardness of an unknown mineral is found by scratching its edge against a flat surface on each reference mineral. If the reference mineral scratches an unknown, the reference mineral is harder than the unknown. If the unknown scratches a reference mineral, the unknown is harder than the reference mineral. If two minerals do not scratch one another, they have the same hardness. Diamond, with a hardness of ten, scratches all other minerals. Talc, with a hardness of one, is scratched by most other minerals. If you are collecting rocks and minerals on a field trip, a field scale like the one shown in Table 16–4 is convenient to use.

Table 16–4

Field Scale of Hardness
1—soft, greasy, flakes on fingers
2—scratched by fingernail
3—cuts easily with knife or nail or scratched by penny
4—scratched easily by knife
5—scratched by knife but with difficulty
6—scratched by steel file
7—scratches steel file
8—scratches quartz
no approximations above 8

3. What is the streak of the mineral? **Streak** is the color of the powdered mineral. Streak is an important clue used to identify minerals that are fairly soft. To find the streak color, rub the unknown mineral across a piece of unglazed porcelain tile. Minerals with a hardness greater than seven do not leave a streak. Streak is also not much help in recognizing minerals that have a colorless or white streak. Appendices M and N list the streak color of many minerals.
4. What is the form or general shape of the mineral? Is the mineral specimen a single crystal, a group of small crystals, or a dense mass? All these features describe the mineral's **form** or appearance. Some crystals may be recognized by their shape. Crystals of halite, galena, and fluorite are cubic. Because perfect crystals are rare, most minerals take other forms.

FIGURE 16–7. The color of a mineral's streak may differ from that of the mineral sample.

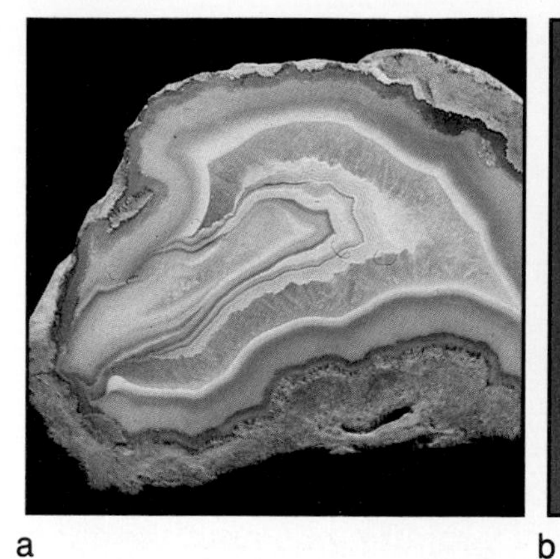

a b

FIGURE 16–8. Agate is a massive specimen (a). Asbestos is described as fibrous (b).

If you cannot see any crystals, the mineral may be massive. Massive samples may be described by terms that suggest their appearance. These terms include compact, fibrous, and granular. Compact minerals are dense and solid. Fibrous minerals are composed of slender, long threads or fibers stacked together. Granulated minerals are composed of tiny sandlike grains.

5. Does the mineral have any broken surfaces? The way a mineral breaks is described either as **cleavage** (KLEE vihj) or as fracture. Minerals cleave if they break along smooth, flat planes. Sometimes cleavage occurs in only one or two directions. Calcite has three directions of cleavage.

   Fracture is breakage along an irregular surface. The surface may be rough, conchoidal (kahn KOYD uhl) or curved, or hackly with thin, jagged points. Cleavage and fracture are both related to a mineral's internal arrangement of atoms.

What is fracture?

6. Is the mineral heavy? Specific gravity (Section 4:4) is useful in recognizing heavy minerals. Specific gravity is the ratio of the mineral's mass to the mass of an equal volume of water. For example, galena (PbS) has a specific gravity of 7.5. One cubic centimeter of galena is 7.5 times more massive than one cubic centimeter of water. Minerals with specific gravities above 4.0 usually contain metals. These dense minerals often can be judged by picking up the mineral and tossing it up and down. The term "heft," meaning density, is used in these cases where the specific gravity is not actually measured. Specific gravities of most nonmetallic minerals usually are lower than 4.0.

FIGURE 16–9. Galena has cubic cleavage.

7. What is the color of the mineral? Color is important in the identification of only a few minerals. Such minerals are always the same color. The color of most

a

b

FIGURE 16–10. Quartz (a) and fluorite (b) can be a variety of colors. Why is color not a good way to identify a mineral?

minerals changes with the chemistry of the mineral, with the presence of impurities, or because of surface tarnish. Pyrite, gold, and chalcopyrite (kal kuh PI rite) are metallic yellow. Azurite is deep blue. In general, streak is more reliable than color to identify a mineral.

8. Does the mineral have some unique characteristics? Some minerals can be identified by special properties. For example, you can taste halite (salt), smell sulfur, and tap jade for a bell-like ring. Some minerals react by giving off a gas when hydrochloric acid is dropped on them. When this reaction occurs, the mineral appears to bubble. Appendices M and N have a column headed "Uses and other properties." "Properties" in this column will help you to determine differences among several minerals that appear the same.

FIGURE 16–11. Some minerals have distinct characteristics. Sulfur (a) has a distinct odor. Calcite (b) has special optical properties.

a

b

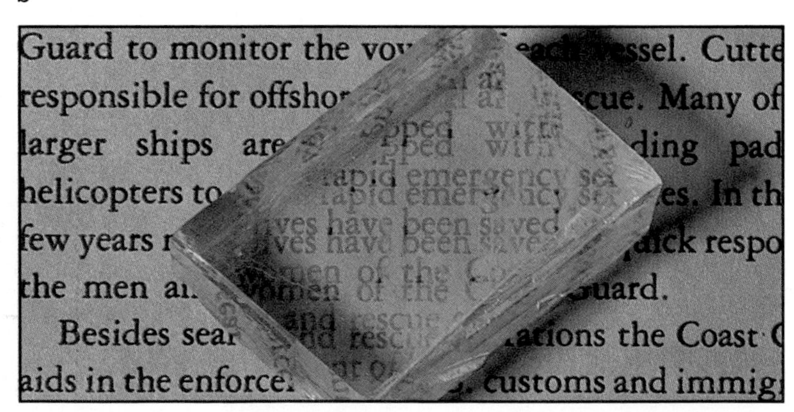

SKILL
Classification

Problem: How can you classify minerals?

Materials

paper
pencil
Appendix G

Background

You have learned that minerals can be classified according to their chemical compositions. Recall from Section 16:2 that these compositions are expressed by a type of shorthand called chemical formulas.

Procedure

1. Study the different minerals and the chemical formulas listed in the Data Table. Use the periodic table in Appendix G to find out how elements are abbreviated in chemical formulas.
2. Classify the minerals in the Data Table into four groups according to their chemical compositions.
3. Compare your mineral classification system with the systems compiled by your classmates.

Questions

1. How did you classify the minerals listed in the Data Table?
2. How were your groupings similar to those of your classmates?
3. How was your classification system different from those compiled by other students?
4. Is there only one way to classify minerals? Explain your answer.

Data Table

Mineral	Formula	Mineral	Formula	Mineral	Formula
acanthite	Ag_2S	enstatite	$MgSiO_3$	pyrolusite	MnO_2
adularia	$KAlSi_3O_8$	galena	PbS	quartz	SiO_2
alabandite	MnS	gold	Au	rhodonite	$MnSiO_3$
albite	$NaAlSi_3O_8$	hematite	Fe_2O_3	rutile	TiO_2
anatase	TiO_2	ilmenite	$FeTiO_3$	sanidine	$KAlSi_3O_8$
andalusite	Al_2SiO_5	iron	Fe	scheelite	$CaWO_4$
anorthite	$CaAl_2Si_2O_8$	magnesite	$MgCO_3$	siderite	$FeCO_3$
aragonite	$CaCO_3$	magnetite	Fe_3O_4	silver	Ag
argentite	Ag_2S	mercury	Hg	sphalerite	ZnS
bornite	Cu_5FeS_4	millerite	NiS	sulfur	S
calcite	$CaCO_3$	olivine	$(Mg,Fe)_2SiO_4$	tenorite	CuO
cassiterite	SnO_2	orpiment	As_2S_3	tin	Sn
cerussite	$PbCO_3$	periclase	MgO	triolite	FeS
cinnabar	HgS	platinum	Pt	zincite	ZnO
diamond	C	pyrite	FeS_2	zircon	$ZrSiO_4$

PROBLEM SOLVING

Is It Real?

Hazel had visited several areas of California. At one stop, a place called Sutter's Mill, where gold was discovered in 1849, Hazel saw some bright yellow metallic objects glistening in the clear water of a fast moving stream. Reaching into the stream, she found what appeared to be four or five small nuggets of gold.

Excitedly, Hazel, who had been working in a jewelry store, tested the nuggets. She found that the gold nuggets left a greenish-black powder when rubbed across a piece of white porcelain. The nuggets scratched a copper penny she had with her. The nuggets had a high specific gravity. Did Hazel hit pay dirt?

16:4 Gems

Some mineral varieties stand apart from the more commonly occurring ones in terms of radiance and color. These rare and highly prized minerals are known as **gems** or gemstones. The physical properties of minerals that make them valuable as gems are color, luster, how light passes through them, and hardness.

The difference between a gem and the common form of the mineral can be very slight. Often, only the presence

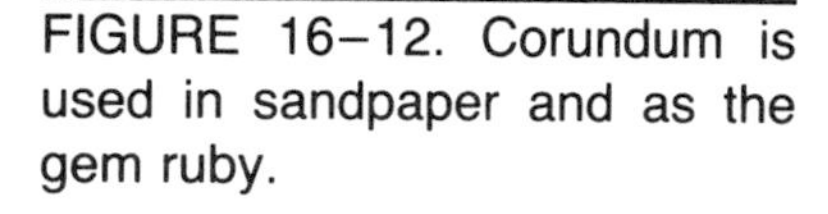

FIGURE 16–12. Corundum is used in sandpaper and as the gem ruby.

a

b

FIGURE 16–13. Rough diamonds (a) are not as brilliant as cut and polished diamonds (b).

of traces of other elements or a slightly different crystalline structure distinguishes one from the other. For example, a gem form of corundum is the ruby. Corundum is also used in emery, which is an abrasive commonly used on fine sand paper. Beryl is rather common; but one of its gem forms is the rarest and most valuable of all gems, the emerald. Table 16–5 lists only a few gems.

Table 16–5

Common Minerals and Rare Gem Forms

Mineral	Composition	Gem Form	Color	Where Found
Diamond	C	Diamond	many colors	Australia, India, Brazil, Uganda
Corundum	Al_2O_3	Ruby Sapphire	red blue	Burma, Thailand, Sri Lanka Thailand, Sri Lanka, India, Australia
Beryl	$Be_3Al_2(Si_6O_{18})$	Emerald Aquamarine	green blue-green	Colombia, Brazil, Siberia Brazil, Madagascar, United States
Tourmaline	$Al_6(BO_3)_3Si_6O_{18}(OH)_4$	Rubellite Indicolite	red blue	Brazil, U.S.S.R., Madagascar, United States
Spinel	$MgAl_2O_4$	Ruby spinel Picoltite	red yellow to greenish brown	Thailand, Burma, Sri Lanka, Madagascar
Garnet	$(Mg, Fe)_3Al_2(SiO_4)_3$	Pyrope Almandite	red red	United States India, Sri Lanka, Brazil
Topaz	$Al_2SiO_4(F, OH)_2$	Topaz	clear, yellow, blue-green	U.S.S.R., Brazil, Mexico, Japan, United States
Quartz	SiO_2	Crystal Amethyst	clear purple	Europe, Brazil, Japan, United States

INVESTIGATION 16–2

Mineral Identification

Problem: How are minerals identified?

Materials   

mineral samples (known and unknown)
hand lens
pan balance
graduated cylinder
water
copper penny
glass slide
steel file
streak plate
5% hydrochloric acid with dropper
goggles
Mohs' minerals (optional)
Appendices M and N

Procedure

Part A: Identifying Known Minerals

1. Use the hand lens to examine the known mineral specimens. Determine and record the luster, hardness, streak, form, cleavage or fracture, color, and specific gravity of each sample.
2. Test the samples for special properties such as reaction to hydrochloric acid. **CAUTION:** *HCl may cause burns. If spillage occurs, rinse with water. Wear your goggles.*
3. Record all observable characteristics in a table like the one shown.

Part B: Identifying Unknown Minerals

1. Examine the unknown minerals and compare them with the known minerals from Part A.
2. Determine the physical properties of each mineral sample and record your observations in the table.
3. Using Appendices M and N, identify each mineral sample.

Questions

1. What property was most useful to you in mineral identification?
2. Which test was most difficult to perform?
3. What property was least helpful in identifying minerals?
4. How can you identify unknown minerals?

Data and Observations

Mineral	Luster	Hardness	Streak	Form	Cleavage/ fracture	Color	Specific gravity	Other properties
quartz	non-metallic, glassy	7	none	massive	conchoidal	gray	2.6	none

16:5 Rock-forming Minerals

Rocks are Earth materials that are made of one or more minerals. Rocks are grouped according to the manner in which they formed and by the minerals that compose them. The three rock groups are igneous, sedimentary, and metamorphic. Igneous rocks form from molten Earth material. Sedimentary rocks form as the result of processes at Earth's surface. Metamorphic rocks are rocks that are changed by internal Earth processes.

The most common minerals are formed from the eight most abundant elements. Silicates make up the largest group of minerals. **Silicates** are compounds of silicon and oxygen combined with other metals and nonmetals.

Another group of minerals is the **carbonates,** composed of carbon, oxygen, and some other element or elements. Calcite is probably the most familiar carbonate. **Oxides**, another group of minerals, are combinations of oxygen and some other element or elements. Sulfates are minerals composed of the sulfate ion and one or more other elements. Halides are a mineral group composed of the halogen ion and another element.

Rocks are composed of one or more minerals. Of the 3000 different minerals, only about a dozen are common rock formers. The common rock-forming minerals are listed in Table 16–6.

BIOGRAPHY

Norman L. Bowen

1887–1956

Norman Bowen studied the way in which molten material crystallizes. He theorized that as magma cools, it passes through stages. Bowen called this the "reaction series."

Table 16–6

Rock-forming Minerals

Chemical classification	Mineral	Rocks in which they are commonly found
Silicates	orthoclase	granite, sandstone, gneiss
	plagioclase	basalt, gneiss, gabbro
	amphibole	rhyolite, some metamorphic rocks
	pyroxene	gabbro, basalt
	olivine	basalt
	micas	gneiss, schist
	clay minerals	shale, some metamorphic rocks
	quartz	granite, sandstone, quartzite
Carbonates	calcite	limestone, dolomite, marble
	dolomite	dolomite, marble
Oxides	hematite	sandstone, shale
	limonite	sandstone, shale
Sulfates	gypsum or anhydrite	rock gypsum
Halides	halite	rock salt

Review

4. What physical properties are most helpful in identifying minerals?
5. What is streak?
6. What is the largest group of minerals?
7. Describe the difference between cleavage and fracture.
★ 8. Identify the following gem: streak—colorless; hardness—7.0; specific gravity—2.6; cleavage—none; fracture—conchoidal; luster—glassy; color—light purple.

TECHNOLOGY: APPLICATIONS

Minerals and Their Liquid Past

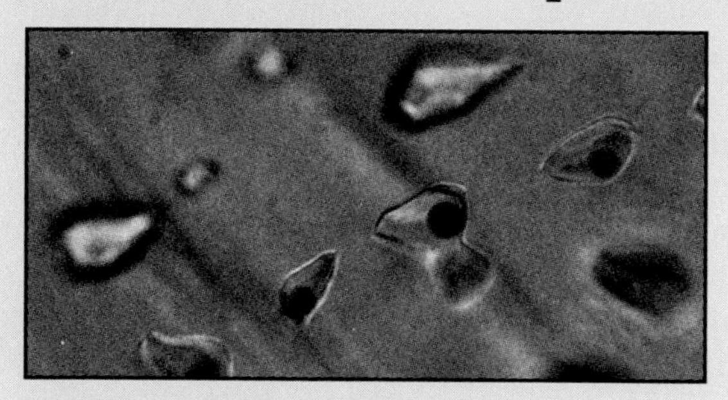

Igneous rocks form from magma within the mantle. Due to impurities, billions of tiny bubbles of fluids often are trapped within minerals as they crystallize. These bubbles, called fluid inclusions, provide valuable information for unravelling geologic processes of the past. Each tiny bubble is a geothermometer that has "preserved" the specific temperature of an event in the history of the mineral. Scientists can determine the temperature at which a particular inclusion, and in turn, the mineral itself, formed. Fluid inclusions also can provide data on the pressure of the environment in which a mineral formed. Densities of fluids that were present during the time of crystallization also can be determined from a study of fluid inclusions.

Fluid inclusions are studied in many areas of geology. "Bubbles" are used to provide an understanding of the physical and chemical environment of some ore deposits. Such information is valuable when exploring for nonrenewable resources. For example, in the search for oil and gas, bubbles can provide information on the formation and migration of oil and natural gas.

Fluid inclusion studies also are important in the development of some of Earth's other energy resources. Fluid inclusions have played an important role in determining the safety of nuclear reactor and waste depository sites. The "bubbles" provide information on whether a region is structurally active and therefore, potentially unsafe for the storage or use of radioactive materials.

Although some jewelers view fluid inclusions as negative features that can reduce the value of a gemstone, the bubbles often are important in identification. Bubbles also aid in distinguishing a mineral from a synthetic compound.

Despite their size—one cubic centimeter of white quartz may contain a billion bubbles—fluid inclusions preserve important data for reconstructing geologic events. In many cases, fluid inclusions are the only clues a geologist has to the history of a mineral.

Chapter 16 Review

SUMMARY

1. Minerals are naturally formed inorganic solids with definite internal atomic patterns and definite chemical compositions. 16:1, 16:2
2. Earth is composed of elements that, in various combinations, form minerals. 16:1
3. Minerals crystallize from magma or are deposited from solutions or gases. 16:1
4. Crystals are solids that have definite shapes due to their atomic structures. 16:2
5. Minerals are identified by their physical properties. These properties include hardness, luster, streak, cleavage, color, form, and specific gravity. 16:3
6. Gems are highly-prized, rare minerals whose color, luster, and hardness produce unique physical properties. 16:4
7. Rocks are Earth materials that are composed of one or more minerals. 16:5
8. Three common groups of minerals in Earth's crust are silicates, carbonates, and oxides. 16:5
9. Although about 2000 minerals have been identified, only a dozen or so are common rock-forming minerals. 16:5

VOCABULARY

a. carbonates
b. cleavage
c. crystal
d. form
e. fracture
f. gems
g. hardness
h. luster
i. magma
j. mineral
k. oxides
l. quartz
m. rocks
n. silicates
o. streak

Matching

Match each description with the correct vocabulary word from the list above. Some words will not be used.

1. the most common group of rock-forming minerals
2. SiO_2
3. molten material located beneath Earth's surface
4. composed of one or more minerals
5. the powder of a mineral
6. the breaking of a mineral along smooth, flat planes
7. the mineral group that contains calcite
8. a naturally occurring inorganic solid with uniform crystalline structure and consistent composition
9. the way light is reflected from the surface of a mineral
10. the resistance of a mineral to being scratched

Chapter 16 Review

MAIN IDEAS

A. Reviewing Concepts

Choose the word or phrase that correctly completes each of the following sentences.

1. The most common minerals are formed from the *(six, eight, twelve)* most abundant elements.
2. One mineral with a distinct smell is *(halite, sulfur, iron)*.
3. Oxides are the combinations of a given element and *(carbon, oxygen, silicon)*.
4. *(Streak, Luster, Cleavage)* is the color of the powdered mineral on a piece of unglazed porcelain.
5. On the Mohs' scale, *(talc, quartz, diamond)* is the softest mineral.
6. *(Halite, Garnet, Quartz)* is a mineral with a distinct taste.
7. The largest group of minerals in Earth's crust are *(silicates, oxides, carbonates)*.
8. Specific gravity is a(n) *(physical, chemical, optical)* property of a mineral.
9. The most abundant element in Earth's crust is *(iron, silicon, oxygen)*.
10. A mineral that will react with hydrochloric acid is *(orthoclase, calcite, quartz)*.
11. A mineral may be an element or *(a rock, magma, a chemical compound)*.
12. The least useful physical property for identifying minerals is *(streak, specific gravity, color)*.
13. Sapphires and rubies are gem forms of *(corundum, quartz, beryl)*.
14. Magma is a *(solid, liquid, gas)*.
15. When the two elements of oxygen and silicon combine, the mineral formed is *(calcite, quartz, pyrite)*.

B. Understanding Concepts

Answer the following questions using complete sentences.

16. Why are only certain forms of some minerals considered to be gems?
17. Distinguish between cleavage and fracture in testing a mineral sample.
18. Which are usually the most useful tests in identifying an unknown mineral?
19. Describe how you would determine the hardness of an unknown mineral.
20. If rocks are composed of minerals, of what are minerals composed?
21. Use Appendices M and N to identify the mineral with the following properties: luster—nonmetallic; color—green; streak—colorless; hardness—6 to 7; specific gravity—3.5.
22. Use Appendix M or N to identify the mineral with the following properties: specific gravity—5.2; color—black; streak—black; hardness—5.5 to 6.5; luster—metallic.
23. Industrial diamonds are produced by people. Are they minerals? Explain.
24. What is a transparent mineral?
25. Why are silicate minerals the most common group of rock-forming minerals on Earth?

C. Applying Concepts

Answer the following questions using complete sentences.

26. Coral is made of $CaCO_3$. Calcite has the same chemical formula. Why is coral not a mineral?
27. Why can quartz be many different colors?
28. Why do most minerals not exist as large, perfect crystals?
29. How could you identify a mineral if you were blindfolded?
30. What color is diamond's streak?

SKILL REVIEW

1. Compare and contrast the gems ruby and sapphire.
2. Use lightweight cardboard to make models of the six crystal systems.

3. You correctly identify a metallic mineral with a specific gravity of 19.3, a yellow streak, and a hardness of 2.5–3 as gold. Is this an observation or an inference?
4. Use Appendix M to compare and contrast pyrite ("fool's gold") and gold.
5. The minerals muscovite and biotite both break or cleave into large, flexible, sheetlike plates. Describe the cause and effect of this cleavage.

PROJECTS

1. Grow crystals of several different substances and display them in your classroom.
2. Start your own mineral collection. Include with each mineral sample the location where it was collected, its chemical formula, and its properties and uses.

READINGS

1. Bains, Rae. *Rocks & Minerals*. Mahwah, NJ: Troll Associates, 1985.
2. Cheney, Glenn Alan. *Mineral Resources*. New York: Franklin Watts, 1985.
3. Lindsten, Don C. "Emerald, the Rarest Beryl." *Lapidary Journal*. August, 1985, pp. 25–32.

IGNEOUS ROCKS
IGNEOUS ACTIVITY

Igneous Rocks

You may recall that in May, 1980, Mount St. Helens erupted, spewing lava and ash over many kilometers. Mount St. Helens is a volcano. Recall from Chapter 3 that a volcano is a type of mountain. The hot material that erupted from the mountain formed deep within Earth. The material cooled and hardened when it reached Earth's surface and formed igneous rocks.

IGNEOUS ROCKS

One type of igneous rock that formed when the lava and ash thrown from Mount St. Helens cooled and hardened is called basalt. It is a dark-colored rock. There are many types of igneous rocks. They differ in the way they are formed and the minerals that compose them.

GOALS

1. You will learn how igneous rocks form.
2. You will learn to identify some igneous rocks.

17:1 Origin of Igneous Rocks

Igneous rocks are rocks formed from molten, or hot, liquid Earth materials. Magma and **lava** are two kinds of molten material. Igneous rocks were named after the Latin word *igneus,* which means fire. Most magma originates 60 to 200 kilometers below Earth's surface, at or near the crust/mantle boundary. At these depths, temperatures are about 1400°C. The pressure and heat that produce magma are caused by the mass of the overlying rocks and by the radioactive decay of minerals.

F.Y.I. Intrusive igneous rock, the most common of all rocks, makes up approximately 95% of Earth's crust to a depth of 10 kilometers.

Magma can remain in place or move upward. Magma trapped beneath the surface cools and solidifies slowly and forms **intrusive igneous rocks.** Intrusive rocks generally are coarse-grained because the slow rate of cooling allows time for large crystal growth. These are the most commonly occurring igneous rocks. Sometimes magma moves upward to Earth's surface. The magma, now called lava, is released during volcanic eruptions or flows out through fractures onto Earth's surface. This lava cools

Why are intrusive igneous rocks generally coarse-grained?

FIGURE 17–1. Lava is molten material that reaches Earth's surface.

quickly to form fine-grained **extrusive igneous rocks.** Minerals of extrusive rocks are often so small that they cannot be identified without magnification.

There are two general types of magma. **Basaltic magma** flows readily and forms thin layers of dark-colored lava that cover large areas. This magma makes up much of the ocean floor. Temperatures of basaltic magma range

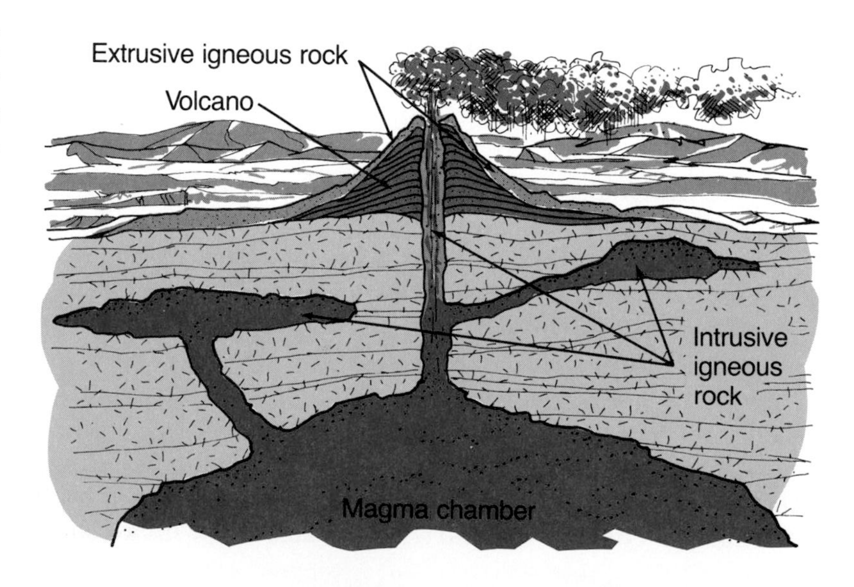

FIGURE 17–2. Extrusive igneous rocks form at or near Earth's surface. Intrusive rocks crystallize beneath Earth's surface.

a

b

FIGURE 17–3. The Columbia Plateau is composed of layers of basaltic lava (a). A plug of granitic magma (b) is exposed as the result of erosion.

from 900 to 1200°C at the surface. **Granitic magma** is thick and stiff due to its high silica content. Its temperature is generally below 800°C at the surface.

Basaltic magma contains a high percentage of iron and magnesium. The presence of these elements produces igneous rocks that are generally quite dense and dark in color. The minerals that form from granitic magma have high percentages of aluminum and potassium or sodium. These produce less dense, lighter-colored igneous rocks.

17:2 Minerals in Igneous Rocks

Magma is a complex mixture of liquids, solids, and gases. The properties of a magma are related to its composition and the temperature and pressure at which it forms. All magmas are silicate melts composed mostly of oxygen and silicon. Other elements include aluminum, calcium, sodium, potassium, iron, and magnesium.

To what are the properties of a magma related?

FIGURE 17–4. Light-colored silicate minerals include quartz (a), feldspar (b), and olivine (c).

a

b

c

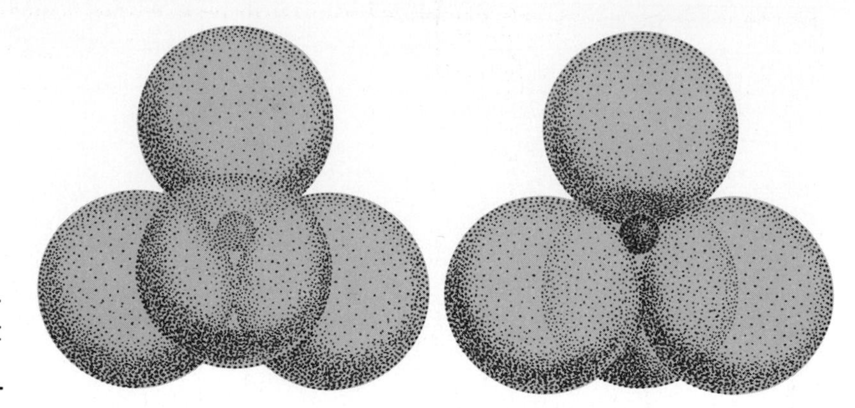

FIGURE 17–5. The silicate tetrahedron is a basic unit of most rock-forming minerals.

What is the silicate tetrahedron?

Oxygen and silicon combine to form the **silicate tetrahedron** (teh truh HEE drun), a unit basic to most of the rock-forming minerals. The silicate tetrahedron is one ion of silicon combined with four ions of oxygen. Thus, it has a negative charge of four. The silicate tetrahedron in turn combines with one or more other elements to form rock-forming minerals.

Norman Bowen, a mineralogist, discovered that minerals form from a melt in a certain order. This order is called Bowen's reaction series, shown in Figure 17–7.

Several minerals are made of iron, magnesium, and the silicate tetrahedron. These minerals include biotite, olivine, and the amphibole and pyroxene families. Muscovite, a mica similar to biotite, contains the silicate tetrahedron plus aluminum and potassium. If the tetrahedron combines with aluminum, calcium, and sodium, the feldspar plagioclase is formed. If the tetrahedron combines with aluminum and potassium, orthoclase is formed.

Quartz forms late and is composed exclusively of silicon and oxygen. Figure 17–7 shows the general order in which minerals crystallize from a cooling magma. Note that iron-magnesium and feldspar minerals may form at the same time.

FIGURE 17–6. Amphibole (a), pyroxene (b), and mica (c) are common igneous rock minerals.

a

b

c

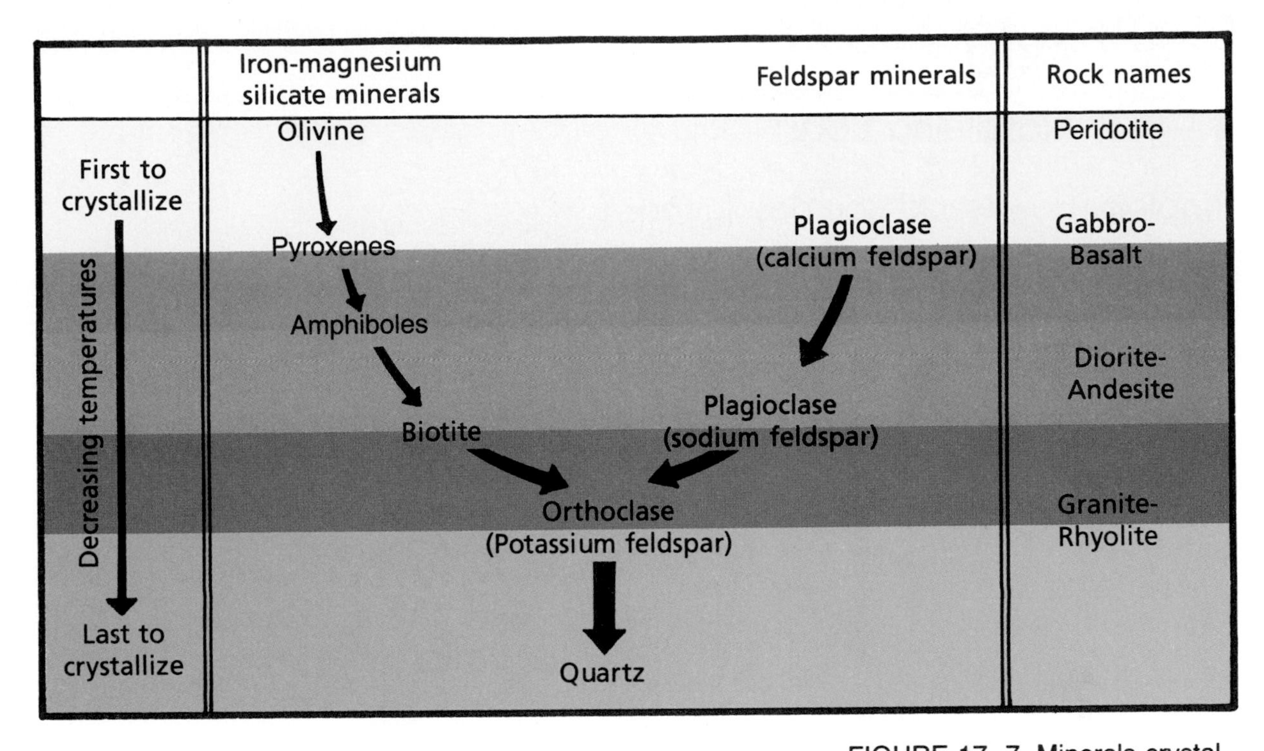

FIGURE 17–7. Minerals crystallize from magma in this order.

17:3 Igneous Rock Classification

Igneous rocks are classified according to texture and composition. When magma cools and hardens, intrusive rocks are formed. No feldspars are present in the early forming rocks. Thus, the rock is named based on the iron-magnesium minerals. Olivine alone or together with a pyroxene is the first intrusive rock to form from a silicate melt. This rock is called peridotite. With more cooling, a rock called gabbro forms. The most abundant minerals in a gabbro are from the pyroxene family and the calcium feldspars. Next to form is the rock diorite. A diorite is made of sodium plagioclase with some iron-magnesium amphibole. The last rock to form is granite. A granite is

What is a peridotite?

FIGURE 17–8. Peridotite (a), diorite (b), and granite (c) are intrusive igneous rocks.

a

b

c

SKILL

Classifying Rocks

Problem: How would you classify rocks?

Materials   

10 rock samples
hand lens or magnifying glass
rock hammer
towel
goggles
metric ruler
glass slide
steel file
streak plate

Procedure

1. Examine the rock samples. Observe characteristics such as minerals present, the size and shape of mineral grains, and the arrangement of grains. Record your observations in a table similar to the one shown in Data and Observations. An example of what you might observe has been done for you.
2. Group all the rocks that have a common characteristic. Make at least three categories. In the last column of the table, record the common characteristic on which you based your groupings.
3. Compare your system of classification with those devised by your classmates.
4. Crush bits of the rocks. **CAUTION:** *Wrap rock samples in a towel before hitting them with the hammer. Always wear goggles when using a rock hammer.*
5. Examine the crushed samples with the hand lens. Record any observable characteristics not seen in the larger specimens.

Questions

1. Was there any characteristic common to all the rock samples?
2. What feature was most useful in grouping the rocks?
3. What feature was least helpful in grouping the rocks?
4. Were there any characteristics observed in the crushed rocks that aided or changed your system of grouping?
5. Was your system of grouping different from those of your classmates? Why might different classification systems be devised?
6. How would you classify rocks?

Data and Observations

Sample	Minerals present	Size/shape of minerals	Arrangement of grains	Other information	Common characteristic
1	quartz feldspar biotite	.5 cm, rectangular			grain size similar to samples 3 and 7
2					
3					
4					
5					
6					

a

b

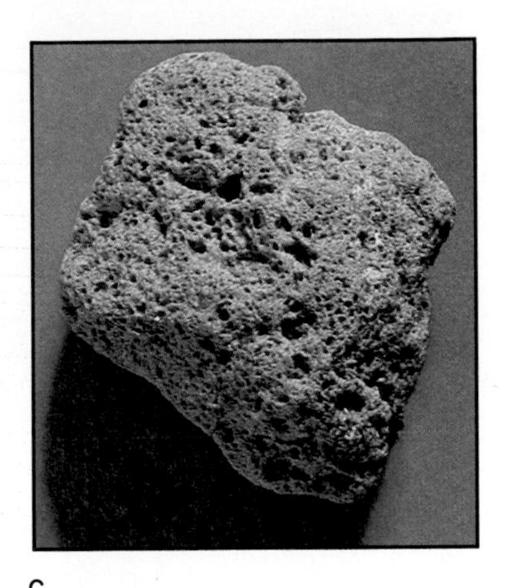

c

FIGURE 17–9. Basalt (a) and andesite (b) are extrusive igneous rocks. Pumice (c) is a volcanic rock formed when gases escape from a felsitic lava.

mostly orthoclase and quartz with smaller amounts of sodium feldspar, amphibole, and biotite.

Fine-grained, or extrusive, rocks include basalt, andesite, and rhyolite. These rocks form from the same kind of silicate melt as coarse-grained intrusive rocks. Extrusive rocks, however, cool more quickly and therefore are made up of smaller grains. Basalts are dark and dense and generally can be recognized even if the minerals cannot be identified. **Felsites** are light-colored, fine-grained rocks in which small grains can be seen. Felsites include andesite and rhyolite. Andesite is the fine-grained equivalent of diorite. Rhyolite is the fine-grained equivalent of granite.

What are felsites?

Cooling at Earth's surface can be so rapid that grains do not have time to form. The resulting rock is obsidian, or volcanic glass. Some magmas have a high percentage

FIGURE 17–10. Obsidian is volcanic glass (a). Compare the size of the grains in a rhyolite and an andesite porphyry (b). The black bar represents 2.54 centimeters. Why is obsidian fine-grained?

a

b

of water vapor and carbon dioxide present. Rapid cooling at the surface may cause gases in the rock to escape, leaving many small cavities. Pumice is formed when gases escape from a felsite. Scoria is formed when gases escape from a basalt. Pumice and scoria are extrusive rocks.

Some rocks begin to crystallize at great depth, and the early forming crystals grow large. Then, magma rises toward Earth's surface where it cools rapidly, forming smaller grains. An igneous rock that has two or more different grain sizes is a **porphyry** (POR fuh ree). Usually a porphyry contains large crystals surrounded by smaller grains. If cooling occurs at the surface, the large crystals may be surrounded by basalt or a felsite.

Review

1. Define igneous rocks.
2. Where does melting of rocks occur?
3. What is the source of heat for rock melting?
4. How do extrusive and intrusive rocks differ?
5. ★ How is the texture of an igneous rock used to identify the conditions under which the rock formed?

CAREER

Thin Section Technician

Mary Hennison is a rock thin section technician. She prepares petrographic sections of rocks. A thin section is made by cutting off a slice of rock about 1 mm thick using a saw embedded with diamonds. One side of the rock chip is polished with silicon carbide grinding powders and then is cemented to a glass slide. Next, Ms. Hennison grinds the open side to a thickness of about 0.03 mm. The last stages of polishing are done by hand.

Minerals are named and identified by using a polarizing microscope, which passes polarized light through the thin section. Geologists use the information obtained by Ms. Hennison to determine what minerals are present in a particular rock. Ms. Hennison has a bachelor of science in geology. Although rocks are hard Earth materials, polishing them to the thickness of this page requires great care and much patience.

For career information, write:
American Geological Institute
4220 King Street
Alexandria, VA 22302

INVESTIGATION 17–1

Classifying Igneous Rocks

Problem: How are igneous rocks classified?

Materials 

igneous rock specimens
hand lens
Appendices M and N
Figure 17–7, 17–8, 17–9, and 17–10

Procedure

1. Examine each rock specimen with a hand lens. Determine the texture of each rock sample. Separate the rocks into groups based on the texture. If the grains or crystals are large and easily seen, the texture is coarse. If the grains are so small that they are not easily distinguished, the texture is described as fine.
2. Answer Questions 1 and 2.
3. Determine the color of each rock. Is the rock light-colored, dark-colored, or intermediate? Regroup your samples based on color.
4. Answer Questions 3 and 4.
5. Examine the coarse-grained rocks to determine the minerals present.
6. Refer to the Figures and give each sample a name according to its composition and grain size.

Analysis

1. Describe the size of the grains in the intrusive igneous rocks.
2. If grain sizes within one sample are noticeably different, what type of rock is it?
3. What minerals must be responsible for the color of sample B?
4. Name at least two other igneous rocks that owe their color to the presence of these minerals.

Conclusions and Applications

5. Why do igneous rocks of the same composition sometimes have different sizes of grains?
6. What two characteristics determine the identity of an igneous rock?
7. How are igneous rocks classified?

Data and Observations

Rock Sample	Texture	Color	Mineral present	Rock Name
A				
B				
C				
D				
E				
F				

IGNEOUS ACTIVITY

GOALS

1. You will learn about the different types of volcanoes on Earth's surface.
2. You will learn about igneous rock structures that result from volcanic activity.

A volcano is a crack or vent through which magma or lava is extruded. Volcanic eruptions are often thought of as destructive events. Each, however, adds new rock to Earth's surface. Volcanic activity at Earth's surface provides clues to Earth's interior.

17:4 Volcanic Mountains

What controls the shape of a volcanic mountain?

Volcanic mountains are present on Earth's continents as well as on the ocean floor. The shapes of these volcanic mountains are controlled by the composition of the extruded material.

How do strato-volcanoes differ from dome volcanoes?

Strato-volcanoes, or composite volcanoes, are cone-shaped structures made of flows of andesite lava and ash, cinders, and rock fragments. Mount Fuji in Japan and Mount Vesuvius in Italy are strato-volcanoes. **Dome volcanoes** form from rhyolite, which is more viscous (VIHS kus), or slower flowing, than andesite. Dome volcanoes are smaller, with steep sides, and have a dome-shaped mass within the crater. **Shield volcanoes** form from basalt. When the basalt reaches the surface, it flows out quietly around a central opening. The best known shield volcanoes are those that form the Hawaiian Islands.

Cinder cones are formed by violent eruptions that blow out fragments of lava in the form of cinders. The cinders collect around a vent and build a steep-sided cone. Examples of cinder cones include Capulin Mountain in New Mexico and Mount Pelée on the island of Martinique.

FIGURE 17–11. Strato-volcanoes are made of layers of lava and ash (a). Masses of rhyolitic lava make up dome volcanoes (b).

a

STRATO-VOLCANO

Lava flows

Cinders

b

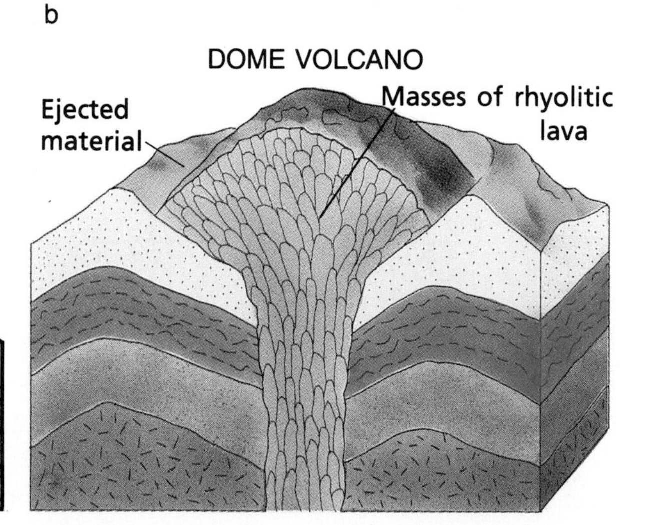

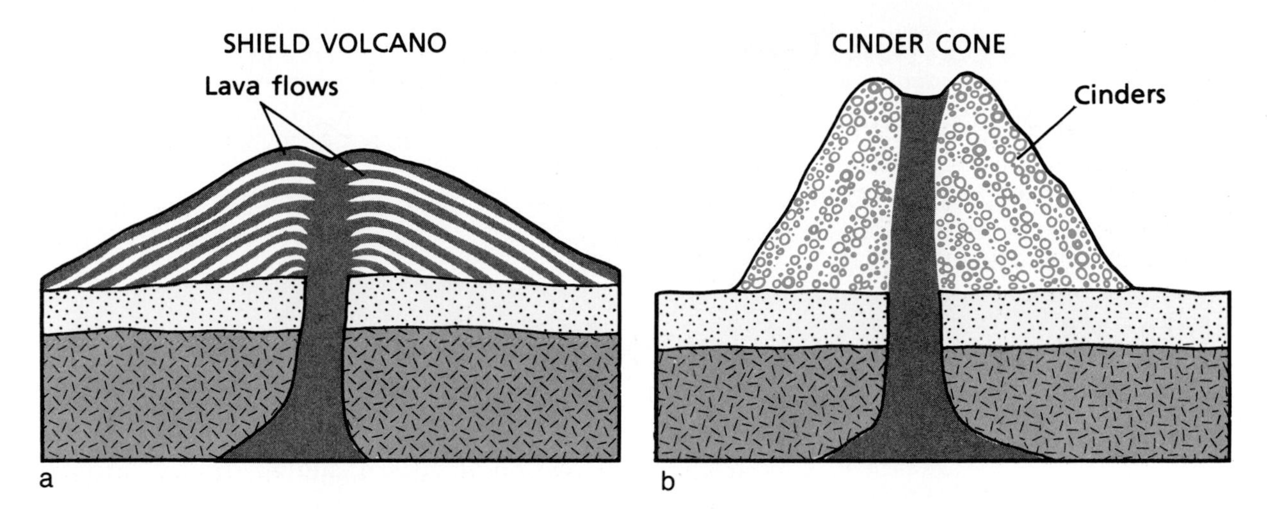

FIGURE 17–12. Shield volcanoes are composed of thin layers of basaltic lava (a). Cinder cones are formed from violent eruptions of lava fragments (b).

All volcanic eruptions emit gas into the atmosphere. In addition to the lava flows, volcanoes eject broken rocks, bombs, cinders, and fine ash and dust. These materials extruded by volcanoes are classified by size. Ash and dust range from a few millimeters to a few centimeters in size. Cinders are the size of peas. Bombs and rocks are much larger.

Table 17–1

15 Selected Volcanic Eruptions in History

Volcano and Location	Year	Type	Eruptive Violence	Magma Content (SiO_2)	Magma Content (H_2O)	Magma Viscosity	Products of Eruption
Etna, Sicily	1669	strato	moderate	high	low	medium	lava, ash, cinders
Tambora, Indonesia	1815	cinder	high	high	high	high	ash, cinders, gas
Krakatoa, Indonesia	1883	cinder	high	high	high	high	ash, cinders, gas
Pelée, Martinique	1902	cinder	high	high	high	high	gas, ash
Vesuvius, Italy	1906	strato	moderate	high	low	medium	lava, ash
Lassen, California	1915	cinder	moderate	high	low	high	ash, cinders
Mauna Loa, Hawaii	1933	shield	low	low	low	low	lava
Paricutín, Mexico	1943	cinder	moderate	high	low	medium	ash, cinders
Surtsey, Iceland	1963	shield	moderate	low	high	low	lava, ash
Kelut, Indonesia	1966	dome	high	high	high	high	gas, ash
Arenal, Costa Rica	1968	cinder	high	high	low	high	gas, ash
Helgafell, Iceland	1973	shield	moderate	low	high	medium	gas, ash
St. Helens, Washington	1980	strato	high	high	high	high	gas, ash
Laki, Iceland	1983	shield	moderate	low	high	medium	lava, ash
Kilauea Iki, Hawaii	1987	shield	low	low	low	low	lava

INVESTIGATION 17–2

Volcanic Eruptions

Problem: How are volcanic eruptions related to properties of magma?

Materials

Table 17–1
paper
pencil

Procedure

1. Copy the graphs shown in Figure 17–13.
2. Using the information from Table 17–1, plot the magma content data for each volcanic eruption by placing an "H" for high quantities of silica (SiO_2) or water (H_2O).
3. Place an "L" for low quantities of silica or water in the appropriate slot on the graph. The data for the 1669 eruption of Mount Etna have already been plotted. (Note: This type of graph is called a scattergram.)
4. When the plotting of the scattergrams has been completed, use the information to answer the questions.

Analysis

1. What relationship appears to exist between the viscosity of the magma and the eruptive violence of the volcano?
2. Which is more liquid in its properties, a magma of high or low viscosity?
3. What relationship appears to exist between the silica or water content of the magma and the nature of the materials ejected from the volcano during the eruptions?

Conclusions and Applications

4. How does the viscosity of the magma appear to be related to its silica or water content?
5. Which of the two variables (SiO_2 or H_2O) has the greatest effect on eruption?
6. What relationship appears to exist between the chemical composition of the magma and the type of volcano that is produced?

FIGURE 17–13.

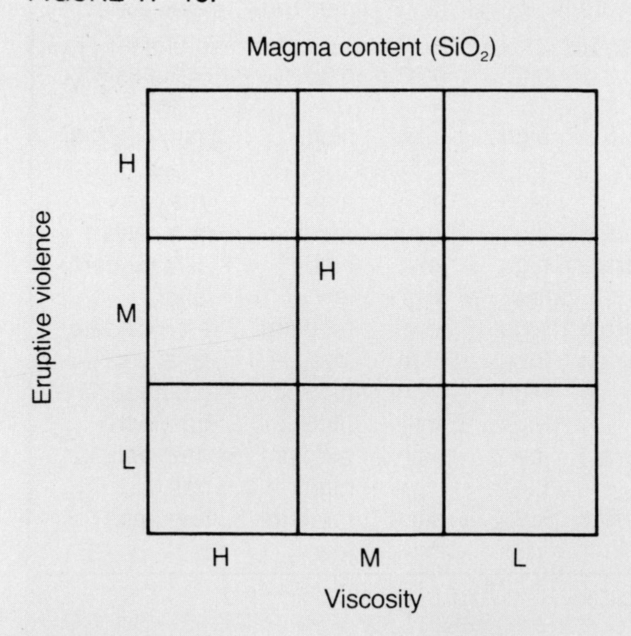

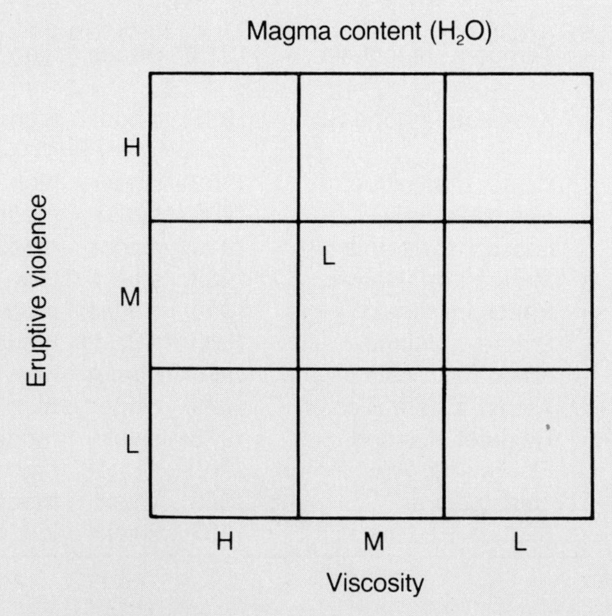

TECHNOLOGY: APPLICATIONS

Predicting Volcanic Eruptions

On May 18, 1980, Mount St. Helens erupted. The violent blast devastated the surrounding 320 square kilometer area. Many people were killed. Could these deaths have been prevented? Could people have been warned?

Volcanoes are difficult to study because they are not commonly found in an active state. When one becomes active, scientists try to gather all the information they can about it. Since an eruption involves the movement of magma or gas beneath Earth's surface, earthquakes are common. Scientists use tiltmeters, seismographs, and other sensitive instruments to monitor the underground movement. Scientists can then pinpoint the depth of the tremors. Rising magma is indicated by rising tremors. When earthquakes become frequent very near the surface, an eruption becomes quite possible. Although technology is improving the ability to predict eruptions, it is still ultimately a guessing game controlled by nature.

17:5 Igneous Rock Structures

Igneous rock structures often form the core of structural mountains. Magma cools and hardens as it rises toward the surface, forming different types of rock structures. These structures are named according to their size, shape, and position in or on surrounding rock.

How are igneous rock structures named?

Magma that cools and hardens below Earth's surface forms intrusive igneous rock structures. **Batholiths** are the largest structures of this type. They are 50 to 80 kilometers across and extend for hundreds of kilometers in length. They may form several kilometers below Earth's surface. Batholiths are too thick for their lower surfaces to be exposed. Their upper surfaces are exposed at the

FIGURE 17–14. Intrusive igneous rock structures form below Earth's surface. Extrusive structures are found on the surface.

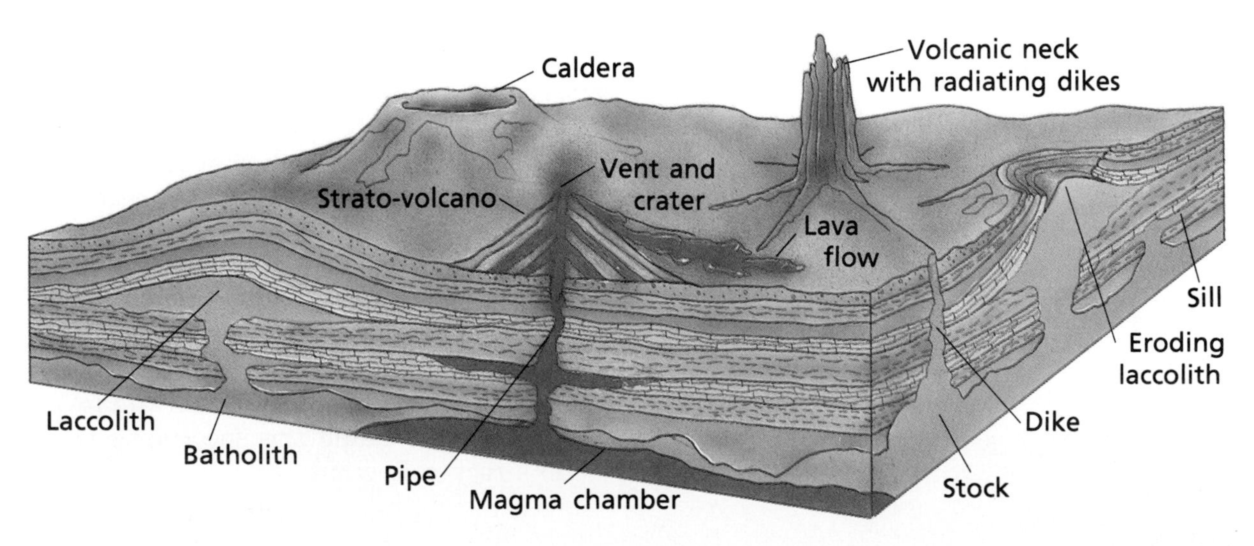

BIOGRAPHY

Thomas A. Jaggar, Jr.
1871–1953

Founder of the Hawaii Volcano Observatory, Jaggar studied volcanoes. He lived in Volcano House perched on the edge of the Kilauea crater.

surface only when the overlying rock is eroded. Most batholiths are composed of granite. They often form the cores of major mountain systems such as the Sierra Nevadas.

Recall that orogeny, or mountain building, causes many faults and fractures in the surrounding rock. **Dikes** are formed when magma enters a vertical fracture and hardens. Sometimes magma squeezes between two rock layers and hardens into a thin horizontal sheet called a **sill.** A **laccolith** (LAK uh lihth) is formed in much the same way as a sill, but the magma forms a thicker sheet that domes upward. An inactive volcano may have magma solidified in its pipe. This structure, called a volcanic neck, is resistant to erosion.

Review

6. Name four types of volcanoes.
7. Name five intrusive igneous rock structures.
8. What is the difference between a dike and a sill?
9. What determines the shape of a volcanic mountain?
★ **10.** What affects the viscosity of molten material?

PROBLEM SOLVING

A Rock by Any Other Name . . .

While on a field trip to the southwestern United States, Peter collected many igneous rocks. Among those in his collection were a light-colored, coarse-grained rock containing quartz, potassium feldspar, and biotite; a green coarse-grained rock; a fine-grained, dark-colored rock with many pores; a black glassy rock; a white porous rock that floats on water; a coarse-grained rock composed of plagioclase feldspar and amphiboles; a coarse-grained rock made of only calcium feldspar; and a rock with large feldspar crystals surrounded by smaller crystals of quartz and biotite. Use what you have learned about minerals and igneous rocks to name the specimens in Peter's rock collection.

Chapter 17 Review

SUMMARY

1. Igneous rocks crystallize from lava or from magma. Fine-grained extrusive rocks are formed at or near Earth's surface. Coarse-grained intrusive rocks harden beneath the surface. 17:1
2. Basaltic magma is less viscous than granitic magma. Basaltic magma produces dark-colored igneous rocks; granitic magma produces light-colored igneous rocks. 17:1
3. Oxygen and silicon combine to form the silicate tetrahedron 17:2
4. Most igneous rock-forming minerals contain silicon and oxygen combined with one or more elements. 17:2
5. Minerals form from a melt in a certain order known as Bowen's reaction series. 17:2
6. Igneous rocks are classified according to texture and composition. 17:3
7. The nature of a volcanic eruption depends on the chemical composition of the magma. 17:4
8. Volcanoes are classified by shape and composition as strato-volcanoes, dome volcanoes, shield volcanoes, or cinder cones. 17:4
9. Igneous rock structures are named according to their size, shape, and position in or on surrounding rock. 17:5

VOCABULARY

a. basaltic magma
b. batholiths
c. cinder cones
d. dikes
e. dome volcanoes
f. extrusive igneous rocks
g. felsites
h. granitic magma
i. igneous rocks
j. intrusive igneous rocks
k. laccolith
l. lava
m. porphyry
n. shield volcanoes
o. silicate tetrahedron
p. sill
q. strato-volcanoes

Matching

Match each description with the correct vocabulary word from the list above. Some words will not be used.

1. extruded magma
2. structures resulting from magma low in SiO_2 and H_2O
3. the basic unit of the rock-forming minerals
4. fined-grained igneous rocks
5. formed when magma enters a vertical crack and hardens
6. an igneous rock with two or more different grain sizes
7. magma with a high silica content
8. the largest intrusive igneous rock structures
9. horizontal sheet of igneous rock
10. violent, steep-sided volcano

Chapter 17 Review

MAIN IDEAS

A. Reviewing Concepts

Choose the word or phrase that correctly completes each of the following sentences.

1. The silicate tetrahedron is made of silicon and *(oxygen, iron, carbon)*.
2. Slow cooling of magma produces *(small, large, no)* crystals.
3. Igneous rocks are named according to composition and *(color, texture, hardness)*.
4. An extrusive igneous rock is *(granite, peridotite, basalt)*.
5. Volcanoes formed by violent eruptions are *(cinder cones, dome volcanoes, shield volcanoes)*.
6. Most batholiths are made of *(basalt, obsidian, granite)*.
7. The largest intrusive igneous structures are *(sills, batholiths, laccoliths)*.
8. Large crystals in an igneous rock indicate that it is a(n) *(extrusive, volcanic, intrusive)* rock.
9. An extrusive igneous rock from which gases have escaped is *(basalt, scoria, felsite)*.
10. Volcanic mountains composed of alternating layers of lava, ash, and cinders are *(strato-volcanoes, shield volcanoes, cinder cones)*.
11. A light-colored, coarse-grained intrusive igneous rock is *(basalt, granite, rhyolite)*.
12. The structural unit basic to the rock-forming minerals is the *(batholith, silicate tetrahedron, caldera)*.
13. A horizontal sheet of igneous rock is a *(sill, batholith, felsite)*.
14. Volcanic mountains resulting from magma low in both silica and water content are *(strato-volcanoes, shield volcanoes, cinder cones)*.
15. Light-colored, fine-grained extrusive igneous rocks within which small grains can be seen are *(felsites, basalts, porphyries)*.

B. Understanding Concepts

Answer the following questions using complete sentences.

16. What are the most commonly occurring Earth rocks?
17. How are igneous rocks classified?
18. How do intrusive and extrusive igneous rocks differ?
19. Distinguish between cinders and bombs.
20. Distinguish between magma and lava.
21. Identify the two general types of magma. How do they differ in appearance?
22. How do the two types of magma differ in chemical composition?
23. What is the difference between a batholith and a laccolith?
24. Where does most magma originate?
25. Describe the characteristics of cinder cone volcanoes.

C. Applying Concepts

Answer the following questions using complete sentences.

26. Why can the minerals in granite and rhyolite, two different igneous rocks, be the same?
27. Why is quartz last to form in a cooling magma?
28. How can you tell scoria and pumice apart without a microscope?
29. Why do volcanoes differ in size and shape?
30. Peridotite, gabbro, diorite, and granite are all igneous rocks and yet are rarely, if ever, associated with volcanic activity. Explain why this is so.

SKILL REVIEW

1. Compare and contrast intrusive and extrusive igneous rocks.
2. Table 17–1 indicates that two recent volcanic eruptions (Mount St. Helens in 1980 and Kilauea Iki in 1987) were quite different in the violence and products of their respective eruptions. List the causes and effects in these two cases from the data provided in the table.
3. Suppose you were given an igneous rock that could be described as follows: black or dark color, very fine-grained, quite dense but with small air spaces throughout. What conclusions could you draw about the origin of this rock?
4. Use data in Table 17–1 to construct a scattergram or graph showing the relationship between the viscosity of magma and the products of the eruption.
5. Construct a simple world map and show the approximate location of each of the volcanoes listed in Table 17–1. Save this map for use in Chapters 19 and 20.

PROJECTS

1. Construct a model of the Mount St. Helens area after the major eruption in 1980. Show the mountain, the destruction of trees, and the nearby river. Include as much detail as possible.
2. Construct a three-dimensional model of the intrusive igneous rock structures and resulting extrusive surface features that are shown in Figure 17–14.

READINGS

1. Lambert, David. *Volcanoes*. New York: Franklin Watts, 1985.
2. Pasachoff, Jay M. "Volcanoes in the Solar System." *Science Digest*. February, 1986, pp. 28-33.
3. Thorpe, Richard, and Geoffrey Brown. *The Field Description of Igneous Rocks*. New York: Wiley, 1985.

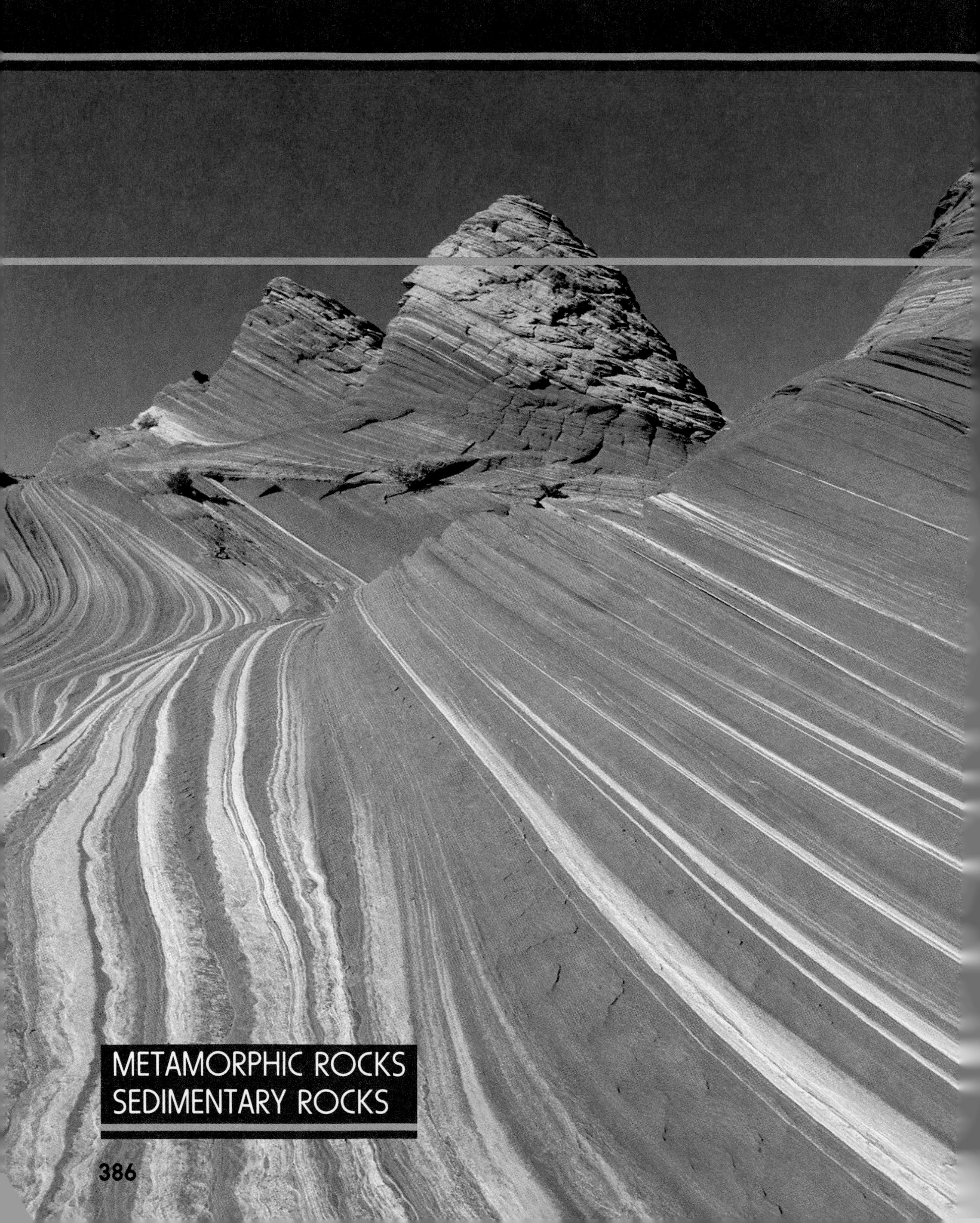

METAMORPHIC ROCKS
SEDIMENTARY ROCKS

Chapter 18

Metamorphic and Sedimentary Rocks

The Grand Canyon is composed of a series of igneous, sedimentary, and metamorphic rocks about 1.6 kilometers thick. Much of the rock sequence is sedimentary and was deposited as an ancient sea gradually advanced and retreated across the North American continent millions of years ago. The Colorado River eroded the sequence to form a spectacular gorge.

METAMORPHIC ROCKS

Metamorphic rocks are rocks changed by temperature and pressure. *Meta* is the Greek word meaning to transform or change. Metamorphic rocks form from preexisting igneous, sedimentary, or other metamorphic rocks.

GOALS

1. You will learn how metamorphic rocks form.
2. You will learn how metamorphic rocks are classified.

18:1 Origin of Metamorphic Rocks

What are metamorphic rocks?

Metamorphic and igneous rocks form in similar environments. During metamorphism, however, temperatures are usually lower and most often melting does not occur. Metamorphic changes usually occur at depths of 12 to 15 kilometers beneath Earth's surface. Temperatures at this depth range from 150 to 800°C.

Three changes may occur as the result of metamorphism: a rearrangement of mineral grains, an enlargement of crystals, and/or a change in the chemistry of the rock. The first two changes result from the rock being subjected to high temperatures and pressures. The change in the chemistry results in the recombination of elements to form different minerals. Chemical changes in a rock occur because of the presence of hot solutions that move through the rock during metamorphism. These changes can occur alone or in any combination.

Several types of metamorphism occur in Earth's crust. Low-pressure, low-temperature changes in the upper crust

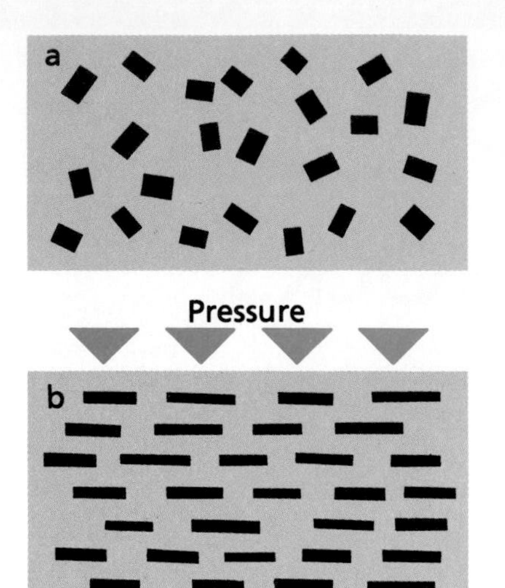

FIGURE 18–1. During metamorphism, mineral grains in the original rock (a) may be reoriented (b) due to pressure.

F.Y.I. Contact metamorphism is also referred to by geologists as thermal metamorphism.

are the result of **burial metamorphism.** Changes occur due to the weight of overlying rocks. This causes a reorientation of mineral grains within the rock.

Regional metamorphism often is associated with mountain building. As mountains form, magma is forced toward the surface where it hardens into the mountain core. Rocks surrounding the core for hundreds of kilometers are subjected to intense temperatures and pressures. Changes in the rocks are most intense near the core. Here, heat and pressure cause new minerals to form and may cause an increase in the size of existing mineral grains. The degree of metamorphism decreases outward from the mountain core.

Contact metamorphism occurs when magma intrudes overlying rock. The rock is changed by heat, solutions, and gases from the magma. New minerals are formed in the surrounding rock as fluids move through cracks and react with the elements present. These kinds of changes occur at temperatures near 800°C. Heat from the magma often causes some changes in the crystal structure of the minerals in the rock. Ions from the magma can be added to rocks already in place. The light-colored rocks in Figure 18–2b were formed when magma intruded a pre-existing rock body.

18:2 Metamorphic Rock Composition

New minerals often are formed during metamorphism. Minerals in metamorphic and igneous rocks are similar because the rocks form under similar conditions. However, minerals that compose sedimentary rocks are formed at low temperatures and pressures at Earth's surface. Thus, during metamorphism, these sedimentary minerals

FIGURE 18–2. During contact metamorphism, rocks are altered by heat, solutions, and gases (a). The light-colored rocks are intrusive igneous rocks (b). Where would you expect metamorphism to have occurred?

a

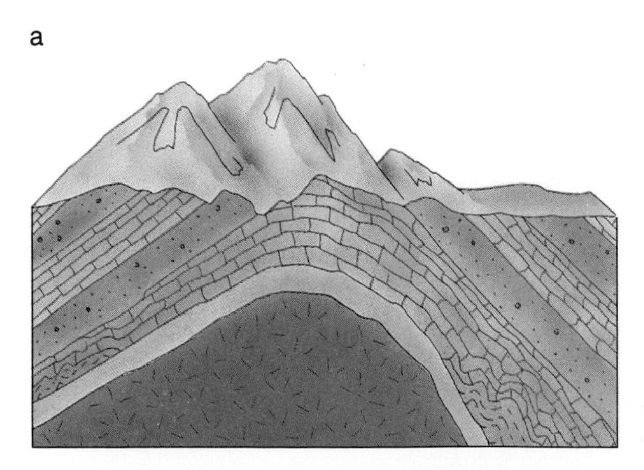

b

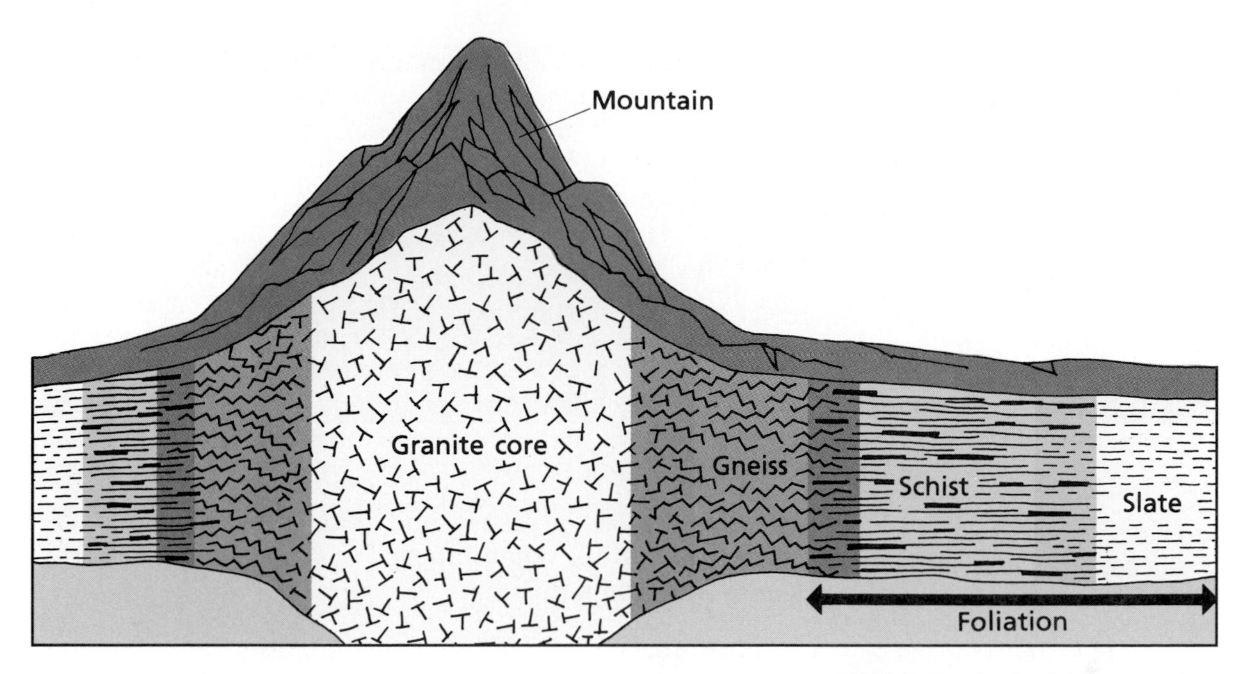

FIGURE 18–3. Metamorphism associated with mountain building decreases outward from the mountain core.

are changed into minerals that are stable at higher temperatures and pressures.

Rocks formed during metamorphism depend on the composition of the parent rock and the temperatures and pressures to which the parent rock is subjected. For example, the sedimentary rock shale contains clay minerals. As temperature and pressure increase, a shale becomes a slate. Biotite and garnet are minerals present in slates.

Table 18–1

Changes During Metamorphism

Parent rock	Low temperature Low pressure	Medium temperature Medium pressure	High temperature High pressure	Higher temperature Higher pressure
shale	slate	phyllite schist	gneiss	MELTING
quartz sandstone		quartzite		
limestone	marble			
peridotite	serpentinite			
basalt		amphibolite or schist		
felsite	slate		schist gneiss	
granite			gneiss	
soft coal (bituminous)	anthracite (hard coal)		graphite	

a

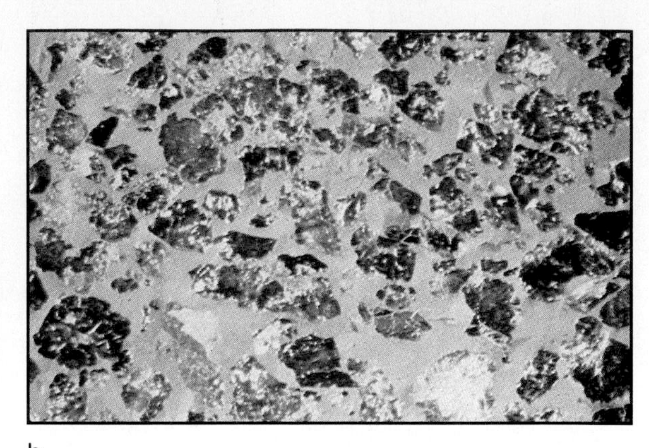

b

FIGURE 18–4. Metamorphism can cause minerals to align (a) or can cause an enlargement of mineral grains (b).

With increased pressure and temperature, a slate becomes a phyllite. What happens to a granite at high pressures and temperatures?

18:3 Classification of Metamorphic Rocks

How are metamorphic rocks classified?

Metamorphic rocks are classified by texture. **Foliated** rocks are banded. At about 150°C, minerals begin to line up in bands, or layers. Banded or foliated rocks include slates, phyllites, schists, and gneisses. *Slates* show no visible internal change. Slates are more compact than their parent rock shale. Grains in slates are extremely small and can be seen only with magnification. However, the arrangement of grains results in visible foliation on most slates. Mica flakes make a *phyllite* surface shiny in contrast to the dull surface of the slate. *Schists* are rocks that show some internal change and definite foliation. They are the most common metamorphic rock. Bands of dark minerals, either biotite or amphibole, alternate with light-colored bands of quartz or feldspar. Schists may form from shales, felsites, or basalts. *Gneisses* are the result of high temperatures and pressures. Some types of gneisses look like granite except for the foliation present.

FIGURE 18–5. Slate (a) and phyllite (b) show little banding.

a

b

a

b

FIGURE 18–6. Banding is visible in schist (a) and gneiss (b).

What are the characteristics of nonfoliated metamorphic rocks?

Nonfoliated rocks are massive and lack banding. These rocks usually contain only one mineral. Nonfoliated rocks include marble and quartzite. All nonfoliated rocks have undergone internal changes. *Marble* is metamorphosed limestone. Marble is harder than limestone and its crystals are much larger. *Quartzite* forms from the sedimentary rock sandstone. *Serpentinite* has marble streaks formed from a parent rock that contains calcium. Soft coal is changed to anthracite at low temperatures and pressures. With an increase in temperature and pressure, it becomes *graphite*.

Review

1. What are the three types of metamorphism?
2. What minerals form when shale is metamorphosed?
3. How are metamorphic rocks classified?
4. ★ Would a gneiss ever form at Earth's surface? Explain your answer.

Table 18–2

Metamorphic Rocks			
	Name	Texture	
		Arrangement of grains	Size of grains
Foliated	slate phyllite	layers almost visible	microscopic microscopic (except for small muscovite flakes)
	schist	layers visible to 1.25 cm apart	recognizable
	gneiss	layers 1.25 cm to 1 m apart	easily recognizable
Nonfoliated	marble quartzite serpentinite	no visible layers	calcite grains recognizable quartz grains recognizable serpentine grains recognizable; calcite may be present

SEDIMENTARY ROCKS

GOALS

1. You will learn about the formation, composition, and classification of sedimentary rocks.
2. You will gain an understanding of the processes involved in the rock cycle.

Sedimentary rocks are products of weathering. They form at or near Earth's surface. The word *sedimentary* comes from the Latin word *sedimentum,* which means to settle out. Sedimentary rocks form in bodies of water and on land. These rocks are classified by the way in which they form and the kinds of materials that compose them.

18:4 Formation of Sedimentary Rocks

Sedimentary rocks are rocks made of loose Earth materials called sediments. These rocks form at Earth's surface. Sedimentary rocks form from weathered igneous, metamorphic, or other sedimentary rocks.

Weathered sediments are deposited by wind, water, and ice. After deposition, the materials are buried, and with time, harden into rock. Many sedimentary rocks are deposited in a series of layers similar to the layers in a cake. These layers are recognized by differences in color, grain size, and composition. The oldest layer or bed is deposited first at the bottom of a series of layers. The youngest bed is at the top. Crustal movements may alter this arrangement.

Materials found in sedimentary rocks include fragments of broken rocks, mineral grains, and substances

FIGURE 18–7. These fossils are preserved in the sedimentary rock limestone (a). Sedimentary rocks are deposited in layers, with the oldest layer deposited on the bottom (b).

a

b

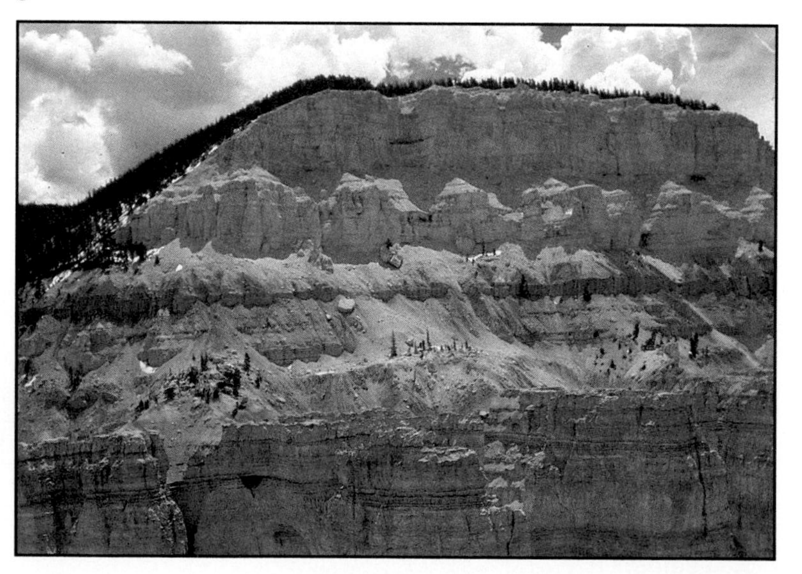

deposited from solution. These sediments are cemented together by calcite, silica, or limonite. Fossils, especially of marine life, are common in sedimentary rocks. **Fossils** are the remains or traces of once-living organisms preserved in Earth's rocks. Bones, shells, and prints of plants and animals may be preserved as fossils. Fossils are most plentiful in limestone, but also occur in shale, coal, and sandstone.

What are fossils?

TECHNOLOGY: APPLICATIONS

Quarrying Rocks

There are basically two types of quarries: quarries that produce blocks or slabs of rock called dimension-stone; and those that produce crushed stone. The first step in the quarrying process is to locate the desired rock. Next, the overburden, or undesirable top layers, are removed. After the overburden has been removed, several methods for quarrying the rock can be used.

One of the most common methods of removing dimension-stone is called plug and feather. In this process, a series of holes about 2 centimeters in diameter is drilled in a row about 15-20 centimeters apart. Two "feathers," or half-round shims, are placed in the hole. A wedge plug is then placed between the feathers. A hammer is used to drive the plugs into the rock. This creates a line of tensile stress between the holes and causes the rock to split.

Another method used to quarry dimension-stone is wire sawing. Very long wire cables are stretched around a pulley system that comes in contact with the stone. Sand and water are constantly placed under the cable as it moves. This increases the abrasive power of the cable and allows it to cut through rock. Wire sawing can cut through rock at the rate of about 5 centimeters per hour, depending on the hardness of the rock.

Two other methods for quarrying dimension-stone are jet piercing and channelling machines. The jet piercing method uses the combustion of fuel oil and oxygen to burn a channel in the rock. A channeling machine is a series of chisels that move along a track on the rock. Each individual chisel removes a small piece of rock during the chopping motion.

The quarrying of crushed stone requires less delicacy and uses explosives throughout most of the quarrying process. Stone is later processed by machine to crush it further or sort it according to size.

INVESTIGATION 18–1

Rock Identification

Problem: How can you identify different rocks?

Materials  

labeled rock set
5% HCl
streak plate
hand lens
goggles
paper towels
water
cloth towel
rock hammer
steel nail
unknown rocks
glass plate
Appendices M and N
magnet
copper penny

Procedure

1. Carefully break off a small piece of each labeled rock sample using the hammer. **CAUTION:** *Be sure to wear goggles when using the hammer. Wrap the rock sample in the cloth towel before striking.*
2. Test a fresh surface of each rock for hardness and reaction to HCl. "Bubbling" indicates the presence of a carbonate mineral. **CAUTION:** *Acid may cause burns. Wear goggles. If spillage occurs, rinse with water.*
3. Observe and record the presence of crystals, layers or openings, texture, odor, and ability of each sample to mark paper.
4. Use the streak plate, steel nail, glass plate, magnet, and penny to identify the minerals you see in the rocks.

FIGURE 18–8.

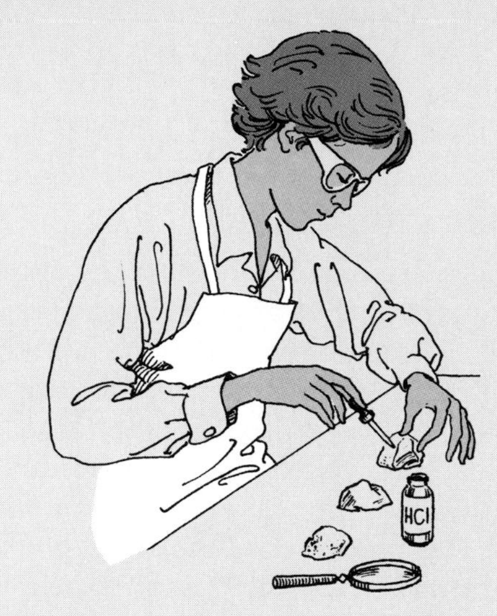

5. Classify each rock as igneous, sedimentary, or metamorphic.
6. Repeat Steps 1 through 4 for the unknown samples.
7. Use the tables in Chapters 17 and 18, as well as Appendices M and N, to give each rock a name based on characteristics you observe.

Data and Observations

Sample	Observations	Rock type	Rock name

Analysis

1. Compare the unknown samples to the known samples. Are any alike?
2. What properties were most helpful in determining the rock type?
3. Which rocks were most similar and thus the most difficult to tell apart?

Conclusions and Applications

4. What processes form metamorphic rocks from sedimentary and igneous rocks?
5. How can you identify different types of rocks?

18:5 Features of Sedimentary Rocks

You may have seen ripples on dunes or on beaches. Sometimes these ripples are preserved in the sand when it is cemented to form sandstone. **Ripple marks** are wavy features of some sandstones.

Mud cracks form along shores or in river beds where mud deposits are completely dried out from time to time. Sand may be blown into the cracks. The cracks may be preserved if they are covered with water and the sediment hardens into rock.

Concretions (kahn KREE shuns) are another feature of some sedimentary rocks. **Concretions** are ball-like objects or irregularly shaped masses of cementing material that collected around a nucleus. Concretions often form in layers and may be centimeters to several meters in diameter.

Geodes (JEE ohds) are ball-like objects found in some sedimentary rocks. Quartz and other minerals grow inward from a hard outer rim of silica. The quartz crystals may be colorless, smoky, or purple.

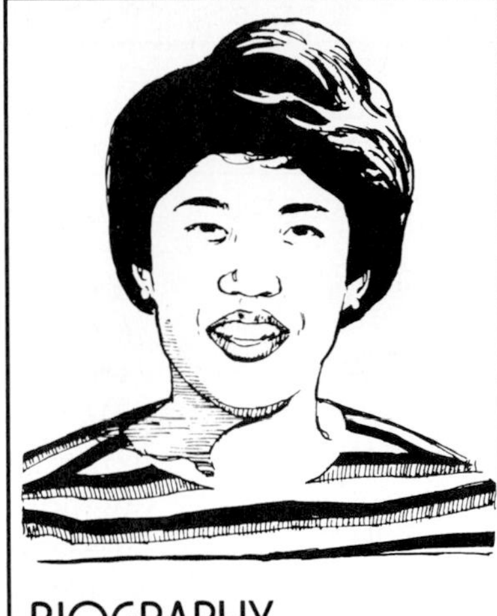

BIOGRAPHY

Marjorie A. Chan
1955–

Marjorie Chan is a geologist at the University of Utah. She studies sedimentary rocks of the Colorado Plateau and reconstructs the way in which the rocks formed millions of years ago. Similar rocks, buried beneath the surface, are of special interest to Chan, because they harbor oil and gas.

18:6 Classification of Sedimentary Rocks

Sedimentary rocks are classified as clastics (KLAS tihks) and nonclastics. **Clastics** are rocks made of fragments of rocks and minerals and broken shells. **Nonclastics** are sedimentary rocks that are deposited from solution or by organic processes.

Clastics contain fragments of rocks and minerals carried by water, wind, or glaciers. They are deposited as

FIGURE 18–9. Ripple marks (a), geodes (b), and concretions (c) are features of some sedimentary rocks.

a

b

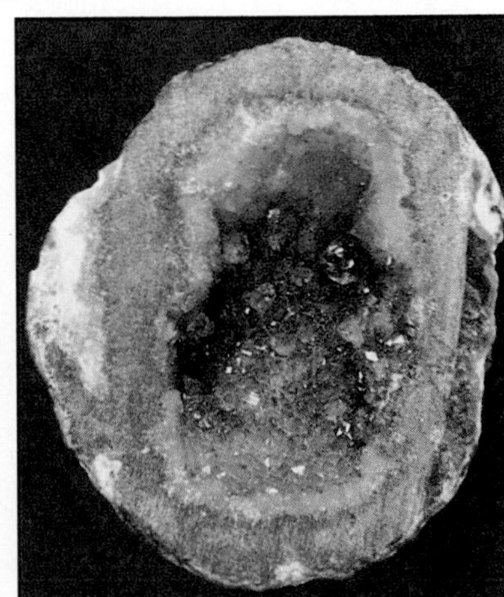

c

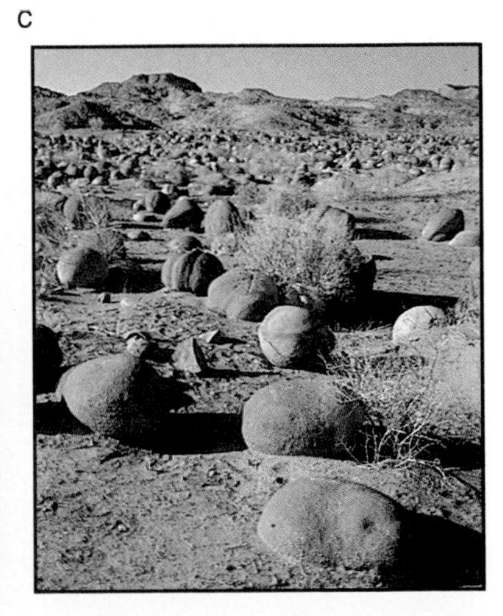

Table 18–3

	Name	Texture	Composition	Comments
Sedimentary Rocks				
Clastics	conglomerate	round pebbles (> 2 mm diameter)	any kind of rock or minerals	pebbles held together with sand, clay, and cement
	breccia	angular pebbles (> 2 mm)		
	sandstone	sand-size grains (0.0625 to 2 mm)	quartz or feldspar and quartz	cement may be calcite, iron oxide or clay
	siltstone	very fine grains (0.004 to .06 mm)	mostly quartz some clay	gritty feel
	shale	microscopic grains and flakes	mostly clay and mud, some mica	no gritty feel
Nonclastics	limestone	coarse to microscopic	calcite or shells and fragments	may have a microscopic shell texture
	chert (flint)	microscopic	chalcedony	precipitate
	alabaster (rock gypsum)	microscopic to coarse	gypsum or anhydrite	evaporite
	rock salt	cubic crystals	halite	evaporite
	peat, coal	coarse to microscopic	plant fragments	fragments of plants or carbon compounds

How are clastics formed?

the velocity of the transporting agent decreases. Large, heavy fragments are dropped first. As speed decreases, smaller and smaller sizes are dropped. When glaciers melt, their entire load is dropped immediately.

Clastic rocks are named according to the size and shape of their fragments. *Conglomerates* (kun GLAHM ruts) are mixtures of rounded pebbles of any kind and shape. Clay and sand grains also may be present. *Breccia* (BRECH uh) is similar to conglomerate except that its fragments are sharp, angular pieces.

Sandstones usually consist of small quartz grains, but may be composed of calcite, feldspar, or pieces of other rocks. Sandstones feel gritty and the grains that compose them are easy to see.

a

b

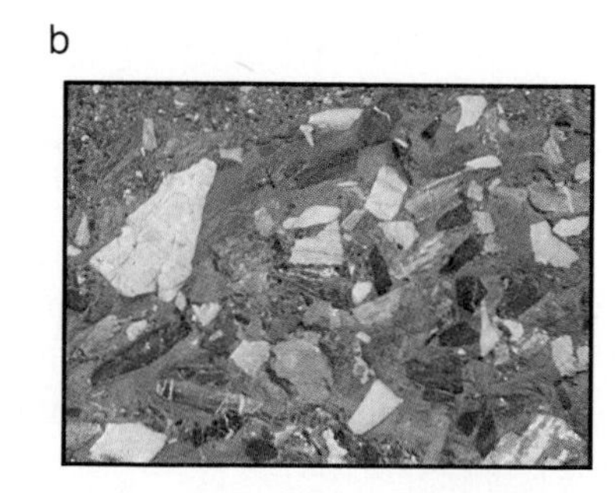

FIGURE 18–10. Conglomerate (a) has rounded fragments. Breccia (b) is composed of sharp, angular fragments.

a

b

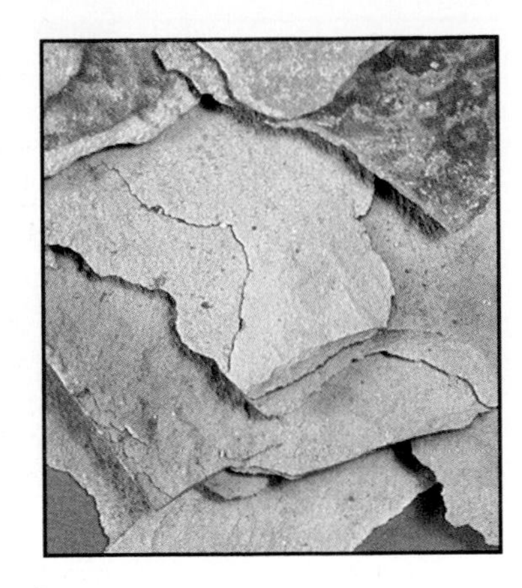

c

FIGURE 18–11. Sandstone (a) has grains that are easily seen. Grains in siltstone (b) can be seen only with magnification. Shale (c) is composed of layers of compacted clay fragments.

Siltstone is like sandstone but with much smaller grains, which can be felt but can be seen only with magnification. Clay usually is present in large amounts in siltstone.

Shale is made of thin layers of clay- and mud-sized particles that are too small to be seen without magnification. The presence of mud suggests that shale is deposited in a quiet, deep body of water where only the finest sediment is transported. Mud also suggests complete decomposition of the original rock or soil. Mica flakes, which are common in shales, generally give the surface of shale a smooth, slippery feel.

How are clastic rocks similar?

All clastic rocks undergo some kind of consolidation process. Shale and siltstone become rocks by compaction (kum PAK shun). Mudballs show how easily clay can be consolidated by squeezing. **Compaction** of mud or silt in nature occurs when it is buried and water and air are squeezed out. Compaction is generally caused by the weight of the overlying sediment. The reduction of volume due to compaction can be as much as 50 percent in some shales. Particles in shale are pressed together so tightly that water cannot move through the rock. Shale is impermeable (ihm PUR mee uh bul), that is, fluids are not able to move through it.

What is cementation?

Sand, pebbles, or mineral grains are consolidated when water, carrying dissolved minerals, moves through these sediments. The minerals are precipitated between the grains and fragments. This process is called **cementation.** The cementing material holds the particles together. Spaces between sediment grains may be completely filled with cement. Some rocks, like sandstones and conglomerates, usually have some openings left that are connected. Such rocks are permeable and liquid can move through them. Oil or water may accumulate in pore spaces in these rocks.

a

b

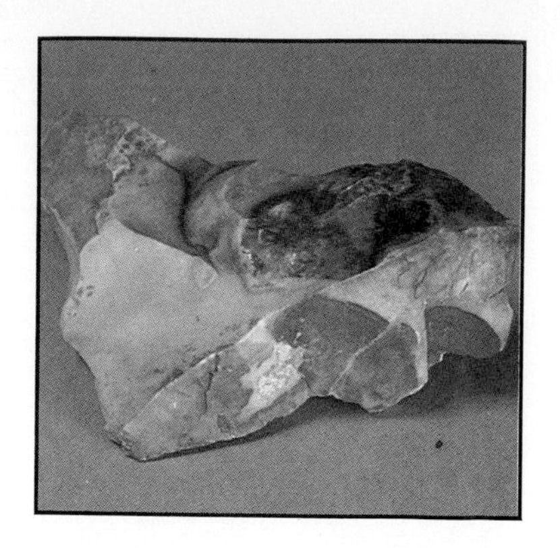

c

FIGURE 18–12. Nonclastics include limestone (a), flint (b), and chert (c).

Nonclastic rocks are either chemically or organically formed. **Precipitates** occur when chemical reactions form solids that settle out of solution. Calcite is the most common precipitate. It is found in caves and as cement in other rocks. Some beds of limestone are precipitates. **Evaporites** form when water evaporates, leaving its dissolved solids behind. Evaporites include beds of salt and gypsum.

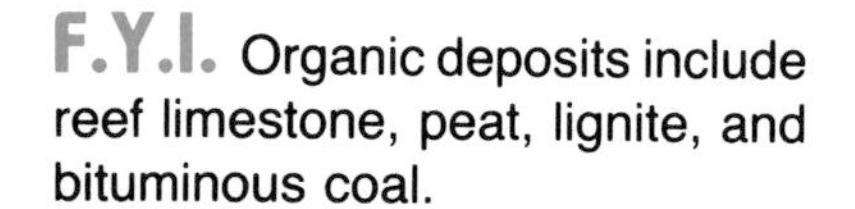

F.Y.I. Organic deposits include reef limestone, peat, lignite, and bituminous coal.

Corals, sea animals that secrete a hard covering of calcite around their bodies, form reefs hundreds or thousands of meters thick. Reefs that are compacted become limestone.

Plant debris such as trees, twigs, and ferns may be buried in a swamp. Over millions of years, this organic material is compacted into peat. With further burial, peat is changed to coal.

Nonclastic rocks are named according to composition. These rocks include *limestone*, composed of calcite; *flint* or *chert,* made of silica; *rock salt*, made of halite; and *rock gypsum* or *alabaster,* made of gypsum. *Chalk* is a kind of limestone that has crystals and shells visible only under a microscope.

FIGURE 18–13. Alabaster forms when water containing certain compounds evaporates (a). These salt flats in the Dead Sea are evaporites (b).

a

b

INVESTIGATION 18–2
Sedimentary Rocks

Problem: How can you classify sedimentary rocks?

Materials  

sedimentary rock samples
marking pen
5% hydrochloric acid (HCl)
dropper
hand lens
goggles
paper towels
water
unknown sedimentary rock samples

Procedure

1. On your paper, make a Data and Observations chart similar to the one shown below.
2. Group your samples into clastics and nonclastics.
3. Give each rock a label and number, S–1, S–2, and so on.
4. Sketch each rock.
5. Put a few drops of HCl on each rock sample. **CAUTION:** *HCl is an acid and can cause burns. Wear goggles. Rinse spills with water.* "Bubbling" on a rock indicates the presence of carbonate minerals.
6. Look for fossils and describe them if any are present.
7. Identify each rock sample using the information in Table 18–3.

Questions

1. How do clastic sedimentary rocks differ from nonclastics?
2. Why did you test the rocks with hydrochloric acid?
3. What mineral reacts with hydrochloric acid?
4. What is needed in order for sedimentary rocks to form from fragments?
5. How can you classify sedimentary rocks?

Data and Observations

Sample	Clastic or Nonclastic	Observations	Rock Name
S–1			
S–2			
S–3			
S–4			
S–5			

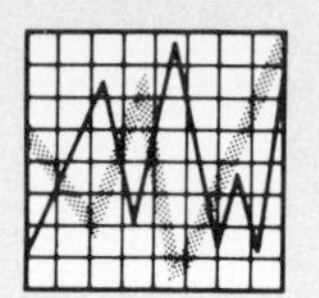

SKILL

Using Tables and Charts

Problem: How do you use tables and charts?

Materials

paper
pencil
Focus on Earth Science textbook

Background

Tables and charts are excellent ways to organize and present data and scientific information. Thus, they are included in most scientific literature. Your textbook contains many tables and charts. Tables and charts serve two purposes. First, they allow you to get specific information quickly and easily. Second, they allow you to arrange a great deal of information for easy reference.

In order to use tables and charts accurately, you must understand the vocabulary used; note the purpose of each table; determine how the table is organized; and read the background information presented in the text. The following questions will help you to better use tables and charts. Below are questions about the tables and charts used in Chapters 16, 17, and 18 on rocks and minerals. Use the tables in those chapters to answer the following questions.

Questions

1. Is each table explained by general background information in your text? If so, how does this information help you read each table? How does the information in each table increase your understanding of the text materials?
2. Where can you find the meaning of each of the following words taken from the tables in Unit 5?

   hardness, shield volcano, nonfoliated

   Give the section number where the definition of each word is found.
3. Explain how Tables 18–1 and 18–2 help you to better understand metamorphic rocks.
4. Study Table 16–6, Rock-forming Minerals. How many chemical classifications are listed? What type of information is given about each classification?
5. Look at the titles of the tables in Chapter 18, Metamorphic and Sedimentary Rocks. What do the titles tell you?
6. Use Table 16–1 to find out the most abundant element in Earth's crust. What is this element's symbol?
7. Use Table 16–3, Mohs' Scale of Hardness, to determine which is harder, diamond or gypsum. Will topaz scratch calcite? Will corundum scratch fluorite? Will quartz scratch gypsum?
8. Table 16–5 lists common minerals and rare gem forms. What are the differences between rubies and sapphires? What one element is common to all red gemstones listed on the chart? What is the chemical formula of spinel?
9. Use Table 16–6 to find out in which rocks the carbonates, calcite and dolomite, are commonly found.
10. Use Table 18–2, Metamorphic Rocks, to distinguish among a phyllite, a schist, and a gneiss.
11. Selected volcanic eruptions are listed in Table 17–1. Which type of volcanic cone is most common in Hawaii?
12. Table 18–1 describes changes during metamorphism. What two metamorphic rocks form from shale, felsite, and granite?
13. Table 18–3 lists sedimentary rocks. Distinguish between a conglomerate and a breccia.

CAREER

Pottery Maker

Julie Meyers works for a tile company that glazes tile and porcelain. Her official title is pottery maker. She designs kitchenware and sketches patterns for plates and wall tiles. Ms. Meyers has learned that it takes a great deal of experience to create the product she desires.

Ms. Meyers chose her career partly because of the knowledge she obtained in high school. She took an earth science class that taught her about the formation of different types of rocks. As she learned about clay, feldspars, and flint in earth science, she shaped and molded them into pottery during art class.

For career information, write:
National Association of Schools of
Art and Design
11250 Roger Bacon Drive
Reston, VA 22090

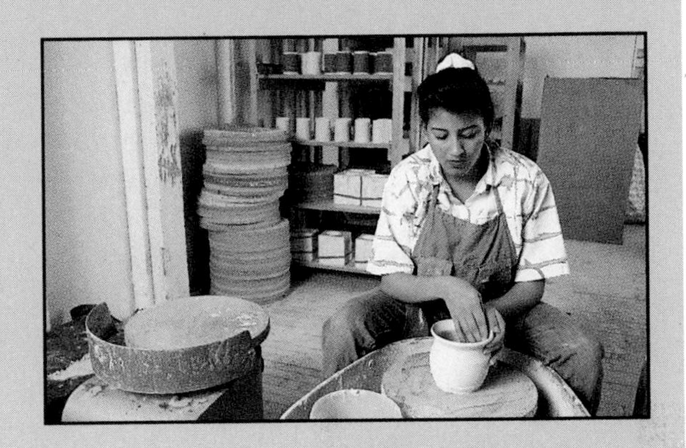

18:7 The Rock Cycle

Earth materials constantly change. Some changes occur below the surface. Others occur at Earth's surface. Some changes are the result of chemical processes, some are due to physical processes. Beneath Earth's surface, heat and pressure change rocks. One source of this heat and pressure is the mass of overlying rocks. Another source is the decay of radioactive elements in rocks.

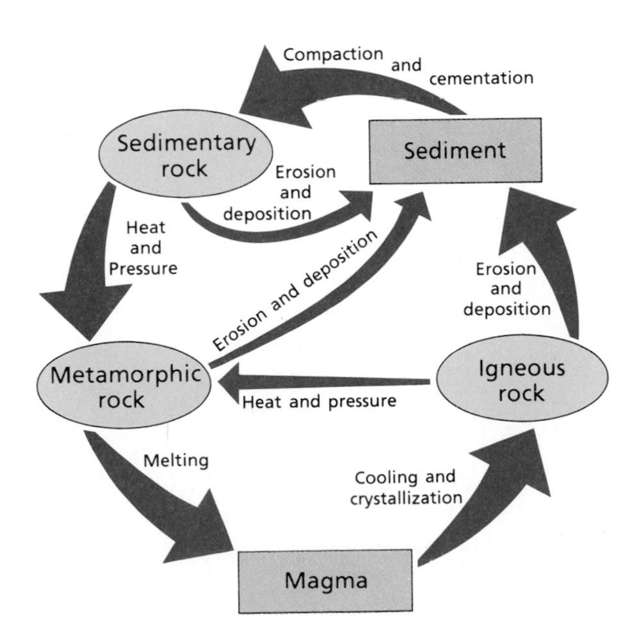

FIGURE 18–14. The rock cycle shows the changes that rocks may undergo.

The processes that occur both within the crust and at Earth's surface not only play an important role in shaping landscapes, but they also are related to the rocks that result from the these processes. Figure 18–14 shows how certain processes play a role in the formation of the three major rock types. The paths that rock materials follow and the processes that affect them along the way is known as the **rock cycle.**

What is the rock cycle?

The rock cycle, like the water cycle, results from the response of Earth materials to various forms of energy. In Figure 18–14, the outer circle represents the complete cycle as it may occur without interruption. The arrows within the circle, however, show that certain conditions may create "short cuts" that miss parts of the cycle. Each rock type can result from one of the other two. Under the right conditions, a certain rock type can eventually provide the materials for the formation of other rock types. The rock cycle shows that Earth is a dynamic system being changed by the forces at its surface and those within.

Review

5. Identify the two types of sedimentary rock.
6. How do metamorphic and igneous rocks become sedimentary rocks?
7. How do igneous and sedimentary rocks become metamorphic rocks?
8. ★ Suppose you found a tan rock made of smooth rounded pebbles cemented together by mud. Name your rock.

PROBLEM SOLVING

That's an Order!

Lucia and Chad took a vacation with their parents to the Appalachian Mountains. While on a hike one afternoon, they observed some tightly folded sedimentary and metamorphic rocks. Lucia had learned about these rocks in her earth science class. She explained the geologic history of the area, and concluded her "lesson" by explaining that the lowest bed in the rock sequence was the oldest. Was Lucia correct? Explain.

Chapter 18 Review

SUMMARY

1. Metamorphic changes include rearrangement of grains, enlargement of crystals, and/or chemical changes in the original rock. 18:1
2. Three types of metamorphism are: burial, regional, and contact. 18:1
3. Minerals formed during metamorphism depend on the composition of the original rock and the temperatures and pressures to which the rock is subjected. 18:2
4. Metamorphic rocks are classified by texture. 18:3
5. Sedimentary rocks are formed from Earth materials that are compacted and cemented to form rocks. 18:4
6. Fossils, ripple marks, and mud cracks may be found in sedimentary rocks. 18:4, 18:5
7. Clastics are named according to the size and shape of their fragments. Nonclastics are named according to their composition.
8. Clastics are consolidated by compaction and cementation. Nonclastics form when minerals precipitate from solutions. 18:6
9. The rock cycle describes the changes that rocks undergo. 18:7

VOCABULARY

a. burial metamorphism
b. cementation
c. clastics
d. compaction
e. concretions
f. contact metamorphism
g. evaporites
h. foliated
i. fossils
j. geodes
k. metamorphic rocks
l. nonclastics
m. nonfoliated
n. precipitates
o. regional metamorphism
p. ripple marks
q. rock cycle
r. sedimentary rocks

Matching

Match each description with the correct vocabulary word from the list above. Some words will not be used.

1. hollow ball-like objects found in some sedimentary rocks
2. the paths and processes that affect all three major rock types
3. rocks that result from heat and pressure without melting
4. remains of once-living organisms preserved in Earth's rocks
5. occurs when magma intrudes overlying rock
6. massive metamorphic rocks that lack banding
7. the consolidation process that forms shale and siltstone
8. beds of salt or gypsum
9. a metamorphic rock with layers or bands
10. low pressure and temperature changes due to the weight of overlying rock

Chapter 18 Review

MAIN IDEAS

A. Reviewing Concepts

Choose the word or phrase that correctly completes each of the following sentences.

1. The metamorphic rock that forms from limestone is *(slate, gneiss, marble)*.
2. A metamorphic rock with a coarse-grained texture is *(slate, phyllite, quartzite)*.
3. The first metamorphic rock to form from shale as temperature and pressure increase is *(schist, gneiss, slate)*.
4. The amount of metamorphic change is greatest in *(gneiss, slate, phyllite)*.
5. The conditions under which metamorphic rocks tend to form occur *(near Earth's surface, 12 to 15 kilometers below Earth's surface, several hundred kilometers below Earth's surface)*.
6. Quartzite is the metamorphic equivalent of *(marble, sandstone, basalt)*.
7. Fossils are most commonly found in *(limestone, shale, sandstone)*.
8. Metamorphic rocks are produced when magma intrudes overlying rock in a process known as *(burial, regional, contact)* metamorphism.
9. A nonclastic rock made of silica is *(limestone, rock gypsum, chert)*.
10. Ball-like objects of cementing material that form in layers are called *(breccias, concretions, conglomerates)*.
11. A sedimentary rock formed from compaction of organic debris is *(sandstone, coal, chert)*.
12. Salt beds often form from a process called *(compaction, evaporation, precipitation)*.
13. An example of a nonclastic sedimentary rock is *(chalk, sandstone, conglomerate)*.
14. Rocks with connected openings are said to be *(permeable, nonfoliated, impermeable)*.
15. The rock cycle represents the response of Earth materials to *(compaction, cementation, energy)*.

B. Understanding Concepts

Answer the following questions using complete sentences.

16. What does the word *metamorphic* mean?
17. What are fossils?
18. What is burial metamorphism?
19. Describe the changes that may occur to the sedimentary rock shale as both temperature and pressure increase to extremely high levels.
20. Distinguish between foliated and nonfoliated metamorphic rocks.
21. Distinguish between a conglomerate and a breccia.
22. What are ripple marks and what do they indicate?
23. How do clastic and nonclastic sedimentary rocks differ?
24. How are clastic sedimentary rocks named?
25. What processes change sedimentary and metamorphic rocks to magma? Metamorphic to sedimentary?

C. Applying Concepts

Answer the following questions using complete sentences.

26. How would you distinguish between marble and quartzite?
27. How are geodes different from concretions?
28. How could quartz that was originally formed in the igneous rock granite eventually end up in a sandstone?
29. Would you expect a schist to form as the result of burial metamorphism? Explain.

30. Why are fossils not found in metamorphic rocks?

SKILL REVIEW

1. You are given three white crystalline rocks. All three appear to have similar physical properties, but you have been informed that one is limestone while the other two are of igneous origin. Describe an experiment you might perform to identify the sedimentary rock.
2. Use Table 18–3 to determine which nonclastic sedimentary rocks are classified as evaporites.
3. Study the photograph on page 432. How might these rocks have formed?
4. Explain the causes and effects of pressure and temperature on a shale.
5. Compare and contrast clastic and nonclastic sedimentary rocks.

PROJECTS

1. Construct a display that demonstrates how various rocks such as marble, slate, gneiss, or granite are obtained from rock quarries. Include the locations of well-known sources of these rocks.
2. Use sand, gravel, mud, clay, and a salt solution to make sedimentary, metamorphic, and igneous "rocks." Simulate the conditions under which the rocks are formed.

READINGS

1. Fry, Norman. *The Field Description of Metamorphic Rock*. New York: Halsted Press, 1984.
2. Lambert, Norman. *Rocks & Minerals*. New York: Franklin Watts, 1986.
3. Selsam, Millicent E., and Joyce Hunt. *A First Look at Rocks*. New York: Walker & Company, 1984.

SCIENCE AND SOCIETY

NEW USES FOR OLD ELEMENTS: SUPERCONDUCTORS

Imagine riding a high-speed train that is flying through the night, held aloft by opposing sets of electromagnets. Think of a submarine gliding silently beneath the sea without the aid of engines or propellers. Imagine carrying a powerful computer in your pocket. These machines may sound like something out of a science fiction novel, but they are not. A train already exists. The submarine and computer will probably be in use within the next few years. The development of these products is being made possible by an incredible scientific breakthrough that took place in 1987. That discovery is called "high-temperature superconductivity."

Background

In normal conductivity, such as that through copper electrical wiring, electrons bump into each other and into imperfections in the copper's molecular structure. Each time the electrons collide, energy is lost in the form of heat. In other words, there is resistance to the flow of electricity. In superconductors, there is no resistance to the flow and, therefore, no loss of energy as the electrons move along.

Superconductivity was discovered in 1911 by Heike Kamerlingh Onnes, a Dutch physicist. He cooled a crystal of mercury to 4 Kelvin (K) using liquid helium and found that the mercury lost all resistance to electricity. Liquid helium, however, is very expensive.

Some superconductors have been in use for several years. Magnetic Resonance Imaging Machines use superconducting magnets that allow doctors to obtain X-raylike images of soft body tissues. The high-speed train discussed earlier that runs on superconducting electromagnets has been tested by Japanese scientists. Similar electromagnets wound with superconducting wires will be used for the Superconducting Super Collider.

Since 1911, scientists have been looking for materials that would become superconductors at certain temperatures so they would no longer need to use helium as the coolant. Nitrogen becomes liquid at 77 K. Nitrogen is cheap, plentiful, and easy to use. By 1973, an alloy of niobium and germanium was found to work at 23 K, but that temperature was too low for practical use.

Things started to heat up in 1983 when two researchers with International Business Machines in Zurich decided to try metal oxides rather than the old metal alloys. For two years, the scientists ground up

FIGURE 1. Bullet trains are a means of mass transit in Japan. Superconductors may revolutionize mass transit.

hundreds of compounds with a mortar and pestle. Finally, in late 1985, the researchers combined barium, copper, oxygen, and lanthanum and found that the transition temperature (the temperature at which a substance will superconduct) of this alloy was as high as 10 K. The race was on in earnest!

In the winter of 1987, Paul C.W. Chu at the University of Houston and his associates experimented with other elements and achieved superconductivity at 93 K. They were able to use liquid nitrogen for the coolant! Other groups have repeated these experiments and achieved the same results. Now researchers are working to find ceramic materials that will become superconductors at room temperature.

Once high transition temperatures are achieved, however, the work of finding practical applications will begin. Several problems exist. Most people associate ceramics with cups and plates or delicate figurines. It is hard to imagine them wrapped in coils carrying electricity. Although they are basically metals, when combined with other substances and oxygen, their structure changes and they become ceramics. One of the problems that has to be overcome is the brittleness of ceramics.

Another problem yet to be solved is that these materials seem to lose their superconductivity if large amounts of electricity move through them at high temperatures. The cause of this is not fully understood and is the subject of current research.

Look to the Future

It is the year 2025 and you have just zipped to school in your SMASH (Superconducting Magnetic Aerodynamic Surface Hopper). Your social studies class has been studying the changes in life-styles in the last 15 years. Your latest assignment has been to look at the changes caused by the discovery of superconductivity. Make a list of the things you can think of that have been affected. When you have finished your list, put a "+" next to the ones you think have been beneficial and a "−" next to the ones you feel have not been good. Compare notes with your classmates.

Developing a Viewpoint

1. Inventors are often secretive about their inventions because they fear that others will steal their ideas and receive the profits from them. Do you think the scientists working in industrial laboratories will benefit from hiding their discoveries in the superconductor field? What benefits could come from sharing information?

2. In many areas of technology, the United States seems to be interested in how it stacks up compared with other countries. The international "space race" began in the 1950s and which country possesses the most up-to-date weapons is always of great concern. These concerns have also begun to surface about the "superconductor race." Do you think the U.S. should be concerned about being first in superconductor technology? What will it mean to be the country that either does or does not "get there" first with commercial uses for superconductors?

Suggested Readings

William J. Cook. "Seeking the Perfect Wire?" *U.S. News and World Report,* Vol. 102. May 11, 1987.

Michael D. Lemonick. "Superconductors: The Startling Breakthrough That Could Change Our World." *Time,* Vol. 129. May 11, 1987.

Gina Maranto. "Superconductivity: Hype vs. Reality." *Discover.* August, 1987.

UNIT 6

Earth's interior is active and changing. Earthquakes remind us of the awesome internal powers of Earth. What causes Earth to shift? Why do continents move? Internal forces enlarge oceans, build mountains, and split continents. The Great Rift Valley of East Africa may be a beginning stage in the eventual separation of East Africa from the mainland.

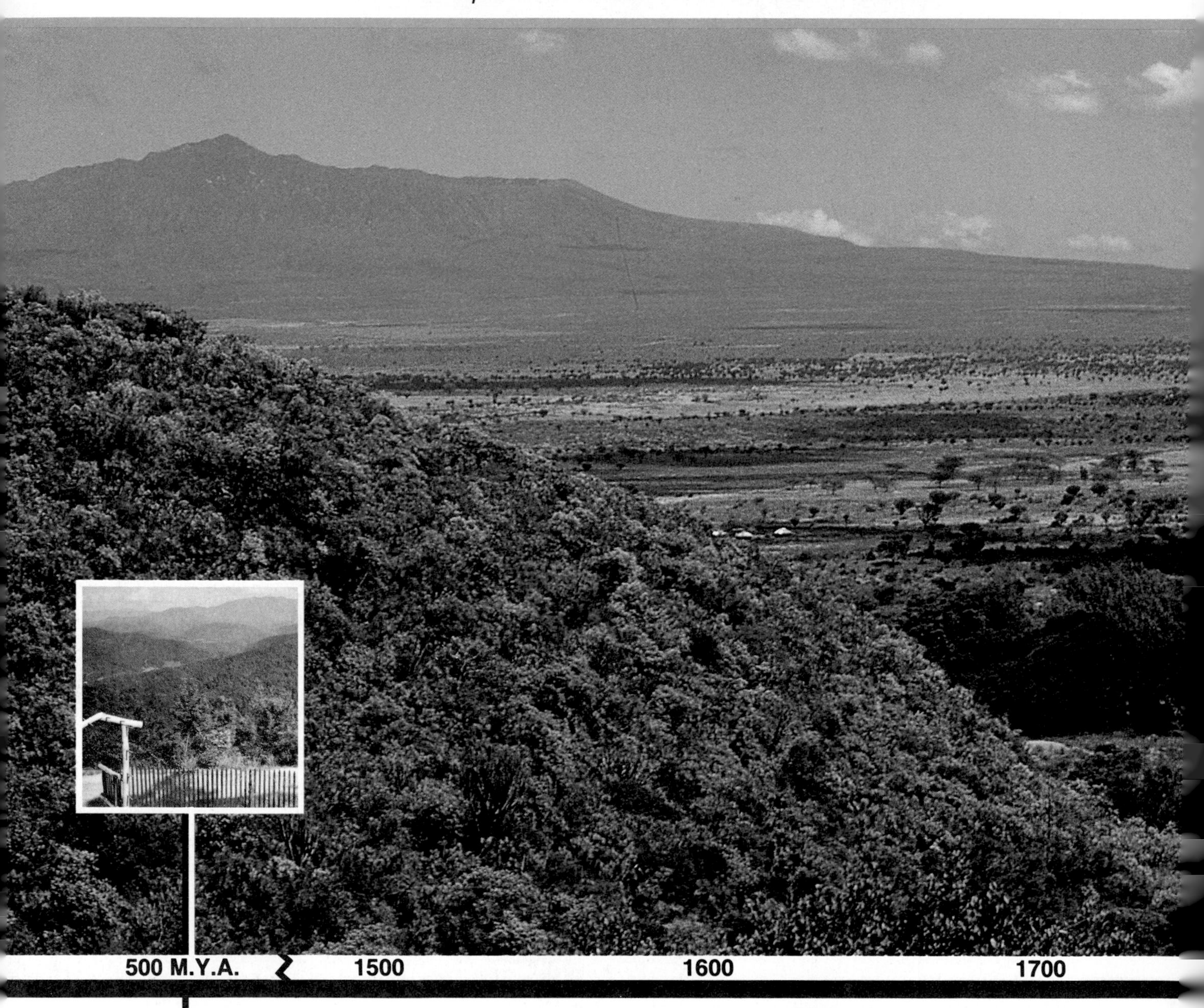

500 M.Y.A. | 1500 | 1600 | 1700

~500 M.Y.A.
Appalachian Mountains begin to form.

1636
Harvard College founded.

INTERNAL PROCESSES

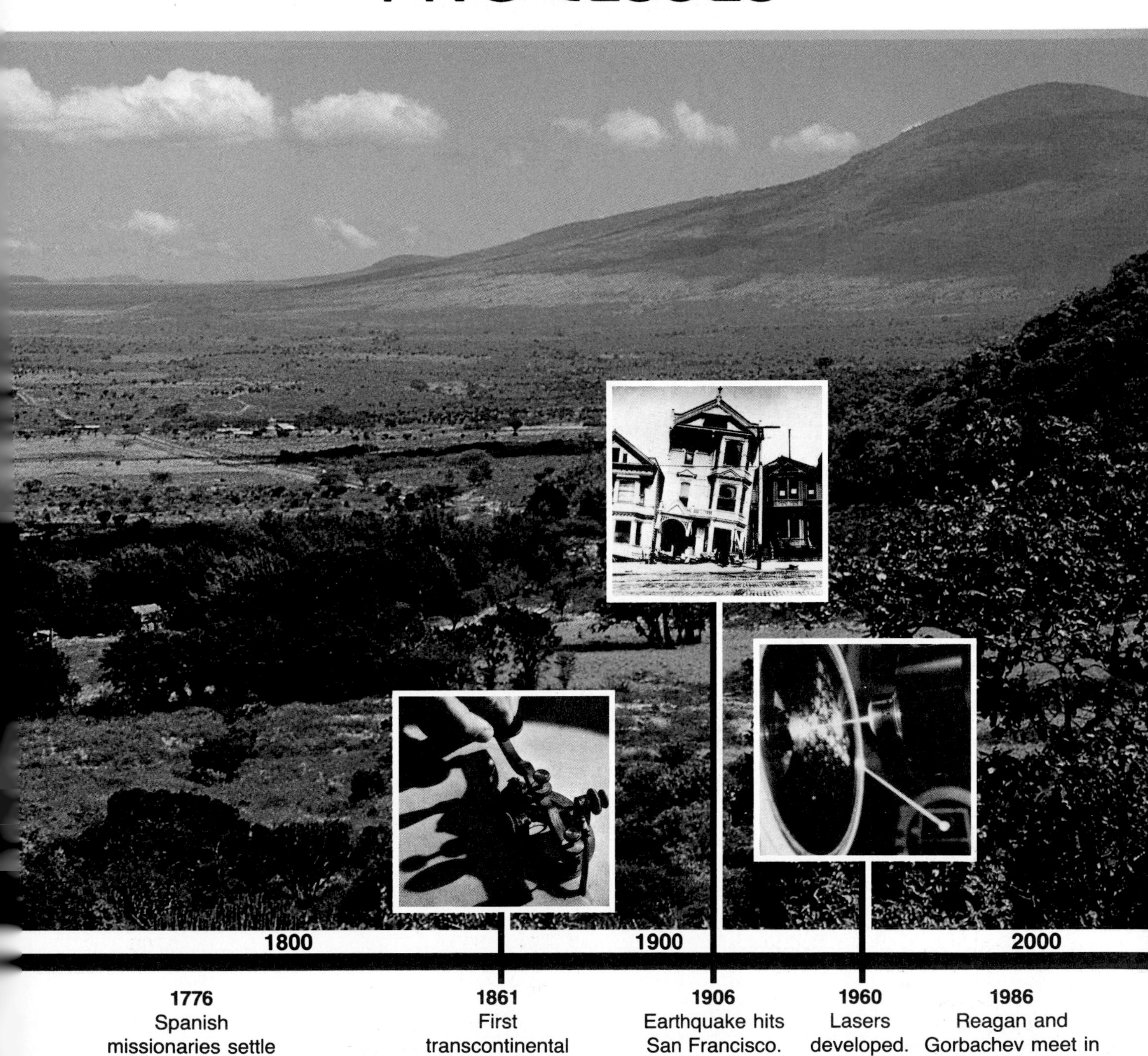

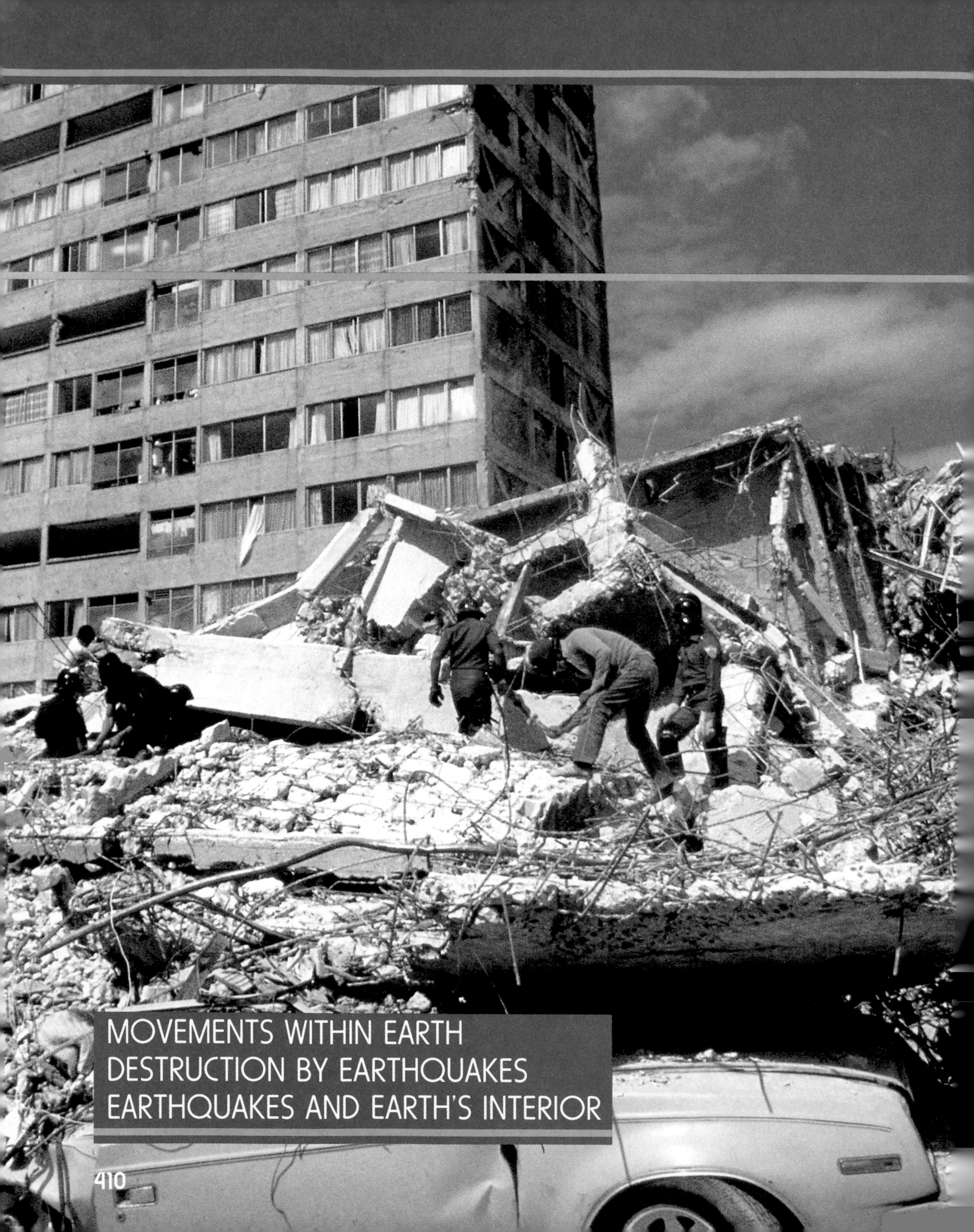

MOVEMENTS WITHIN EARTH
DESTRUCTION BY EARTHQUAKES
EARTHQUAKES AND EARTH'S INTERIOR

Earthquakes

Tremendous forces exist within Earth. Over time, these forces exert so much pressure on rocks that they break, sometimes resulting in earthquakes. Can earthquakes be prevented? Probably not. Can they be controlled or predicted? Some scientists think so.

MOVEMENTS WITHIN EARTH

GOALS

1. You will gain an understanding of the forces that cause earthquakes.
2. You will learn how faults and folds occur in rocks.

The energy and forces that exist inside Earth often affect those who live at its surface. Knowledge of these forces and effects may help us understand what is happening on our dynamic planet.

19:1 Earthquakes

Just after 7 A.M. on a clear, sunny Thursday in September, 1985, most of Mexico City's residents had begun to prepare for the day. Suddenly, the ground began to move. Four minutes later, at least 250 buildings had collapsed, over 100 more were badly damaged, and thousands of people were dead or injured.

For those directly affected, this earthquake was a once-in-a-lifetime event. But for Earth's crust, it was one of many adjustments that occurs every day due to internal forces. Each year, hundreds of earthquakes are felt and recorded worldwide. **Earthquakes** are vibrations caused by the sudden movement of surface rocks. These vibrations, called **seismic waves,** result when rock suddenly breaks and moves, releasing large amounts of energy.

Where do most earthquakes occur?

Scientists know that these seismic events do not happen by chance, but in definite patterns. Most earthquakes occur along well-known belts. One such belt circles the Pacific Ocean. Eighty percent of all earthquakes occur in this belt. Other belts of activity run through central Asia, the Mediterranean Sea, and the Caribbean Sea. Earthquakes also are common along the mid-ocean ridges.

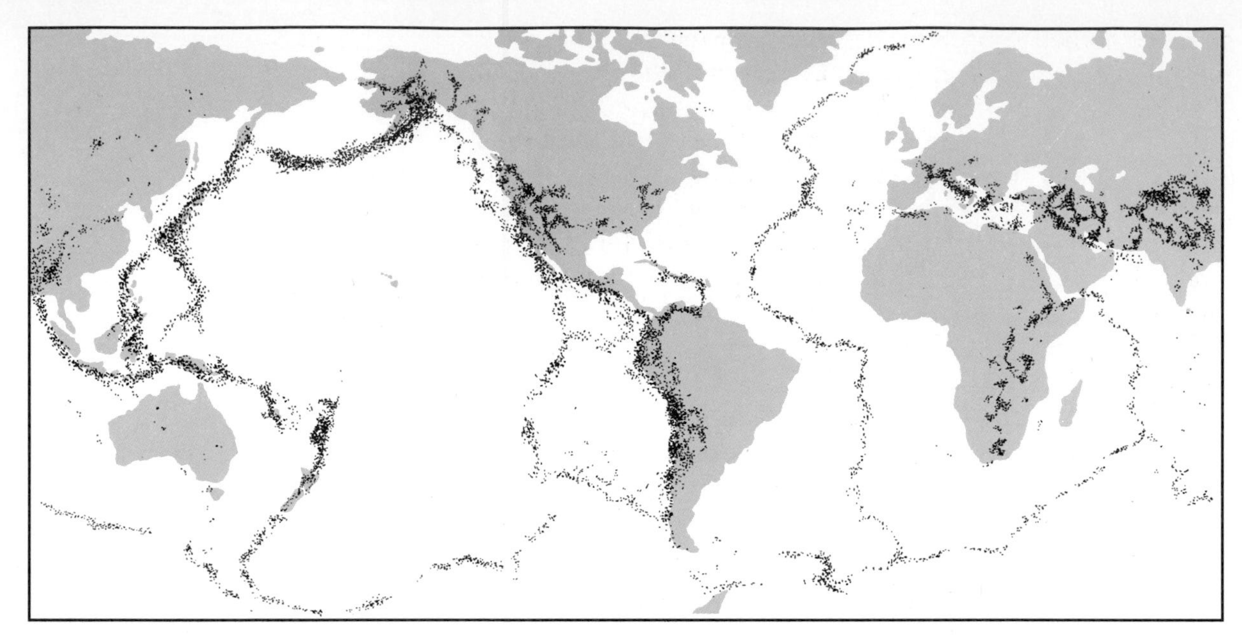

FIGURE 19–1. Most earthquakes occur in belts of tectonic activity, some of which coincide with mid-ocean ridges.

19:2 Forces Within Earth

What types of forces are applied to solids within Earth?

Earth's crust and a part of its upper mantle are solid. Solids hold together because their molecules attract one another strongly. Only a force acting from outside a solid can overcome this attraction. Forces applied to solids are of three kinds: tension, compression, and shearing. **Tension** is a stretching or "pulling-apart" force. **Compression** is a system of forces that pushes against a body from directly opposite sides and tends to squeeze it into folds. **Shearing** is a system of forces that pushes against a body from different sides, producing twisting and tearing. Shearing forces are not directly opposite one another.

Crustal rocks are subjected to forces that produce many landforms and rock structures. Structures, as used here, refer to the shape and position of rock layers that have

FIGURE 19–2. Tension (a), compression (b), and shearing (c) are three forces that act on solids.

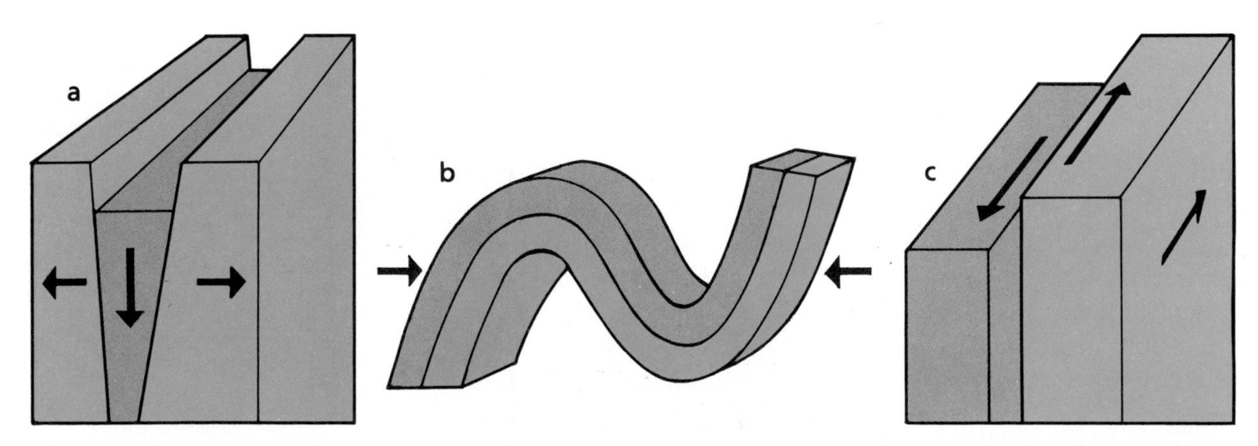

been deformed by forces in Earth's crust. Factors that control the formation of rock structures include the amount of force, the rate at which the force is applied, and the kind of rock under stress. Shearing and compression tend to bend rocks if the force is applied slowly. If these forces are extremely great or are applied rapidly, however, rocks tend to break. Deep within the crust, rocks under pressure tend to bend under forces that would cause breaking nearer the surface.

These forces within Earth's crust are very strong. Solids with elastic properties recover their original shape and size unless the force is too great. Then the solid passes its elastic limit and breaks. Stretch a rubber band until it breaks. You have exerted a force that passed the elastic limit of the rubber band. Rocks also show elastic properties. When the force on a rock passes its elastic limit, it breaks, and stored energy is released as seismic waves.

What happens when a rock passes its elastic limit?

Breaks in rocks are called **fractures.** Many uplifted rocks are fractured in somewhat rectangular patterns. If the fracture pattern covers a large area, it is called a joint system. Joints are important to weathering because they increase the surface area of the rock mass.

Faults are fractures along which movement takes place. Movement may occur in any direction along the surface of the break. One side may move upward or downward, or both sides may move in opposite directions. Movement may also occur in a horizontal direction with rocks sliding past one another in opposite directions or in the same direction at different rates.

In what directions may movement occur along a fault?

The San Andreas fault in California is a transform fault. It is nearly 1000 km in length. It runs from the northern Coast Range to the Mexican border. The San Andreas fault is a fracture zone composed of many faults that are more or less parallel. Because of its length, it has been the site of several major earthquakes.

F.Y.I. Refer to Chapter 20 for more information on transform faults.

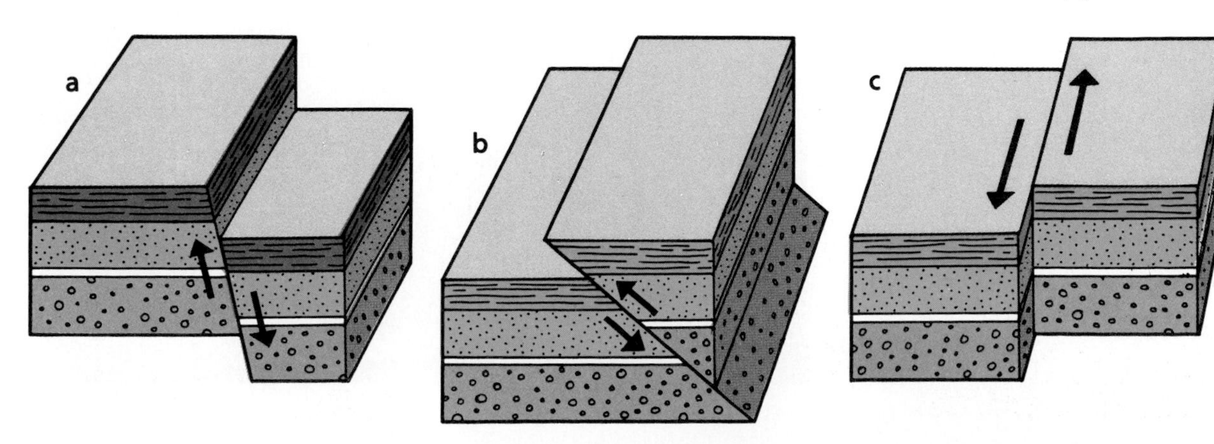

FIGURE 19–3. Faults show both vertical movement (a, b) and horizontal movement (c).

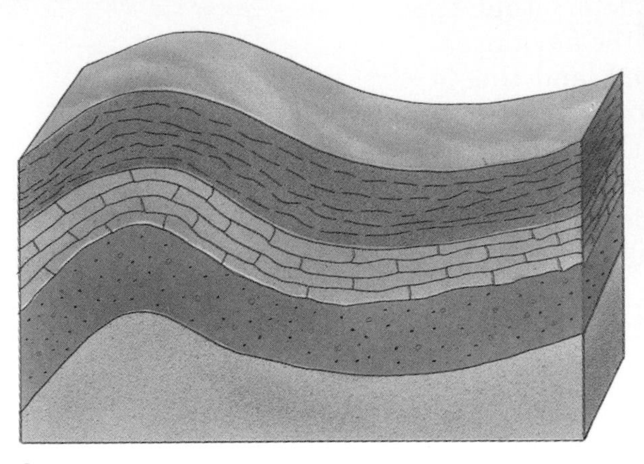

a

b

FIGURE 19–4. Anticlines and synclines form when rock layers are subjected to compressional forces (a). Synclines (b) are concave folds.

Folds are bends in rock layers. Compressional forces may move rock layers from horizontal positions into alternating ridges, or anticlines (ANT ih klinez), and troughs, or synclines (SIHN klinez). These folds may be compressed vertically or pushed over. Folds, fractures, and faults accompany orogeny.

Review

1. Where are Earth's earthquake belts?
2. What are the three kinds of forces that are applied to Earth's rocks?
3. Distinguish between a fracture and a fault.
4. What are the two types of folds in rock?
5. ★ What type of force formed the feature in the photograph on page 432?

FIGURE 19–5. Identify the anticline and the syncline in the photograph.

INVESTIGATION 19–1

Internal Forces

Problem: How can you identify the three types of forces that affect rocks?

Materials  

metric ruler
rubber band
wood molding
bricks (2)
goggles
gloves
newspaper
slabs of modeling clay (gray, yellow, blue, green)

Procedure

1. Measure the length of the rubber band without stretching it.
2. Stretch the rubber band until it breaks and measure its length. **CAUTION:** *Wear goggles and gloves.*
3. Stand a piece of wood molding on end and brace it with the bricks so that it cannot move. Try to bend the molding in an arc.
4. Grasp one end of the molding in each hand and break it. **CAUTION:** *Wear goggles and gloves.*
5. Place the two bricks on a table so that they touch one another.
6. Push one brick in one direction and the other brick in the opposite direction so they slide past one another.
7. Cover your desk with newspaper. Construct a sedimentary deposit using the four slabs of clay.
8. Fold the layers to make an anticline and a syncline. Make the folds very distinct.

FIGURE 19–6.

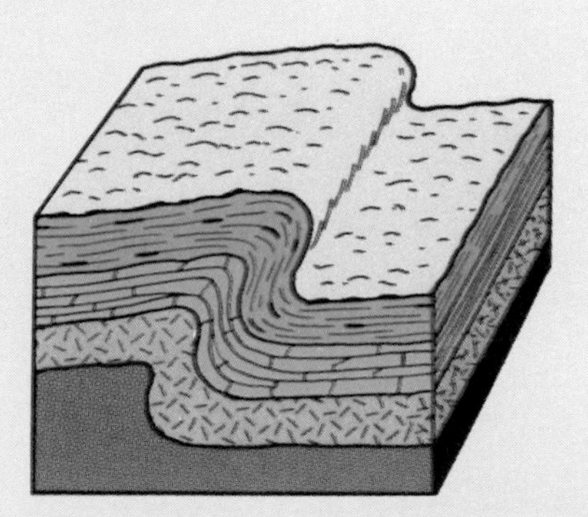

Analysis

1. Does the stretching and breaking of the rubber band change its length?
2. What type of force breaks the rubber band?
3. What kind of force are you applying to the wood molding?
4. Describe the sensation in your hands when the molding broke.
5. What kind of force are you applying to the bricks?
6. What type of force formed the anticline and the syncline?

Data and Observations

Material	Force Applied	Result
rubber band		
wood molding		
bricks		
clay slabs		

Conclusions and Applications

7. What is the relationship between the rubber band and faulting?
8. What would happen to folds in rocks below Earth's surface if compression were applied too quickly?
9. What three types of forces affect rocks?

DESTRUCTION BY EARTHQUAKES

GOALS

1. You will learn how the effects of earthquakes are described on a quantitative scale.
2. You will learn how earthquakes are measured.

One of the most destructive earthquakes to occur in the United States was the San Francisco earthquake of 1906. Movement occurred along the San Andreas fault as the coastal side of the fault moved toward the northwest. The amount of shift on opposite sides of the fault was about five to six meters. Much of the damage was due to fires caused by broken gas mains. Water lines also were broken, which hampered fire-fighting efforts.

19:3 Earthquake Damage

The actual point on the fault where movement occurs and vibrations begin is called the **focus.** Usually the focus is located deep beneath Earth's surface. The effects of seismic activity at the focus are first felt at a point on the surface directly above the focus. This point is called the **epicenter.** Earthquake damage is often greatest at or near the epicenter.

FIGURE 19–7. The Mexico City earthquake caused extensive damage.

Table 19–1

Modified Mercalli Scale of Earthquake Intensity

I. Not felt except by a very few under especially favorable circumstances. Birds and animals uneasy. Delicately suspended objects may swing.

II. Felt only by a few persons at rest, especially on upper floors of buildings.

III. Felt noticeably indoors, especially on upper floors of buildings, but many people do not recognize it as an earthquake. Parked cars may rock slightly. Vibrations like the passing of light trucks. Duration of shaking can be estimated.

IV. Felt indoors by many, outdoors by few. If at night, some awakened. Dishes, windows, doors disturbed. Walls creak. Sensation like the passing of heavy trucks. Parked cars rock noticeably.

V. Felt by nearly everyone. Some dishes, windows, etc., broken. A few instances of cracked plaster. Unstable objects overturned. Disturbances of trees, poles, and other tall objects sometimes noticed. Pendulum clocks may stop.

VI. Felt by all. Many frightened and run outdoors. Some heavy furniture moved. Books knocked off shelves, pictures off walls. Small church and school bells ring. A few instances of damaged chimneys. Otherwise damage is slight.

VII. Everybody runs outdoors. Difficult to stand up. Negligible damage in buildings of good design and construction; slight to moderate in well-built ordinary structures; considerable in poorly built or badly designed structures; some chimneys broken.

VIII. Damage slight in specially designed structures; partial collapse in ordinary buildings; great damage to poorly built structures. Panel walls thrown out of frame structures. Chimneys, factory stacks, columns, monuments, and walls fall. Heavy furniture overturned. Some sand and mud ejected from cracks in the ground. Changes in well water.

IX. Damage considerable in specially designed structures; well-designed frame structures thrown out of plumb; partial collapse of substantial buildings. Buildings shifted off foundations, ground cracked. Serious damage to reservoirs and underground pipes. General panic.

X. Some well-built wooden structures destroyed; most masonry and frame structures destroyed. Ground badly cracked. Rails bent slightly. Considerable landslides from river banks and steep slopes. Shifted sand and mud. Water splashed over banks.

XI. Few masonry structures remain standing. Bridges destroyed. Broad fissures in ground. Underground pipelines out of service. Earth slumps and land slips in soft ground. Rails bent severely.

XII. Damage total. Waves seen on ground surfaces. Lines of sight and level distorted. Objects thrown upward into the air.

A measure of how much damage a quake caused at the surface is called the **intensity** of the earthquake. Intensity is expressed in terms of physical damage or geologic change that occurs. Under certain conditions, it is possible for a moderate earthquake to cause severe damage.

In the past, the Mercalli scale has been used to indicate the intensity of earthquakes. This scale uses a series of roman numerals to express different intensities. See Table 19–1. The Mexico City earthquake shown in Figure 19–7 had an intensity of X to XI.

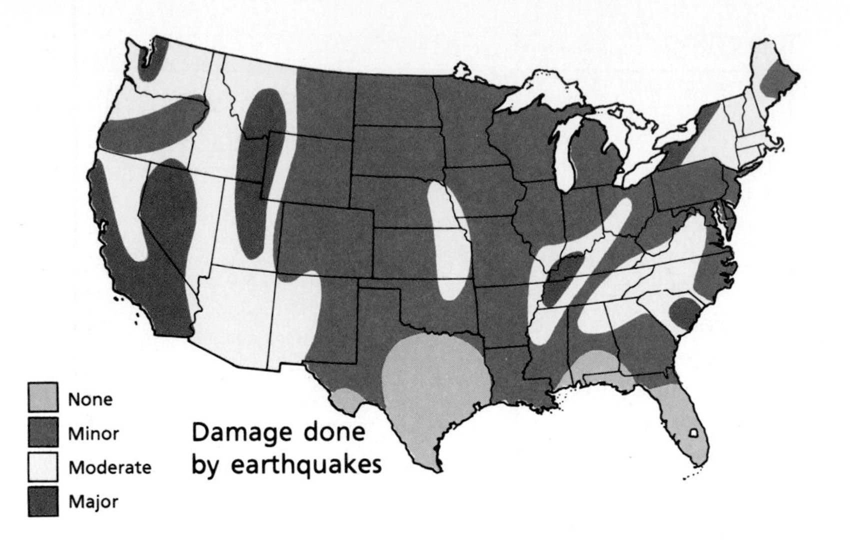

FIGURE 19–8. Most earthquake damage in the United States occurs near the west coast. Why?

In 1976, perhaps the most devastating earthquake in history occurred at Tangshan in northern China. In one of the strongest seismic events ever recorded, an estimated 750 000 people died when a series of severe tremors completely destroyed 32 square kilometers of the city. This quake had an intensity of XII on the Mercalli scale.

For coastal cities, the greatest danger associated with earthquakes is the seismic sea wave. This wave, resulting from an earthquake that displaces the ocean floor, may have a wavelength of 150 kilometers. These waves, or tsunamis, can reach velocities of 500 kilometers per hour and heights of 30 meters near shore. During the Alaskan quake of 1964, a seismic sea wave caused heavy damage to villages near the Gulf of Alaska. It also was responsible for twelve deaths in Crescent City, California, where a six-meter wave washed ashore.

F.Y.I. Parts of the seafloor off the coast of Alaska were raised more than 15 meters during the 1964 Alaskan quake.

19:4 Earthquake Safety

More than 900 000 earthquakes occur each year. Less than one percent are severe enough to cause extensive damage. Most tremors occur in areas associated with earthquake belts. However, two of the most damaging quakes to hit the United States happened in Missouri and South Carolina, areas far removed from any active belt. Thus, seismic risks must be determined in all geographic regions to minimize damage and injury.

Although it is difficult to predict earthquakes, geophysicists constantly monitor sensitive instruments that measure changes in Earth's crust. Magnetometers detect changes in the magnetic field of an area. Tiltmeters record changes in the slope of the surface. Radon detectors mea-

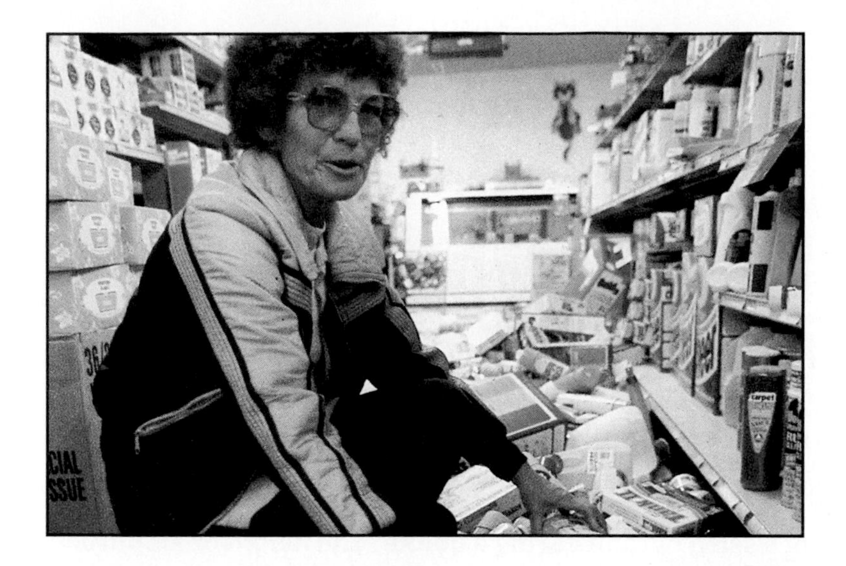

FIGURE 19–9. During an earthquake, objects may topple from shelves.

sure the level of radon gas in deep well water. Radon is a radioactive gas that is produced when uranium decays. Increases in the amount of this gas in wells often precede earthquakes. Slight changes in the readings of these instruments may indicate that a quake is likely.

Awareness is the key to earthquake safety. Drills should be conducted in cities to test the response of medical personnel. People should be taught how to react quickly but safely in the event of an earthquake. Water heaters and gas appliances should be properly secured. Fire damage can result from broken gaslines and appliance connections. Large, heavy objects on bookshelves should be placed on the lowest shelves to avoid damage and injury in case of toppling. During a quake, position yourself in a strong doorway away from windows and furniture. If outside, avoid tall objects that could fall. If possible, move to an open area away from all hazards.

19:5 Measuring an Earthquake

Seismologists are scientists who study earthquakes. They operate instruments, called **seismographs,** that record tremors traveling through Earth. Although tremors may only be felt near the epicenter of a quake, modern seismographs record even the faintest seismic vibrations. Seismograph stations use at least three instruments to obtain data about earthquakes. In order for the seismograph to record vibrations, some point within the instrument must remain at rest. A fixed frame is fastened to Earth. A pendulum or heavy mass is suspended from the frame as a reference point. During a quake, the frame moves, but the pendulum remains at rest.

BIOGRAPHY

Waverly J. Person

1926–

Chief of the National Earthquake Information Service in Golden, Colorado, Waverly Person began his career as a geophysicist for the United States Geological Survey. At NEIS, Person uses computers to interpret seismograph data and find earthquake centers. The seismograph information comes from recording stations all over the world.

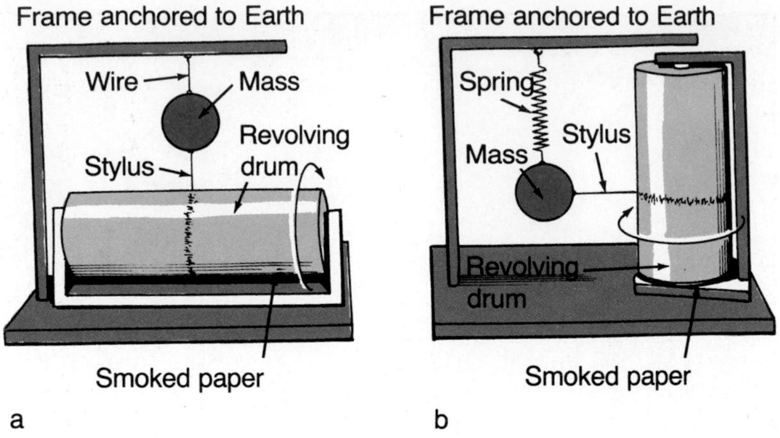

FIGURE 19–10. Horizontal movement of Earth causes the drum to move from side to side as it rotates (a). Vertical movement causes the drum to move up and down as it rotates (b).

On what does the height of the line on a seismogram depend?

Seismograms are lines traced on the recording tape of a seismograph during a quake. The height or amplitude of the lines is a measure of the strength or **magnitude** of the earthquake. The Richter scale expresses earthquake magnitude as measured by seismographs. Readings of 7.0 or higher indicate a major earthquake. The Richter scale is open ended. There is no "highest" number because the strongest quake ever may not yet have occurred. Presently, the strongest quakes ever recorded had magnitudes of 8.9 on the Richter scale. These quakes occurred in Columbia and Equador and off the coast of Japan. The 1985 Mexico City quake was rated 8.1.

Table 19–2

Earthquake Magnitude

Magnitude at focus	Approximate distance away from the epicenter that tremors are felt	Average number expected per year
3.0 to 3.9	24 km	49 000
4.0 to 4.9	48 km	6 200
5.0 to 5.9	112 km	800
6.0 to 6.9	200 km	120
7.0 to 7.9	400 km	18
8.0 to 8.9	720 km	1

Review

6. Distinguish between focus and epicenter.
7. How is the intensity of an earthquake measured?
8. What is a seismograph?
★ 9. Seismic sea waves have been described as tidal waves. Is this description correct? Explain.

a

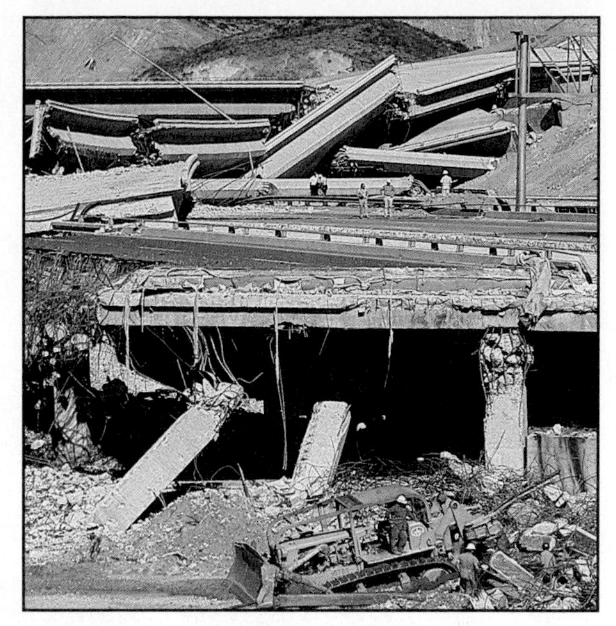

b

FIGURE 19–11. Waverly Person examines a seismogram (a). Energy released by earthquakes can cause widespread destruction in populated areas (b).

EARTHQUAKES AND EARTH'S INTERIOR

Seismic data are used not only to study particular features of an earthquake, but also to gain an understanding of the materials and structure of Earth's interior. Seismolcgists use these data to develop and modify models of Earth's interior.

GOALS

1. You will learn how body waves are used to determine the location of an earthquake.
2. You will learn how data gathered about earthquakes have led to the modern model of Earth's interior.

19:6 Locating an Earthquake

The focus of a seismic event may occur at any depth between Earth's surface and about 700 kilometers. Waves traveling outward from the focus in all directions through Earth's interior are called **body waves.** Body waves are of two types. The **P-wave,** or primary wave, travels forward, moving individual rock particles back and forth along its travel path. As P-waves advance, rock particles are compressed, move apart, and then return to their original positions. **S-waves,** or secondary waves, move forward but vibrate at right angles to the direction of movement. Rock particles move up and down as the S-waves pass through.

In what direction does the P-wave travel?

Because of its higher velocity, the P-wave always arrives at the surface before the S-wave. You can compare a lightning bolt and thunder with P- and S-waves.

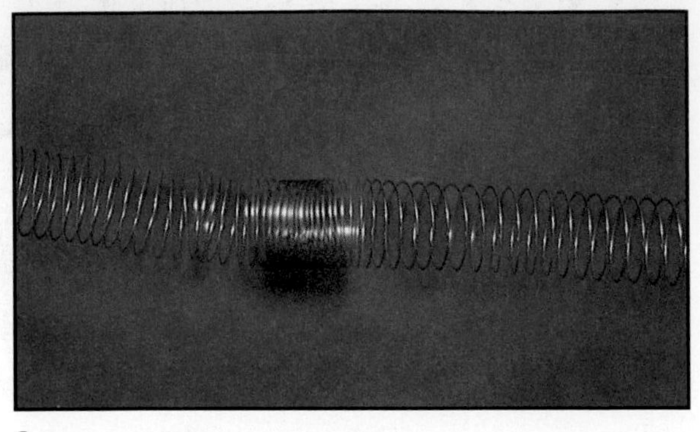

a

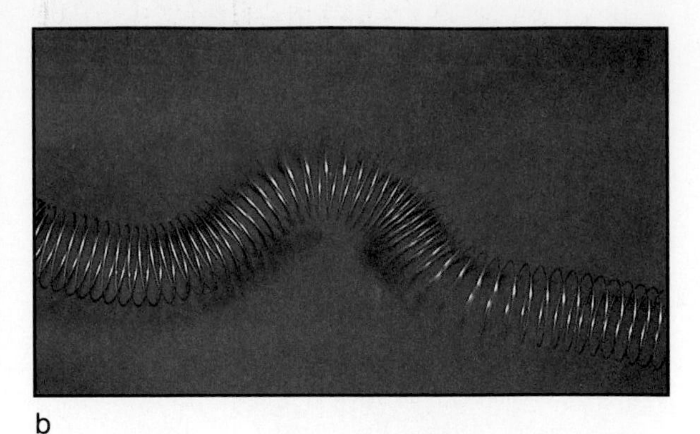

b

FIGURE 19–12. P-waves compress particles together (a). S-waves vibrate at right angles to the direction of movement (b).

P-waves arrive at Earth's surface before S-waves, and can be likened to the lightning. Thunder follows lightning just as S-waves follow P-waves. Distance from a seismic station to the epicenter is calculated from the difference between the arrival time of the P- and S-waves.

Figure 19–13 is a typical record of earthquake waves recorded by a seismograph. Note that the seismic data indicate a difference in arrival times of the P- and S-waves of about five minutes. Figure 19–14 is a simplified P- and S-wave travel time graph. Use Figure 19–14 to find the point where the two travel times are five minutes apart. Read the value from the x-axis. This distance is 3600 km. Thus, the vibrations recorded on the seismogram originated about 3600 km from the seismic station. To determine the precise location of the epicenter, data from at least three seismograph stations are needed.

Waves that travel along the surface outward from the epicenter are called **surface waves.** These waves, also called L-waves, travel parallel to Earth's surface. Surface waves are the last to reach the seismic station. L-waves are the most destructive waves and have a motion similar to that of ocean waves.

FIGURE 19–13. This is a typical record of earthquake waves made on a seismograph. Note the time interval between the arrival of each wave type.

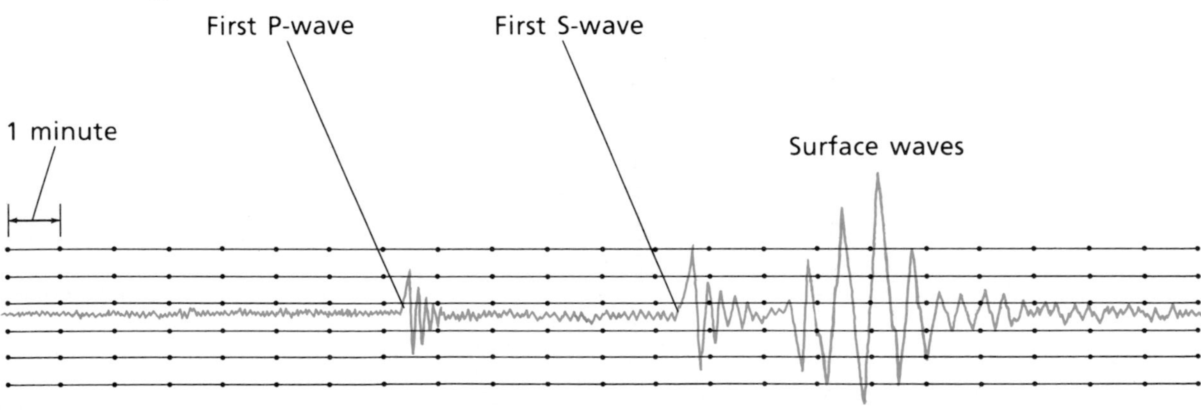

FIGURE 19–14.

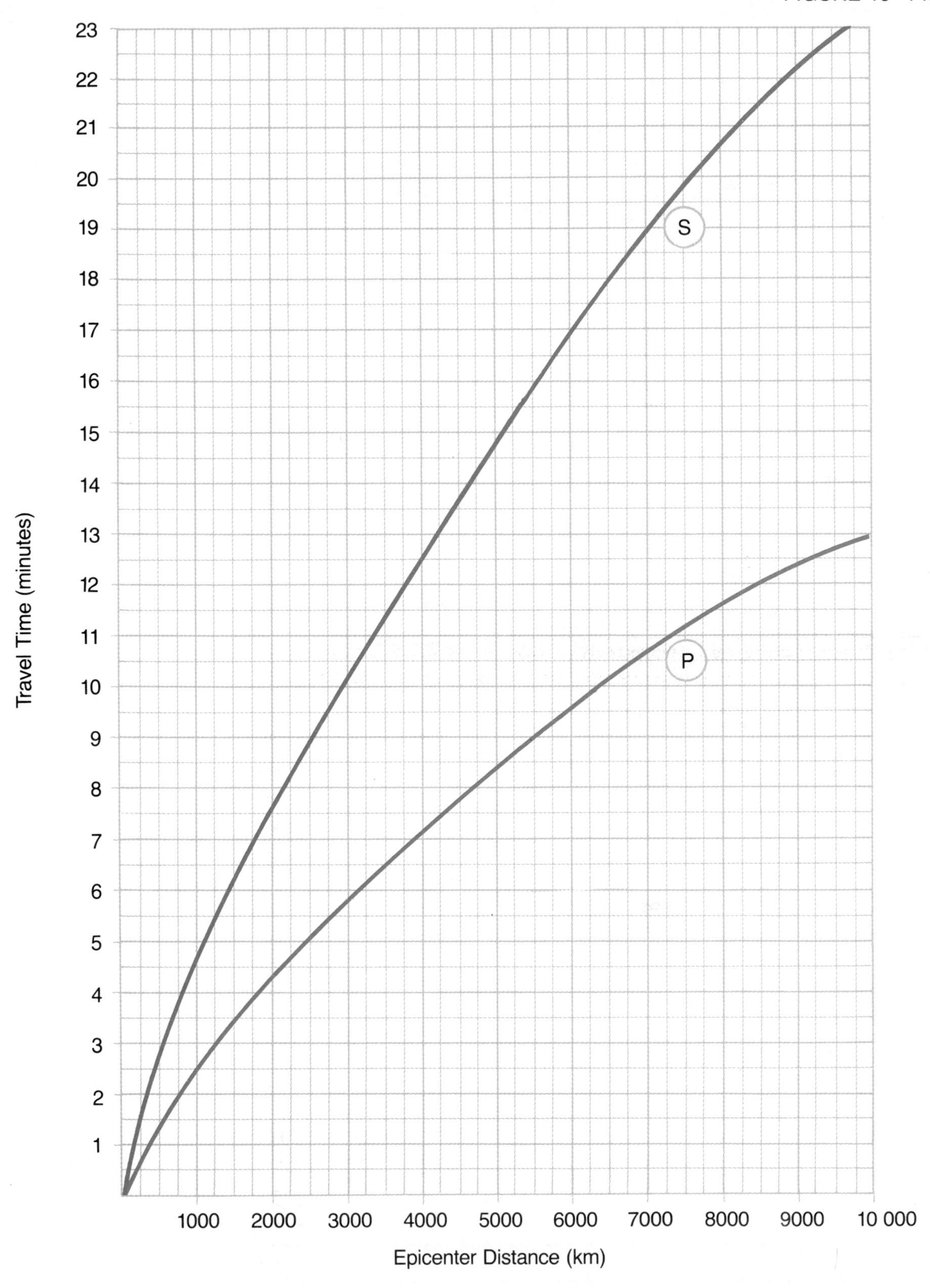

23
22
21
20
19
18
17
16
15
14
13
12
11
10
9
8
7
6
5
4
3
2
1
Travel Time (minutes)
S
P
1000
2000
3000
4000
5000
6000
7000
8000
9000
10 000
Epicenter Distance (km)

SKILL
Graphing Data

Problem: How do you construct a travel time graph?

Materials

pencil
graph paper
Figure 19–14

Procedure

1. Use Figure 19–14 to determine the difference in arrival times for P- and S-waves at the distances listed in the Data and Observations table below. Two examples have been done for you.
2. Start your graph three lines up from the bottom of the page and three lines from the left edge.
3. You will be plotting two variables: the difference in arrival times of P- and S-waves at a seismic station and the distance from that station to the earthquake.
4. Decide which is the independent variable and which is the dependent variable.
5. Label the horizontal axis with the name of the independent variable, and the vertical axis with the name of the dependent variable. Title your graph.
6. Determine the range of each variable. Number the squares along the appropriate axis for each variable.
7. Plot the distance to the earthquake against the difference in arrival times of the seismic waves.
8. Draw a smooth line that best fits the pattern of points that you have plotted. This line shows the relationship between the difference in arrival times of P- and S-waves at a seismic station and the distance from that seismograph to the earthquake.

Questions

1. What values are plotted on the horizontal axis? On the vertical axis?
2. What is the difference in P- and S-wave arrival time at 3000 km? 5000 km? 7000 km?
3. What happens to the difference in P- and S-wave arrival times as the distance to the earthquake increases?

Data and Observations

Distance from seismograph to earthquake (km)	Difference in P- and S-wave arrival time (min and s)
1500	2 minutes 45 seconds
2250	
2750	
3000	
4000	5 minutes 35 seconds
5000	
7000	
9000	

INVESTIGATION 19–2

Locating an Epicenter

Problem: How are epicenters located?

Materials

Figure 19–14
string
metric ruler
globe
paper
chalk or water-soluble marker

Procedure

1. Determine the difference in arrival time between the P- and S-waves at each station for each quake from the Data Table below.
2. Use Figure 19–14 to determine the distance in kilometers of each seismograph from the epicenter of each earthquake. Record these data in the Data and Observations table.
3. Using the string, measure the circumference of the globe. Determine a scale of centimeters of string to kilometers on Earth's surface. (Earth's circumference = 40 000 km)
4. For each earthquake (A through E), place one end of the string at each seismic station location. Use the chalk or marker to draw a circle with a radius equal to the distance to the epicenter of the quake.
5. Identify the epicenter for each earthquake.

Data and Observations

	Calculated distance to epicenter (km) from each seismograph location				
Quake	(1)	(2)	(3)	(4)	(5)
A					
B					

Analysis

1. How is the distance of a seismograph from the earthquake related to the arrival time of the body waves?
2. What is the location of the epicenter of each earthquake?
3. How many stations were necessary in order to accurately locate each epicenter? Should more data have been provided? Explain.

Conclusions and Applications

4. Predict why some seismographs did not receive body waves from some tremors.
5. How are epicenters located?

Location of Seismograph	Wave	Body wave arrival times (Greenwich Mean Time)				
		A	B	C	D	E
(1) New York	P S	2:24:05 pm 2:28:55 pm	1:19:00 pm 1:24:40 pm	1:00:30 pm 1:09:05 pm	3:44:40 am 3:51:45 am	11:07:10 am 11:12:50 am
(2) Seattle	P S	2:24:40 pm 2:30:00 pm	1:14:37 pm 1:16:52 pm	1:02:20 pm 1:12:30 pm	3:40:08 am 3:43:40 am	11:09:18 am 11:16:45 am
(3) Rio de Janeiro	P S	2:29:00 pm 2:38:05 pm	xxx xxx	1:02:35 pm 1:13:05 pm	xxx xxx	11:12:45 am 11:23:20 am
(4) Paris	P S	2:30:15 pm 2:40:29 pm	1:24:05 pm 1:34:05 pm	12:53:00 pm 12:55:30 pm	3:47:00 am 3:56:05 am	11:04:08 am 11:07:40 am
(5) Tokyo	P S	xxx xxx	1:23:30 pm 1:33:05 pm	1:02:45 pm 1:13:20 pm	3:45:00 am 3:52:20 am	11:12:00 am 11:21:55 am

CAREER

Geophysicist

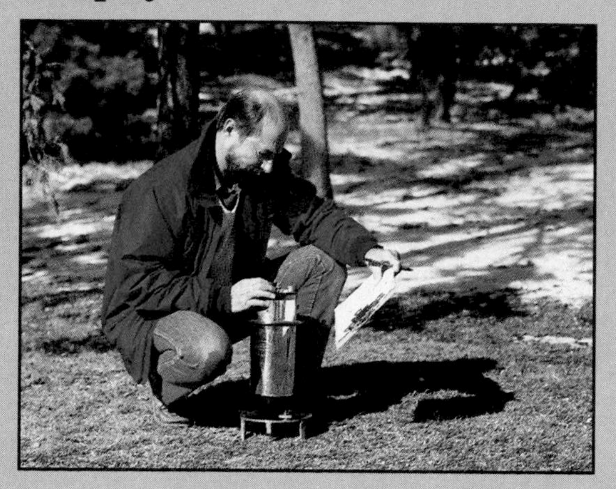

Hallan Noltimier is a geophysicist who studies how Earth responds to seismic waves. While in the field, Dr. Noltimier calculates the arrival times of seismic waves and is able to determine the types of rocks through which the waves traveled. From these data, models of the area are constructed and the locations of deposits of oil and natural gas are pinpointed.

Dr. Noltimier uses his knowledge of geology, mathematics, and physics to analyze geophysical data. He received a doctorate in geophysics.

For career information, write:
Society of Exploration Geophysicists
P.O. Box 70240
Tulsa, OK 74170

19:7 Earth's Interior

Most of what we know about Earth's interior comes from the study of earthquake waves. Seismograms provide clues to the density and physical states of materials within Earth. Seismic stations located near the epicenter and at an angular ground-surface distance up to 103° from the focus record waves that pass only through Earth's upper layers. Both P- and S-wave reach these stations. Stations located between 103° and 143° from the epicenter do not receive any body waves. This zone is the **shadow zone.** Stations located beyond 143° in either direction from the epicenter receive only P-waves.

Where is the shadow zone?

F.Y.I. The Moho received its name from its discoverer, Andrija Mohorovičić, a seismologist in Yugoslavia, in 1909.

What is the Moho?

From these data, seismologists have made a model of Earth's structure. The outermost layer or **crust** extends from the surface to a depth of about 35 km or more beneath continents, and about 10 km or less beneath ocean basins. Below this boundary, both P- and S-waves increase in velocity. This boundary is called the Mohorovičić Discontinuity, or **Moho.** The Moho marks the point where Earth's middle layer or **mantle** begins. From this point, the upper mantle seems to be rigid down to about 100 km, where another change in wave velocity occurs. This change suggests that the mantle is close to its melting point. This zone of near-fluid mantle or asthenosphere extends to about 250 km, and in it both P- and S-waves are slowed,

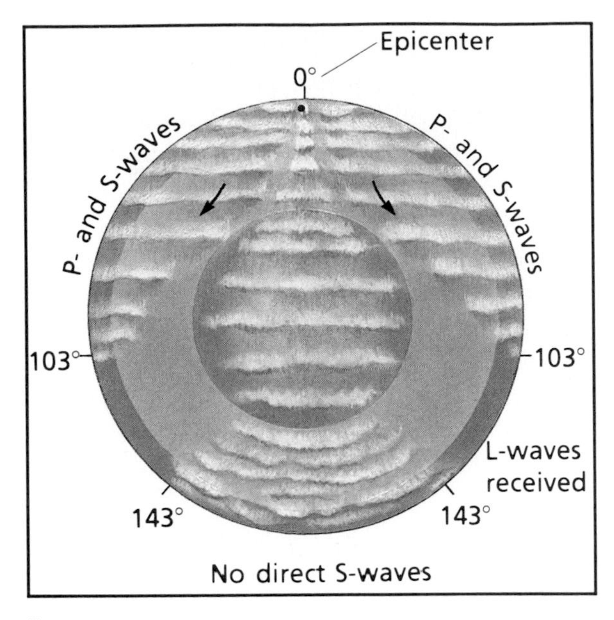

a

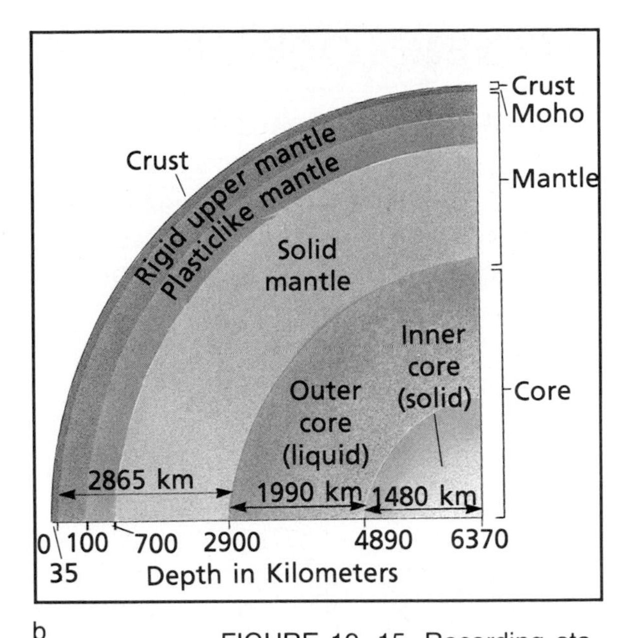

b

FIGURE 19–15. Recording stations between 103° and 143° away from the epicenter receive only L-waves (a). Analysis of seismic data has led to the present model of Earth's interior (b).

but not completely lost. The mantle at this depth probably has a plasticlike consistency. Between 250 and 2900 km, the mantle probably is solid. At 2900 km, a decrease in velocity bends the P-waves. This marks the boundary of Earth's innermost layer or **core.** This bending causes P-waves to miss the shadow zone and only reach stations

PROBLEM SOLVING

What Is In the Box?

Maria returned home from school one day to find a box, approximately 30 cm × 18 cm × 12 cm, sitting on the kitchen table. Her aunt had come to visit and brought Maria a present. In order for Maria to get the present, she had to be able to tell her aunt at least three facts about it. Maria could do anything she needed to do to the box except open it to look at the gift directly. Of course, Maria would be careful not to damage the gift. How many facts do you think Maria can learn about the gift? How is Maria's challenge related to earthquakes and Earth's interior?

TECHNOLOGY: APPLICATIONS

Earthquake-proof Buildings

Scientists are experimenting with building designs in earthquake-prone areas of the United States. In California, a four-story building is being anchored to flexible moorings. These flexible, circular moorings are made of steel plates filled with alternating layers of cured rubber and thin slices of steel. When an earthquake occurs, the lower portion of the mooring will stretch back and forth with the waves. The rubber portions of the moorings will absorb most of the wave motion of the quake. Thus, the building itself only sways gently. Tests have shown that buildings supported with these moorings should be able to withstand an earthquake measuring up to 8.3 on the Richter scale without major damage.

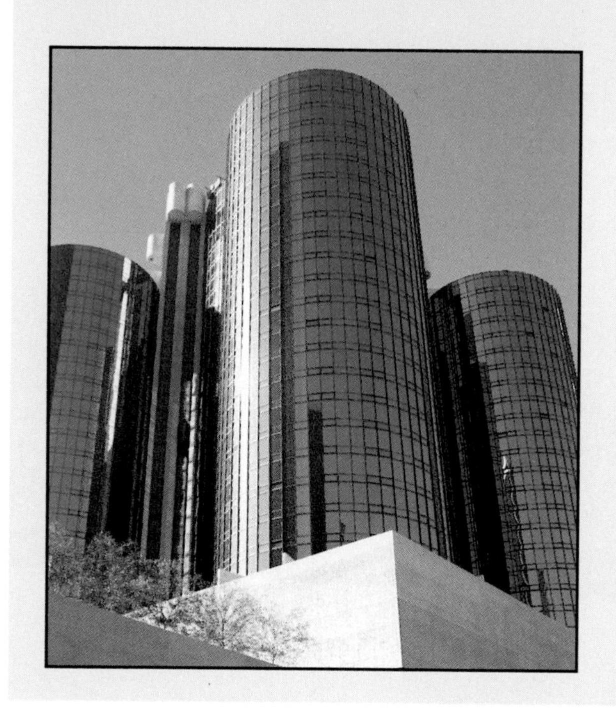

Engineers are working on other structural designs using computer-operated cables attached to hydraulic pistons. These models react to external pressure. Some designs being tested include using separation joints and beams that deform rather than break. Also being studied is the connection of buildings with elevated walkways that have a steadying effect on the buildings during a quake.

beyond 143° from the epicenter. S-waves cannot pass through liquids. Their disappearance indicates that the core is in two parts and that the outer core must be a fluid. At a depth of 4890 km, the velocity of the P-waves increases, suggesting that the inner core is solid. Earth's inner core extends to a depth of 6370 km.

Review

10. What are the two main types of body waves that radiate from the focus of an earthquake?

11. Describe the relative rates of motion of these two types of waves.

★ **12.** The P-wave from an earthquake arrives at a seismic station 4000 km away at 1:10 P.M. At what time will the S-wave first be detected by that same seismograph?

Chapter 19 Review

SUMMARY

1. Earthquakes are vibrations caused by sudden movements of Earth's rocks. 19:1
2. Tension, compression, and shearing within Earth's crust often cause rocks to break or to bend. Faults are fractures in Earth's crust along which there has been movement. 19:2
3. Earthquakes originate at the focus. The epicenter is the point on the surface directly above the focus. 19:3
4. Awareness of procedures to follow in the event of an earthquake will lower seismic risk to individuals. 19:4
5. Seismographs record earthquake vibrations. 19:5
6. The Richter scale is used to express earthquake magnitudes. 19:5
7. Body waves travel through Earth's interior. P-waves are compressional waves. S-waves are shear waves that vibrate at right angles to the direction of movement. L-waves, or surface waves, travel parallel to Earth's surface. 19:6
8. The shadow zone is the area between 103° and 143° from the epicenter. Earth has three layers: crust, mantle, and core. 19:7

VOCABULARY

a. body waves
b. compression
c. core
d. crust
e. earthquakes
f. epicenter
g. faults
h. focus
i. folds
j. fractures
k. intensity
l. magnitude
m. mantle
n. Moho
o. P-wave
p. seismic waves
q. seismograms
r. seismographs
s. seismologists
t. shadow zone
u. shearing
v. surface waves
w. S-waves
x. tension

Matching

Match each description with the correct vocabulary word from the list above. Some words will not be used.

1. the boundary between Earth's crust and mantle
2. L-waves
3. breaks in rocks along which movement takes place
4. a stretching or "pulling apart" force
5. no P- or S-waves are received in this area
6. the first body wave detected on a seismogram
7. instruments used to analyze seismic waves
8. vibrations caused by sudden movements of Earth's rocks
9. the point on Earth's surface directly above the focus of an earthquake
10. a measure of the surface damage caused by an earthquake

Chapter 19 Review

MAIN IDEAS

A. Reviewing Concepts

Choose the word or phase that correctly completes each of the following sentences.

1. Vibrations caused by sudden movement of surface rocks are *(earthquakes, folds, fractures)*.
2. The force that causes stretching of rocks is *(compression, tension, shearing)*.
3. A break in rocks along which there is movement is a *(fault, fold, fracture)*.
4. The earthquake wave with the fastest speed is the *(L-, S-, P-)* wave.
5. Earth's *(outer core, inner core, crust)* is probably fluid.
6. Serious structural damage to buildings caused by seismic activity would be indicated by a Mercalli rating of *(III, 2.3, X)*.
7. The earthquake wave that causes the greatest surface damage is the *(L-, P-, S-)* wave.
8. The Richter scale number of a major quake might be *(III, 8.3, 2.5)*.
9. The *(mantle, crust, core)* can be divided into an inner and outer layer.
10. *(S-, L-, P-)* waves cannot pass through liquids.
11. Anticlines and synclines are the products of *(compression, tension, shearing)*.
12. S-waves do not travel through the *(outer core, mantle, crust)*.
13. A seismograph measures the *(intensity, magnitude, damage)* of an earthquake.
14. The point on Earth's surface located directly above the fault where an earthquake occurs is the *(focus, shadow zone, epicenter)*.
15. The Moho is the boundary between the crust and the *(core, mantle, shadow zone)*.

B. Understanding Concepts

Answer the following questions using complete sentences.

16. Distinguish between a seismograph and a seismogram.
17. In addition to body and surface seismic waves, what other effects are produced by earthquakes that may seriously threaten coastal areas?
18. What will happen to most rocks when the forces of shearing and compression are rapidly applied?
19. What happens to seismic body waves at the Moho?
20. Distinguish between earthquake intensity and magnitude.
21. Compare and contrast P-waves and S-waves.
22. Which earthquake waves are the most destructive during or immediately after a major quake?
23. How long does it take a P-wave to travel 4000 km? How long for an S-wave to travel the same distance?
24. What is the difference in the arrival times of P- and S-waves for an earthquake that occurs 1500 km from the seismic station?
25. Describe Earth's shadow zone.

C. Applying Concepts

Answer the following questions using complete sentences.

26. How do seismographs work?
27. How have seismologists used seismic data to develop the current model of Earth's interior?
28. Why do major earthquakes seem to be more destructive today than 100 years ago?
29. How do scientists locate the epicenter of an earthquake?

30. Why might an earthquake with a large magnitude have a low intensity?

SKILL REVIEW

1. Compare and contrast the Richter and Mercalli scales for measuring earthquakes.

2. Compare and contrast compression and tension.

3. Construct a line graph from the data that are provided in Table 19–2. Explain what your completed graph indicates.

4. Use the travel time graph for earthquake waves in Figure 19–14 and describe what it indicates about the relative velocities of those waves as they travel through Earth's interior.

5. Construct a diagram showing how each of the three types of earthquake waves travel from both the focus and epicenter of an earthquake to a seismograph located at least several thousand kilometers away.

PROJECTS

1. Use a large flat pan half-filled with water, flexible cardboard, and a dropper to investigate how waves are reflected from flat and curved surfaces. Produce the ripples with water from the dropper. Observe the waves with an unshaded light source.

2. Make a three-dimensional model of Earth's interior. Include all the layers and sublayers discussed in Section 19:7. What do scientists believe is the composition of each layer?

READINGS

1. Brownlee, Shannon. "Waiting for the Big One." *Discover*. July, 1986, pp. 52–71.

2. Morris, Charles. *The San Francisco Calamity by Earthquake*. Secaucus, NJ: Citadel Press, 1986.

3. Winner, Peter. *Earthquakes*. Morristown, NJ: Silver Burdett, 1986.

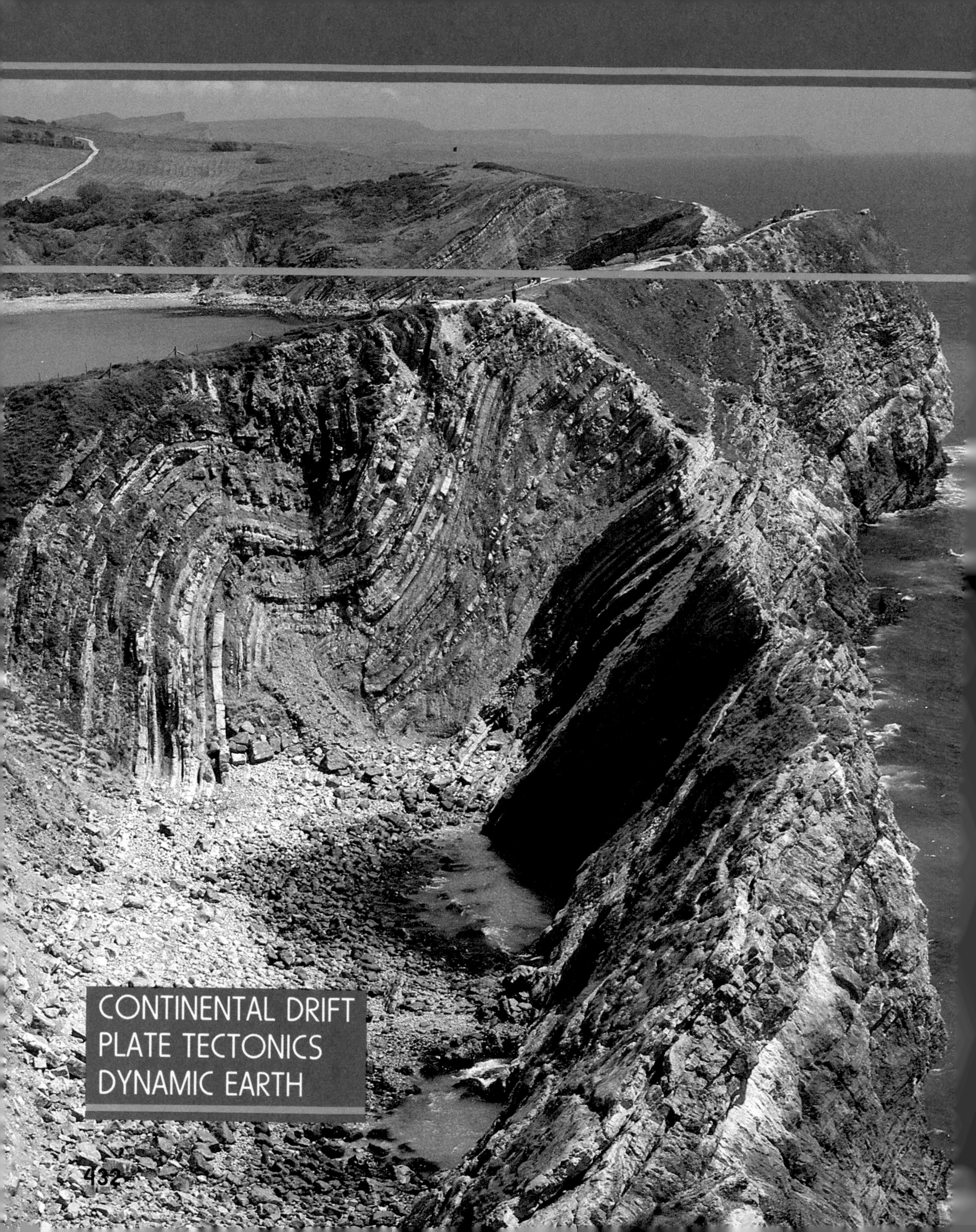

CONTINENTAL DRIFT
PLATE TECTONICS
DYNAMIC EARTH

Plate Tectonics

Earth is always changing. These tightly folded, highly deformed layers of rocks resulted from tensional, compressional, and shearing forces deep within Earth. These changes occurred as segments of Earth's crust and upper mantle collided with one another over millions of years. What causes such movements?

CONTINENTAL DRIFT

GOALS

1. You will learn about the idea of continental drift.
2. You will learn why the idea that continents have moved across Earth's surface has only recently been accepted.

In the past, people have believed the positions of the continents and oceans have been constant. When accurate world maps were developed, people noted that the continents, especially South America and Africa, fit together like a puzzle. Until recently, the idea that the continents could move across Earth's surface was thought foolish.

20:1 Continents in Motion

In 1911, a meteorologist named Alfred Wegener was completing research on an interesting hypothesis. Wegener believed that world climates had once been quite different in some areas than present climate patterns. He thought existing life forms as well as fossils would support this idea. Wegener proposed that land bridges had once connected the major continents.

In 1915, Wegener published *The Origin of Continents and Oceans,* in which he described continental drift. **Continental drift** is the hypothesis that continents had once been one or more larger landmasses that had separated and moved apart. Wegener cited as evidence the fit of the continents, similar fossils, rock structures, and ancient climates on different continents. Wegener named his original "supercontinent" **Pangaea,** meaning *all Earth.* To him, all of the present continents were once a part of Pangaea, and had since drifted to their present positions.

What evidence led to the development of the idea of continental drift?

FIGURE 20–1. The discovery of *Glossopteris* in Africa, Australia, India, and Antarctica was evidence for continental drift.

Although Wegener recorded all of the research and data that supported his idea, he could not explain how, when, or why these changes in the position of the continents had taken place. Since other scientists at that time could not provide these explanations either, Wegener's idea of drifting continents was generally ignored.

20:2 Fossil, Climate, and Rock Clues

What is *Lystrosaurus?*

Despite the fact that many scientists could not believe that the continents moved, evidence in support of continental drift continued to be found. *Lystrosaurus* (li struh SAWR us), a small, squat, doglike reptile, lived about 200 million years ago. Scientists have found bones and teeth of the animal in South Africa and a complete skeleton in Antarctica. *Lystrosaurus* lived around rivers and swamps. In addition, the fossil fern *Glossopteris* has been found in Africa, Australia, India, and Antarctica. At the time of these discoveries, scientists assumed that a land bridge once connected the continents. But no remnants of such a bridge exist below sea level. Wegener believed that the discovery of these fossils in rocks on different continents indicated that the landmasses were once joined.

Further evidence of continental drift was found in glacial deposits and the effects of glaciation on five continents in the Southern Hemisphere. At the same time in Earth's past, tropical swamps existed in the Northern Hemisphere. Thus, Wegener and other scientists concluded that these southern continents must have been nearer the South Pole 250 million years ago. At about the same time as the glaciation, salt beds were deposited in areas now located in west Texas and Germany. Today, such deposits form only in arid areas near the equator. Could these beds have formed nearer the equator and drifted north?

FIGURE 20–2. A major episode of glaciation occurred 200 million years ago when the continents were joined (a). Evidence of this event exists in areas in which glaciation is now unlikely (b).

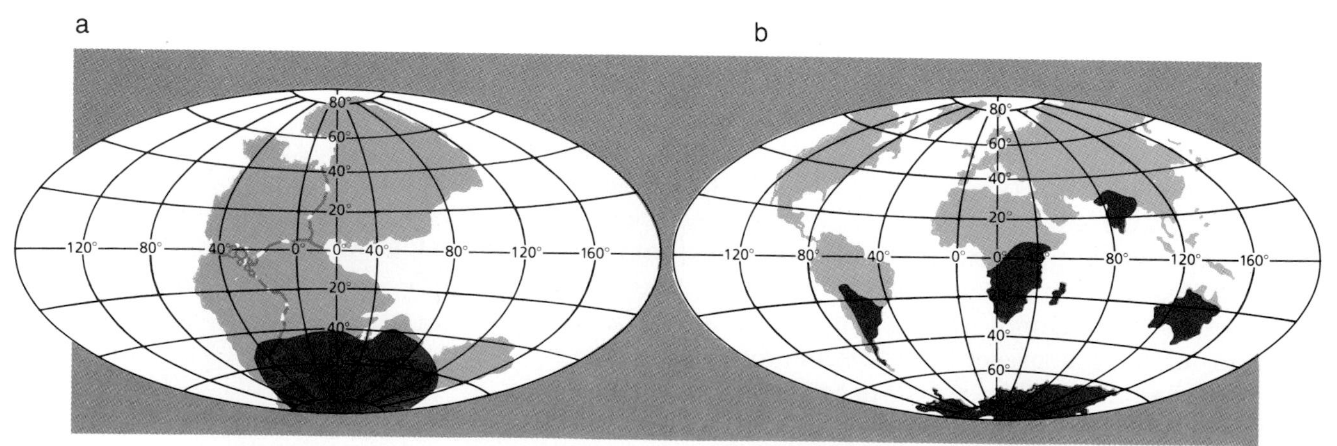

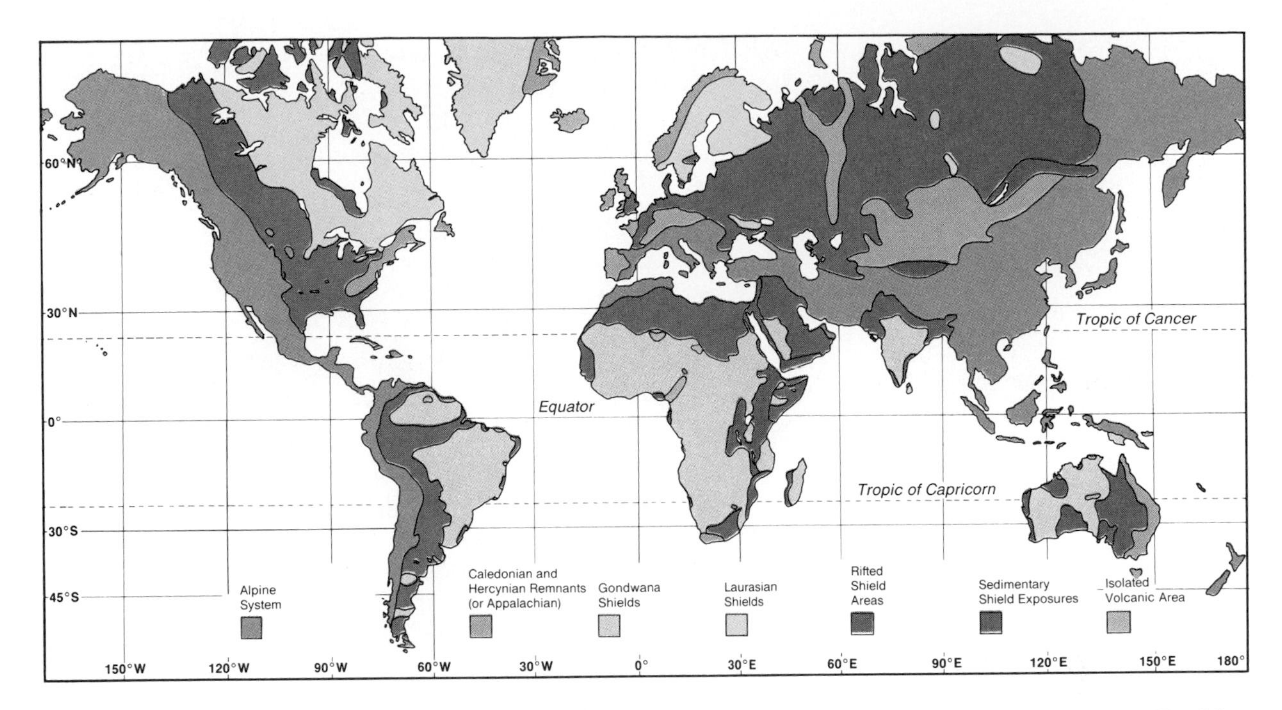

FIGURE 20–3. Many of Earth's rock structures cross continental borders.

Discovery of rock structures that end abruptly at the edges of continents also supported Wegener's hypothesis. For example, the mountain belt that includes the Appalachians runs through the northeastern United States and off the coast of Newfoundland. Mountains of similar age and structure are found in Greenland and in Scandinavia. When the edges of the continents are fitted together on a map, the mountains are nearly continuous from one continent to the other. Similar structures can be matched when South America and Africa are joined.

PROBLEM SOLVING

A Perfect Fit?

While looking at a map of Earth, William had noticed what many others before him had suggested about the continents. It seemed that they might all fit together like pieces of a puzzle. One day he took an old map and cut out each of the continents. He laid them all on a table top and tried to fit them into a single large continent.

Much to his surprise, he found that the pieces did not fit together very well. Yet there were several areas where the fit was almost perfect.

William thought about this problem for a short while and had an idea. He took one look at another map and made one small adjustment in his procedure. Cutting another old map he found that almost all of the pieces would fit together with the one slight change in what he had done. How do you suppose William solved this problem?

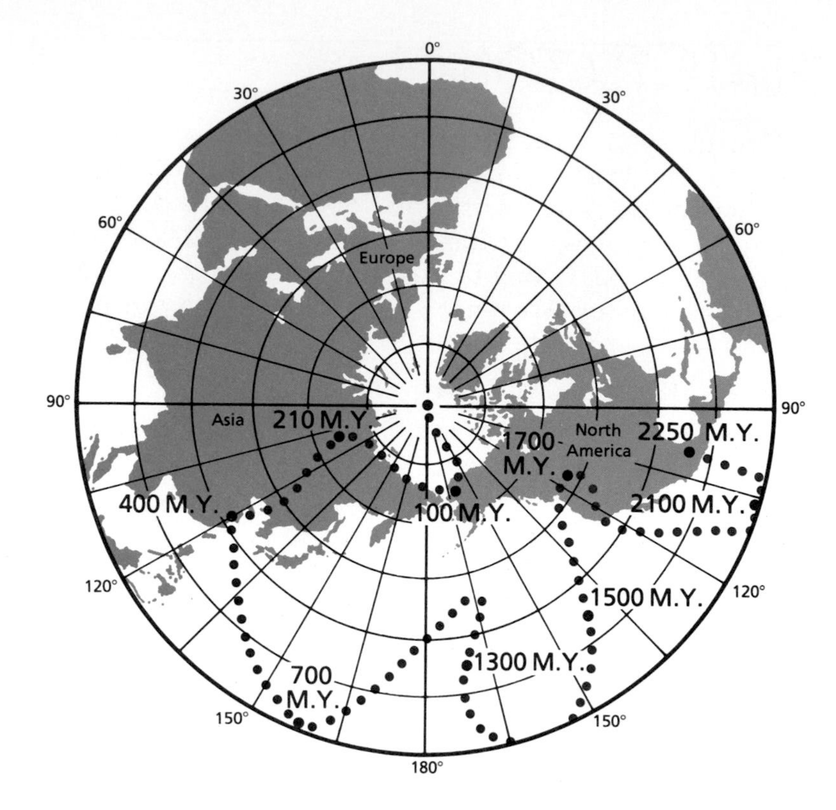

FIGURE 20–4. This polar-wandering curve shows the apparent positions of the pole determined from North American rocks. "M.Y." means millions of years ago.

How are the polar-wandering curves for Europe and North America similar?

20:3 Magnetic Clues

Long after Wegener's death, more evidence in support of continental drift came from the study of magnetism in ancient rocks, or **paleomagnetism.** Some rocks, like basalt, contain iron-bearing minerals that become magnetized as they form. These mineral grains line up, pointing toward the magnetic pole. From the directions of magnetism indicated by these rocks, scientists determined the pole position for European rocks of many different ages. When these data were plotted on a map, the location of the magnetic pole appeared to have migrated through time. This idea is known as **polar wandering.** When a polar-wandering path was later plotted for North America, both the American and European plots had similar paths. However, the paths were separated by about 30° of longitude. Scientists thought it unlikely that two different North Poles once existed. They found that the paths are nearly the same if continents are rotated next to one another, as in Wegener's idea of continental drift.

Review

1. Explain continental drift.
★ 2. Discuss three findings that indicate the continents may once have been joined.

PLATE TECTONICS

The ideas of Wegener and other scientists were not accepted by most geologists because no one could figure out how or when the drift could have happened. It was not until technology produced highly precise and sensitive instruments, beginning in the early 1950s, that the data needed to support Wegener's ideas could be obtained.

GOALS

1. You will learn about seafloor spreading and how it supports the idea of continental drift.
2. You will learn about the plate tectonics theory.

20:4 Data from the Seafloor

The geology of Earth's ocean basins has been studied for over 100 years. It was not until the invention of the precision depth recorder in 1953, however, that oceanographers were able to collect accurate data about the ocean floors. Using this instrument, scientists discovered the mid-ocean ridge systems and deep-sea trenches.

In the early 1960s, *Glomar Challenger,* a research ship, began its existence as a floating drilling platform. The ship has a hole in its bottom so that scientists can drill into the seafloor for samples of ocean crust. After many years of data collection, scientists found that the seafloor is composed of bands of rock. The youngest rocks are at the mid-ocean ridge. On either side of the ridge the bands of rocks become increasingly older. Scientists also found no sea-floor deposits older than 200 million years. In contrast, some continental rocks are more than 3 billion years old. Why is the ocean floor so young? What is happening to the old seafloor?

Where on the seafloor are the youngest rocks?

FIGURE 20–5. Ocean trenches usually are found close to continental margins (a). The *Glomar Challenger* (b) continues to gather data on ocean basin geology.

a

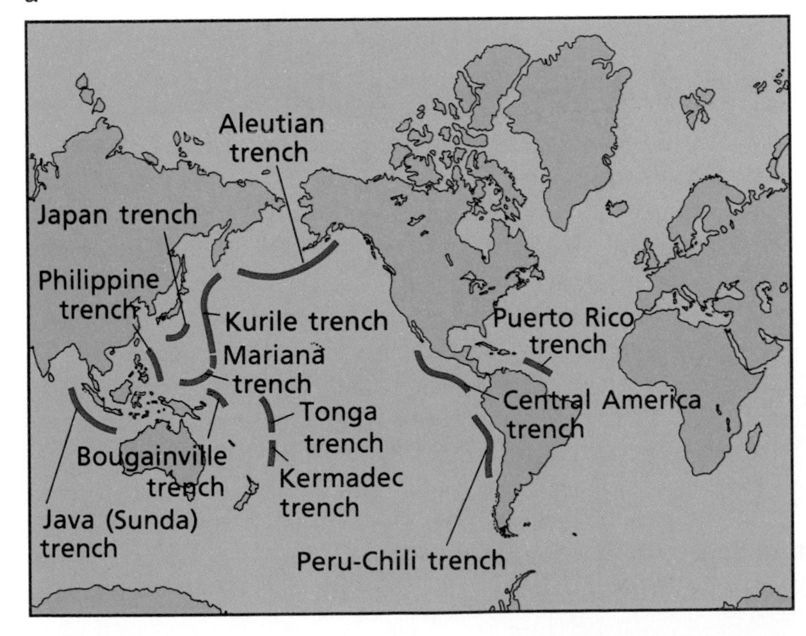

b

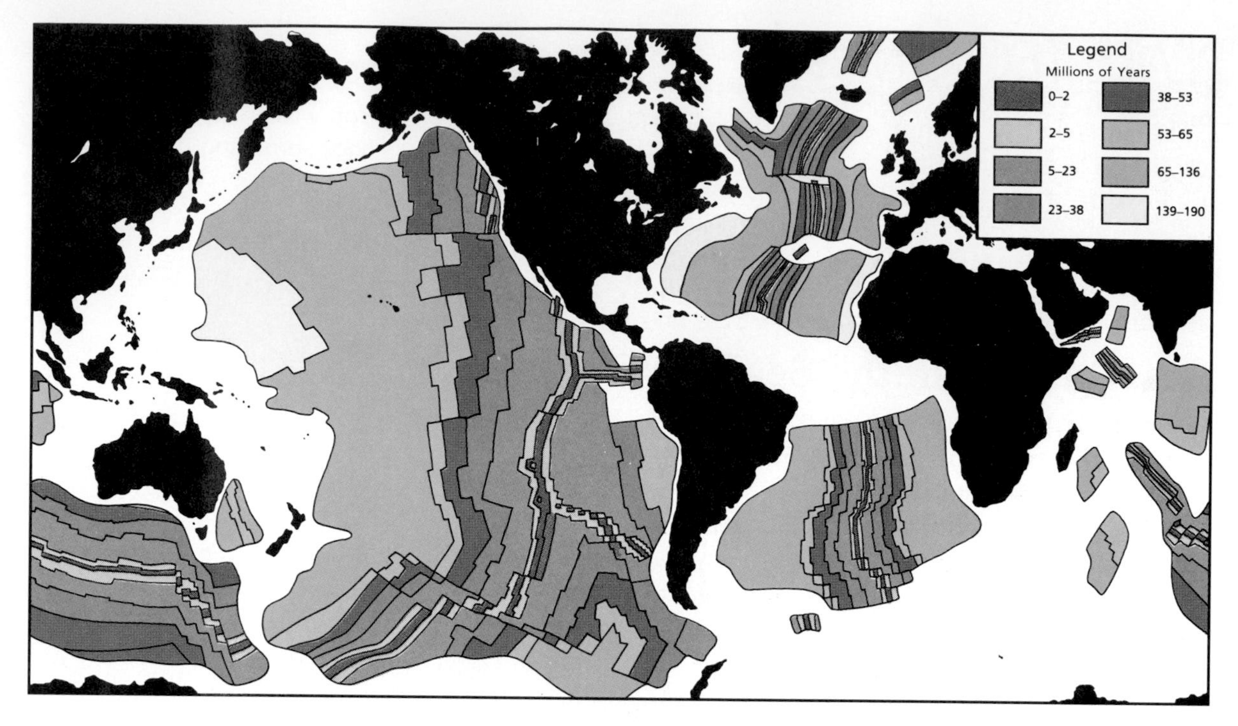

FIGURE 20–6. The seafloor is composed of increasingly older material as one moves away from the mid-ocean ridges. Why?

A second clue obtained from sea-floor studies involved the magnetism of basaltic rocks. When scientists towed magnetometers across segments of the ocean floor, alternating bands of magnetism were discovered. **Magnetometers** detect and measure the presence of weak magnetic fields in crystalline rocks. Scientists already knew that Earth's magnetic field had reversed repeatedly in the past. These bands reflected the magnetic reversals. But how were parallel bands of basalt deposited across the ocean basin floors?

20:5 The Shifting Plates

Data obtained from many sources led scientists to develop the theory of **plate tectonics.** This theory is the most current model that explains not only the movement of the continents, but also the changes in Earth's crust caused by internal forces.

What are plates?

Plates are rigid blocks of Earth's crust and upper mantle. These rigid blocks make up the **lithosphere** (LITH uh sfihr). Plates extend from Earth's surface to a depth of about 100 km. Below 100 km, mantle material is near its melting point and acts similar to a plastic material. This weak, plasticlike zone is called the **asthenosphere.** It is about 200 km thick. The less dense lithosphere floats on the more dense asthenosphere the way ice floats on water in a lake.

According to plate tectonics, Earth's lithosphere is broken into nine large sections and several small ones. The major plates are named after the continents they are "transporting." However, the continents show little relationship to the shape or size of the plate.

Where is oceanic crust formed?

Recall that oceanic crust is being formed at rift zones. Thus, it seems to be in a constant state of change. Continental crust, on the other hand, seems to be a relatively permanent part of Earth's surface. This permanence may be a result of the lower density and greater thickness of continental crust as compared to oceanic crust.

Plates move over Earth's surface. Plates may bump into one another, may spread apart, or may move horizontally past one another. Earthquakes, volcanoes, and mountain building commonly occur along boundaries where plates interact with one another. The geology of these areas varies with the type of interaction occurring between plates.

Divergent boundaries occur where two plates are pulling apart. The mid-ocean ridges are sites of divergent plate boundaries. At these ridges, vertical walls plunge 1.5 km toward a central crack, or rift zone. These rifts generally average from 35 to 45 km wide. Along the ridge, oceanic crust is being upwarped. From time to time, the crust bulges to the point where it cannot withstand the stress. Then the rift opens and allows basaltic magma to

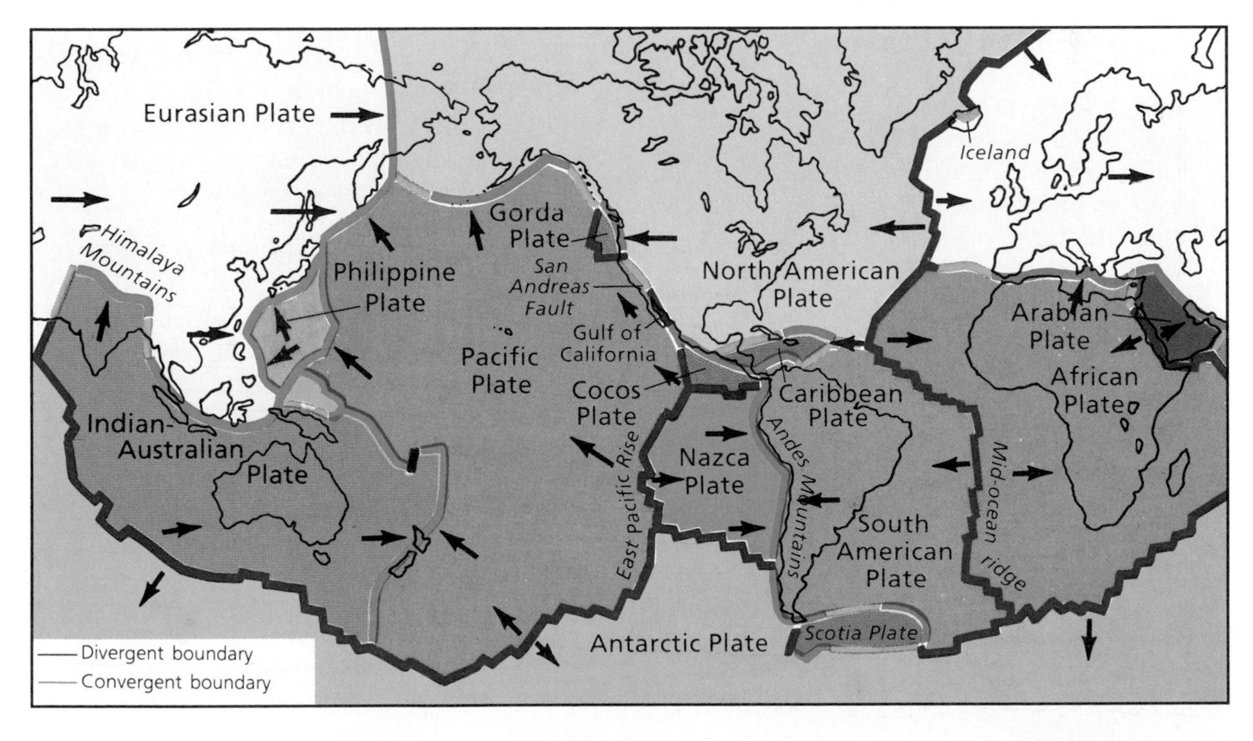

FIGURE 20–7. Earth's crust and upper mantle are broken into nine major plates and several small ones. What do you think the arrows indicate?

INVESTIGATION 20–1

Sea-floor Spreading

Problem: How do blocks of Earth's crust and upper mantle move?

Materials 

food coloring
water
hot plate
deep metal pan
thermal mitts
metric ruler
paper
tape
small magnetic compass (2)
bar magnet
pen or marker

Procedure

Part A

1. Fill the pan with water to 5 cm from the top.
2. Center the pan on the hot plate and heat gently. **CAUTION:** *Pan will become hot. Wear thermal mitts to protect your hands.*
3. Add a few drops of food coloring to the water directly above the hot plate.
4. Watch for currents to form in the water. Record your observations in a data table.

Part B

1. Tape several sheets of paper together to produce a strip from 40 to 60 cm in length.
2. Fold the strip of paper and place it between two close desks or piles of books as shown in Figure 20–8. The paper represents oceanic crust on either side of a mid-ocean ridge.
3. Place the magnet as in Figure 20–8.

FIGURE 20–8.

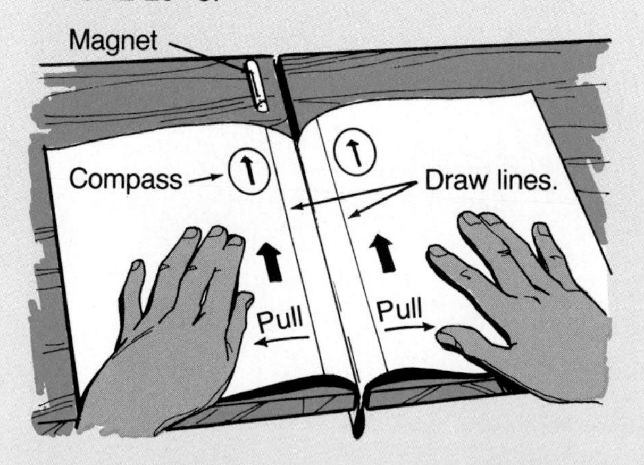

4. Place the two compasses next to each other on either side of the space between the desks.
5. Draw a line along each side of the space to represent the edges of the ocean ridge.
6. Beside the line, draw arrows showing the direction the compass needles are pointing.
7. Split the "seafloor" by moving the paper away from the center approximately 3 cm on each side. Reverse the magnet by turning it 180°.
8. Return the compasses to their original positions along the side of the space between the desks. Draw new arrows on the paper to represent the direction that the compass needles are now pointing.
9. Repeat this procedure several times.

Data and Observations

Part A	Part B

Analysis

1. What does the water in Part A represent?
2. Why did you add food coloring to the water?
3. What happens to the current when it reaches the sides of the pan?
4. In Part B, where are the "oldest" marks on the strip of paper?
5. Compare your completed strip to the patterns in Figure 20–6. What are the similarities and differences?

Conclusions and Applications

6. How does this investigation compare to the movement of crustal and mantle material?
7. How do the plates move?
8. What is a magnetic reversal?
9. How does this model answer the question of why the ocean basins have younger crustal rocks than the continents?

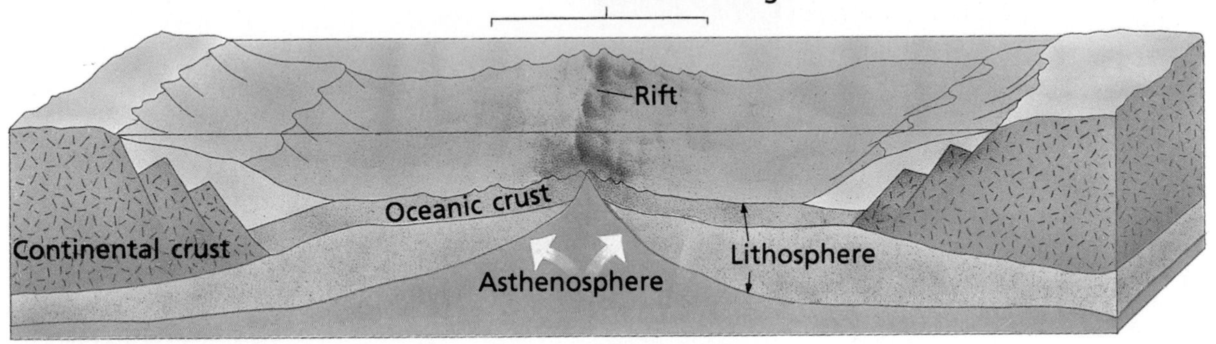

well up through the cracks. Each new addition of lava hardens to form new ocean crust. Thus, the width of the ocean basin is increased.

FIGURE 20–9. Mid-ocean ridges are divergent plate boundaries. Locate these boundaries in Figure 20–7.

When first recognized, this process was called **sea-floor spreading.** The average rate of spreading from a typical ridge is about six cm per year. Sea-floor spreading explains the bands of alternating directions of magnetism. It also explains the increasing age of the bands of lava away from the ridges. Volcanic activity associated with divergent boundaries also may occur on landmasses, such as the Rift Valley of East Africa. Here, hot material from the mantle moves upward, upwarps, and stretches the crust. As the crust pulls apart, blocks of crust sink and form the Rift Valley. If this process continues, the Rift Valley will lengthen and deepen, and eventually extend to the ocean. Then this area will resemble the Red Sea, another divergent boundary.

F.Y.I. The mid-ocean ridge is called a spreading zone because the adjacent plates slowly move away carrying new seafloor with them.

FIGURE 20–10. The East African Rift Valley is a divergent plate boundary forming today (a). The photo shows the area outlined on the map (b) in red. Note how the Arabian Peninsula could fit against Africa.

a

b

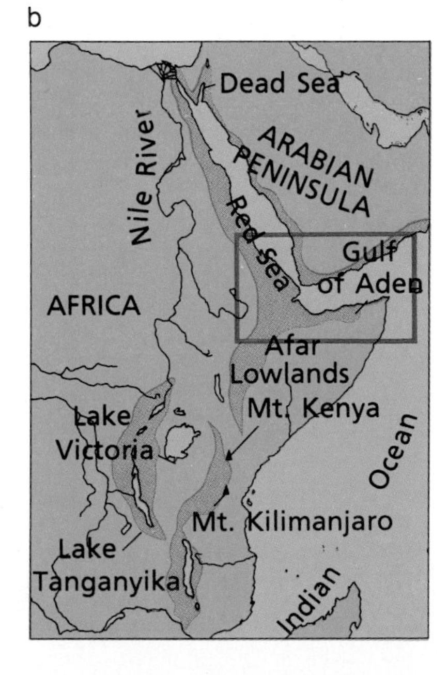

FIGURE 20–11. The San Andreas fault is a transform fault.

What is a transform fault?

Transform faults are boundaries at which plates move past one another in opposite directions or in the same direction but at different rates. Earthquake activity is stronger in these areas than it is along divergent margins, but little volcanic activity occurs. The San Andreas fault is a transform fault. Numerous transform faults offset the mid-ocean ridges. Transform faults are shown in gray on Figure 20–7.

20:6 Convergent Boundaries

Where are deep-sea trenches produced?

Convergent boundaries are boundaries between two colliding plates. Earthquakes are common in these areas. When two oceanic plates collide, the edge of one is bent downward. As the leading edge of this plate descends into the asthenosphere, it melts. The region where the plate descends is the **subduction zone.** A deep-sea trench is

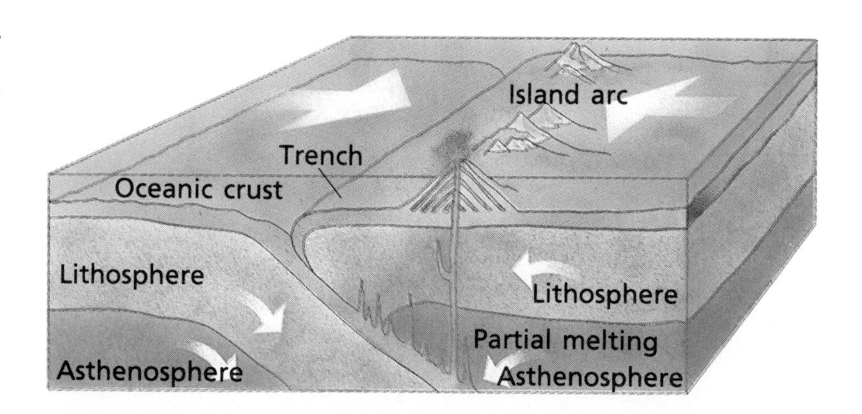

FIGURE 20–12. Two converging oceanic plates may cause the formation of an island arc.

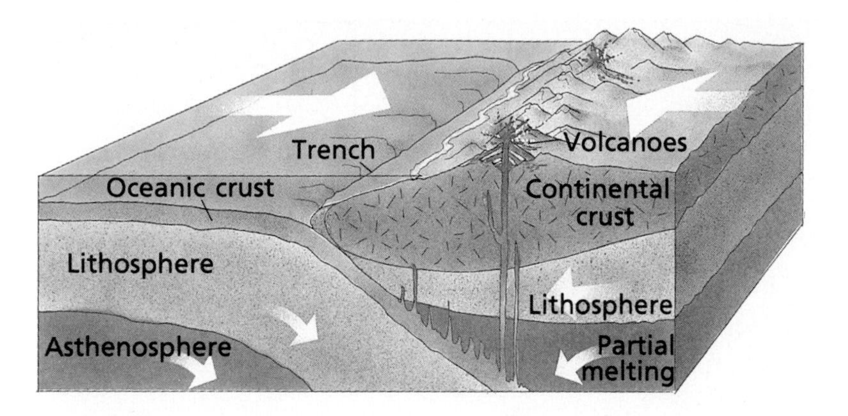

FIGURE 20–13. When oceanic and continental plates converge, volcanoes may form.

produced on the ocean floor at the point where the plate is subducted. Magma is formed as the descending plate melts. It is less dense than the surrounding mantle material. Thus, this magma rises and forms a chain of volcanoes on the ocean floor parallel to the trench. In time, these volcanoes may form a chain of islands, such as the Japanese Islands, called an **island arc.** As the islands erode, the debris collects in the trench. Any ocean sediment that reaches the trench is trapped there also. The sediments are carried into the subduction zone where they are melted or metamorphosed.

What is an island arc?

When oceanic and continental plates collide, the denser oceanic plate descends into the asthenosphere. Magma created as the sinking plate melts may form a chain of volcanic mountains along the edge of the continent. Volcanoes such as the Cascade Range are believed to have been formed this way.

F.Y.I. The Pacific plate is made of oceanic crust and occupies all of the Pacific Ocean west of the East Pacific Rise. It is overridden by the North American plate.

When plates carrying continental crust collide, the continents eventually meet. Continental rocks have low densities. Thus, they do not sink into the asthenosphere.

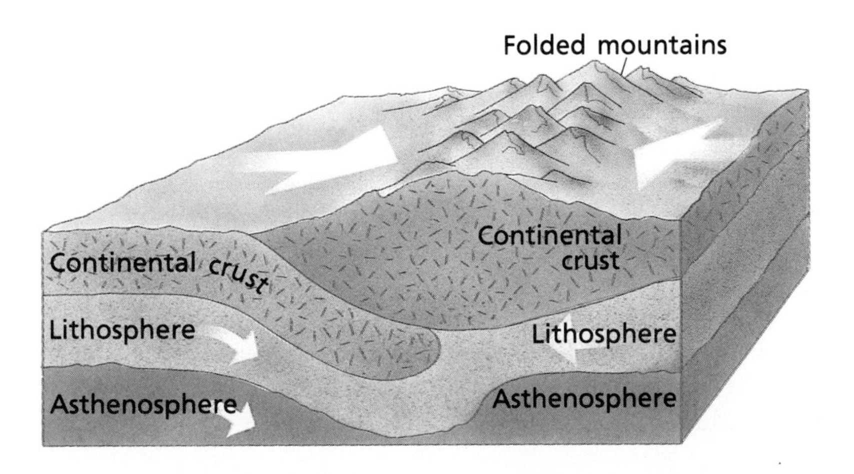

FIGURE 20–14. The collision of two continental plates produces folded mountains.

FIGURE 20–15. The Himalayas formed when two continental plates collided.

How do folded mountains form?

Instead, these rocks buckle and rise. The sediments along continental margins, as well as volcanic material produced by subduction, are lifted high above sea level. This "crumpling" effect forms the folded mountains common along many continental margins. Little volcanic activity occurs when continents collide, because the rocks are not thrust deeply into the mantle. Earthquakes, however, are common as the folded rocks break under the compressive forces of colliding plates. The Himalaya Mountains are being formed as the Indian plate collides with the Asian plate. The Alps, Appalachians, and Urals also are thought to have formed in this way.

20:7 What Causes Plate Motion?

Energy for the uplift of folded mountains comes from the convergence of plates. Heat is produced at subduction zones into which the leading edges of the plates are dragged. A mechanism is needed, however, for the movement of plates.

Along the mid-ocean ridges, scientists think basaltic magma from the mantle is rising due to convection currents. Recall from Chapter 9 that convection currents are due to density differences. Although some of the magma wells up through the mid-ocean ridges, much of it is turned horizontally by the rigid crust. As it moves away from the rift zone, the current of basalt carries the overlying lithosphere with it. Eventually the basalt cools and becomes more dense or comes in contact with a thick continental plate. Then the current turns downward and sinks into the mantle, carrying the overlying oceanic crust with it. The amount of oceanic crust subducted into the mantle seems to be about equal to the amount of new crust added at the rift zone. Thus, convection currents in the mantle appear to provide the mechanism necessary to move Earth's plates.

How do plates move?

Review

3. What invention allowed accurate data to be collected from the ocean floor?
4. Define lithosphere and asthenosphere.
5. What are divergent boundaries? Where are they found?
6. What is a transform fault?
★**7.** Locate the Andes Mountains in Figure 20–3. How might these mountains have formed?

CAREER

Geologist

Cynthia Dusel-Bacon is a project chief geologist for the Branch of Alaskan Geology in the United States Geological Survey's Western Region. Ms. Dusel-Bacon studies the metamorphic rocks in east-central Alaska. She is also heading a project that compiles a map and text giving information on metamorphism, particularly the temperature and pressure conditions under which the rocks recrystallized in all of Alaska. Ms. Dusel-Bacon spends much of her time outdoors. As a field geologist, she observes and collects data at rock outcrops. An outcrop is the part of a geologic formation or structure that appears at the surface of Earth. She makes measurements of the thicknesses of formations and records these data in a field notebook.

In her work, Ms. Dusel-Bacon not only uses field research, but also radiometric dating, microscopic examination of thin sections of rock, and geochemical data. She does field work on alternate years, gives talks to professional groups, and writes scientific papers.

For career information, write:
American Geological Institute
5202 Leesburg Pike
Falls Church, VA 22041

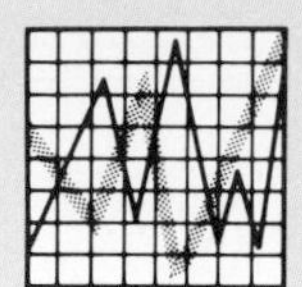

SKILL

Interpreting Data

Problem: How can you interpret data to determine the rate of sea-floor spreading?

Materials

Figure 20–16 metric ruler pencil

Procedure

1. Study the magnetic field profile in Figure 20–16. You will be working with six major peaks east and west of the mid-Atlantic ridge for both normal and reversed polarity.
2. Place the ruler through the first peak west of the main rift. Determine and record the distance in km to the mid-Atlantic ridge.
3. Repeat Step 2 for each of the six major peaks east and west of the main rift, for both normal and reversed polarity.
4. Find the average distance from peak to ridge for each pair of corresponding peaks on either side of the ridge. Record these values.
5. Use the normal polarity readings to find the age of the rocks at each average distance.
6. Using normal polarity readings, calculate the rate of movement in cm/year. Use the formula (distance = rate × time) to calculate the rate. You must convert km to cm.

FIGURE 20–16.

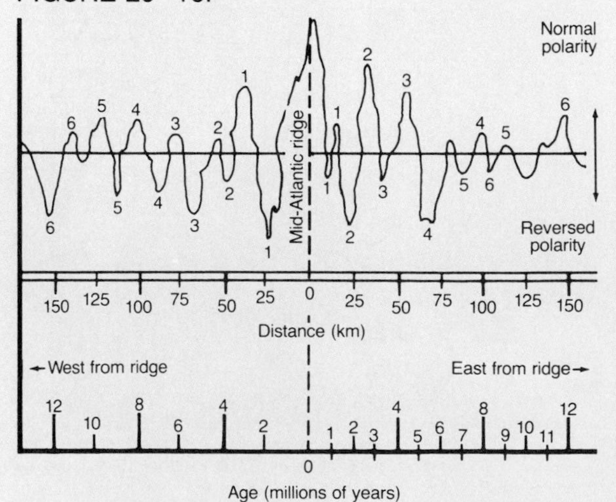

Data and Observations

Peak	1	2	3	4	5	6
Distance west normal polarity						
Distance east normal polarity						
Average distance						
Distance west reversed polarity						
Distance east reversed polarity						
Average distance						
Age from scale (millions of years)						
Rate of movement (cm/year)						

Questions

1. Compare the age of the igneous rock found near the mid-ocean ridge to that of the rock found farther away from the ridge.
2. In what way does the information shown in the graph in Figure 20–16 relate to the procedure for Part B in Investigation 20–1?
3. On your paper, draw a line that would represent the amount of total movement that would occur between a point east of the mid-Atlantic ridge and a point west of the ridge in one year.
4. If the distance from a point on the coast of Africa to the mid-Atlantic ridge is approximately 2400 km, how long ago was that point in Africa at or near that mid-ocean ridge?

DYNAMIC EARTH

It was not until the mid-1960s that scientists began to realize that a new model of a dynamic Earth had to be developed. In 1968, seismologists charted the epicenters of all earthquakes that had been recorded between 1961 and 1967. The resulting map identified the sites of each mid-ocean ridge, transform fault, and convergent plate boundary. The pieces of the puzzle were now falling into place.

GOALS

1. You will learn how the occurrence and location of earthquakes may be used as supporting evidence for the present plate tectonics model.
2. You will learn what Earth will look like if plate motions continue as they are.

20:8 Earthquakes and Plate Tectonics

Plate tectonics can be used to explain the distribution and occurrence of earthquakes. Many quakes are associated with deep-ocean trenches wherever oceanic plates descend into the mantle. **Shallow-focus earthquakes** are produced at the outer edge of trenches where plates are bent downward and scrape against the overriding plate. **Deep-focus earthquakes** occur as the plate is subducted deeper into the mantle. These earthquakes only occur in the rigid, descending plate. Few earthquakes have foci below 700 km. This is because most of the plate has melted by the time this depth is reached.

Where do shallow-focus earthquakes occur?

When continents collide, shallow-focus earthquakes occur as rocks are folded into mountains. Shallow-focus quakes also occur as plates move past one another at transform faults and at mid-ocean ridges.

Table 20–1

Depth and Location of Earthquakes (Foci) Along the Coast of a Continent

Quake	Focus depth	Distance and direction from coast (km)	Quake	Focus depth	Distance and direction from coast (km)
A	−55 km	0	L	−45 km	95 east
B	−295 km	100 east	M	−305 km	495 east
C	−390 km	455 east	N	−480 km	285 east
D	−60 km	75 east	O	−665 km	545 east
E	−130 km	255 east	P	−85 km	90 west
F	−195 km	65 east	Q	−525 km	205 east
G	−695 km	400 east	R	−85 km	25 west
H	−20 km	40 west	S	−445 km	595 east
I	−505 km	695 east	T	−635 km	665 east
J	−520 km	390 east	U	−55 km	95 west
K	−385 km	335 east	V	−70 km	100 west

INVESTIGATION 20–2

Tracking Plates

Problem: What do earthquakes tell us about plate boundaries?

Materials

Table 20–1
graph paper
pencil
Figures 20–12, 20–13, 20–14

Procedure

1. The data in Table 20–1 show the location and depth of earthquakes along the coast of a major continent.
2. Construct a graph similar to Figure 20–17 on your graph paper. Place "Distance east or west of the coast" on the horizontal axis beginning with 100 km west and moving east. The coast of the continent is represented by 0 km.

FIGURE 20–17.

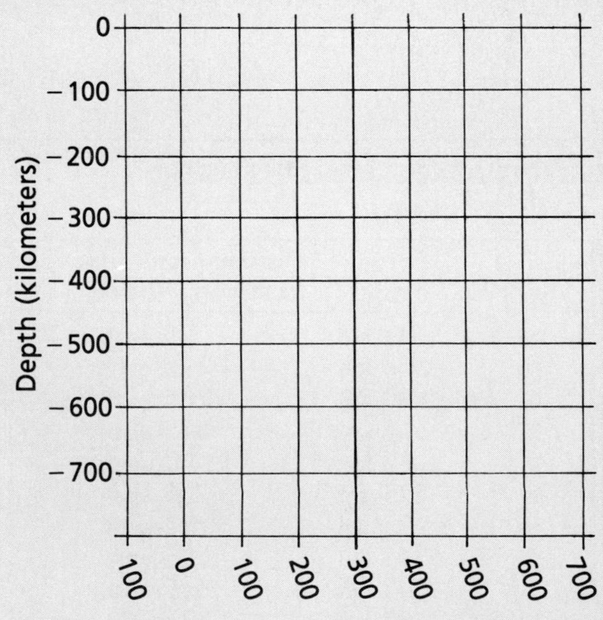

3. The vertical axis of the graph is "Depth below the surface" in kilometers. The top of the graph should represent 0 km, or the surface. The bottom of the vertical axis should be −800 km.
4. Plot the focus depth in kilometers against the distance and direction from the coast in kilometers for each earthquake from Table 20–1 on your graph. Use your graph and Figures 20–12, 20–13, and 20–14 to answer the questions.

Analysis

1. Describe the relationship between the location of the epicenters (east or west of the coast) and the depth of the earthquakes.
2. Does this appear to be a converging plate boundary or a diverging plate boundary? Explain your answer.

Conclusions and Applications

3. Describe a major sea-floor feature that probably can be found near this plate boundary. Would this feature be located east or west of the coast? How do you know?
4. Predict what landforms you would see along this continental margin. Explain your prediction in terms of plate interactions.
5. Is the continent located east or west of the coast? How do you know?
6. Using Figure 20–7 of the plates and plate boundaries, predict which continental coast is being studied in this investigation. Give reasons for your answer.
7. Why do earthquakes not occur below a depth of −700 km?
8. What do earthquakes tell us about plate boundaries?

20:9 Testing the Plate Tectonics Model

As with any theory, scientists are constantly searching for new data that will either support or change the current model of a dynamic Earth. One of the most convincing pieces of evidence for plate tectonics is the age of the rocks within the Pacific Ocean basin. New ocean floor is created at mid-ocean ridges and returned to the mantle in the trenches. Thus, the rate of plate movement should determine the maximum age of the rock in the plate.

The average rate of spreading at the East Pacific rise is 4 cm/year. The maximum distance from the ridge to any trench is 10 000 km. If the spreading rate is constant, then no existing Pacific Ocean rock should be older than 250 million years. To date, no rocks older than 200 million years have been found in the Pacific Ocean basin. Thus, the evidence supports the plate tectonics model. Should rocks older than 250 million years be found in the Pacific basin, the theory will have to be revised.

Another "proof" of the theory would be to visit Earth 100 million years from now to observe the positions of the continents. Figure 20–18 shows where the continents will be located if present rates and directions of plate movement continue. Part of California will be separated from North America by a shallow sea. A new ocean will form between two sections of Africa. Australia will begin to collide with Asia, piling the islands of the East Indies onto the continental margin.

BIOGRAPHY

Kenneth Jinghwa Hsü
1929-

Kenneth Hsü has studied plate tectonics, specializing in the study of geosynclinal sediments that accumulate on the edges of Earth's plates. Hsü's work helps us understand the origins of mountains.

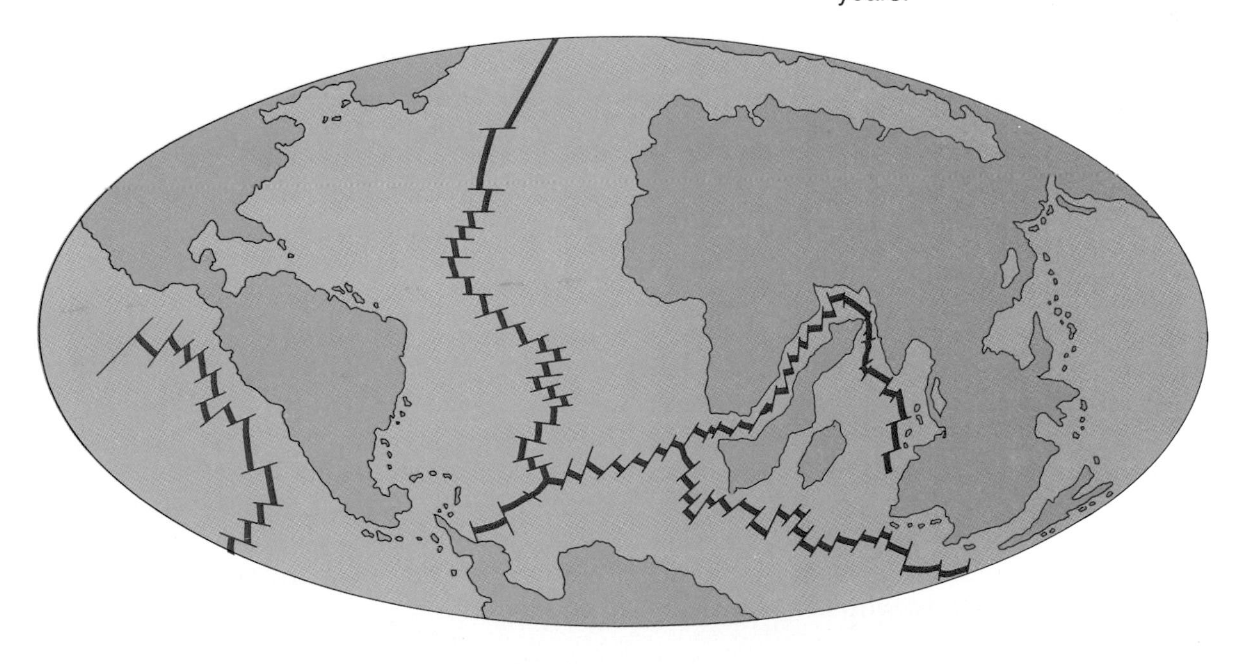

FIGURE 20–18. If current plate movement continues, this is the way Earth will look in 100 million years.

It certainly would be easy, 100 million years from now, to know just how accurate the plate tectonics model really is. By comparing predictions of continental locations with their actual positions, the theory could be accepted or revised in light of new facts. Geologists have done essentially that, however, in arriving at the present version of the theory. You will learn in Chapter 21 that by studying the apparent positions of the continents over the past 200 million years, scientists can better understand the processes shaping Earth today.

Review

8. Why are deep-focus earthquakes only associated with the subduction of oceanic plates into the asthenosphere?

9. The belt of earthquakes and volcanoes around the Pacific Ocean is called the "Ring of Fire." Explain this belt in terms of the plate tectonics theory.

10. Why do few earthquakes occur below 700 km?

11. What similarities in earthquake foci patterns occur between divergent plate boundaries and transform faults?

★ **12.** Explain how the fact that no rocks older than 200 million years have been found in the Pacific Ocean basin supports the model of plate tectonics.

TECHNOLOGY: APPLICATIONS

The Alaskan Puzzle

According to the theory of plate tectonics, continents are believed to grow at a slow but irregular rate along their outer margins. Researchers in Alaska, using fossil and magnetic data, have found evidence that suggests that sometimes continents grow by the addition of large blocks of crust called suspect terranes. Some scientists believe that Alaska is a jumble of these terranes. Many of the Alaskan terranes are of oceanic origin and are composed of oceanic crust, island arc material, or mid-ocean ridge material. These terranes have been "scraped off" oceanic plates as they were subducted into the mantle.

Other terranes are fragments of continents, some of which originated at or near the equator. These lithospheric blocks have been transported 9000 kilometers north during the past 200 million years. Paleontologists have found fossils, coal deposits, and dinosaur footprints in these terranes that suggest a tropical origin for these lithospheric blocks. Paleomagnetic data also indicate that several of the blocks formed in areas different from their present locations. As geologists continue to study the Alaskan terranes, the history of Earth's surface and the movement of plates will become clearer. Perhaps one day the puzzle will be solved.

Chapter 20 Review

SUMMARY

1. The idea of continental drift stated that continents had moved and had been in different positions in the geologic past. 20:1
2. Fossils, climate, and rock structures are all clues to past positions of continents. 20:2
3. The study of paleomagnetism led to the idea of polar wandering. Now continents are believed to have moved, not the poles. 20:3
4. Two important features of the ocean floor are the mid-ocean ridges and the trenches. The age of sea-floor rocks gets increasingly older away from the mid-ocean ridges. 20:4
5. Plate tectonics states that Earth's lithosphere is broken into several large plates that move on the plasticlike asthenosphere. 20:5
6. Divergent boundaries occur where plates are pulling apart. Transform faults occur where plates move past one another. 20:5
7. Convergent boundaries occur where two plates collide. 20:6
8. Convection currents appear to transport plates. 20:7
9. Shallow-focus earthquakes occur along mid-ocean ridges and transform faults, at trench boundaries, and when two continents collide. Deep-focus earthquakes occur in subduction zones. 20:8
10. The relative youthfulness of rocks in the Pacific Ocean basin supports the plate tectonics model. 20:9

VOCABULARY

a. asthenosphere
b. continental drift
c. convergent boundaries
d. deep-focus earthquakes
e. divergent boundaries
f. island arc
g. lithosphere
h. magnetometers
i. paleomagnetism
j. Pangaea
k. plates
l. plate tectonics
m. polar wandering
n. sea-floor spreading
o. shallow-focus earthquakes
p. subduction zone
q. transform faults

Matching

Match each description with the correct vocabulary word from the list above. Some words will not be used.

1. according to Wegener, the original continent
2. boundaries where plates move past each other horizontally
3. the study of ancient magnetism
4. rigid blocks of crust and upper mantle
5. the process by which ocean crust is formed and basins expand
6. a region where a plate descends into the mantle
7. a chain of volcanoes parallel to a deep-ocean trench
8. sites where plates collide
9. the idea that Earth's magnetic poles have moved with geologic time
10. the partially melted part of Earth's mantle

Chapter 20 Review

MAIN IDEAS

A. Reviewing Concepts

Choose the word or phrase that correctly completes each of the following sentences.

1. Plates moving in the same direction at different rates are separated by a *(convergent, divergent, transform fault)* boundary.
2. Lithosphere returns to the mantle at a *(subduction zone, mid-ocean ridge, transform fault)*.
3. The oldest parts of the seafloor usually are found near *(mid-ocean ridges, trenches, volcanic arcs)*.
4. A(n) *(volcanic mountain, island arc, rift zone)* forms when two oceanic plates collide.
5. An instrument used by oceanographers to detect magnetic changes in crustal rocks is a *(precision depth recorder, magnetometer, seismograph)*.
6. Plates are composed of *(crust, crust and mantle, mantle)*.
7. An island arc may form near a *(mid-ocean ridge, continental margin, deep-sea trench)*.
8. The continents of *(Africa and South America, India and Australia, Europe and Antarctica)* seem to fit together like pieces of a puzzle and first gave rise to the idea of drifting continents.
9. The most current explanation for the movement of the continents is called *(continental drift, plate tectonics, sea-floor spreading)*.
10. The oldest rocks found in ocean basins are generally about *(200 million, 500 million, 3.5 billion)* years old.
11. The scientist who first provided supporting evidence for the idea of continental drift, Alfred Wegener, was a *(meteorologist, geophysicist, seismologist)*.
12. A common feature of diverging plate boundaries is a *(deep-focus earthquake, subducting plate, mid-ocean ridge)*.
13. A feature that often is the result of the collision of two continental plates is a(n) *(mid-ocean ridge, folded mountain range, ocean trench)*.
14. Generally, as one moves away from a mid-ocean ridge, the age of the crustal rock in the ocean basin floor *(increases, decreases, remains the same)*.
15. Current theories suggest that convection currents in Earth's *(asthenosphere, lithosphere, plates)* are the driving mechanism for plate transport.

B. Understanding Concepts

Answer the following questions using complete sentences.

16. Describe where the youngest rocks tend to be found in any of the major ocean basins on Earth's surface.
17. What are the three types of plate boundaries?
18. Describe the "history" of Pangaea.
19. Describe the basic differences between oceanic and continental crust.
20. What generally appears to occur when oceanic plates collide with continental plates?
21. Explain how fossils such as *Lystrosaurus* and *Glossopteris* support the concepts of sea-floor spreading and plate tectonics.
22. Identify three scientific instruments that have provided data that support the theory of plate tectonics.
23. The Appalachian Mountains run through the northeastern United States and appear to end abruptly off the coast of Newfoundland. Where else on Earth's surface are mountains of almost identical age and structure found? What might this indicate?

24. How many plates make up Earth?
25. What are rift valleys and what do they indicate?

C. Applying Concepts

Answer the following questions using complete sentences.

26. How are convection currents important to plate tectonics?
27. Why are there few volcanoes in the Himalayas, but many earthquakes?
28. Glacial deposits often form at high latitudes and high altitudes. Glacial deposits have been found in Africa. Explain.
29. How will the continents appear 100 million years from now if the present rate and direction of plate movement continues?
30. The Ural Mountains run north and south between Europe and Asia. How might they have been formed?

SKILL REVIEW

1. Refer to the graph you completed for Investigation 20–2. If the plate boundary being plotted is located between the Nazca and South American plates, in what direction are you looking when analyzing the graph? You may want to use Figure 20–7.
2. Construct a diagram of Earth at a diverging plate boundary. Show a major feature that can be expected to form there and the ages of the crustal rocks as one moves away from the divergent plate boundary.
3. Compare and contrast divergent and convergent plate margins.
4. Assuming a constant rate and direction for plate movement, make a map of Earth's surface one hundred million years from today.
5. Use a globe to plot the locations of convergent and divergent plate boundaries as shown in Figure 20–7. Use either colored chalk or water soluble markers to show the locations of the major and minor plates. Draw arrows, such as those in Figure 20–7, to show the direction of plate movement.

PROJECTS

1. Construct a three-dimensional model of each type of plate boundary showing typical earthquake foci depths and the structures that tend to result at each type of boundary.
2. Construct a model that shows how the size and shape of the ocean basins will appear through time, given the present rates of change.

READINGS

1. Cox, Allan, and Brian R. Hart. *Plate Tectonics: How It Works*. Boston: Blackwell, 1986.
2. Gallant, Ray. *Our Restless Earth*. New York: Franklin Watts, 1986.
3. Yulsman, Tom. "Our Restless Earth." *Science Digest*. March, 1986, pp. 48–51, 82–83.

SCIENCE AND SOCIETY

EARTHQUAKES

What do San Francisco, California, Tokyo, Japan, Managua, Nicaragua, and Mexico City, Mexico have in common? They are all metropolitan areas that have been hit by major earthquakes. These cities are located on or near the edges of active plate margins. A plate is a piece of Earth's crust and upper mantle. Earth can be broken into nine large plates and several smaller plates.

Background

San Francisco is located along the San Andreas fault. Along this transform fault, the North American plate is moving past the Pacific plate. The city of San Francisco is "riding" northwest on the Pacific Plate. Tokyo, as well as the rest of Japan, is an island arc in the Circum-Pacific Ring of Fire, an area well known for its seismic activity. This ring is made up of the volcanic mountains of North, Central, and South America and the volcanic islands of the north and west Pacific Ocean. Managua is in an area affected by the movement of several small plates including the Nazca, Cocos, and Caribbean plates. The most recent earthquake to hit Mexico City was the result of movement along the Cocos Plate. Even though these cities are built in areas that have particular economic values that may include a good harbor, abundant natural resources, or a strategic location, the people living in these cities must learn to cope with the potential dangers of earthquakes.

Faults are fractures in Earth's crust along which there has been movement. Some faults are described as "creeping." They move constantly in a series of light vertical and horizontal movements that can result in frequent small tremors. Other faults are described as "locked." No movement occurs along these fractures until extreme pressure accumulates. When slip suddenly occurs, extreme destructive force is released.

Seismic gap refers to the time span between movements along faults. The longer the gap, the greater the likelihood of a major earthquake. Based on its seismic gap, it has been predicted that southern California will have a major earthquake within the next 30 years.

FIGURE 1. Earthquakes often originate along plate boundaries such as this one in California.

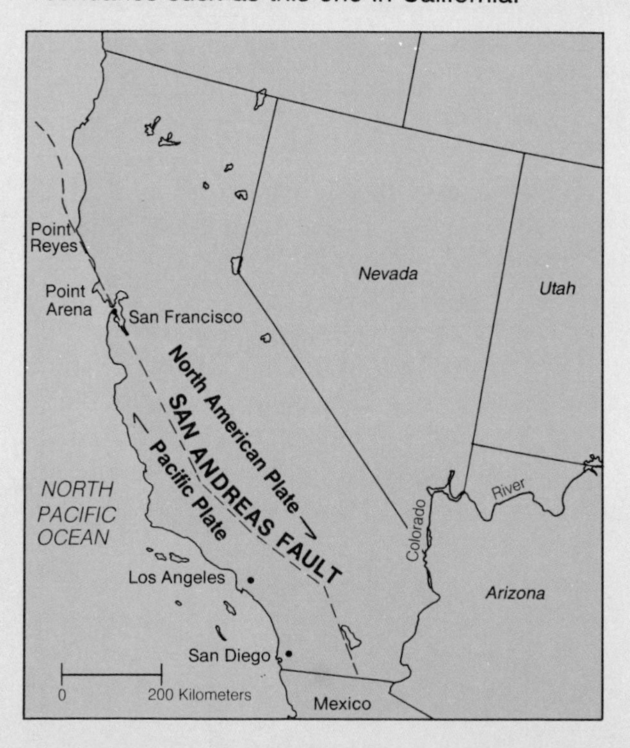

Case Studies

1. Prediction techniques have improved with ongoing research, but prediction is generally reliable only on a long-term basis. Modern techniques for earthquake prediction use computer modeling based on the cyclic nature of some earthquakes. Information is gathered with equipment such as tilt meters, creep meters, and laser beams that measure changes in the positions of Earth's rocks. Animals also have been thought to give a warning of impending earthquakes. Their behavior often changes just before an earthquake.

2. New building codes and techniques have been designed to help the residents of earthquake-prone areas. Some of the steps taken involve reinforcing joints in buildings, removing ornaments or sills that might fall off buildings during a quake, and limiting the heights of structures. Buildings undergo stresses during an earthquake and sway and vibrate with the frequency of the ground waves. Designs have been devised that allow for this sway. Another factor to consider is placement of a building. A building with the long axis of its foundation parallel to the ground motion will move less than one that is perpendicular to the motion. Architects in Tokyo have constructed some new buildings on rollers. Residents who have "rolled" through moderate earthquakes report that it is a rather alarming experience, but there is little damage.

3. Several cities, including San Francisco and Tokyo, run annual preparedness drills. In Tokyo, every citizen has an assigned role that varies from firefighter to grave digger. When the alarm goes off, normal activity stops and everyone reports to a designated station. Homeowners and tenants shut off gas lines, TVs, and electrical appliances to minimize the danger of fire.

Developing a Viewpoint

1. Your family has just been offered a very high paying job in a town located on a major fault. What information would you want to obtain in order to make a wise decision about whether or not to accept the job?

2. You have inherited 10 000 hectares of scrub land. It is not suited for farming or even grazing cattle. You decide the best thing to do with it is to build a retirement community. However, when the deed arrives you find that a fault runs diagonally through the property. There has not been a tremor or an earthquake in the past fifty years. At first you are apprehensive, but then you realize that if your plans were laid out carefully, you might be able to proceed. You must present your plans for the community to the state planning board. How will you convince them that a retirement community could be safe even if there were a major tremor? What parts of the community could safely be located along the fault? Draw the plans of your community.

Suggested Readings

Jerry Adler. "Forecasting Future Shock: Experts Know Where Quakes will Strike—But Not When." *Newsweek*. Vol. 106. September 30, 1985.

Allen Boraiko. "Earthquake in Mexico." *National Geographic*. May, 1986.

Golden, Frederick. *The Trembling Earth*. New York: Charles Scribner and Sons, 1983.

Morris, Charles. *The San Francisco Calamity by Earthquake and Fire*. Secaucus: Citadel Press, 1986.

Walker, Bryce, et al. *Planet Earth Series: Earthquake*. Virginia: Time-Life Books, 1982.

UNIT 7

Many changes have occurred on Earth over millions of years. Earth's landscape has been subjected to processes of building and breaking down. Many organisms have come and gone. Some have left traces in rock records. Scientists study Earth's history in places like the Grand Canyon. Where do you find evidence of the changes Earth has undergone?

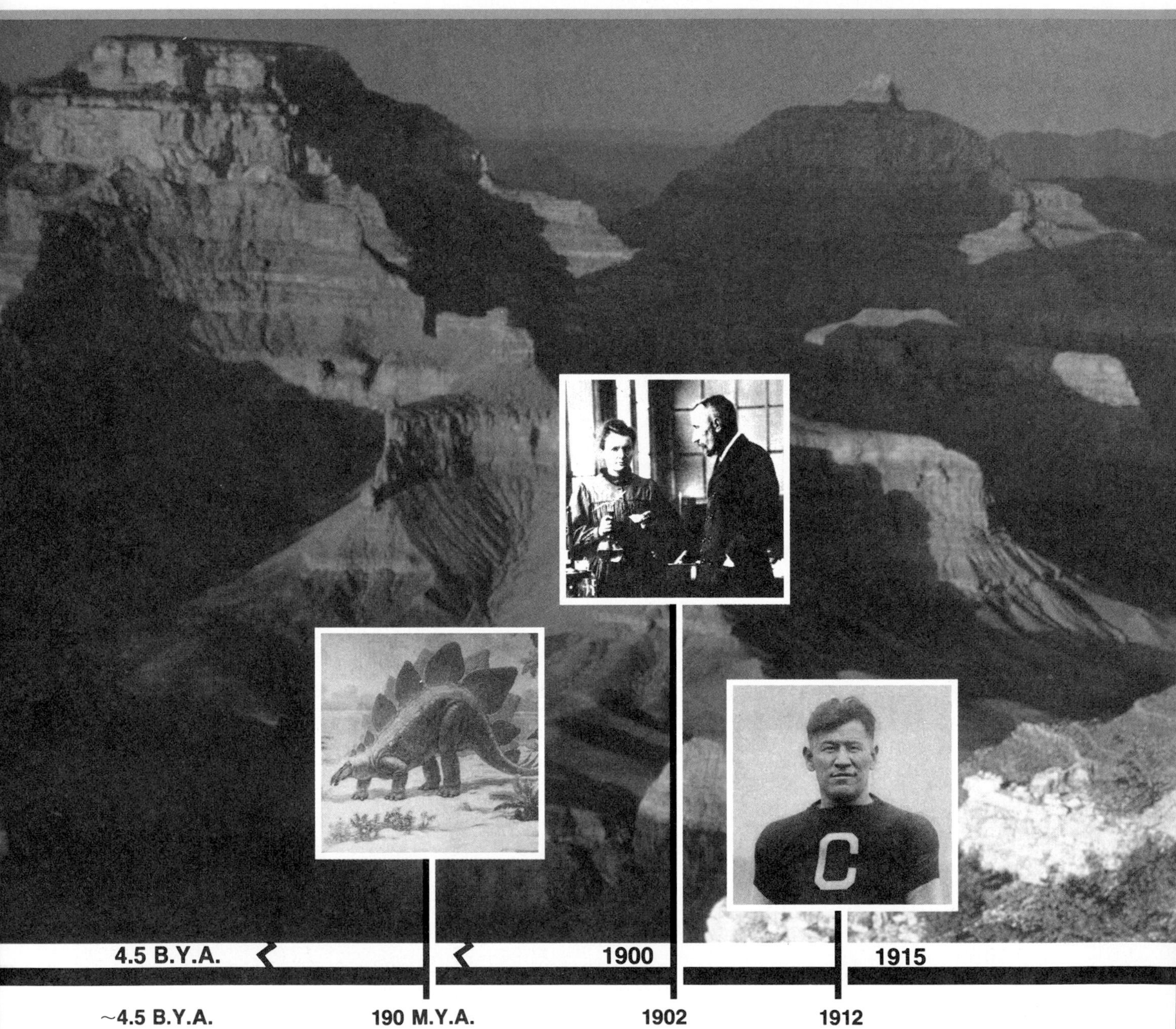

~4.5 B.Y.A.
Earth forms.

190 M.Y.A.
Dinosaurs roam Earth.

1902
Marie and Pierre Curie discover radium.

1912
Jim Thorpe wins two Olympic gold medals.

EARTH'S HISTORY

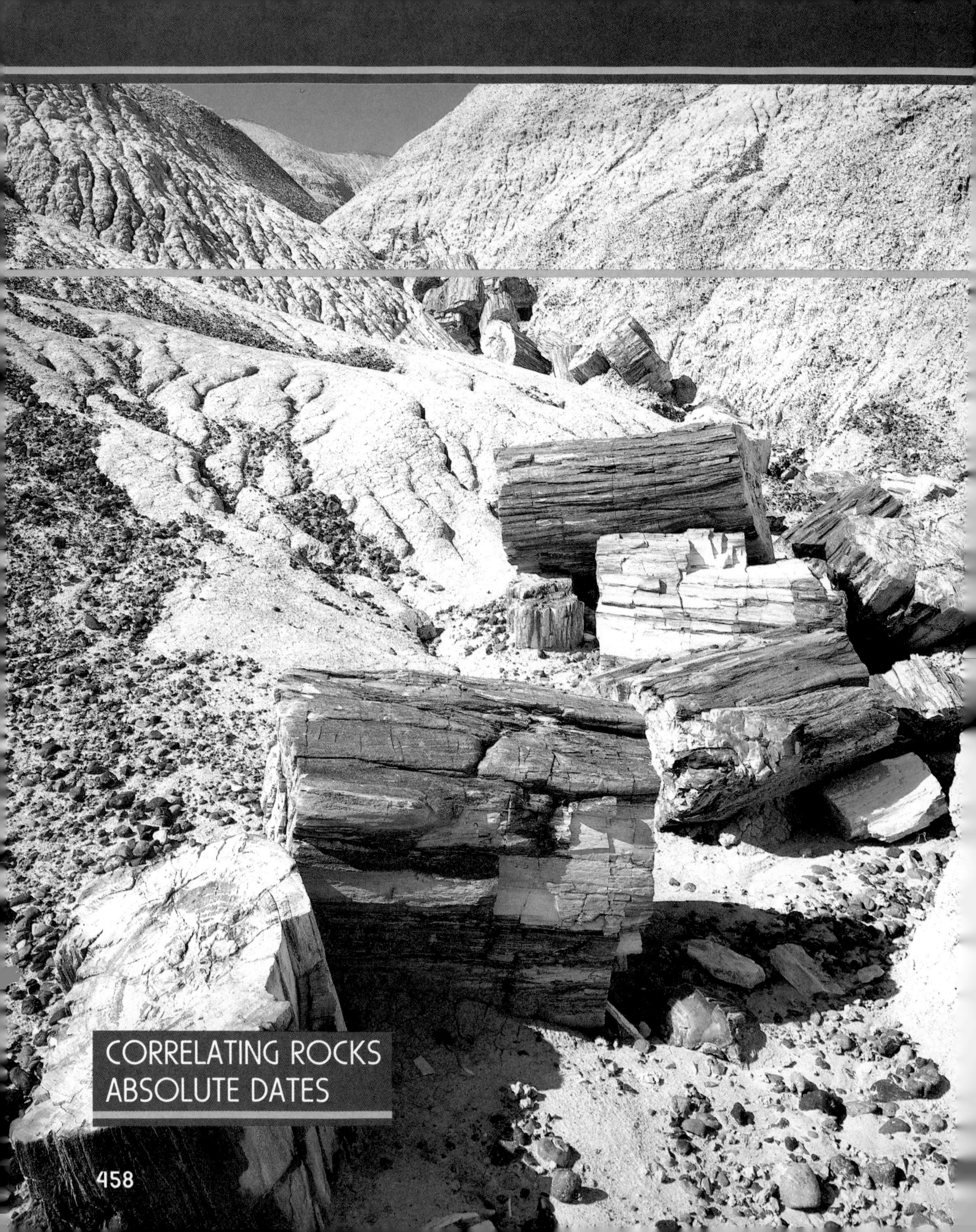

CORRELATING ROCKS ABSOLUTE DATES

Chapter 21

Clues to Earth's Past

Geologists use many tools to try to unravel Earth's history. They study Earth processes that are occurring today, and then determine if and how these processes occurred in the past. Geologists also study fossils to learn about ancient life forms. Unraveling Earth's complex history involves research in many fields of earth science.

CORRELATING ROCKS

How do we tell geologic time? Geologists use many different methods to date Earth's rocks. Some methods result in relative ages while others give absolute ages. The ages are then used to correlate or match up rocks in different regions of Earth.

GOALS

1. You will learn about methods used to determine the relative ages of rocks.
2. You will learn how fossils are used to determine the age of rocks.
3. You will learn how geologists interpret clues to determine the geologic history of an area.

21:1 Age Relationships Among Rocks

Many attempts have been made to measure the age of Earth materials and periods of geologic time. Much of our understanding of Earth's past has come from the study of sedimentary rocks. Most sedimentary rocks are laid down in horizontal layers. The oldest beds are at the bottom and the youngest beds are at the top of a rock sequence. This order of layering is called the **law of superposition.** Geologists apply this principle to determine relative ages of beds in a rock sequence. Reconstructing Earth's history would be simple if all sedimentary rocks remained horizontal. However, many rock layers have been deformed by folding and faulting associated with plate movements. This deformation makes it more difficult to unravel Earth's history. Thus, geologists have to use other principles in their investigations of Earth and its past.

FIGURE 21–1. Applying the law of superposition, conglomerate is the youngest rock shown and limestone is the oldest rock in this sequence.

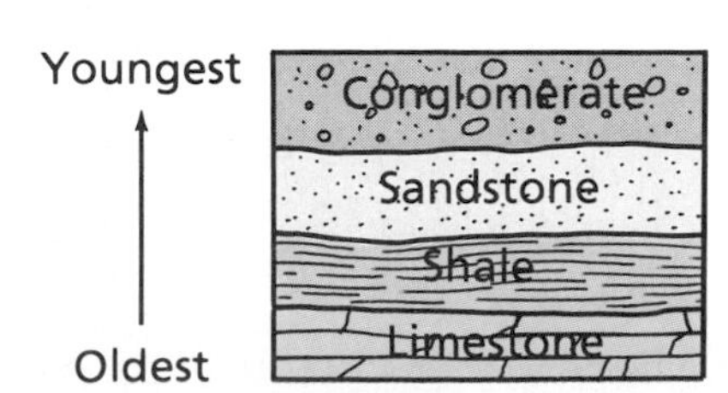

F.Y.I. The law of superposition only applies at sites where the rock sequence is undisturbed by folding or faulting.

One such principle uses fossils to identify beds in a sedimentary sequence. Fossils are remains or evidence of ancient life preserved in Earth's rocks. Fossils usually are found only in sedimentary rocks. These rocks can be identified by the fossils that they contain. **Fossil correlation** has been used to reconstruct Earth's history in areas where relationships among beds are complex. It also can be used to match beds that are widely separated.

Early in the 19th century, William Smith, an English engineer, found that he could identify rock layers in England and Wales by the kind of fossils they contained. By using the law of superposition, he was able to determine the relative ages of strata from the particular types of fossils found in each layer. He was able to match similar rock layers in France, Holland, and Belgium with those in England and Wales. Smith was the first to apply the principle of fossil correlation and to use fossil data to construct a geologic map. A **geologic map** shows the outcrop areas of various rock units and the structural features and ages of surface rocks.

What is a geologic map?

Another principle used to determine relative age in layered rocks uses crosscutting relationships. This method states that faults and igneous intrusions are younger than the rocks they cut across. In an area where crosscutting has occurred, the relative age of rocks, intrusions, and faults can be determined by applying this principle.

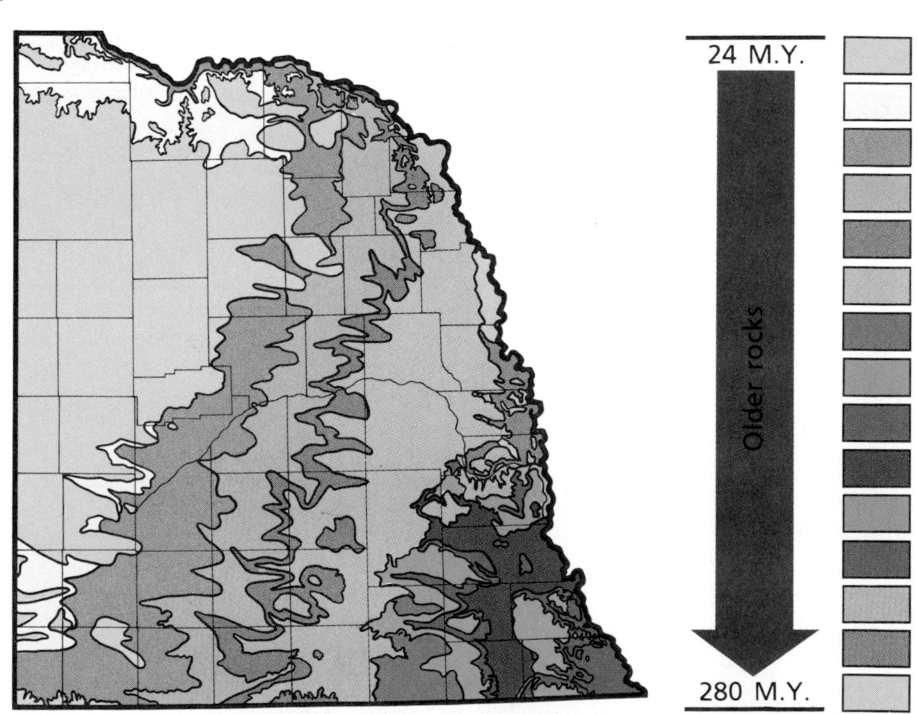

FIGURE 21–2. This geologic map of eastern Nebraska shows the relative age of rocks exposed at the surface.

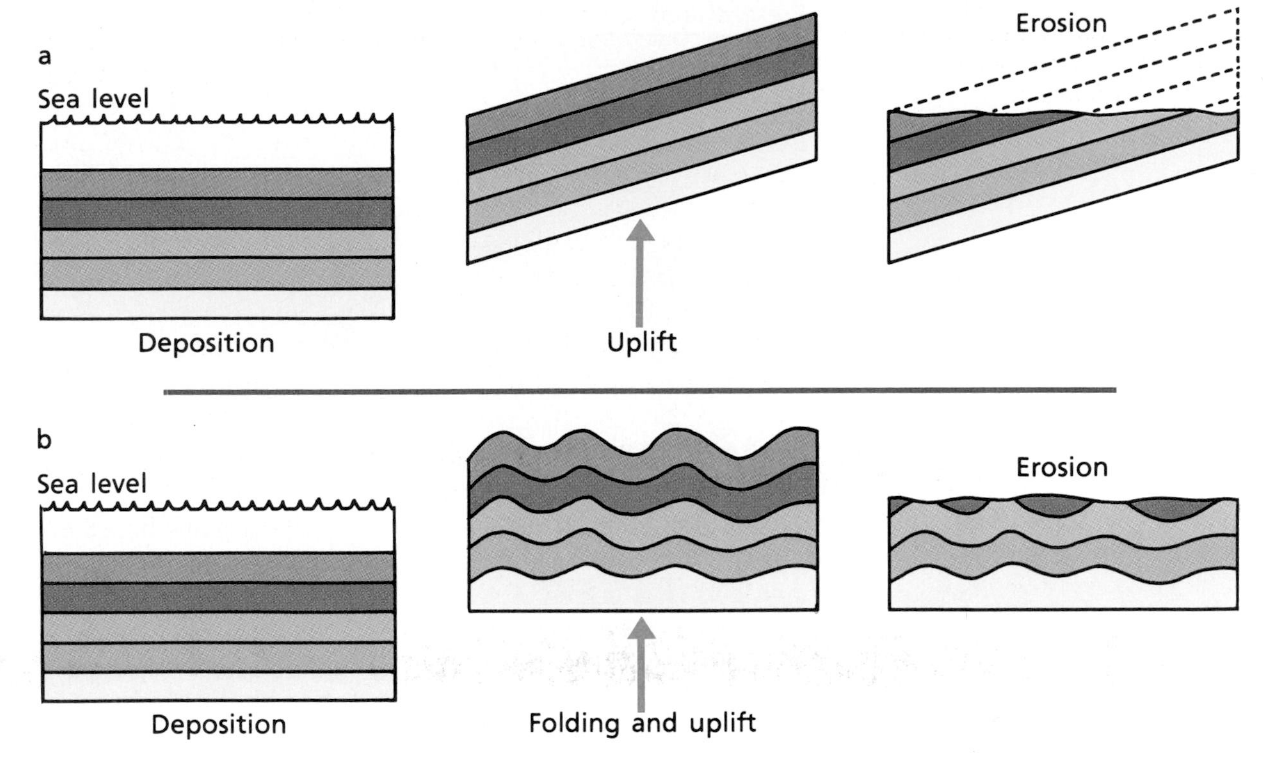

FIGURE 21–3. Relationships among rock layers become more complex as tilted layers (a) and folded layers (b) are eroded.

Another principle used to study Earth's history is **uniformitarianism.** This principle states that all Earth processes occurring today are similar to those that took place in the past. Although Earth processes have been essentially the same throughout time, their rates have undoubtedly varied.

Accurate age estimates of Earth and its rocks have been made using radiometric dating methods. Some minerals in most igneous rocks contain radioactive elements that can be used to determine when the rock formed. The relative amounts of radioactive material present in a rock will reveal its age.

21:2 Relative Dating

Early geologists used the law of superposition to determine relative ages of layered rock. This method can be applied to sedimentary beds and to some layered igneous rocks, such as volcanic ash and lava flows. **Relative dating** places events in their proper sequence or order. This method will not tell how long ago something happened, but it will indicate that one event took place before or after another event. For example, you may have a friend who has an older sister and a younger brother. You may not know their birth dates, but you do know the

What is relative dating?

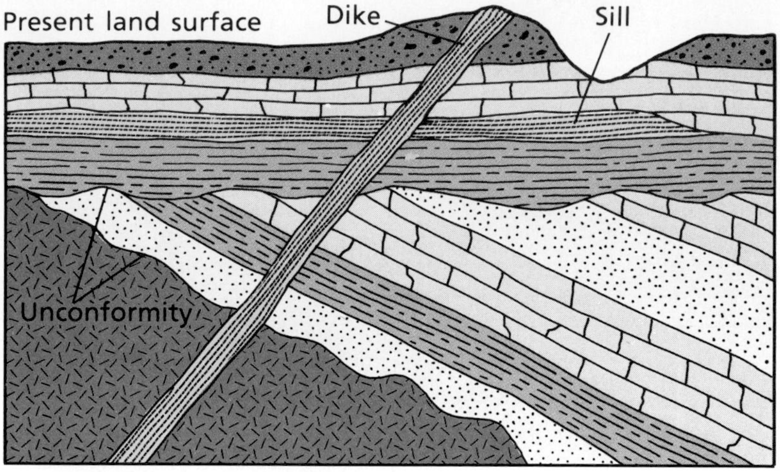

FIGURE 21–4. A dike that cross cuts several beds is younger than the beds themselves. What is the youngest rock in this geologic cross section?

relative ages of all three. Relative rock ages are based on the relationship among the various rock layers. A volcanic ash bed in a series of sedimentary layers indicates that the ash is younger than the beds below it and older than the beds above it. We do not know the age of the layers in terms of years, but we do know their relative ages.

Relationships among most igneous rocks are more difficult to determine than among sedimentary rocks, because igneous rocks usually lack layering. Sometimes when igneous rocks are in contact with sedimentary rocks, the crosscutting principle can be applied to determine their relative ages. If an igneous rock cuts across sedimentary layers, the sedimentary rocks are older. If sedimentary beds have been uplifted and deformed by an igneous intrusion, the intrusion is younger.

How is a geologic column constructed?

To determine relative ages of rocks, geologists construct a local history of geologic events from layered rocks in a particular area. They then attempt to match this history with layered rocks of nearby areas, and then with strata of more distant areas. From this work, geologists are able to arrange all rock layers in the order in which they were deposited. This sequence of layers, or **geologic column,** represents a certain amount of time during which the rocks were deposited.

21:3 Gaps in the Rock Record

To determine the history of an area, geologists must identify places in the geologic column where certain layers are missing. These breaks in the rock record are called unconformities (un kun FOR muh teez). **Unconformities** are surfaces that separate a series of rock layers. These gaps represent periods of erosion, nondeposition, or both.

SKILL

Sequencing Events

Problem: How are relative dates determined?

Materials

pencil
paper

Procedure

1. List any four events that have occurred in your lifetime, beginning with the most recent. Events can include personal occasions like a special party or a broken ankle, or public events such as a presidential election or an earthquake. The most recent event could have occurred yesterday, last week, or last year. The second event you list must have occurred before the first event. Do not include the dates of the events on your list.
2. Working with two or three other students, describe each event and compile a single list of everyone's events in the order you think they occurred. You may *not* use any dates to place the events in the correct order.
3. Complete the Data and Observations chart before determining the actual dates (or times) of the events.
4. Check the accuracy of your combined list by adding the dates of occurrence of each event. Make corrections to your list where necessary.
5. Save your completed Data and Observations chart. It will be used in Investigation 21–1.

Questions

1. Compare the processes you followed in this skill with the processes used to construct a geologic column. How are they similar? How are they different?
2. How are relative dates determined?

Data and Observations

Event	Description	Student	Date/time*	Event	Description	Student	Date/time*
1				9			
2				10			
3				11			
4				12			
5				13			
6				14			
7				15			
8				16			

*To be determined after Steps 1–3 have been completed

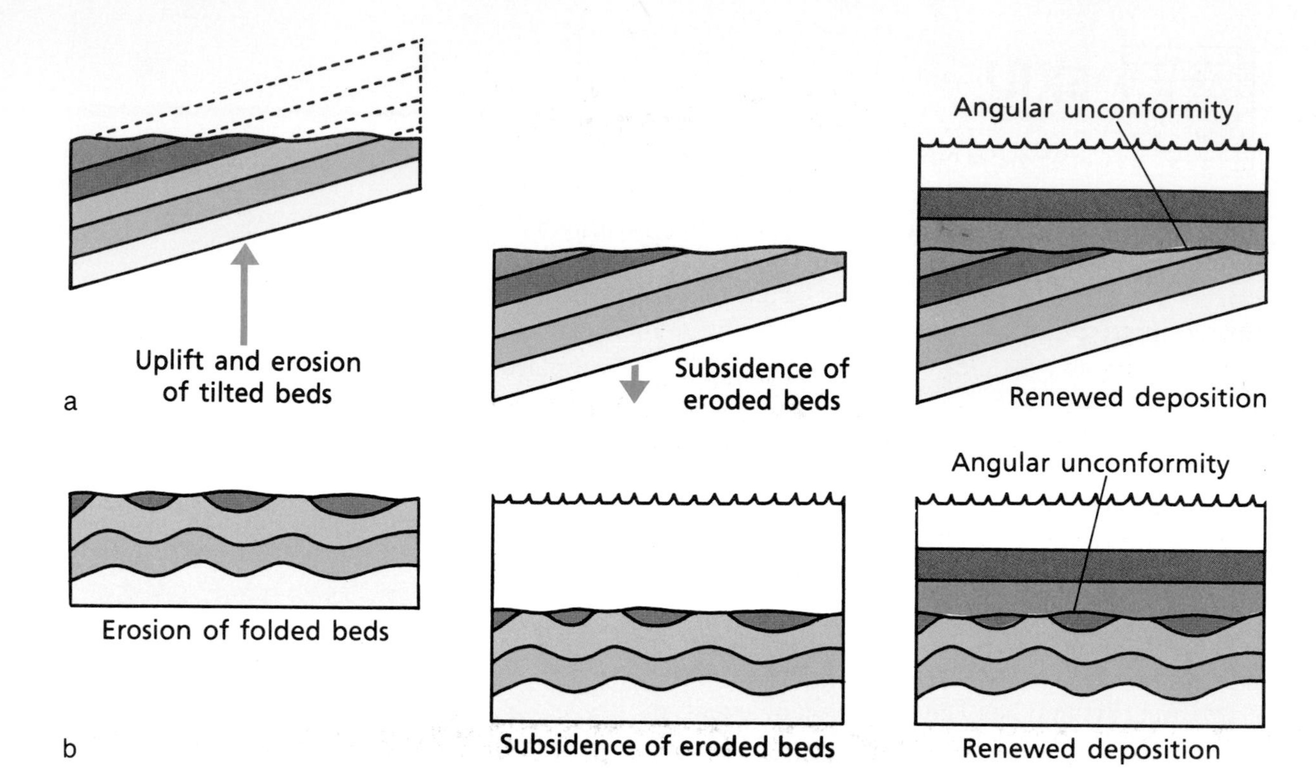

FIGURE 21–5. Unconformities occur where deposition follows erosion of tilted (a) and folded (b) rock layers.

These breaks represent periods of time in which no rock record is present in the geologic column. Unconformities may be useful clues in determining the relative ages.

Angular unconformities consist of tilted or folded rocks overlain by horizontal layers. These unconformities reveal a series of events including deposition, folding or tilting, uplift, and erosion of the lower layers. These events were followed by deposition of the overlying layers.

Unconformities also may separate two sets of horizontal beds. In such areas, deposition occurred, and then the area was uplifted. Erosion or nondeposition followed. Later, the region was submerged again, and the younger beds were deposited. Gaps that separate two horizontal series of beds are called **disconformities.** Study Figure 21–4. Find an angular unconformity and a disconformity.

FIGURE 21–6. Disconformities are unconformities that occur between layers of horizontal strata. What do disconformities represent?

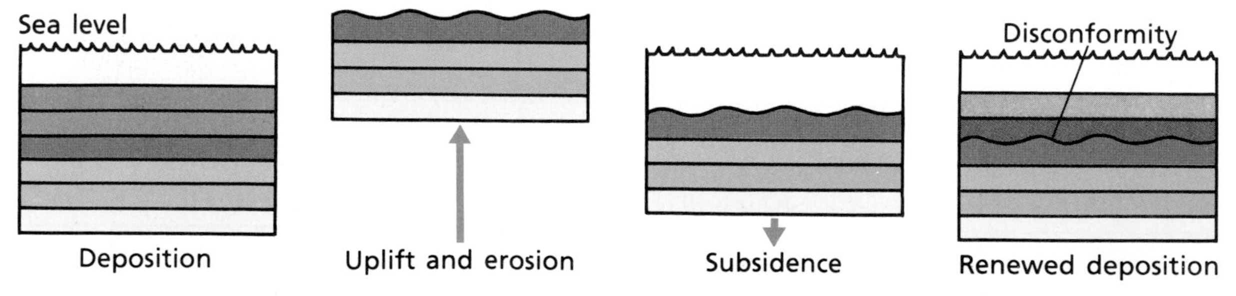

21:4 Fossil Clues

Paleontologists (pay lee ahn TAHL uh justs) study fossils. Fossils form when the hard parts of an organism are rapidly covered by sediment. Organic remains are most often preserved in marine sediments, in floodplain materials, or in lake deposits where prompt burial is most likely. Occasionally, plants, soft-bodied animals, and animal tracks are preserved in soft sand or mud.

There are several types of fossil preservation. Sometimes the original substance, such as wood or bone, is replaced by minerals that precipitate from groundwater. Minerals also may fill spaces in the original substance. These fossils are said to be **petrified.**

What is a mold?

Some fossils are molds or casts. A **mold** is a cavity left when groundwater dissolves a shell or bone. If a mold later is filled with mud or minerals, the filling is called a cast. A **cast** has the same shape as the original organism, but no inner structure.

PROBLEM SOLVING

Closing the Gap

Lana and Geoff spent part of their summer vacation on a field trip through western Colorado and eastern Utah. They had observed many rock outcrops and had recorded what they saw in notebooks. The geologic column on the left was drawn by Lana from observations made in Green River, Utah. The column on the right was made by Geoff from data he had collected in Westwater, Colorado. Help them reconstruct the geologic history of the area by answering the following. What type of rock is found at the base of each column? How many unconformities occur in each column? Describe the locations of these unconformities. Explain the geologic history of the Green River area in terms of erosion and deposition. Why are some formations missing from the Westwater column? What does the comparison of geologic columns tell about the history of an area?

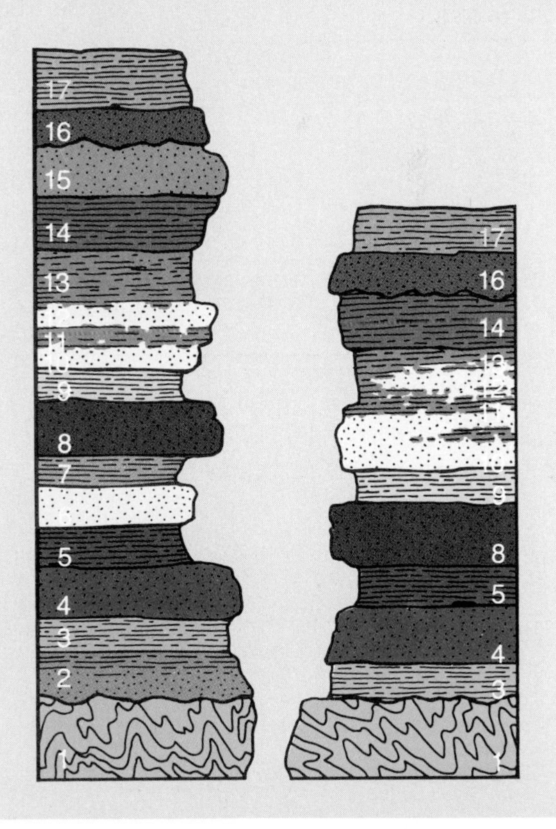

Step 1

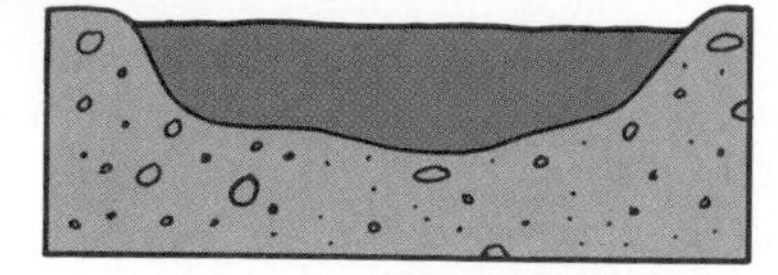

Depression filled with mud

Step 2

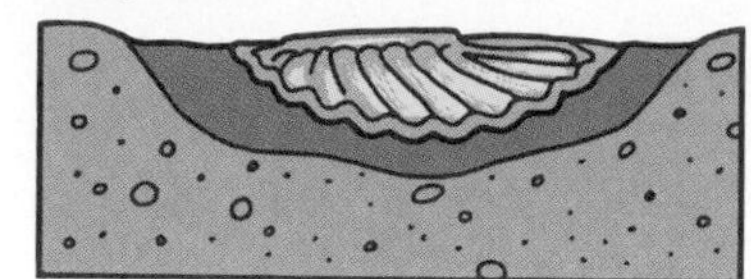

Shell pressed into mud

Step 3

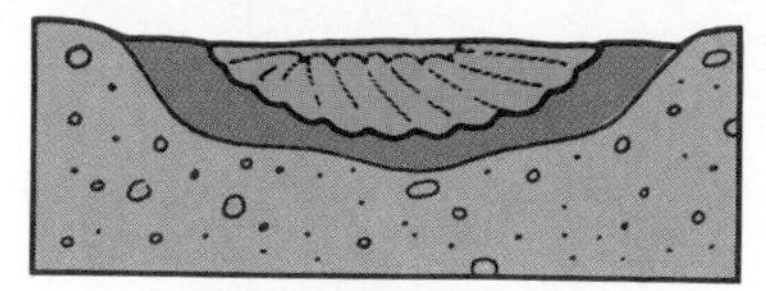

Shell is gone,
but impression is left

a

Step 1

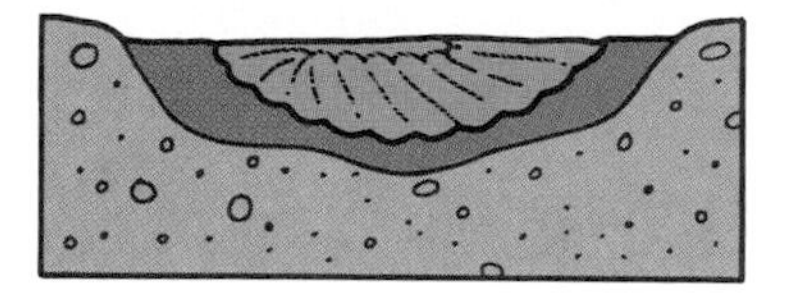

b Mold of shell

Step 2

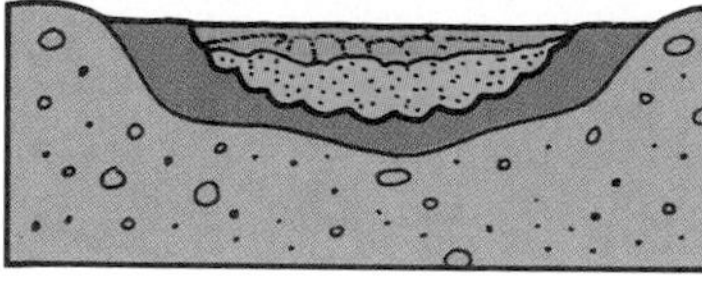

Filling of mold
with sediment

Step 3

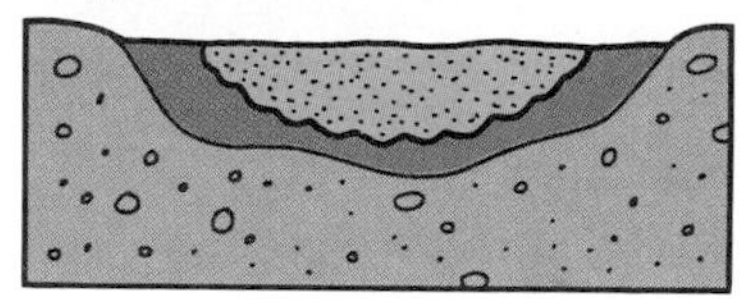

Resulting cast fossil
in original mold

FIGURE 21–7. When the original material is dissolved out of hardened mud, a mold is formed (a). Casts form when sediments fill a mold and then harden (b).

Carbon impressions may form when leaves or soft-bodied animals are buried in soft mud. When the plant or animal matter decays during compaction of the mud, a thin film of carbon is produced. The carbon film preserves the impression of the leaf or animal.

How is Earth history divided into time units?

The fossil record has many examples of the disappearance, or **extinction** (ihk STINGK shun), of some forms of life and the appearance of others. Earth history is divided into time units based on these changes in the fossil record. In other words, divisions of time are based on the appearance and disappearance of certain fossils. These divisions of geologic time are recognized on a worldwide basis. Fossils that are used to identify specific units of geologic time are called guide fossils or **index fossils.** These fossil forms existed for only a short period of geologic history, but their remains are widely distributed.

FIGURE 21–8. How were each of these fossils preserved?

FIGURE 21–9. Ammonoids can be used as index fossils due to the evolution of the suture patterns on their shells.

Ammonoids are extinct shelled organisms that lived in shallow seas from about 395 million to about 136 million years ago. These fossils are good index fossils because of their rapid but stable evolution, widespread occurrence, and ease of recognition. The calcareous shells of ammonoids are highly ornamented by suture lines.

There are three major groups of ammonoids based on the type of sutures present. Goniatites (GO nee uh tites) are the oldest type of ammonoid. They have very gently curved sutures. As these ammonoids evolved, the sutures became more complex. Sutures of ceratites (SER uh tites) are more "wrinkled" than those of ancestral forms. The most advanced form of ammonoid is the ammonite. Ammonites have extremely wrinkled and complex sutures.

The distinct differences in ammonoid shell structures have allowed paleontologists to recognize ammonoid zones that are present worldwide. Thus, relatives ages of rocks in which the fossils are found can be determined.

Facies (FAY sheez) **fossils** are the remains of organisms that existed for long periods of geologic time, but only in certain stable environments. These organisms died when the environment changed. Facies fossils help scientists to determine past environments.

FIGURE 21–10. Ammonoid sutures are lines on the outer surface of the shells.

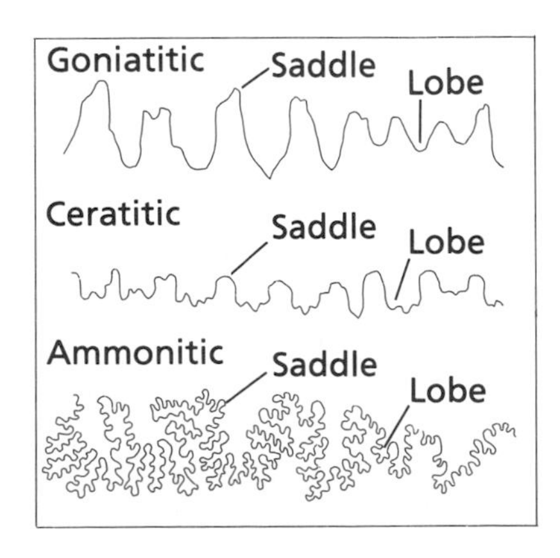

Review

1. List six methods used in geologic dating.
2. How can igneous rocks be used to determine the relative ages of sedimentary rocks?
3. What is a geologic column?
4. What is an unconformity?
5. ★ How can fossils be used to correlate rock layers between two geologic columns?

CAREER

Biostratigrapher

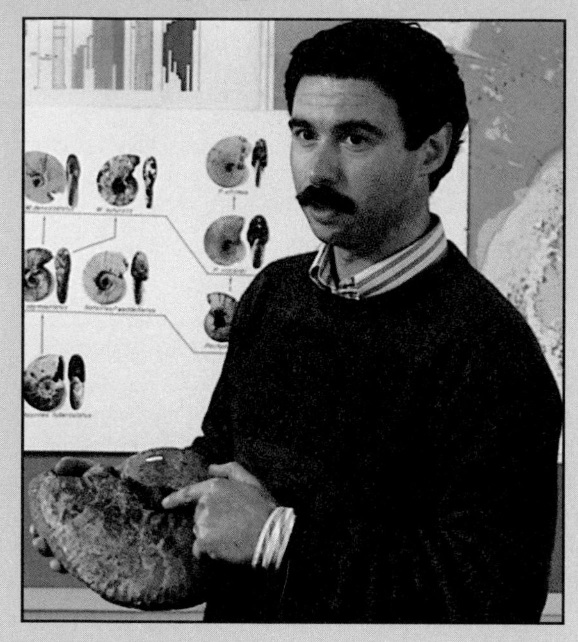

Dr. Carlos Macellari is a biostratigrapher doing research on Earth's history at the Cretaceous/Tertiary boundary.

Biostratigraphers study fossil records from all parts of the world in an attempt to correlate and date events in Earth's past. Based on the fossils present within strata, ages can be assigned to rock units.

Dr. Macellari is studying the evolutionary patterns of ammonites and bivalves during the 70 million year time span of the late Mesozoic Era. He has examined these marine invertebrates from Antarctica and South America to try to explain the mass extinctions that occurred at the end of the Cretaceous.

Dr. Macellari received a masters and doctorate degree in geology. His dissertation research was done in South America and Antarctica.

For career information, write:
The Paleontological Society
U.S. Geological Survey
E/501 National Museum Building
Smithsonian Institution
Washington, DC 20506

GOALS

1. You will learn about radioactive decay.
2. You will learn how scientists determine the absolute age of some rocks.

ABSOLUTE DATES

Geologists use various clues to determine the order of events in Earth's history. Relative dating does not give an exact age of a rock or an event. To determine an exact age, scientists use rocks that contain radioactive elements. These rocks may be used as the "clocks" that tell the absolute time of events in Earth's history.

21:5 Absolute Dating

The rocks in Figure 21–11 show undisturbed layers at the top that are younger than the layers near the bottom. But how much older are the bottom layers? 100 years? 1000 years? 1 000 000 years? Geologists face these same questions when trying to determine the age of rocks using relative dating techniques. How can scientists determine the exact age of these rocks? **Absolute dating** is a method that determines the actual length of time between events.

FIGURE 21–11. The absolute ages of some rock layers can be determined by studying radioactive elements.

Measurement of absolute time must be based on an accurate scale. Units of time are based on actions that are repeated at regular intervals. Earth's time units of days and years are based on the regular motions of rotation and revolution of our planet. But, how do we measure the lifetime of the planet? Is there an activity that regularly repeats itself that we can use to determine time?

Scientists found that there is a repeating activity that spans billions of years. Atoms of radioactive elements form new elements or different isotopes of the same element as their nuclei break apart or decay. During this process, called **radioactive decay,** changes take place in the nucleus of the atom. Some radioactive elements change into other elements by emitting a stream of alpha particles composed of two protons and two neutrons. Other elements decay into different isotopes as neutrons break apart and electrons, or beta particles, are released. Radioactive decay occurs spontaneously. Thus, the fixed rate of radioactive decay can be used as a unit of time to measure the age of Earth and its rocks.

F.Y.I. The first suggestion that natural radioactivity might provide the means for measuring the age of rocks was made by an American physicist and chemist, Bertram Boltwood, in 1907.

FIGURE 21–12. Alpha particles are given off as uranium-238 decays to thorium-234 (a). The decay of thorium-234 to protactinium-234 produces beta particles (b).

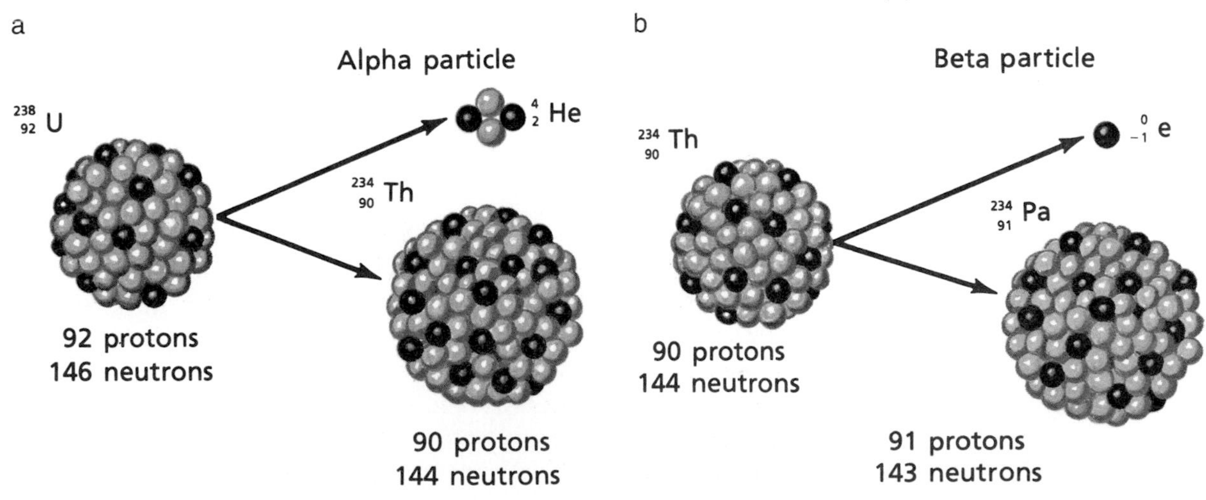

INVESTIGATION 21–1

Constructing an Absolute Time Scale

Problem: How is an absolute time scale constructed?

Materials

list of relative events from the Skill, "Sequencing Events"
metric ruler
paper
pencil

Procedure

1. Determine the age of the oldest student you worked with in the Skill, "Sequencing Events."
2. Draw a straight line on your paper.
3. Divide the line into the same number of equal segments as that student's age. For example, if the oldest student was 13, you should have 13 segments on your line.
4. Place each event from your combined list in the proper place along the time line according to when each actually occurred. If several events occurred on the same day, determine the time of day each event occurred to place it on the line.
5. You have now constructed an absolute time scale that shows exactly how long ago each event occurred.

Analysis

1. How does an absolute time scale compare to the relative list you compiled in the Skill on sequencing events?
2. In what situations is a relative scale useful?
3. Under what conditions is an absolute time scale useful?
4. Describe the location of events on your absolute scale. Do any of the events seem to be grouped together in one area of the line?

Conclusions and Applications

5. Study your absolute scale. Think about what you have learned about Earth's history. Predict where most events will be clustered on a geologic time line.
6. Do more events close together mean that more events actually are occurring? Explain.
7. How is an absolute time scale constructed?

21:6 Radiometric Dating

The method of using the decay of radioactive elements to determine the absolute age of rocks is called **radiometric dating**.The rate of decay has been determined for each radioactive element. The amount of time needed for one-half of the atoms of an element to decay is called the **half-life** of the element. The half-life of uranium-238 is 4.5 billion years. In 4.5 billion years, one-half of the atoms of a mass of this isotope of uranium will have changed to another element. The half-life of radioactive elements varies widely. The half-life of carbon-14, for example, is only 5730 years. The half-life of oxygen-19 is only 19 seconds. New elements formed during the decay process are called **daughter elements.** Daughter elements frequently are also radioactive and decay into still other elements. Eventually, a stable element forms that is not radioactive and the process ends. Lead-206 is the stable daughter element that results from decay of radioactive uranium-238.

In order to use radioactive decay to estimate the age of a rock, the rock must contain measurable amounts of both the radioactive parent element and the stable daughter element. Granite contains minerals such as feldspars that have small amounts of radioactive uranium-238, thorium-232, potassium-40, or rubidium-87. If uranium-238 is present in a rock, the age of the rock can be determined using the following equation.

$$\text{age} = \frac{\text{amount of lead-206}}{\text{amount of uranium-238}} \times (7.6 \times 10^{-9})$$

One gram of uranium-238 produces 7.6×10^{-9} grams of lead-206 per year. Radioactive elements are found in most igneous rocks. Thus, radiometric dating can be used to establish their actual dates of formation.

Recall that sedimentary and metamorphic rocks are made of recycled material. The radioactive elements in

BIOGRAPHY

Marie Curie
1867-1934

Marie Curie went to Paris in 1891 to study physics at the Sorbonne. There, she and her husband, Pierre Curie, began work on separating an unknown element from pitchblende. After eight years, they were able to announce the discovery of a new element, a radioactive element they named "radium." For this discovery they received the Nobel Prize. Curie was the first woman to receive this award.

Time		
0 years	Original amount of carbon-14	
After 5730 years	$\frac{1}{2}$ remaining	$\frac{1}{2}$ decayed
After 11 460 years	$\frac{1}{4}$	$\frac{3}{4}$ decayed
After 17 190 years	$\frac{1}{8}$	$\frac{7}{8}$ decayed
After 22 920 years	$\frac{1}{16}$	$\frac{15}{16}$ decayed

FIGURE 21–13. Carbon-14 has a half-life of 5730 years. After one half-life passes, one-half of the original amount of carbon-14 is decayed. After two half-lives, one-fourth of the original amount of carbon-14 remains.

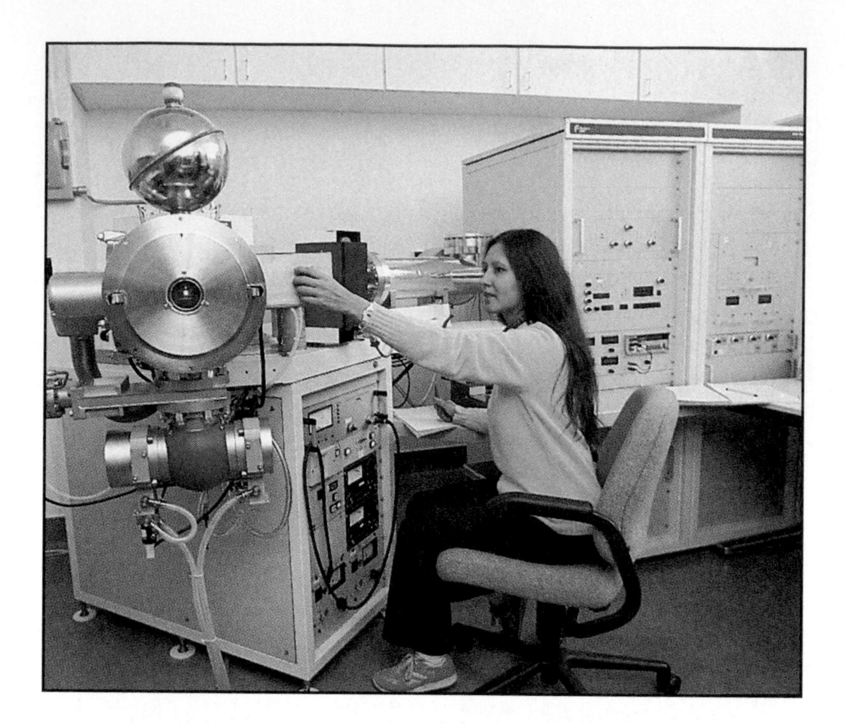

FIGURE 21–14. Instruments that measure the isotopic ratios of a given element help scientists estimate when the rock was formed.

Why is it not possible to determine the date of formation of a metamorphic rock using radiometric dating?

sedimentary rocks may be used to determine the age of mineral grains contained in these rocks, but not the actual time when the minerals were deposited. Slight changes in temperature due to metamorphism can affect the amount of parent and daughter material that remains in a rock. Thus, a date obtained in a metamorphic rock may be the date of the metamorphism that affected the rock rather than the actual age of the rock.

What is radiocarbon dating?

Carbon-14 is a radioactive element that is present in extremely small amounts in all living matter. Carbon-14 is absorbed by plants from carbon dioxide. Plants, in turn, are eaten by animals. After death, the amount of carbon-14 in an organism is no longer renewed by life processes. The carbon-14 decays into nitrogen-14. The approximate time an organism died is found by comparing the amount of carbon-14 in the fossil remains with the amount of carbon-14 in the same amount of living matter. This type of radiometric dating is **radiocarbon dating.**

If organisms and sediments are deposited at the same time, the carbon-14 in the fossils can be used to date the rock. But if an animal dies in a cave or falls into a canyon, the carbon-14 in the fossil remains would be useful only in dating the fossil, not the age of the cave or canyon. Only a relative date could be determined for the cave or canyon.

Radiometric dating has provided a way to calibrate the time scale of Earth history with absolute values. Ages

INVESTIGATION 21–2

Radioactive Decay

Problem: How can absolute age be determined from a model of radioactive decay?

Materials

shoe box with lid
paper clips (100)
pennies (100)
brass fasteners (100)
graph paper
colored pencils (2)

Procedure

1. Place the pennies into the shoe box with all heads up.
2. Place the lid on the box and shake it one time.
3. Remove the lid. Replace the pennies that are now tails up with paper clips. Record the number of pennies remaining in the box in a data table similar to the one shown.
4. Repeat Steps 2 and 3 until all the pennies have been removed.
5. Remove the paper clips from the box. Put an "X" on one of the shorter sides of the box. Place the fasteners in the box.
6. Repeat Steps 2 and 3 until all of the fasteners have been removed. Remove only the fasteners that point toward the "X". Be sure to replace them with paper clips.
7. Plot both sets of data on the same graph. Graph the "shake number" on the horizontal axis and the "number of pennies or fasteners remaining" on the vertical axis. Be sure to use a different color pencil for both sets of data.

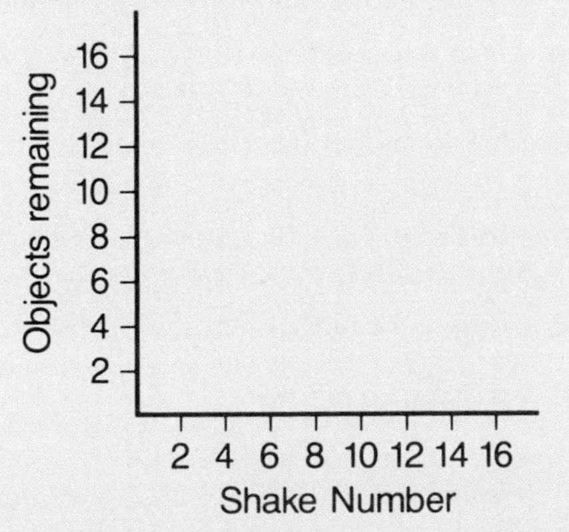

Data and Observations

Shake number	Objects remaining	
	Pennies	Fasteners
0	100	100
1		
2		
12		
13		
14		
15		

Analysis

1. In this model of radioactive decay, what do the coins and fasteners represent? The paper clips? The box? Each shake?
2. What was the half-life for each set of objects?
3. How does the difference between the two objects affect the half-life? Compare the objects to the differences among radioactive elements.

Conclusions and Applications

4. Suppose that you could make only one shake in 100 years. How many years would it take to have 25 objects and 75 paper clips remaining?
5. How can absolute age of rocks be determined?

have been determine for hundreds of thousands of rock masses on Earth's surface. Through the analysis of these data, scientists have determined that Earth is about 4.5 billion years old.

Review

6. Describe how absolute dating is different from relative dating.
7. What are the half-life and stable daughter element of uranium-238?
8. What is the equation used to determine the age of a rock that contains uranium-238?
9. ★ Explain why carbon-14 is not used to determine the absolute age of fossils more than 100 000 years old.

TECHNOLOGY: APPLICATIONS

Recovering Fossils

Earth's history is a very large, complex puzzle. Scientists are working to answer questions about the formation of Earth and the development and evolution of life. This complicated puzzle is different from most because it has pieces missing. The missing pieces are fossils. Scientists rely on fossils to provide evidence about Earth's past.

If fossils are thought to be buried under rock or soil, scientists may use large earth-moving equipment or explosives to remove the unwanted material. Progressively smaller tools are used as the excavation draws nearer to the fossil. Eventually, tiny picks and brushes will be used to remove soil from the fossils.

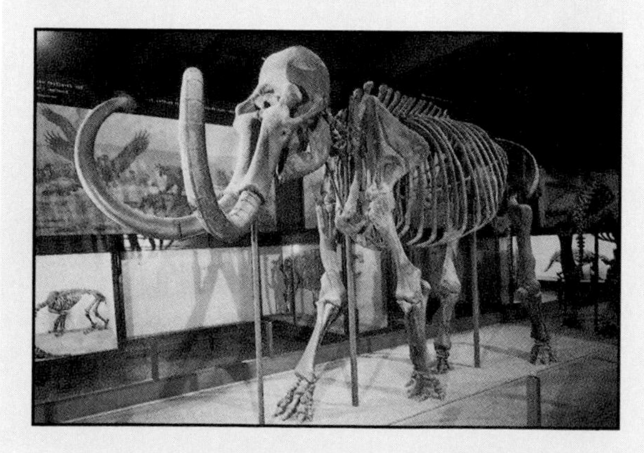

In order for a fossil to be removed or transported, it must be strengthened and protected. Many large, brittle bones are first covered with a layer of shellac. Then, strips of wet newspaper are molded to the fossil. Next, a plaster mixture is used to coat burlap straps, which are then applied to the fossil. This process produces a kind of cast that protects and supports the fossil. This cast can be made even stronger with wooden sticks. The fossil specimen is now ready to be transported to a museum, where it undergoes its final preparation.

Once a fossil is cleaned, it can be prepared in different ways for display in a museum. One method of display for fossils is called the free mount. Skeletons are sometimes used in free mounts to create spectacular displays. The bones are separated and strengthened individually. Using a model of the skeleton, a steel framework is welded to support the skeleton. The bones are then assembled to hide the steel framework, and the skeleton appears to stand by itself.

Chapter 21 Review

SUMMARY

1. The law of superposition and the principles of fossil correlation, cross-cutting, and uniformitarianism help scientists determine the relative ages of rock layers. 21:1, 21:2
2. A sequence of all rock layers in the order in which they were deposited is a geologic column. 21:2
3. Unconformities are important clues to uplift, erosion, and/or non-deposition in a rock sequence. 21:3
4. A fossil is the remains or trace of an organism that has been preserved in Earth's crust. Types of fossils include molds, casts, carbon impressions, and petrified substances. 21:4
5. Index fossils are used to divide geologic time into units. 21:4
6. Absolute dating of geologic time is based on radioactive decay of certain elements. 21:5
7. Radiometric dating is most useful for dating igneous rocks, and is possible if some of the parent material and some of the stable daughter element are present in measurable amounts. 21:6

VOCABULARY

a. absolute dating
b. ammonoids
c. angular unconformities
d. carbon impressions
e. cast
f. daughter elements
g. disconformities
h. extinction
i. facies fossils
j. fossil correlation
k. geologic column
l. geologic map
m. half-life
n. index fossils
o. law of superposition
p. mold
q. paleontologists
r. petrified
s. radioactive decay
t. radiocarbon dating
u. radiometric dating
v. relative dating
w. unconformities
x. uniformitarianism

Matching

Match each description with the correct vocabulary word from the list above. Some words will not be used.

1. method by which certain rocks are identified by fossils they contain
2. unconformities between tilted and horizontal rocks
3. unconformities between two horizontal beds of rock
4. elements formed during the process of radioactive decay
5. the actual length of time between geologic events
6. remains of organisms that existed for relatively long periods of geologic time in stable environments
7. form when soft-bodied animals are buried in soft sand or mud
8. a cavity left in rock when a shell or bone is dissolved
9. scientists who study fossils
10. extinct shelled organisms

Chapter 21 Review

MAIN IDEAS

A. Reviewing Concepts

Choose the word or phrase that correctly completes each of the following sentences.

1. An arrangement of rock layers showing the sequence in which they were deposited is a *(geologic column, law of superposition, geologic map)*.
2. Fossils are most commonly found in *(igneous, metamorphic, sedimentary)* rocks.
3. Uranium is a radioactive element most useful in dating *(igneous, sedimentary, metamorphic)* rocks.
4. People who specialize in the study of fossils are called *(physicists, paleontologists, chemists)*.
5. Bone or wood sometimes can be used for an age determination because it contains *(potassium-40, uranium-238, carbon-14)*.
6. Relative ages for igneous rocks are determined by *(their relationship to sedimentary rocks, radioactive elements, minerals present)*.
7. Fossils used to divide geologic time into small units are *(facies fossils, index fossils, species)*.
8. The age and structural features of surface rocks are shown on a(n) *(geologic map, geologic column, unconformity)*.
9. Casts are identified as being different from the original organism by *(shape, lack of surface characteristics, lack of internal structure)*.
10. An erosional surface between two horizontal rock layers is a(n) *(disconformity, angular unconformity, facies)*.
11. *(Absolute ages, Relative ages, Index fossils)* are determined by studying the positions of layered rocks.
12. The result of the process of radioactive decay is a *(daughter element, half-life, disconformity)*.
13. "The present is the key to the past" is a statement that best represents the principle of *(crosscutting, superposition, uniformitarianism)*.
14. The most accurate method of determining the age of a sedimentary rock is the use of *(radiometric dating, facies fossils, index fossils)*.
15. A fossil that has 1/16 of its original carbon-14 remaining is approximately *(5700, 23 000, 75 000)* years old.

B. Understanding Concepts

Answer the following questions using complete sentences.

16. Describe the principle of crosscutting.
17. What are index fossils?
18. What is the ratio of uranium-238 to lead-206 in a rock that is 4.5 billion years old? Explain how you determined the answer.
19. What type of dating techniques are used most effectively with igneous rocks? Explain your answer.
20. How many half-lives have occurred in a rock containing 1/8 of the original radioactive mineral and 7/8 of the daughter element?
21. What is an unconformity?
22. Identify an unconformity that occurs between two horizontal layers of sedimentary rock.
23. Describe the law of superposition.
24. What is uniformitarianism?
25. What is a geologic column?

C. Applying Concepts

Answer the following questions using complete sentences.

26. How do relative and absolute dating methods differ?
27. Why is the fossil record incomplete?
28. Why is uranium not suitable for dating rocks formed during the last two million years?

29. How would a lava flow between two series of sedimentary layers be useful in determining approximate dates for the layers in contact with the flow?

30. How could you use a geologic map to help you make a fossil collection?

SKILL REVIEW

1. Compare and contrast relative and absolute dates.
2. Construct a simple graph that shows the radioactive decay rate for an element with a half-life of one million years.
3. Describe a sequence of geologic events that led to the formation of the angular unconformity in Figure 21–4.
4. Use the data from Figure 21–13 to construct a graph of amount of carbon-14 versus half-life.
5. Hypothesize why there are no rocks older than about 24 million years in the western part of the geologic map shown in Figure 21–2.

PROJECTS

1. Construct a model of a geologic column for the rock formations in your local area. Show how this column correlates to several other geologic columns in your state or region.
2. Start your own fossil collection. Label each fossil as to type, age, and location found.

READINGS

1. Baylor, Byrd. *If You Are a Hunter of Fossils*. New York: Macmillan, 1984.
2. Colbert, Edwin H. *Great Dinosaur Hunters and Their Discoveries*. New York: Dover, 1984.
3. Revkin, Andrew C. "Heirlooms and Old Air." *Science Digest*. February, 1986, p. 24.

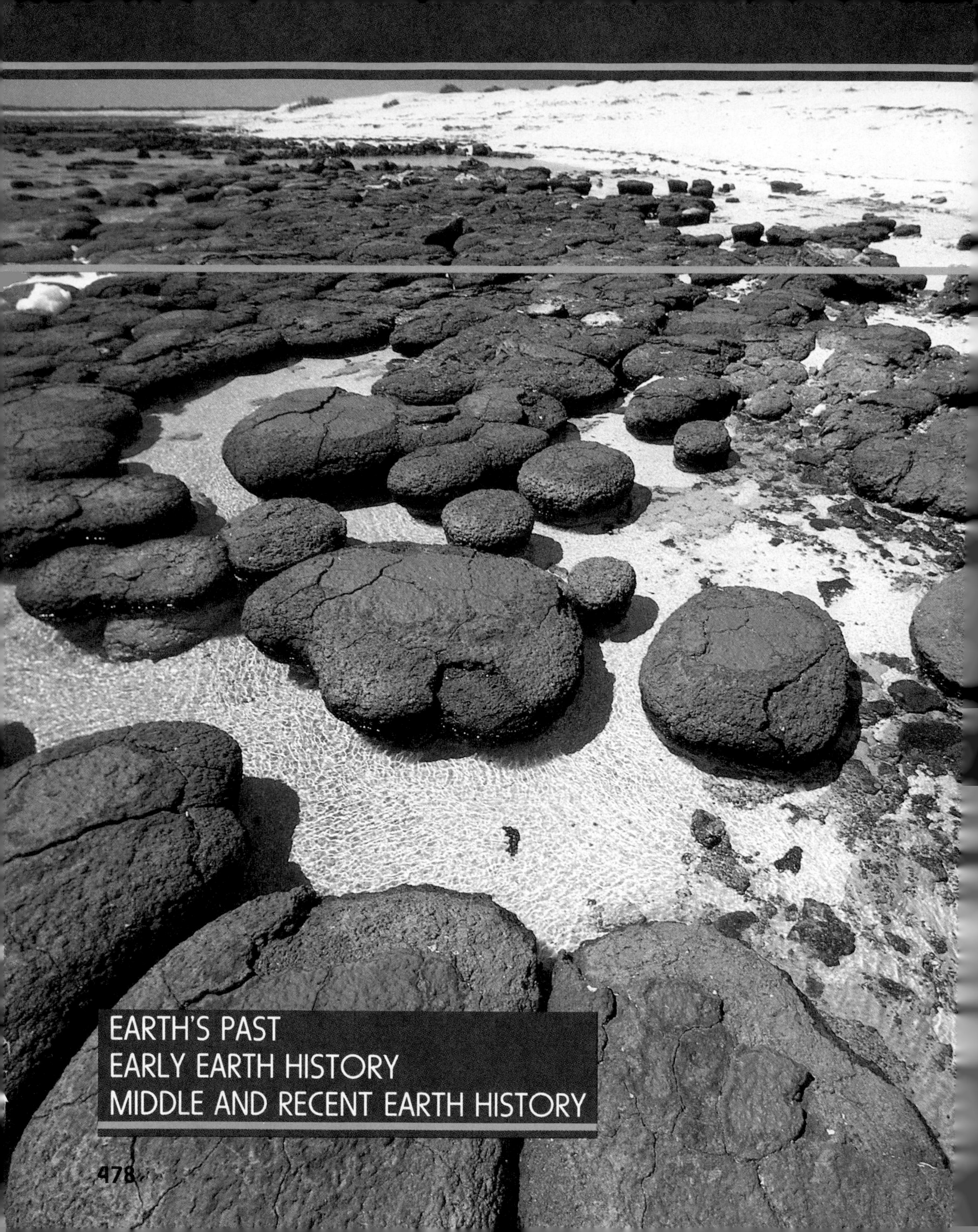

EARTH'S PAST
EARLY EARTH HISTORY
MIDDLE AND RECENT EARTH HISTORY

Geologic Time

Some of the oldest fossils found on Earth are 3.5 billion-year old stromatolites. These structures were formed by the life processes of cyanobacteria. Today, stromatolites are forming in Australia's Shark Bay.

F.Y.I. Cyanobacteria were once known as blue-green algae. Stromatolites are also called algal mounds or columns.

EARTH'S PAST

GOALS

1. You will gain an understanding of the length of geologic history.
2. You will learn how scientists subdivide the span of Earth's history.

Early paleontologists recognized that many of the fossils they studied represented life forms that disappeared from Earth hundreds of million of years before. As these scientists became aware of the vast spans of time represented by Earth history, they recognized a need to subdivide Earth time into units.

22:1 The Geologic Time Scale

Geologists use many relative dating techniques to construct geologic columns. By combining data from all the columns, scientists have been able to outline Earth's history in the geologic time scale. The **geologic time scale** is a chronologic sequence of events in Earth's history.

How is the geologic time scale subdivided?

The time scale is subdivided according to changes in life forms. The largest subdivisions of the scale are **eras.** Most life forms have evolved in the three most recent eras—the Paleozoic (ancient life), the Mesozoic (middle life), and the Cenozoic (recent life). Each era is further divided into **periods.** Periods are divided into smaller units called **epochs.** The period of time between Earth's formation about 4.5 billion years ago and the beginning of the Paleozoic is known as **Precambrian** time. The Precambrian includes two eras, the Archean and the Proterozoic. Little is known about the evolution of life during this time, even though it represents the largest segment of Earth history. Thus, we will not refer to these eras in our discussion of Precambrian time.

Table 22–1

Geologic Time Scale					
ERA	PERIOD	EPOCH	M.Y.B.P.*	MAJOR EVENTS	
				FOSSIL RECORD	PHYSICAL GEOLOGY
CENOZOIC	Quaternary	Recent	0.005	humans appear	ice ages, formation of Grand Canyon
		Pleistocene	2.5		
	Tertiary	Pliocene	7	mammals abundant angiosperms dominant	Alps and Himalayas begin (and continue) to rise
		Miocene	26		
		Oligocene	38		
		Eocene	54		
		Paleocene	65		
MESOZOIC	Cretaceous		136	dinosaurs extinct angiosperms appear	beginning of Rocky Mountains
	Jurassic		190	first birds dinosaurs dominant	orogeny in western North America
	Triassic		225	first mammals conifers and cycads	Atlantic Ocean begins forming, Pangea breaks up
PALEOZOIC	Permian		280	many marine invertebrates extinct	end of Appalachian orogeny, glaciation
	Pennsylvanian		325	coal swamps, first reptiles, insects	shallow seas begin to withdraw
	Mississippian		345	amphibians dominant	glacial advances occur
	Devonian		395	fish dominant first amphibians	continents exist as two landmasses
	Silurian		430	first land plants, corals dominant	warm, shallow seas cover much of North America
	Ordovician		500	first fish, invertebrates dominant	beginning of Appalachian orogeny
	Cambrian		570	trilobites, brachiopods, other marine invertebrates abundant	thick rocks deposited on continental margins
PRECAMBRIAN			4 500 +	bacterialike forms, microfossils	several episodes of mountain building

(*Million years before present)

INVESTIGATION 22–1

Geologic Time Line

Problem: How is an absolute time line constructed?

Materials

adding machine tape
meter stick
pencil
scissors

Procedure

1. Using a scale of 1 millimeter equals one million years (1 mm = 1 000 000 years), measure and cut a piece of adding machine tape equal to the approximate age of Earth (4.5 billion years).
2. Mark one end of the tape "today."
3. Using Table 22–1 as a reference, measure and mark the places on the tape that represent the time when each era began.
4. Examine the events and ages listed in the Data Table. Measuring carefully, include each event on your adding machine tape in the proper place in time. Note that the dates are provided in years B.P. (before present), not *millions* of years B.P.

Analysis

1. Which events were most difficult to plot?
2. How does the existence of humans on Earth compare with the duration of geologic time?
3. Approximately what percent of geologic time occurred during the Precambrian?

Conclusions and Applications

4. Form a hypothesis as to why more is known about recent history than about the Precambrian. How could you test this hypothesis?
5. What can be determined from your time line about the rate at which events have occurred on Earth's surface? Does this rate reflect what has actually happened on Earth? Explain.

Data Table

Earth History Events	
Event	**Years B.P.**
1. today	0
2. astronauts land on moon	20
3. American Civil War	130
4. Columbus discovers America	500
5. Pompeii destroyed	1 900
6. Eratosthenes calculates Earth's circumference	2 100
7. continental ice retreats from North America	10 000
8. beginning of ice age	1 million
9. first humans	2.3 million
10. first elephants	40 million
11. first horse	50 million
12. beginning of Paleocene	65 million
13. dinosaurs become extinct and Rocky Mountains rise	80 million

Earth History Events	
Event	**Years B.P.**
14. beginning of Cretaceous	136 million
15. first birds	150 million
16. beginning of Jurassic	190 million
17. first mammals and dinosaurs	225 million
18. beginning of Permian	280 million
19. first reptiles	325 million
20. coal forests; Appalachians rise	330 million
21. beginning of Mississippian	345 million
22. first amphibians	400 million
23. first land plants; Silurian starts	430 million
24. earliest vertebrate	480 million
25. beginning of Ordovician	500 million
26. early sponges	600 million
27. oldest microfossils (algae)	3 300 million
28. oldest known rocks	3 800 million

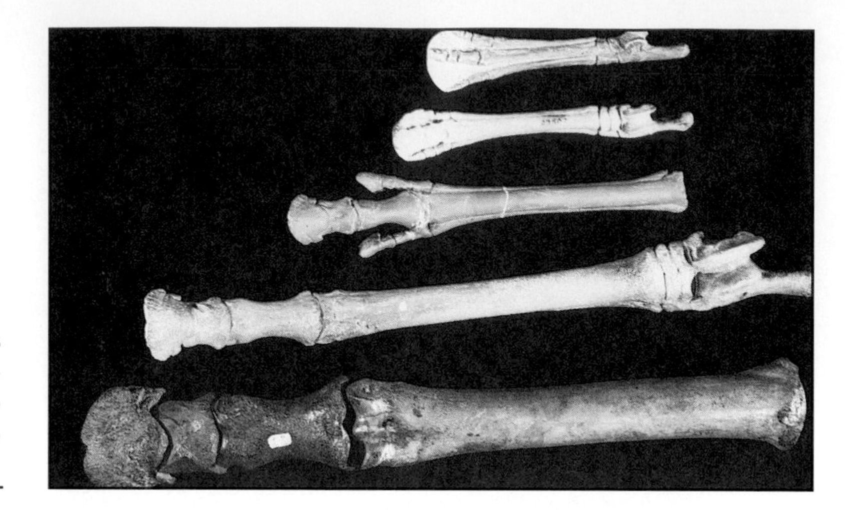

FIGURE 22–1. These fossils show the changes that have occurred as the leg structure of the horse evolved. Which fossil is the oldest?

22:2 Changing Patterns of Life

Life forms and environments on Earth have changed through geologic time. The process of change of life forms through time is known as **organic evolution.** Similar theories of evolution were proposed by both Charles Darwin and Alfred Wallace. Recall that a theory is an explanation backed by supporting evidence. According to this model, species of organisms change through time in direct response to changes in the environment.

F.Y.I. The investigations and observations of Charles Darwin that led to his theory of evolution took place during the cruise of the ship *Beagle* through islands in the South Pacific Ocean from 1831 through 1833.

Many species seem to have evolved slowly through geologic time as Darwin's explanation suggests. A **species** is a group of organisms that can produce fertile offspring in nature. Members of a species are never identical. Some individuals have traits that give them a better chance of surviving long enough to reproduce. These members of a species then pass these advantageous traits to their offspring. Individuals of a species that have less advantageous traits have less of a chance of surviving to reproduce. Thus, the advantageous traits will be concentrated in subsequent generations of a species. The disadvantageous traits will be gradually eliminated. In this way, a species changes through time to become better adapted to its environment, and eventually a new species arises. Thus, Darwin's theory is that the mechanism of evolution is both a natural process and a selective process. This is known as **natural selection.**

What is natural selection?

In an isolated group of organisms, small individual differences, such as variations in size, color, bone structure, or eye position, will be passed from generation to generation. These traits may spread throughout the group and lead to the development of new species. As the horse evolved from the Eocene Epoch to the present, it gradually changed from a four-toed animal to a one-toed, or hoofed, animal. At the same time, it became larger and swifter. This evolution occurred as the species adapted to changes in the environment over 40 million years.

Recently, scientists have discovered evidence that many groups evolve very quickly and then remain relatively unchanged for long periods of time. This spasmodic pattern of evolution is called **punctuated equilibria.** The fossil record of numerous groups of organisms, from dinosaurs to clams, exhibits this type of non-uniform evolutionary pattern.

What is punctuated equilibria?

22:3 Earth History and Plate Tectonics

Many areas of biology and geology provide evidence of evolution. Movements of continents may have affected the evolution of species. Data show that through time, ocean basins open and close. This reduces the different environments in which species may evolve. For example, corals experienced a period of abundance in the Silurian when warm shallow seas covered much of Earth's surface. Later, during the Pennsylvanian, seas retreated from the land. These new environments enabled reptiles to evolve rapidly.

What factors other than biological causes may influence the evolution of a species?

Mountain ranges that form as the result of plate collisions may cause groups of organisms to be isolated from

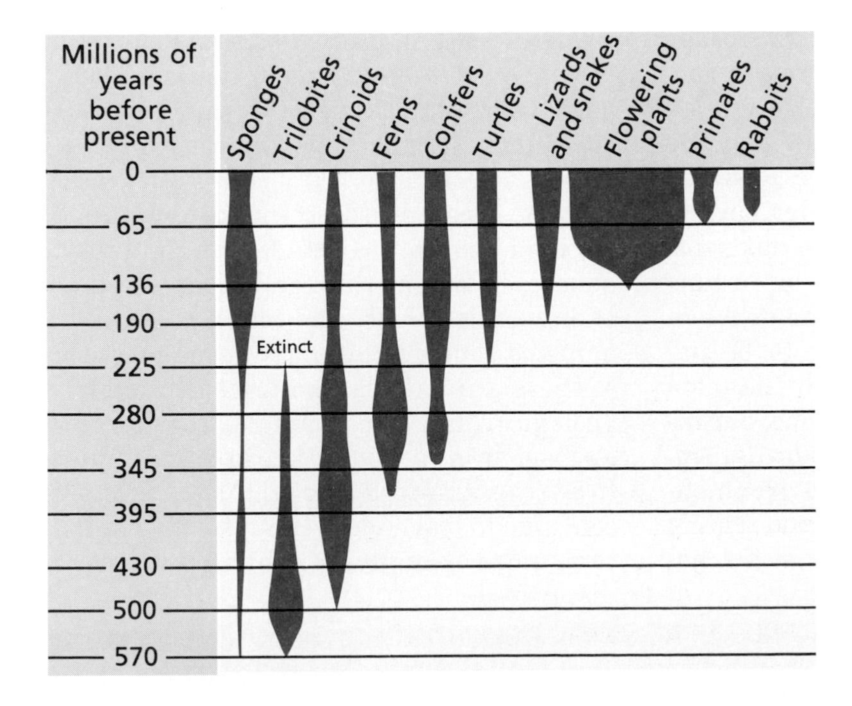

FIGURE 22–2. As continents moved and environments changed, different groups of organisms developed and flourished. The relative numbers of a group living at a certain time is indicated by the width of the band below the group's name.

TECHNOLOGY: ADVANCES

Clues to Evolution

Many areas of biology and geology provide examples of evolution. One area of scientific evidence suggesting that organisms on Earth have evolved is comparative biochemistry. Red blood cells contain a complex protein called hemoglobin. If the hemoglobin of a chimpanzee is compared to that of a human, many similarities are found. If human hemoglobin is compared to the hemoglobin of a dog, fewer similarities in hemoglobin show that humans and chimpanzees are more closely related than humans and dogs, and that they have a more recent common ancestor from which they evolved in their own different ways. Other chemical similarities between chimpanzees and humans suggest a close evolutionary relationship between the two organisms.

Comparative anatomy also offers evidence for evolution. In the study of comparative anatomy, scientists look for similarities between two organs or other body structures that result from similar development from the same or similar embryonic tissues. Homologous structures include the human arm, the foreleg of a horse, the wing of a bat, and the flipper of a porpoise. All are modified limbs that develop from similar embryonic tissues, even though each has a different function. Structures or organs must have both structural and developmental similarities in order to be homologous. Structures with only the same function may not be homologous, but instead indicate evolutionary parallelism, which is caused by similar environmental pressures. Wings of birds and insects are functionally similar but are not homologous.

Comparative embryology is the study of the development of the embryos of different kinds of organisms. Scientists have discovered that embryos of certain animals in the early stages of development are so similar that it is difficult to tell them apart. For example, the embryos of sharks and chickens look almost identical in their early stages of development. It is only in the later stages of development that differences begin to appear. These similarities indicate that sharks and chickens probably had a common ancestor. The differences indicate that they have evolved differently from that ancestor through geologic time.

Within many organisms, scientists have discovered organs that seem to have no function. These organs, called vestiges or vestigial organs, provide evidence that evolution has occurred. For example, the appendix in humans is a vestigial organ that has no known function in a human body. At one time, some ancestor of humans had an appendix that had an important function. As the species evolved, the function carried out by the appendix was no longer needed. The basic body plan stayed the same, however, and thus the appendix remains as a vestigial organ.

In the fossil record, scientists have discovered complete evolutionary sequences in Eocene through Recent horses, in Paleozoic corals, and many other organisms. The evolutionary development of these animals can be traced through geologic time. For example, the horse evolved from a short, four-toed animal to a large, one-toed animal. As scientists continue to learn more about Earth's history, more examples of evolution will be discovered.

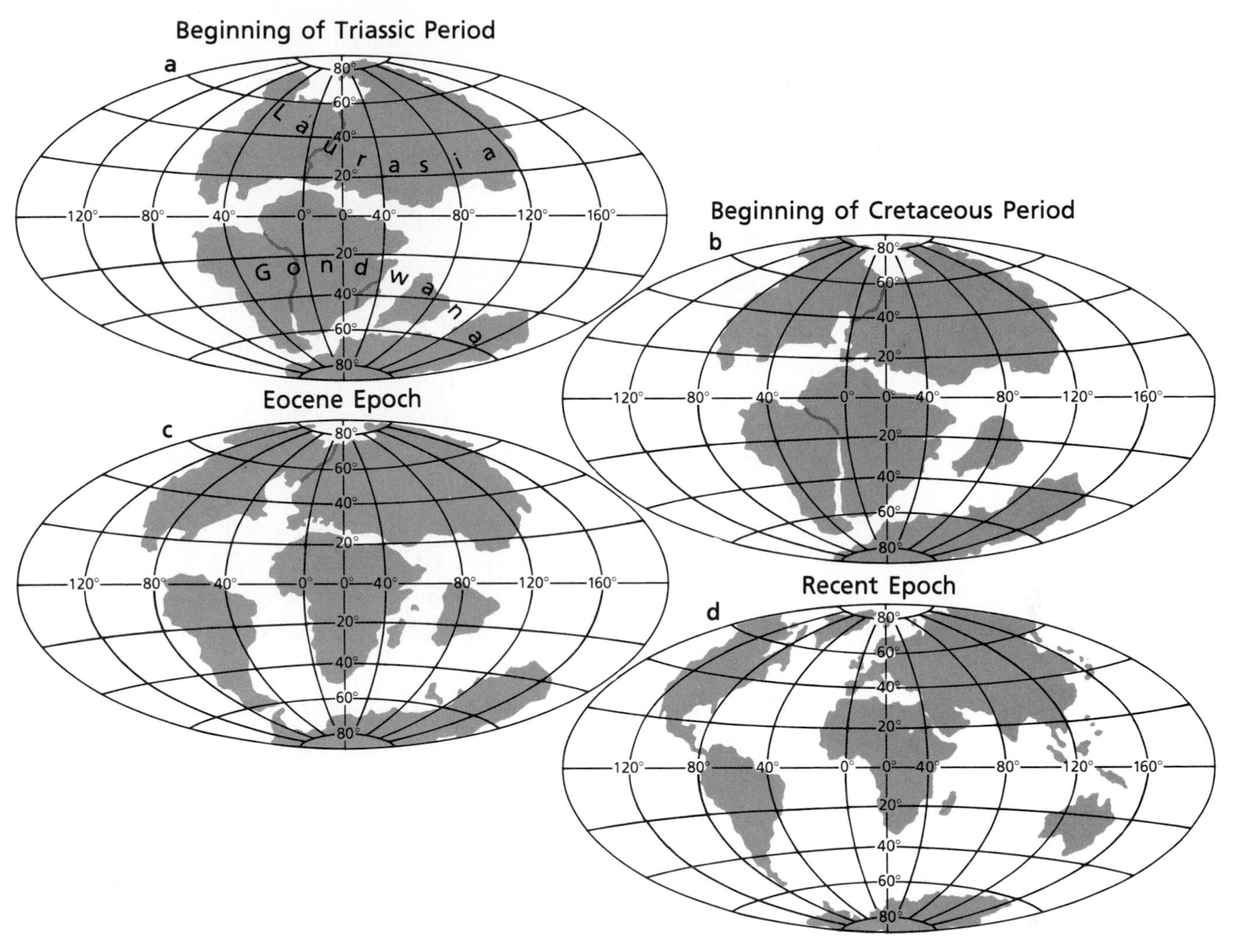

FIGURE 22–3. The breakup of Pangaea caused isolation of different populations of organisms. How did this affect organisms?

one another. This barrier would allow individual differences to become widespread in a population. Eventually, isolated populations of the same species may become so different that they can no longer produce offspring with one another. At this point, the original species will have evolved into two new species.

How has the movement of continents influenced evolution?

At several points during Earth's history, the continents have been joined into one or more landmasses. As these landmasses broke apart, populations of organisms were isolated and thus evolved separately. When continents later were joined, better adapted species caused the extinction of less adapted species. At the beginning of the Mesozoic Era, reptiles were just beginning to dominate other forms of life. At that time, present-day continents formed a single landmass called Pangaea. Pangaea began to break apart, forming Laurasia and Gondwana. **Laurasia** was composed of what is now North America, Europe, and Asia. **Gondwana** was made of what is now South America, Africa, Australia, Antarctica, and part of India. Further movement led to the present locations.

Review

1. Define organic evolution.
2. What is punctuated equilibria?
3. How did the breakup of Pangaea affect evolution?
4. Identify the major subdivisions of Earth history.
5. ★ The breakup of Pangaea took over 200 million years. Is it likely that there have been other landmasses at other times in the past? Explain your answer.

EARLY EARTH HISTORY

GOALS

1. You will learn about early Earth history.
2. You will learn how fossils are used to interpret Earth's past.

Precambrian time represents almost 90% of Earth's history. Life was abundant in Precambrian oceans. However, it was not until about 570 million years ago that organisms began to develop hard body parts and internal support structures. This marks the beginning of the Cambrian Period, the first period of the Paleozoic Era.

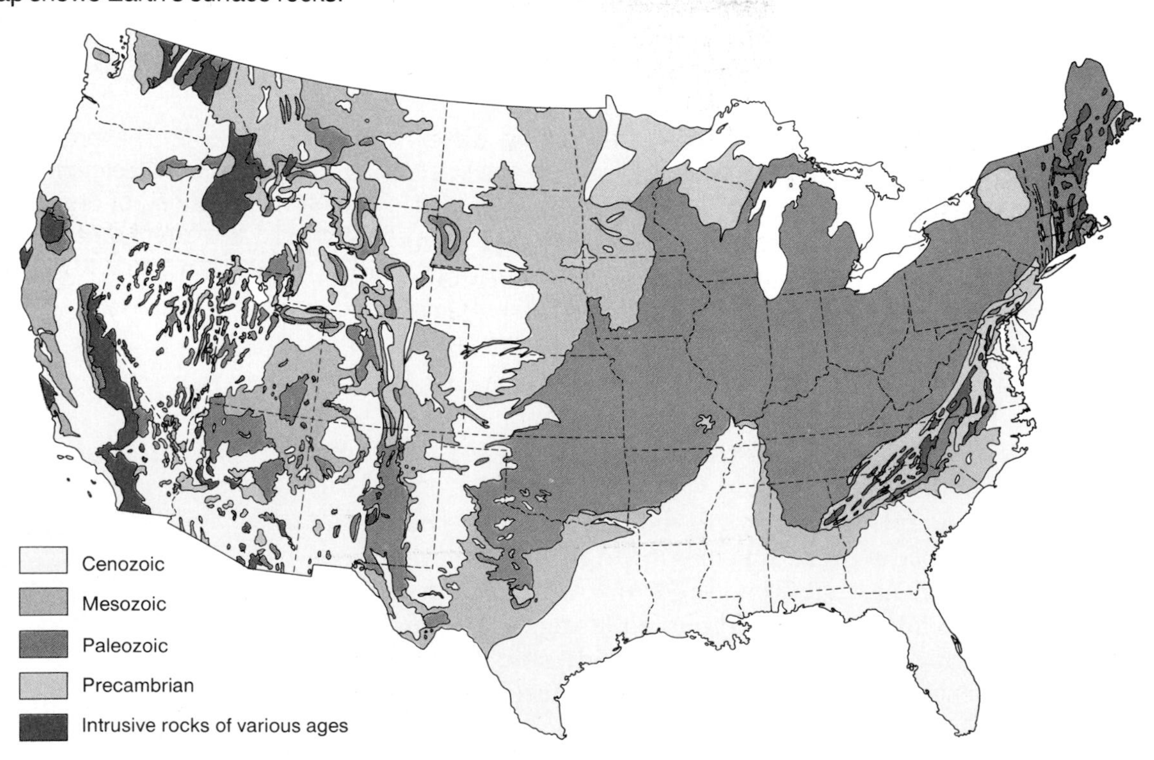

FIGURE 22–4. This geologic map shows Earth's surface rocks.

FIGURE 22–5. This Landsat image shows the shield areas of North America.

22:4 Precambrian Time

What is the span of Precambrian time?

Precambrian time lasted from Earth's beginning about 4.5 billion years ago to about 600 million years ago. However, because many Precambrian rocks have been buried, eroded, or metamorphosed, the first billion years of Earth's history is not well known.

The nucleus of each continent consists of Precambrian rocks, most of which are metamorphic or igneous. These areas of the continents are known as **shields.** Earth's oldest dated rocks, found in Greenland and Africa, are about 3.5 to 3.8 billion years old. In North America, the Canadian Shield is about 2.5 billion years old.

Radiometric dating indicates that there were many small landmasses in the Precambrian. Plate movements were more rapid than in later geologic time. There is also evidence of glaciation during Precambrian time.

The earliest records of life on Earth are buried in Precambrian rocks. The oldest fossils discovered to date are microfossils of ancient cyanobacteria. Microfossils are fossils that are so small they must be studied with a microscope. The fossil cyanobacteria found in western Australia are nearly 3.5 billion years old. The discovery of these single-celled organisms led scientists to propose that even simpler life forms may have formed on Earth as early as 4.0 billion years ago.

Cyanobacteria, such as those discovered in Australia, have been preserved as fossils called **stromatolites** (stroh MAT uh lites). Oxygen released by these plantlike organisms combined with iron in ocean water to produce Earth's great reserves of iron ore. Precambrian stromatolites have been identified in shield rocks located in Africa, Asia, Europe, and North America as well as Australia.

FIGURE 22–6. Organisms similar to cyanobacteria were abundant billions of years ago.

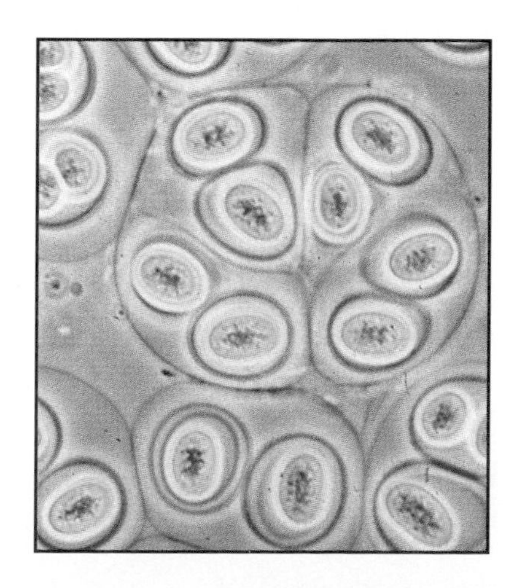

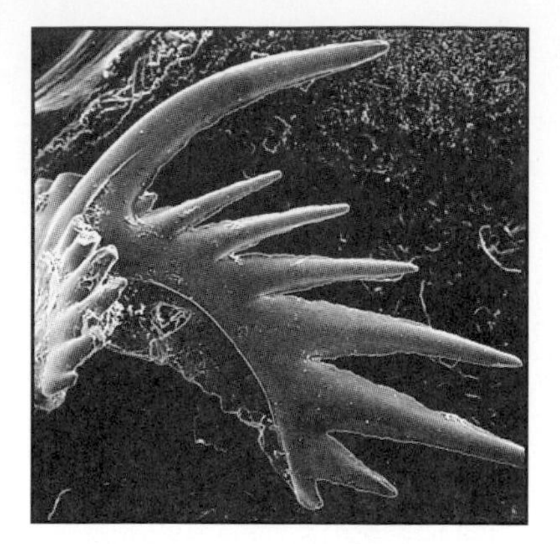

FIGURE 22–7. Conodonts are microfossils that were abundant during the Paleozoic Era.

F.Y.I. The Paleozoic Era is subdivided into seven periods. Each period is named for the geographic area where its rocks were first studied.

22:5 The Paleozoic Era

The **Paleozoic Era** began about 570 M.Y.B.P when species with hard parts that are easily fossilized first evolved. This abrupt appearance of invertebrate marine life in the fossil record included animals with body coverings like those of modern shrimp and crabs. Other Paleozoic invertebrates had hard outer coverings. **Trilobites** (TRI luh bites) and **brachiopods** (BRAY kee uh pahdz) are characteristic fossils of the Paleozoic. Trilobites are extinct, but some brachiopods still exist today. In fact, most of today's basic animal groups began to evolve during the Paleozoic.

During the Paleozoic, the existing landmass broke into at least three continents in the Northern Hemisphere and one continent in the Southern Hemisphere. Shallow seas invaded the continents and provided environments in which marine invertebrates flourished.

Life of the Cambrian Period has been well documented in a rock formation known as the Burgess Shale. This deposit formed at the base of a steep reef where mud quickly buried and preserved dead organisms. Carbon impressions of soft-bodied arthropods, jellyfish, sea anemones, and worms were preserved.

Uplift of the Appalachians began as the European and North American plates joined during the Ordovician and continued to the Permian. Fish appeared during the Ordovician and became the dominant occupants of the sea by the Devonian. One kind of fish evolved a lung that allowed it to survive out of water. This type of fish also had fins that would propel its mass on land. Because the

FIGURE 22–8. Crinoids (a) flourished during the Paleozoic. These invertebrates had flowerlike bodies supported on long stalks. Trilobites (b) were extinct by the end of the era. Some species of brachiopods (c) exist today.

a

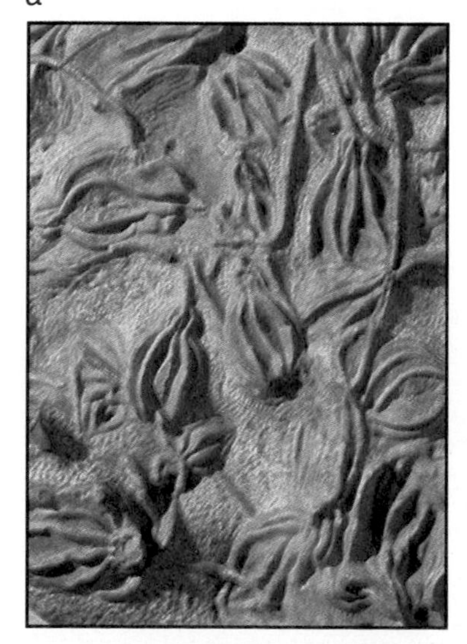

b

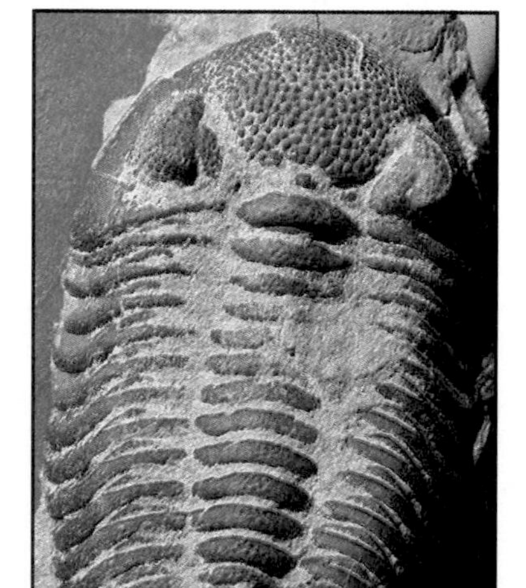

c

PROBLEM SOLVING

Paleozoic Puzzle

Neila enjoyed finding and collecting fossils. She had investigated many of the rock outcrops in her city and had begun an excellent collection of local fossil types. Her favorite fossil was a particular species of brachiopod known as a *Mucrospirifer*. She identified her fossil using pictures and descriptions from a book on Paleozoic fossils.

While on a trip with her family to visit relatives in another state, Neila found what seemed like the same type of *Mucrospirifer* fossil in a rock formation on her aunt's farm. What could Neila say about the rocks in which she found both fossils?

fish could breathe air and move on land, it is considered to be the ancestor of amphibians (am FIHB ee unz). Amphibians first appeared during the Devonian. Most **amphibians** live on land, but they must return to water to reproduce. Their eggs, which have no protective covering, must be laid in water. Even adult amphibians must spend part of their time in water, otherwise they lose their body fluids. Modern amphibians include species of toads, frogs, and salamanders.

Why is an ectotherm called "cold blooded?"

During the Mississippian, amphibians had become numerous and dominated life on land. One variety of amphibian evolved an egg with a hard covering. By the Pennsylvanian, this type of animal had evolved into what we call a reptile. **Reptiles** do not need to return to water to produce their young. Reptiles have a skin composed of hard scales that prevents loss of body fluids. Thus, they are adaptable to a variety of environments. Like fish and amphibians, reptiles are ectotherms. The body temperature of an **ectotherm** changes with that of the environment. Such animals are often said to be "cold blooded."

During the Pennsylvanian, inland seas were cut off from the ocean. Swamps similar to the Florida Everglades covered what is now Pennsylvania, Ohio, Michigan, Illinois, Oklahoma, and part of Kansas. When the thick

FIGURE 22–9. Swamp vegetation such as *Lepidodendron* flourished during the Pennsylvanian. This vegetation later became coal. Shown here is part of *Lepidodendron's* trunk.

FIGURE 22–10. The rock layers exposed by the meanders of the San Juan River were deposited during the Paleozoic.

swamp vegetation died and was buried, it became the coal fields that are mined today.

What mountains formed when Europe collided with Asia?

Near the end of the Permian, continental plates again drifted together. Laurasia once more joined Gondwana, forming mountains. The Hercynian (hur SIHN ee un) Mountain belt rose in Europe. The Ural Mountains formed as the European and Asian plates collided. The uplift of mountains caused the seas to drain away and interior deserts spread over much of the United States and parts of Europe. Vast salt beds were deposited in what is now Germany and in west Texas as the seas evaporated. In the Southern Hemisphere, glaciers covered the southern parts of South America and Africa, as well as India and Antarctica. These events affected the oceans and the continents. The events also caused mass extinctions of both land and sea animals at the end of the Paleozoic.

Review

6. Why is information about Precambrian time sparse?
7. What types of life forms existed during Precambrian time?
8. What major change separates Precambrian time from the Paleozoic Era?
9. How do amphibians and reptiles differ?
10. Explain how fish may be considered ancestors of amphibians.
11. What is an ectotherm?
★ **12.** Describe the events at the end of the Permian that ended the Paleozoic. What effect did these events have on Paleozoic life forms?

MIDDLE AND RECENT EARTH HISTORY

The mass extinctions near the end of the Paleozoic Era were due to drastic changes in environments. Shallow seas contracted. Climates changed. Pangaea brought together formerly separated organisms that now competed with one another. Thus, life forms changed drastically as Earth grew older.

GOALS

1. You will learn about some of the different species that evolved during the Mesozoic and Cenozoic Eras.
2. You will learn about some of the major landform changes that occurred during the Mesozoic and Cenozoic Eras.

22:6 The Mesozoic Era

The **Mesozoic Era** began about 200 M.Y.B.P. At this time, Laurasia and Gondwana again became separated by a sea. Land animals appear to have moved from one continent to another, so a land bridge must have existed. Species that had survived the massive extinctions of the late Paleozoic began to adapt to Mesozoic environments.

F.Y.I. The Mesozoic Era includes three geologic periods—the Triassic, Jurassic, and Cretaceous.

By the Jurassic, reptiles were the dominant life form in land, air, and sea. One flying reptile, the pterosaur, discovered preserved as a fossil in the Big Bend area of west Texas, had a wingspan of about 15.5 meters. Dinosaurs are probably the best-known reptiles. Some dinosaurs were huge. The largest ones had masses of at least 17 metric tons! *Tyrannosaurus* (tuh ran uh SOR us) was one of the largest carnivorous, or meat eating, land animals that ever lived. It stood about six meters tall and was 15 meters long. Dinosaurs have always been considered to be ectotherms, like all reptiles. A current theory

What was the dominant life form in the Jurassic?

FIGURE 22–11. These Mesozoic leaves look much like present-day tree leaves.

FIGURE 22–12. Reptiles first appeared in the Paleozoic.

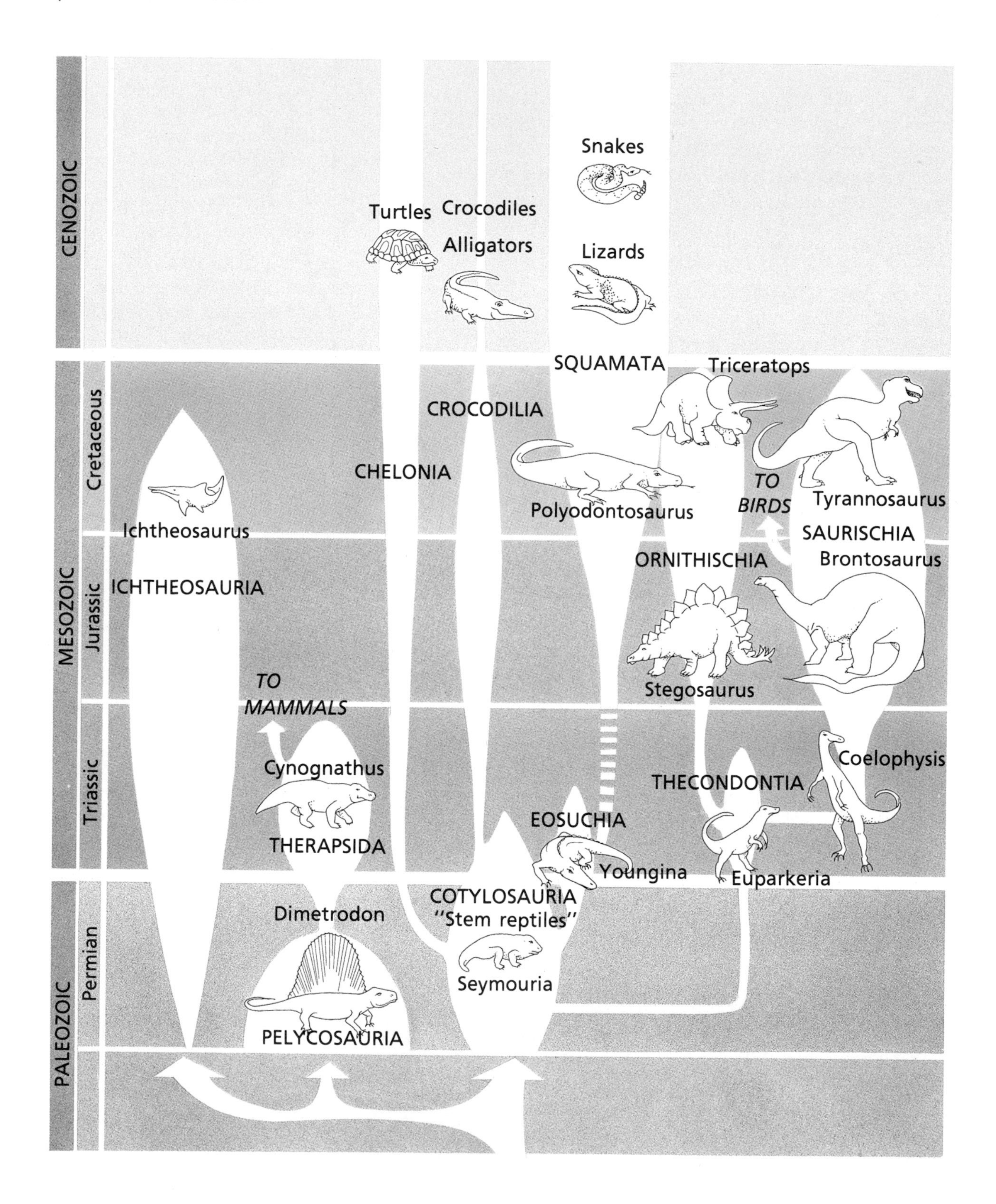

however, suggests that they may have been endothermic ancestors of birds. **Endotherms** have a constant body temperature and are sometimes said to be "warm blooded." A Jurassic fossil with birdlike feathers and a reptilelike skeleton is one of the clues that suggests a close relationship may have existed between dinosaurs and birds. Birds and mammals had also appeared by the middle of the Mesozoic. The species in these two groups are endothermic.

By the Cretaceous, seas spread inland and covered many areas of Europe and the Americas. These changes in environments caused great changes in the plant and animal life. Plants of the Mesozoic had to adapt to an unstable environment. **Angiosperms** (AN jee oh spurmz), the flowering plants, developed during the Cretaceous. These plants had seeds that could survive changing environments because of a hard outer covering. Ammonoids had evolved during the Mesozoic. They could not adapt to the changing environments and became extinct by the end of the Cretaceous. Recall that ammonoids had distinct markings on their shells that make them excellent index fossils of the Mesozoic Era.

By the end of the Mesozoic, the Americas began to separate from Gondwana. Eurasia survived as a unit. North and South America, Antarctica, Australia, Africa, and India became separate plates. As the breakup progressed, seas drained off the land, volcanic activity increased, and mountain building waxed and waned. Dinosaurs and many other life forms became extinct or greatly reduced in numbers. Today, only turtles, snakes, lizards, crocodiles, and alligators survive among the reptiles. It is estimated that at the close of the Mesozoic Era,

BIOGRAPHY

James Hutton
1726–1797

James Hutton, a Scottish scientist, and his followers were called "Plutonists." They were the first to recognize that some of Earth's rocks were formed under intense heat. Today, we call these rocks igneous rocks. Hutton also set forth the principle of uniformitarianism, which stated that Earth had changed in the past and would continue to change slowly in the future.

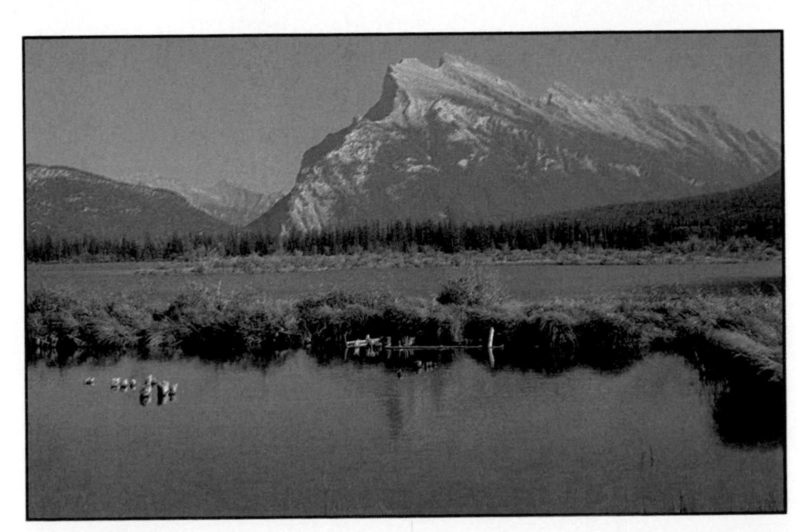

FIGURE 22–13. Mt. Rundle in Alberta, Canada, was uplifted in the late Mesozoic.

about 15 percent of all animal species that existed up to that time became extinct.

What may have caused the mass extinctions that occurred at the end of the Mesozoic?

No one is certain what caused the mass extinctions that ended the Mesozoic. Certainly, major changes in environments and climates were involved. One recent hypothesis, supported by the discovery of a rare element in late Mesozoic rocks, is that a meteorite struck Earth. The impact is believed by some geologists to have led to severe climatic changes. These changes caused the extinction of all but the most adaptable species. However, there is fossil evidence that dinosaurs on different parts of Earth became extinct at different times. These various hypotheses, among others, are being tested by paleontologists today.

22:7 The Cenozoic Era

F.Y.I. The Cenozoic Era is comprised of two major periods—the Tertiary (65-2.5 M.Y.B.P.) and the Quaternary (the last 2.5 million years).

The **Cenozoic Era** began about 65 M.Y.B.P. Early in the Cenozoic, great flows of basalt formed the Columbia River Plateau in the northwestern United States. The Rocky Mountains were uplifted. At about the same time, the Coast Ranges in North America, the Andes Mountains in South America, and the volcanoes of the Cascade Range, including Mount St. Helens, formed. Greenland, part of the American plate, began to drift away from Norway. This opened a water passage between the Arctic Sea and the Atlantic Ocean.

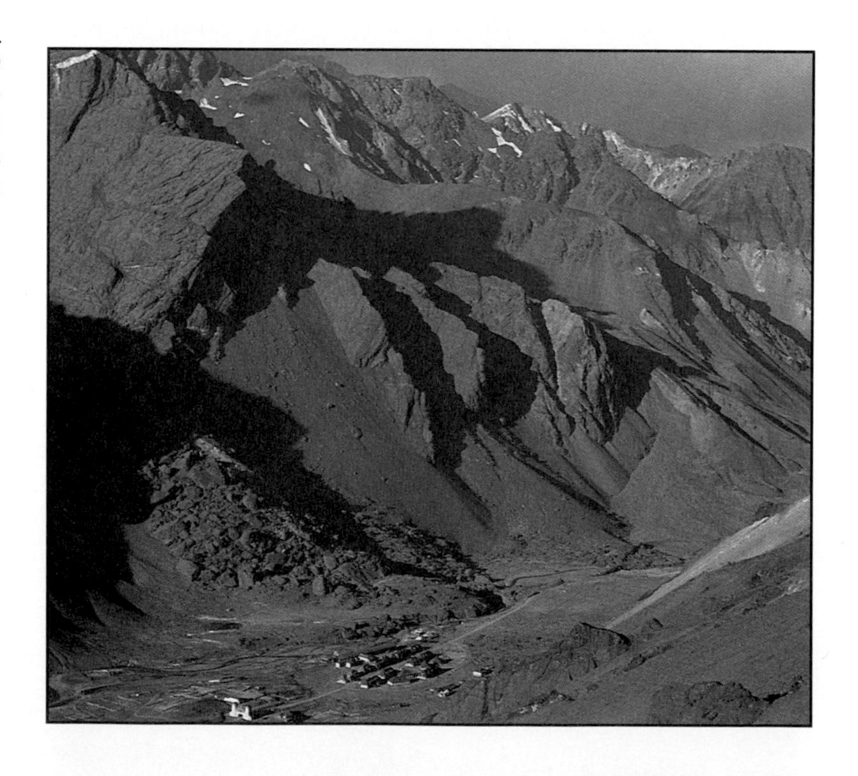

FIGURE 22–14. The rocks of the Andes Mountains were deposited during the Mesozoic and Cenozoic Eras. These layers were uplifted to form mountains late in the Cenozoic.

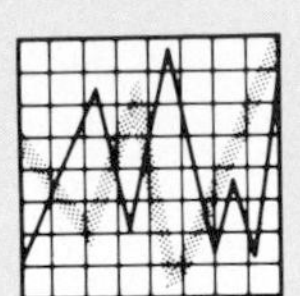

SKILL

Interpreting a Cross Section

Problem: How do you interpret a geologic cross section?

Materials

Figure 22–15
paper
pencil

Background

A geologic cross section is a drawing of a vertical or side view of a rock outcrop.

Procedure

1. Carefully study the geologic cross section shown in Figure 22–15.
2. Answer the questions based on your knowledge of earth science, the information on Earth history provided in Table 22–1, and the relationships among the rocks shown in Figure 22–15.

Questions

1. During what era and period did rock unit A form? What type of rock is unit A?
2. Which of the following fossils are most likely to be found in rock unit B—earliest reptiles, earliest fish, earliest amphibians, or earliest land plants?
3. Which two rock layers appear to be deposited on unconformities?
4. During which era did rock unit D most likely form? Explain your answer.
5. Which of the following fossils are most likely to be found in rock unit H—first birds, trilobites, or earliest land plants?
6. Which rock unit was forming as Pangaea began to break up?
7. During which geologic era and period did rock unit E form?
8. Which of the following events occurred most recently—deposition of rock layer A; deposition of rock layer G; erosion of rock layer F; or erosion of rock layer J?
9. In which rock unit are fossils least likely to be found? Explain your answer.
10. Which of the following is the only probable age of rock unit I—30 million years, 150 million years, 200 million years, or 340 million years?

FIGURE 22–15.

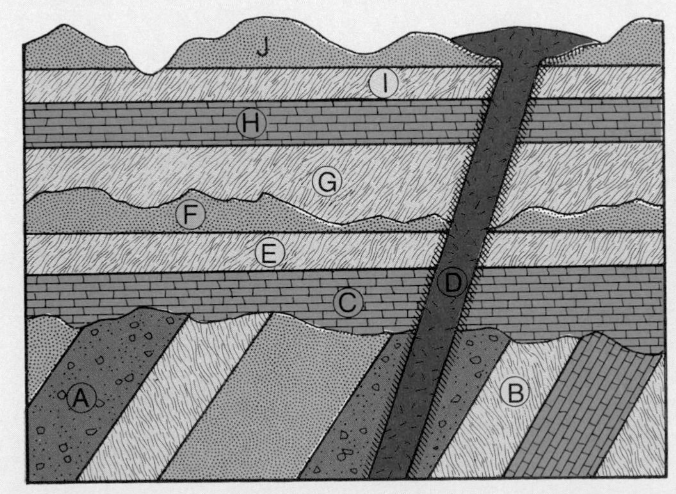

Rock type

Sedimentary Rock

Igneous Rock

Contact Metamorphism

Rock unit	Age (m.y.)
A	320
B	480
C	220
E	189
J	70

Key

INVESTIGATION 22–2

Making a Fossil

Problem: How can you make a "fossil"?

Materials

paper towels
several heavy books
baking soda
stems, leaves, flowers, or other organic material

Procedure

1. Make a stack of 4 or 5 squares of paper toweling.
2. Sprinkle the top of the pile with baking soda.
3. Arrange a leaf, a flower, or any other organic material so that it lies flat on the baking soda.
4. Cover the organic material with a thin layer of baking soda, and place another 4 or 5 squares of paper toweling over it.
5. Place a pile of heavy books on the paper toweling.
6. Leave the books undisturbed on the toweling for 2 or 3 days.
7. Remove the books, the upper layers of toweling, and the top layer of baking soda.

Analysis

1. Describe the color of the "fossilized" leaves or flowers. How does the color compare to fresh organic matter?
2. How are these "fossils" different from imprints?
3. If the materials had been buried in mud, what would eventually have happened to the plant tissue?
4. What is the purpose of the baking soda?

Conclusions and Applications

5. What two elements are missing to make these true fossils?
6. How can you make a "fossil"?

FIGURE 22–16. Strata exposed in the White River Badlands of South Dakota were deposited during the Cenozoic.

The Alps of Europe were uplifted during the Cenozoic. At the same time, Africa drifted into Eurasia. The collision of the Indian and Arabian plates with the eastern edge of the Eurasian plate formed the Himalayas. Australia and South America were separated from Antarctica and drifted into their present positions.

What formed the Himalaya Mountains?

Isolation during the Cenozoic seems to have changed the development among the animals found in Australia and South America. Marsupials (mar SEW pee ulz), like the kangaroo, became the dominant type of animal. They are still common in Australia. For a while, marsupials were common in South America. But in the late Cenozoic, the Isthmus of Panama formed a land bridge between North and South America. Animals from these continents could migrate from one place to the other. More advanced mammals from North America moved south and occupied South America. The unusual groups of animals living in South America disappeared. The armadillo (ar muh DIHL oh), however, moved north out of South America. It now inhabits much of the southwestern United States.

The dominant marine invertebrates in the Cenozoic include members of the starfish, snail, and clam groups. Mammals expanded into niches left vacant by the extinct reptiles. Mammals and a few other vertebrates bear their young alive. Early protection of their young helps mammals survive in many different environments.

A major event of the Pleistocene Epoch was the invasion of the Northern Hemisphere by continental glaciers. The glaciers repeatedly advanced, melted, and then readvanced during a two million year period. It was during the Quaternary Period that evidence of a new species,

FIGURE 22–17. The armadillo migrated from South America during the Cenozoic and now inhabits the southwestern United States.

FIGURE 22–18. The fjords of New Zealand were carved by glaciers during the Cenozoic.

Homo sapiens, appeared. *Homo sapiens* are mammals that are highly adaptable. Humans are *Homo sapiens.*

Review

13. When did angiosperms, or flowering plants, first appear in the fossil record?
14. When did the Cenozoic Era begin?
15. During what period and epoch did the North American ice ages occur?
★ **16.** What might have caused the mass extinctions at the end of the Mesozoic?

CAREER

Museum Technician

Much work must be done from the time a fossil is found to the day it goes on display in a museum. Transforming scientific information and fossils into an understandable and visible form is the job of Lisa Hartford. She is a museum technician. Ms. Hartford works with designers, artists, and paleontologists to create fossil displays. Constructing these displays is a very complex and detailed process. Ms. Hartford uses her artistic abilities and bachelors degree in geology to aid her in her fascinating career.

For career information, write:
American Association of Museums
1055 Thomas Jefferson St. NW
Washington, DC 20007

Chapter 22 Review

SUMMARY

1. The geologic time scale is a chronologic sequence of events in Earth's history. 22:1
2. The process of change of life forms through time is called organic evolution. 22:2
3. The location and the shape of continents have changed throughout Earth's history, affecting the evolution of organisms. 22:3
4. Precambrian time lasted for 3.9 billion years. 22:4
5. During the late Paleozoic Era, amphibians dominated life on land, the earliest reptiles developed, and many mountain ranges formed on Earth. 22:5
6. Great changes in plant and animal life marked the Mesozoic Era. Modern flowering plants, birds, and mammals first appeared in the Mesozoic. 22:6
7. The Cenozoic Era began about 65 million years ago. The late Cenozoic saw the advance of continental glaciation over parts of North America and Europe, as well as the first records of humans. 22:7

VOCABULARY

a. amphibians
b. angiosperms
c. brachiopods
d. Cenozoic Era
e. ectotherm
f. endotherms
g. epochs
h. eras
i. geologic time scale
j. Gondwana
k. Laurasia
l. Mesozoic Era
m. natural selection
n. organic evolution
o. Paleozoic Era
p. periods
q. Precambrian
r. punctuated equilibria
s. reptiles
t. shields
u. species
v. stromatolites
w. trilobites

Matching

Match each description with the correct vocabulary word from the list above. Some words will not be used.

1. change of species through time in response to changes in the environment
2. chronologic sequence of events in Earth's history
3. landmass composed of what is now North America, Europe, and Asia
4. organisms with skins made of hard scales
5. land-dwelling organisms that must return to water to reproduce
6. era during which the Ural Mountains formed
7. flowering plants
8. accounts for about 90% of Earth's history
9. major units of geologic time
10. spasmodic or irregular rate of evolutionary change

Chapter 22 Review

MAIN IDEAS

A. Reviewing Concepts

Choose the word or phrase that correctly completes each of the following sentences.

1. Earth's oldest rocks are *(Precambrian, Paleozoic, Cenozoic)* in age.
2. Trilobites and *(reptiles, mammals, brachiopods)* are Paleozoic index fossils.
3. Flowering plants developed during the *(Cenozoic, Mesozoic, Paleozoic)*.
4. Turtles, snakes, lizards, crocodiles, and alligators are survivors of the once-dominant group of *(amphibians, reptiles, angiosperms)*.
5. Early Paleozoic life included *(invertebrates, angiosperms, dinosaurs)*.
6. During the late Cenozoic, there is evidence for *(the presence of humans, the last of the dinosaurs, ammonites)*.
7. Ice ages that affected Europe and North America occurred during the *(Mesozoic, Paleozoic, Cenozoic)*.
8. *(Dinosaurs, Ammonites, Trilobites)* were probably reptiles.
9. *(Eras, Epochs, Periods)* are the smallest divisions of geologic time.
10. During most of the Paleozoic *(deep seas, mammals, shallow seas)* covered much of Earth.
11. Animals with relatively constant body temperatures are *(endotherms, ectotherms, angiosperms)*.
12. Most of geologic time occurred during the *(Precambrian, Mesozoic, Paleozoic)*.
13. *(Trilobites, Angiosperms, Mammals)* are common Paleozoic fossils.
14. The Paleozoic Era began about *(600, 200, 65)* million years ago.
15. Groups of organisms that can produce fertile offspring are *(trilobites, dinosaurs, species)*.

B. Understanding Concepts

Answer each of the following questions using complete sentences.

16. List the four major units of Earth history in order of occurrence.
17. Describe how geologic time is subdivided into units.
18. How old are Earth's oldest rocks?
19. Describe the theory of punctuated equilibria.
20. Describe the geologic and biologic changes that occurred during the Paleozoic Era.
21. Identify the single most significant event that separates Precambrian time from the Paleozoic Era.
22. Describe the theory of evolution.
23. What is natural selection?
24. What is the significance of stromatolites in Earth history?
25. Describe the geologic and biologic changes that occurred during the Mesozoic Era.

C. Applying Concepts

Answer each of the following questions using complete sentences.

26. Why do you think the development of land plants preceded the development of land animals?
27. Why do trilobites make excellent index fossils?
28. How do reptiles differ from amphibians?
29. Why are the oldest rocks generally found near the center of each continent, while younger rocks are found nearer the edges?
30. How has the movement of continents caused changes in the types of animals found on Earth?

SKILL REVIEW

1. Study the drawing below. In an undisturbed area, how would the fossils appear in sedimentary rock layers?

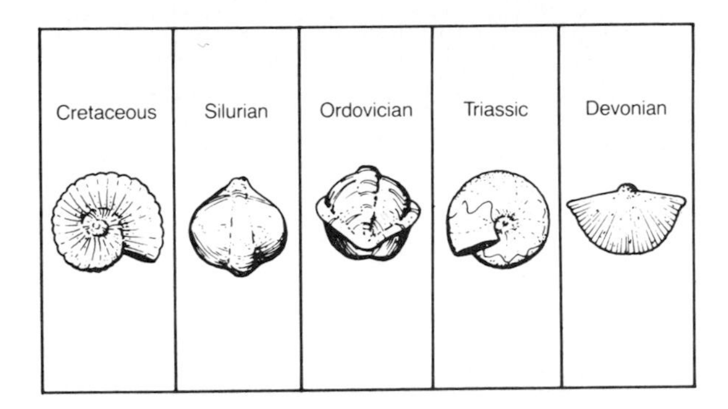

2. Using the outlines of present-day continents, make a sketch of the Mesozoic supercontinent Pangaea.
3. Hypothesize why trilobites became extinct.
4. Jonathan found what he believed was a small coral in a piece of coal. What conclusions can you draw about Jonathan's discovery?
5. Compare and contrast endotherms and ectotherms.

PROJECTS

1. Choose one part of a continent and trace it through geologic time. Include drawings of the different positions on Earth's surface and summaries of the climate, environment, and topography of this part of Earth throughout geologic time.
2. Research the most recent theories on the mass extinctions that occurred at the end of the Mesozoic. Include in your project an explanation of how a single meteorite could lead to the extinction of so many species. Construct a chart that demonstrates the events that may have led to these extinctions.

READINGS

1. Boy Scouts of America. *Geology*. Irving, TX: BSA, 1985.
2. Boyer, Robert E., and P. B. Snyder. *Geology,* 2nd ed. Northbrook, IL: Hubbard Scientific, 1986.
3. Rossbocher, Lisa A. *Recent Revolutions in Geology*. New York: Franklin Watts, 1986.

SCIENCE AND SOCIETY

THE GREENHOUSE EFFECT

A greenhouse is a glass-covered enclosure where plants are grown in a warm environment. The glass roof lets in some forms of radiation and keeps out other forms. Visible light penetrates the glass, but when it is reflected as infrared waves or heat, it cannot leave the greenhouse. Earth's atmosphere works very much like a glass roof on a greenhouse. The atmosphere lets in visible light, but when the energy is reflected back from Earth's surface, the atmosphere traps heat. It has recently been discovered that an increase in carbon dioxide in the atmosphere has upset the delicate balance of gases that protect Earth. The increased production of carbon dioxide is causing an increase in the greenhouse effect that may lead to changes in Earth's climate.

Background

Scientists believe that the increase of carbon dioxide in the atmosphere is due to the burning of fossil fuels such as coal, oil, and natural gas. These fuels are called "fossil fuels" because they are composed of the remains of plants and animals. When fossil fuels are burned, stored carbon is given off as carbon dioxide. People have been using fossil fuels for many years. However, the effects of additional carbon dioxide in the atmosphere are just now being realized.

Many theories have been proposed concerning the consequences of the increase in atmospheric carbon dioxide. A global warming of as little as 1°C could melt the ice caps. This melting could result in a rise in sea level as much as six meters, which would flood many of Earth's coastal cities and lowlands. Temperate and tropical climate zones could extend poleward. Some scientists predict that hurricanes will have increased intensity due to the higher temperature of the ocean water over which they form. Others say the amount of land covered by deserts would increase. Warmer temperatures would increase the rate of evaporation and could cause an increase in precipitation in some locations. Since warm air holds more water vapor than cold air, cloud cover may increase with higher tem-

FIGURE 1. Carbon dioxide enters the atmosphere from a variety of sources.

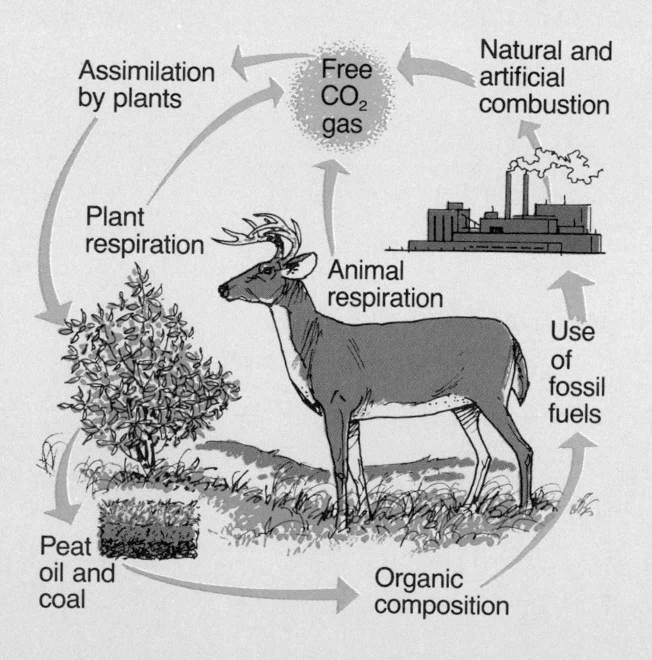

peratures. This increased cloud cover could offset a further rise in temperature by blocking some of the sunlight that would reach Earth.

Although the consequences listed are just hypotheses, most scientists agree on one point—the increasing level of carbon dioxide in Earth's atmosphere is a problem that needs to be addressed. Disagreements arise over the methods of solving the problem. Should people stop using fossil fuels? If we stop using fossil fuels for the production of electricity, can we find alternatives that are safe, available, and economical?

Many alternative energy sources have disadvantages that may be as bad as those of fossil fuels. Solar power is still too expensive to use on a large scale. Nuclear power plants produce thermal pollution and radioactive wastes that pose safety hazards. Rivers suitable for hydroelectric power plants are limited and the dams can cause harm to the river environment.

Debate

There are differing opinions about the role people play in altering Earth's atmosphere. Some scientists believe that increases in carbon dioxide levels will increase the greenhouse effect. Others believe that increases in carbon dioxide may be offset by other factors and no global warming will occur. Choose ten classmates to form two debate teams. Research the differing opinions about the role of carbon dioxide and the greenhouse effect. Prepare to debate the topic, "Should the United States government limit the burning of fossil fuels to prevent an increase in the greenhouse effect?"

Developing a Viewpoint

1. Solar power is an example of an alternative energy source that is clean and efficient. One reason that it is not used more is because solar cells are very expensive to produce. What role should the United States government play in the development of alternative energy sources? Explain your viewpoint.

2. Nuclear energy is a very controversial subject. Nuclear accidents have occurred in the past, causing many people to oppose the development and use of nuclear power plants. What are the pros and cons of using nuclear power plants? Do the benefits outweigh the problems and possible hazards? Explain your viewpoint.

3. The problems associated with an increase in global temperatures include melting of the polar ice caps and a rise in sea level. These seem to be possibilities for our future if something is not done. Do you believe that people have the power to make a difference in the future? Explain. Would you be willing to pay for more expensive sources of energy so that carbon dioxide levels could be controlled? Explain.

Suggested Readings

"Are Stronger Hurricanes in the Offing?" *New York Times*. April 7, 1988.

"Eocene Greenhouse." *Scientific American,* Vol. 252. April, 1985.

Stanley N. Wellborn. "Facing Life in a Greenhouse." *U.S. News & World Report*. September 29, 1986.

"Probing the Permafrost" (for evidence of global warming to prove greenhouse effect theory). *Scientific American,* Vol 258. February, 1987.

UNIT 8

Earth materials that are useful are called natural resources. These resources include fuel, air, wood, water, and minerals. Resources may be renewable or nonrenewable. Some resources are being used faster than they can be recycled or replaced. What are some resources you use? Can they be renewed?

~2000 B.C.
Textiles used in South America.

1880
First electric street lights are used in New York.

1933
Tennessee Valley Authority is established.

EARTH'S RESOURCES

1940 1950 1960 1970 1980 1990

1954
United States explodes first hydrogen bomb.

1966
Department of Transportation created.

1980
Gold reaches over $850 per ounce.

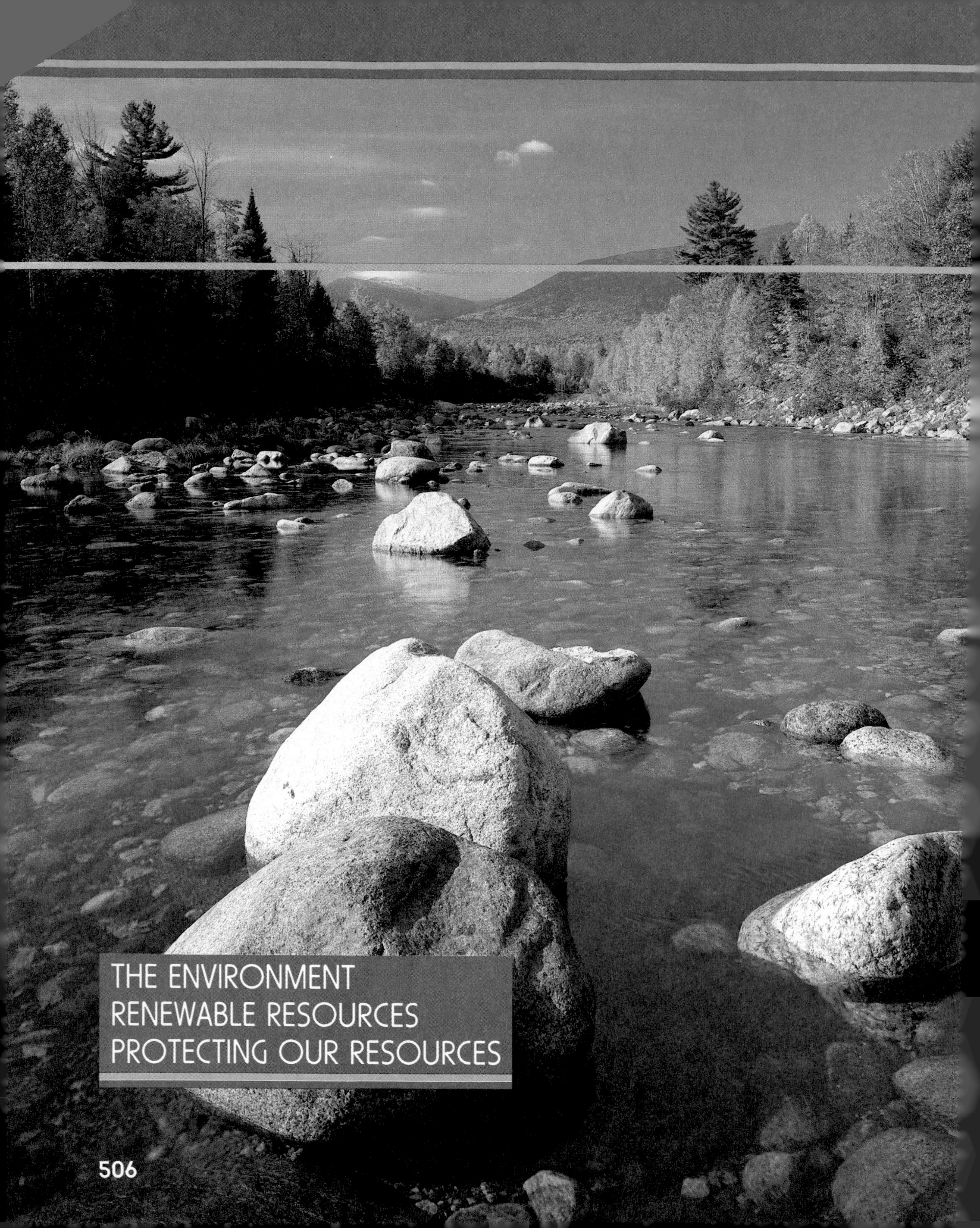

THE ENVIRONMENT
RENEWABLE RESOURCES
PROTECTING OUR RESOURCES

Renewable Resources

All the materials you use are produced using Earth's resources. Water is one resource that is constantly recycled. Water is used for drinking, home use, industrial processes, and recreation. How is the water supply renewed? How do pollutants affect water supplies? What other resources are renewable?

THE ENVIRONMENT

GOALS

1. You will learn how human activities alter the environment.
2. You will learn which natural resources are renewable.

Climate, soil, water, mineral resources, and landforms determine where people live. In turn, as people live and work they change their surroundings.

23:1 Humans and the Environment

The **environment** is the sum of all external conditions that act upon or influence an organism or community. We are a part of our environment and are influenced by it. The environment changes in response to both natural events and human activities. Thunderstorms, floods, volcanic eruptions, and earthquakes are natural processes that affect the environment. Human activities include construction of roads and buildings, farming, and mining. Human activities usually change the environment more rapidly and in ways different from natural activities.

F.Y.I. Nine out of every ten babies born today are born in a developing country.

The human population doubled between the years 1650 and 1850. It is predicted that human population will double in only 41 years based on the current growth rates. Growth rates vary from country to country, but a million people are added to the world every four or five days—85 million each year. World population is predicted to reach 6 billion by the year 2000. Where will all these people live? What will they eat? This rapidly increasing

Table 23–1

Population Doubling Time

Country	Years
Kenya	17
Mexico	27
Afghanistan	28
Ethiopia	33
China	65
U.S.S.R.	71
Canada	90
United States	100
United Kingdom	630

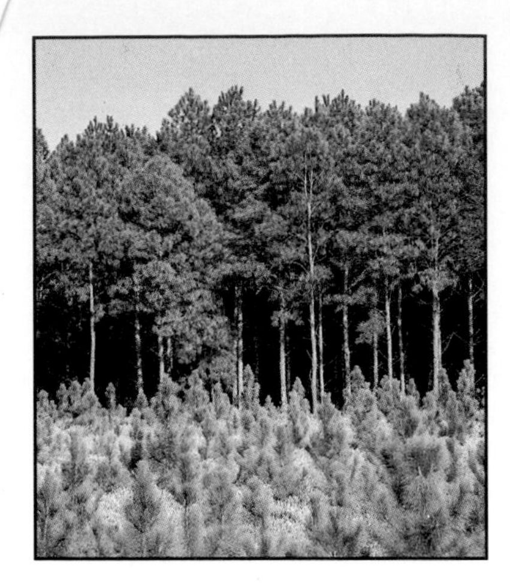

FIGURE 23–1. Trees are renewable natural resources. What do they supply?

human population has greatly altered the natural environment in recent times. Scientists predict even more drastic environmental changes in the future as more people draw on Earth's limited resources.

23:2 Earth's Resources

People use many materials from Earth. This book is made of paper from trees. The dyes used to color the paper fiber came from Earth. A polyester knit shirt or blouse is a product of petroleum; cotton and linen come from plants. Materials from Earth's crust that we use are called **natural resources.** They include water, air, minerals, trees, and fuels such as coal, oil, and natural gas. Minerals are used to make useful objects such as pencils, machines, and cooking utensils. Fuel is used for cooking and heating. Some of Earth's resources are renewable; others are not. **Renewable resources** are those that are recycled either by nature or by humans within a useful period of time. They include forests, water, and air. We can use these resources without exhausting them completely for future generations.

List three renewable natural resources.

Forests can be replanted by humans after trees are removed. Abandoned lands also can be allowed to reforest as a natural process. Water and air are other renewable resources. As water cycles through the hydrosphere, sources of water such as lakes, streams, and groundwater are replenished by precipitation. Air is another resource that is recycled by natural processes. Oxygen is constantly being added to the atmosphere through photosynthesis by plants. It is then used by plants and animals in respiration. Impurities may be washed from air as precipitation forms and falls.

Review

1. List ten parts of your environment that are natural. List ten parts of your environment that are made by humans.
2. What are renewable natural resources and why are they important?
3. List three renewable natural resources.
4. Why are water and air considered renewable resources?
5. ★ Use Table 23–1 to determine which country doubles its population in half the time of China.

RENEWABLE RESOURCES

GOALS

1. You will learn about four renewable resources.
2. You will learn about pollution problems associated with these resources.

Forests, air, and water are all very important resources. Through personal, agricultural, and industrial activities, people not only use, but often abuse these important natural resources.

23:3 Land Use

Earth's crust contains about 13 billion hectares of land. Only 1.5 billion hectares presently are being farmed. Another 200 000 to 600 000 hectares could be farmed. However, vast regions of Earth are either too cold, rocky, wet, or dry to farm.

What methods do farmers in agriculturally developed nations use to increase food production?

In agriculturally developed areas, such as the United States, Europe, and Canada, the same fields are planted each year. The nutrients used by the plants are replaced by fertilizers and by crop rotation. Modern machinery, pesticides, and improved plant types have greatly increased food production. About ten million farmers produce enough food to meet the needs of the United States and to export surpluses.

Agriculturally developing countries cannot feed their populations. Many farmers in parts of Africa and Central and South America cannot afford machinery, improved strains of seeds, pesticides, or fertilizers. Locusts and other insects may damage up to two-thirds of their crops. Because fertilizers are unavailable or too expensive, some

a

b

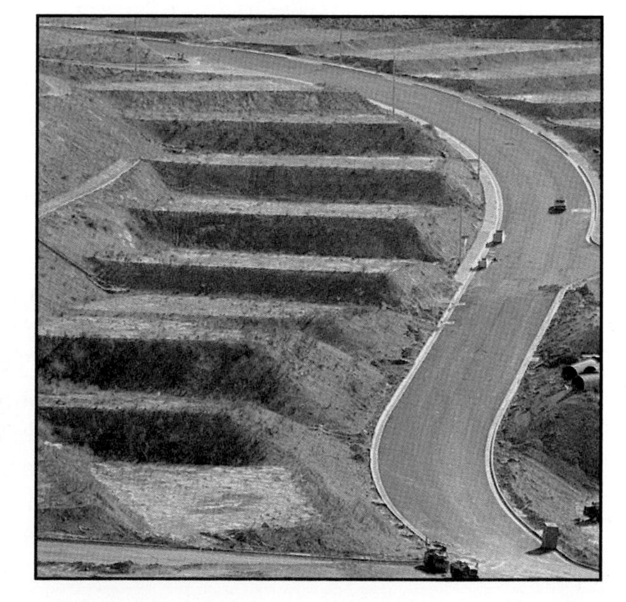

FIGURE 23–2. Erosion results when vegetation is removed from the land surface (a). Humans speed the process of erosion by removing vegetation during construction (b).

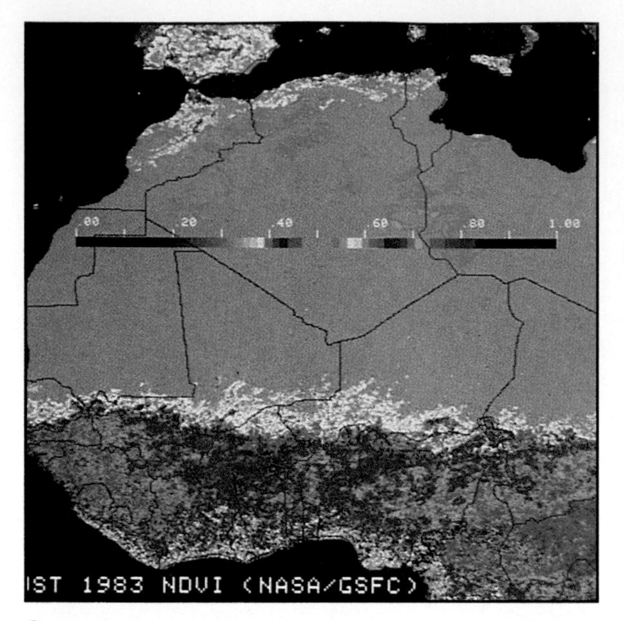

a

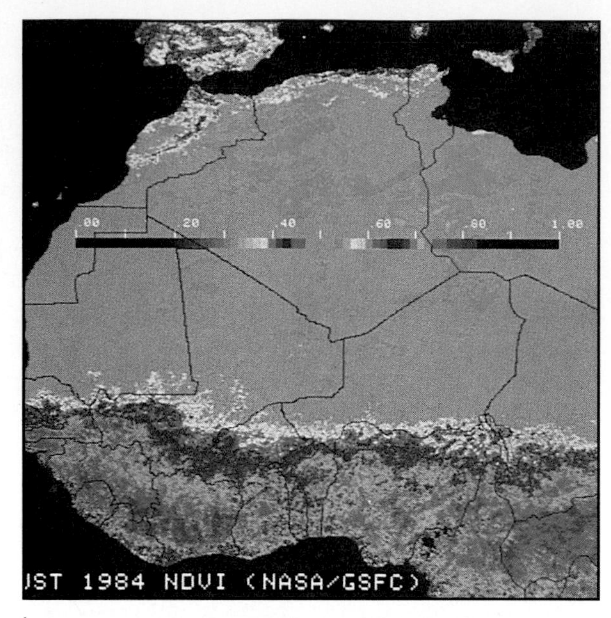

b

FIGURE 23–3. These satellite images of northwestern Africa show the extension of the desert southward between 1983 and 1984.

farmers move to new sites when soil nutrients are exhausted. The abandoned land has no plant cover and is eroded easily when flooding occurs. Landslides and dust storms may follow the floods. Some land that has been productive in the past is being converted to desert due to overgrazing of animals, deforestation, drought, excessive erosion, and overuse of available water supplies. Millions of people in Africa and India are suffering from malnutrition and starvation as a result of these changes in land use.

What results if ground cover is removed?

Grazing land for livestock is another land use that requires large areas. Rangeland may be planted with crops for animal food or animals may eat native plants. Rangeland that is not fertile enough or that is too hilly for farm crops may provide good grazing for cattle, sheep, or goats. During periods of drought, these areas may be overgrazed as plant life becomes scarce. Erosion may result if too much plant cover is removed by grazing animals.

Forests and tree farms are other valuable uses for land. These resources supply lumber for buildings, pulpwood for paper products, and fuel for woodburning stoves and fireplaces. Trees also protect land from erosion. Runoff is slowed enough in wooded areas for the water to seep into the ground. Trees also reduce the carbon dioxide content of the air and produce oxygen. Today, the forests of the Amazon River region of South America produce much of the world's oxygen. But because these forests are now being cleared, there is growing concern about the stability of the oxygen cycle.

Another use of land is for garbage disposal. As the world's population keeps increasing, so does the amount of garbage created. Where can we dispose of all this garbage? Each day, the people in the United States throw away 362 880 metric tons of garbage. Ninety percent is buried in landfills. Some cities have run out of space at their landfills and send their trash elsewhere. In the spring of 1987, several tons of garbage from Islip, New York, was put aboard a barge and shipped south for burial. However, during a 9654-kilometer journey that lasted 155 days, the garbage was rejected by six states and three foreign countries. It was finally shipped back to Brooklyn, New York, where it was incinerated. The ash that resulted was buried in a landfill.

Some landfills have been designated as hazardous-waste sites. Toxic materials from improperly contained wastes can leach into the soil and eventually into the groundwater. Many communities fight having these types of landfills near their homes and schools.

CAREER

Sanitary Landfill Operator

Pamela Brown is a sanitary landfill operator. She performs many tasks while disposing of solid waste materials at the landfill site and operates heavy equipment such as bulldozers, front-end loaders, and compactors. She also transports solid waste materials, spreads and compacts layers of waste and Earth cover, and directs incoming vehicles to the dumping areas. Ms. Brown sprays chemicals over the waste materials in order to control disease-carrying pests, and drives a truck to distribute oil and water over the landfill to control dust. Ms. Brown's job includes weighing the vehicles that enter and leave the landfill site, and collecting dumping fees. It is very important that she examine the cargo that is brought to the site and prohibit the disposal of any caustic wastes according to government regulations.

For career information, write:

The National Solid Wastes
Management Association
Suite 1000
1730 Rhode Island Avenue N.W.
Washington, DC 20036

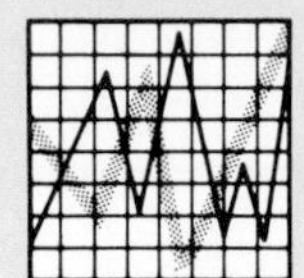

SKILL

Designing an Experiment

Problem: How do you design an experiment?

Background

The first step in designing an experiment is to choose a topic to study. Suppose you want to find out how the hardness of water in your home compares to the hardness of water from other sources in your community. How would you begin? The following steps will aid you in designing this and other experiments.

Procedure

1. A good place to begin is in the library. You will need to find out what causes hard water and what tests have been designed to test for hard water. Perhaps you may come across the following information: *Hand soaps and detergents do not lather well in hard water.*
2. The next step is to state a hypothesis. Based on your library research and on experiences with your water, you may hypothesize that your water is harder than water from other sources in your community. If you wanted to test this hypothesis, what water sources might you use? What would you use to collect samples?
3. An experiment must be designed to test your hypothesis. Recall that experiments involve variables and controls. Decide the method you will use. List the materials you will need. It is often possible to use simple materials you have around the house. Based on information in Step 1, what household materials might you use to test the hypothesis you formed in Step 2?
4. The final step is to decide the procedures you will use. Think logically. What should you do first? Be sure that you use a control. Design the experiment in such a way that quantities are measured. Decide how many times the experiment should be conducted to ensure accurate results. Use the materials from Step 3 to complete the design of a simple experiment to test the hardness of water in your home compared with two other sources in your community. Rank the three samples from hardest to softest.

FIGURE 23–4.

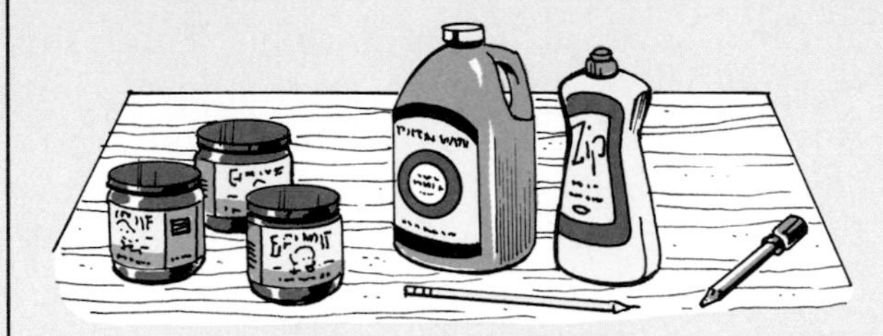

Analysis

1. What were the water sources you used? What variables did you have to take into consideration when you gathered your samples?
2. How did you label your containers?
3. What materials did you use to check the hardness of the water samples? What variables did you control?
4. What was the source of the softest water? The hardest? How did you determine the ranking?
5. If you wanted to present these results to others, what format would you use?

Conclusions and Applications

6. What steps should be taken in designing an experiment?
7. What factors should you consider when deciding on the equipment you will need for an experiment?
8. What factors should you consider when listing the procedures you will use during an experiment?

23:4 Air

Most resources are not distributed equally. Air is everywhere, but its quality varies. The more industries a region has, the harder it is to keep air free of dust and chemical pollution. Industries also may pollute the air with sulfur dioxide, nitric oxide, other gases, and soot. Although industries add impurities to the environment, some of the products and services they provide improve the quality of life and make technology and conveniences possible.

F.Y.I. Deadly dioxins, found in some fish in the Great Lakes, are thought to be transported over great distances in the air.

Volcanic eruptions, forest fires, and grass fires pollute the air with dust, ash, and gases. Plowing of farmland, construction projects, and mining also put dust in the air. Small amounts of dust in the air are important because drops of water collect around these nuclei and fall as rain. However, when the particles contain chemicals, or when they are too plentiful, they may be considered pollutants.

Smog is a fog that is made darker and more dense by smoke and chemical fumes. There are different types of smog. In areas like Los Angeles, a chemical reaction between the sun and materials such as nitric oxide, sulfur dioxide, and hydrocarbons produces a photochemical smog. **Hydrocarbons** are compounds containing carbon and hydrogen. These smogs are usually brownish in color and are associated with the burning of coal, oil, and natural gas. In addition, industry, cars, and planes produce byproducts such as carbon dioxide, water, smoke, and ash. Carbon monoxide is formed if burning is incomplete. This

What are the characteristics of a photochemical smog?

FIGURE 23–5. Air pollution is caused by both natural means and human actions.

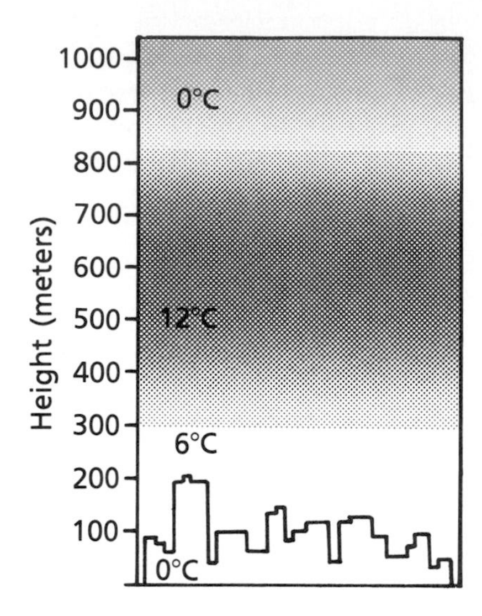

FIGURE 23–6. Sometimes polluted air is trapped near the surface by a temperature inversion.

gas is deadly in large quantities. Even in amounts as small as 100 parts per million, carbon monoxide makes many people ill. When the concentration of the gas rises to 1000 parts per million in heavy smog, people may die.

Another smog type is a sulfurous smog, which used to be common in London and New York. The burning of coal or oil in large power plants produced sulfur oxides and particulates in the air. These combined with a layer of stable air or cloud cover to form a thick, grayish fog that lasted several days. In 1952, a London smog was responsible for the deaths of over 4000 people. In 1953, a New York smog is believed to have caused 200 deaths.

Mountains may contribute to the development of smog by restricting the movement of air. When weather conditions are right, warm air overlying cooler air acts as a lid and prevents upward movement of air near the surface. The dense, cool air remains near the surface, along with the pollutants trapped in it. This smog can cause respiratory illnesses in people living in the area. Eventually, the weather changes and the smog is diluted with cleaner air blown in from other areas. People in Los Angeles and Denver suffer from atmospheric conditions made worse by nearby mountains.

Sulfur dioxide is an air pollutant that is produced when fuels with sulfur impurities are burned. Coal from some mines contains large amounts of sulfur. When smoke from the burning of such coal combines with moisture in the air, sulfuric acid forms. The acid then falls with rain or

FIGURE 23–7. Smog can trap harmful materials near the ground. People with respiratory problems need to stay indoors.

a

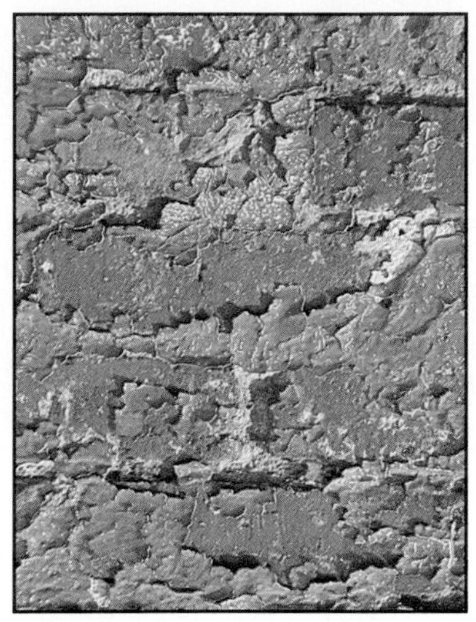

b

FIGURE 23–8. Tall smokestacks help disperse local air pollutants, but may contribute to acid precipitation in other areas (a). Acid precipitation may cause painted surfaces to deteriorate quickly (b).

snow. This **acid precipitation** damages the surfaces of buildings, changes the acidity of lakes, and may reduce crop yields.

Many industries are working to recover sulfur for reuse and to prevent air pollution. The gases produced from the burning of sulfur coal are processed through a scrubber in an attempt to reduce sulfur emissions. Some industries use tall smokestacks to help disperse their gases. However, these tall stacks vent some of the sulfur emissions higher in the atmosphere where they are carried hundreds of kilometers from the pollution source to fall as acid precipitation. The extent of damage from acid precipitation depends on rain and snowfall patterns, winds, and soil sensitivity. Alkaline soils of the midwest plains neutralize much of the acid. However, some northeastern states, eastern Canada, and Scandinavian countries do not have these neutralizing soils and have suffered damage. People in these areas often hold highly industrialized regions responsible for their acid precipitation problems.

What kind of soil neutralizes acid rain?

Some industries are experimenting with bacteria to aid in reducing sulfur emissions. Coal with a high sulfur

List three methods that can be used to remove sulfur from coal.

content is treated with the bacteria, which eat the sulfur in the coal and produce a harmless by-product. Crushing and washing the coal also may remove up to one-third of the pyrite that produces the acid emissions. The remaining sulfur can be removed through a gasification process or through scrubbing the gases produced by burning. As the cost of energy rises, more coal high in sulfur content will be burned. People in every community must work to find feasible ways to produce energy while maintaining air quality.

PROBLEM SOLVING

What Is Happening to the Fish?

Joe and Scott grabbed their fishing gear after school and hiked to their favorite fishing spot on a nearby farm. They had been fishing this same pond for seven years. However, they had noticed that they seemed to be catching fewer fish during the past two years than they had when they first starting fishing there.

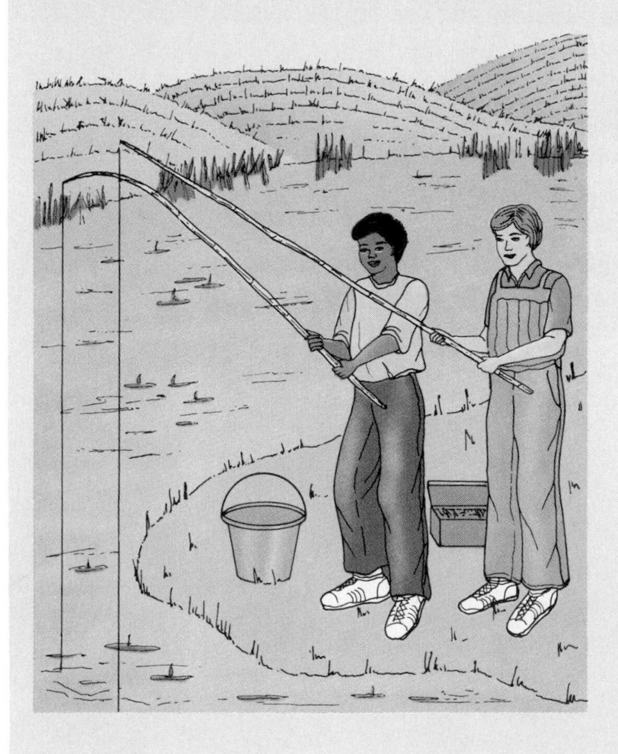

They were also aware that the algae cover seemed to be getting thicker each year.

Joe and Scott were learning about air and water pollution in their science class. They learned that high levels of phosphates and nitrates in water can cause an increase in algae populations. As these algae populations die, they sink to the bottom of the body of water. Then, the decomposer populations increase. Oxygen in the water is used up by the large numbers of decomposers. Fish and other aquatic populations do not have enough oxygen to survive.

Joe and Scott decided to make some careful observations near their fishing spot. They found that the farm had a septic tank system for waste disposal. They questioned the owners about the types of soaps and detergents that were used for doing laundry. They also asked about the types and amounts of fertilizers that were being used on the crops that were being raised. Why were Joe and Scott interested in this information? What other questions do you think they will ask? If they determine that the pond is being polluted, what could be done to decrease the pollution?

INVESTIGATION 23–1

Dust in the Air

Problem: What types of dust or particles are in the air in your area?

Materials

small box of plain gelatin
hot plate
pan of water
plastic lids (4)
hand lens
binocular microscope
marker
refrigerator
thermal mitt

Procedure

1. Follow the mixing directions on the box of gelatin. Pour a thin layer of gelatin into each lid. **CAUTION:** *Wear a thermal mitt while working with the hot plate and while pouring the gelatin from the pan into the lids.*
2. Place the lids in a refrigerator until the gelatin is set.
3. Place the lids in four different locations in your community. Make sure that you choose places where the gelatin will not be disturbed.
4. After one week, collect the lids. Label each lid with its location.
5. Examine each lid with a hand lens.
6. Record your observations in a data table similar to the one shown in Data and Observations. Record whether the material on each lid is dust, large particles, or other materials.
7. Sketch some of the material found on each of your lids.
8. Sort the particulate matter (large pieces of dust, plant pieces, seeds, parts of insects, and so on) from each sample site.
9. Arrange these materials on microscope slides.
10. Label the slides with the location from which each sample came.
11. Examine each slide with the microscope. Draw any structures you see.
12. Try to identify what type of particulate matter you have collected.

Data and Observations

Location	Type of material	Description of material
1.		
2.		
3.		
4.		

Analysis

1. What sort of materials collected on each lid?
2. Rank the lids in order from the most solid particles to the least solid particles.
3. Which of the materials that were collected might be a result of human activities?
4. Explain why some areas had more and larger particles in the air.

Conclusions and Applications

5. How does your body filter solid particles from the air you breathe?
6. Do any of the seeds that you might have collected suggest how some plants migrate from one area to another? Explain.
7. Try to identify what plants are responsible for the plant material on your slides. Try to identify the insect material. Were there any particles you could not identify?
8. Do you think any of the material you collected might be harmful to humans? Explain your answer.
9. Which areas in your study should be given more attention in terms of air quality?

23:5 Water

F.Y.I. Americans use more than 1703 billion liters of water daily.

Water is necessary for all living organisms. People need large amounts of fresh water. In the United States, home water use includes drinking, bathing, flushing toilets, sprinkling lawns, and washing clothes, dishes, and cars. Daily home water use in the United States averages 265 to 568 liters per person per day. Farmers need water for irrigation. Water is also used in many industrial processes. It takes more than 360 000 liters of water to make a car. When we consider all of these uses, each person in the United States uses about 7100 liters per day. This is a much higher average than in 1900, when Americans used less than 379 liters per day. Projections are that by the year 2000, Americans will use 22 712 liters per day.

Unlike air, water is not distributed equally at all times nor to all places. Although Earth has a total water supply that is far greater than current demands, much of the water is not available for use. Ninety-seven percent of the supply is ocean water. We can fish in it, swim in it, and use it for transportation, but we cannot drink it unless the water is processed and the salt is removed. However, desalination is currently too expensive to provide large quantities of fresh water.

Where is most of Earth's fresh water supply?

Globally, fresh water is distributed unequally. Each continent has desert areas. Some areas have high annual rainfalls. Ninety-eight percent of the freshwater supply is frozen in the glaciers of the polar regions. Usable water resources include rivers, lakes, and groundwater. Surface water provides about ¾ of the water needed for homes, industry, and irrigation. Groundwater supplies the rest of our needs. People usually try to settle in areas where fresh water is available. Sometimes, though, climates change and the supply of water fails. Many early civili-

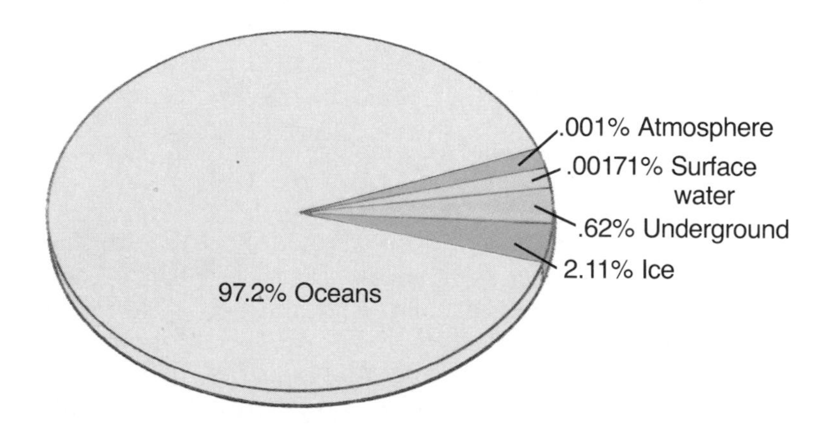

FIGURE 23–9. Most of Earth's water supply is found in the oceans.

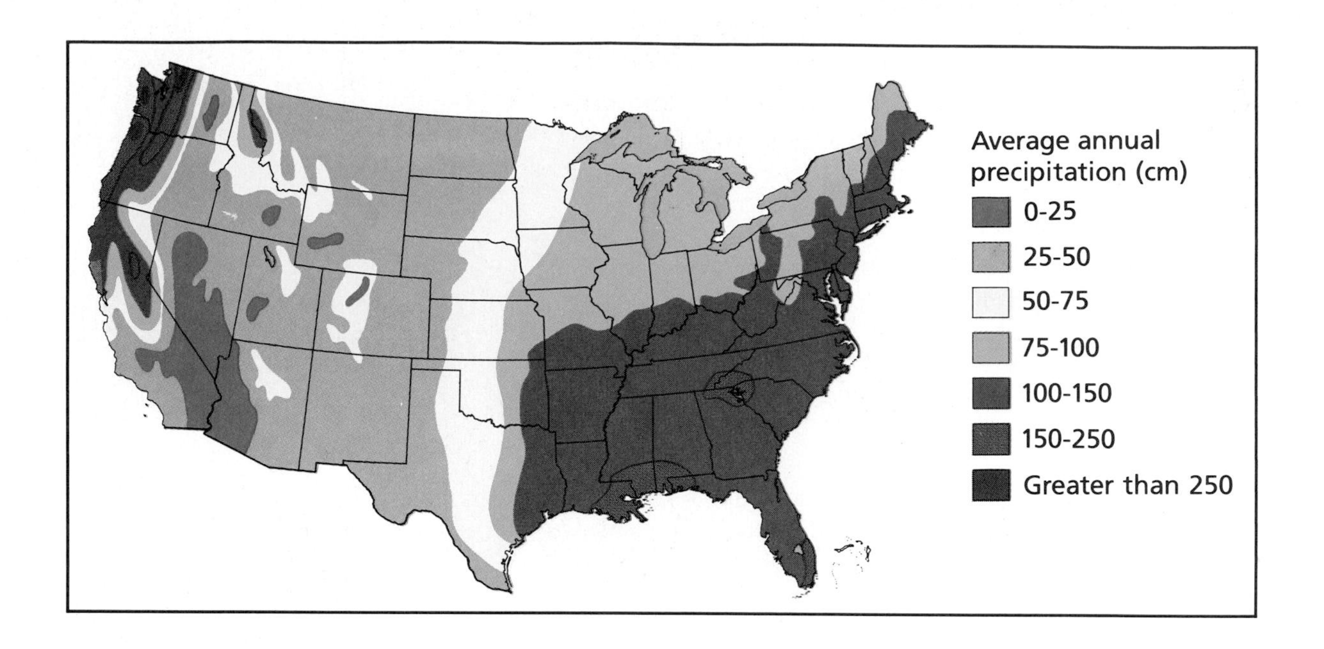

FIGURE 23–10. Rainfall is not distributed evenly over the United States. Where does most rain fall?

zations moved to new locations during periods of drought, just as people do today in parts of Africa. However, moving an industry to a new location is expensive. Today, people usually try to find ways to bring water to an area where it is in short supply.

Water is found beneath Earth's surface almost everywhere. Humans are using larger and larger amounts of groundwater to meet their daily needs. When too much water is used, the water table falls and some wells may become dry. Surface water may eventually seep back into the aquifers, but the renewal time may be too long to be useful to humans. Buried wastes, including radioactive substances, chemical wastes, and septic tanks, can pollute the groundwater, which then may be pumped to the surface before it has been naturally filtered by the rocks in the aquifer. Most groundwater moves about 10 to 20 meters per day through limestone or coarse gravel aquifers, and about 1.5 meters per day through most sandstone aquifers. Sandstone is a more efficient filter and the water removed from sandstone aquifers tends to be freer of pollutants than water that is pumped from limestone or gravel aquifers.

What type of aquifer tends to provide pollution-free water?

Surface water may be polluted by organic wastes, excessive nutrients, sediments, chemicals, and excess heat. Organic wastes come from human, plant, and other animal life and from some manufacturing processes. Oxygen is needed to digest the organic wastes in water environments. If too much organic waste enters the water for the

FIGURE 23–11. If they are not properly stored, toxic chemical wastes may seep into groundwater supplies.

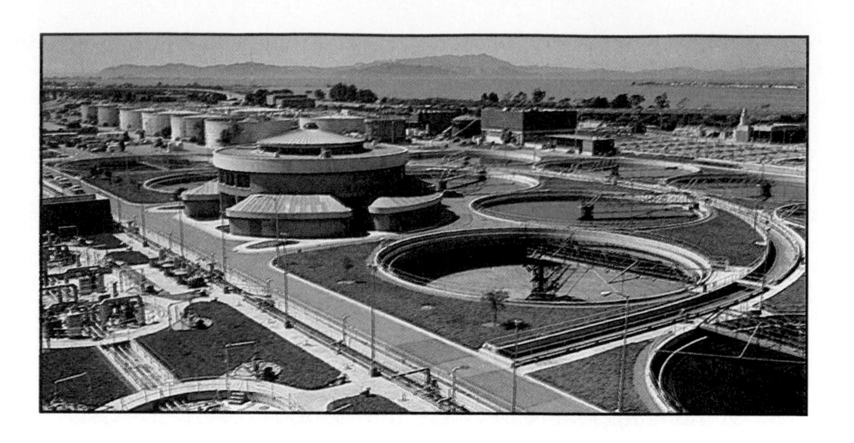

FIGURE 23–12. Sewage treatment plants remove impurities and some bacteria from wastewater before it is returned to the environment.

amount of oxygen present, some of the waste will not be digested. Chlorine added to sewage minimizes the disease threat posed by excess wastes, but these pollutants still must be carefully controlled.

Problems develop in water supplies when too many plant nutrients, such as nitrates and phosphates, enter lakes and rivers. Plant nutrients come from detergents in sewage, from industrial wastes, and from the drainage of fertilizers from farmlands. An excess of these nutrients causes plant life to increase rapidly. When large numbers of these plants die, bacteria that decompose them quickly use up the oxygen supply. When the oxygen is gone, the fish suffocate.

Some pesticides, cleaning agents, and inorganic materials from industries and farmland also pollute water. Some of these materials are toxic even in small quantities. Large amounts of sediments and certain minerals may affect water quality. **Hard water** results when minerals such as calcium and magnesium carbonates or sulfates are dissolved in the water. When heated, these compounds form scale in tea kettles, hot water tanks, and industrial boilers.

Sand and soil settle to the bottom of rivers and lakes, gradually filling them in and covering the food supply used by aquatic organisms. Sediment must be dredged from rivers and lakes to keep the channels usable for navigation.

In some areas, thermal pollution of rivers or lakes may kill organisms and cause evaporation that concentrates minerals and metals in the water. **Thermal pollution** is caused by the discharge of water used for cooling purposes in many industrial processes and power plants. The temperature of this heated water may be lowered through

BIOGRAPHY

Lorna J. Mike

1955-

Lorna Mike is a fisheries manager for the Lower Elwha Fisheries for the Point No Treaty Council of Kingston, Washington. She also serves as Chairwoman of the Tribal Fisheries Committee for the Lower Elwha Klallam Tribe. Mike is concerned that the waters near Kingston, Washington, remain free from pollution.

the use of cooling towers or ponds before the water is returned to its source.

Review

6. Describe at least three human activities that can cause rapid erosion.
7. How is smog produced?
8. Which type of rock is best for filtering impurities from groundwater? Explain.
9. What happens to organisms when excess nutrients are added to a lake?

★ 10. Explain why people living in eastern Canada blame highly industrialized regions to the west for their problems with acid precipitation.

TECHNOLOGY: APPLICATIONS

Wash-Away Plastics

Plastics are used by nearly every major industry in the world. The United States produces close to 17 million metric tons of plastics each year. Plastics are used to make everything from appliance parts to packaging materials. They are waterproof, durable, and chemically stable. However, after these materials have been used, the plastics create a tremendous environmental problem. They do not decompose when buried in landfills. Burning them produces toxic smoke.

Concern for these problems has caused a small polymer-processing company in Switzerland to invent a new type of plastic that repels water during use, but decomposes into harmless residues when discarded. It can be used on cars in place of wax, sprayed on statues to prevent the effects of acid rain, and made into disposable plastic bags and containers.

The plastic can be made into pellets, films, tubes, adhesives, and sheets. In order to produce a particular type of product, a computer is used to choose the appropriate processing procedures.

INVESTIGATION 23–2

Water Use

Problem: How much water do you use?

Materials

home water meter

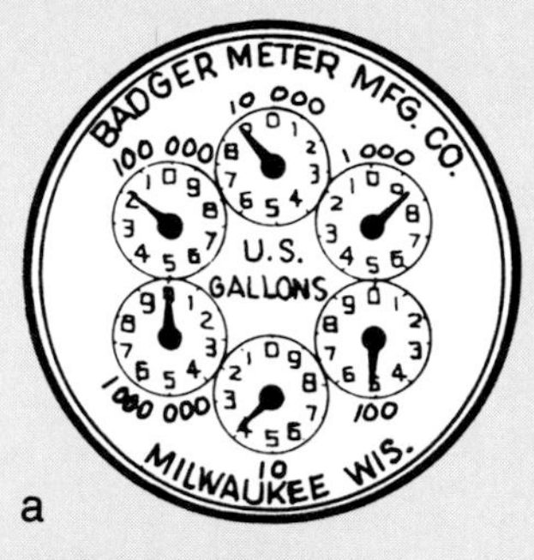

a

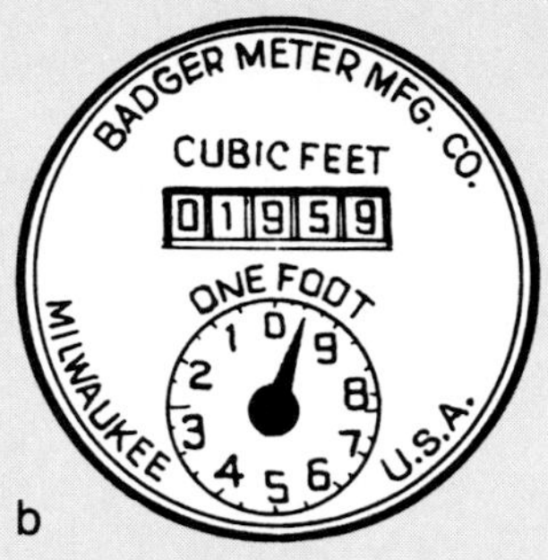

b

c

FIGURE 23–13.

Background

There are several different types of water meters. Meter *a* has six dials. As water moves through the meter, the pointers on the dials rotate. To read a meter similar to *a*, find the dial with the lowest denomination indicated. The bottom dial is labeled 10. Record the last number that the pointer on that dial has passed. Continue this process for each dial. Meter *a* shows 18 854 gallons. Meter *b* is read like a digital watch. It indicates 1959.9 cubic feet. Meter *c* is similar to meter *b*, but indicates water use in cubic meters.

Procedure

1. Use Figure 23–13 to determine how to read your water meter.
2. Record your home water meter reading at the same time of the day for eight days.
3. Subtract the previous day's reading to determine the actual amount of water used each day.
4. Record how water is used in your home each day.
5. Plot your data on a graph like the one shown. Label the vertical axis with the units used by your meter.

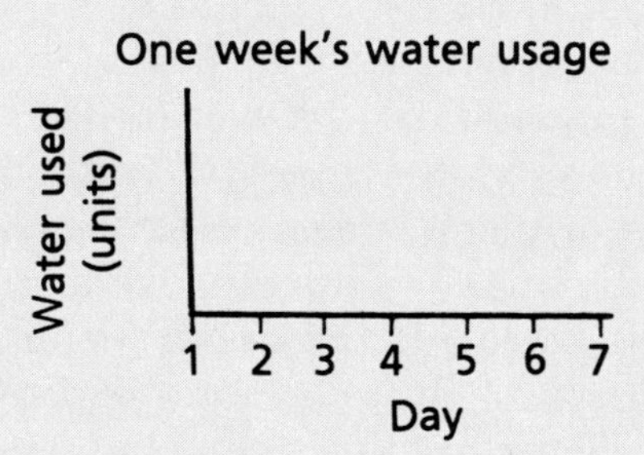

Analysis

1. When was more water used? Why?
2. Calculate the total amount of water used by your family during the week.
3. Calculate the average amount of water each person used during the week by dividing the total amount of water used by the number of persons. Calculate a monthly average.

Conclusions and Applications

4. Why is your answer to question 3 only an estimate of the amount of water used?
5. Would your family's water use vary with the time of year? Explain.
6. How can your family conserve water?

PROTECTING OUR RESOURCES

GOALS

1. You will learn several ways to conserve resources.
2. You will learn what is being done to reduce pollution in the United States.

With the increase in population discussed in Section 23:1, there is a greater demand than ever for Earth's natural resources. Pollution of these resources also has become a serious problem.

23:6 Conservation of Earth's Resources

Industrialization and **urbanization**, or the expansion of cities, are on the rise. As a result, much land is covered with buildings and asphalt, a tremendous amount of water is being used, and soils and air are being polluted. There is a real need to conserve our natural resources. **Conservation** is the planned and wise management to prevent destruction, exploitation, and neglect of Earth's resources.

What is conservation?

In order to conserve, people must reduce waste. They must practice responsible ways of using minerals, forests, water, air, and fuel. All people can help with the conservation effort. For example, you can help conserve water by reducing the time you spend in the shower. A ten-minute shower uses 190 liters of water. By reducing your time to five minutes, you can save 95 liters of water.

You can improve the quality of soil by adding compost, a mixture of decayed organic matter. You can improve the quality of air and water by planting trees and other vegetation. A grass lawn 15 meters by 15 meters uses carbon dioxide and produces enough oxygen for four people. The plants also slow the erosion process, helping to keep sediments from clogging lakes and streams. Thus, vegetation aids in reducing or preventing pollution while providing a pleasant environment for people to enjoy.

F.Y.I. See Section 24:6 for suggestions on recycling natural resources.

FIGURE 23–14. Solid wastes are a pollution problem you can help solve.

FIGURE 23–15. EPA investigators work to locate the sources of pollution.

23:7 Pollution Control

What is the EPA?

Most cities have health departments that constantly monitor air and water quality. In the United States, the **Environmental Protection Agency,** or EPA, is the federal agency that acts as a watchdog for the environment. Due to government regulation in the past 25 years, air and water quality are carefully monitored. The 1970 **Clean Air Act** requires that air pollutants be reduced to safe levels. Air cleaners, scrubbers, and tall stacks are being used by industry to help remove and disperse air pollutants. Cities where smog is a serious problem have installed systems to warn people when pollution reaches unsafe levels.

In 1987, the **Clean Water Act** was established by the United States Congress. It earmarked money through 1994 for states to build sewage and wastewater treatment facilities and to curb storm-water runoff from streets, mines, and farms. Storm-water runoff was the cause of up to one-half of the water pollution in the United States before 1987.

Our natural environment is far too valuable a resource to let pollution destroy it. We cannot sacrifice our environment for industry or any other reason. We simply cannot survive without ample clean water and air. International, national, state, and local communities need to work together to keep pollution to a minimum. We are learning that we must be careful with the environment.

Review

11. What is the purpose of the EPA?
12. How has the Clean Water Act helped to reduce water pollution?
13. What is conservation?
14. What caused more than one-half of the water pollution in the United States before 1987?
★ **15.** Describe the results of an increase in urbanization on Earth's natural resources.

Chapter 23 Review

SUMMARY

1. The natural environment includes conditions of climate, organisms, air, water, landforms, and soil. Changes in the environment caused by human activities often occur more rapidly than most natural changes. 23:1
2. Forests, water, and air are renewable resources. 23:2
3. Uses of land resources include farming, grazing, forests, tree farms, and landfills. 23:3
4. Atmospheric pollution comes from both natural and human sources. Smog is a fog made darker and more dense by smoke and chemical fumes. 23:4
5. When sulfur dioxide combines with moisture in the atmosphere, acid precipitation is formed. 23:4
6. Much of Earth's water is not available for many of our uses. 23:5
7. Sandstone filters many of the impurities from groundwater. 23:5
8. An excess of nitrogen and phosphorus in lakes and rivers causes plants to increase rapidly and exhausts the oxygen supply. 23:5
9. To conserve Earth's resources, people must reduce waste. 23:6
10. The Clean Air Act requires that pollutants in the air be reduced to safe levels. The Clean Water Act requires that sewage and wastewater treatment facilities be built. 23:7

VOCABULARY

a. acid precipitation
b. Clean Air Act
c. Clean Water Act
d. conservation
e. environment
f. Environmental Protection Agency
g. hard water
h. hydrocarbons
i. natural resources
j. renewable resources
k. smog
l. thermal pollution
m. urbanization

Matching

Match each description with the correct vocabulary word from the list above. Some words will not be used.

1. caused when industries dump heated water into lakes and rivers
2. may be created when some air pollutants react with sunlight
3. formed when sulfur reacts with moisture in the air
4. materials from Earth's crust that we use
5. compounds that contain carbon and hydrogen
6. results when much calcium carbonate is dissolved in water
7. planned and wise management to prevent destruction of resources
8. requires that air cleaners, scrubbers, and tall stacks be used by industries
9. sum of all external conditions that influence an organism
10. appropriated money for states to build sewage and wastewater treatment facilities

Chapter 23 Review

MAIN IDEAS

A. Reviewing Concepts

Choose the word or phrase that correctly completes each of the following sentences.

1. *(Limestone, Sandstone, Gravel)* is the most effective filter for groundwater.
2. An excess of *(nutrients, chlorine, hydrocarbons)* causes plant life to increase rapidly and exhaust the oxygen supply in a lake.
3. Erosion is not a problem on *(abandoned farmland, overgrazed rangeland, tree farms)*.
4. *(Scrubbers, Cooling towers, Chlorine)* may be used to reduce sulfur emissions from power plants.
5. Incomplete burning of hydrocarbons forms *(oxygen, hydrogen, carbon monoxide)*.
6. Groundwater may be contaminated by *(septic tanks, smog, air pollution)*.
7. *(Groundwater, Surface water, The Antarctic glacier)* supplies 3/4 of the water needed for human use.
8. Heated water may cause *(hard water, thermal pollution, smog)* when returned to rivers.
9. *(Seventy, Eighty-five, Ninety-eight)* percent of the world's fresh water is in glacial ice.
10. *(Fifty, Seventy-five, Eighty-five)* million people are born each year.
11. Renewable natural resources include air, water, and *(minerals, forests, dams)*.
12. Trees reduce the *(sulfur dioxide, carbon dioxide, oxygen)* in the air.
13. A chemical reaction between the sun and nitric oxides, sulfur dioxides, and hydrocarbons produces a *(sulfurous, photochemical, dense)* smog.
14. Human activity in the United States uses about *(380, 5208, 7100)* liters of water per person per day.
15. Acid precipitation forms when *(sulfur, gold, dust)* from burning coal combines with moisture in air.

B. Understanding Concepts

Answer the following questions using complete sentences.

16. What happens to a lake or pond if an abundance of plant nutrients is added?
17. Describe why oxygen is considered a renewable resource.
18. Explain why much of Earth's crust cannot be farmed.
19. Explain why many developing countries cannot feed their populations.
20. Describe pollutants that industries may add to the air.
21. Describe why much of the water on Earth is not available for many of our uses.
22. Why is water that is removed from sandstone aquifers free of most pollutants?
23. Describe five ways that surface water may be polluted.
24. Discuss the function of the Environmental Protection Agency.
25. Explain why the addition of heat to lakes or rivers is harmful.

C. Applying Concepts

Answer the following questions using complete sentences.

26. What do cities with smog problems try to do to lessen the dangers to their inhabitants?
27. How are industries both helpful and harmful to the environment?
28. How can you help control pollution?
29. Why is the discharge from sewage treatment plants monitored? What types of pollutants may be present in this discharge and how could they harm the environment?

30. What steps might a community in a desert area take to cope with the water supply problem?

SKILL REVIEW

1. What are the steps you need to take when designing an experiment?
2. Describe the effect that an increase in population will have on our natural resources.
3. If smog is brownish, is it sulfurous or photochemical?
4. Use Table 23–1 to determine the doubling time for the population in the United States.
5. Infer what will happen to the desert regions of Africa if drought and poor farming practices continue.

PROJECTS

1. Design an experiment for testing the effect of acid rain on vegetation.
2. Determine some problems you think will arise as population continues to increase. Discuss these problems and some possible solutions in a report.

READINGS

1. Brown, Michael H. "Toxic Wind." *Discover*. November, 1987, pp. 42-49.
2. Cajacob, Thomas, and Teresa Burton. *Water for Life*. Minneapolis: Carolrhoda Books, 1986.
3. Congressional Quarterly Inc. *Energy and Environment: The Unfinished Business*. Washington: Congressional Quaterly, 1985.

ORES
MINING EARTH'S RESOURCES

Mineral Resources

Renewable resources are recycled by nature or by humans. However, many of Earth's natural resources cannot be replaced in the foreseeable future. These resources include the minerals and other materials removed from Earth's crust. What can you do to help conserve these resources?

ORES

GOALS

1. You will learn the characteristics of metallic and nonmetallic ore deposits.
2. You will learn how these deposits are used.

Resources that are not renewable include fossil fuels such as oil, coal, and natural gas; and minerals such as salt, gold, silver, copper, and clay. Sand and gravel are other resources that are not renewable. Most of these resources are found underground. Before they can be used, they must be found, mined, and processed.

24:1 Nonrenewable Resources

Nonrenewable resources are accumulations or concentrations of useful materials from Earth's crust that cannot be recycled or replaced within a useful period of time. The first Earth materials used by ancient people were flint, chert, obsidian, and limestone. These rocks and minerals were used as weapons, utensils, and for carving. As early as 25 000 B.C., burned clay figurines discovered in Moravia and clay pottery found in Egypt may represent the first large-scale mining industries. One of the great pyramids of Egypt contains 2 300 000 blocks of stone averaging 2.5 metric tons apiece!

People have used a variety of minerals for thousands of years. Gems such as amethyst, turquoise, jade, and garnet were prized by ancient Egyptians, Babylonians, and Assyrians. Precious metals like gold and silver were mined by the Greeks as early as 2500 B.C. Copper was being used by the Egyptians at least 7000 years ago.

What is an ore?

Ores are deposits in which a mineral or minerals exists in large enough amounts that they can be mined at a profit. The noneconomic material that must be removed before a mineral can be used is called waste rock, or **gangue** (GANG). Some nonmetallic deposits, such as sand and gravel, have little or no gangue and require little processing before they can be used.

What factors determine if an ore can be mined at a profit?

Changing economic factors affect the mining industry. In order for a deposit to be mined at a profit, the resource must be in demand. Also, enough of the mineral must be present in a deposit to pay for the work of getting the mineral to market. Finally, the price paid for the rock or mineral must allow a profit. Both the cost of mining and the price paid for a mineral change from time to time. These changes may make it possible to mine a deposit that once was not profitable. In the past few years, some abandoned gold mines have been reopened because the price of this metal has risen greatly. On the other hand, some copper and silver mines have closed because the price of these metals has dropped. Although the minerals are still in the deposits, it was not profitable to mine them.

24:2 Types of Ore Deposits

The original source of all minerals and rock is magma. But as magma rises toward the surface, changes occur. High-temperature gases and certain liquids escape from the main body of magma and invade the surrounding rocks. When these hot fluids cool, they may form **high-temperature vein deposits,** which often are important

FIGURE 24–1. Diamonds sometimes are found in pipes of igneous rock called kimberlite (a). Mines such as these (b) have produced millions of diamonds, mostly for use in industry.

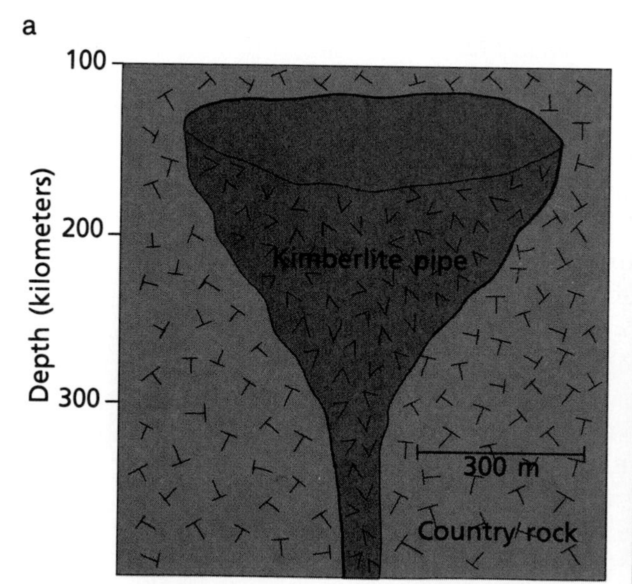

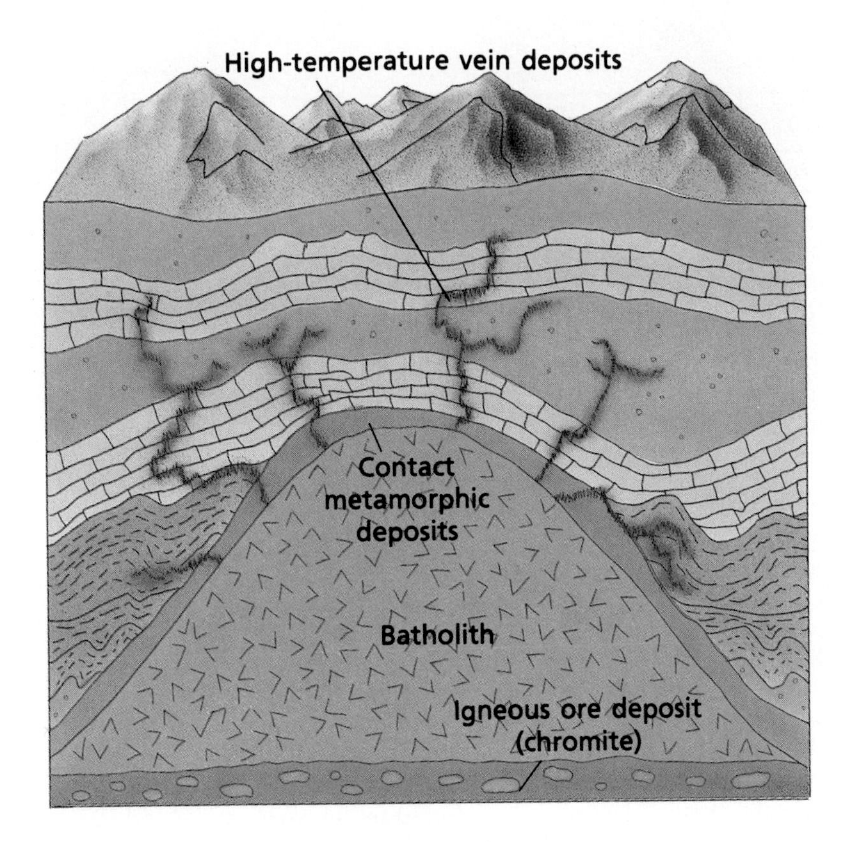

FIGURE 24–2. Igneous ore, high-temperature vein, and contact metamorphic deposits are associated with igneous intrusions.

sources of lead, silver, copper, and gold. Some famous vein deposits include Cripple Creek, Colorado (gold), Sunshine, Idaho (silver), and Butte, Montana (copper).

Igneous ore deposits are concentrations of minerals within the intrusive body itself. Some minerals form while the magma is still liquid. Heavy minerals sink to the bottom of the magma chamber. Igneous ore deposits of magnetite are found in the Adirondack Mountains. Ore deposits of chromite are found in the Stillwater Complex in Montana.

Where do contact metamorphic deposits form?

Contact metamorphic deposits are formed at the contact between an igneous intrusion and the surrounding rock. High-temperature fluids from the magma seep into the rock. These fluids then trade ions with the rock. Minerals formed during this ion exchange include silver and gems such as rubies and sapphires. Deposits of rubies and sapphires are found in Burma, Thailand, and India.

Sedimentary ore deposits result from surface processes. Copper ores often are enriched by circulating groundwater. In the zone of weathering, copper is changed to a water-soluble form. The copper solution is carried downward by groundwater. Near the water table, the copper is changed back to an insoluble form and is deposited in the enriched zone. With time, concentrated

Table 24–1

Types of Ores and Their Uses		
Type of deposit	**Ore**	**Use**
High-temperature vein deposits	galena (PbS) sphalerite (ZnS) cinnabar (HgS) chalcopyrite $(CuFeS)_2$	lead zinc mercury copper
Igneous bodies	kimberlite chromite ($FeCr_2O_4$) magnetite (Fe_3O_4) pentlandite $(Fe, Ni)_9S_8$	diamond chromium iron nickel
Contact metamorphic	corundum (Al_2O_3) graphite (C) garnet (varies) magnetite (Fe_3O_4)	abrasive lubricant, pencil "lead" abrasive iron
Sedimentary	gypsum ($CaSO_4{\cdot}2H_2O$) calcite ($CaCO_3$), limestone bauxite (varies) salt (NaCl) clay (varies) sulfur (S) sand and gravel	plaster cement, building stone aluminum food preparation brick, pottery fertilizers, explosives construction

copper is gradually added to earlier deposits. Copper ores of this type are actively being mined in New Mexico, Oklahoma, and Bolivia.

24:3 Metallic Ores

Metals are valuable for many reasons. Metals are malleable, which means they can be shaped by pounding

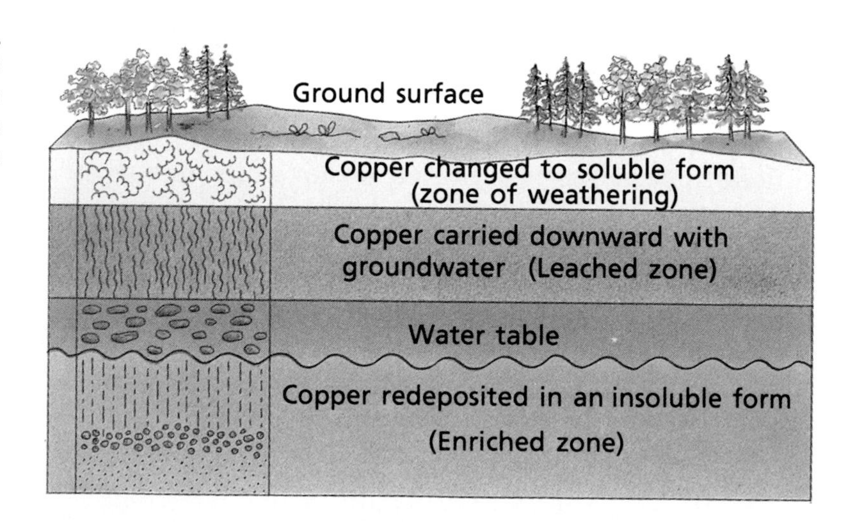

FIGURE 24–3. Copper may be concentrated by groundwater into a sedimentary ore deposit. Locate the different soil horizons in this profile.

a

b

FIGURE 24–4. Taconite (a) is an ore that contains iron. Pure copper (b) is rarely found in nature.

without losing strength. Some metals, like copper and aluminum, are good conductors of electricity. Silver and gold have long been valued for their beauty. Silver is in great demand for electronics and photography, and gold is used in dentistry and medicine. Some metals are used in combination. When two or more metals are combined, an **alloy** is produced. Steel is an alloy of iron and other metals. Steel is stronger than iron alone.

Metallic ores are deposits that may yield one or more metals. Ores that generally are worked for a single metal are those of iron, aluminum, tin, and some ores of copper. Gold ores may yield only gold, but silver is a common by-product. Ores of nickel, copper, lead, and zinc often yield two or more metals. Common ore associations include gold-silver, lead-zinc-copper, nickel-copper, and iron-manganese.

Metallic ores are removed from Earth's crust. The ore then is crushed and the gangue is removed. Some gangue has small amounts of ore present. This waste rock may become valuable if the price of the metal increases or improvements in mining and processing methods are developed.

When may waste rock become valuable?

After removal of the gangue, the ore is taken to the smelter where it is refined. During the smelting process, the metal is separated from unwanted elements by chemical processes. When the metal is free from elements such as silicon, oxygen, or sulfur, it is ready for use.

Although gold occurs in high-temperature vein deposits, it is often concentrated by weathering and erosion.

FIGURE 24–5. Iron deposits such as this one in the Nimba Mountains of Liberia were formed by the weathering of Precambrian rock.

BIOGRAPHY

Jo Davison

1935-

Jo Davison is a former high school teacher. She is now President and Research Director of a biotechnology firm in Columbus, Ohio. As a teacher, she knew that microorganisms existed about 300 million years ago that decomposed organic matter containing metals and sulfur to form coal. Davison believes these organisms are released when coal is mined. She has developed a way to use them to correct acid water pollution from mines naturally.

When the veins are exposed to surface processes, the surrounding rock weathers and releases the gold. Gold is not chemically active and does not react easily with other elements. Thus, it can survive weathering and erosion better than many metals. Gold grains often are picked up by a stream and deposited as the stream's velocity decreases at the foot of a slope or where the stream curves. These deposits are called **placers** (PLAS urs). Placer gold deposits are found in Alaska, California, Australia, and the U.S.S.R.

Iron is a metallic ore that sometimes accumulates in sedimentary ore deposits. Some of the large iron deposits of North America are the result of concentration by weathering of former iron-rich sedimentary beds. These thick series of beds were weathered and metamorphosed to form alternating layers of slate or chert and iron oxides. Most of these iron ores are found in Precambrian rocks. These banded iron formations are believed to have formed 1900 to 2500 million years ago. These large deposits of iron are found in the United States in Minnesota, Wisconsin, and Michigan. Iron is an important industrial material, and oxides of iron such as limonite and hematite are used in the preparation of pigments and dyes.

Zinc and lead are metals that often occur together as vein deposits in limestone and dolomite. Lead water pipes have been found in Roman ruins. The Chinese used lead as money. Zinc bracelets date back as far as 500 B.C. Zinc is also used to coat steel to prevent it from rusting. This process is called galvanizing. In modern industry, zinc and lead rank next to copper as essential metals. Lead is used in batteries. Brass is a zinc and copper alloy.

Manganese is used in the making of steel. Most manganese deposits are sedimentary in origin. Recall from Chapter 12 that manganese nodules on the ocean floor may be one of the world's largest resources of manganese and other metals. Estimated reserves of these resources total billions of metric tons.

24:4 Nonmetallic Deposits

Nonmetallic deposits are useful products from Earth's crust that contain no metals. Some of these ores are rocks, others are minerals. Many of the nonmetallic materials are mined and used without much processing. At least 35 nonmetallic materials are necessary to industry.

Where do nonmetallic minerals occur?

Nonmetallic minerals may occur in vein, contact metamorphic, igneous, or sedimentary ore deposits. Nonmetallic minerals such as travertine ($CaCO_3$) may become concentrated by circulating groundwater. Graphite and talc are nonmetallic minerals associated with metamorphic deposits. Common uses for graphite are as pencil

PROBLEM SOLVING

Golden Opportunity?

Bethany had inherited a great deal of money from her wealthy aunt. She had never had this much money before. She wondered what she should do with all of it. She could buy many things she had always wanted or she could invest it. After all, she was planning on going to college in four years and would need to have money for tuition, room, board, and books. Maybe she should put the money in a savings account to earn interest. Perhaps she should invest it in the stock market.

That evening Bethany was looking through the business section of the newspaper and came across an article about investing in the Golden Glory Mine in California. She decided to discuss this idea with her mother. Her mother was very cautious about the mine-investment idea. She warned Bethany that there were many questions she should research before investing in this, or any mine. If you were Bethany, what questions would you research before making such an investment?

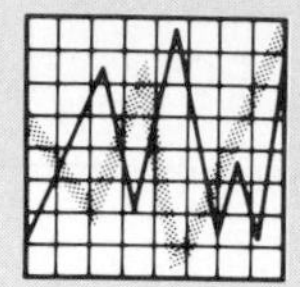

SKILL

Interpreting Tables

Problem: How can tables be interpreted?

Background

As you learned in Chapter 18 in the Skill, "Using Tables and Charts," tables are widely used by scientists to organize data. In this skill you will learn how to interpret information from tables.

Materials

data table
pencil and paper

Procedure

1. Tables are organized into columns and rows. The characteristics that are to be compared are across the top of the table. Each row describes the characteristics of a particular item. In the data table, what four mineral characteristics are being compared?
2. Count the items in the first column. How many minerals are being compared?
3. Look at each column carefully. Which of the following best describes the order used to list the minerals? They are arranged by (A) increasing specific gravity; (B) increasing hardness; (C) type of deposit; (D) alphabetical order.
4. Look at the rows for galena and quartz, and column 2. Which has the greater specific gravity?
5. Look at the last column. How many of the minerals form as high-temperature vein deposits? Look across the rows and name these minerals.
6. Using columns 4, 5 and 1, in that order, determine which mineral has both a metallic luster and forms as a contact metamorphic deposit. Record your answer.
7. Compare all of the information in the rows and columns and determine which of the following statements is true about talc. Of the minerals listed, talc (A) is the hardest and has the lowest specific gravity; (B) has the lowest hardness and is a high-temperature vein deposit; (C) has the lowest hardness and has a metallic luster; (D) has the lowest hardness and has the second lowest specific gravity.

Data Table

A Comparison of Common Minerals

Mineral	Specific Gravity	Hardness	Luster	Deposit Type
Talc	2.7	1.0	Pearly	Sedimentary ore
Galena	7.5	2.5	Metallic	High-temperature vein
Magnetite	5.8	6.0	Metallic	Contact metamorphic
Tourmaline	3.1	7.0	Vitreous	High-temperature vein
Quartz	2.65	7.0	Vitreous	High-temperature vein and igneous ore

a

b

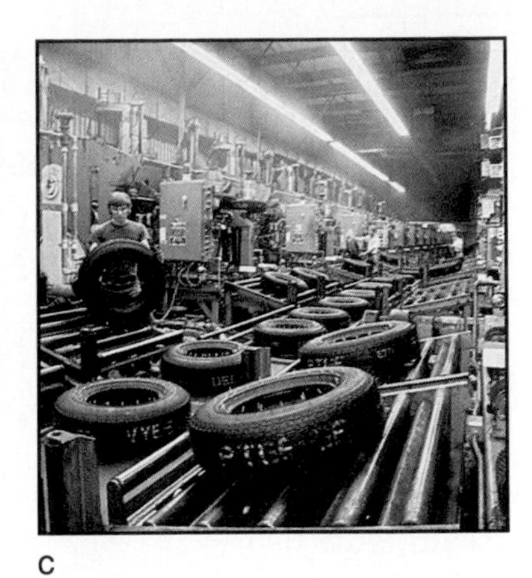
c

FIGURE 24–6. Uses of nonmetallic materials include phosphate rock in fertilizers (a), clay in bricks (b), and sulfur in the manufacture of tires (c).

"lead" and as industrial lubricants. Talc is used in powders and face creams, and in dinnerware.

Nonmetallic high-temperature vein minerals include quartz, mica, and fluorite. However, depending on the composition of the deposit, quartz and fluorite may be mined as ore, or disposed of as gangue. Quartz crystals are used in optics, radio and telephone operations, and in watches and clocks. Fluorite is used in making high-power microscope lenses and in toothpastes.

What nonmetallic minerals are found in high-temperature vein deposits?

Sulfur is another nonmetallic mineral deposit. It is associated with volcanic eruptions in Japan and Mexico, with certain sulfur-reducing bacteria in Sicily, and with cap rocks of salt domes along the Gulf Coast of the United States. Sulfur is used in fertilizers, pigments, explosives, insecticides, and rubber and steel making.

Other nonmetallic deposits include sandstone, limestone, gypsum, and marble. These materials are used in building and construction. Millions of metric tons of sand, gravel, limestone, and clay are used by the construction industry. Sand, gravel, and limestone are used to make concrete and cement. Sand also is used to make glass and computer parts. Clay is used for brick, tile, pottery, ceramics, and dishes.

What nonmetallic materials are used in construction?

Phosphates, nitrates, and potash are nonmetallic deposits used in fertilizers. The phosphate-rich sedimentary beds of Utah, Idaho, Wyoming, and Colorado are believed to have formed from precipitation in an ancient sea. Oceans also contain many valuable nonmetallic materials. Halite (table salt), gypsum, and chlorine are among the resources of the sea. Halite has been recovered from seawater since ancient times. In San Francisco Bay, evaporating basins are used to separate halite from ocean water.

F.Y.I. Halite is made from sodium and chlorine, the two most common elements in seawater. It can be removed from seawater by desalination. See Section 11:2.

INVESTIGATION 24–1

Placer Deposits

Problem: How do miners locate and recover placers?

Materials  

iron filings
plastic buckets (2)
stream table
water
wood block
pie pan
ball bearings (10)
dishpan
sand

Procedure

1. Use the wood block and one bucket to set up the stream table.
2. Make a stream bed with two definite curves in the sand. Form a hollow at the mouth of the stream.
3. Pour some iron filings near the top of the stream.
4. Using a second bucket, pour water into the stream bed above the filings. Keep pouring water until some of the filings reach the mouth of the stream.
5. Sketch the stream and indicate where the filings collect.
6. Half fill the pie pan with sand. Put the 10 ball bearings in the sand.
7. Pour water into the pan until it is almost to the top.
8. Over the dishpan, shake the pan from side to side with a rotary motion. See Figure 24–7.
9. Carefully pour off some of the sand and water, making sure that the ball bearings do not escape.
10. Put more water into the pan and continue the shaking process until all of the ball bearings and only a small amount of sand are left in the pan.

FIGURE 24–7.

Analysis

1. Where do the most iron filings seem to collect in the stream?

Conclusions and Applications

2. What caused the iron filings to be deposited where you found them?
3. Where might a placer miner locate a gold deposit?
4. Why did the ball bearings stay behind when the sand was poured away?
5. Gold is heavier than steel ball bearings. Would gold be easier to pan? Explain.
6. Do you think panning is a profitable way to recover gold? Explain your answer.
7. How do miners locate and recover a placer deposit?

CAREER

Miner

Bob Billingsworth mines coal in a shaft mine in West Virginia. Deep within the mine, he uses a cutting machine and a pick to cut channels under the working face of the mine. He operates a mounted power drill to bore blasting holes in the working face. For some coal seams, it is necessary to position charges and set off explosives to blast out the coal. The shattered coal fragments are shoveled into mine cars or onto a conveyor. Mr. Billingsworth and his fellow workers install timber and roof bolts to help support the walls and roof. Other coal miners create new passageways, air vents, and shafts, and lay track to accommodate mine cars and track-mounted equipment.

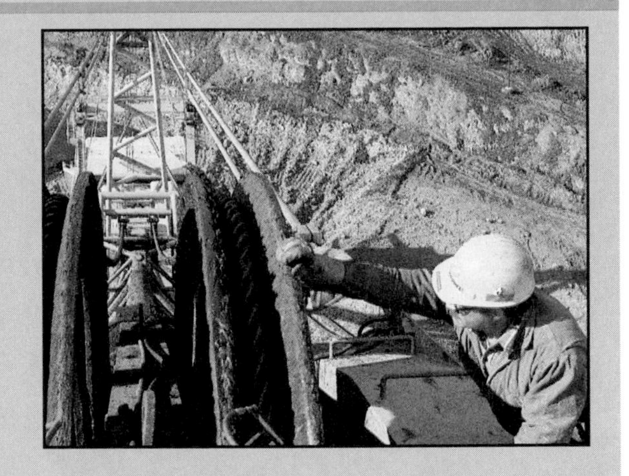

For career information, write:
International Union United Mine
Workers of America
900 15th St., N.W.
Washington, DC 20005

Diamonds are found associated with igneous deposits and as placers. Diamond, the hardest natural substance, is resistant to erosion and weathering. This mineral is pure carbon. It forms only at the extremely high temperatures and pressures found deep within Earth. Diamonds probably are formed at plate boundaries near the base of Earth's crust. Most diamonds are found in Africa, but some occur in Brazil. Many of the famous gem diamonds occurred as placers in India and Borneo. Most diamonds, however, are not of gem quality and are used as industrial abrasives.

Review

1. What is the difference between a metal and an alloy?
2. List five nonmetallic ores and their uses.
3. What is an ore?
4. Describe how contact metamorphic deposits form.
5. ★ Explain how changing economic factors affect the mining industry.

MINING EARTH'S RESOURCES

GOALS

1. You will learn about shaft, strip, and quarry mining.
2. You will learn about land reclamation and recycling.

Rocks and minerals are mined in several different ways. There may be environmental problems created from mining. However, land reclamation projects have been effective in restoring damaged land. Pollution is reduced when people recycle Earth's resources.

24:5 Mines

What are the advantages of backfilling?

Some mines are underground or **shaft mines.** These mines do not disturb Earth's surface very much. Rock is brought to the surface where it is crushed and the ore is removed. Then the waste rock is mixed with water. This mixture is pumped back into the mine passages from which the ore was taken. After it settles, the water is pumped out and the passages are closed. This process of pumping waste materials back into a mine is called **backfilling.** Backfilling has two advantages. It disposes of the gangue, and it supports the rock layers above the mine passages so that they do not collapse. However, backfilling is expensive and some miners cannot afford the costs. Instead, the operators of these mines may pile the gangue on the surface. These wastes may then pollute the area when eroded by streams or blown by the wind. Sometimes gangue contains elements that are toxic to organisms in the environment.

FIGURE 24–8. Tremendous amounts of waste rock are produced as a result of mining and refining many resources.

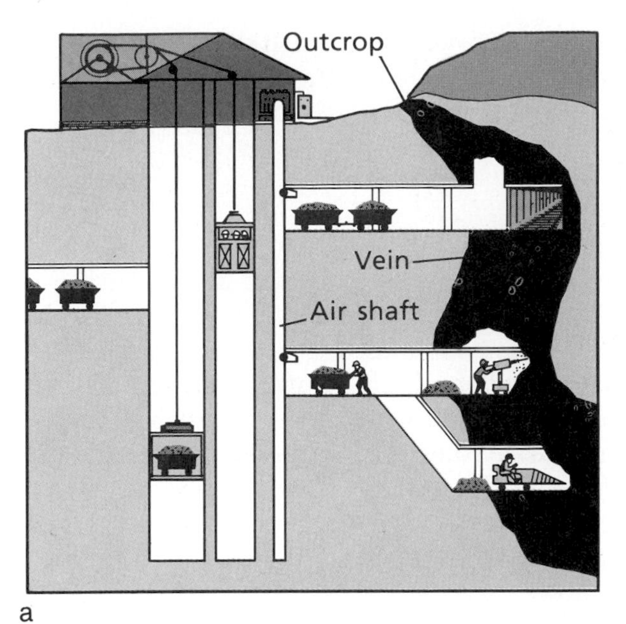

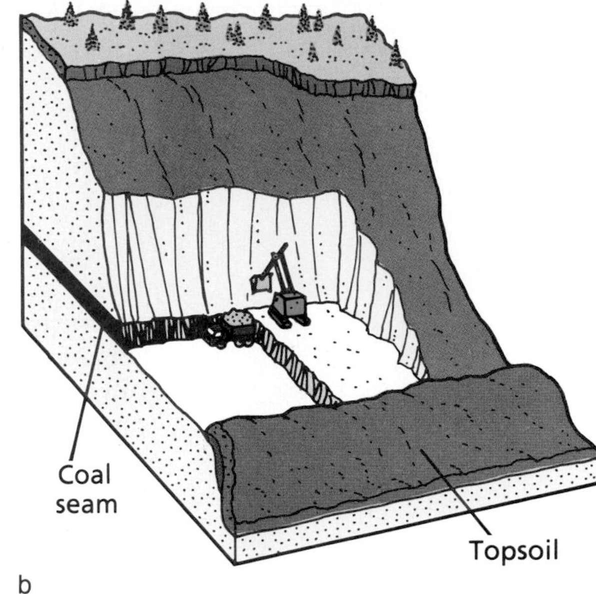

FIGURE 24–9. Shaft mines (a) are used when valuable resources are oriented vertically. Strip mines (b) are used when valuable resources are present in horizontal layers close to Earth's surface.

Strip mines are mines in which resources are removed through an opening, or pit, at the surface. Coal and metals such as iron and copper are recovered by stripping surface materials and overburden to reach the ore. Where strip mines are properly handled, the topsoil is removed carefully and saved. After the ore is mined, overburden is returned to the open pit and the topsoil is spread over the surface and replanted. In this way, the area is restored. Some beautiful parks and forests have resulted from restoration of strip mines. Most states have strict rules that require strip mine operators to restore the land surface to its original contour.

What is a strip mine?

Some mining activities move only small amounts of rock and produce little gangue. **Quarries** are open pit mines from which blocks of sandstone, limestone, or marble are removed. Clay, sand, or gravel pits usually are small quarries that do not produce much waste rock. The holes left when the quarry operation closes may fill with water or vegetation.

F.Y.I. Some strip mines produce up to 50 000 metric tons of coal a day.

24:6 Effects of Mining

Besides the problem of disposing of waste rock produced by mining, area streams and groundwater may be affected. Particulate matter and dissolved sediments from eroding waste piles clog streams. Trace elements such as cadmium, cobalt, copper, lead, molybdenum, and others may leach from wastes and be concentrated in water, soil, plants, or fish. These trace elements may be toxic or may cause disease in people and animals exposed to high concentrations of these substances. The presence of acid water is a problem in some mines. The mineral pyrite (FeS_2) occurs with some coals and certain ores. When groundwater comes in contact with pyrite, sulfuric acid forms. Circulating groundwater may move the acid water to a stream or lake where it may kill aquatic organisms. In some areas, the effects of acid water do not extend far downstream because the acid is diluted as it mixes with rain or stream water. In areas where acid water persists, the water can be treated to remove the acid. Abandoned mines also can be drained.

How does acid water form?

24:7 Land Reclamation

It is easy to see the damage that has been done as the result of mining, especially strip mining. Alpine meadows, deserts, and forests have been scarred by mining activities. But how is it possible to obtain much-needed materials and still preserve the land?

One of the largest and most successful programs of land reclamation is SEAM (**Surface Environment and Mining Program**). SEAM was established by the U.S. Forest Service in 1973 to help land managers cope effectively with the complex problems of mine management. After more than 15 years, the experts feel confident that their

FIGURE 24–10. Areas used for strip mining (a) can be reclaimed as useful land (b) after mining operations cease.

a

b

INVESTIGATION 24–2

Pollution from Mining

Problem: How could coal mining affect groundwater, streams, and lakes?

Materials  

goggles
hammer
towel
pyrite
sulfur coal
glass jars (3)
water
dropper
blue litmus paper
glass marker

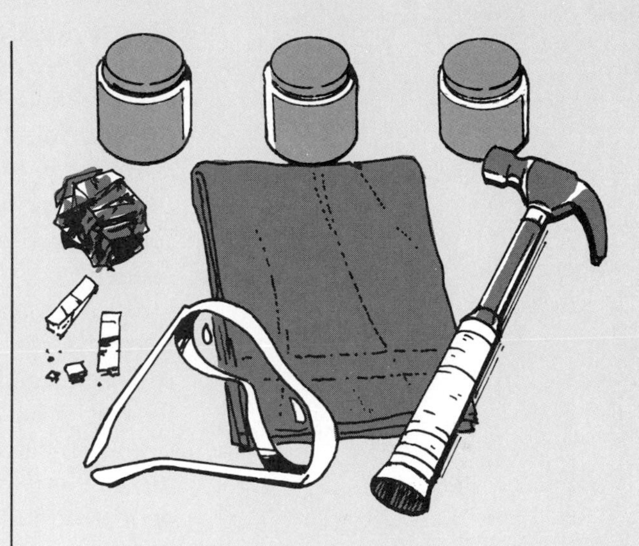

Procedure

1. Wrap a sample of pyrite in a towel. Crush the sample with a hammer. **CAUTION:** *Be sure to wear your goggles. Keep hands and fingers away from the pyrite as you crush the sample.*
2. Repeat Step 1 using a coal sample.
3. Place each sample in a separate glass jar. Drip water over the samples until they are covered.
4. Pour tap water in the third jar. Label each jar with its contents.
5. Test the water in each jar with a piece of blue litmus paper. Record any color changes.
6. Set the jars in the sun or a warm place for three days.
7. After three days, retest the water in each jar with blue litmus paper. Record any color changes.
8. Observe the samples in the jars. Record any changes that have occurred.

Data and Observations

Jar	Litmus test		Observations
	Before sitting	After sitting	
pyrite and water			
coal and water			
water			

Analysis

1. In which of your samples did the color of the litmus paper change after the jar sat for 3 days?
2. Did you observe any changes in the water color or samples after 3 days?
3. Which jar was your control?

Conclusions and Applications

4. Why did you add water to the jars?
5. Consider the litmus color change. What type of substance formed in the pyrite jar?
6. Does the coal mixture contain any acid?
7. How could coal mining affect groundwater, streams, and lakes?

reclamation efforts are succeeding. People and equipment follow in the wake of the bulldozers and draglines, reshaping and restoring mined areas. Grasses, herbs, shrubs, and trees are planted to stabilize the soil and to provide wildlife habitats.

One successful SEAM project is the reclamation of Gulf Oil's McKinley Mine, located in the Four Corners area of New Mexico. More than four million tons of coal come from McKinley Mine each year. Every year approximately 500 hectares are reclaimed at the site. Reclamation follows closely behind mining operations. Bulldozers grade the mined land to its original shape and topsoil that was removed during mining is returned. The topsoil is then prepared with fertilizer and the area is seeded.

Because the land is arid, moisture is important in the revegetation plans. Water control must be maintained in the reclaimed areas. The land around McKinley Mine was formerly used for grazing, so much of the reclamation plan focuses on returning the area to pasture. With the aid of New Mexico State University, Gulf Oil has developed seed mixtures specifically for grazing. Scientists say that in some areas the pasture has improved as much as 30 percent.

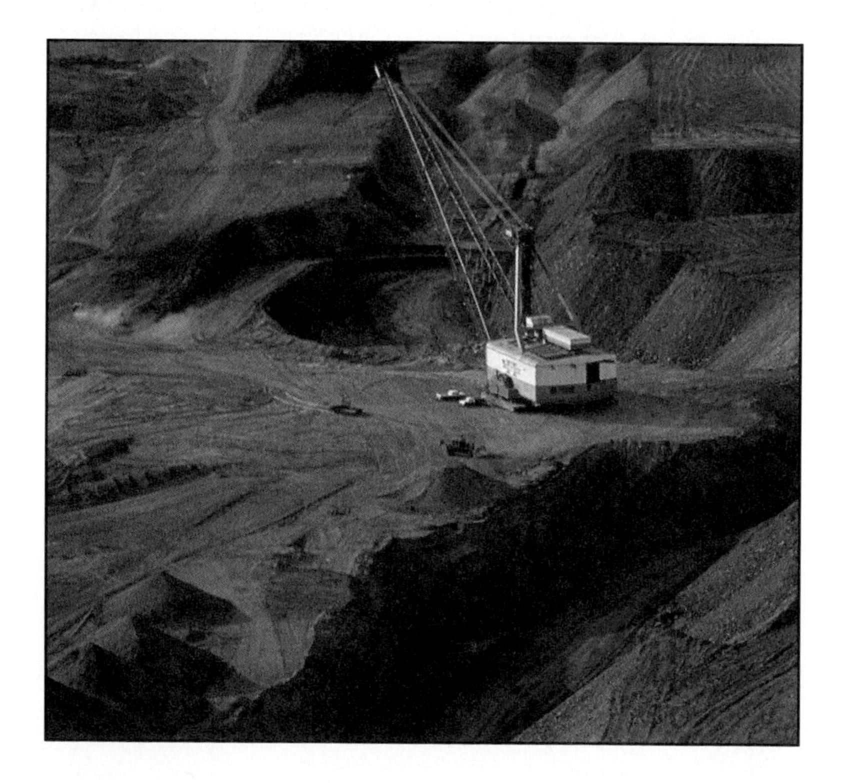

FIGURE 24–11. Reclamation of mined land is a successful effort of the Surface Environment and Mining Program.

TECHNOLOGY: APPLICATIONS

Mining Manganese Nodules

Manganese nodules are solid materials that accumulate on the ocean floor. They are composed of about 20 percent manganese dioxide (MnO_2) and 15 percent iron oxide (Fe_2O_3), plus clay minerals, calcium carbonate, and volcanic fragments. Present in minor amounts are copper, nickel, and cobalt. The nodules grow by the accumulation of material around a nucleus. Often, crystalline material enriched in manganese, copper, nickel, and cobalt alternates with iron-rich layers. The nodules range in size from less than one millimeter to several decimeters in diameter. Their growth rate averages a few millimeters in one million years.

Manganese nodules are being mined from the ocean floor primarily because of their cobalt content. Cobalt is a rare element important to several industries, including the production of heat-resistant steel. Many different techniques have been proposed to recover the nodules. In one method, a television camera is lowered from a ship and is positioned a few meters above the ocean floor. Any nodules seen are dredged from the seafloor by a series of buckets attached to long cables. The buckets are taken aboard by raising the cables. A much more efficient technique employs a giant "vacuum cleaner" that sweeps the ocean floor.

Mining manganese nodules may cause environmental problems. The effects of mining on the water, the sediment, and the organisms of the deep sea are unknown.

24:8 Recycling Earth's Natural Resources

Think back to yesterday or last week. Did you throw away anything solid? There may soon be a shortage of many natural resources, including metals such as lead, tin, chromium, nickel, and manganese. These materials are now being buried as solid wastes. One way to make our natural resources go further is to recycle them. Recycling reduces the amount of new natural resources needed each year and helps reduce pollution from waste disposal and resource processing. Recycling also reduces the amount of energy needed to produce products containing natural resources.

What are some advantages of recycling?

Many industries are involved in recycling usable materials. About 360 000 metric tons of copper, or about 20

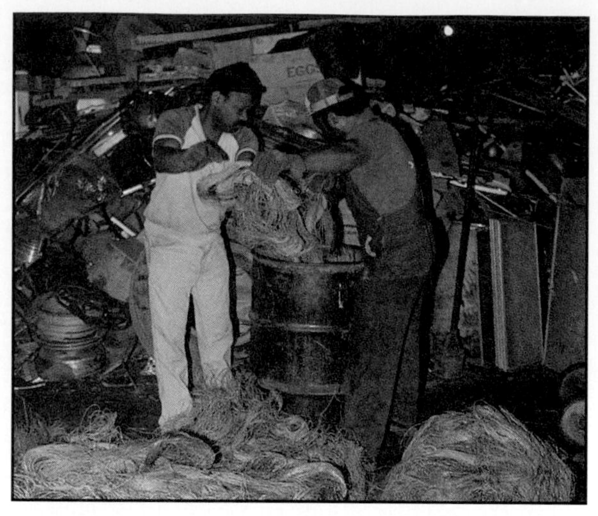

FIGURE 24–12. Recycling reduces the amount of new natural resources needed each year. It also reduces pollution from waste disposal and resource processing.

percent of the annual United States' production of copper, are recycled each year. Nearly 90 percent of the automobiles that are discarded in the United States are recycled. Auto wrecking and salvage companies break them apart for the metal they contain. Nearly one-third of our iron production comes from recycled autos and scrap iron. Lead, mercury, and aluminum are also recycled. Many beverage companies pay for the return of aluminum cans. It is expected that larger amounts of materials will be reused in the future because of higher prices and increasing worldwide shortages.

Other materials that are fairly easy to recycle include newspaper and glass. Recycled newspapers can be used to make cardboard, paper bags, and writing paper. Recycling a stack of newspaper only one meter high saves one tree. Many states encourage the recycling of glass soft drink bottles by requiring that a small deposit be paid when the beverage is purchased. This deposit is returned when the empty bottles are returned to the store. What materials do you recycle?

Review

6. Describe two advantages of backfilling.
7. List five types of rocks and sediments that are mined from quarries.
8. Why are some trace elements found near mines dangerous?
9. Describe the cause of acid water in some mines.
★ **10.** Contrast shaft mines and strip mines.

Chapter 24 Review

SUMMARY

1. Ores are deposits that can be mined at a profit. 24:1
2. Ores are found in high-temperature vein, igneous ore, contact metamorphic, and sedimentary ore deposits. 24:2
3. Combining metals forms an alloy. 24:3
4. Metallic ores are refined at a smelter. 24:3
5. Gold and diamonds may be found as placer deposits. 24:3, 24:4
6. Thirty-five nonmetallic materials are used in industry. 24:4
7. Shaft mines, strip mines, and quarries are areas where materials are removed from Earth's crust. 24:5
8. Gangue from mining of nonrenewable resources may cause pollution problems. 24:5
9. Acid water from pyrite decomposition may cause pollution. 24:6
10. A successful program of land reclamation is SEAM. 24:7
11. Recycling reduces the amount of new natural resources needed per year, pollution from wastes, and the amount of energy needed to process resources. 24:8

VOCABULARY

a. alloy
b. backfilling
c. contact metamorphic deposits
d. gangue
e. high-temperature vein deposits
f. igneous ore deposits
g. metallic ores
h. nonmetallic deposits
i. nonrenewable resources
j. ores
k. placers
l. quarries
m. sedimentary ore deposits
n. shaft mines
o. strip mines
p. Surface Environment and Mining Program

Matching

Match each description with the correct vocabulary word from the list above. Some words will not be used.

1. local concentrations of minerals formed at the contact between an igneous intrusion and the surrounding rock
2. may be used to dispose of waste rock from shaft mines
3. waste rock from mining
4. open pits from which limestone, gravel, and sand are removed
5. accumulations of useful materials from Earth's crust that cannot be recycled or replaced in a useful period of time
6. deposits in which a mineral exists in large enough quantity that it can be mined at a profit
7. deposits formed where a stream's velocity decreases
8. underground mines
9. forms when two or more metals combine
10. mines where materials are removed at Earth's surface

Chapter 24 Review

MAIN IDEAS

A. Reviewing Concepts

Choose the word or phrase that correctly completes each of the following sentences.

1. High-temperature vein deposits are formed by *(contact metamorphism, fluids from magma, alloys)*.
2. Heavy minerals sink to the bottom of the magma chamber in *(igneous, contact metamorphic, sedimentary)* ore deposits.
3. *(Gold, Silver, Graphite)* is a nonmetallic mineral.
4. The hardest known natural substance is *(iron, diamond, gold)*.
5. *(Graphite, Sandstone, Silver)* is associated with contact metamorphic deposits.
6. Marble and sandstone are removed from *(strip mines, quarries, shaft mines)*.
7. Phosphates, nitrates, and potash are used in *(ceramics, fertilizers, building materials)*.
8. Coal, gold, and gravel are *(nonrenewable resources, renewable resources, minerals)*.
9. *(Metals, Crystals, Nonmetals)* can be shaped by pounding without loss of strength.
10. At least *(20, 8, 35)* nonmetallic materials are necessary to industry.
11. Nickel, copper, and lead are *(nonmetallic, metallic, sedimentary)* ore deposits.
12. The process used to refine metals is *(smelting, recycling, backfilling.)*
13. The original source of all Earth's rock and minerals is *(contact metamorphic deposits, metallic ore, magma)*.
14. Steel is a(n) *(alloy, nonmetal, high-temperature vein deposit)*.
15. When groundwater comes in contact with pyrite, *(an alloy, sulfuric acid, gangue)* is formed.

B. Understanding Concepts

Answer the following questions using complete sentences.

16. How do high-temperature vein deposits form?
17. Discuss the three considerations that determine if an ore deposit is profitable for mining.
18. How do sedimentary ore deposits form?
19. Explain why metals are valuable.
20. Explain why gangue can become valuable.
21. Why are most diamonds used as industrial abrasives?
22. Describe the advantages of recycling.
23. Describe how rubies and sapphires form.
24. What is brass?
25. Describe three uses for sulfur.

C. Applying Concepts

Answer the following questions using complete sentences.

26. How does a prospector know where to look for gold placers?
27. How do shaft and strip mines differ?
28. Why are some ores unprofitable to mine?
29. How is coal mined? Why do many coal mining operations require restoration of the mining area?
30. How can gold be concentrated by weathering and erosion?

SKILL REVIEW

1. Why did you have the third jar of tap water in Investigation 24–2?

2. Use Table 24–1 to determine which type of ore formed as a contact metamorphic deposit can be used as a lubricant.
3. Use Table 24–1 to determine one use for the sedimentary deposit with the chemical formula NaCl.
4. Compare and contrast renewable and nonrenewable resources.
5. What are three effects of mining on the environment?

PROJECTS

1. Repeat Investigation 24–2, "Pollution from Mining." This time add a few drops of ammonia to the water in each jar. Describe what the addition of a base to the water does to your results.
2. Design a better way to recover placers than by panning.

READINGS

1. Barnhardt, Wilton. "The Death of Ducktown." *Discover*. October, 1987, pp. 34-43.
2. Cheney, Glenn A. *Mineral Resources*. New York: Franklin Watts, 1985.
3. Law, Dennis L. *Mine-Land Rehabilitation*. New York: Van Nostrand Reinhold, 1984.

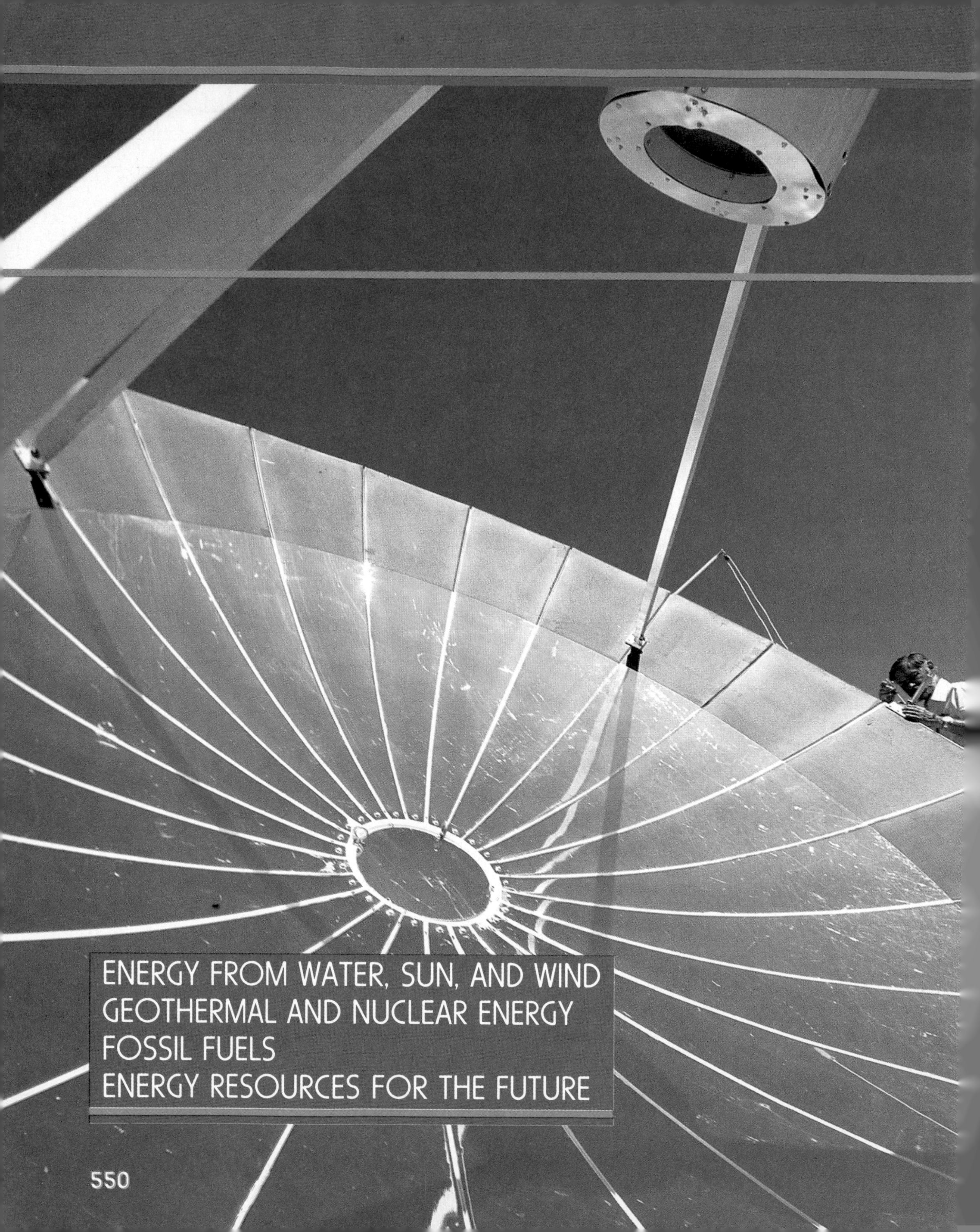

ENERGY FROM WATER, SUN, AND WIND
GEOTHERMAL AND NUCLEAR ENERGY
FOSSIL FUELS
ENERGY RESOURCES FOR THE FUTURE

Energy Resources

We live in an energy-dependent world. Think about your kitchen at home. Everywhere you look you find objects that require energy. The stove, microwave oven, electric can opener, and the mixer are just a few examples. Now think about the world outside your house. Energy is needed for cars and factories, to heat businesses and schools, and for many other activities we often take for granted. Where do we get all this energy?

ENERGY FROM WATER, SUN, AND WIND

GOALS

1. You will learn how water, the sun, and wind are used to produce energy.
2. You will learn advantages and disadvantages of these energy sources.

Energy is the ability or the capacity to do work. In order for work to be done, energy must be used. We use energy sources such as the sun, wind, and fossil fuels. Fossil fuels are limited and are nonrenewable energy resources. Resources such as water, the sun, and wind are alternatives to the use of fossil fuels.

25:1 Water Energy

Using water to do work is not a new concept. Tidal mills for grain processing existed on both sides of the English Channel and in Spain as early as the eleventh century. Water wheels six meters in diameter were installed in 1580 under the arches of the London Bridge. They were still pumping a portion of the city's water supply in 1824. Today, energy from running water is used to turn turbines to make electricity. Many large rivers in the Pacific Northwest have dams that are used to generate hydroelectric power. In Europe and New England, many small streams provide electricity for local areas.

Water power is fairly clean and cheap. It produces no air pollution or waste and is an efficient means of making

a

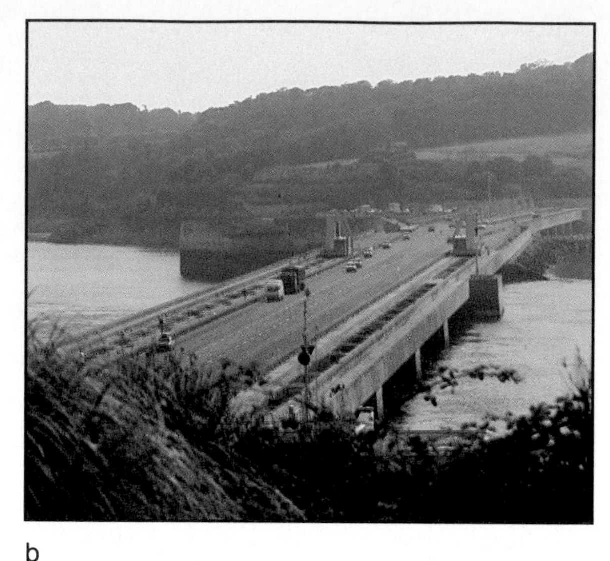

b

FIGURE 25–1. Dams produce electricity from the energy of falling water (a). A dam across the Rance River in France harnesses the power of tides with 24 special turbines (b).

electricity. However, water falling over high dams may acquire nitrogen that can kill fish. The lakes formed behind the dams may cover farmland or scenic areas. Dams also disrupt sediment flow and cause silting of lakes and reservoirs upstream and erosion downstream.

What are some disadvantages of water power?

The ocean is an enormous energy resource that can be tapped in some areas of the world. **Tidal power** is the use of ocean tides to produce electricity. Water flows in and out of bays and river openings as tides rise and fall. If a dam is constructed across the narrow openings, the passage of water can be used to turn electric generators. Tidal power is possible in only a few areas of the world due to low tidal ranges, lack of narrow openings for the water to flow through, cost of operation, and effects on ocean life. The area of the Rance River in France has the conditions necessary for tidal power. Tidal power has been used there for over 20 years. There are possible sites in eastern Canada, the coasts of Maine and Alaska in the United States, small areas of the Soviet Union and China, and the Kimberley Coast of Australia. While the use of tidal power is limited, it can be useful in reducing the amount of fossil fuels used for the world's energy needs.

Where can tidal power be used?

Another source of energy based on solar technology but involving ocean temperature differences between surface and deep waters is the **Ocean Thermal Energy Conversion (OTEC)** method. This method holds a fluid such as ammonia under low pressure. Warm surface waters are used to change liquid ammonia to ammonia vapors, which then drive a turbine. Cold, deep ocean water is pumped to the surface and is used to condense the am-

What is OTEC?

monia vapors to liquid again. The process of heating and cooling continues to power the turbine that generates electricity. Small OTEC plants have been operating since 1979. Larger OTEC plants are in the planning stages. It is unclear how effective OTEC energy sources will be because of high construction costs and limited numbers of acceptable sites.

25:2 Solar Energy

Many of our energy sources are related to solar energy. Solar energy constantly reaches Earth, where some of it is stored by green plants. When wood is burned, some of this stored energy is released. Animals receive solar energy when they eat plants. **Fossil fuels** such as oil, natural gas, and coal were formed during the decay of organisms that lived millions of years ago. When fossil fuels are burned, stored solar energy is released. Wind and wave energy are the result of solar heating. OTEC is possible because the sun heats surface water.

Most of the direct solar energy that strikes Earth is unused. Many attempts are being made to find ways to use solar energy. A **solar cell** collects energy from the sun and transforms it into electricity. Solar cells were first developed in the 1950s for generating electric power

BIOGRAPHY

Rene Dubos

1901–1982

Rene Dubos began his career as a microbiologist. Late in his career he began to write about the relationship between life and a healthy environment. In 1969, Dubos won a Pulitzer Prize for his book on the interrelationship between environment and humans.

TECHNOLOGY: ADVANCES

Solar Plane

Researchers for NASA have designed a solar plane capable of staying aloft for an entire year. Although it will have a wingspan of 98 meters and a 12-meter propeller, the plane will have a low mass because most of it will be made of light, strong wood. Solar energy will be collected by 3948 square meters of solar cells that will cover vertical arrays and hinged wing tips. During the day, the wing tips will swing to a 90-degree angle to receive maximum solar radiation. Excess electricity produced by the cells will be channeled to a fuel cell. At night, the wing tips will become horizontal to increase aerodynamic lift and the fuel cell will provide electricity. Uses for this craft, due to be built in 1992, include providing surveillance of coasts, monitoring of fires and blights, and serving as a communications relay.

FIGURE 25–2. This passive solar home has south-facing windows that collect the sun's heat.

on satellites. A major project at the Phoenix, Arizona, airport uses 7200 solar cells to supply electricity. This project tests the possibility of using solar cells for large-scale electricity generation.

What are some ways solar power is being used?

Solar energy also is used to heat and cool homes and buildings, to heat water, and to run radio stations. Flat-plate thermal collectors used in homes allow solar energy to be transferred by air or a liquid such as water. Reflective-dish tracking concentrators collect solar energy as they follow the sun across the sky. Some passive solar buildings are designed with south-facing vertical windows that absorb shortwave solar energy, which is converted to heat.

As more technology becomes available, solar energy will become a more common energy resource. Giant solar energy farms, each with thousands of square kilometers of collectors, could be located in desert areas that receive a large amount of sunlight. Or, solar energy might be collected by a large array of solar panels in space and transmitted to Earth in the form of microwaves. The receiving stations would convert the microwaves into usable electric power.

What are some problems associated with using direct solar power?

Problems with using direct solar power include the time of day, weather conditions, seasons, and latitude. These greatly affect the amount of solar energy that is available. Also, most of the devices used to collect solar energy are presently very expensive.

INVESTIGATION 25–1

Solar Energy

Problem: At what rate do materials absorb solar energy? How does the color of a material affect its ability to absorb energy?

Materials  

dry black soil
dry brown soil
dry white sandy soil
clear glass or plastic dishes (3)
thermometers (3)
200-watt lamp with reflector and clamp
watch or clock with second hand
ring stand
glass marker
graph paper
colored pencils (3)
metric ruler

Procedure

1. Use the glass marker to label the dishes A, B, and C.
2. Arrange the dishes close together on your desk.
3. Fill dish A with dry black soil to a depth of 2.5 cm.
4. Fill dish B to the same depth with dry brown soil.
5. Fill dish C to the same depth with dry white sandy soil.
6. Place a thermometer in each dish. Be sure to cover the thermometer bulb in each dish completely with the material.
7. Record the temperature of each dish in a table similar to the one shown.
8. Clamp the lamp to the ring stand and position over all three dishes. See Figure 25–3.
9. Turn on the lamp. Be sure the light shines equally on each dish.
10. Read the temperature of each material every 30 seconds for 25 minutes and record in your data table.
11. Use the data to construct a graph. Time should be plotted on the horizontal axis and temperature on the vertical axis. Use a different colored pencil to plot the data and draw the line for each material.

FIGURE 25–3.

Data and Observations

Time minutes	Temperature		
	Dish A	Dish B	Dish C
0			
0.5			

Analysis

1. Which material had the greatest temperature change?
2. Which material had the least change?

Conclusions and Applications

3. Why do the curves on the graph flatten?
4. Why do you think flat-plate solar collectors have black plates behind the water pipes?
5. How does the color of a material affect its ability to absorb energy?

25:3 Wind Power

How was wind power first used?

Wind power was first used to power sailing ships. The Dutch used windmills to grind corn and to pump water. In the United States during the early 1900s, metal windmills were designed to pump water in rural areas. Today, as the cost of electricity rises, windmills again are being used to generate energy.

F.Y.I. The blade on the Solano County, California, wind turbine is 100 meters long and weighs 90 tons.

In 1982, a wind turbine was installed on a hilltop in Solano County, California. The windmill is expected to provide enough power to generate electricity for 2500 average homes per year. For each 12 hours that this turbine operates at full power, one barrel of oil (152.5 L) is saved. The Solano area is bounded by the San Francisco Bay area and the Sacramento Valley. The temperature

PROBLEM SOLVING

Hot Spot

It was a hot summer day. Christie decided she wanted to make iced tea. She had heard that "sun tea" was really good. Her friend, Will, had told her that you use a large glass jar with a lid, tea bags, water, and the sun. She decided to make some.

She filled the jar up to the brim with water and put in four small tea bags. Then she screwed the lid tightly on the jar and placed the jar on the patio in the sun. She had forgotten to ask Will how long it took to make it, so she decided to check the jar in a half hour. At the end of the half hour, Christie found that only the top 1/4 of the jar had become tea-colored. The bottom 3/4 was still clear water. Also, she noticed when she touched the jar that the top was warmer than the bottom. She checked it again 30 minutes later to find that the top 1/2 was tea-colored, but the bottom was still clear. Finally, at the end of one more hour, the whole jar was tea-colored.

Christie added ice cubes to the warm tea and drank a tall glass. As she sipped the long-awaited beverage, she was puzzled by several questions. Why did only the top part of the water become tea at first? What made the tea so warm? Why was the top part of the jar warm before the bottom part?

FIGURE 25–4. Wind power has been used for centuries. How are these wind turbines used?

and pressure differences between these two areas cause large masses of air to move from cooler to warmer zones. This movement creates a good wind power environment. When the wind blows at 19 kilometers per hour or more, the rotor blade begins to turn. The spinning blade transfers energy to a generator that makes electricity. Technicians use computers to control the windmill. The computers keep the blade facing into the wind so that the maximum amount of energy can be obtained. If the wind speeds exceed 96 km/h, the turbine shuts itself down. It has endured hurricane force winds and fierce storms that toppled trees and electrical towers.

The Altamont Pass area is the wind energy capital of California. About 5000 wind turbines are located there on wind farms. The generating capacity for this area on a windy day may be as high as 500 000 kilowatts.

What is an advantage of using wind power?

Wind energy is a clean source of energy that does not cause pollution. However, using wind to produce electricity is limited to certain geographical areas. The best locations for wind turbine generators in the United States are the Great Plains, mountainous areas, and some coastal regions where strong winds are common.

Unfortunately, wind is not always available and sometimes it blows too hard to be useful. Research is currently being done on how to reduce the costs of storing and transporting electricity generated by wind power. New technologies may make wind a more useful resource.

Review

1. What is energy?
2. What are some of Earth's energy alternatives to the use of fossil fuels?
3. Why are tidal and ocean thermal power presently limited means of providing energy?
4. What are the advantages and disadvantages of wind power?
5. ★ Explain how tides are used today as an energy resource. Why is this a limited resource in the United States?

GEOTHERMAL AND NUCLEAR ENERGY

GOALS

1. You will learn why geothermal power is only available in certain geographic locations.
2. You will learn the advantages and disadvantages of using fission and fusion processes to produce energy.

Because fossil fuels are nonrenewable resources, people continue to search for and refine the use of alternative energy sources. Some sources are limited by their availability and the cost of the technology needed to tap them. Others are limited by the human health and environmental risks connected with their development and use. Geothermal and nuclear energy sources hold promise for the future in spite of some problems.

25:4 Geothermal Energy

What is geothermal energy?

Geothermal energy is the energy that comes from the internal heat of Earth. Oceanic ridges, plate boundaries, and volcanoes are areas of high heat flow. In some regions, heat from igneous rocks warms circulating groundwater. The water then escapes to the surface as steam in geysers or in hot springs. In northern California, steam taken from underground wells drives turbines to generate about fifty percent of the electricity for San Francisco. In some small western towns, steam is produced through drilled bore holes and is then piped to homes and industries. Italy has used geothermal energy since 1904 to generate electricity at Larderello. Iceland, Mexico, New Zealand, and Japan, all areas of volcanic activity, also use geothermal power.

Not all geothermal wells are related to igneous activity. Wells can be drilled down to rocks containing hot water rather than steam. The heat in these wells is the product of the decay of radioactive elements in Earth's crust. This

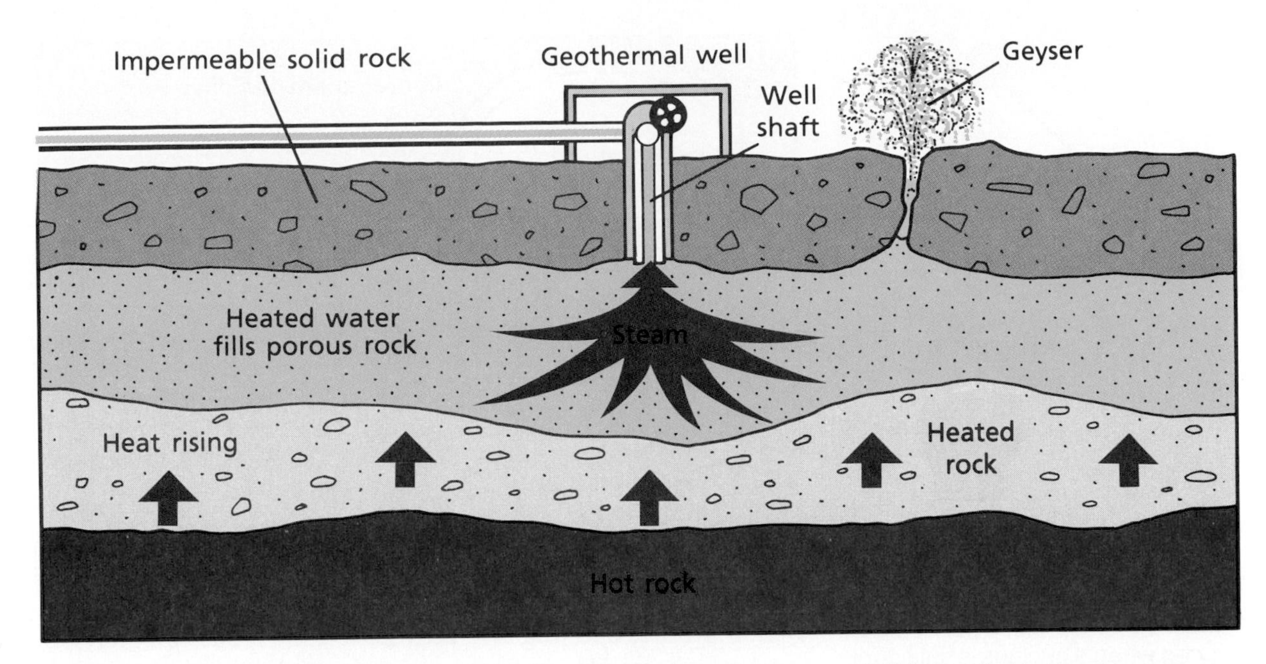

FIGURE 25–5. Geothermal energy in the form of steam is used in areas of igneous activity or where radioactive decay heats rocks near Earth's surface.

energy source is currently being used along the Gulf Coast of the United States. It may someday be used to supply hot water to homes and industries if the cost of drilling to the necessary depths can be lessened.

Environmental problems associated with geothermal energy do exist. Hot water under pressure dissolves and carries many minerals in solution. Thus, geothermal pollution includes the release of saline wastewater and the precipitation of salts in pipes and valves. Also, thermal pollution may occur from hot wastewater. Subsidence of the land may eventually take place if large quantities of groundwater are removed.

Presently, due to cost and the limited number of available places, geothermal sources supply only very small amounts of energy. This resource will become more important as fossil fuel resources become depleted.

Why is geothermal power not more widely used?

25:5 Nuclear Energy

Nuclear fuels provide more than six percent of the energy used in the United States. Nuclear energy is produced by **fission,** the splitting of the nuclei of atoms. The fuel for fission power plants comes from uranium-235, which is found concentrated in sandstones. Once the ore is mined, the uranium is processed into fuel rods or pellets. The fuel is loaded into the core of the reactor in a power plant. Neutrons hit the nuclei of the uranium-235 atoms. They break apart easily and release neutrons that hit

F.Y.I. The first commercial fission reactor was developed in Shippingport, Pennsylvania, in 1957.

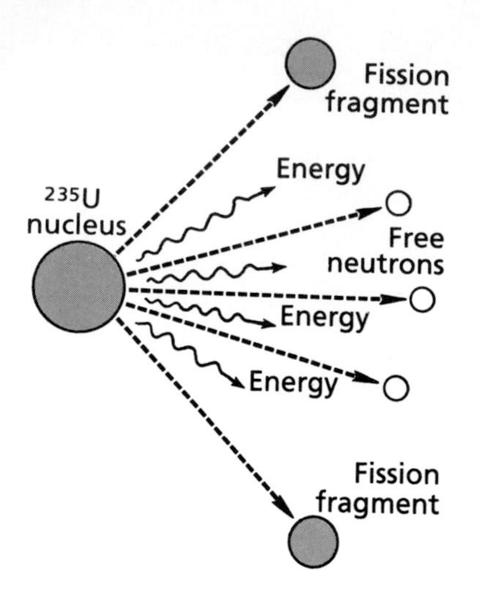

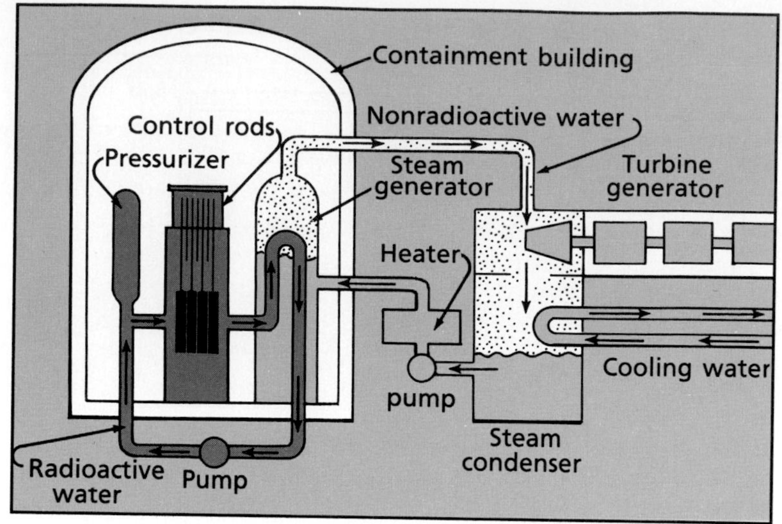

FIGURE 25–6. Energy is produced when a nucleus is split during fission (a). Control rods determine the speed at which nuclei split in a fission reactor (b).

other atoms, causing a chain reaction. Heat is released and may be used to produce steam, which powers a turbine to generate electricity.

Another type of reaction, called **breeding,** produces more fuel than it uses. In the breeding process, uranium-235 atoms undergo fission and release neutrons. These neutrons are absorbed by uranium-238, which changes to the element plutonium-239. Plutonium-239 undergoes fission and can be used to produce energy. Currently, France, Great Britain, and the U.S.S.R. are testing and using forms of breeder reactors.

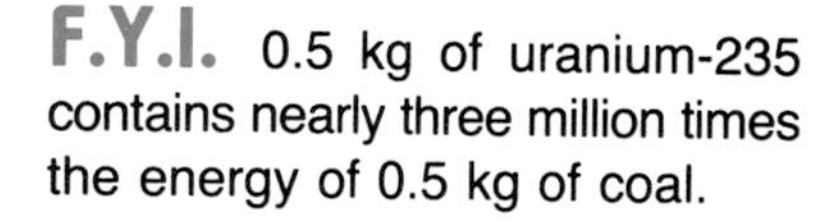

F.Y.I. 0.5 kg of uranium-235 contains nearly three million times the energy of 0.5 kg of coal.

Why do wastes from nuclear plants pose an environmental problem?

Relatively small amounts of uranium are needed to produce electricity compared to the amounts of coal or oil needed to produce the same amount of electricity. However, nuclear power generation has problems of safety and waste storage. For every 100 grams of uranium used, 99 grams remain as waste. These waste products release radiation, which destroys cells and changes or destroys genetic material. If these wastes are not recycled, they must be stored until they are no longer harmful. This storage period ranges from 600 to tens of thousands of years. Some waste has been stored in containers that have been dumped into the sea. Scientists are not sure how safe the present storage methods are. Tight safety and environmental regulations and increasing public opposition have caused a decrease in the number of new nuclear power plants being built.

Another nuclear process is **fusion.** In this process, the nuclei of hydrogen atoms join to release energy and form helium. The reaction is much the same as reactions in the sun. Fusion produces a waste that needs to be stored a shorter length of time than fission wastes before becoming harmless. The hydrogen needed as fuel can be obtained from certain types of water. Fusion is not practical at present because temperatures in the millions of degrees are needed for reactions to occur. Fusion processes require more research to solve the temperature problem, and will not be economically possible until sometime in the future.

Why is fusion impractical at the present time?

Review

6. List several sources of geothermal energy.
7. What are the pros and cons of nuclear energy?
8. What environmental problems are associated with geothermal energy?
9. Describe a breeding reaction. What are the advantages of breeder reactors?
10. Contrast fission and fusion reactions.
★ 11. Some nuclear wastes have been put in containers and dumped in the oceans. What problems could this cause?

CAREER

Powerhouse Supervisor

As a supervisor of powerhouse mechanics, Tamara Reno's job is to supervise and coordinate all of the activities of the electricians and mechanics that work in the Beaver Valley Hydroelectric Plant. She teaches workers how to repair and maintain hydroelectric turbines, generators, water-inlet and floodgate controls, valves, pumps, compressors, and pipes. As a supervisor, her ultimate responsibility is to be sure that the hydroelectric plant runs smoothly. Ms. Reno began working at the plant as a mechanic and was promoted to supervisor after several years. Her keen understanding of the workings of the machinery and her ability to work well with other people make her very effective.

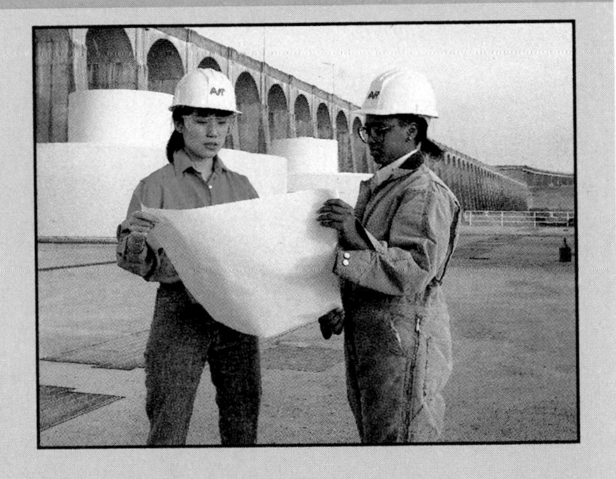

For career information, write:
National Hydropower Association
Suite 500
1133 21st N.W.
Washington, DC 20036

FOSSIL FUELS

GOALS

1. You will learn how fossil fuels are formed.
2. You will learn the advantages and disadvantages of using fossil fuels.

Fossil fuels are hydrocarbons formed during plant and animal decay millions of years ago. They are considered nonrenewable resources even though these fuels may be forming in some areas of the world today. Coal, oil, and natural gas are familiar fossil fuels.

25:6 Coal

Coal forms in swampy areas as the result of the decay of plants in the absence of oxygen. Certain bacteria attack the plants and release oxygen and hydrogen. Carbon builds up over a period of time. Coal goes through several changes during its formation. With time and increased pressure, impurities and moisture are removed. In swamps where coal forms, other sediments, such as sand, clay, and silt, are also deposited. The weight of the sediment presses on the organic matter beneath it. During this process, moisture and other materials are squeezed out, leaving a high carbon concentration.

F.Y.I. North America has 29% of the world's coal reserves.

Peat, the first stage in coal formation, is composed of about 75 to 90 percent water plus twigs, leaves, branches, and other plant parts. Although peat itself is not coal, it is an important fuel used in Ireland and the U.S.S.R. **Lignite,** the second stage in coal formation, is a soft, brown coal composed of compressed woody matter that

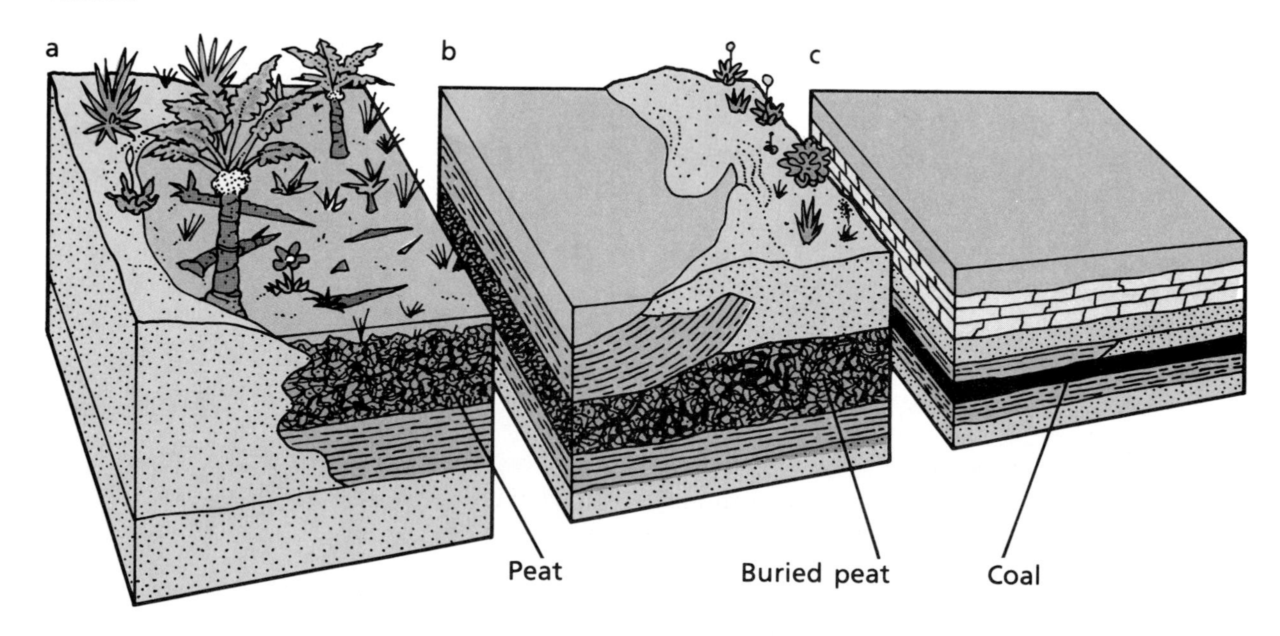

FIGURE 25–7. If peat (a) at the bottom of a swamp is buried and compressed (b), coal (c) may be formed.

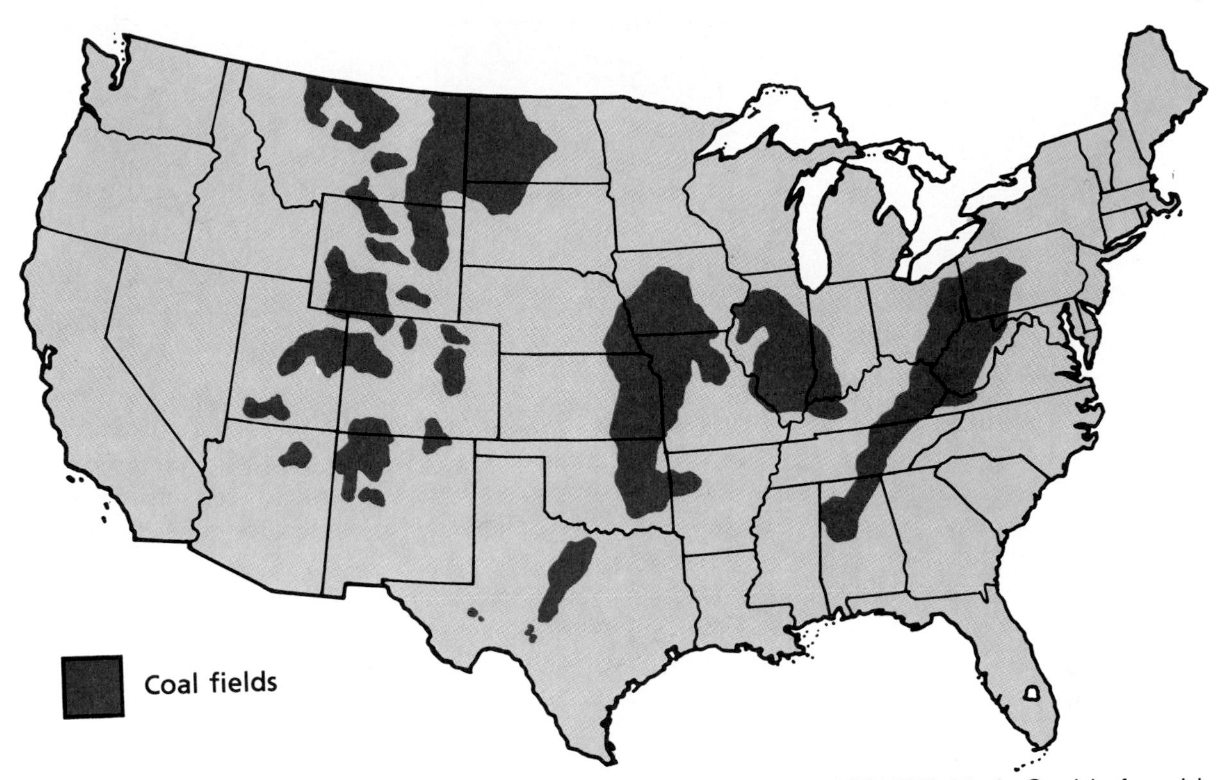

FIGURE 25–8. Coal is found in many areas throughout the United States.

has lost most of its moisture. It is used for local fuels in homes and industries. Germany uses its lignite to produce synthetic petroleum. **Bituminous coal,** or soft coal, is the third stage in coal formation. It is a dense, dark, brittle material that has lost all its moisture and most other impurities. It is ignited easily by a flame. Although bituminous coal is an efficient heating material, it produces a smoky yellow flame, ash, and sulfur compounds when it is burned. Strict emission laws have limited the amount of pollutants industries can release when this coal is burned. Bituminous coal is mined throughout the United States, with major fields in the Appalachians, the Great Plains, and the Colorado Plateau.

What problems are associated with the burning of bituminous coal?

Anthracite, or hard coal, is the final stage in coal formation. Lignite and bituminous coal are sedimentary rocks. Anthracite is a metamorphic rock. It is found only in areas of mountain building where heat and pressure were great. Anthracite is the cleanest of all coals with the fewest impurities because it is mostly carbon. It does not produce as much heat as bituminous coal, but it is preferred because it burns cleaner and longer. Anthracite fields occur in northeastern Pennsylvania, Great Britain, and parts of the U.S.S.R.

b

c

d

a

FIGURE 25–9. Stages in coal formation include peat (a), lignite (b), bituminous coal (c), and anthracite (d).

Problems associated with coal use include proper reclamation of mining sites, the greenhouse effect caused by burning, acid precipitation, and a disease of miners called black lung. Coal use has changed greatly in the last 100 years. Coal overtook wood as the major source of fuel in the 1880s. By 1925, coal provided 70 percent of the energy in the United States. Today coal provides only 20 percent of the energy needs for home heating, manufacturing, and generating electricity. As supplies of oil and natural gas decrease, coal again is becoming an important source of energy in many parts of the world. Coal reserves are estimated to last for several hundred years.

What problems are associated with coal use?

25:7 Petroleum and Natural Gas

Petroleum, or crude oil, and **natural gas** are important hydrocarbons that are found in nature within pores and fractures of rocks. Oil and gas form over millions of years as the result of the decay of marine organisms. These organisms die and collect on the ocean floor. Sediments such as clay and mud are deposited above these organisms. During burial and compaction, the organic matter becomes heated and chemical reactions occur. Petroleum and natural gas are formed. The fluids are forced out of the source rock into permeable beds such as sandstones.

Because oil and gas are not very dense, they migrate upward through water-saturated rock layers. In some cases, this movement is stopped by a cap rock such as shale or rock salt. The cap rock stops the upward movement and traps the hydrocarbons. The oil and natural gas form a reservoir in the rock. This type of hydrocarbon reservoir requires a source rock, a reservoir rock, and a cap rock. Most of the world's reservoirs are in sandstones, limestones, and dolomites.

Common oil traps are folds, faults, and salt domes. Oil and gas can be trapped in an anticline. The uppermost material found when drilling an oil well is usually natural gas. This gas is less dense than the oil. Water is the densest fluid and is found at the bottom of a reservoir.

Why are oil and natural gas found at the top of a reservoir?

When a hole is drilled through a cap rock, the natural gas and oil are forced from the reservoir rock into the drill hole. Sometimes there is enough natural pressure to force the oil from the drill hole. In most cases, a pump is used to extract the oil from the reservoir.

Natural gas was once burned at wells as waste. Today, natural gas is a very important fuel because it is the easiest fossil fuel to transport in pipelines and the cleanest to burn. Although natural gas often occurs with oil, some fields produce only natural gas. Both natural gas and oil supplies are limited and the cost of using these fuels is rapidly increasing. Each person in the United States uses the equivalent of a barrel of oil every six days. In 1986 the United States imported about four million barrels of oil each day. Before 1970, a barrel of oil cost $3.00. In 1973-74, the Arab oil embargo caused the price to rise to $12.00 per barrel. In 1981, the price rose to $37.00. However, in August of 1987, the price dropped to below $20.00 per barrel because the Organization of Petroleum Exporting Countries (OPEC) was overproducing oil. This drop in price came with the oversupply.

FIGURE 25–10. Gas and oil occur in a reservoir rock according to their densities (a). Reservoirs formed where permeable rocks grade into impermeable layers are called pinch outs (b). Where does oil collect?

a

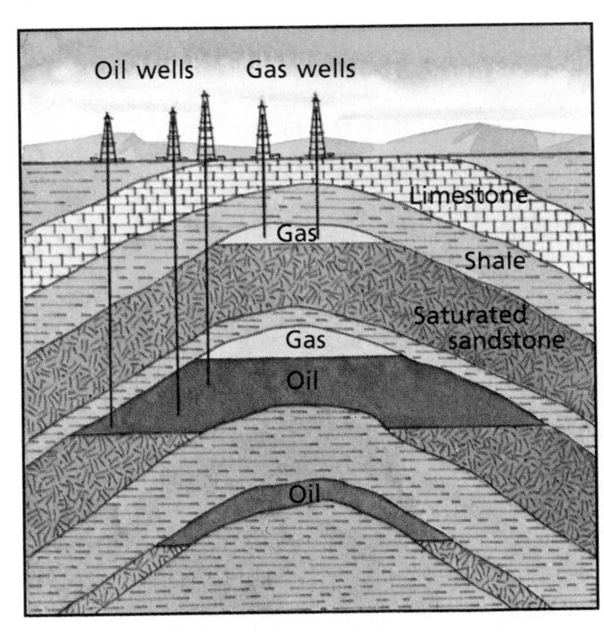

b

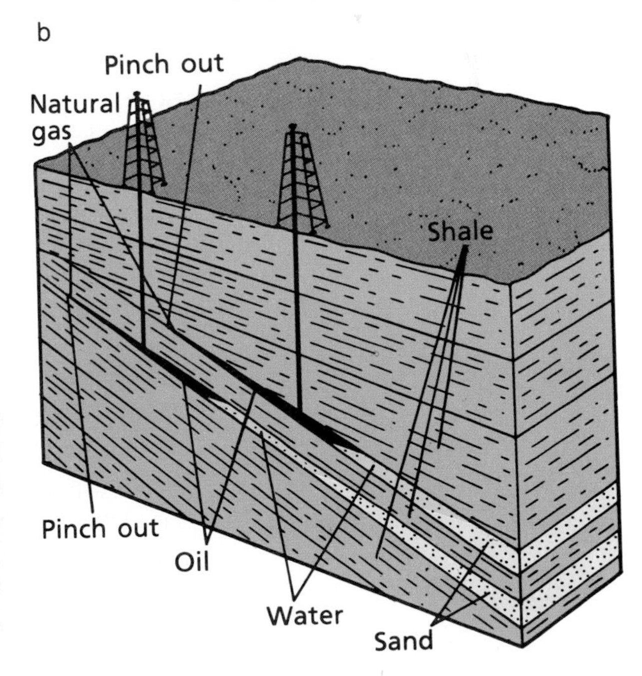

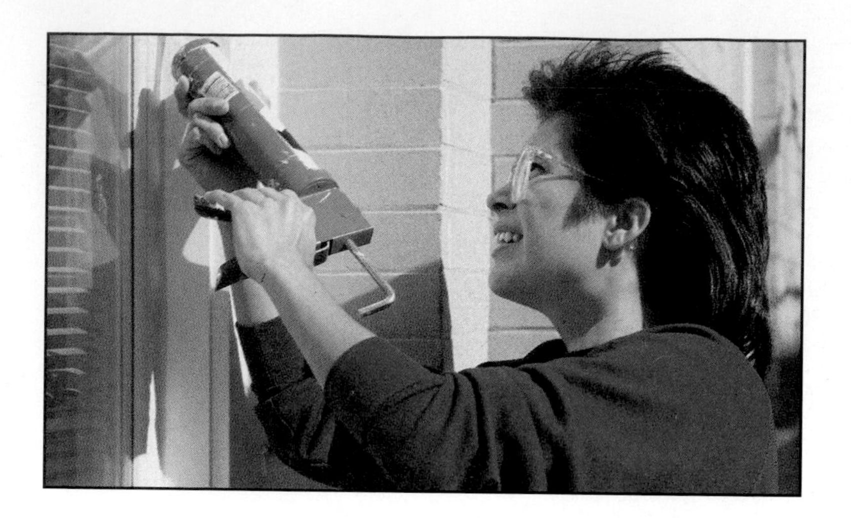

FIGURE 25–11. Insulating buildings is one way to conserve fossil fuel reserves.

How can more oil be removed from a well?

Presently, only about 30 percent of the crude oil in a well can be recovered. Secondary recovery methods can be used to increase the amount of crude oil that can be pumped from wells. Steam, carbon dioxide, and detergents can be used to force out the heavy oil that normally remains in the ground.

Because oil and natural gas are nonrenewable natural resources, each person must take an active part in the conservation of these valuable energy resources. One important way to conserve energy is to reduce the amount of energy needed for the heating and cooling of buildings. Use of weather-stripping and double-paned windows will aid in reducing the amount of heated or cooled air that is lost. Insulation can be added to walls and ceilings to make homes more energy efficient. Carpooling, operating fuel-efficient cars and furnaces, and the lowering of home thermostats during the winter months are just a few examples of energy-saving measures. How can you assure adequate supplies of oil and gas for the future?

Review

12. Why is coal a nonrenewable resource?
13. Why is anthracite considered a metamorphic rock?
14. Describe how oil and natural gas are formed and trapped beneath Earth's surface.
15. Describe some secondary methods used to increase the amount of oil recovered from a well.
★ **16.** How has the importance of coal as an energy resource changed in the last 100 years?

SKILL
Predicting Outcomes

Problem: How can outcomes be predicted?

Materials

Figure 25–12
pencil and paper

Background

Many times scientists must be able to predict outcomes. One way of predicting outcomes is to analyze trends. Trends are general movements or directions that have happened in the past. In this skill, you will learn ways of predicting outcomes.

Procedure

1. Examine Figure 25–12. It shows the billions of cubic meters of natural gas in U.S. reserves (identified deposits) and in marketed production from 1925 to 1985. Answer the following questions.
   A. Which best describes the trend in reserves between 1925 and 1965? Between 1965 and 1985?
      a. Reserves increased.
      b. Reserves decreased.
      c. Reserves remained constant.
   B. Which best describes the trend in marketed production between 1945 and 1975? Between 1975 and 1985?
      a. Marketed production increased.
      b. Marketed production decreased.
      c. It remained constant.
2. When analyzing trends, you need to understand why they happened the way they did. Answer the following questions.
   A. Why do you think the U.S. gas reserves increased from 1925 to 1965, but have decreased since that time?
   B. Why did marketed production increase between 1945 and 1975 and then drop?

FIGURE 25–12.

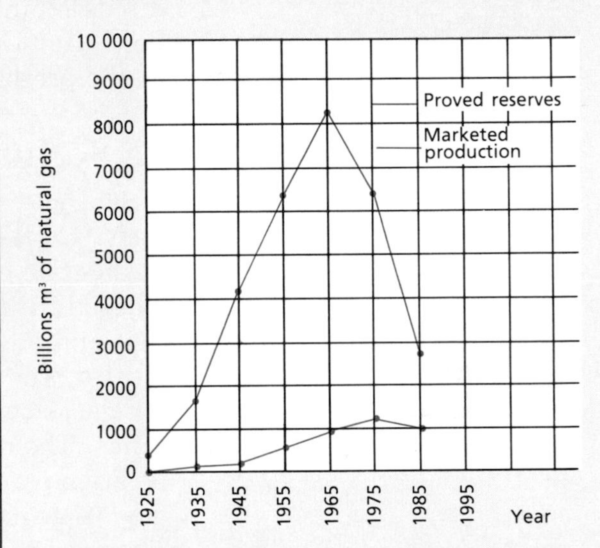

3. In order to predict the future when examining a graph, a method called extrapolation is used. Extrapolation assumes that the slopes of the curves will not change a great deal and that the latest trend will continue. Use this method to answer the following questions.
   A. If trends continue in reserves and marketed production, about how many billion cubic meters of gas will remain in 1990?
   B. If the trends continue in both areas, about when will the gas reserves run out?
4. Sometimes when predicting outcomes you must take into account that some factors will change and make your predictions while considering these changes. Predict what could happen to the reserves if people really started conserving gas and the marketed production dropped to where it was in 1945. Choose the best answer.
   A. Natural gas reserves will last longer.
   B. Gas reserves will disappear faster.

ENERGY RESOURCES FOR THE FUTURE

GOALS

1. You will lean about synfuels and their advantages and disadvantages.
2. You will learn about biomass fuels and how they can be produced.

The human population is making more and more demands on Earth's energy resources. Many fuel reserves are becoming exhausted. We must look to new and improved ways of getting energy. Some of the energy sources discussed in this section are not new, but they have not been used because they are more costly than fossil fuels at the present time.

25:8 Synfuels

Synthetic fuels, or synfuels, are an alternative to fossil fuels. **Coal gasification** is a process in which coal is changed to other products. This technology was first developed in 1909. Hydrogen is added to a paste of ground coal and oil and the mixture is heated with a catalyst. Heating the mixture produces a liquid similar to natural oil, gases such as natural gas and hydrogen, and sulfur solids. The only unwanted product is ash. One coal gasification plant converts coal into 55 000 barrels of oil each day. In a similar process, coal is ground to a fine powder. Then it is treated with steam and oxygen. A gas forms

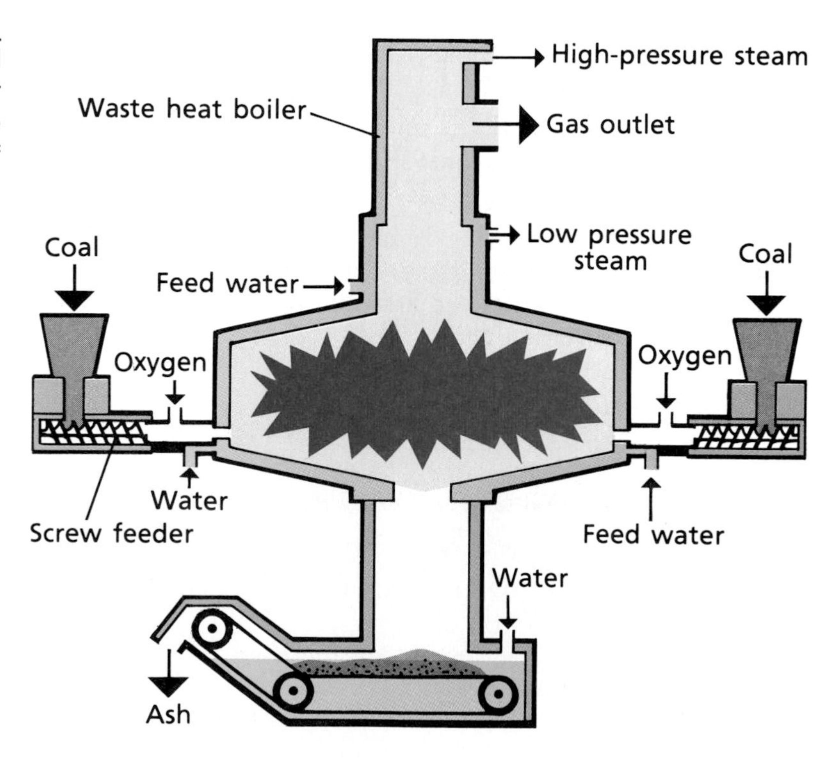

FIGURE 25–13. Coal is mixed with steam and oxygen under controlled conditions to produce natural gas during the process of coal gasification.

a

b

FIGURE 25–14. Oil shales are a potential source of energy (a), but mining them produces much waste rock (b).

that is purified to yield natural gas or methane. Coal gasification processes are complex and expensive. Research is underway to make them less costly. Coal synfuels used to cost 1.5 times as much as oil, but projections show that they may be economical by 1990.

Another synfuel is crude oil released from certain shales by heating. **Oil shales** are rocks that contain a waxy hydrocarbon material capable of producing at least 38 liters of oil per ton. The shale is crushed and heated until the hydrocarbon material changes into heavy crude oil. This process of extracting oil from shale is costly. However, oil shales may be used in the near future. Some scientists expect that by 1992, the United States will be producing two million barrels of oil daily from oil shales. Scientists estimate that between 600 billion and one trillion barrels of oil from shales are recoverable in the United States.

What problems are involved with the mining and processing of oil shales?

Shales produce a large amount of waste rock, which is an environmental problem. Also, large amounts of water are needed for the processing of shales. Water is usually scarce where these shales are located. The mining and transport of the shales create dust. Ore processing also emits pollutants into the air. Therefore, shale processing is being closely watched by the EPA.

Oil can also be removed from tar sands. **Tar sands** are sandstones saturated with a heavy oil that looks like tar. About 550 tar sand deposits have been found in the United States, most of them in Utah. Geologists believe Canada has the world's largest deposits of tar sands. The oil is removed from the sand by heating it or by injecting steam,

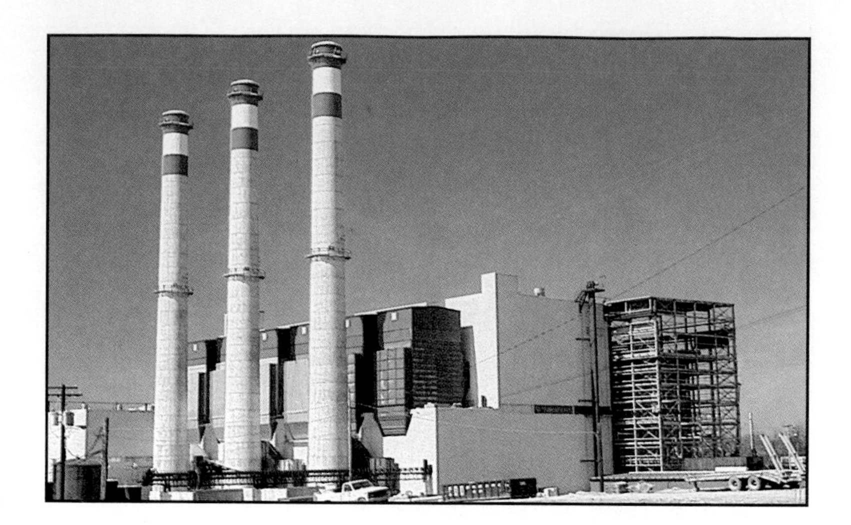

FIGURE 25–15. This trash-burning power plant uses municipal waste mixed with coal to produce electricity.

which liquefies the oil. Geologists estimate that at least 300 billion barrels of oil can be recovered from tar sands, but the recovery cost is high. Scientists predict that at least 60 000 barrels of oil per day could be mined by 1990, and 200 000 barrels per day by the year 2000.

25:9 Biomass Fuels

What are biomass fuels?

Biomass fuels are other important energy sources for the future. These burnable fuels are made from organic matter. Wood, one of the oldest known biomass fuels, is again gaining popularity as an important energy resource. A utility in Vermont is building an electric power plant that will use wood chips as fuel. Scientists are experimenting with making methane by fermenting kelp in vats that have been seeded with bacteria. In some cities, methane is being extracted from municipal sewage. One biological laboratory in Utah is using weeds to make crude oil. Biomass fuels can be added to fossil fuels to extend the supplies. In Columbus, Ohio, municipal garbage and trash are burned along with coal to heat water to make steam. The steam is used to generate electricity. Wastes from these processes can be a problem. Ash deposits must be disposed of properly to prevent possible groundwater contamination. If the processes for converting solid wastes to energy can be made profitable and a way to dispose of wastes can be found, cities will have an alternative to simply filling more and more landfills.

Another type of biomass fuel is gasohol. **Gasohol** is a combination of 10 percent alcohol and 90 percent unleaded gasoline. Alcohol can be produced from any plant material, including corn and sugarcane. One bushel of

INVESTIGATION 25–2

Insulation

Problem: Which material is the best insulator?

Materials  

soft drink cans (4)
thermometers (4)
clock or watch with second hand
tepid water
200-watt lamp with reflector and clamp
ring stand
newspaper
aluminum foil
large polystyrene cup with lid
tape
clay
funnel
graph paper
colored pencils (4)

Procedure

1. Wrap can 1 with a thick layer of newspaper. Use tape to secure the newspaper. Be sure to cover the top of the can.
2. Wrap can 2 with the aluminum foil, shiny side out. Cover the top of the can.
3. Place can 3 inside the polystyrene cup. Do not put on the lid.
4. Leave can 4 uncovered.
5. Punch a small hole through the newspaper and foil on the tops of cans 1 and 2. Use the funnel to carefully fill each can with tepid water.
6. Fill cans 3 and 4 with tepid water.
7. Place the lid on can 3.
8. Insert a thermometer into the hole in the top of each can and secure with clay. You will first need to punch a hole in the lid of can 3.
9. Make a table for recording the data for Steps 11 and 13.
10. Attach the lamp to the ring stand and turn it on. Be sure that the light shines equally on all cans. **CAUTION:** *The lamp may cause burns.*

FIGURE 25–16.

11. Record the temperature in each can every two minutes for 20 minutes.
12. Turn off the lamp.
13. Record the temperature every two minutes for another 20 minutes.
14. Construct a graph to plot both the heating and cooling phases. Use a different colored pencil to plot the data for each can.

Analysis

1. Which can increased in temperature most quickly when the light was turned on?
2. Which can increased least quickly?
3. Which can decreased in temperature most rapidly when the light was turned off?
4. Which can decreased the slowest?

Conclusions and Applications

5. Of the materials you tried, which were the best insulators against heat absorption? Against heat loss?
6. Does insulation prevent heat gain or loss? Explain.

FIGURE 25–17. One way to aid fossil fuel conservation is to develop transportation alternatives such as mass transit or electric-powered or solar-powered cars.

corn yields 9.5 liters of alcohol, which is then mixed with gasoline. Cars can use fuel with as much as 20 percent alcohol content. There is conflict among some people over the use of grains for fuel production instead of for food. The issue continues to be debated.

As we progress into the 21st century, many of Earth's energy resources described in this chapter will become more important. Improved energy efficiency and conservation, more fuel-efficient cars, better insulating materials, and the development of fossil fuel alternatives will help extend our limited fossil fuel resources. It is important that technologies continue to advance the use of alternative sources of energy. Different sources will be needed for different geographic areas.

Review

17. Describe two coal gasification processes and the products they yield.
18. What are the major problems with producing oil from oil shales?
19. What are tar sands?
20. How is oil removed from tar sands?
21. What is biomass fuel?
★ **22.** Why will energy from synthetic and biomass fuels become more important in the future?

Chapter 25 Review

SUMMARY

1. Dams, tidal power, and OTEC are uses of water resources. 25:1
2. Fossil fuels, such as oil, coal, and natural gas, contain stored solar energy. Direct solar energy is being tested as a potential energy source. 25:2
3. Wind can be used to turn turbines and produce electricity. 25:3
4. Geothermal energy comes from the internal heat of Earth. 25:4
5. Electricity can be produced from nuclear energy. Fission is now in use, but fusion may be possible in the future. 25:5
6. Stages in coal formation are peat, lignite, bituminous coal, and anthracite. 25:6
7. Oil and natural gas form as the result of the decay of marine organisms. These energy sources, like coal, are limited. 25:7
8. Steam, carbon dioxide, and detergents can be pumped into oil wells to force out oil from a well. 25:7
9. The natural gas and oil produced during coal gasification and the oil extracted from oil shale and tar sands are synfuels. 25:8
10. Biomass fuels are made from organic matter. Gasohol is a biomass fuel composed of gasoline and alcohol. 25:9
11. Energy efficiency and conservation are important factors in extending Earth's energy resources. 25:9

VOCABULARY

a. anthracite
b. biomass fuels
c. bituminous coal
d. breeding
e. coal gasification
f. fission
g. fossil fuels
h. fusion
i. gasohol
j. geothermal energy
k. lignite
l. natural gas
m. Ocean Thermal Energy Conversion (OTEC)
n. oil shales
o. peat
p. petroleum
q. solar cell
r. tar sands
s. tidal power

Matching

Match each description with the correct vocabulary word from the list above. Some words will not be used.

1. a metamorphic rock
2. the first stage in coal formation
3. energy produced by splitting uranium-235 atoms
4. process where coal is changed to natural gas and oil
5. the nuclear process that occurs in the sun
6. burnable fuel made from organic matter
7. fuel that is a combination of alcohol and gasoline
8. reaction that uses uranium-238 as a major fuel source
9. energy that comes from the internal heat of Earth
10. collects energy from the sun and transforms it into electricity

Chapter 25 Review

MAIN IDEAS

A. Reviewing Concepts

Choose the word or phrase that correctly completes each of the following sentences.

1. A substitute for natural gas can be produced from *(coal, limestone, sandstone)*.

2. Within an oil-bearing layer of rock, the densest fluid is *(water, oil, gas)*.

3. Solar energy collected in space could be transmitted to Earth by *(radio waves, seismic waves, microwaves)*.

4. The hardest type of coal is *(anthracite, lignite, bituminous coal)*.

5. A problem with producing electricity from wind turbines is *(air pollution, lack of or too much wind, water pollution)*.

6. *(Coal, Wind, Synfuel)* is a clean source of energy that produces little pollution.

7. An example of a biomass fuel is *(wind, water, wood)*.

8. The most abundant fossil fuel, which should last for hundreds of years, is *(natural gas, coal, crude oil)*.

9. About 50 percent of San Francisco's electricity is produced from *(fusion, solar, geothermal)* energy.

10. When fossil fuels are burned, *(solar, geothermal, fission)* energy that was stored by ancient organisms is released.

11. The fuel for fission power in the United States is *(lignite, uranium-235, plutonium-239)*.

12. *(Natural gas, Coal, Oil)* is a source of energy that comes from tar sands.

13. Black lung disease is associated with the removal of *(natural gas, crude oil, coal)* from Earth's crust.

14. The easiest fossil fuel to transport and the cleanest to burn is *(natural gas, coal, crude oil)*.

15. The process that uses cold, deep ocean water to condense ammonia vapor is *(OPEC, OTEC, BREEDING.)*

B. Understanding Concepts

Answer the following questions using complete sentences.

16. Explain the disadvantages of using the Ocean Thermal Energy Conversion method of power.

17. Describe the origin of fossil fuels.

18. Why are alternative methods of producing energy being sought?

19. What environmental problems develop when rivers are dammed?

20. Describe how power can be obtained from the tides.

21. How are passive solar buildings designed?

22. Describe why Solano County, California, is a good place to install a wind turbine.

23. Why is geothermal energy a limited source of energy?

24. Describe the differences between bituminous coal and anthracite.

25. Explain why oil prices can change so drastically.

C. Applying Concepts

Answer the following questions using complete sentences.

26. How can biomass fuels be used to help extend fossil fuel supplies?

27. How can large coal reserves be used to help supply natural gas and gasoline?

28. Why is fusion the safest and best way to produce nuclear energy?

29. Why are energy efficiency and conservation so very important today?

30. Why is only a small amount of oil being extracted from oil shales at present?

SKILL REVIEW

1. Use the table below to describe why it is difficult to make short term predictions for the cost of a barrel of oil.

Date	Price per barrel of oil
Before 1970	$3.00
1973-1974	$12.00
1981	$37.00
1987	< $20.00

2. Refer to Section 25:7. Why is it reasonable to make the long-term prediction that the prices of oil and natural gas will increase?
3. Arrange the following steps in the order in which they occur.
   A. Bituminous coal forms.
   B. Plants die and decay.
   C. Anthracite forms.
   D. Plants grow in swamp.
   E. Lignite is formed.
   F. Peat forms.
4. Which can was the control in Investigation 25–2, "Insulation"?
5. Compare and contrast the fission and fusion processes.

PROJECTS

1. Design and make a three-dimensional model of a futuristic energy-efficient building.
2. Conduct an experiment using a silicon solar cell and a milliammeter to measure available solar energy.

READINGS

1. Hawkes, Nigel. *Oil.* New York: Franklin Watts, 1985.
2. Johnson, Gary L. *Wind Energy Systems.* Englewood Cliffs, NJ: Prentice-Hall, 1985.
3. Petersen, David. *Solar Energy at Work.* Chicago: Childrens Press, 1985.

SCIENCE AND SOCIETY

HYDROELECTRIC POWER AND THE ENVIRONMENT

Harnessing the energy of falling water is one of the oldest known techniques for generating power. Before the advent of electrical power, our ancestors used water power to grind their grain, run their mills and factories, and power their ships. Now it is used to turn the blades of turbines to generate electricity. Dams are placed across rivers or streams, creating a lake behind the dam. The water is then released at a controlled rate over the dam to turn the turbines. The fuel—water—is free and renewable, and there is relatively no pollution. The only major costs involved are the construction and maintenance of the dam.

Background

In the early days of this country, most of the hydroelectric projects were small. Now, however, giant, federally funded hydroelectric projects dot the country's rivers. The best known of these in the eastern United States is probably the Tennessee Valley Authority or T.V.A. There are many others in the western and southwestern parts of the country. These large projects do more than just provide power. They also provide flood control to the local area and water to major cities located kilometers away from the sites. The water can be stored in reservoirs and released when needed.

The environmental cost of these large projects is high. As early as 1913, John Muir and other early conservationists fought the construction of the Hetch Hetchy reservoir that flooded one of the most beautiful valleys in Yosemite National Park. The impact on an area can be seen by examining Arizona's Glen Canyon Dam on the Colorado River. The dam, started in the 1950s and completed in 1963, formed Lake Powell behind it. The water from the lake flows past the dam, cuts through the Grand Canyon, and eventually flows into Lake Mead behind the Hoover Dam. The complex of dams has affected the whole area. The most obvious change is that there are now two lakes in what used to be valleys. These lakes are indeed scenic recreation areas but the price for these was the destruction of kilometers of natural canyons.

Case Studies

1. There are more subtle or less obvious changes from the Hoover Dam system. Although one purpose of the dams is to provide water for dry areas, there is actually

FIGURE 1. Glen Canyon Dam on the Colorado River provides hydroelectric power for parts of the southwest.

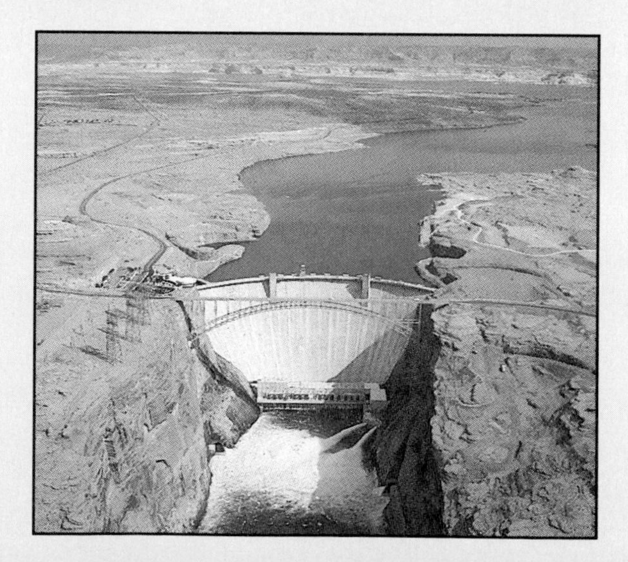

a net loss of water through increased evaporation from the large lake surfaces. The evaporation from Lake Powell alone amounts to about 270 000 cubic meters per year. Not only is this water lost to the area, the salts it leaves behind increase the salinity of the lake.

2. Another problem, common to all large hydroelectric projects, is that of silting. As a stream flows, it carries particles of small rocks, sand, and silt eroded from its banks and bed. In a lake, however, the water is relatively still and these particles settle out, building layers of sediments on the bottom. After many years, the lake can fill up with these materials. Before the Glen Canyon Dam was built, approximately 140 metric tons of sediments traveled through the Grand Canyon. These sediments acted as an abrasive to scour the banks and river bottom. Now most of that sediment load is deposited in Lake Mead. Below the lake, the water continues to erode the river banks, but it carries no material to replace what it wears away. The cycle of erosion and deposition has been altered.

3. Nutrients leached from the soil are also normally carried downstream by a river. This cycle has also been interrupted, leaving these nutrients in lakes rather than carrying them to the sea at the end of the river. Many species of plants and animals are deprived of their benefits.

4. Aquatic life is also affected. Water falling over a dam contains high amounts of nitrogen. This gas expands in the bloodstream of fish and kills them. Water temperature can also be a problem. In a natural situation, stream water released into the river below Glen Canyon comes from depths up to 60 meters below the surface of Lake Powell. The water temperature in August remains at 7°C because no sunlight penetrates to those depths. Fish adapted to breed in the warmer summer waters may not reproduce at all due to the colder water temperatures.

Developing a Viewpoint

Before any large, federally funded project is approved, a cost-benefit analysis is done. This study determines the value of the project by comparing the costs to the benefits. Costs are not limited to monetary costs, but can also include loss of land use, loss of one or more plant or animal species, and environmental damage. You are going to prepare a cost-benefit analysis for a large hydroelectric dam. Work in groups of four or five to complete the cost-benefit analysis for a proposed new hydroelectric power plant.

Divide a sheet of paper into two columns. Label them COSTS and BENEFITS. In the COSTS column, list all the problems or disadvantages your group can imagine. In the second column, list the BENEFITS. When you have finished your list, look at each item and rank its importance on a scale of 1 to 10, with 1 being low. When you have finished, total each column. Compare the results. Compare your group's results with others in the class and discuss the differences.

Few other energy sources are as clean and efficient as hyroelectric power. However, adequate sites are limited and there are environmental drawbacks. If you were to live in the area of a proposed new plant, what questions would you want the builders to answer? Would you be willing to give up your home and land to the project? Explain.

Suggested Readings

Daniel Deudney. "Hydropower: An Old Technology for a New Era." *Environment*. September, 1981.

William A. Loeb. "How Small Hydro is Growing Big." *Technology Review*. August-September, 1983.

"After 85 years, the Era of Big Dams Nears End." *New York Times*. January 24, 1987.

APPENDIX A

International System of Units

The International System (SI) of Units is accepted as the standard for measurement throughout most of the world. Three base units in SI are the meter, kilogram, and second. Frequently used SI units are listed below.

Table A–1

Frequently Used SI Units
LENGTH
1 millimeter (mm) = 1000 micrometers (µm)
1 centimeter (cm) = 10 millimeters (mm)
1 meter (m) = 100 centimeters (cm)
1 kilometer (km) = 1000 meters (m)
1 light-year = 9 460 000 000 000 kilometers (km)
AREA
1 square meter (m^2) = 10 000 square centimeters (cm^2)
1 square kilometer (km^2) = 1 000 000 square meters (m^2)
VOLUME
1 milliliter (mL) = 1 cubic centimeter (cc) (cm^3)
1 liter (L) = 1000 milliliters (mL)
MASS
1 gram (g) = 1000 milligrams (mg)
1 kilogram (kg) = 1000 grams (g)
1 metric ton = 1000 kilograms (kg)
TIME
1 s = 1 second

Temperature measurements in SI are often made in degrees Celsius. Celsius temperature is a supplementary unit derived from the base unit kelvin. The Celsius scale (°C) has 100 equal graduations between the freezing temperature (0°C) and the boiling temperature of water (100°C). The following relationship exists between the Celsius and kelvin temperature scales:

$$K = {}^{\circ}C + 273$$

Several other supplementary SI units are listed below.

Table A–2

Supplementary SI Units			
Measurement	**Unit**	**Symbol**	**Expressed in base units**
Energy	Joule	J	$kg \cdot m^2/s^2$
Force	Newton	N	$kg \cdot m/s^2$
Power	Watt	W	$kg \cdot m^2/s^3$ (J/s)
Pressure	Pascal	Pa	$kg/m \cdot s^2$ ($N \cdot m$)

APPENDIX B

Safety in the Science Classroom

1. Always obtain your teacher's permission before beginning an investigation.
2. Study the procedure. If you have questions, ask your teacher. Be sure you understand any safety symbols shown on the page.
3. Use the safety equipment provided for you. Goggles and a safety apron should be worn when any investigation calls for using chemicals.
4. Always slant test tubes away from yourself and others when heating them.
5. Never eat or drink in the laboratory, and never use laboratory glassware as food or drink containers. Never inhale chemicals. Do not taste any substance or draw any material into a tube with your mouth.
6. If you spill any chemical, wash it off immediately with water. Report the spill immediately to your teacher.
7. Know the location and proper use of the fire extinguisher, safety shower, fire blanket, first aid kit, and fire alarm.
8. Keep all materials away from open flames. Tie back long hair and loose clothing.
9. If a fire should break out in the classroom, or if your clothing should catch fire, smother it with the fire blanket or a coat, or get under a safety shower. **NEVER RUN.**
10. Report any accident or injury, no matter how small, to your teacher.

Follow these procedures as you clean up your work area.

1. Turn off the water and gas. Disconnect electrical devices.
2. Return all materials to their proper places.
3. Dispose of chemicals and other materials as directed by your teacher. Place broken glass and solid substances in the proper containers. Never discard materials in the sink.
4. Clean your work area.
5. Wash your hands thoroughly after working in the laboratory.

Table B–1

FIRST AID	
Injury	**Safe response**
Burns	Apply cold water. Call your teacher immediately.
Cuts and bruises	Stop any bleeding by applying direct pressure. Cover cuts with a clean dressing. Apply cold compresses to bruises. Call your teacher immediately.
Fainting	Leave the person lying down. Loosen any tight clothing and keep crowds away. Call your teacher immediately.
Foreign matter in eye	Flush with plenty of water. Use eyewash bottle or fountain.
Poisoning	Note the suspected poisoning agent and call your teacher immediately.
Any spills on skin	Flush with large amounts of water or use safety shower. Call your teacher immmediately.

APPENDIX C

Safety Symbols

 **DISPOSAL ALERT** This symbol appears when care must be taken to dispose of materials properly.	 **ANIMAL SAFETY** This symbol appears whenever live animals are studied and the safety of the animals and the students must be ensured.
 **BIOLOGICAL HAZARD** This symbol appears when there is danger involving bacteria, fungi, or protists.	 **RADIOACTIVE SAFETY** This symbol appears when radioactive materials are used.
 **OPEN FLAME ALERT** This symbol appears when use of an open flame could cause a fire or an explosion.	 **CLOTHING PROTECTION SAFETY** This symbol appears when substances used could stain or burn clothing.
 **THERMAL SAFETY** This symbol appears as a reminder to use caution when handling hot objects.	 **FIRE SAFETY** This symbol appears when care should be taken around open flames.
 **SHARP OBJECT SAFETY** This symbol appears when a danger of cuts or punctures caused by the use of sharp objects exists.	 **EXPLOSION SAFETY** This symbol appears when the misuse of chemicals could cause an explosion.
 **FUME SAFETY** This symbol appears when chemicals or chemical reactions could cause dangerous fumes.	 **EYE SAFETY** This symbol appears when a danger to the eyes exists. Safety goggles should be worn when this symbol appears.
 **ELECTRICAL SAFETY** This symbol appears when care should be taken when using electrical equipment.	 **POISON SAFETY** This symbol appears when poisonous substances are used.
 **PLANT SAFETY** This symbol appears when poisonous plants or plants with thorns are handled.	 **CHEMICAL SAFETY** This symbol appears when chemicals used can cause burns or are poisonous if absorbed through the skin.

APPENDIX D

Earth Data

Mean distance to sun	1.496×10^8 kilometers
Average velocity around sun	29.77 kilometers/second
Age	4.6×10^9 years
Mass	5.96×10^{27} grams
Average density	5.52 grams/cubic centimeter
Volume	1.08×10^{27} cubic centimeters
Surface area	5.1×10^{18} square centimeters
Percent surface area of oceans	71%
Percent surface area of land	29%
Polar diameter	12 714 kilometers
Equatorial diameter	12 756 kilometers
Polar circumference	40 008 kilometers
Equatorial circumference	40 075 kilometers
Continents	
Average elevation	623 meters
Highest elevation	8848 meters above sea level
Lowest elevation	300 meters below sea level
Oceans	
Average depth	3.8 kilometers
Deepest part	11 kilometers
Mass	1.4×10^{24} grams
Temperature	
Average surface	14°C
Highest recorded	58°C
Lowest recorded	−88°C
Atmosphere	
Height	~ 1600 kilometers above surface
Mass	5.1×10^{21} grams
Percent of nitrogen	78%
Percent of oxygen	21%
Percent of other gases	1%
Crust	
Average depth	35 kilometers
Average mass	2.5×10^{25} grams
Average temperature	870°C
Mantle	
Average depth	2900 kilometers
Average mass	4.05×10^{27} grams
Average temperature	2200°C
Core	
Average depth	6370 kilometers
Average mass	1.90×10^{27} grams
Average temperature	5000°C

APPENDIX E

Topographic Map Symbols

Primary highway, hard surface
Secondary highway, hard surface
Light-duty road, hard or improved surface
Unimproved road
Railroad: single track and multiple track
Railroads in juxtaposition

Buildings
School, church, and cemetery — cem
Buildings (barn, warehouse, etc.)
Wells other than water (labeled as to type) — o oil o gas
Tanks: oil, water, etc. (labeled only if water) — water
Located or landmark object; windmill
Open pit, mine, or quarry; prospect

Marsh (swamp)
Wooded marsh
Woods or brushwood
Vineyard
Land subject to controlled inundation
Submerged marsh
Mangrove
Orchard
Scrub
Urban area

Spot elevation — ×7369
Water elevation — 670

Index contour
Supplementary contour
Intermediate contour
Depression contours

Boundaries: National
State
County, parish, minicipio
Civil township, precinct, town, barrio
Incorporated city, village, town, hamlet
Reservation, National or State
Small park, cemetary, airport, etc.
Land grant
Township or range line, United States land survey
Township or range line, approximate location

Perennial streams
Elevated aqueduct
Water well and spring
Small rapids
Large rapids
Intermittent lake
Intermittent streams
Aqueduct tunnel
Glacier
Small falls
Large falls
Dry lake bed

APPENDIX F

Solar System Information

Planet	Mercury	Venus	Earth	Mars	Jupiter	Saturn	Uranus	Neptune	Pluto
Diameter (km)	4880	12 100	12 750	6800	142 800	120 660	51 800	49 500	~2400
Diameter (E = 1.0)*	0.38	0.95	1.00	0.53	11.23	9.46	4.06	3.88	0.19
Mass (E = 1.0)*	0.05	0.82	1.00	0.11	317.9	95.2	14.6	17.2	0.002
Density (g/cm^3)	5.2	5.3	5.5	3.8	1.3	0.7	1.3	1.7	2.1
Period of rotation days hours minutes	58 15 00	243 4 00	00 23 56	00 24 37	00 9 55	00 10 39	00 17 14	00 17 00	6 9 17
Surface gravity (E = 1.0)	0.39	0.90	1.00	0.38	2.58	1.11	1.07	1.40	?
Average distance to sun (AU)	0.387	0.723	1.000	1.524	5.20	9.54	19.18	30.06	39.44
Period of revolution	87.96 d	224.7 d	365.26 d	687 d	11.86 y	29.46 y	84.01 y	164.79 y	248 y
Eccentricity of orbit	0.206	0.007	0.0167	0.093	0.048	0.056	0.047	0.009	0.250
Average orbital speed (km/s)	47.8	35.0	29.8	24.2	13.1	9.7	6.8	5.4	4.7
Number of known satellites	0	0	1	2	16	~20	15	2	1

*Earth = 1.0

APPENDIX G

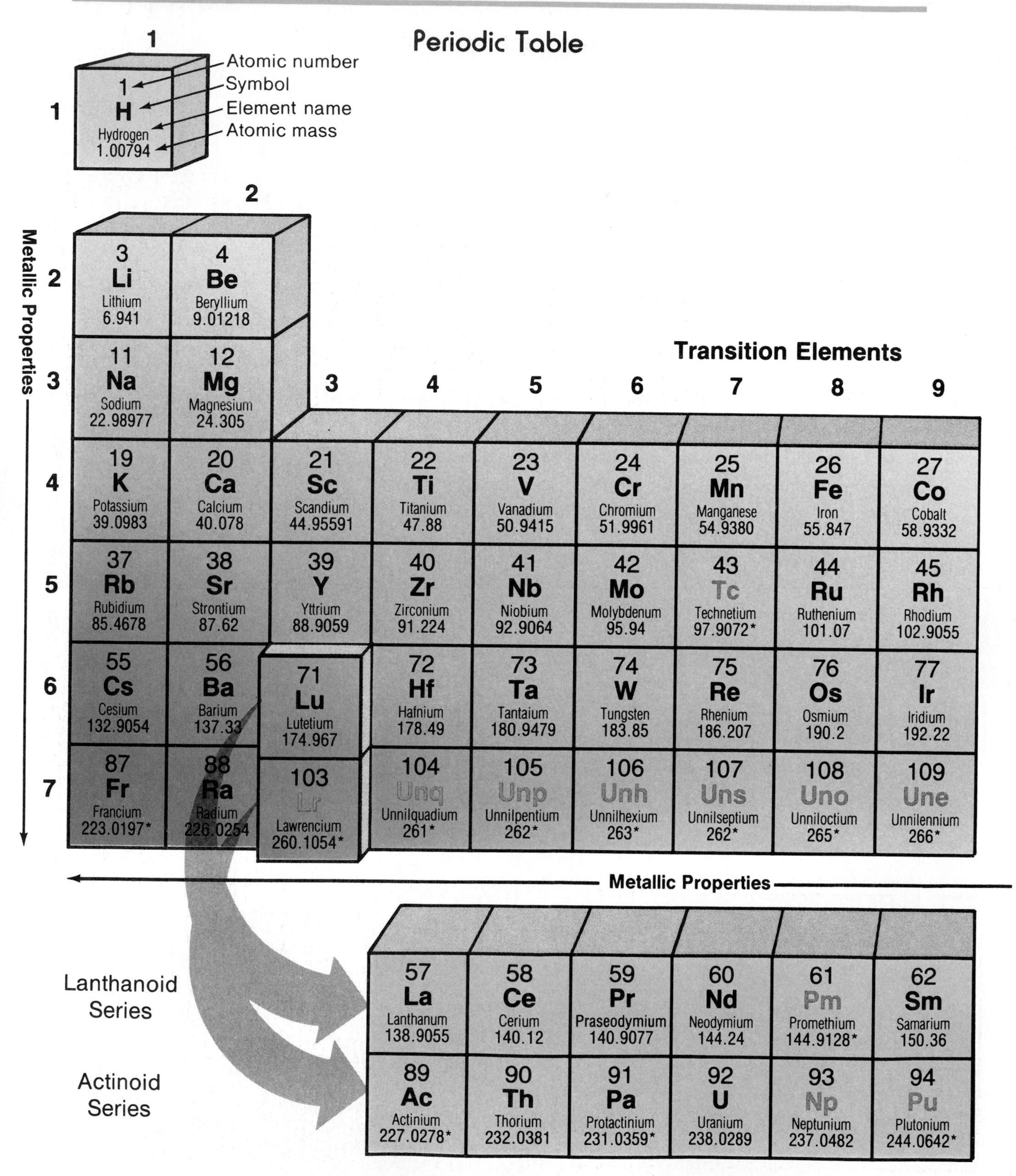

Periodic Table
1
Atomic number
Symbol
Element name
Atomic mass
1 H Hydrogen 1.00794
2
Metallic Properties
3 Li Lithium 6.941
4 Be Beryllium 9.01218
11 Na Sodium 22.98977
12 Mg Magnesium 24.305
Transition Elements
3 4 5 6 7 8 9
19 K Potassium 39.0983
20 Ca Calcium 40.078
21 Sc Scandium 44.95591
22 Ti Titanium 47.88
23 V Vanadium 50.9415
24 Cr Chromium 51.9961
25 Mn Manganese 54.9380
26 Fe Iron 55.847
27 Co Cobalt 58.9332
37 Rb Rubidium 85.4678
38 Sr Strontium 87.62
39 Y Yttrium 88.9059
40 Zr Zirconium 91.224
41 Nb Niobium 92.9064
42 Mo Molybdenum 95.94
43 Tc Technetium 97.9072*
44 Ru Ruthenium 101.07
45 Rh Rhodium 102.9055
55 Cs Cesium 132.9054
56 Ba Barium 137.33
71 Lu Lutetium 174.967
72 Hf Hafnium 178.49
73 Ta Tantaium 180.9479
74 W Tungsten 183.85
75 Re Rhenium 186.207
76 Os Osmium 190.2
77 Ir Iridium 192.22
87 Fr Francium 223.0197*
88 Ra Radium 226.0254
103 Lr Lawrencium 260.1054*
104 Unq Unnilquadium 261*
105 Unp Unnilpentium 262*
106 Unh Unnilhexium 263*
107 Uns Unnilseptium 262*
108 Uno Unniloctium 265*
109 Une Unnilennium 266*
Metallic Properties
Lanthanoid Series
Actinoid Series
57 La Lanthanum 138.9055
58 Ce Cerium 140.12
59 Pr Praseodymium 140.9077
60 Nd Neodymium 144.24
61 Pm Promethium 144.9128*
62 Sm Samarium 150.36
89 Ac Actinium 227.0278*
90 Th Thorium 232.0381
91 Pa Protactinium 231.0359*
92 U Uranium 238.0289
93 Np Neptunium 237.0482
94 Pu Plutonium 244.0642*

APPENDIX G

Periodic Table

10	11	12	13	14	15	16	17	18 (Noble Gases)
								2 **He** Helium 4.002602
			5 **B** Boron 10.811	6 **C** Carbon 12.011	7 **N** Nitrogen 14.0067	8 **O** Oxygen 15.9994	9 **F** Fluorine 18.998403	10 **Ne** Neon 20.179
			13 **Al** Aluminum 26.98154	14 **Si** Silicon 28.0855	15 **P** Phosphorus 30.97376	16 **S** Sulfur 32.06	17 **Cl** Chlorine 35.453	18 **Ar** Argon 39.948
28 **Ni** Nickel 58.69	29 **Cu** Copper 63.546	30 **Zn** Zinc 65.39	31 **Ga** Gallium 69.723	32 **Ge** Germanium 72.59	33 **As** Arsenic 74.9216	34 **Se** Selenium 78.96	35 **Br** Bromine 79.904	36 **Kr** Krypton 83.80
46 **Pd** Palladium 106.42	47 **Ag** Silver 107.8682	48 **Cd** Cadmium 112.41	49 **In** Indium 114.82	50 **Sn** Tin 118.710	51 **Sb** Antimony 121.75	52 **Te** Tellurium 127.60	53 **I** Iodine 126.9045	54 **Xe** Xenon 131.29
78 **Pt** Platinum 195.08	79 **Au** Gold 196.9665	80 **Hg** Mercury 200.59	81 **Tl** Thallium 204.383	82 **Pb** Lead 207.2	83 **Bi** Bismuth 208.9804	84 **Po** Polonium 208.9824*	85 **At** Astatine 209.98712*	86 **Rn** Radon 222.017*

Nonmetallic Properties →

Metallic Properties

Nonmetallic Properties

Metalloids

Synthetic Elements

63 **Eu** Europium 151.96	64 **Gd** Gadolinium 157.25	65 **Tb** Terbium 158.9254	66 **Dy** Dysprosium 162.50	67 **Ho** Holmium 164.9304	68 **Er** Erbium 167.26	69 **Tm** Thulium 168.9342	70 **Yb** Ytterbium 173.04
95 **Am** Americium 243.0614*	96 **Cm** Curium 247.0703*	97 **Bk** Berkelium 247.0703*	98 **Cf** Californium 251.0796*	99 **Es** Einsteinium 252.0828*	100 **Fm** Fermium 257.0951*	101 **Md** Mendelevium 258.986*	102 **No** Nobelium 259.1009*

APPENDIX H

Spring Star Chart

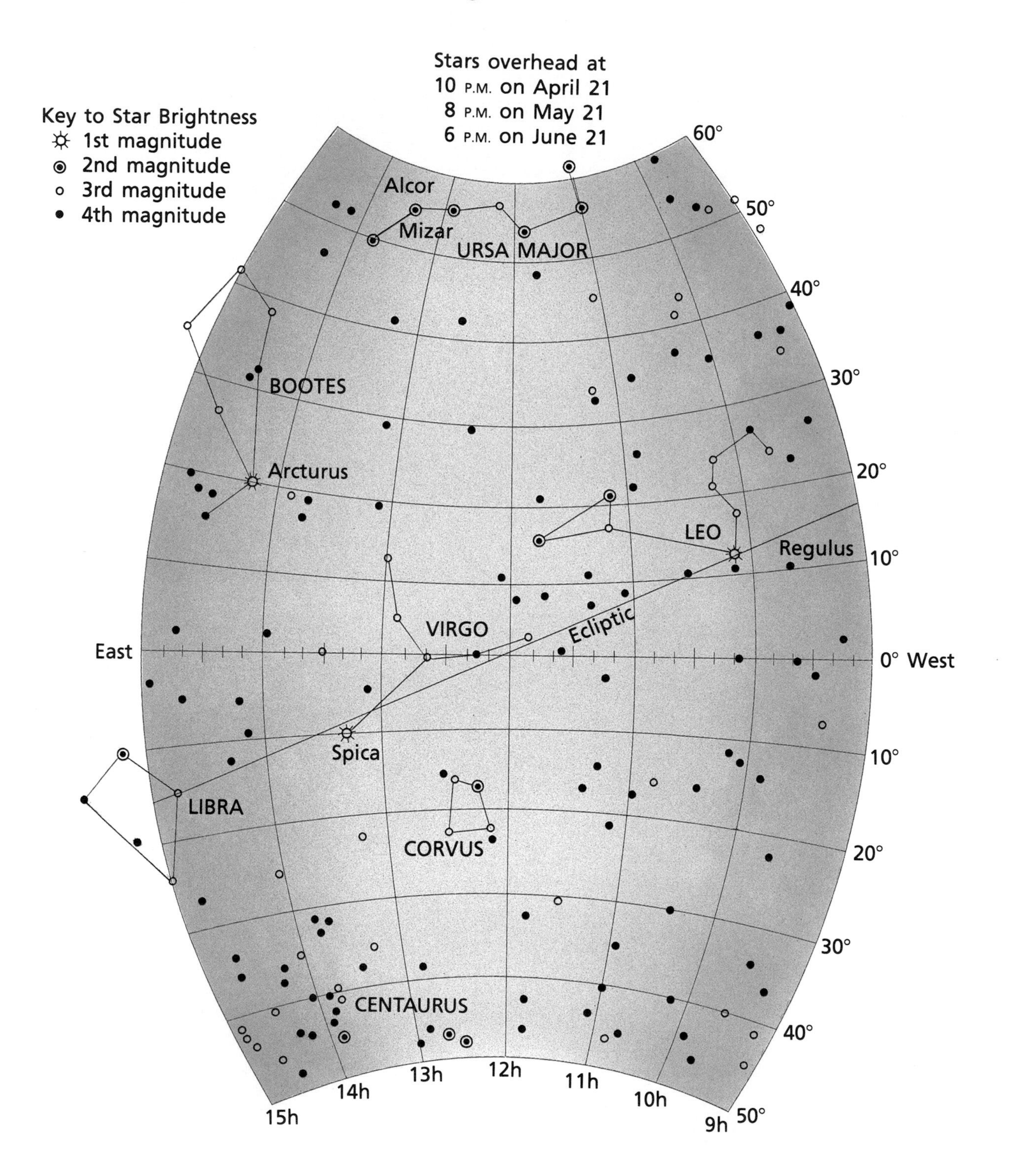

APPENDIX I

Summer Star Chart

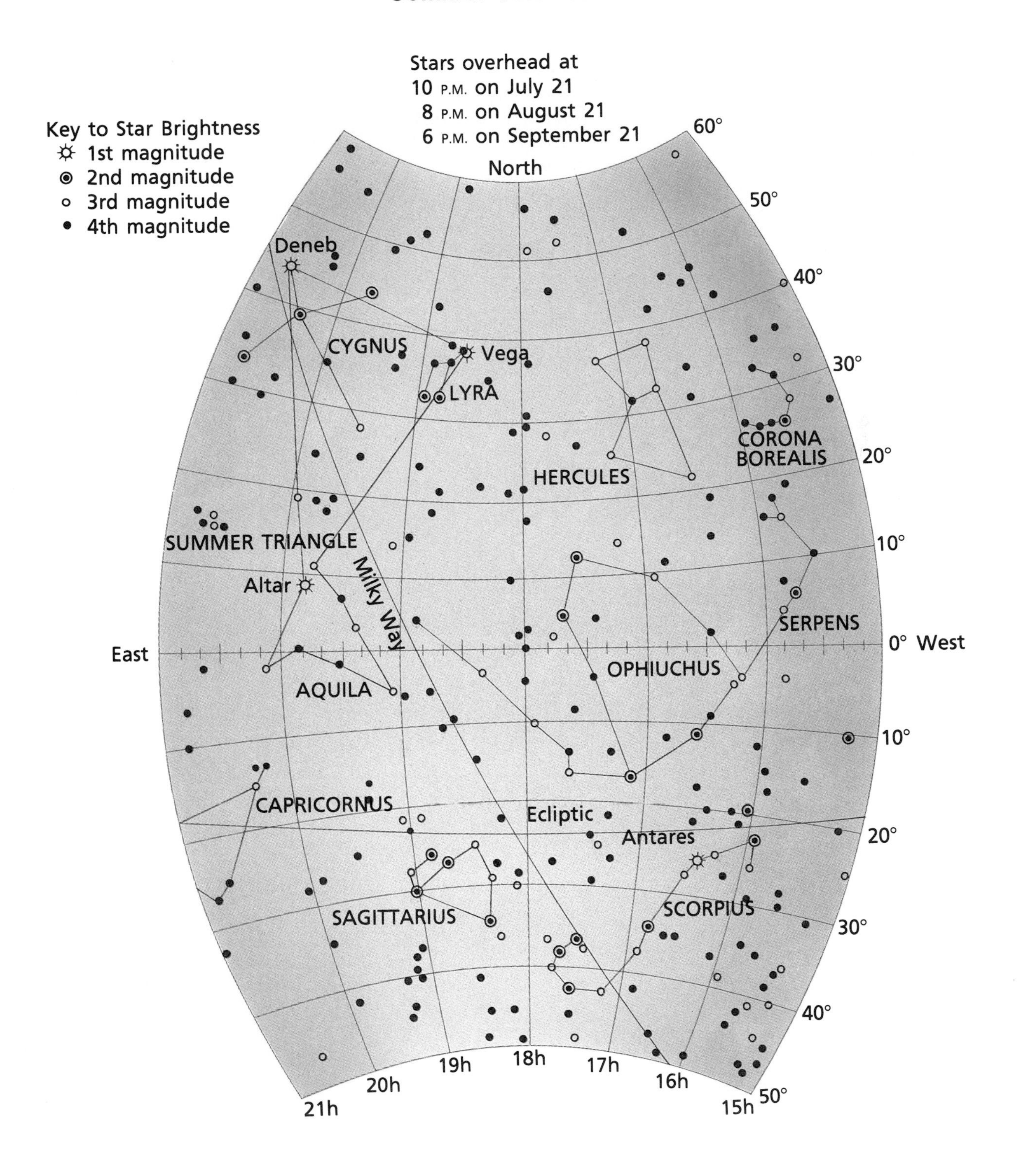

APPENDIX J

Fall Star Chart

Stars overhead at
10 P.M. on October 21
8 P.M. on November 21
6 P.M. on December 21

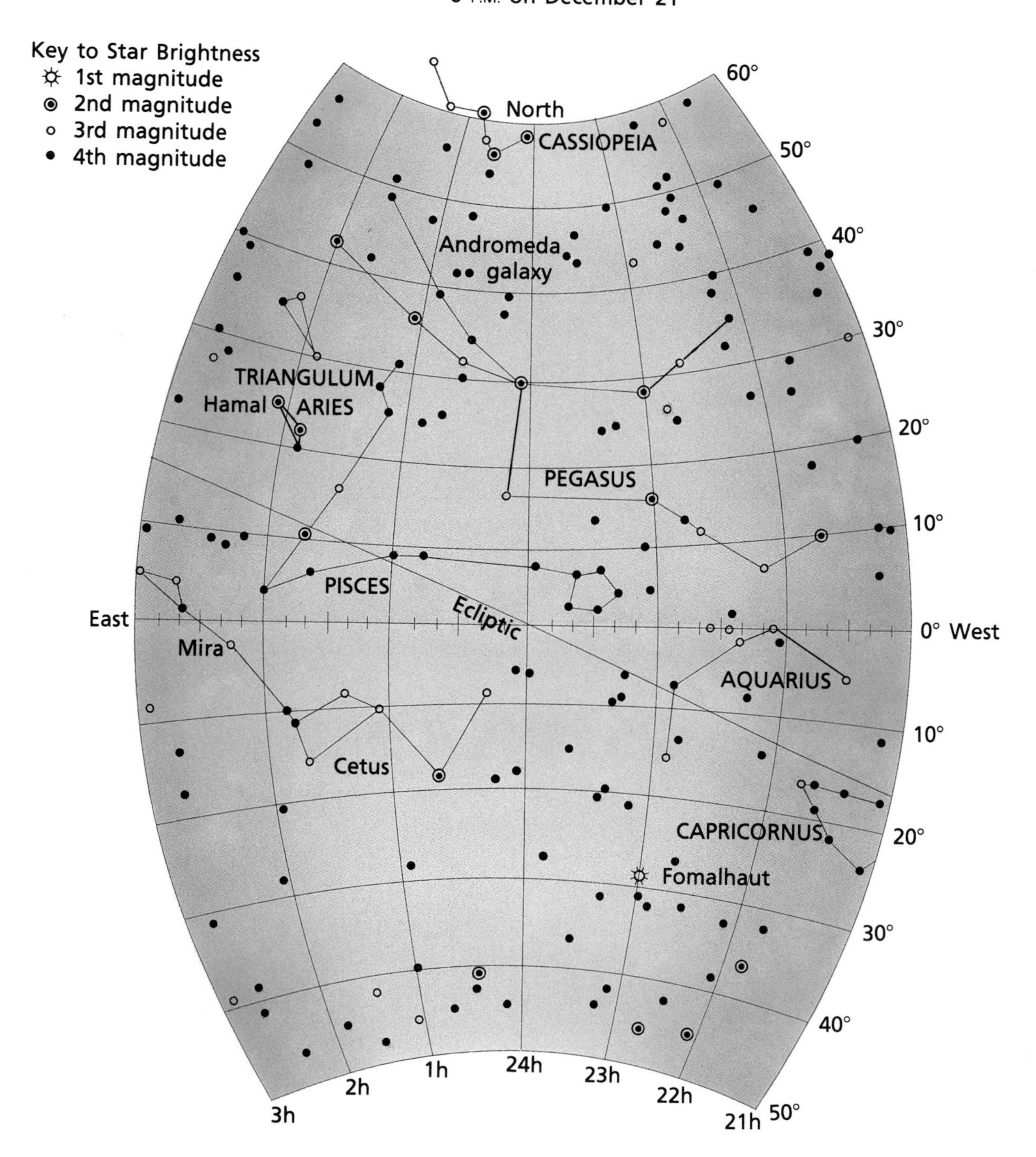

APPENDIX K

Winter Star Chart

Stars overhead at
10 P.M. on January 21
8 P.M. on February 21
6 P.M. on March 21

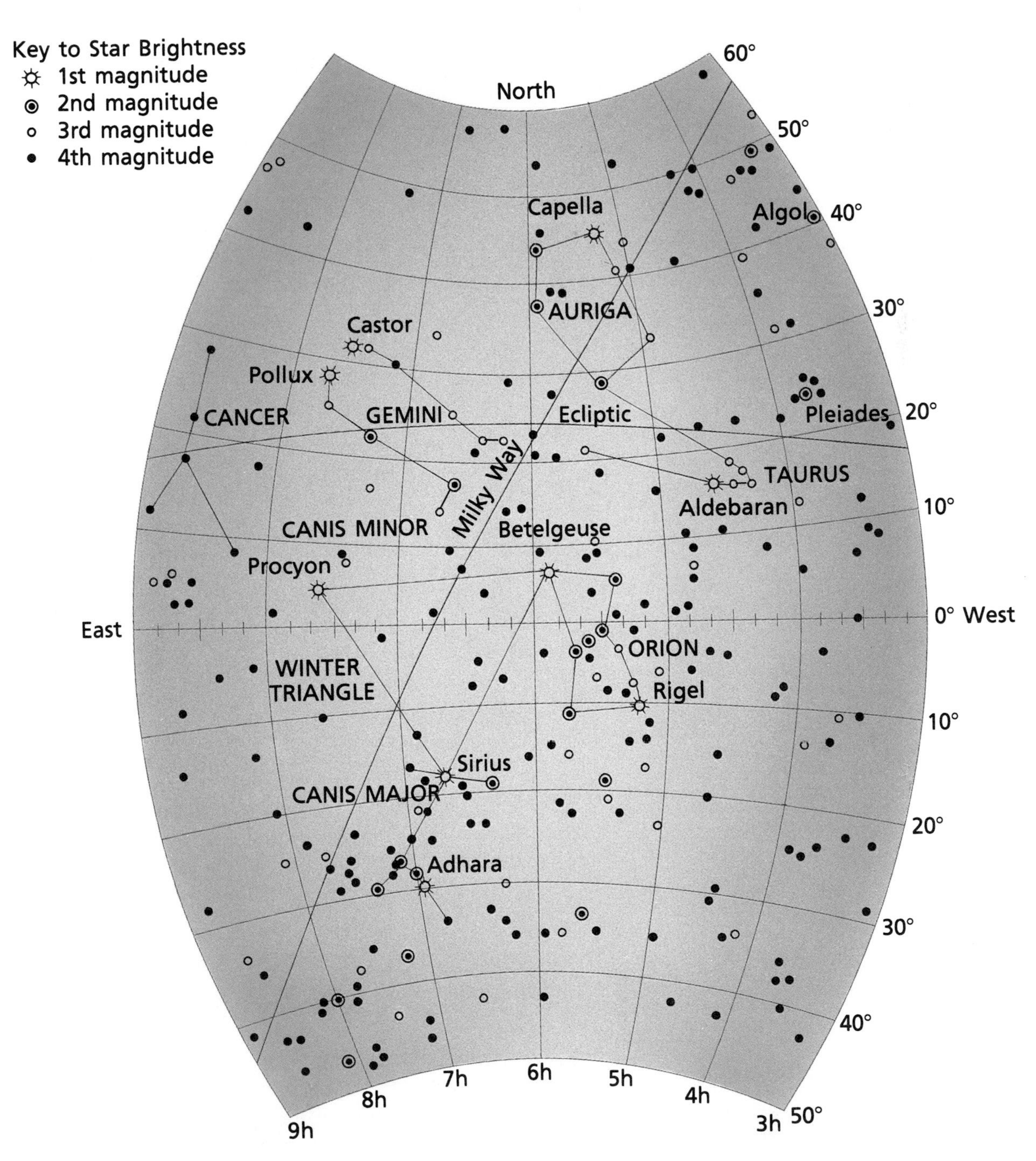

APPENDIX L

Weather Map Symbols

SAMPLE PLOTTED REPORT AT EACH STATION

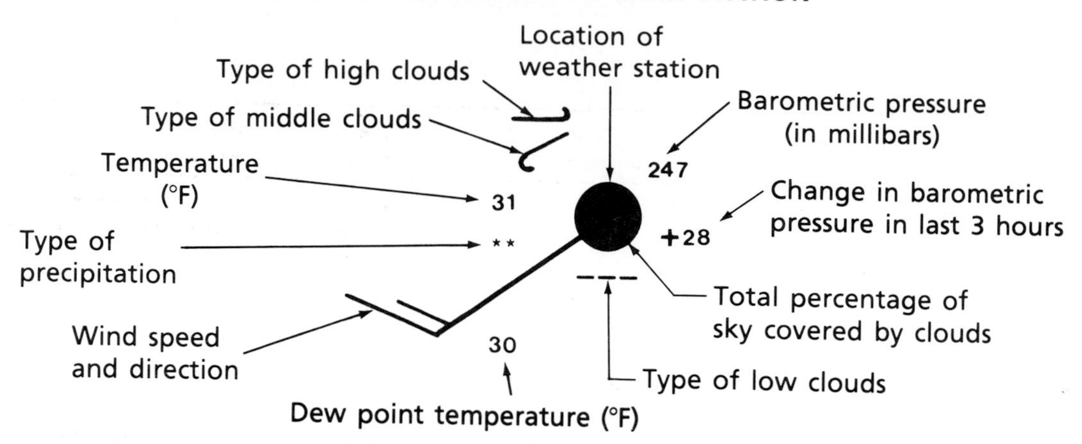

SYMBOLS USED IN PLOTTING REPORT

Precipitation	Wind speed and direction	Sky coverage	Some types of high clouds
Fog	0 calm	No cover	Scattered cirrus
Snow	1–2 knots	1/10 or less	Dense cirrus in patches
Rain	3–7 knots	2/10 to 3/10	Veil of cirrus covering entire sky
Thunder-storm	8–12 knots	4/10	Cirrus not covering entire sky
Drizzle	13–17 knots	½	
Showers	18–22 knots	6/10	
	23-27 knots	7/10	
	48-52 knots	Overcast with openings	
	1 knot = 1.852 km/h	Complete overcast	

Some types of middle clouds	Some types of low clouds	Fronts and pressure systems	
Thin altostratus layer	Cumulus of fair weather	(H) or High (L) or Low	Center of high or low pressure system
Thick altostratus layer	Stratocumulus		Cold front
Thin altostratus in patches	Fractocumulus of bad weather		Warm front
Thin altostratus in bands	Stratus of fair weather		Occluded front
			Stationary front

APPENDIX M

Minerals with Metallic Luster

Mineral (formula)	Color	Streak	Hardness	Specific gravity	Crystal system	Breakage pattern	Uses and other properties
graphite (C)	black to gray	black to gray	1-2	2.3	hexagonal	basal cleavage (scales)	pencil lead, lubricants for locks, rods to control some small nuclear reactions, battery poles
silver (Ag)	silvery white, tarnishes to black	light gray to silver	2.5	10-12	cubic	hackly	coins, fillings for teeth, jewelry, silverplate, wires; malleable and ductile
galena (PbS)	gray	gray to black	2.5	7.5	cubic	cubic cleavage perfect	source of lead, used in pipes, shields for X rays, fishing equipment sinkers
gold (Au)	pale to golden yellow	yellow	2.5-3	19.3	cubic	hackly	jewelry, money, gold leaf, filling for teeth, medicines; does not tarnish
bornite (Cu_5FeS_4)	bronze, tarnishes to dark blue, purple	gray-black	3	4.9.-5.4	tetragonal	uneven fracture	source of copper; called "peacock ore" because of the purple shine when it tarnishes
copper (Cu)	copper red	copper red	3	8.5-9	cubic	hackly	coins, pipes, gutters, wire, cooking utensils, jewelry, decorative plaques; malleable and ductile
chalcopyrite ($CuFeS_2$)	brassy to golden yellow	greenish black	3.5-4	4.2	tetragonal	uneven fracture	main ore of copper
chromite ($FeCr_2O_4$)	black or brown	brown to black	5.5	4.6	cubic	irregular fracture	ore of chromium, stainless steel, metallurgical bricks
pyrrhotite (FeS)	bronze	gray-black	4	4.6	hexagonal	uneven fracture	often found with pentlandite, an ore of nickel; may be magnetic
hematite (specular) (Fe_2O_3)	black or reddish brown	red to reddish brown	6	5.3	hexagonal	irregular fracture	source of iron; roasted in a blast furnace, converted to "pig" iron, made into steel
magnetite (Fe_3O_4)	black	black	6	5.2	cubic	conchoidal fracture	source of iron, naturally magnetic, called lodestone
pyrite (FeS_2)	light, brassy, yellow	greenish black	6.5	5.0	cubic	uneven fracture	source of iron, "fool's gold," alters to limonite

APPENDIX N

Minerals with Nonmetallic Luster

Mineral (formula)	Color	Streak	Hardness	Specific gravity	Crystal system	Breakage pattern	Uses and other properties
talc ($Mg_3(OH)_2 Si_4O_{10}$)	white, greenish	white	1	2.8	monoclinic	cleavage in one direction	easily cut with fingernail; used for talcum powder; soapstone; is used in paper and for table tops
bauxite (hydrous aluminum compound)	gray, red, white, brown	gray	1-3	2.0-2.5	—	—	source of aluminum, used in paints, aluminum foil, and airplane parts
kaolinite ($Al_2Si_2O_5 (OH)_4$)	white, red, reddish brown, black	white	2	2.6	triclinic	basal cleavage	clays; used in ceramics and in china dishes; common in most soils; often microscopic-sized particles
gypsum ($CaSO_4{\cdot}2H_2O$)	colorless, gray, white, brown	white	2	2.3	monoclinic	basal cleavage	used extensively in the preparation of plaster of paris, alabaster, and dry wall for building construction
sphalerite (ZnS)	brown	pale yellow	3.5-4	4	cubic	cleavage in six directions	main ore of zinc; used in paints, dyes, and medicine
sulfur (S)	yellow	yellow to white	2	2.0	ortho-rhombic	conchoidal fracture	used in medicine, fungicides for plants, vulcanization of rubber, production of sulfuric acid
muscovite ($KAl_3Si_3O_{10} (OH)_2$)	white, light gray, yellow, rose, green	colorless	2.5	2.8	monoclinic	basal cleavage	occurs in large flexible plates; used as an insulator in electrical equipment, lubricant
biotite ($K(Mg, Fe)_3 AlSi_3O_{10} (OH)_2$)	black to dark brown	colorless	2.5	2.8-3.4	monoclinic	basal cleavage	occurs in large flexible plates
halite ($NaCl$)	colorless, red, white, blue	colorless	2.5	2.1	cubic	cubic cleavage	salt; very soluble in water; a preservative
calcite ($CaCO_3$)	colorless, white, pale tints	colorless, white	3	2.7	hexagonal	cleavage in three directions	fizzes when HC1 is added; used in cements and other building materials
dolomite ($CaMg(CO_3)_2$)	colorless, white, pink, green, gray, black	white	3.5-4	2.8	hexagonal	cleavage in three directions	concrete and cement, used as an ornamental building stone

APPENDIX N

Minerals with Nonmetallic Luster

Mineral (formula)	Color	Streak	Hardness	Specific gravity	Crystal system	Breakage pattern	Uses and other properties
fluorite (CaF_2)	colorless, white, blue, green, red, yellow, purple	colorless	4	3-3.2	cubic	cleavage	used in the manufacture of optical equipment; glows under ultraviolet light
limonite (hydrous iron oxides)	yellow, brown, black	yellow, brown	5.5	2.7-4.3	—	conchoidal fracture	source of iron; weathers easily, coloring matter of soils
hornblende ($CaNa(Mg, Al, Fe)_5 Al, Si)_2 Si_6O_{22}(OH)_2)$)	green to black	gray to white	5-6	3.4	monoclinic	cleavage in two directions	will transmit light on thin edges; 6-sided cross-section
feldspar (orthoclase) ($KAlSi_3O_8$)	colorless, white to gray, green, and yellow	colorless	6	2.5	monoclinic	two cleavage planes meet at 90° angle	insoluble in acids, used in the manufacture of porcelain
feldspar (plagioclase) ($NaAlSi_3O_8$) ($CaAl_2Si_2O_8$)	gray, green, white	colorless	6	2.5	triclinic	two cleavage planes meet at 86° angle	used in ceramics; striations present on some faces
augite ($(Ca, Na) (Mg, Fe, Al) (Al, Si)_2 O_6$)	black	colorless	6	3.3	monoclinic	2-directional cleavage	square or 8-sided cross-section
olivine ($(Mg, Fe)_2 SiO_4$)	olive green	colorless	6.5	3.5	ortho-rhombic	conchoidal fracture	gemstones, refractory sand
quartz (SiO_2)	colorless, various colors	colorless	7	2.6	hexagonal	conchoidal fracture	used in glass manufacture, electronic equipment, radios, computers, watches, gemstones
garnet ($(Mg, Fe, Ca)_3$ $(Al_2Si_3O_{12})$)	deep yellow-red green, black	colorless	7.5	3.5	cubic	conchoidal fracture	used in jewelry, also used as an abrasive
topaz (Al_2SiO_4 $(F, OH)_2$)	white, pink yellow, pale blue, colorless	colorless	8	3.5	ortho-rhombic	basal cleavage	valuable gemstone
corundum (Al_2O_3)	colorless, blue, brown, green, white, pink, red	colorless	9	4.0	hexagonal	fracture	gemstones: ruby is red, sapphire is blue; industrial abrasive

GLOSSARY

The glossary contains all of the major science terms of the text and their definitions. Below is a pronunciation key to help you use these terms. The word or term will be given in boldface type. If necessary, the pronunciation will follow the term in parentheses.

a . . . **b**a**ck** (bak)
ay . . . **day** (day)
ah . . . **father** (fahth ur)
ow . . . **flower** (flow ur)
ar . . . **car** (car)
e . . . **less** (les)
ee . . . **leaf** (leef)
ih . . . **trip** (trihp)
i (i + con + e) . . . **idea** (i dee uh), **life**, (life)
oh . . . **go** (goh)
aw . . . **soft** (sawft)
or . . . **orbit** (or but)
oy . . . **coin** (coyn)
oo . . . **foot** (foot)
ew . . . **food** (fewd)
yoo . . . **pure** (pyoor)
yew . . . **few** (fyew)
uh . . . comm**a** (cahm uh)
u (+ con) . . . **flower** (flow ur)
sh . . . **sh**elf (shelf)
ch . . . na**t**ure (nay chur)
g . . . **g**ift (gihft)
j . . . **g**em (jem)
ing . . . s**ing** (sing)
zh . . . vi**s**ion (vihzh un)
k . . . **c**a**k**e (kayk)
s . . . **s**eed, **c**ent (seed, sent)
z . . . **z**one, rai**s**e (zohn, rayz)

A

abrading: scouring of the bedrock by a glacier

abrasion (uh BRAY shun): a scouring action of wind-carried particles resulting in the erosion of rock surfaces

absolute dating: a method that determines the actual length of time between events

absolute magnitude: a measure of a star's actual brightness

absolute zero: a temperature at which all molecular movement within atoms stops, equivalent to −273°C

abyss: ocean depths of 2000 to 6000 meters

abyssal plain: the flat, almost level area of the ocean basin

acid precipitation: rain or snow that forms as the result of mixture of air pollutants with moisture in the atmosphere

air mass: a body of air that has the same properties as the region over which it develops

alloy: a substance composed of two or more metals

alluvial (uh LEW vee ul) **fans:** apron-shaped deposits at the foot of a slope

ammonoids: extinct shelled organisms that lived from about 395 to 136 million years ago

amphibians (am FIHB ee unz): ectothermic vertebrates intermediate between fish and reptiles that spend part of their lives on land and part in water

amplitude: the distance a wave rises or falls as it travels

andesite: the extrusive equivalent of diorite

aneroid (AN uh royd) **barometer:** an instrument used to measure air pressure

angiosperms (AN jee oh spurmz): flowering plants having seeds in a protective covering, which enables them to survive changing environments

angular unconformities: unconformities in which tilted or folded rocks are overlain by more horizontal layers

anthracite: the final stage in coal formation

anticline (ANT ih kline): a convex upward fold in which the oldest strata are in the middle

anticyclone (ant ih SI klohn): an air mass in which air circulates away from the center in a clockwise motion in the Northern Hemisphere

aphelion: the point in an orbit when an object is farthest from the sun

Apollo: a series of spaceflights that landed astronauts on the moon

apparent magnitude: the brightness of a star observed at its actual distance from Earth

apparent motion: the motion of an object relative to the position of its observer

aquifers (AK wuh furz): permeable rocks containing water

area: the amount of surface within a set of boundaries

artesian (ahr TEE zhun) **well:** a well in which water is under pressure due to the weight of the overlying column of water in the aquifer

asteroids: fragments of matter similar to planetary matter that orbit between Mars and Jupiter

asthenosphere (as THEN uh sfihr): the upper, plasticlike portion of the mantle

astronomical unit (AU): the average distance between Earth and the sun; used to measure distances within the solar system

astronomy: the study of objects in space

atoll: a coral reef that surrounds a lagoon

atom: the smallest unit of an element that still keeps the properties of the element

atomic number: the total number of protons present in the nucleus of an atom

authigenic deposits: deposits that form in place

B

backfilling: the process of putting waste materials back into a mine

bar graph: a graph that uses thick bars to display data

barchans: crescent-shaped sand dunes

barometer: an instrument used to measure air pressure

barrier islands: sand deposits that parallel the shore but are separated from the mainland

barrier reef: a coral reef that is separated from a landmass by water

bars: ridges of sand built up by currents in coastal waters

barycenter: the common center of gravity of two objects that orbit the sun

basal slip: a type of glacial flow that occurs when meltwater acts as a lubricant

basalt: a dark-colored extrusive igneous rock

basaltic magma: magma that originates in the mantle and flows readily

batholiths: large masses of intrusive igneous rock extending to unknown depths; often form cores of mountains

bed load: sediment that is rolled along a river bottom

bedrock: the solid rock found under soil

benthos (BEN thohs): bottom-dwelling organisms in the ocean

Big Bang: a theory stating that the universe originated from the explosion of a huge mass of matter, which cooled, collected into clouds, and formed galaxies

binary star: two stars revolving around a common center

biology: the study of living organisms

biomass fuels: burnable fuels made from organic matter

bituminous coal: the third stage in coal formation

black hole: a star in which matter is condensed and its gravity field so strong that light cannot escape

body waves: waves traveling outward from the focus of an earthquake in all directions through Earth's interior

boiling point: the temperature at which a liquid boils

brachiopods (BRAY kee uh pahdz): marine invertebrates characteristic of the Paleozoic Era

breeding: a nuclear reaction that produces more fuel than it uses

buoyancy (BOY un see): the upward lift exerted by water upon an immersed or floating body

burial metamorphism: low pressure, low temperature changes in rocks due to the weight of overlying rocks

butte (BYEWT): a flat-topped hill with steep slopes

C

Cambrian: a period of geologic time lasting from about 570 to 500 M.Y.B.P., during which marine invertebrates were the dominant life forms

cap rock: a rock such as shale or rock salt that stops the upward movement of oil and gas

carbonates: a group of minerals composed of carbon, oxygen, and some other element or elements

carbon impressions: impressions of buried leaves or soft-bodied animals preserved by a thin film of carbon

cast: a method of fossil preservation that occurs when a mineral fills a fossil mold

Celsius: a unit of temperature

cementation: the process by which loose particles are cemented to form clastic sedimentary rocks

Cenozoic Era: an era of geologic time that began about 65 million years ago

centi-: prefix meaning one-hundredth

Ceres: the largest asteroid, discovered in 1801

Challenger: a space shuttle launched in January, 1986

chemical property: a characteristic that depends on the reaction of a substance to form other substances

chemistry: the study of the properties and composition of matter

chemosynthesis: a process by which bacteria produce energy and nutrients necessary for survival

chert: nonclastic sedimentary rock made of silica

chromosphere: a bright, hot layer of gases surrounding the photosphere of the sun

cinder cones: steep-sided volcanoes formed by violent eruptions

circumference: the distance around a circle

cirque: the original hollow in which snow accumulates to form a valley glacier

cirrus: a high, white, feathery cloud composed of ice crystals or supercooled water; associated with fair weather

clastics (KLAS tihks): sedimentary rocks made of rock fragments, minerals, and broken shells

clay: a fine sediment formed from chemical weathering of either iron-magnesium minerals or feldspars

Clean Air Act: a law that requires that pollutants be reduced to safe levels in the atmosphere

Clean Water Act: a law that provides funds for states to build sewage and wastewater treatment facilities

cleavage (KLEE vihj): a physical property that describes the way a mineral breaks

climate: the average of all weather conditions of an area over a long period of time

coal gasification: a process in which coal is converted to gasoline

cold front: the boundary developed when a cold air mass meets a warm air mass

coma: a cloud of dust and gas that forms a halo around the nucleus of a comet

comet: a mass of frozen gases, cosmic dust, and small rock particles that orbits the sun

compaction (kum PAK shun): a process by which sediments become consolidated into rocks

compounds: substances in which two or more elements combine chemically

compression: a system of forces that pushes against a body from directly opposite sides

conclusion: an answer to a question based on analyzing data and observations gathered in an experiment

concretions (kahn KREE shuns): irregularly shaped masses of cementing material that collect around a nucleus

condensation: process by which water vapor becomes a liquid or a solid

conduction: the transfer of heat through matter by the actual contact of molecules

conductivity: the power of conducting or transmitting heat

conservation: the planned and wise management of Earth's resources

consolidation: process in which materials are compressed into a compact mass

constellations: groups or patterns of stars

contact metamorphic deposits: local concentrations of minerals formed at the contact point between an intruding igneous rock and the surrounding rock

contact metamorphism: changes in rocks that occur as the result of an intruding magma body

continental drift: a hypothesis that suggests the continents have been in different positions through geologic time

continental glacier: a large ice sheet whose movement is outward due to pressure from overlying ice and snow

continental shelf: the relatively flat part of a continent that is covered by seawater

continental slope: the steeply dipping surface between the outer edge of the continental shelf and the ocean basin

contour interval: the difference in elevation between two adjacent contour lines

contour line: a line drawn on a map to join all points of the same elevation

control: a standard for comparison in an experiment

convection: the transfer of heat due to density differences

convergent boundaries: boundaries between two colliding plates

corals: animals that attach to the ocean bottom and join to form a colony or reef

core: the inner part of the sun where fusion occurs; the innermost part of Earth, consisting of an inner solid portion surrounded by an outer liquid region

Coriolis (kohr ee OH lus) **effect:** a phenomenon that causes a body in motion to be deflected from its initial path due to Earth's rotation

corona: the transparent zone beyond the chromosphere of the sun visible only during a total solar eclipse

craters: depressions on the surfaces of some planets and their satellites

creep: a slow, downhill movement of material

crest: the highest point of a wave

Cretaceous: a period of geologic time that occurred from about 136 to 65 M.Y.B.P., during which angiosperms appeared, and many forms of life became extinct

crude oil: a nonrenewable resource; petroleum

crust: the outermost layer of Earth that extends from the surface to a depth of about 35 kilometers

crystal: a solid bounded by plane surfaces that has a definite shape due to its internal arrangement

cumulus: a thick, puffy cloud that develops when rising columns of moist air are cooled to the dew point

cyanobacteria: single-celled, plantlike organisms

cyclones (SI klohnz): low-pressure systems in which air circulates toward the center in a counterclockwise motion in the Northern Hemisphere

D

daughter elements: new elements formed during radioactive decay

daylight saving time: a plan in which clocks are set one hour ahead of standard time for a certain number of months

deci-: prefix meaning one-tenth

decomposition (dee kahm puh SIHSH un): the chemical process of weathering

deep-focus earthquake: earthquake that occurs in the mantle in the rigid, descending plate

deep-sea trenches: parts of the ocean floor where an oceanic plate has been subducted and descends into the mantle

deep-water wave: a wave moving in water that is deeper than one-half its wavelength

deflation (dih FLAY shun): the removal of loose material from the ground surface by wind

delta: a fan-shaped deposit that accumulates when a moving body of water loses its velocity

dendritic (den DRIHT ik): a treelike stream pattern

density: the mass per unit volume of any substance

density current: a current formed by the movement of more dense water toward an area of less dense water

deposition: the process by which products of erosion are laid down

desalination: a process by which salts are removed from seawater to produce fresh water

desert: a hot, dry area of land receiving less than 25 centimeters of rainfall annually

desert pavement: boulders and pebbles left by deflation

Devonian: period of geologic time that lasted from about 395 to 345 M.Y.B.P., during which fish were the dominant marine life forms and the first amphibians developed

dew point: the temperature at which condensation occurs

dikes: rock structures formed when magma enters a vertical fracture in rock and hardens

diorite: an intrusive igneous rock

disconformities: gaps that separate two horizontal series of beds

disintegration (dis ihnt uh GRAY shun): the physical or mechanical process of weathering

diurnal: an event occurring once a day; usually refers to the tides

divergent boundaries: boundaries where two plates are pulling apart

divide: an area of high ground that separates drainage basins

doldrums: a windless zone at the equator

dome volcanoes: small, steep-sided volcanoes made from rhyolitic lava

drainage basin: an area separated by divides and drained by the major water body of the region

drainage system: a network of channels that joins to form the main water body

drought: a long period of little or no precipitation

drumlins: streamlined hills composed of glacial till

dust: the finest sediment carried by wind

Dust Bowl: the name given to the Great Plains during a period of drought

E

Earth: the third planet from the sun having an atmosphere in which life forms can exist

earthquakes: vibrations caused by the sudden movement of surface rocks

earth science: the study of planet Earth and its place in space

eclipse: the passing of one object into the shadow of another

ectotherm: an animal whose body temperature changes with that of the environment

electromagnetic spectrum: forms of radiant energy arranged in the order of their wavelengths and frequencies

electron: a negatively-charged particle that moves around the nucleus of an atom

electron cloud: the space occupied by the electrons in an atom

elements: basic substances that cannot be broken down or changed into simpler substances

elevation: height above sea level

ellipse: an elongated, closed curve

elliptical galaxies: systems of stars that contain little dust and gas and may be composed of millions to trillions of stars

endotherms: animals with constant body temperatures

energy: the ability or the capacity to do work

environment: the sum of all external conditions that affect an organism or community

Environmental Protection Agency: a governmental agency that monitors the various levels of pollutants in the environment in the United States

epicenter: the point at Earth's surface that is directly above the focus of an earthquake

epochs: geologic time units smaller than periods

equator: an imaginary line that lies halfway between the North and South Poles

equinox: either of the two times each year when the sun crosses the equator, and day and night are of equal length

eras: the largest subdivisions of geologic time

erosion: the process by which Earth materials are carried away and redeposited by wind, water, gravity, or ice

eskers: long, winding ridges of glacial outwash

evaporation: process by which a liquid becomes a gas

evaporites: nonclastic sedimentary rocks that form when water evaporates, leaving dissolved solids behind

exosphere: the upper zone of the thermosphere that begins at about 500-700 kilometers above Earth and extends into interplanetary space

extinction (ihk STINGK shun): the total disappearance of a species

extrusive igneous rocks: rocks formed when lava cools at or near Earth's surface

F

facies (FAY sheez) **fossils:** the remains of organisms that existed for a long period of geologic time but in very limited types of environments

fault-block mountains: mountains formed when an area between two parallel faults is uplifted

faults: fractures in a rock body along which movement has taken place

felsites: light-colored, fine-grained igneous rocks

first aid: emergency treatment given to an ill or injured person before regular medical aid can be obtained

first quarter: a phase of the moon in which the sun, Earth, and moon form a 90° angle, making the moon appear half bright and half dark

fission: the splitting of the nuclei of atoms

flat-plate thermal collector: a device that allows solar energy to be transferred by a liquid, such as air

floodplain: an area of fertile soil composed of fine sediment deposited during floods

fluorocarbon: a chemical compound used in refrigerants and propellants that can destroy ozone molecules

focal plane: the point on a telescope where the image is formed

focus: the actual point on a fault where movement occurs and vibrations begin

fog: a stratus cloud close to Earth's surface formed from the condensation of water vapor

folded mountains: mountains that form as the result of compression

folds: bends in a rock layer due to force

foliated: a metamorphic rock texture in which minerals align in bands or layers

force: any push or pull on an object

form: the general appearance or shape of a mineral

fossil correlation: the principle that uses fossils to identify rock beds in a sedimentary sequence

fossil fuels: fuels such as coal and oil formed from the decay of organisms that lived millions of years ago

fossils: the remains or traces of once-living organisms preserved in Earth's rocks

fracture: a distinct manner of breaking in a mineral other than along planar surfaces; a break in a rock body due to force

freezing point: the temperature at which liquids change to solids

frequency: the number of waves that pass a given point per second

friction: resistance to motion between two bodies in contact

fringing reef: a coral reef that develops close to the shore of a continent or island

full moon: a phase of the moon that occurs when Earth is between the sun and the moon, and the entire side of the moon facing Earth is bright

fusion: the nuclear reaction in which hydrogen atoms join to form helium, releasing large amounts of energy

G

gabbro: an intrusive igneous rock

galaxies: large systems of stars, gases, and dust

gamma rays: forms of radiant energy given off by a radioactive substance

gangue (GANG): the noneconomic material that is removed before a mineral can be used; waste rock

gas: a state of matter in which atoms or molecules move freely and independently to fill all the available space

gasohol: a biomass fuel consisting of alcohol and gasoline

Gemini: a space program that preceded the Apollo program

gems: rare minerals that are highly-prized due to their color, luster, and hardness

geocentric universe: first concept of the universe in which Earth was thought to be the central object

geodes (JEE ohds): hollow, ball-like objects found in some sedimentary rocks

geologic column: a sequence of rock layers that represent a certain amount of time during which the rocks were deposited

geologic map: a map showing the outcrop areas of rock units, structural features, and ages of surface rocks

geologic time scale: a chronological arrangement or sequence of events of Earth's history

geologists: scientists who study the processes that form and change Earth

geology: the study of all processes affecting Earth

geophysicist: a scientist who studies internal processes of Earth

geothermal energy: energy that comes from the internal heat of Earth

geysers (GI surs): hot springs that erupt water and steam at regular intervals

glaciers: thick masses of ice in motion

globe: a model of Earth

Glomar Challenger: a research ship used to drill for ocean crust samples

Gondwana: the Late Paleozoic continental assemblage of the Southern Hemisphere

gram: the standard unit of mass

granitic magma: thick, stiff magma that forms just below the crust

gravimeter: an instrument used to measure variations in gravitational force

gravitational forces: attractions between objects due to their masses

gravity: the attraction between two objects due to their masses; an agent of erosion

Great Red Spot: the gaseous, hurricane-like mass on the planet Jupiter

greenhouse effect: the process by which water vapor and carbon dioxide in the atmosphere absorb and reflect infrared waves

ground moraine: debris deposited from the base of an ice sheet as it melts

groundwater: water that sinks into the porous areas of the crust

guyot: a submerged volcanic island with a flat top

H

hachure lines: short, straight lines drawn perpendicular to contour lines to indicate a surface in relief, such as a depression

half-life: the amount of time needed for one-half of the atoms of a radioactive element to decay

halite: table salt

hanging valleys: U-shaped scars that form when a glacial tributary enters the main valley above its base

hardness: a mineral's resistance to being scratched

hard water: water in which excessive amounts of calcium and magnesium carbonates are dissolved

headward extension: the lengthening and cutting upstream of a valley by a body of water or a glacier

heliocentric universe: a model of the universe in which the planets revolve about the sun, which was thought to be the center of the universe

Hertzsprung-Russell (H-R) diagram: a classification system of stars based on absolute magnitude and temperature

high-temperature vein deposits: deposits formed when hot gases and liquids from magma invade the surrounding rocks

Homo sapiens: species to which humans belong

horizons: soil layers of different colors and textures

horn: a high, sharpened peak formed by the headward extension of cirques by glaciers

hot springs: water heated within Earth's crust

humus: material formed from decayed organic matter; found in topsoil

hurricanes: storms that develop when warm, moist air carried by trade winds rotates around a low-pressure "eye"

hydrocarbons: compounds that are composed of carbon and hydrogen

hydroelectric power: the use of water to produce electricity

hydrogen: an element having an atomic number of one

hydrosphere: the waters of Earth including oceans, lakes, rivers, groundwater, snow, ice and glaciers, and water vapor in the atmosphere

hypothesis: a proposed answer to a question that can be tested

I

ice age: a period of time when ice sheets and alpine glaciers are far more extensive than they are today

igneous ore deposits: concentrations of minerals within a body of magma

igneous rocks: rocks formed from molten Earth materials

impermeable (ihm PUR mee uh bul): a condition of a rock or soil in which fluids cannot move through the tightly-packed particles

index fossils: the widely-distributed remains of certain organisms that are used to divide geologic time into major units

inertia: the tendency of matter to remain at rest or in motion unless acted upon by an outside force

inference: a conclusion based on both observations and knowledge of a subject

inner planets: the planets relatively close to the sun—Mercury, Venus, Earth, and Mars

inorganic: not formed by life processes

intensity: a measure of the surface damage caused by an earthquake

interglacial period: a length of time characterized by warm or mild climates that separates periods of widespread glaciation

International Date Line: the 180° meridian directly opposite the prime meridian

International System of Units (SI): a system of units used by scientists and other people in most countries of the world

intrusive igneous rocks: rocks formed when magma cools slowly beneath Earth's surface

ionosphere (i AHN uh sfihr): the lower zone of the thermosphere that contains ions and free electrons

ions: electrically charged particles

irregular galaxies: galaxies that contain clouds of gas and dust and stars that are in the early stages of their life cycles

island arc: a chain of islands which is parallel to an oceanic trench, formed from volcanoes on the ocean floor

isobars (I suh barz): lines drawn to connect points of equal air pressure

isotherms (I suh thurmz): lines drawn to connect points of equal temperature

isotopes: atoms of the same element having the same number of protons but different number of neutrons

J

jet streams: narrow belts of wind near the tropopause that form when warm tropical air meets cold polar air

Jupiter: the largest planet in the solar system

Jurassic: a period of geologic time occurring from about 190 to 136 M.Y.B.P., during which the first birds appeared, and dinosaurs dominated land life forms

K

kames: small knobby hills of sand and gravel that commonly form between lobes of ice advancing from different directions

Kelvin: a unit of measurement for temperature in the International System of Units; 0 K = −273°C

kettle lakes: small basins formed when blocks of ice, surrounded by debris, melt very slowly

kilo-: prefix meaning thousand

kilograms: the base unit for measuring mass in SI

L

L-waves: surface waves

laccolith (LAK uh lihth): a mushroom-shaped body of intrusive igneous rock smaller than a batholith

lagoon: a shallow sound near a larger body of water

land breeze: a circulation pattern in which warm air over water rises and is replaced by cooler air from land; occurs at night

land reclamation: restoration of land after mining

landslides: rapid downhill movements of large amounts of Earth materials

last quarter: a phase of the moon in which the moon is waning

latitude: a parallel line that identifies locations or distances north or south of the equator

Laurasia: the Late Paleozoic continental assemblage of the Northern Hemisphere

lava: magma that is released at Earth's surface during volcanic eruptions

law: a theory that has been sufficiently tested and validated

law of superposition: a principle that states that in an undeformed rock sequence, the oldest beds are at the bottom, and the youngest beds are at the top

leaching: the process by which some soil components are dissolved and carried downward by water

leeward: away from the wind

legend: an explanation of each symbol used on a map

levee (LEV ee): a low ridge of coarse sediment deposited by a river along its margins, parallel to the channel

light-year: a measure of the distance light travels in one year, equal to 9.5×10^{12} km

lignite: the second stage in coal formation

line graph: a graph consisting of a vertical axis and a horizontal axis used to plot points to show the relationship between two sets of data

liquid: a state of matter in which atoms or molecules are free to move; state of matter that changes shape but not volume

liter: the standard unit of volume in SI

lithosphere (LITH uh sfihr): rigid blocks of crustal and mantle material of which the continents are a part

loess (les): a wind-blown deposit of fine dust particles gathered from deserts, dry riverbeds, or old glacial lakebeds

longitude: a line perpendicular to the equator used to identify distances or locations east or west of the prime meridian

longshore current: a current that flows parallel to the shore and forms when waves approach the shore at an angle

lunar eclipse: an eclipse that occurs when the moon passes into Earth's shadow

luster: a physical property that refers to the way light is reflected from a mineral's surface

M

magma: molten material found beneath Earth's surface

magnetic pole: either of two regions located in the polar areas and toward which a compass needle points

magnetometers: instruments that detect and measure Earth's magnetic field and its changes

magnitude: a measure of the strength of an earthquake

main sequence stars: stars that lie along a diagonal line from upper left to lower right on the Hertzsprung-Russell diagram; use up their hydrogen fuel at a steady rate

malleable: able to be shaped by pounding without losing strength

mantle: the middle layer of Earth located between the crust and the core

map scale: a fixed ratio between a unit of measure on a map and the distance it represents on land

maria: low-lying regions on the moon's surface

Mars: the fourth planet from the sun

mass: a measure of the amount of matter in an object

mass number: the sum of the protons and neutrons in each atom of a given element

matter: anything that has mass and occupies space

meander (mee AN dur): a curve developed from the side-to-side wandering of a river across its valley or floodplain

measurement: a comparison of an object to a standard that includes a unit of measure and a number stating how many of the units are present

melting point: the temperature at which a solid changes to a liquid

meltwater: water resulting from the melting of glacial snow and ice

mental models: ideas of how objects look

Mercalli scale: a scale that indicates the intensity of earthquakes expressed between I and XII

Mercury: the closest planet to the sun; a space program that provided data and experience in the basics of spaceflight

mercury barometer: an instrument used to measure air pressure

meridians: lines of longitude that pass through the poles

mesas (MAY suz): flat-topped hills covered by a resistant rock layer protecting soft rock beneath

mesosphere (MEZ uh sfihr): the coldest zone of the atmosphere that extends upward from Earth from approximately 50 kilometers to about 85 kilometers

Mesozoic Era: an era of geologic time lasting from about 200 to 65 M.Y.B.P.

metallic: having a shiny luster; resembling a metal

metallic ores: ores that contain one or more metals

metals: materials valuable because of their malleability and conductivity

metamorphic rocks: rocks that form from pre-existing rocks as the result of temperature and pressure changes

meteorites: small fragments of matter that strike Earth

meteoroids (MEET ee uh royds): small fragments of matter moving in space that vaporize upon entering Earth's atmosphere

meteorologist (meet ee uh RAHL uh just): a scientist who studies storm patterns and climates in order to predict daily weather

meteorology: the study of weather and the forces and processes that cause it

meteors: meteoroids that burn up in Earth's atmosphere

meter: the standard unit of distance or length in SI

methane: a biomass fuel produced from sewage or from the fermentation of kelp

micro-: prefix meaning one-millionth

mid-ocean ridge: an underwater mountain chain that rises from the ocean basins

Milky Way: the spiral galaxy to which Earth and the rest of our solar system belong

milli-: prefix meaning one-thousandth

milliliters: measures of liquid volume

mineral: a naturally occurring, inorganic, crystalline solid, with a definite chemical composition

Mississippian: a period of geologic time that occurred from about 345 to 325 M.Y.B.P., during which amphibians were the dominant life form

mixture: a physical combination of different substances that keep their own properties

model: a representation of an object or an idea of how an object looks

Moho: the boundary between Earth's crust and mantle

Mohorovičić discontinuity: the point where Earth's mantle begins; Moho

Mohs' scale: a scale of hardness used to describe minerals

mold: a cavity or impression left by an organism

molecule: a particle in which atoms composing it are held together by covalent bonds

mudflows: rapid, downhill movements of materials occurring after heavy rains

N

NASA: National Aeronautics and Space Administration; an agency that oversees the space program in the United States

native elements: elements that occur uncombined in nature

natural gas: a clean-burning fossil fuel formed from decayed marine organisms

natural resources: materials from Earth's crust that people use, including air, water, minerals, trees, and fuels

natural selection: an evolutionary process that results in the survival of individuals best adapted to their environments

neap tides: low tides that occur when the sun, Earth, and moon form a right angle

nebula: a low-density cloud of gas and dust in which a star is born

nekton (NEK tun): all swimming forms of life that inhabit the ocean

Neptune: a gaseous, outer planet similar in size and composition to Uranus

neutron: a particle in the nucleus of an atom that has no electrical charge

neutron star: a dense mass that results from the collapse of a supernova

new moon: a phase of the moon in which the side of the moon facing Earth is dark

newtons: the basic SI unit for measuring force

nimbus: a dark gray cloud with ragged edges from which rain or snow continually falls

nonclastics: sedimentary rocks that are deposited from solution or by organic processes

nonfoliated: a metamorphic rock texture used to describe rocks that lack banding

nonmetallic: having a dull, pearly, silky, glassy, or brilliant luster

nonmetallic deposits: ore deposits that contain no metals

nonrenewable resources: materials from Earth's crust that cannot be recycled or replaced within a useful period of time

nova: an ordinary star that suddenly increases in brightness and then fades slowly back to its original brightness

nucleus: the core of an atom that is made up of protons and neutrons

O

oasis (oh AY sus): a fertile, green area in an arid region

oblate spheroid (SFIHR oyd): a sphere that bulges at the equator and is flattened at the poles

observation: the act of gathering information using the senses

occluded front: a boundary that forms when two cool air masses merge, forcing the warmer air between them to rise

Ocean Thermal Energy Conversion (OTEC): a method that uses temperature differences in ocean water to generate electricity

oceanic trench: a deep trough on the ocean basin floor where oceanic crust is being forced below continental or other oceanic crust

oceanographers: scientists who study the physical and chemical properties of ocean water and the processes that occur within oceans

oceanography: the study of Earth's oceans

oil shales: rocks that contain a hydrocarbon material from which oil is extracted

oil spill: a form of pollution in which oil from various sources leaks into the ocean

ooze: sediment that contains at least thirty percent organic fragments

optical telescope: an instrument that collects light waves and is used to observe objects in space

orbit: the path followed by a planet

orbital plane: an imaginary plane formed by a planet in orbit

Ordovician: a period of geologic time occurring from about 500 to 430 M.Y.B.P., during which fish and invertebrates were the dominant organisms

ores: minerals or rock deposits that can be mined for a profit

organic: refers to living organisms

organic evolution: the process in which life forms change through time

orogeny: the processes involved in mountain building

outer planets: the giant, gaseous planets—Jupiter, Saturn, Uranus, Neptune—and Pluto

outwash: sorted, layered deposits of glacial debris

overburden: waste rock overlying a rock or mineral deposit that must be removed prior to mining

oxbow lake: a cut-off meander

oxides: a group of minerals composed of combinations of oxygen and some other element or elements

ozone: a form of oxygen that absorbs most of the ultraviolet radiation that enters the atmosphere

P

P-wave: a body wave that travels forward in a horizontal direction

paleomagnetism: the study of magnetism in ancient rocks

paleontologists (pay lee ahn TAHL uh justs): scientists who study fossils

Paleozoic Era: an era of geologic time lasting from about 570 to 225 M.Y.B.P.

Pangaea (pan JEE uh): a supercontinent that existed during the Mesozoic and included most of Earth's present-day continents

parabolic dunes: crescent-shaped sand dunes held down by vegetation

passive solar building: a building with south-facing vertical windows that absorb solar energy, which is then converted to heat

peat: the first stage in coal formation

Pennsylvanian: a period of geologic time lasting from about 325 to 280 M.Y.B.P., during which coal forests and swamps were abundant, and the first insects and reptiles developed

penumbra: a partial shadow formed during an eclipse

peridotite: an intrusive igneous rock

perihelion: the point in an orbit when an object is closest to the sun

periods: geologic time units that are subdivisions of eras

permeable: a condition of some rocks in which pore spaces exist within the rock thus allowing fluids to move through the rock

Permian: a period of geologic time that occurred from about 280 to 225 M.Y.B.P., during which many marine invertebrates became extinct

petrified: a form of fossil preservation in which organic material is replaced by minerals from groundwater

petroleum: crude oil

photochemical smog: smog produced by the chemical reaction between the sun and materials such as nitric oxide, sulfur dioxide, and hydrocarbons

photosphere: the visible surface of the sun

photosynthesis: the process by which plants use carbon dioxide, nutrients, and sunlight to produce food

physical models: models that represent actual objects

physical properties: characteristics of matter that can be observed and measured, such as color, length, and density

physics: the science that deals with forces, motion, and energy and their effects on matter

phytoplankton: the plant portion of plankton

pie graph: a graph that uses parts of a circle to show how each part is related to the whole

piedmont (PEED mahnt) **glacier:** a glacier that forms at the foot of a mountain when valley glaciers advance onto the plains

placers (PLAS urs): deposits containing valuable minerals that accumulate at the foot of a slope or where a stream curves

plains: vast flat areas of low elevation

planet: an object in space that reflects light from a nearby star around which it revolves

plasma: a state of matter that exists only at very high temperatures in the form of ions and free electrons

plastic flow: a type of internal glacial flow in which layers of ice slide over one another

plateaus: high, relatively flat areas next to mountains

plates: rigid blocks of Earth's crust and upper mantle

plate tectonics: a theory that explains movements of continents and changes in Earth's crust caused by internal forces

playa (PLI uh): a low area between mountains in which water may collect to form a lake

plucking: an erosional process whereby glacial meltwater freezes and removes surrounding rock fragments as it advances

Pluto: the ninth planet from the sun

plutonium-239: an element used in the breeding process

polar easterlies: cold, dry, dense air located between the poles and 60° latitude

Polaris: the North Star

polar wandering: a concept based on magnetic data that suggests that the continents have been moving throughout geologic time

polar zone: a cold climate zone that extends from the poles to 66½° north and south latitudes

pollutants: substances that produce harmful changes in the environment

pollution: the introduction of substances into an environment that produce harmful changes in the environment

porphyry: an igneous rock that has two or more different grain sizes

Precambrian: the time between Earth's formation and the beginning of the Paleozoic Era

precipitates: nonclastic sedimentary rocks that form when solids settle out of solution

precipitation (prih sihp uh TAY shun): water that falls to Earth's surface from the atmosphere as rain, snow, hail, or sleet

prevailing westerlies: winds located between about 30° and 60° latitude

primary wave: P-wave

prime meridian: the meridian that passes through Greenwich, England (0° longitude), from which distances east and west on Earth's surface are measured

problem solving: a process for finding solutions to problems

profile of equilibrium (ee kwuh LIHB ree um): a state of balance along a river in which erosion and deposition are balanced

projections: models used to produce convenient representations of Earth's surface

prominences: trapped gases that appear to shoot outward from the chromosphere of the sun

proton: a positively charged particle in the nucleus of an atom

pterosaur: a flying reptile

pulsar: a rapidly spinning neutron star that emits radio waves

punctuated equilibria: an evolutionary pattern in which certain groups evolve very quickly and then remain relatively unchanged for long periods of time

Q

quarries: open pit mines from which sandstone, limestone, gravel, or marble is removed

quartz: a mineral composed of oxygen and silicon

R

radial patterns: drainage patterns in which streams flow outward from a central location

radiant energy: energy that is transferred from one place to another by means of waves

radiation: the transfer of energy by means of electromagnetic waves

radioactive decay: a spontaneous process by which nuclei of atoms change and emit energy as the result of this change

radiocarbon dating: a method of determining age by measuring the amount of carbon-14 present in a substance

radiometric dating: a method that uses the decay of radioactive elements to determine the absolute age of rocks and minerals

radio telescope: an instrument consisting of a reflector, a receiver, and an antenna used to collect radio waves from space

rays: bright streaks that extend outward from some large moon craters

real motion: the actual movement of an object

rebounding: the rising of land surfaces as a glacier melts and pressure is removed

rectangular pattern: a drainage pattern in which the main stream makes right angle turns, and its tributaries enter at right angles

red giants: stars near the end of their existence whose cores collapse due to an exhaustion of hydrogen

reflecting telescope: an instrument in which light is collected on a concave mirror and the image formed is reflected and magnified in the eyepiece

refracting telescopes: instruments that use an objective lens to bend light toward the focal plane where the image is formed

regional metamorphism: intense pressure and temperature changes that occur in rocks as the result of mountain building

regolith: the outermost layer of the moon or Earth that is made of dust, rock fragments, and boulders

relative dating: a method that places an event in its proper order by comparing the event with other chronological events

relative humidity: the ratio of the amount of water vapor in a given volume of air to the total amount of moisture that that volume of air could hold at a given temperature

relative motion: the motion of one object as compared to the motion of another object

relief: the variation in the elevation of an area

renewable resources: materials from Earth that can be recycled within a useful period of time

reptiles: ectothermic vertebrates that live on land and lay hard-shelled eggs

residual soils: soils formed in place by the gradual weathering of parent rock

retrograde: having direction contrary to the general direction of similar bodies

revolution: the circling of one object about another

rhyolite: a fine-grained igneous rock that is the intrusive equivalent of granite

Richter scale: an expression of earthquake magnitudes as measured by seismographs

rift zone: a system of cracks in Earth's crust through which molten material rises

rills: a network of small channels that result from a sheet of water flowing downhill

rip currents: narrow currents that flow seaward at a right angle to the shoreline

ripple marks: wavy features of some sandstones

rock cycle: the changes that rocks undergo from the magmatic state, to rocks, and back to magma

rockets: action-reaction engines based on Newton's third law of motion

rockfalls: downhill movements of large masses of rock or small pebbles

rocks: Earth materials made of one or more minerals

rotation: the turning motion of an object on its axis

runoff: precipitation that flows across a land surface as drainage

S

S-waves: body waves that vibrate at right angles to a primary wave

safety symbols: symbols in this textbook that warn of possible laboratory dangers

salinity: the total quantity of dissolved solids in seawater

sand dunes: hills of sand deposited by the wind when it slows down or meets an obstacle

satellite: an object that revolves around a large primary body

Saturn: the second largest planet in our solar system

scale: a fixed ratio between the size of a real object and the size of a model of the same object

science: a process that produces knowledge about the physical world

scientific method: a process in which a series of steps are followed in order to solve problems

scientist: a person who uses scientific methods to learn about and explain natural events

sea breeze: a circulation pattern in which warmer air over land rises and is replaced by cooler air from water; occurs during the day

sea-floor spreading: a process by which new oceanic crust is being formed at mid-ocean ridges

sea level: the level of the surface of the sea midway between the average high and low tides

seamount: a volcano that does not rise above sea level

secondary wave: S-wave

sedimentary ore deposits: concentrations of ores resulting from weathering at or near Earth's surface and the movement of groundwater

sedimentary rocks: rocks formed as the result of weathering and redeposition of loose Earth materials

sediments: loose Earth materials or debris resulting from weathering

seismic sea waves: shallow water waves caused by a sudden shift of the ocean floor; also called tsunamis

seismic waves: vibrations that result when rock suddenly breaks and moves, releasing large amounts of energy

seismograms: lines traced on the recording tape of a seismograph

seismographs: instruments that record Earth tremors

seismologists: scientists who study earthquakes

semidiurnal: having two high tides and two low tides each day

shadow zone: a region of Earth between 103° and 143° from an earthquake epicenter that does not receive body waves

shaft mines: underground mines

shale: a clastic sedimentary rock composed of thin layers of clay- and mud-sized particles

shallow-focus earthquakes: earthquakes produced at the outer edge of ocean trenches, as rocks are folded into mountains, at transform faults, and at mid-ocean ridges

shallow-water wave: a wave moving in water that is shallower than one-half its wavelength

shearing: a system of forces that pushes against a body from different sides not directly opposite each other

shield volcanoes: basaltic volcanoes with gently sloping sides

shields: areas of continents composed of Precambrian rocks

shore zone: the area lying between high and low tides

silicate tetrahedron (teh truh HEE drun): a structure composed of one ion of silicon and four ions of oxygen

silicates: a group of minerals composed of silicon, oxygen, and one or more other elements

sill: a horizontal sheet of rock formed when magma intrudes two rock layers and hardens

silt: sediment smaller than sand

Silurian: a period of geologic time occurring from about 430 to 395 M.Y.B.P., during which corals dominated the seas, and the first land plants developed

sinkholes: funnel-shaped depressions that result from the dissolution of limestone along cracks and joints

slump: a type of mass movement in which loose material or rock layers slip downward as a unit

smog: a foglike pollution made darker and more dense by smoke and chemical fumes

soil: a mixture of weathered rock and decayed organic material

soil profile: a vertical section of all horizons that make up a soil

solar cell: a device that collects the sun's energy and changes it into electricity

solar day: the period in which Earth makes one complete rotation on its axis with respect to the sun; 24 hours

solar eclipse: an eclipse that occurs when Earth is in the moon's shadow

solar energy: energy from the sun

solar flares: sudden increases in brightness of the sun's chromosphere

solar system: a system of objects in orbit around our sun

solar wind: a flow of ions and electrons that moves outward from the sun at high speeds

solid: a state of matter in which each atom or molecule has a fixed position; resists change in shape and volume

solstice: either of two points where the sun reaches its maximum distances north and south of the equator

solution: a mixture containing two or more substances, one of which is dissolved in the other

spacelab: a workshop and laboratory located in the cargo bay of a space shuttle

space probes: rocket-launched vehicles that carry data-gathering equipment above Earth's atmosphere

space shuttle: a reusable craft designed to transport astronauts, materials, and satellites to and from space

space station: a living and work space that contains all the equipment and life-support systems necessary for astronauts in space

species: a group of organisms that can produce fertile offspring

specific gravity: the ratio between the mass of a given substance and the mass of an equal volume of water

spectrograph: an instrument used to photograph a spectrum

spectroscope: an instrument that separates visible light into its various wavelengths

spiral galaxies: disk-shaped galaxies with arms that rotate around a dense center

spit: a long ridge of sand deposited by a longshore current when the current loses velocity

spring tides: tides that occur when the sun, moon, and Earth align

stack: an island of resistant rock left after weaker rock is worn away by waves and currents

stalactites: elongated structures of calcium carbonate that hang from cave ceilings

stalagmites: upward growths of calcium carbonate from a cave floor

star dunes: sand dunes with four or five arms of sand extending outward from a central point

Starr's model: a representation of the major wind systems of Earth

stars: hot, bright spheres of gas

stationary front: a boundary that forms when either a warm front or a cold front stops moving forward

station models: the part of a weather map that describes the local weather in an area and includes wind speed and direction, atmospheric pressure, temperatures, dew point, and other data

strategies: plans or ways to solve problems

stratosphere (STRAT uh sfihr): a layer of the atmosphere that contains the ozone layer

strato-volcanoes: cone-shaped structures made of lava and other volcanic debris; also called composite volcanoes

stratus: a layered cloud, often covering the whole sky, associated with light drizzle

streak: the color of a powdered mineral

striation: a glacial scratch or groove left in bedrock as ice moves over it

strip mines: mines in which resources are removed through an opening or pit at Earth's surface

stromatolites (stroh MAT uh lites): microfossils similar to cyanobacteria

subduction: the process whereby an oceanic plate descends into the mantle and melts

subduction zone: the region where an oceanic plate descends into the upper mantle

subsidence: sinking of the land

sulfurous smog: smog produced when sulfur oxides and particulates in the air combine with a layer of stable air or cloud cover

sun: a star that produces large amounts of radiant energy each second; Earth's star

sunspots: relatively cool, dark areas on the sun's surface

supergiants: red giant stars

supernova: a star whose sudden increase in brightness indicates a greater outburst of energy than a nova

surf: the result of the forming and breaking of many waves

surface currents: movements of water as a result of wind

Surface Environment and Mining Program: a land reclamation program established by the U.S. Forest Service to reshape and restore mined areas

surface waves: tremors that travel along Earth's surface

surges: sudden movements of glaciers after long periods of little movement

suspended load: sediment picked up and carried by a river

syncline (SIHN kline): a fold that is convex downward in which the youngest strata are in the middle

synfuel: synthetic fuels produced by coal gasification and the extraction of hydrocarbons from shales and tar sands

T

talus (TAY luhs): material that accumulates at the foot of a steep slope or cliff

tar sands: porous sandstones saturated with thick hydrocarbons

technology: applied science

temperate zone: a climate zone between the tropics and the polar zones where weather changes with the seasons

temperature: a measure of the amount of heat in an object

tension: a stretching or pulling force

terminal moraines (muh RAYNZ): ridges of glacial till left at the margins of an ice sheet

theory: an explanation supported by facts

thermal pollution: the result of returning heated water to its source

thermohaline currents: currents that are synonymous with density currents and refer to the temperature and salinity of ocean water

thermosphere (THUR muh sfihr): the outermost layer of the atmosphere that extends from about 80 kilometers upward into space

thrust: a force produced by the expansion of hot gases that propels a rocket forward

tidal power: the use of ocean tides to produce electricity

tidal range: the difference between high tide and low tide

tides: shallow water waves caused by the gravitational attraction among Earth, moon, and sun

till: an unsorted, unlayered glacial deposit of boulders, sand, and clay

time: the measurement of the span between two events

tombolo: a sand or gravel deposit that connects an island to the mainland

topographic maps: maps that show the location, landscape, and cultural features of a part of Earth's surface

topography (tuh PAHG ruh fee): the surface features of an area

topsoil: the uppermost layer of soil

tornado: a violent, whirling wind that moves in a narrow path over land

trade winds: winds that blow toward the equator from about 30° north and south of the doldrums

transform faults: boundaries at which plates move past one another horizontally

transpiration: the process by which water escapes from the leaves of plants back to the atmosphere

transported soil: soil that has been removed from its place of origin by erosion and deposited in another location

trellis patterns: drainage patterns in which the main stream cuts across resistant rock layers, while the tributaries follow valleys of weak rock

tremors: seismic vibrations

trilobites (TRI luh bites): fossil invertebrates found in some Paleozoic rocks

tropics: a climate zone lying between 23½° north and south latitude that receives the greatest concentration of sunlight

tropopause: the boundary near the top of the troposphere that acts as a ceiling to the weather zone

troposphere (TROHP uh sfihr): a layer of the atmosphere nearest Earth, containing 75 percent of the gases of the atmosphere

trough: the lowest point of a wave

tsunamis (soo NAHM eez): seismic sea waves

Tyrannosaurus (tuh ran uh SOR us): the largest carnivorous land animal that lived during the Mesozoic Era

U

U-shaped valley: a valley eroded by a valley glacier

umbra: an inner complete shadow formed during an eclipse

unconformities (un kun FOR mut eez): gaps or breaks in the rock record due to erosion, nondeposition, or both

uniformitarianism: the principle that states that all Earth processes occurring today are similar to those that took place in the past

upwarped mountains: mountains formed as a result of crustal uplifting

upwelling: the rising of cold, deep ocean water toward the surface, especially along continental coasts

uranium-235: a radioactive element used as fuel for nuclear power plants

Uranus: a gaseous planet with a retrograde revolution

urbanization: the expansion of cities

V

V-shaped valley: a valley eroded by a river

valley glaciers: glaciers that form at high elevations where snow remains year-round in hollows or abandoned river valleys; also called alpine glaciers

variables: changeable factors that are tested in experiments

variable stars: stars that change brightness

velocity: the speed and direction of a moving object

Venus: the second planet from the sun

volcano: a mountain formed by the accumulation of material that has been forced out of Earth's interior onto its surface

volume: the space that an object occupies

Voyager: a space probe

W

warm front: a boundary that develops when a less dense, warm air mass meets a denser, colder air mass

warning: a weather advisory issued when severe weather conditions exist

watch: a weather advisory issued when conditions are such that severe weather could occur

water table: the upper surface of a zone of saturation

wave base: the depth of water equal to one-half the wavelength of a wave

wave height: the vertical distance between the crest and trough of a wave

wavelength: the horizontal distance between successive wave crests or troughs

wave period: the time it takes two successive wave crests to pass a given point

weathering: all processes by which rocks change at or near Earth's surface

weight: a measure of the gravitational force exerted on an object by another body

white dwarf: the dying core of a giant that radiates heat into space as light waves

windward: facing into the wind

X

x-axis: the horizontal axis on a line or bar graph

X rays: invisible forms of radiant energy

Y

y-axis: the vertical axis on a line or bar graph

year: the time needed for Earth to make one revolution about the sun; 365¼ days

Z

zone of aeration: the area above the water table in which pores are filled with air between rainstorms

zone of saturation: the area above an impermeable rock layer in which all pores are filled with water

zooplankton: the animal portion of plankton

INDEX

D

E

F

G

H

I

J

K

L

M

N

O

P

T

U

V

W

Y

Z

PHOTO CREDITS

2–3, Larry Burton, (insets l to r) James Jenning, Museum of the American Indian, File Photo, Courtesy: The Singer Co., Doug Martin; **4,** Spencer Swanger/Tom Stack & Assoc.; **6,** Pictures Unlimited; **7,** File Photo; **8,** (l) Bogart Photography, (r) Doug Martin; **13,** Steve Lissau; **15,** NASA/JPL; **17,** (l) Lawrence Migdale, (r) National Optical Astronomy Observatories; **18,** (l) Tim Courlas, (c) Bogart Photography, (r) Cobalt Photography; **19,** David Frazier; **20,** Tim Courlas; **21,** (l) NASA, (r) Woods Hole Oceanographic Institution; **26,** Chip Clark; **28,** (t) Susan Leach, (b) Bogart Photography; **29,** (l) Brian Parker/Tom Stock & Assoc., (r) Tim Courlas; **30,** (l) Bogart Photography, (r) File Photo; **32,** File Photo; **33,** (l) Tim Courlas, (r) Frank Lerner; **35, 36,** Tim Courlas; **38,** Latent Image; **46,** The Granger Collection; **49,** Doug Martin; **54,** (l) Earth Scenes/Lynn M. Stone, (r) Carlos Elmer/FPG; **55,** (l) Bill & Joan Popejoy, (r) Larry Burton; **56,** (t) Maxwell Mackenzie/Uniphoto; (b) Breck Kent; **57,** (t) Breck Kent, (b) James Mason/Black Star; **58,** U.S.G.S.; **59,** (t) File Photo; (b) U.S.G.S.; **60,** Courtesy: ETAK; **61, 63,** U.S.G.S.; **66,** (l) Roger K. Burnard, (r) Chris Kramer; **67, 68, 69,** U.S.G.S.; **74,** Jon Riley: Click/Chicago Ltd.; **76,** Farrell Goreham/Photo Researchers; **79,** University of Tokyo/ Discover Magazine © 1987, Family Media, Inc.; **80,** Photographs by Carolina Biological Supply Company; **81,** Russ Lappa; **83,** (t) Michael Abramson/Woodfin Camp & Assoc., (c) Steve Ogden/Tom Stack & Assoc., (r) Smithsonian Institution; **86,** File Photo; **87,** Breck Kent; **88,** Tim Courlas; **92,** Latent Image; **94–95,** © Gary Ladd, (insets l to r), Mount Wilson & Palomar Observatory, Tersch, NASA; **96,** EROS; **99,** Pictures Unlimited; **100,** NASA; **102,** Courtesy: Replogle Globes, Inc.; **105, 110, 111, 112,** NASA; **116,** Tim Courlas; **118,** Courtesy: Morton Thiokol, Inc.; **120,** Doug Martin; **126, 127, 128, 129, 130,** NASA; **131,** (t) NASA, (b) JPL; **133,** NASA; **138,** JPL; **143,** (t) NASA, (b) Tiara Observatory; **144,** Doug Martin; **145,** (l) Journal of Geological Education, (r) JPL/California Institute of Technology; **147, 148,** JPL; **151,** (t) File Photo, (b) NASA; **152,** (t) John Sanford/Science Photo Library/Photo Researchers, (bl) Dave Davidson/Tom Stack & Assoc., (br) Verlag/ZEFA; **158,** Michael Carroll; **161,** (r) Tersch, (others) NASA; **162,** JPL/California Institute of Technology; **164,** Kitt Peak National Observatory; **168,** California Institute of Technology/Carnegie Institute of Washington; **172,** (l) Hale Observatory, (r) U.S. Naval Observatory; **173,** (l) U.S. Naval Observatory, (r) Kitt Peak National Observatory; **175,** (t) Big Bear Solar Observatory/ California Institute of Technology, (b) California Institute of Technology/Carnegie Institute of Washington; **176,** Kitt Peak National Observatory; **180,** NASA; **182–183,** John Pearce, (inserts l to r) U.S. Navy, File Photo, File Photo, Library of Congress, British Airways; **184,** John Pearce; **192,** Cobalt Productions; **202,** John Anderson/University of Wisconsin, Madison; **206,** Edi Ann Otto; **208,** File Photo; **210,** NCAR; **211,** (l) File Photo, (r) Bill Popejoy; **212,** (t to b) Betty Crowell, David Dennis, Betty Crowell, William Tucker/Uniphoto, David Dennis, David Frazier, James Fullmer; **213,** Larry Burton; **218,** (t) F. Mendonca/Tom Stack & Assoc., (b) Ohio Disaster Service Agency; **219,** File Photo; **221,** National Severe Storm Laboratory; **223,** Hank Morgan/Photo Researchers; **228,** Steve Lissau; **232,** K. Benser/ ZEFA; **234,** (l) Woods Hole Oceanographic Institution, (r) U.S. Navy; **236,** Pacific Gas & Electric; **238,** Courtesy H.S. Bell, California Institute of Technology/Sedimentation Laboratory; **240,** (t) J.R. Berintenstein/Photo Researchers, (b) NOAA; **244,** Ward's Natural Science Establishment; **246,** Courtesy: Nova Scotia Power Corp.; **250,** Nicolas de Vore III/Bruce Coleman, Inc.; **252,** (l) Russ Kinne/Photo Researchers, (r) Steve Lissau; **253,** Steve Lissau; **254,** EROS; **255,** Geri Murphy; **257,** (l) Scripps Institution of Oceanography, (r) Lawrence Sullivan/Lamont-Doherty Geological Observatory; **260,** U.S.G.S.; **261,** File Photo; **262,** Lamont-Doherty Geological Observatory; **264,** (l) Hugh Spencer/ Photo Researchers, (r) Jerry Sieve; **266,** NASA; **267,** Geri Murphy; **268,** Woods Hole Oceanographic Institution; **269,** Doug Martin; **270,** PHOTRI; **274,** NASA; **276–277,** Ed Cooper, (insets l to r) Charles R. Knight and American Field Museum, John Hancock Mutual Life Insurance Co., The New York Convention & Visitors Bureau; **278,** Tony Morrison, Suffolk; **280,** (l) Caroline Coleman/William E. Ferguson, (r) S.J. Krasemann/Peter Arnold, Inc.; **281,** (t) James Westwater, (b) Breck Kent; **282,** (l) James Westwater, (r) Tom McHugh/Photo Researchers; **283,** James Patterson; **288,** Tim Courlas; **289,** (l) Tom Bean, (r) John Elk III/Photo Researchers; **290,** Englehert/Photo Researchers; **291,** (l) Michael Collier, (r) G.R. Roberts; **293,** Doug Martin; **294,** James Westwater; **295,** (l) K.L. Tompkins/Tom Stack & Assoc., (r) E.S. Ross; **296,** Collier/Condit Aerial Photo; **297,** ARAMCO; **298,** Peter Pearson: Click/Chicago; **300,** Larry Hamill; **304,** Ed Cooper; **307,** Cyril Toker/Photo Researchers; **310,** Betty Crowell; **312,** (l) EROS, (r) Doug Martin; **315,** (t) EROS, (b) Masachika Suhara; **317,** (l) Larry Burton, (r) Michael Collier; **319,** Timothy O'Keefe/Tom Stack & Assoc.; **320,** Chip Clark; **324,** Tom Bean; **326,** James Patterson; **327,** Betty Crowell; **328,** Roger K. Burnard; **329,** (l) Roger K. Burnard, (r) G.H. Matchneer; **332,** William E. Ferguson; **336, 338,** Tom Bean; **340,** Spitzburg/Gamma-Liaison; **344,** Steele Perkins/Magnum; **346–347,** Dr. Georg Gerster, (insets l to r) File Photo, File Photo, Northwind Pictures Archives, Gerard Photography; **348,** Chip Clark; **350,** Tim Courlas; **351,** Doug Martin; **352,** (t) Hickson-Bender, (others) Tim Courlas; **354,** (l) Eric Hoffhines, (r) Elaine Comer; **355,** Tim Courlas; **356,** (t) Tim Courlas, (b) Doug Martin; **357,** (t) Lorie Kramer, (b) Tim Courlas; **358,** (br) Craig Kramer, (others) Tim Courlas; **360,** Tim Courlas; **361,** (l) Tom McHugh/Photo Researchers, (r) Pictures Unlimited; **364,** R.W. Sheets/Ohio State University; **368,** Stern/Black Star; **370,** Cecil Stoughton/U.S. Department of the Interior; **371,** (tl) Betty Crowell, (tr) Collier/Condit Aerial Photo, (bl) Doug Martin, (bc) Craig Kramer, (br) Cobalt Productions; **372,** (r) Doug Martin, (others) Eric Hoffhines; **373,** Young/ Hoffhines; **375,** (tl, tc) Tim Courlas, (tr) Cobalt Productions, (bl) W.C. Bradley Collection, (br) Eric Hoffhines; **376,** Tim Courlas; **386,** Tom Bean; **388,** File Photo; **390,** (t) Carolina Biological Supply Co., (b) Young/Hoffhines; **392,** (l) Elaine Comer, (r) Michael Collier; **393,** Paola Koch/Photo Researchers; **395,** (l) Elainer Comer, (c) Young/Hoffhines, (r) Betty Crowell; **396,** (l) Russ Lappa, (r) Paul Nesbit; **397,** (l) Roger K. Burnard, (others) Elaine Comer; **398,** (t) Elaine Comer, (bl) Michael Collier, (br) Bjorn Bolstad/Peter Arnold, Inc.; **401,** Tim Courlas; **406,** Blair Seitz/Photo Researchers; **408–409,** M. Philip Kahl Jr./Bruce Coleman, Inc., (insets l to r) Tenn. Conservation Dept. Photo, File Photo, U.S.G.S., Courtesy: General Electric Development Center; **410,** David Walters/Miami Herald/Black Star; **414,** (t) Jerome Wyckoff, (b) Van Bucker/Photo Researchers; **416,** Wesley Bocxe/Photo Researchers; **419,** Andrew Rafkind: Click/Chicago; **421,** (l) U.S.G.S., (r) Steve Martin/Tom Stack & Assoc.; **422,** Doug Martin; **426,** Tim Courlas; **428,** Elisa Leonelli/Bruce Coleman, Inc.; **432,** G.R. Roberts; **434,** File Photo; **437,** Kenneth Gareth/Woodfin Camp; **441,** NASA; **442,** James Balog/Black Star; **444,** Gunnar/Tom Stack &

Assoc.; **445,** © Elliot Varner Smith/Courtesy: Cynthia Dusil-Bacon; **456–457,** Tom Bean, (insets l to r) American Field Museum, The Granger Collection, File Photo, Herbert Hoover Presidential Library; UPI; **458,** Tom Algire/Tom Stack & Assoc.; **466,** (l) Roger K. Burnard, (r) Michael Collier; **467,** Dr. Takeo Suzuki; **468,** Rick Pantonial; **469,** William E. Ferguson; **472,** American Field Museum; **474,** Tim Courlas; **478,** Rick Smolan/Woodfin Camp & Assoc.; **482,** Michael Collier; **487,** (t) EROS, (b) Ward's Natural Science Establishment; **488,** (t) W. Britt Leatham/Ohio State University, (bl) Tim Courlas, (bc) Craig Kramer, (br) Doug Martin; **489,** Tim Courlas; **490,** Breck Kent; **493,** Peter Baker/FPG; **494,** Betty Crowell; **497,** (t) Breck Kent, (b) Stephen J. Krasemann; **498,** (t) Ted Spiegel/Black Star; (b) Ted Rice; **504–505,** William E. Ferguson, (insets l to r) The Granger Collection, Tenn. Valley Authority, United Nations/Photo by Sygma; **506,** Tom Algire/Tom Stack & Assoc.; **508,** William E. Ferguson; **509,** (l) AID, (r) Alpha; **510,** NASA/Goddard Space Flight Center; **511,** Tim Courlas; **513,** Brian Brake/Photo Researchers; **514,** William E. Ferguson; **515,** (l) File Photo, (r) Larry Hamill; **519,** Tim Courlas; **520,** Lawrence Migdale; **521,** Ted Rice; **523,** Doug Martin; **524,** Tim Courlas; **528,** Paola Koch/Photo Researchers; **530,** Weldon King/FPG; **533,** (l) Tim Courlas, (r) Ward's Natural Science Establishment; **534,** Jacques Janquox/Peter Arnold, Inc.; **537,** (l) John Zoiner/Peter Arnold, Inc., (c) Tim Courlas, (r) Courtesy: Morton Salt Co.; **539,** Earl Roberge/Photo Researchers; **540,** Glenn Stiener/FPG; **542,** Ohio Department of Natural Resources/Division of Reclamation; **544,** Collier/Condit Aerial Photo; **545,** Scripps Institution of Oceanography; **546,** (l) Institute of Scrap Iron & Steel, Inc., (r) Doug Martin; **550,** Dan McCoy/Rainbow; **552,** (l) Mickey Gibson/Tom Stack & Assoc., (r) Cameramann International; **553,** NASA; **554,** David Brownell; **557,** David Frazier; **561,** Latent Image; 564, David Cameron; **566,** Latent Image; **569,** Courtesy: Shell Oil Company; **570,** Eric Hoffhines; **572,** Doug Martin; **576,** Tom Bean.

6 7 8 9 10 11 12 13 14 15—98 97 96 95 94 93 92 91 90